Modern Mathematical Methods
for Scientists and Engineers

a street-smart introduction

Modern Mathematical Methods
for Scientists and Engineers
a street-smart introduction

Athanassios Fokas

University of Cambridge, UK &
University of Southern California, USA

Efthimios Kaxiras

Harvard University, USA

 World Scientific

NEW JERSEY · LONDON · SINGAPORE · BEIJING · SHANGHAI · HONG KONG · TAIPEI · CHENNAI · TOKYO

Published by

World Scientific Publishing Europe Ltd.

57 Shelton Street, Covent Garden, London WC2H 9HE

Head office: 5 Toh Tuck Link, Singapore 596224

USA office: 27 Warren Street, Suite 401-402, Hackensack, NJ 07601

Library of Congress Cataloging-in-Publication Data

Names: Fokas, A. S., 1952–   author. | Kaxiras, Efthimios, author.

Title: Modern mathematical methods for scientists and engineers : a street-smart introduction /
   Athanassios Fokas, University of Cambridge, UK & University of Southern California, USA,
   Efthimios Kaxiras, Harvard University, USA.

Description: New Jersey : World Scientific, [2023] | Includes bibliographical references and index.

Identifiers: LCCN 2022013695 | ISBN 9781800611801 (hardcover) | ISBN 9781800611832 (paperback) |
   ISBN 9781800611818 (ebook for institutions) | ISBN 9781800611825 (ebook for individuals)

Subjects: LCSH: Mathematical analysis.

Classification: LCC QA300 .F66 2023 | DDC 518/.6--dc23/eng20220517

LC record available at https://lccn.loc.gov/2022013695

British Library Cataloguing-in-Publication Data

A catalogue record for this book is available from the British Library.

Cover design by Lali Abri

Image: M.C. Escher's *Relativity* © 2021 The M.C. Escher Company – The Netherlands. All rights reserved. www.mcescher.com

For any available supplementary material, please visit
https://www.worldscientific.com/worldscibooks/10.1142/Q0348#t=suppl

Desk Editors: Nimal Koliyat/Shi Ying Koe

Typeset by Stallion Press
Email: enquiries@stallionpress.com

Printed in Singapore

Dedicated to

Regina and Eleni

with immeasurable
affection
and gratitude

Preface

Philosophy of the book

Mathematics is the language of science. One can learn a new language in a formal way and have a deep knowledge and appreciation of its many aspects (grammar, semantics, literature, poetry, etc.). This takes a great deal of effort and at the beginning can be rather tiresome. On the other hand, one can pick up the essentials of a language and develop the ability to communicate effectively without an extensive knowledge of its richness. This book is a guide for a quick familiarization with essential tools of applied mathematics for solving a variety of problems in science and engineering. Its approach is similar to learning a language at a street-smart level, picking it up by talking to friends and exploring it by making new friends. Of course, in this way, some of the richness of the language is lost. But the satisfaction of being able to use the language quickly and the accompanying wealth of new experiences compensate for this loss.

Consistent with this viewpoint, the text maintains some of the less-than-formal character of a lecture, with more attention paid to examples and applications rather than to rigorous statements and proof of theorems. Nevertheless, throughout the book proofs of all the important statements are provided or at least sketched in some detail, so that the reader can understand the logic behind each statement.

How it came to be

The seed that became this book was planted three decades ago when the second author, then a fresh junior faculty, inherited an undergraduate course on complex analysis from the well-known applied mathematician Richard E. Kronauer (1925–2019). The inheritance comprised a set of hand-written notes totaling 40 pages — one page per lecture plus a one-page syllabus. Kronauer's original notes were augmented with detailed derivations and examples to become an early version of what is now Part II of the book. This material, typed in LaTeX and enhanced with figures drawn in xfig, was distributed as the course Lecture Notes for many years. Additional material was developed later in order to provide the background for a more thorough understanding of complex analysis. Another motivation was to present some topics that typically are not covered in an introductory mathematics course, for instance, special functions or probability theory, which students only encounter in more specialized courses. This material forms Part I and Chapter 13.

A major change in the scope and the style of the book was introduced around 2013, following an extended visit by the first author to Harvard University in the fall of 2012. During this visit, it dawned on us that a marriage of the material on complex analysis, with the basic ideas of the powerful "unified transform" method for solving partial differential equations, would make an excellent introduction to this method, which up to that point had been presented only in advanced graduate-level texts. Close collaboration between the two authors during the past few years produced the exposition of the method and its applications that form Part III of the book, as well as innumerable additions and improvements in the other parts. Finally, Chapters 14 and 15 were added recently, as a complement to the rest of the material, covering some topics that are of increasing importance in modern scientific and engineering applications.

We hope that the final product is a tightly integrated text, that covers the unified transform with all the necessary background as well as a wide range of other important topics in an accessible and comprehensive manner.

Subject matter

We provide here an overview of the contents of the book and the rationale for the choices we have made.

In Part I we present a discussion of single-variable calculus (Chapter 1), followed by multi-variable calculus (Chapter 2) and series expansions of functions (Chapter 3). We assume the reader has some familiarity with these subjects, so their treatment is more at a review level rather than an exhaustive presentation. We discuss in some detail only certain topics of relevance to later chapters; for instance, we focus on topics like the θ-function and the δ-function that are often not adequately covered in textbooks but are widely employed in science and engineering applications. At the end of each chapter, we also include applications. Some of these applications are common, like signal filtering (which illustrates the use of the convolution integral) but others are less conventional, such as neural networks (which illustrate the use of a function depending on many variables), and wavelets (which illustrate series expansion in a basis beyond polynomials, powers and trigonometric functions). The less conventional applications are meant to provide students with an understanding of the basic ideas behind the methodology, so that they are better prepared when they encounter such methods in more advanced courses.

In Part II we introduce functions of complex variables (Chapter 4) and discuss how their singularities can be employed to calculate integrals by contour integration (Chapter 5), as well as the mappings produced by complex functions (Chapter 6). These topics are treated more thoroughly and assume no previous familiarity. We also discuss the Fourier expansion and the Fourier transform (Chapter 7), tools that employ complex functions and are of utmost importance in many applications.

In Part III we start with a review of traditional methods for solving ordinary and partial differential equations (Chapter 8), and then proceed to discuss at length a new method for solving PDEs, the "unified transform" (also known as the "Fokas method"). The method lends itself to efficient analytical or numerical solution of a variety of important problems. We apply the method to solve representative problems for evolution equations (Chapters 9 and 10), the wave equation (Chapter 11), the Laplace, Poisson, Helmholtz and modified Helmholtz equations (Chapter 12). Our investment in a detailed description of this method is motivated by the hope that this alternative approach to partial differential equations will equip students with a powerful tool for solving interesting and challenging problems.

Lastly, in Part IV we discuss the theory of probabilities (Chapter 13), provide an elementary introduction to numerical methods (Chapter 14), and use concepts from those areas to present stochastic simulation and optimization methods (Chapter 15). Stochastic methods are being used ever more broadly in a variety of complex problems, so early exposure of STEM students to such methods is indispensable.

We have made the decision not to include any discussion of linear algebra and related topics in our book, because there already exist several textbooks that treat this subject at a satisfactory level. Moreover, even though many of the topics covered in this book can be handled by methods of linear algebra and vector spaces, we have not found it necessary to invoke these methods in our treatment.

Teaching plan suggestions

The book can be used in various ways for teaching a course, in the timeframe of either one or two semesters (we assume a semester consists of 37–39 one-hour lectures).

(a) A two-semester course can cover most of the book's content, with the first semester devoted to Parts I and II and the second semester to Parts III and IV. The material in Parts I and II can be comfortably covered in one semester, and provides a thorough training in single-variable and multi-variable calculus and complex analysis. Chapters 1, 2 and 3 can be covered at a slightly faster pace, assuming some familiarity with the material from an AP-level or introductory college-level calculus course. However, the contents of Parts III and IV amount to more than what can be comfortably covered in one semester, because the material is more conceptually dense and technically challenging, especially the five chapters of Part III. Some judicious choices must be made in what to include: One choice, which emphasizes the unified transform method and its applications, is to cover Part III in its entirety and possibly some topics from numerical methods (Chapter 14), skipping Chapters 13 and 15. A different choice is to cover Part III selectively, for instance Chapters 8 (Sections 8.1–8.4), 9 and 11 (Sections 11.1 and 11.2), and most of Part IV, thus giving roughly equal weight to partial differential equations, the unified transform method and probabilities and stochastic approaches.

(b) A one-semester course assumes that the students are already familiar with single-variable and multi-variable calculus, thus Part I can be skipped altogether, perhaps with only a quick discussion of unfamiliar topics like the generalized functions (Section 1.6). From Part II, the essential chapters are 4, 5 and 7, taking roughly the first half of the semester. The second half of the semester can be devoted to partial differential equations and the unified transform method, with the essential chapters being 8 (Sections 8.1–8.4), 9 and 11 (Sections 11.1 and 11.2).

Alternatively, the four different parts of the book can be used in conjunction with other sources to cover a particular subject more thoroughly. This is the reason the book was constructed in a modular form, with each part being thematically self-contained. Specifically:

(i) Part II can be used to build a one-semester course on complex and Fourier analysis, assuming that the students have already taken a college-level calculus course.

(ii) Part III can be used to build a one-semester course on partial differential equations, assuming that the students have already taken a course in complex and Fourier analysis.

(iii) Part I or Part IV, by themselves, do not contain enough material to build a semester-long course, but either one can be used as a component of a general math course, in combination with other topics. For instance, Part I can be paired with linear algebra topics to provide a useful background for many applications in physical science and engineering; this part could be comfortably covered in approximately half a semester. Part IV could be covered in approximately half a semester and can be paired with either linear algebra topics or topics on discrete mathematics.

Related sources

There are several widely used textbooks that discuss the conventional approaches to the topics covered in our book. These are useful sources for topics that we treat very briefly, and for providing a different perspective. They also contain a large number of examples and problems. Finally, several of these textbooks offer a good coverage of linear algebra, a topic not addressed in the present book. Some notable examples are:

1. George B. Arfken and Hans J. Weber, *Mathematical Methods for Physicists*, Sixth Ed. (Elsevier Academic Press, 2005).

2. K.F. Riley, M.P. Hobson, S.J. Bence, *Mathematical Methods for Physics and Engineering*, Third Ed. (Cambridge University Press, 2006).

3. D.W. Jordan and P. Smith, *Mathematical Techniques, An Introduction for the Engineering, Physical, and Mathematical Sciences*, Fourth Ed. (Oxford University Press, 2008).

4. Michael D. Greenberg, *Advanced Engineering Mathematics*, Second Ed. (Prentice Hall, 1998).

5. Alan Jeffrey, *Advanced Engineering Mathematics* (Harcourt/Academic Press, 2002).

6. Erwin Kreyzig, *Advanced Engineering Mathematics*, Tenth Ed. (Wiley, 2011).

7. Peter V. O'Neil, *Advanced Engineering Mathematics*, Eighth Ed. (Cengage Learning, 2017).

The order of presentation of the various topics in the last four textbooks mentioned above is roughly the inverse of what we have adopted here, namely starting with differential equations and ending with functions of complex variables.

About the Authors

 **Athanassios Fokas** has been the newly inaugurated Chair of Nonlinear Mathematical Sciences at the University of Cambridge, UK since 2002 (prior to this appointment he was Chair of Applied Mathematics at Imperial College London, UK). In addition, since 2015 he has been Adjunct Professor at the departments of Civil and Environmental Engineering and of Biomedical Engineering at the University of Southern California, USA. Fokas holds a BSc in Aeronautics from Imperial College (1975), a PhD in Applied Mathematics from Caltech (1979), and an MD from the University of Miami (1986).

Fokas has made broad scientific contributions with seminal results in several areas in mathematics, including the "Fokas method", which features in the textbook as a method of greatly simplifying calculations for undergraduates.

In 2000, Fokas was awarded the Naylor Prize, the most prestigious prize in Applied Mathematics and Mathematical Physics in the UK (in 1999 this prize was awarded to Stephen Hawking). He is also a recipient of a Guggenheim Fellowship, and has been elected in several academies, including the prestigious Academy of Athens, the only applied mathematician to be elected at this academy. His contributions to the areas of medicine and biology were recently recognized by his election to the American Institute of Medical and of Biological Engineering (one of only a handful of mathematicians to be members of this Institute). His overall contributions were recognized in 2015 with a Senior Fellowship by EPSRC, UK, which covers his salary at the University of Cambridge, so that he can concentrate on research, released from all teaching and administrative responsibilities.

Professor Fokas has published close to 400 papers. He has appeared in the list of the most highly cited researchers in mathematics on Web of Science. He has co-edited eight books and has authored or co-authored five monographs. His book *Complex Variables and Applications* with M. J. Ablowitz, published by Cambridge University Press, has sold nearly 20,000 copies.

Efthimios Kaxiras was educated at the Massachusetts Institute of Technology where he received a PhD in theoretical condensed matter physics. He joined the faculty of Harvard University in 1991, where he currently holds the title of John Hasbrouck Van Vleck Professor of Pure and Applied Physics in the Department of Physics and Professor of Applied Mathematics in the School of Engineering and Applied Sciences (SEAS). He served as Chair of Applied Mathematics, SEAS, from 2017 to 2020 and was appointed Chair of the Department of Physics in 2020. He is the Founding Director of the Institute for Applied Computational Science and served as its Director for two terms (2010–2013 and 2017–2020). He also served as Director of the Initiative on Innovative Computing (2007–2009). He has held faculty appointments and administrative positions in Switzerland (Ecole Polytechnique Federal de Lausanne) and Greece (University of Crete, University of Ioannina, Foundation for Research and Technology Hellas). He is a Fellow of the American Physical Society, and Chartered Physicist and Fellow of the Institute of Physics (UK).

His research interests encompass a wide range of topics in the physics of solids and fluids, with recent emphasis on materials for renewable energy, especially batteries and photovoltaics, and on the physics and applications of two-dimensional materials. Professor Kaxiras serves on the editorial board of several scientific journals and has published over 400 papers in refereed journals and several review articles and chapters in books. His group has developed several original methods for efficient simulations of solids using high-performance computing as well as multiscale approaches for the realistic modeling of materials.

During the 30 years of his tenure as a faculty member at Harvard University, Professor Kaxiras has taught more than a dozen different courses at both the graduate and undergraduate levels, covering diverse topics from Applied Mathematics ("Complex and Fourier Analysis"), to Condensed Matter Physics ("Introduction to Quantum Theory of Solids") and High Performance Computing ("Extreme Computing"). More recently, he has developed novel pedagogical methods and corresponding content for introductory courses in the physical sciences.

He has published two graduate-level textbooks: *Atomic and Electronic Structure of Solids* (Cambridge University Press, 2003), and *Quantum Theory of Materials*, co-authored with J. D. Joannopoulos (Cambridge University Press, 2019).

Acknowledgments

While as authors we bear all responsibility for the book's shortcomings, the contributions of some former students, postdocs, colleagues and collaborators have made a notable difference in its richness. We mention some of these contributions briefly here, with gratitude.

AF is deeply grateful to Dionysis Mantzavinos, Dave Smith and Matt Colbrook and especially Kostis Kalimeris, for their assistance regarding Part III. He also expresses his sincere thanks to the numerous collaborators who contributed to the development of the unified transform and in particular to Bernard Deconinck, Jonathan Lennels and Beatrice Pelloni. Lastly, he wishes to acknowledge the Alexander Onassis Foundation for funding the visit to Harvard University in the fall of 2012, when the plan to write this book was formulated.

EK is deeply grateful to Martin Bazant and Dionisios Margetis, who helped teach and enrich the early versions of the complex analysis course, and to George Tsironis and Georgios Neofotistos, for their significant contributions to the expanded version of the course in more recent years; all four have left their mark on Parts I, II and IV. He is also grateful to Isaac Lagaris for providing an initial draft of Section 14.7 and for many thoughtful comments on numerical and stochastic methods, and to Stefanos Trachanas for helpful suggestions on the pedagogical presentation of several topics and many insightful discussions on the deeper meaning of various concepts in mathematics (and beyond).

Finally, both of us are grateful to the students at our home institutions who took courses based on early versions of the material contained in this book. Their innumerable and often penetrating questions helped us sharpen the presentation of various ideas, which vastly improved the book (they also made better teachers of us). We sincerely hope that many future students will benefit from this experience.

Contents

Part I

Functions of Real Variables

Chapter 1

Functions of a Single Variable

Everyone knows what a curve is until he has studied enough mathematics to
become thoroughly confused by the countless exceptions.

Felix Klein (German mathematician, 1849–1925)

In this chapter, we provide a broad overview of functions of *real variables* while
in later chapters we shall investigate the properties of functions of *complex variables*. A real variable represents a *real number*, and a complex variable represents a *pair of real numbers*, of which the first is called the "real part" and the
second is called the "imaginary part". As a preamble and in order to introduce
some notation used throughout the book, the first section of the chapter provides a brief review of the various types of numbers, starting with the simplest
and most intuitive (the natural numbers) and extending to complex numbers
and beyond.

1.1 Introduction: The various types of numbers

Natural numbers: These are the numbers $1, 2, 3, 4, \ldots$ that one could use to
count, for example, physical objects or concepts. This set of numbers is denoted by the symbol $\mathbb{N}$.

With natural numbers, we can define the familiar operations of addition and
multiplication. The set or natural numbers is closed with respect to these operations, meaning that the sum (the result of addition) of two natural numbers
is another natural number and the product (the result of multiplication) of two
natural numbers is another natural number. An extension of multiplication is
the nth power, which is the product of a number with itself n times:

$$5^n = \underbrace{5 \times 5 \times \cdots \times 5}_{n \text{ times}}.$$

A natural number can often be expressed as the product of smaller natural
numbers, for example

$$105 = 7 \times 5 \times 3 \times 1.$$

Including the 1 at the end of the product is redundant, but we use it by convention to indicate that we have arrived at the smallest possible natural number
in this product expansion.

A subset of the natural numbers, called **prime numbers**, *cannot* be expressed as products of smaller natural numbers (other than 1); in other words, a prime number can only be expressed in product form as itself times 1. The prime numbers smaller than 100 are as follows:

$$2, 3, 5, 7, 11, 13, 17, 19, 23, 29, 31, 37, 41, 43, 47, 53, 59, 61, 67, 71, 73, 79, 83, 89, 97.$$

Having defined the prime numbers, we can then express any natural number as a product of prime numbers (some of them appearing in powers higher than 1): this is called *factorization*.

Once thought to be merely mathematical oddities, prime numbers are actually fascinating and can be quite useful, through their role in factorization, for computation and in many applications, including cryptography. Euclid, in his work *Elements*, first proved that the set of prime numbers is infinite. They are also at the heart of one of the most challenging unsolved problems in pure mathematics, known as the "Riemann hypothesis".

Integers: These are the natural numbers and their opposites, including zero:

$$\ldots, -3, -2, -1, 0, 1, 2, 3, \ldots$$

This set of numbers is denoted by the symbol $\mathbb{Z}$. The symbols $\mathbb{Z}^+$ and $\mathbb{Z}^-$ are used to denote the positive and the negative integers, respectively, while the symbol $\mathbb{Z}^{0+}$ is used to denote the non-negative integers (positive integers and zero).

With the introduction of integers, we can expand the range of operations that can be applied to any two numbers to include subtraction, which is the opposite of addition: The difference (result of subtraction) of two integers is another integer.

Rational numbers: A rational number is any number that can be written as a fraction of two integers, for example:

$$\frac{1}{11}, \quad \frac{23}{139}, \quad \frac{-44}{1051}.$$

With the introduction of rational numbers, we can expand the range of operations that can be applied to any two numbers to include division, which is the inverse of multiplication. The ratio (the result of division) of two rational numbers is another rational number. When a rational number is expressed in decimal form, it has either a finite number of digits after the decimal point, or an infinite number of digits but with a finite repeating sequence:

$$\frac{1}{11} = 0.090909090909\ldots, \quad \frac{23}{139} = 0.16546762589\ldots, \quad \frac{-44}{1051} = -0.04186489058\ldots$$

The finite repeating sequence can be very long in some cases, as the last two examples above indicate.

Irrational numbers: These are numbers that *cannot* be represented as a fraction of two integers, for example, expressions that involve the square root of a prime number:

$$\sqrt{2} = 1.414213562373095\ldots, \quad \frac{1+\sqrt{5}}{2} = 1.618033988749894\ldots$$

These two examples are the roots of certain polynomials with rational coefficients: $\sqrt{2}$ is one of the two roots of $x^2 - 2 = 0$ and $(1+\sqrt{5})/2$ is one of the two roots of $x^2 - x - 1 = 0$ (polynomials and their roots are discussed in Section 1.2.1). The second number is also known as the "golden ratio", and plays an important role in art and architecture. Irrational numbers always have an infinite number of digits after the decimal point, with no repeating sequence.

Two special irrational numbers are: e, the base of the natural logarithms, and π, the ratio of the perimeter to the diameter of a circle; their values to 15 decimal places are:

$$e = 2.718281828459045\ldots, \quad \pi = 3.141592653589793\ldots$$

e and π are called "transcendental" numbers because they cannot be obtained as the roots of polynomials with rational coefficients. As we will see in later chapters, these two numbers play a central role in complex numbers and the functions of complex numbers.

With the introduction of irrational numbers, we can expand the range of operations that can be applied to a number to include the *n*th root, which is the inverse operation of the *n*th power.

Real numbers: Putting together the integers, rational and irrational numbers, we obtain the set of **real numbers**. This set is represented by the symbol $\mathbb{R}$. The symbols $\mathbb{R}^+$ and $\mathbb{R}^-$ are used to denote the positive and the negative real numbers, respectively. We can represent the real numbers as points on a straight line stretching from $-\infty$ to $+\infty$, and call the arbitrary point on this line the x value: this is what we refer to as the *real variable*. Thus, real numbers are represented by the one-dimensional infinite line.

Properties of the basic operations: We note that the two basic operations we have defined are addition and multiplication. For each of these operations, there is an element of the set which, when involved in the operation, gives back the other number in the operation. For addition, this element is 0: $a + 0 = a$; for multiplication, it is 1: $a \times 1 = a$.

The complementary operations, namely subtraction and division, can be thought of as the operations which, given a number a, define the "opposite" of this number, $-a$, for the addition, or the "inverse" of this number, $(1/a)$, for the multiplication. When the opposite is added to the number it gives 0: $a + (-a) = 0$; when the inverse is multiplied by the number it gives 1: $a \times (1/a) = 1$ (for $a \neq 0$).

For all the sets of numbers defined above, addition and multiplication have the following two fundamental properties:

- commutativity: $a + b = b + a$ and $a \times b = b \times a$.

- associativity: $a + (b + c) = (a + b) + c$ and $a \times (b \times c) = (a \times b) \times c$.

In the following, and in the main text of the book, we will no longer use the symbol '$\times$' for multiplication; this symbol will be reserved for a special type of product between vectors, the "cross product" (see Chapter 2, Section 2.3). The usual product of two numbers a and b will simply be denoted as ab.

Complex numbers: By introducing the imaginary unit i, which is defined by

$$i \equiv \sqrt{-1} \Rightarrow i^2 = -1,$$

we can construct the complex number z using a pair of real numbers, (x, y), as follows:

$$z = x + iy.$$

x is referred to as the "real part" of the complex number z, denoted by Re[z]=x, and y is referred to as the "imaginary part" of z and denoted by Im[z]=y. The set of complex numbers is represented by the symbol $\mathbb{C}$. The notation $\mathbb{C} \setminus \{0\}$ is used to denote all the complex numbers excluding the value $(x, y) = (0, 0)$.

Given the definition of the complex number z in terms of the pair (x, y) and the definition of the imaginary unit i, and applying the usual rules of addition and multiplication including commutativity and associativity, we can obtain the sum and the product of two complex numbers as follows: given the two complex numbers z_1, z_2, with

$$z_1 = x_1 + iy_1 \quad \text{and} \quad z_2 = x_2 + iy_2,$$

their sum is given by

$$z_1 + z_2 = (x_1 + x_2) + i(y_1 + y_2),$$

and their product is given by

$$z_1 z_2 = (x_1 x_2 - y_1 y_2) + i(x_1 y_2 + x_2 y_1),$$

where in calculating the product we have made use of the fact that $i^2 = -1$. A more detailed look at the properties of operations with complex numbers is provided in Chapter 4 (see Section 4.1).

We can use two orthogonal axes to represent the values of the real variable x (the horizontal axis) and y (the vertical axis), so that the complex number z is represented by a point on the plane spanned by these two orthogonal axes. Thus, a complex number is represented by a point on the two-dimensional complex plane, see Fig. 1.1.

The inclusion of complex numbers represents a milestone because it leads to an *algebraically closed* number system. The essence of this statement is that all operations with complex numbers produce other complex numbers. This is captured by the "fundamental theorem of algebra", which can be stated as follows.

Theorem: *Every polynomial of the single variable x of degree n, which is not constant ($n > 0$), and has complex coefficients, has n complex roots.*

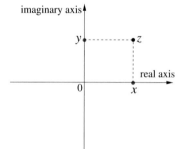

Figure 1.1: The complex plane, on which each point z represents a complex number with real part x and imaginary part y.

Beyond complex numbers: A natural question that arises is whether or not one can use higher-order multiplets of numbers to represent higher-dimensionality spaces, for instance, a triplet of real numbers to represent a three-dimensional space, by analogy to the use of pairs of real numbers (the complex numbers) to represent two-dimensional space. The answer to this question is that, in order to maintain the feasibility of the basic operations (addition and multiplication) and their complementary operations (subtraction and division), this can be done only for certain multiplets. Specifically, we can use quartets of real numbers to represent four-dimensional spaces; these numbers are called **quaternions**. We can also use octets of real numbers, called **octonions**, to represent eight-dimensional spaces. This exhausts the possibilities, that is, no higher-order sets of numbers can be defined while maintaining the basic operations and their complementary ones. In fact, going to higher dimensions necessitates giving up fundamental properties of the basic operations: for the quaternions, we have to relinquish commutativity of multiplication and for the octonions we have to relinquish both commutativity and associativity of multiplication.

Quaternions and octonions are not often encountered in applications. Real numbers and complex numbers, on the other hand, are the foundation for a plethora of applications in science, engineering, and increasingly in the social sciences and life sciences.

1.2   Elementary functions

A real function $f(x)$ of the real number x is defined as the mathematical object which for a given value of the variable x takes some other real and *unique* value. The variable x is called the *argument* of the function. A function can be thought of as an *unambiguous* rule so that for a given value of the variable x this rule computes a unique $f(x)$.

1.2.1   Polynomials

A simple type of functions are those that involve integer powers of the variable x, called "polynomial functions":

$$f_0(x) = a, \quad a \in \mathbb{R}, \tag{1.1a}$$

$$f_1(x) = ax + b, \quad a, b \in \mathbb{R}, \tag{1.1b}$$

$$f_2(x) = ax^2 + bx + c, \quad a, b, c \in \mathbb{R}, \tag{1.1c}$$

$$f_3(x) = ax^3 + bx^2 + cx + d, \quad a, b, c, d \in \mathbb{R}, \tag{1.1d}$$

$$f_n(x) = a_0 + a_1 x + a_2 x^2 + \cdots + a_n x^n, \quad a_0, a_1, \ldots, a_n \in \mathbb{R}. \tag{1.1e}$$

The first is a constant function, the second is a linear function whose graph represents a line, the third is the quadratic polynomial (it represents a parabola), the last one is a general polynomial in powers of the variable x, called nth-order polynomial (from the largest power of x that appears in it).

It is very instructive, as will be discussed in more detail below, to plot functions with their values $f(x)$ on a vertical line (the "y axis") for a range of values of the variable x on a horizontal line (the "x-axis"); see Figs. 1.2 and 1.3 for the plots of the linear (first degree) and quadratic (second degree) polynomials, respectively. For the line, when $x = 0$, the value of the function $f_1(0)$ is $y = b$; this is the point where the line that represents the function $f_1(x)$ intersects the vertical axis.

The quadratic polynomial can also be written as

$$f_2(x) = a(x - x_0)^2 + y_0, \tag{1.2}$$

which gives the following relations between x_0, d and the parameters a, b, c:

$$x_0 = -\frac{b}{2a}, \quad y_0 = c - \frac{b^2}{4a} = -\frac{b^2 - 4ac}{4a}. \tag{1.3}$$

It is assumed that $a \neq 0$; otherwise $f_2(x)$ becomes a linear instead of a quadratic function. From the above relations, it follows that x_0 is the value where the function attains its extremum (minimum for $a > 0$ and maximum for $a < 0$), which is equal to $f_2(x_0) = y_0$.

Roots of polynomial functions: It is often useful to know the specific values of the argument x for which the function $f(x)$ vanishes. These values of x are called the "roots" of the function. For example, the root of the linear polynomial $f_1(x) = ax + b$ can be found by replacing $f_1(x)$ with zero; it is $x_0 = -b/a$. Here we must assume that $a \neq 0$, because otherwise we would have a constant function. The root is the point where the line that represents the function $f_1(x)$ intersects the horizontal axis on the xy-plane, and as long as $a \neq 0$ the linear function always has a root, as can be seen from the plot of the function in Fig. 1.2.

For the quadratic polynomial $f_2(x) = ax^2 + bx + c$, the roots are given by

$$x_1 = \frac{-b + \sqrt{b^2 - 4ac}}{2a}, \quad x_2 = \frac{-b - \sqrt{b^2 - 4ac}}{2a}. \tag{1.4}$$

Indeed, completing the square we obtain

$$ax^2 + bx + c = a\left(x^2 + \frac{b}{a}x + \frac{c}{a}\right) = a\left[\left(x + \frac{b}{2a}\right)^2 + \frac{c}{a} - \frac{b^2}{4a^2}\right].$$

Hence, setting the last expression to zero (with $a \neq 0$), we get

$$\left(x + \frac{b}{2a}\right) = \pm\left[\frac{b^2}{4a^2} - \frac{c}{a}\right]^{1/2},$$

and solving for x we obtain the expressions of Eq. (1.4). From Fig. 1.3, we see that whether or not the curve of $f_2(x)$ intersects the horizontal axis depends on the value of y_0, which is given by

$$y_0 = -\frac{b^2 - 4ac}{4a} = -\frac{\Delta}{4a},$$

where we have defined $\Delta = b^2 - 4ac$; this quantity, called the "discriminant", plays an important role, as we elaborate below. For $y_0 < 0$ or $(\Delta/a) > 0$, the curve intersects the horizontal axis at two points (the two roots) while for

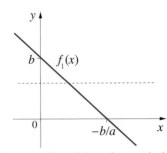

Figure 1.2: Plot of the polynomial of first degree, $f_1(x) = ax + b$ (line).

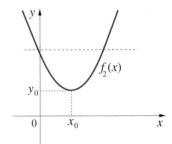

Figure 1.3: Plot of the polynomial of second degree, $f_2(x) = ax^2 + bx + c$ (parabola).

$y_0 > 0$ or $(\Delta/a) < 0$, it does not intersect the horizontal axis (there are no real roots) and for $y_0 = 0$ or $\Delta = 0$ it only touches the horizontal axis at one point (the two roots collapse to one). The first case is illustrated in Fig. 1.4. Thus, the discriminant Δ determines the type of roots of the quadratic equation: from Eq. (1.4), we see that the roots of the quadratic function are real numbers if $\Delta > 0$ and complex numbers if $\Delta < 0$, while for $\Delta = 0$ the two roots collapse into one value, a "double root". It is straightforward to establish the following relations between the two roots of the quadratic function:

$$x_1 + x_2 = -\frac{b}{a}, \quad x_1 x_2 = \frac{c}{a}. \tag{1.5}$$

For the cubic polynomial, $f_3(x) = ax^3 + bx^2 + cx + d$, the roots are given by the expressions:

$$x_1 = Q^{(+)} + Q^{(-)} - p, \tag{1.6a}$$

$$x_2 = -\frac{1}{2}\left[Q^{(+)} + Q^{(-)}\right] + \frac{i\sqrt{3}}{2}\left[Q^{(+)} - Q^{(-)}\right] - p, \tag{1.6b}$$

$$x_2 = -\frac{1}{2}\left[Q^{(+)} + Q^{(-)}\right] - \frac{i\sqrt{3}}{2}\left[Q^{(+)} - Q^{(-)}\right] - p, \tag{1.6c}$$

where we have defined the following quantities in terms of the coefficients a, b, c, d:

$$p \equiv \frac{b}{3a}, \quad q \equiv \frac{c}{2a}, \quad r \equiv \frac{d}{2a}, \tag{1.6d}$$

$$Q^{(\pm)} \equiv \left[(pq - p^3 - r) \pm \sqrt{(pq - p^3 - r)^2 + \left(\frac{2}{3}q - p^2\right)^3}\right]^{1/3}, \tag{1.6e}$$

and i is the imaginary unit, $i^2 = -1$. From the plot of this polynomial, Fig. 1.5, it is obvious that it always has one real root, since the function extends to $\pm\infty$, so that it always intersects the horizontal axis at least once. Depending on the values of the parameters, it may intersect the horizontal axis at three different points in which case it has three real roots (the case illustrated in Fig. 1.5), or at one point only, in which case it has one real and two complex roots. Working with complex numbers is the subject of Chapter 4, and underlies much of the rest of this book.

1.2.2 The inverse function

The inverse of a function is not necessarily a function: For example, for the function $f_0(x) = a$, where a is a constant, every value of the variable x gives the same value for the function, namely a; if we know the value of the argument we know the value of the function, which is consistent with the definition. However, if we know the value of the function $f_0(x)$ we cannot tell from which value of the argument it was obtained, since in this case all values of the argument give the same value for the function. In other case, this relation *can* be inverted, that is, if we know the value of the function $f(x) = y$, we can uniquely determine the value of the argument that produced it. This is exemplified by $f_1(x)$, Eq. (1.1b):

$$y = ax + b \Rightarrow x = \frac{y - b}{a}, \tag{1.7}$$

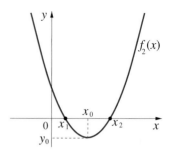

Figure 1.4: Plot of the polynomial of second degree, $f_2(x) = ax^2 + bx + c$, with parameter values that produce two real roots, labeled x_1, x_2.

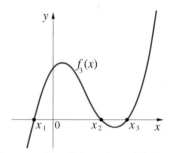

Figure 1.5: Plot of the polynomial of third degree, $f_3(x) = ax^3 + bx^2 + cx + d$, with parameter values that produce three real roots, labeled x_1, x_2, x_3.

assuming $a \neq 0$ (otherwise the function becomes a constant, just like Eq. (1.1a)). Thus, a function takes us **unambiguously** from the value of x to the value of $y = f(x)$, for all values of x in the *domain* of the function (the domain being the set of x values where the function is well defined); but this is not necessarily true in the opposite direction, that is, from y to x. If the relation can be inverted, then we can obtain x from the *inverse function*, denoted by f^{-1}, evaluated at the value of the original function y:

$$f(x) = y \Rightarrow x = f^{-1}(y). \tag{1.8}$$

This motivates the notion of a *mapping*. A function represents the mapping of the set of real values (denoted by the argument x) to another set of real values. A given value of the argument x gets mapped to a specific value denoted by $y = f(x)$. If each value of x gets mapped to one and only one value y, then the relation can be inverted (as in the second example, Eq. (1.1b)), otherwise it cannot (as in the first example, Eq. (1.1a)). Mapping point by point yields the familiar curves produced by several simple functions. These plots provide a straightforward way of determining whether the underlying relation can be inverted or not: Draw a horizontal line (parallel to the x axis, y: constant) that intersects the vertical axis at an arbitrary value within the set of values that the function can take; this value of the vertical axis corresponds to some value of the function. If this line intersects the plot of the function at only one point then the function can be inverted; otherwise it cannot. For example, such a horizontal line intersects the line at only one point, assuming its slope is not zero (see Fig. 1.2); so we can always find the inverse of the line. For the parabola, a horizontal line cuts it at one point for $y = y_0 = c - b^2/4a$ but at two points for $y > y_0$, and at no points for $y < y_0$; so the inverse of the parabola does not exist.

1.2.3   Geometric shapes: Circle, ellipse, hyperbola

Functions can also be interpreted as describing geometric shapes with specific properties. An example is the parabola: for the parabola defined by the quadratic function of Eq. (1.2), with $y = f_2(x)$, there is a special point F on the xy-plane called the **focus**, with coordinates $(x_f, y_f) = (x_0, y_0 + \alpha)$, where $\alpha = 1/4a$, and a special line called the **directrix**, described by the linear equation $y_d = y_0 - \alpha$. The parabola consists of all the points on the plane for which the distance from the focus is the same as the distance from the directrix, as shown in Fig. 1.6. This can be verified as follows: the square of the distance of the point $P = (x, y)$ from the focus is given by

$$(x - x_f)^2 + (y - y_f)^2 = (x - x_0)^2 + (y - y_0 - \alpha)^2.$$

The square of the distance of the point $P = (x, y)$ from the directrix is given by

$$(y - y_d)^2 = (y - y_0 + \alpha)^2.$$

Equating these two quantities and canceling the common terms on the two sides we find

$$(x - x_0)^2 - 2(y - y_0)\alpha = 2(y - y_0)\alpha \Rightarrow (y - y_0) = a(x - x_0)^2,$$

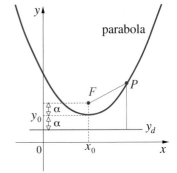

Figure 1.6: The parabola as a set of points P equidistant from the focus, a point at $F = (x_f, y_f)$ and the directrix, a horizontal line described by $y_d = y_0 - \alpha$.

which is the same as the quadratic equation for the parabola, Eq. (1.2), once we have substituted the value $\alpha = 1/4a$ and set $y = f_2(x)$.

Another familiar example of a geometric shape on a plane is the set of points (x, y) that are equidistant from a special point O, the "center"; these points constitute a circle of radius ρ, the distance of each point from the center, as shown in Fig. 1.7. If we assume that the center is situated at the point $O = (x_0, y_0)$, then the equation describing the circle of radius ρ is

$$\text{circle}: \quad \frac{(x - x_0)^2}{\rho^2} + \frac{(y - y_0)^2}{\rho^2} = 1. \tag{1.9}$$

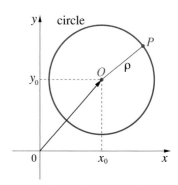

Figure 1.7: The circle as a set of points P at fixed distance ρ from the center at $O = (x_0, y_0)$.

This relationship does not correspond to a real function, because, solving for y from the above expression we get

$$y = y_0 \pm \sqrt{\rho^2 - (x - x_0)^2},$$

thus, for every value of x there are two possible values of y, not a unique one as the definition of a function requires (except at the points $x = x_0 \pm \rho$, for which there is only one solution, $y = y_0$).

Two other equations which involve similar relations between the x and y variables are the **ellipse** and the **hyperbola**. In the case of the ellipse, the relation between x and y takes the form

$$\text{ellipse}: \quad \frac{(x - x_0)^2}{a^2} + \frac{(y - y_0)^2}{b^2} = 1, \tag{1.10}$$

which is a generalization of the circle equation. The ellipse is defined as the set of points $P = (x, y)$ for which the sum of the distances from the two "foci", situated at $(x_f, y_f) = (\pm c - x_0, -y_0)$, is equal to $2a$. For the ellipse, we have the following relation between the parameters that describe its shape:

$$\text{ellipse}: \quad c^2 = a^2 - b^2.$$

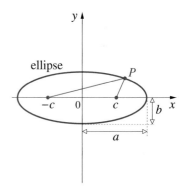

Figure 1.8: The ellipse as a set of points P for which the sum of distances (red lines) from the two foci which are located at $(\pm c, 0)$, is equal to $2a$ [the origin has been shifted to (x_0, y_0) for simplicity].

The corresponding plot for this curve is shown in Fig. 1.8. Evidently, for $a = b$ the ellipse becomes a circle of radius $\rho = a$. In this case, $c = 0$ and the two foci coincide into one point, the center of the circle. For this reason, it is convenient to define the "eccentricity" of the ellipse, which is a measure of how much it differs from a circle. The eccentricity e is defined as

$$e = \sqrt{1 - \frac{b^2}{a^2}}.$$

The circle has eccentricity $e = 0$.

In the case of the hyperbola, the relation between x and y is

$$\text{hyperbola}: \quad \frac{(x - x_0)^2}{a^2} - \frac{(y - y_0)^2}{b^2} = 1. \tag{1.11}$$

The hyperbola is defined as the set of points $P = (x, y)$ for which the difference of the distances from the two foci, situated at $(x_f, y_f) = (\pm c - x_0, -y_0)$, is equal to $2a$. For the hyperbola, we have the following relation between the parameters that describe its shape:

$$\text{hyperbola}: \quad c^2 = a^2 + b^2.$$

The corresponding plot for this curve is shown in Fig. 1.9.

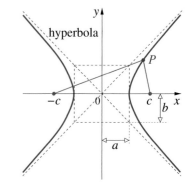

Figure 1.9: The hyperbola as a set of points P for which the difference of distances (red lines) from the two foci which are located at $(\pm c, 0)$ is equal to $2a$; there are two branches for the hyperbola [the origin has been shifted to (x_0, y_0) for simplicity].

1.2.4   Trigonometric functions

For an arbitrary point on the xy-plane which is defined by the values of two coordinates x (the "abscissa") and y (the "ordinate"), we can also define its position by its distance from the origin, which is equal to $\rho = \sqrt{x^2 + y^2}$ and the angle ϕ that the line from the origin to the point makes with the horizontal axis. By convention, we measure ϕ starting at the horizontal axis (corresponding to $\phi = 0$) and going around the circle in the counter-clockwise direction. The angle ϕ measured in radians can take values $0 \leq \phi < 2\pi$. Values beyond this range give the same result as values within the range that differ by 2π. This is the reason that one end of the interval needs to be included in the domain of values of ϕ, which by convention is the value 0, as shown in Fig. 1.10; ρ and ϕ are referred to as "polar" coordinates.

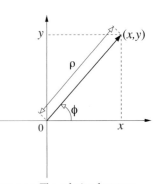

Figure 1.10: The relation between Cartesian coordinates (x, y) and polar coordinates (ρ, ϕ).

Two very important functions of this angle ϕ, called the "trigonometric" functions "cosine" and "sine", are defined as

$$\cos(\phi) \equiv \frac{x}{\sqrt{x^2 + y^2}} = \frac{x}{\rho}, \quad \sin(\phi) \equiv \frac{y}{\sqrt{x^2 + y^2}} = \frac{y}{\rho}. \tag{1.12}$$

Instead of dealing with the general situation as above, we consider a circle of radius $\rho = 1$ which is centered at the origin (the "unit circle"). The projections of a radius of the unit circle making an angle ϕ with respect to the horizontal axis onto the vertical and horizontal axes are equal to the cosine and the sine of the angle, respectively, as shown in Fig. 1.11.

Additional functions can be defined by the line segments where the ray at angle ϕ with respect to the horizontal axis intersects the two axes that are parallel to the vertical and horizontal axes and tangent to the unit circle; these are called the "tangent" ($\tan(\phi)$) and "cotangent" ($\cot(\phi)$) functions, respectively. These definitions are shown in Fig. 1.11. From the definitions, we can deduce

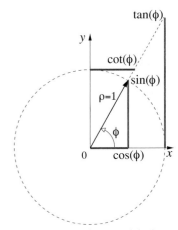

Figure 1.11: The definition of the basic trigonometric functions, $\sin(\phi)$, $\cos(\phi)$, $\tan(\phi)$, $\cot(\phi)$, in the unit circle.

$$\tan(\phi) = \frac{\sin(\phi)}{\cos(\phi)}, \quad \cot(\phi) = \frac{\cos(\phi)}{\sin(\phi)}. \tag{1.13}$$

The sine and cosine functions satisfy the important relation

$$\sin^2(\phi) + \cos^2(\phi) = 1, \tag{1.14}$$

as can be verified from their general definition, Eq. (1.12). The values that the sine and cosine take for the full range of their argument ϕ are shown in Fig. 1.12.

From their definition, it is evident that the trigonometric functions are periodic in the argument ϕ with a period of 2π, that is, if the argument of any of these functions is changed by an integer multiple of 2π, the value of the function does not change:

$$\sin(\phi + 2k\pi) = \sin(\phi), \quad \cos(\phi + 2k\pi) = \cos(\phi), \quad k \text{ integer}. \tag{1.15}$$

Moreover, using geometric arguments a number of useful relations between trigonometric functions of different arguments can be established, from which additional relations are readily derived, for example:

$$\cos(\phi_1 + \phi_2) = \cos(\phi_1)\cos(\phi_2) - \sin(\phi_1)\sin(\phi_2) \tag{1.16a}$$

$$\Rightarrow \quad \cos(2\phi) = \cos^2(\phi) - \sin^2(\phi), \tag{1.16b}$$

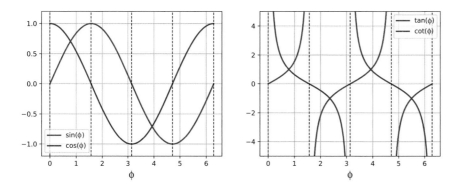

Figure 1.12: Plots of the trigonometric functions $\cos(\phi), \sin(\phi), \tan(\phi), \cot(\phi)$ for $0 \leq \phi < 2\pi$; the vertical dashed lines indicate the values $\phi = 0, \pi/2, \pi, 3\pi/2, 2\pi$.

and

$$\sin(\phi_1 + \phi_2) = \cos(\phi_1)\sin(\phi_2) + \sin(\phi_1)\cos(\phi_2) \qquad (1.16c)$$
$$\Rightarrow \quad \sin(2\phi) = 2\cos(\phi)\sin(\phi). \qquad (1.16d)$$

Instead of providing proofs for these relations here in geometric terms, which can be quite cumbersome, we postpone these proofs to Chapter 4, when the use of complex numbers makes them simpler and elegant. Specifically, it will be shown there that $\sin(\phi)$ and $\cos(\phi)$ can be expressed in terms of exponential functions of a complex argument. Using these expressions, it follows that both algebraic and differential relations involving trigonometric functions can be established using manipulations of exponential functions.

1.2.5 Exponential, logarithm, hyperbolic functions

Another very useful function is the exponential $\exp(x) = e^x$, defined as a limit of a simpler function that involves powers of x:

$$f(x) = e^x = \lim_{N \to \infty} \left[1 + \frac{x}{N}\right]^N, \quad N \in \mathbb{N}. \qquad (1.17)$$

It is obvious from the plot of the exponential function, Eq. (1.17), that e^x can be inverted to produce another function. In fact, the inverse of the exponential function is the familiar logarithm:

$$f(x) = e^x \Rightarrow f^{-1}(x) = \ln(x), \qquad (1.18)$$

as shown in Fig. 1.13. Actually, it is easy to see from these plots that the two functions correspond to the same curve, with the variables $x \leftrightarrow y$ interchanged. The domain of the exponential is the entire real axis (all real values of x),

$$\text{domain of } e^x : \quad x \in (-\infty, +\infty).$$

The function $\exp(x)$ takes only real *positive* values, thus the domain of the logarithm is only the positive real axis

$$\text{domain of } \ln(x) : \quad x \in (0, +\infty),$$

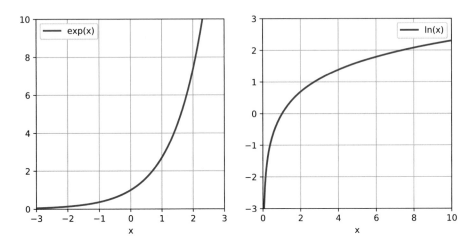

Figure 1.13: Plots of the exponential $\exp(x) = e^x$ and logarithm $\ln(x)$ functions. The graphical representation makes it evident that $\ln(x)$ is the inverse function of e^x, with the roles of the abscissa and the ordinate interchanged.

with the lower bound corresponding to the value of the exponential at the lowest value of its argument $(-\infty)$ and the upper bound corresponding to the value of the exponential at the highest value of its argument $(+\infty)$.

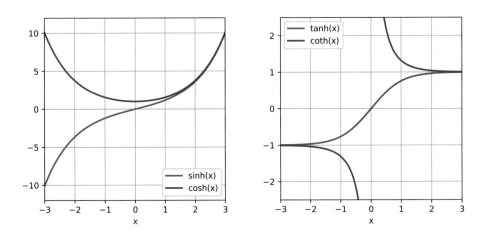

Figure 1.14: Plots of the hyperbolic functions $\cosh(x), \sinh(x), \tanh(x), \coth(x)$ for $-3 \le x \le 3$.

Other important functions that are defined through the exponential, are the hyperbolic sine ($\sinh(x)$) and hyperbolic cosine ($\cosh(x)$) functions:

$$\cosh(x) \equiv \frac{e^x + e^{-x}}{2}, \quad \sinh(x) \equiv \frac{e^x - e^{-x}}{2}. \tag{1.19}$$

Plots of the hyperbolic sine and cosine functions are shown in Fig. 1.14. These definitions imply that the hyperbolic sine and cosine functions satisfy the relation

$$\cosh^2(x) - \sinh^2(x) = 1, \tag{1.20}$$

which is reminiscent of the relation satisfied by the sine and cosine functions, Eq. (1.14). We can also define the hyperbolic tangent ($\tanh(x)$) and cotangent ($\coth(x)$) functions by analogy to the trigonometric sine and cosine functions:

$$\tanh(x) \equiv \frac{\sinh(x)}{\cosh(x)} = \frac{e^x - e^{-x}}{e^x + e^{-x}}, \quad \coth(x) \equiv \frac{1}{\tanh(x)}. \tag{1.21}$$

Plots of the hyperbolic tangent and cotangent functions are shown in Fig. 1.14. In contrast to the trigonometric functions, the exponential and the hyperbolic functions are *not* periodic functions of x.

1.3   Continuity and derivatives

In order to understand the behavior of a function $f(x)$, it is important to compute how fast or how slowly it changes when the variable x changes values. This is captured by the notion of the *derivative* of a function which precisely captures the rate of change of a function at a specific value of its argument.

For the rate of change to make sense, the function must be *continuous* at the value of the argument where we are trying to calculate this rate. The concept of *continuity* is an important one more generally, so we discuss it in some detail before discussing derivatives.

1.3.1   Continuity

Broadly speaking, a function is called *continuous* if its values do not make any jumps as the values of its argument are varied smoothly within its domain; otherwise it is called *discontinuous*. A jump in the value of the function is called a "discontinuity". The jump can be finite or infinite.

To make this notion more precise, we require that as the value of the argument x approaches some specific value x_0, the value of the function $f(x)$ approaches the value which it takes at x_0, that is $f(x_0)$. In mathematical terms, this is expressed as follows: For every value of $\epsilon > 0$, no matter how small, there exists a value of $\delta > 0$ such that when x is closer than δ to the specific value x_0, namely $x_0 - \delta < x < x_0 + \delta$, then $f(x)$ is closer than ϵ to the corresponding value $f(x_0)$, namely $f(x_0) - \epsilon < f(x) < f(x_0) + \epsilon$. This is expressed more compactly as

$$|x - x_0| < \delta \Rightarrow |f(x) - f(x_0)| < \epsilon. \tag{1.22}$$

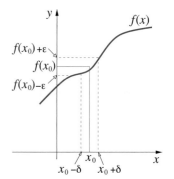

Figure 1.15: Illustration of the concept of continuity: when x is closer than δ to x_0, $f(x)$ gets closer than ϵ to $f(x_0)$.

For a function which is continuous near x_0, we can always find values of $f(x)$ arbitrarily close to $f(x_0)$, provided that the value of the argument x is sufficiently close to x_0. The essence of this argument is the following: we want to make sure that $f(x)$ is within a very narrow range of $f(x_0)$ as x approaches x_0, so we try to make sure that the difference $|f(x) - f(x_0)|$ is bounded by a very small number ϵ as x gets very close to x_0. All we need to show is that for x close enough to x_0, that is, $|x - x_0| < \delta$, then indeed $|f(x) - f(x_0)| < \epsilon$; namely, we must show that we can choose the values of δ and ϵ so that this last condition is satisfied. These ideas are illustrated schematically in Fig. 1.15.

Example 1.1: An example of a function which is is not continuous at some point is the function

$$f(x) = \frac{1}{x},$$

which is discontinuous in the neighborhood of $x_0 = 0$, since $f(x)$ is not defined at $x_0 = 0$. Actually, when $x \to 0^+$ the value of $f(x) \to +\infty$, and when $x \to 0^-$ the value of $f(x) \to -\infty$. Therefore, very close to 0, if we change the argument of the function by some amount δ we can jump from a very large positive value of $f(x)$ to a very large negative value. Therefore, this function is discontinuous at $x_0 = 0$ with an infinite discontinuity. Note that it is continuous at every other real value of x.

To show this explicitly, we take the values of δ and ϵ to be related by

$$\delta = \frac{\epsilon|x_0|^2}{1+\epsilon|x_0|} \Leftrightarrow \epsilon = \frac{\delta}{(|x_0|-\delta)|x_0|},$$

that is, once we have defined the value of ϵ that bounds the value of $f(x)$ to the value $f(x_0)$, we obtain the value of δ from the above expression: for x bounded by δ near x_0, $f(x)$ will be bounded by ϵ near $f(x_0)$. Notice that for a finite value of x_0, we can always assume that $0 < \delta < |x_0|$ since δ can be arbitrarily small, so that both δ and ϵ are positive. We then consider what happens for $|x - x_0| < \delta$: First, we have the relations:

$$|x_0| = |x_0 - x + x| \le |x_0 - x| + |x| < \delta + |x|$$
$$\Rightarrow |x| > |x_0| - \delta \Rightarrow \frac{1}{|x|} < \frac{1}{|x_0| - \delta},$$

where we have used the triangle inequality (see Problem 1) to obtain the first inequality above. Using the last relationship together with $|x - x_0| < \delta$, we obtain:

$$\epsilon = \frac{\delta}{(|x_0|-\delta)|x_0|} > \frac{|x-x_0|}{|x||x_0|} = \left|\frac{x-x_0}{xx_0}\right| = \left|\frac{1}{x} - \frac{1}{x_0}\right| = |f(x) - f(x_0)|.$$

Thus, with our choice of δ and ϵ we have managed to satisfy the continuity requirement, Eq. (1.22).

1.3.2   Definition of derivatives

The derivative of $f(x)$ at the point x_0 denoted by $[df/dx](x_0)$ or $f'(x_0)$ is defined by

$$f'(x_0) = \frac{df}{dx}(x_0) \equiv \lim_{x \to x_0} \frac{f(x) - f(x_0)}{x - x_0}. \tag{1.23}$$

When the function is continuous at $x = x_0$ this limit gives a finite value, that is, $f'(x_0)$ is well defined. If this holds for all values of x in a domain, then the derivative is itself a proper function in this domain. Conversely, if the function is discontinuous, the numerator in the limit can be finite (or even infinite) as the denominator tends to zero, which makes it impossible to assign a precise value to the limit, so that the derivative does not exist. As an example, the derivative of the function $1/x$ is

$$\frac{d(1/x)}{dx}(x_0) = \lim_{x \to x_0} \frac{(1/x - 1/x_0)}{x - x_0} = \lim_{x \to x_0} \frac{(x_0 - x)}{(x - x_0)xx_0} = -\frac{1}{x_0^2}, \tag{1.24}$$

which holds for every value of x_0, except $x_0 = 0$, where the function is discontinuous. Thus, the derivative of $f(x) = 1/x$ is $f'(x) = -1/x^2$ for all $x \ne 0$.

The derivatives of some important functions are given below.

Powers of x:

$$f(x) = x^a \quad \to \quad f'(x) = ax^{a-1}, \quad a \in \mathbb{R}. \tag{1.25}$$

The logarithm and the exponential functions:

$$f(x) = \ln(x) \quad \to \quad f'(x) = \frac{1}{x}, \quad x > 0, \tag{1.26a}$$

$$f(x) = e^x \quad \to \quad f'(x) = e^x. \tag{1.26b}$$

The trigonometric sine and cosine functions:

$$f(x) = \cos(x) \quad \rightarrow \quad f'(x) = -\sin(x), \tag{1.27a}$$

$$f(x) = \sin(x) \quad \rightarrow \quad f'(x) = \cos(x). \tag{1.27b}$$

The hyperbolic sine and cosine functions:

$$f(x) = \cosh(x) \quad \rightarrow \quad f'(x) = \sinh(x), \tag{1.28a}$$

$$f(x) = \sinh(x) \quad \rightarrow \quad f'(x) = \cosh(x). \tag{1.28b}$$

Interestingly, the exponential function is its own derivative, Eq. (1.26b), and the trigonometric functions $\sin(x)$ and $\cos(x)$ are derivatives of each other, with a minus sign involved, Eq. (1.27).

The notion of the derivative can be extended to higher-order derivatives. For instance, in the definition of the second derivative, denoted as $f''(x)$, the role previously played by the function in now played by the first derivative.[1] Specifically, the second derivative of the function $f(x)$ at x_0 is given by

$$f''(x_0) = \frac{d^2 f}{dx^2}(x_0) = \lim_{x \to x_0} \frac{f'(x) - f'(x_0)}{x - x_0}. \tag{1.29}$$

Using the first derivatives of the trigonometric functions $\sin(x)$ and $\cos(x)$, Eq. (1.27), it follows that their second derivatives are the negative of themselves:

$$f(x) = \cos(x) \quad \rightarrow \quad f''(x) = -\cos(x), \tag{1.30a}$$

$$f(x) = \sin(x) \quad \rightarrow \quad f''(x) = -\sin(x). \tag{1.30b}$$

For higher derivatives, it becomes awkward to keep adding primes, so the notation $f^{(n)}$ is often used for $d^n f / dx^n$ for the nth derivative.

[1] Using the prime symbol for derivatives is convenient for the first and second derivatives but becomes unwieldy in more complicated situations like higher-order derivatives and partial derivatives of functions of many variables. In Chapter 2, we introduce a more compact and convenient notation for these situations, which is employed extensively in the rest of the book.

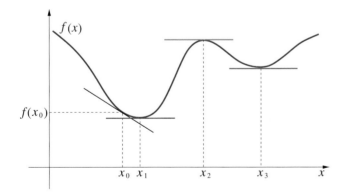

Figure 1.16: Geometric interpretation of the derivative as the slope of the tangent of $f(x)$ at $x = x_0$ (black line), at two minima at $x = x_1$, and $x = x_3$, where $f'(x_1) = 0, f''(x_1) > 0$, and $f'(x_3) = 0, f''(x_3) > 0$ (red lines), and at a maximum at $x = x_2$ where $f'(x_2) = 0, f''(x_2) < 0$ (green line).

1.3.3   Geometric interpretation of derivatives

The geometric interpretation of the derivative of the function $f(x)$ is the following: $f'(x)$ provides the slope of the tangent of the curve that corresponds to $f(x)$ at the point x. This is justified by the fact that the line described by the equation

$$y(x) = f(x_0) + f'(x_0)(x - x_0) = f'(x_0)x + [f(x_0) - x_0 f'(x_0)] = ax + b,$$

with $a = f'(x_0)$ and $b = [f(x_0) - x_0 f'(x_0)]$, passes through the point $(x = x_0, y = f(x_0))$, that is, $y(x_0) = f(x_0)$, and has a slope of

$$a = \frac{y(x) - y(x_0)}{x - x_0} = f'(x_0),$$

as illustrated in Fig. 1.16. Based on this argument, we conclude that for x sufficiently close to x_0, the dominant behavior of the function $f(x)$ is linear in the quantity $(x - x_0)$, with the constant multiplying this quantity being equal to $f'(x_0)$:

$$f(x) \approx f(x_0) + f'(x_0)(x - x_0), \quad \text{for} \quad x \to x_0. \tag{1.31}$$

Indeed, all other terms involve higher powers of $(x - x_0)$ thus are negligible compared to the first power, provided that $|x - x_0|$ is sufficiently small. This is based on the assumption that only powers of $(x - x_0)$ may be used to approximate the function for $x \to x_0$, when the function is continuous and smooth near this point.

What happens if the slope is zero at some value $x = x_0$? In this case, the value of the function does not change as x moves infinitesimally away from x_0 to either higher or lower values of x, since, as argued above, for x close enough to x_0 the function must be proportional to $f'(x_0)(x - x_0)$ and we are assuming here that $f'(x_0) = 0$. Therefore, at this point the function must have an *extremum*, that is, a minimum or a maximum. To determine the type of this extremum one must examine the second derivative. Employing the same logic used to establish that the function $f(x)$ for $x \to x_0$ becomes linear in $(x - x_0)$ with a coefficient $f'(x_0)$, it follows that when this coefficient is zero, the function must be quadratic in $(x - x_0)$,

$$\text{if } f'(x_0) = 0 \Rightarrow f(x) \approx f(x_0) + c(x - x_0)^2, \quad \text{for } x \to x_0.$$

To establish the value of c, we consider the second derivative of the function at this point, $f''(x_0)$, and the second derivative of its approximate value, which must be equal. This leads to

$$2c = f''(x_0) \Rightarrow c = \frac{1}{2}f''(x_0).$$

The net result is that, if the first derivative vanishes at some point x_0, then sufficiently close to this point, the function is approximately given by the expression below:

$$\text{if } f'(x_0) = 0 \Rightarrow f(x) \approx f(x_0) + \frac{1}{2}f''(x_0)(x - x_0)^2, \quad \text{for} \quad x \to x_0. \tag{1.32}$$

We can use this result with the earlier result for the case of $f'(x_0) \neq 0$ to write the function, close enough to x_0, as

$$f(x) \approx f(x_0) + f'(x_0)(x - x_0) + \frac{1}{2}f''(x_0)(x - x_0)^2, \quad \text{for} \quad x \to x_0, \tag{1.33}$$

because, sufficiently close to x_0, the first power of $(x - x_0)$ dominates over the second one, so if $f'(x_0) \neq 0$ the function is linear in $(x - x_0)$ and the second term (the quadratic one) can be neglected; while if $f'(x_0) = 0$, only the quadratic term survives, as already derived above.

We can now use the sign of the second derivative to decide what kind of an extremum is the value that corresponds to $f'(x_0) = 0$. If $f''(x_0) > 0$ then the value of the function *increases* when we go away from x_0 in either the positive, $x > x_0$, or the negative, $x < x_0$, direction; hence at x_0 the function attains a *minimum* value. Similarly, if $f''(x_0) < 0$, then the value of the function *decreases* when we go away from x_0, in either the positive, $x > x_0$, or the negative, $x < x_0$, direction; hence at x_0 the function attains a *maximum* value. These features are illustrated in Fig. 1.16.

If the second derivative also happens to be zero at x_0, then we can extend the same logic to write the function in terms of its third derivative times a cubic term, and so on. However, this quickly becomes a tedious procedure. The generalization of this idea gives rise to the so-called Taylor series representation of the function, a topic discussed in detail in Chapter 3 and revisited throughout the book.

1.3.4 Finding roots by using derivatives

Finding the roots of a function can be difficult. For instance, as the discussion of Section 1.2.1 made clear, the roots of a polynomial function become increasingly complicated expressions of the coefficients as the degree of the polynomial increases. Things can be even more complicated for functions that are not polynomials but involve other elementary functions. In many cases, the only recourse for finding roots is a numerical approach. The conceptual basis for such approaches is the geometric interpretation of the derivative as the slope of the tangent to the function.

A powerful numerical approach for finding simple roots, known as "Newton's method" (also referred to as "Newton–Raphson" method), relies on making a guess for the root and then improving it iteratively using the derivative of the function in the neighborhood of the root. We call our first guess for the root x_1 and assume that for this value the function is close to zero but not exactly zero: $f(x_1) \neq 0$. If $f(x_1) > 0$, we should follow the function to lower values until we find a zero; conversely, if $f(x_1) < 0$ we should follow the function to higher values until we find a zero. However, this is difficult to do, either numerically or analytically. The next best thing we can do is to take advantage of the linearity of the function near x_1, which gives a good approximation for the function itself, provided that $f'(x_1) \neq 0$. We know that if we are *close* but not *at* the root of the function, then following the direction of the *tangent* will get us closer to the root, by moving downward along the tangent (to lower values) if $f(x_1) > 0$ or upward (to higher values) if $f(x_1) < 0$. This has now become an easy operation because the tangent is a straight line. We therefore define the line, denoted as $y_1(x)$, that passes through the point $(x_1, f(x_1))$ and has slope $f'(x_1)$:

$$y_1(x) = f(x_1) + (x - x_1)f'(x_1),$$

which is the tangent of $f(x)$ at $x = x_1$. The line $y_1(x)$ intersects the horizontal axis at the point x_2, that is,

$$y_1(x_2) = 0 \Rightarrow f(x_1) + (x_2 - x_1)f'(x_1) = 0 \Rightarrow x_2 = x_1 - \frac{f(x_1)}{f'(x_1)}.$$

If $f(x_2) = 0$ then we have succeeded in finding the root of $f(x)$, in other words, the desired value of x is $x_0 = x_2$. If $f(x_2) \neq 0$ and $f'(x_2) \neq 0$, then we use x_2 as our improved guess for the root, and repeat the above procedure: we consider the line $y_2(x)$ that passes through the point $(x_2, f(x_2))$ and has slope $f'(x_2)$,

$$y_2(x) = f(x_2) + (x - x_2)f'(x_2),$$

which intersects the horizontal axis at the point x_3, that is,

$$y_2(x_3) = 0 \Rightarrow f(x_2) + (x_3 - x_2)f'(x_2) = 0 \Rightarrow x_3 = x_2 - \frac{f(x_2)}{f'(x_2)},$$

with x_3 our improved guess for the root of $f(x)$. We see that our process consists of successive approximations of the root by

$$x_{k+1} = x_k - \frac{f(x_k)}{f'(x_k)}, \quad k = 1, 2, 3, \ldots \tag{1.34}$$

If we start with a reasonable guess for the root, this procedure quickly converges to the correct value for the root, as shown schematically in Fig. 1.17. The iterations stop when $|f(x_k)| < \epsilon$, for a predetermined tolerance ϵ. The above procedure fails if any of the points $x_k, k = 1, 2, 3, \ldots$ is an extremum because in this case $f'(x_k) = 0$. If the extremum is also a root, then it must be a higher order root.

1.3.5 Chain rule and implicit differentiation

Two useful relations in taking derivatives are the product rule for two functions of x, and the chain rule for a function of a function of x, also referred to as "implicit differentiation":

$$\text{product rule} : \frac{d}{dx}[f(x)g(x)] = f'(x)g(x) + f(x)g'(x), \tag{1.35}$$

$$\text{chain rule} : \frac{d}{dx}f(g(x)) = f'(g(x))g'(x). \tag{1.36}$$

In the last expression, the notation $f'(y)$ implies the derivative of the function f with respect to its argument y, ignoring for this differentiation the fact that y is itself a function of x.

The chain rule can be memorized as follows: differentiate $f(g(x))$ treating $g(x)$ as if it were x, and then multiply by $g'(x)$, the derivative of $g(x)$ with respect to x. For example, in order to compute the derivative of $(\sin(x))^3$, treating $\sin(x)$ as if it were x, we obtain $3(\sin(x))^2$; then we multiply by the derivative of $\sin(x)$ which is $\cos(x)$. Hence,

$$\frac{d}{dx}(\sin(x))^3 = 2(\sin(x))^2\cos(x).$$

This technique can be used to differentiate any function composed of functions that we know their derivatives. For example,

$$\frac{d}{dx}\left(\cos(e^{x^2})\right)^4 = 4\left(\cos(e^{x^2})\right)^3\left[-\sin(e^{x^2})\right]\left[e^{x^2}2x\right],$$

where we have used the derivatives of x^4, $\cos(x)$, e^x and x^2, which are $4x^3$, $-\sin(x)$, e^x and $2x$, respectively.

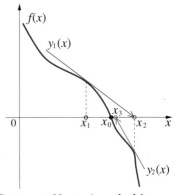

Figure 1.17: Newton's method for finding the value x_0 at which the function $f(x)$ vanishes, $f(x_0) = 0$, by starting from an initial guess x_1, and moving along the tangent at each step, $y_1(x), y_2(x), \ldots$, in successive iterations based on Eq. (1.34).

A useful application of the chain rule is the calculation of the derivative of $1/g(x)$. Since the derivative of x^{-1} is $-x^{-2}$, using the chain rule we find

$$\frac{d}{dx}\left(g(x)\right)^{-1} = -\frac{1}{(g(x))^2}g'(x). \tag{1.37}$$

Another useful application that employs both the chain rule and the product rule is the calculation of the derivative of $h(x)/g(x)$. From the product rule, Eq. (1.35), we get

$$\frac{d}{dx}\left[\frac{h(x)}{g(x)}\right] = h'(x)\left(g(x)\right)^{-1} + h(x)\frac{d}{dx}\left(g(x)\right)^{-1},$$

and using Eq. (1.37) the above expression becomes

$$\frac{d}{dx}\left[\frac{h(x)}{g(x)}\right] = h'(x)\frac{1}{g(x)} - h(x)\frac{1}{[g(x)]^2}g'(x) = \frac{h'(x)g(x) - h(x)g'(x)}{[g(x)]^2}. \tag{1.38}$$

1.4   Integrals

The notion of the *integration* of the values of function $f(x)$ over a range of values of its argument involves the summation of the values of the function at certain values of the variable x, multiplied by the infinitesimal step by which the values of the variable differ, denoted by dx, with the understanding that this interval tends to zero.

1.4.1   The definite and indefinite integrals

We define a function $F(x)$ such that its infinitesimal increment at x, denoted by $dF(x)$ is equal to $f(x)$ multiplied by dx, the infinitesimal increment in the variable x:

$$dF(x) = f(x)dx \Rightarrow F'(x) \equiv \frac{dF}{dx}(x) = f(x). \tag{1.39}$$

From the last relation, it is evident that finding the function $F(x)$ is equivalent to determining a function whose derivative is equal to $f(x)$; for this reason, $F(x)$ is also referred to as the *antiderivative* of $f(x)$. Note that because the derivative of a constant function is zero, $F(x)$ is defined up to an unspecified constant.

Now consider the summation of all the values that $dF(x)$ takes starting at some initial point a and ending at a final point b. Since these are successive infinitesimal increments in the value of $F(x)$, this summation will give the total difference ΔF of the values of $F(x)$ between the endpoints:

$$\Delta F = F(b) - F(a) = \sum_{x=a}^{b} dF(x).$$

Because the last expression involves an infinite number of infinitesimal quantities, it is not a proper summation; for this reason, we refer to it by a different name, the **definite integral**, and we denote it by the symbol "$\int$" instead of the summation symbol:

$$\int_a^b dF(x) = \int_a^b f(x)dx = F(b) - F(a). \tag{1.40}$$

The relation between $f(x)$ and the values of $F(x)$ at $x = a$ and $x = b$ is referred to as the "fundamental theorem of calculus". If we omit the limits on the integration symbol, we call the resulting quantity the **indefinite integral**. This simply involves finding the antiderivative of the function $f(x)$ that appears under the "$\int$" symbol:

$$F(x) = \int f(x) \mathrm{d}x. \tag{1.41}$$

In the definite integral, the difference between the values at the endpoints of integration cancels the unspecified constant involved in $F(x)$. The antiderivatives of some important functions (omitting the unspecified constant) are given below.

Powers of x, except x^{-1}:

$$f(x) = x^a \rightarrow F(x) = \frac{1}{a+1}x^{a+1} \; (x > 0, \, a \text{ real}, \, a \neq -1). \tag{1.42}$$

The x^{-1} power and the exponential functions:

$$f(x) = \frac{1}{x} \quad \rightarrow \quad F(x) = \ln(x) \; (x > 0), \tag{1.43a}$$

$$f(x) = e^x \quad \rightarrow \quad F(x) = e^x. \tag{1.43b}$$

The trigonometric sine and cosine functions:

$$f(x) = \sin(x) \quad \rightarrow \quad F(x) = -\cos(x), \tag{1.44a}$$

$$f(x) = \cos(x) \quad \rightarrow \quad F(x) = \sin(x). \tag{1.44b}$$

The hyperbolic sine and cosine functions:

$$f(x) = \sinh(x) \quad \rightarrow \quad F(x) = \cosh(x), \tag{1.45a}$$

$$f(x) = \cosh(x) \quad \rightarrow \quad F(x) = \sinh(x). \tag{1.45b}$$

The above equations can be obtained by inverting the relations mentioned earlier, Eqs. (1.25)–(1.27).

We can express the integral in terms of a usual sum of finite quantities if we break the interval $b - a$ into N equal parts and then take the limit of $N \rightarrow \infty$:

$$x_j = a + j\Delta x, \; j = 0, \dots, N, \; x_0 = a, x_N = b, \; \Delta x = \frac{b-a}{N}, \tag{1.46}$$

$$\int_a^b f(x)\mathrm{d}x = \lim_{N \to \infty}\left[\sum_{j=0}^{N} f(x_j)\Delta x\right]. \tag{1.47}$$

From this definition we conclude that the definite integral is the area below the graph of the function from a to b; this is illustrated in Fig. 1.18. From this relation we can also infer that the definite integral will have a finite value if there are no infinities in the values of the function $f(x)$ in the range of integration, or if these infinities are of a special nature so that they cancel. Moreover, if the upper or lower limit of the range of integration is $\pm\infty$, then in order to obtain a finite value for the integral, the absolute value of $f(x)$ should vanish faster than $1/|x|$, that is, it should behave as $1/|x|^a$ with $a > 1$, as this limit is approached. A function $f(x)$ appearing under the integral sign is often referred to as the "integrand" of the integral.

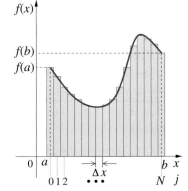

Figure 1.18: Geometric interpretation of the integral as the area below the graph of the function $f(x)$ from $x = a$ to $x = b$, in the limit of the spacing Δx between successive values of x going to zero; the function is evaluated at $x_j = a + j\Delta x, \; j = 0, \dots, N$, with $x_0 = a$, and $x_N = b$.

Finally, we consider a situation in which the limits of an integral are functions of some variable, t:

$$g(t) = \int_{a(t)}^{b(t)} f(x)dx.$$

We are interested in calculating the derivative of $g(t)$ with respect to t. Using the definition of the antiderivative, Eq. (1.41), we conclude that

$$g(t) = F(b(t)) - F(a(t)).$$

Taking the derivative of $g(t)$ with respect to its argument t and using the chain rule of differentiation, we obtain

$$g'(t) = F'(b)\frac{db}{dt} - F'(a)\frac{da}{dt}.$$

But from the definition of the antiderivative, we have:

$$F'(b) = \left[\frac{dF}{dx}\right]_{x=b} = f(b), \quad \text{and} \quad F'(a) = \left[\frac{dF}{dx}\right]_{x=a} = f(a),$$

which gives the formula

$$\frac{d}{dt}\left[\int_{a(t)}^{b(t)} f(x)dx\right] = f(b)\frac{db}{dt} - f(a)\frac{da}{dt}. \tag{1.48}$$

If one of the limits is a value independent of the variable t, then the corresponding derivative with respect to t vanishes.

1.4.2   Integration by change of variables

Often, an integral can be recast into a different one by changing the variable of integration into a new variable. In this type of procedure, care must be taken to introduce the proper changes in the infinitesimal under the integral and the limits of integration, all of which depend on the choice of the new variable. Specifically, suppose we want to compute the definite integral

$$\int_a^b f(x)dx,$$

and we introduce a change of variables from the original variable x to the new variable t, which are related by

$$x = g(t) \Rightarrow dx = \frac{dg(t)}{dt}dt.$$

Then the integral takes the form

$$\int_a^b f(x)dx = \int_{\tilde{a}}^{\tilde{b}} f(g(t))\frac{dg(t)}{dt}dt, \quad \tilde{a} = g^{-1}(a), \quad \tilde{b} = g^{-1}(b). \tag{1.49}$$

We illustrate this process by an example.

Example 1.2: Consider the definite integral

$$I = \int_0^\infty \frac{1}{x^2 + c^2}\, dx, \quad c > 0.$$

We note that the integrand is always finite over the range of integration and it vanishes as $\sim 1/x^2$ for $x \to \infty$, which suggests that this integral should give a finite value. The change of variables from x to t related by

$$x = c\tan(t) = c\frac{\sin(t)}{\cos(t)} \Rightarrow t = \tan^{-1}\left(\frac{x}{c}\right),$$

gives the following expressions in terms of the new variable t for the infinitesimal $\mathrm{d}x$:

$$\mathrm{d}x = c\frac{\mathrm{d}}{\mathrm{d}t}\left[\frac{\sin(t)}{\cos(t)}\right]\mathrm{d}t = c\left[\frac{\cos^2(t) + \sin^2(t)}{\cos^2(t)}\right]\mathrm{d}t = \frac{c}{\cos^2(t)}\,\mathrm{d}t,$$

where we have used the rules of differentiation, Eq. (1.38), and the derivatives of the trigonometric functions, Eq. (1.27), to obtain the final result. As for the limits of integration, we find

$$x = 0 \Rightarrow t = \tan^{-1}\left(\frac{0}{c}\right) = 0, \quad x \to \infty \Rightarrow t = \lim_{x\to\infty}\left[\tan^{-1}\left(\frac{x}{c}\right)\right] = \frac{\pi}{2},$$

as can be seen from the definition of the tangent function, see Fig. 1.12. We thus find that the above integral can be computed in closed form:

$$I = \frac{1}{c^2}\int_0^\infty \frac{1}{(x/c)^2 + 1}\,\mathrm{d}x = \frac{1}{c^2}\int_0^{\pi/2} \frac{1}{\tan^2(t) + 1}\frac{c}{\cos^2(t)}\,\mathrm{d}t = \frac{\pi}{2c}.$$

1.4.3   Integration by parts

A general technique that uses the rules of differentiation and integration discussed earlier is "integration by parts". Suppose that the integrand can be expressed as a product of two functions, one of which is a derivative. Then the associated integral can be written as follows:

$$\int_a^b f'(x)g(x)\mathrm{d}x = \int_a^b \left(\frac{\mathrm{d}}{\mathrm{d}x}[f(x)g(x)] - f(x)g'(x)\right)\mathrm{d}x$$

$$= [f(x)g(x)]_a^b - \int_a^b f(x)g'(x)\mathrm{d}x, \tag{1.50}$$

where we have used primes to denote first derivatives, we have employed the rule for differentiating a product, Eq. (1.35), and we have introduced the notation

$$[h(x)]_a^b = h(b) - h(a).$$

Taking into consideration that $f(x)$ is the integral of $f'(x)$, integration by parts can be memorized as follows: *the integral of a product of two functions is equal to the integral of the first function times the second evaluated at the integral limits, minus the integral of the function obtained earlier, times the derivative of the second function.*

   As we remarked in Section 1.3.2, the exponential function is its own derivative and the trigonometric functions $\sin(x)$ and $\cos(x)$ are derivatives of each other (with a minus sign). This makes them particularly useful functions for doing integration by parts. The following Example 1.3 illustrates this for the case of powers of x multiplied by the sine or the cosine function. For other useful examples, see Problems 10 and 11.

Example 1.3: An example of the integration by parts formula involves products of powers of x and sines or cosines of multiples of x. Consider the following definite integrals:

$$\int_0^\pi \cos(nx)x^k dx, \quad \int_0^\pi \sin(nx)x^k dx,$$

with k, n positive integers. Since the integral of $\cos(nx)$ is $\sin(nx)/n$ whereas the derivative of x^k is kx^{k-1}, for the first case above we find

$$\int_0^\pi \cos(nx)x^k dx = \left[\frac{\sin(nx)}{n}x^k\right]_0^\pi - \int_0^\pi \frac{\sin(nx)}{n}kx^{k-1}dx$$
$$= -\frac{k}{n}\int_0^\pi \sin(nx)x^{k-1}dx.$$

Since the integral of $\sin(nx)$ is $-\cos(nx)/n$ and the derivative of x^{k-1} is $(k-1)x^{k-2}$, the right-hand side becomes

$$-\frac{k}{n}\left[-\frac{\cos(nx)}{n}x^{k-1}\right]_0^\pi - \frac{k}{n}\int_0^\pi \frac{\cos(nx)}{n}(k-1)x^{k-2}dx,$$

which leads to the result

$$\int_0^\pi \cos(nx)x^k dx = \frac{k\pi^{k-1}(-1)^n}{n^2} - \frac{k(k-1)}{n^2}\int_0^\pi \cos(nx)x^{k-2}dx. \quad (1.51)$$

By similar steps we obtain:

$$\int_0^\pi \sin(nx)x^k dx = -\frac{\pi^k(-1)^n}{n} - \frac{k(k-1)}{n^2}\int_0^\pi \sin(nx)x^{k-2}dx. \quad (1.52)$$

The usefulness of these results lies in the fact that the final expression contains an integrand which is similar to the original one with the power of x reduced by 2. Applying these results recursively, we can evaluate any integral containing powers of x and sines or cosines.

1.4.4 The principal value integral

In the general case where the integrand has an infinite value at some point $x = a$ in the interval of integration $a \in [b, c]$, it is possible to define an expression called the "principal value integral" that exists despite the occurrence of infinity at $x = a$; this is expressed as follows:

$$\fint_b^c f(x)\,dx = \lim_{\epsilon \to 0}\left[\int_b^{a-\epsilon} f(x)\,dx + \int_{a+\epsilon}^c f(x)\,dx\right]. \quad (1.53)$$

Through this expression, we evaluate the function $f(x)$ in the neighborhood of $x = a$ *symmetrically* from the left (at $x = a - \epsilon$) and from the right (at $x = a + \epsilon$) and include these contributions in the limit $\epsilon \to 0$; note that the small quantity ϵ is taken to be always a *positive* number. The principal value integral is denoted by the special symbol used above, namely an integral sign with a dash. The infinities in the value of the integrand are referred to as "singularities"; an example of such a singularity is shown in Fig. 1.19 for the

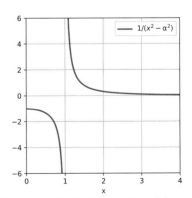

Figure 1.19: Plot of the function $f(x)$ defined in Eq. (1.54), whose integral is studied in Example 1.4; for this plot we have chosen $a = 1$.

function

$$f(x) = \frac{1}{x^2 - a^2},$$ (1.54)

which blows up in the neighborhood of $x = \pm a$: the function goes to $-\infty$ for $x \to a$ from below ($x = a - \epsilon$) and to $+\infty$ for $x \to a$ from above ($x = a + \epsilon$). In the following chapters, we deal in some detail with these types of situations with the use of tools from complex analysis.

Example 1.4: To demonstrate that the principal value integral only works when the limits are taken symmetrically at the singularity, we consider the following definite integral:

$$I = \fint_0^\infty \frac{1}{x^2 - a^2} \, dx, \quad a > 0.$$

The integrand has a singularity at $x = a$, as shown in Fig. 1.19, so we have written the integral as a principal value. Taking into consideration that

$$\frac{1}{x^2 - a^2} = \frac{1}{2a}\left(\frac{1}{x - a} - \frac{1}{x + a}\right),$$

we can write the above integral as

$$I = \frac{1}{2a}\fint_0^\infty \left(\frac{1}{x - a} - \frac{1}{x + a}\right) dx.$$

The second part of the above integrand is not singular, thus we only have to consider the principal value integral of the first part. However, each of the above integrands gives rise to integrals which diverge at infinity; thus, we must treat these integrals carefully, so that the contributions at infinity cancel. We define the integrals I_1, I_2

$$I = \frac{1}{2a}(I_1 - I_2), \quad I_1 = \lim_{b \to \infty}\left[\fint_0^b \frac{dx}{x - a}\right], \quad I_2 = \lim_{b \to \infty}\left[\int_0^b \frac{dx}{x + a}\right].$$

Computing the principal value integral with respect to $x = a$ we find:

$$I_1 = \fint_0^b \frac{dx}{x - a} = \lim_{\epsilon \to 0}\left[\int_0^{a - \epsilon} \frac{dx}{x + a} + \int_{a + \epsilon}^b \frac{dx}{x + a}\right]$$

$$= \lim_{\epsilon \to 0}\left[\ln(\epsilon) - \ln(a) + \ln(|b - a|) - \ln(\epsilon)\right]$$

$$= \ln(|b - a|) - \ln(a).$$

Furthermore,

$$I_2 = \int_0^b \frac{dx}{x + a} = \ln(|b + a|) - \ln(a).$$

Hence,

$$I = \frac{1}{2a}(I_1 - I_2) = \frac{1}{2a}\lim_{b \to \infty}\left[\ln\left(\frac{|b - a|}{b + a}\right)\right] = 0.$$

The above example is an illustration of difficulties that may arise in the evaluation of real definite integrals using conventional approaches. Contour integration on the complex plane, a method to be developed in following chapters, provides an elegant way of circumventing many of these difficulties.

1.5   Norms and moments of a function

Derivatives are very helpful in exploring the *local* properties of a function $f(x)$, that is, its behavior in the neighborhood of a specific value of the variable x, usually denoted as x_0. A manifestation of this is the Taylor expansion which expresses a function $f(x)$ in the neighborhood of x_0 in terms of powers of $(x - x_0)$ and the values of the function and its derivatives x at x_0, as mentioned in Section 1.3. We will discuss this concept in detail in later chapters and encounter it repeatedly throughout the book.

In contrast, integrals are very useful in exploring the *global* behavior of a function, that is, how it behaves over the entire range of the variable x. Two types of integrals that are helpful in exploring the global behavior of a function are the so-called "norms" and "moments" of the function.

1.5.1   The norms of a function

A function can be viewed as a "signal", in the sense that its values contain useful information at each value of the independent variable which may indicate position in space (typically denoted as x) or moment in time (typically denoted as t). In this sense, a useful piece of information is the sum of all of the function's *absolute values*, which gives a measure of the total signal "strength". The absolute value is needed because otherwise the positive and negative values of the function would cancel when summed, thus giving an erroneously low (possibly vanishing) value for the strength of the signal. Alternatively, we may consider the sum of an *even power* of the function's values, which accomplishes the same purpose in terms of avoiding the cancellation of positive and negative values, but gives a different measure of the signal's strength.

The different ways of measuring the total information contained in a function's values are referred to as "norms". These can be defined over the entire range of the independent variable or over a subset of values. The so-called L_1-norm of the function $f(x)$ over the range $x \in [a, b]$ is defined as:

$$\|f(x)\|_{L_1(a,b)} \equiv \int_a^b |f(x)| \, \mathrm{d}x. \tag{1.55}$$

Other norms of the function involve the lowest even powers of $f(x)$,

$$\|f(x)\|_{L_n(a,b)} \equiv \int_a^b [f(x)]^n \, \mathrm{d}x, \quad n \in \mathbb{N}, \, n : \text{ even.} \tag{1.56}$$

The most commonly used norms (other than the L_1 norm) are the L_2-norm ($n = 2$) and the L_4-norm ($n = 4$).

In certain situations, it is desirable to make sure that the norm of a function takes a certain value, which is referred to as the "normalization" of the function. This can be achieved by including a parameter in the definition of the function whose value is adjusted to satisfy the normalization condition. We describe a particularly useful example next.

1.5.2   *The normalized Gaussian function*

The exponential of $-x^2$, known as the "Gaussian function" appears in many different contexts, including the "normal" probability distribution (see Chapter 13) and the solution of the diffusion or heat equation (see Chapter 8). The domain of the Gaussian function is the entire real axis, but the function takes non-negligible values only for a relatively narrow range of values of $|x|$ near $x = 0$, because for larger values of $|x|$ the argument of the exponential becomes very large and negative, making the value of the function negligibly small. We can include a constant multiplicative factor in the exponential, usually chosen to be $1/2\sigma^2$,

$$f(x) = e^{-x^2/2\sigma^2}.$$

The inclusion of this factor makes it possible to adjust the range of the domain in which the Gaussian takes non-negligible values.

Since all values of the Gaussian function are non-negative, its L_1-norm is the same as its integral. It is often desirable to multiply the Gaussian with a "normalization factor" in order to make its L_1-norm over the entire domain exactly equal to unity. To obtain this normalization factor, we calculate the integral of the Gaussian over all values of the variable x. This integral is most easily computed by using Gauss' trick of taking the square root of the square of the integral:

$$\int_{-\infty}^{\infty} e^{-x^2/2\sigma^2} dx = \left[\int_{-\infty}^{\infty} e^{-x^2/2\sigma^2} dx \int_{-\infty}^{\infty} e^{-y^2/2\sigma^2} dy \right]^{\frac{1}{2}}$$

$$= \left[\int_{-\infty}^{\infty} \int_{-\infty}^{\infty} e^{-(x^2+y^2)/2\sigma^2} dx dy \right]^{\frac{1}{2}}.$$

The double integral in the square brackets can be calculated by turning the integration over the Cartesian coordinates (x, y) into an integration over the polar coordinates (ρ, ϕ), where

$$\rho = \sqrt{x^2 + y^2}, \quad \phi = \tan^{-1}\left(\frac{y}{x}\right)$$

(these transformations between coordinate systems will be studied systematically in Chapter 2). The relation between the infinitesimal area element in Cartesian and polar coordinates is given by $dx\, dy = \rho d\rho\, d\phi$. With these transformations, the original integral becomes:

$$\int_{-\infty}^{\infty} e^{-x^2/2\sigma^2} dx = \left[\int_{0}^{2\pi} \int_{0}^{\infty} e^{-\rho^2/2\sigma^2} \rho d\rho d\phi \right]^{\frac{1}{2}} = \left[2\pi \int_{0}^{\infty} e^{-\rho^2/2\sigma^2} \frac{1}{2} d\rho^2 \right]^{\frac{1}{2}}.$$

The remaining integral can then be calculated by the change of variables $t = \rho^2/2\sigma^2$, which yields:

$$\int_{-\infty}^{\infty} e^{-x^2/2\sigma^2} dx = \left[2\pi\sigma^2 \int_{0}^{\infty} e^{-t} dt \right]^{\frac{1}{2}} = \sqrt{2\pi}\sigma. \tag{1.57}$$

We notice that if we shift the variable x by a constant μ, that is, we consider the function

$$f(x) = e^{-(x-\mu)^2/2\sigma^2},$$

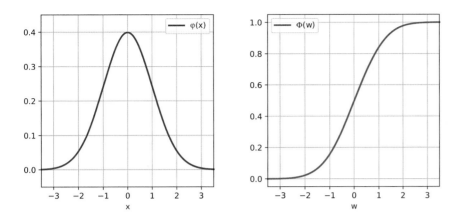

Figure 1.20: **Left**: The normalized Gaussian function, $f_g(x; \mu, \sigma)$, defined in Eq. (1.58), with peak at $\mu = 0$ and width $\sigma = 1$; this function is also known as the "standard normal distribution", denoted by $\varphi(x)$, Eq. (1.59). **Right**: The "cumulative normal distribution" function $\Phi(w)$, defined in Eq. (1.60a).

the integral over all real values x does not change; this shift only affects the position of the maximum of the function, which occurs at $x = \mu$.

Thus, the normalized Gaussian function whose peak occurs at μ and width is σ, denoted here as $f_g(x; \mu, \sigma)$, is given by

$$f_g(x; \mu, \sigma) = \frac{1}{\sqrt{2\pi}\sigma} e^{-(x-\mu)^2/2\sigma^2}. \tag{1.58}$$

The normalized Gaussian function for $\mu = 0$ and $\sigma = 1$, is referred to as the "standard normalized Gaussian" or "standard normal distribution". This function, which is shown in Fig. 1.20, is usually denoted by $\varphi(x)$:

$$\varphi(x) \equiv f_g(x; 0, 1) = \frac{1}{\sqrt{2\pi}} e^{-x^2/2}. \tag{1.59}$$

The function $\varphi(x)$ plays a central role in probability theory (see Chapter 13). In the limit of $\sigma \to 0$ the function $f_g(x; \mu, \sigma)$ tends to an infinitely sharp spike of infinite height which integrates to unity. This limit gives rise to a useful mathematical construct called a δ-function, which we will examine in detail later.

A function derived from the standard normalized Gaussian is its integral over a domain from $-\infty$ to a finite value w, defined as:

$$\Phi(w) = \int_{-\infty}^{w} \varphi(x) dx = \frac{1}{\sqrt{2\pi}} \int_{-\infty}^{w} e^{-x^2/2} dx. \tag{1.60a}$$

From the properties of the normalized Gaussian it is easy to show that an equivalent definition of the function $\Phi(w)$ is given by

$$\Phi(w) = \frac{1}{\sqrt{2\pi}} \int_{-w}^{\infty} e^{-x^2/2} dx, \tag{1.60b}$$

and that the function $\Phi(w)$ has the following properties:

$$\Phi(-w) = 1 - \Phi(w), \quad \Phi(0) = 0.5, \quad \Phi(\infty) = 1, \quad \Phi(-\infty) = 0. \tag{1.60c}$$

$\Phi(w)$ is an important function that represents the cumulative normal probability distribution of the standard normal distribution, as we will discuss in detail in Chapter 13. This function cannot be calculated analytically, so its values are usually obtained from tables (see Appendix B); a plot of $\Phi(x)$ is shown in Fig. 1.20. A very good approximation of its values can be obtained by using the concept of Taylor series expansion, as shown in Chapter 3, Section 3.2.1.

1.5.3 Moments of a function

A different way of gaining insight into the global behavior of a function is to calculate the so-called "moments" of the function. The moments provide information on how the values of the function $f(x)$ are *distributed* over the range of the variable x. For example, moments appear often in the context of probabilities, when the function of interest is the so-called "probability density function" (PDF).

The nth moment of the function $f(x)$ is defined by

$$M_n \equiv \int_{-\infty}^{\infty} x^n f(x) \mathrm{d}x. \tag{1.61}$$

The zeroth moment is simply the integral of the function,

$$M_0 = \int_{-\infty}^{\infty} f(x) \mathrm{d}x,$$

and is a measure of the **average** value $\bar{f}$ of the function.

In the case of a PDF, the function must satisfy the condition $f(x) \geq 0$, and the zeroth moment is equivalent to the L_1-norm of the function over its entire range. Since this represents the total probability, the so-called normalization condition, $M_0 = 1$, must also be satisfied, as in the case of the normalized Gaussian function discussed earlier.

The first moment,

$$\mu = M_1 = \int_{-\infty}^{\infty} x f(x) \mathrm{d}x, \tag{1.62}$$

is related to the **mean** value of the variable x.

The second moment,

$$M_2 = \int_{-\infty}^{\infty} x^2 f(x) \mathrm{d}x = \sigma^2 + \mu^2, \tag{1.63}$$

is related to the **variance**, σ^2, but is not equal to the variance (except for $\mu = 0$). For a normalized function, we find the following:

$$\int_{-\infty}^{\infty} (x - \mu)^2 f(x) \mathrm{d}x = \int_{-\infty}^{\infty} x^2 f(x) \mathrm{d}x + \mu^2 \int_{-\infty}^{\infty} f(x) \mathrm{d}x - 2\mu \int_{-\infty}^{\infty} x f(x) \mathrm{d}x = \sigma^2,$$

where we have used the facts that $\bar{f} = 1$ and the mean value is μ. In other words, the second moment *relative to the mean value μ*, is equal to the variance.

In the same spirit, other interesting moments are defined relative to the mean value: The third moment relative to the mean value is called the **skewness** of the function,

$$s = \int_{-\infty}^{\infty} (x - \mu)^3 f(x) \mathrm{d}x. \tag{1.64}$$

The fourth moment relative to the mean value is called the **kurtosis** of the function,

$$\kappa = \int_{-\infty}^{\infty} (x - \mu)^4 f(x) \mathrm{d}x. \tag{1.65}$$

The larger the number of moments of the function we know, the more information we have about the behavior of this function. If all moments ($n \to \infty$) are known, we can reconstruct the function everywhere in its domain.

Example 1.5: Moments of the normalized Gaussian

As an example, we calculate the moments of the normalized Gaussian function, discussed earlier, Eq. (1.58). We notice that the Gaussian is an even function of the variable x, namely $g(-x) = g(x)$, which implies that all *odd* moments of the Gaussian are zero, because the product of and odd power of x with $g(x)$ is an odd function. To facilitate the calculation of the even moments, we change the notation, namely, we define the parameter $\alpha = 1/2\sigma^2$ which allows us to write the Gaussian integral of Eq. (1.57) as:

$$\int_{-\infty}^{\infty} e^{-\alpha x^2} dx = \sqrt{\pi} \alpha^{-1/2},$$

and then simply take derivatives with respect to α of both sides of this last expression (note that x on the left-hand side of this expression is simply the variable of integration, and therefore *not* a variable on which the integral depends):

$$\int_{-\infty}^{\infty} x^2 e^{-\alpha x^2} dx = -\frac{d}{d\alpha} \int_{-\infty}^{\infty} e^{-\alpha x^2} dx = \frac{1}{2} \sqrt{\pi} \alpha^{-3/2},$$

$$\int_{-\infty}^{\infty} x^4 e^{-\alpha x^2} dx = -\frac{d}{d\alpha} \int_{-\infty}^{\infty} x^2 e^{-\alpha x^2} dx = \frac{3}{4} \sqrt{\pi} \alpha^{-5/2},$$

$$\int_{-\infty}^{\infty} x^{2n} e^{-\alpha x^2} dx = (-1)^n \frac{d^n}{d\alpha^n} \int_{-\infty}^{\infty} e^{-\alpha x^2} dx = \frac{(2n-1)!!}{2^n} \sqrt{\pi} \alpha^{-(2n+1)/2},$$

where the symbol $(2n-1)!!$ denotes the product of all odd integers up to $(2n-1)$. Substituting in these expressions the value of α in terms of σ, namely $\alpha = 1/2\sigma^2$, and using the definition of the normalized Gaussian $f_g(x; \mu, \sigma)$ from Eq. (1.58), we obtain the general formula:

$$\int_{-\infty}^{\infty} x^{2n} f_g(x; \mu, \sigma) dx = (2n-1)!! \sigma^{2n}, \quad n \in \mathbb{N}. \tag{1.66}$$

1.6   Generalized functions: θ-function and δ-function

As a final topic we discuss the concept of the so-called "generalized functions", or "distribution functions". The behavior of these functions is unusual, compared to the common smooth and continuous functions. A simple example of unusual behavior is a finite discontinuity. It is often useful to represent such functions by smooth and continuous functions in a certain limit of a parameter involved in the definition of the function. In this manner, we can apply all the familiar operations, like differentiation and integration, with the same ease as for smooth and continuous functions and then take the limit of the parameter. First, we provide a motivation why generalized functions can be extremely useful tools, and then discuss in some detail two specific examples, the θ-function, or step function, and the δ-function.

In many physical problems one is interested only in specific values that a function takes for certain values of its argument. A simple, intuitive example is a filter. Suppose we are dealing with a function $s(\omega)$ that represents a signal s which is a function of the frequency ω; for instance, $s(\omega)$ could represent the intensity of the sound produced by a musical instrument as a function of the frequency. Suppose also that we want to focus on the signal for only certain values of the frequency, associated with a particular range of notes. We can accomplish this "filtering" of the signal by multiplying $s(\omega)$ with another function that is zero everywhere else except when the frequency takes the particular value or range of values we are interested in. For instance, we might want to filter out the values of the signal below a certain frequency, like in a *high-pass filter* which allows only high frequencies to pass. In this case we want the filter function to take the value 0 for values of $\omega < \omega_0$ and to take the value 1 for $\omega > \omega_0$. This function is called "theta function" or Heaviside function, denoted as $\theta(x - x_0)$. Another possible situation is that we are interested in the signal at a single frequency $\omega = \omega_0$. In this case, the "filtering" could also be accomplished by multiplying the signal with a special function that vanishes everywhere except at ω_0; however, this multiplication alone is not very useful because it does not cover the whole range of frequencies. Better yet, it would be useful to take the *integral* of the product of $s(\omega)$ with a suitable function over all frequencies and get the value $s(\omega_0)$ only; this would be a very powerful tool. The generalized function that accomplishes this task is called the "δ-function" or "Dirac function".

1.6.1   *The θ-function or step-function*

We begin with the $\theta(x - x_0)$ function which is simpler to define. This function, also called the step-function, is defined as follows:

$$\theta(x - x_0) = \begin{cases} 0 & \text{for } x < x_0, \\ 1 & \text{for } x \geq x_0. \end{cases} \tag{1.67}$$

Indeed, the θ-function is rather boring, being constant everywhere, except for the discontinuity of size 1 at the special value of its argument $x = x_0$, as shown in Fig. 1.21.

As is evident from this plot, the θ-function when multiplied by a function $f(x)$ gives back $f(x)$ but only for the range of values of its argument $x \geq x_0$:

$$\theta(x - x_0)f(x) = f(x) \text{ for } x \geq x_0,$$

and gives the value 0 otherwise. In other words, the θ-function picks up a set of values of the function $f(x)$ that corresponds to the values of x for which the argument of the θ-function is positive.

We can define the θ-function through the limit of smooth and continuous functions. For example, a useful representation of the θ-function is given by

$$\theta(x - x_0) = \lim_{\alpha \to 0} \left[\frac{1}{1 + e^{-(x - x_0)/\alpha}} \right]. \tag{1.68}$$

Examples of how this function behaves for different values of the parameter α are shown in Fig. 1.22. This function actually appears in several contexts. In physics, it is known as the "Fermi distribution" function, because it appears in the description of the occupation of energy levels by quantum mechanical particles known as "fermions". In statistics, this function is known as the "logistic sigmoid" function, because of its s-like shape. The sigmoid function can also be written as

$$\theta(x - x_0) = \lim_{\alpha \to 0} \left[\frac{1}{2} \tanh\left(\frac{x - x_0}{2\alpha} \right) + \frac{1}{2} \right], \tag{1.69}$$

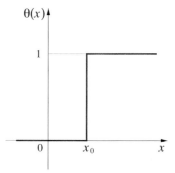

Figure 1.21: Definition of the θ-function, $\theta(x - x_0)$.

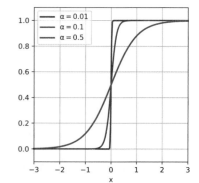

Figure 1.22: The θ-function as represented by the sigmoid function defined in Eq. (1.68), for $\alpha = 0.5$ (green), 0.1 (red) and 0.01 (blue) at $x_0 = 0$.

based on the definition of the hyperbolic tangent function, $\tanh(x)$, Eq. (1.21). A useful property of the sigmoid function is that its first derivative takes the following simple form:

$$\theta'(x - x_0) = \frac{1}{\alpha}\theta(x - x_0)\left[1 - \theta(x - x_0)\right]. \tag{1.70}$$

1.6.2   The δ-function

To introduce the δ-function, we start with a simpler situation, when we deal with an expression, denoted by a_i, that depends on the values of an *integer* index. Next, suppose we want to pick out the value of a_i for a certain value of the index i, specified as j. This is accomplished by introducing the so-called "Kronecker δ", which is defined as

$$\delta_{ij} = \begin{cases} 1 & \text{for } i = j, \\ 0 & \text{for } i \neq j, \end{cases} \tag{1.71}$$

with i, j integers. The Kronecker δ is a very useful symbol in expressions that involve sums. For example, when the Kronecker δ is multiplied by an expression a_i that depends on the integer index i and is summed over all values of i, it picks up the value of this expression for the value of the index $i = j$:

$$\sum_i a_i \delta_{ij} = a_j. \tag{1.72}$$

The δ-function is a generalization to continuous variables of the Kronecker δ symbol. In this case, the continuous variable x plays the role of the index i and the special value x_0 plays the role of the special value j of the index, and of course the summation over all values of the discrete index i becomes an integral over all the values of the continuous variable x. Thus, we may assume that the δ-function takes the value 0 if $x \neq x_0$. Accordingly, we write the δ-function as

$$\delta(x, x_0) = \delta(x - x_0),$$

where in the second expression given above we have taken advantage of the fact that what matters in obtaining a non-zero value for the δ-function is the *difference* between the values x and x_0. However, since the integral over the continuous variable x that replaces the summation over the discrete index i must involve the infinitesimal quantity $\mathrm{d}x$, which multiplies the δ-function for every value of x, the value of the δ-function itself for $x = x_0$ *cannot* be a finite number, because this would lead to a zero value for the integral. Thus we are led to assign the value ∞ to the δ-function at the point x_0. The typical behavior of the δ-function is illustrated in Fig. 1.23. Finally, the integral of the δ-function over all its values cannot be an arbitrary quantity, but must be set equal to 1, so that it is properly normalized. The full definition of the δ-function is

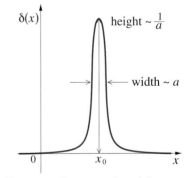

Figure 1.23: Representation of the δ-function, $\delta(x - x_0)$: the limit of $a \to 0$ is implied.

$$\delta(x - x_0) \to \infty \quad \text{for } x = x_0, \tag{1.73a}$$

$$\delta(x - x_0) = 0 \quad \text{for } x \neq x_0, \tag{1.73b}$$

$$\int_{-\infty}^{\infty} \delta(x - x_0)\, \mathrm{d}x = 1. \tag{1.73c}$$

From its definition, it follows that the product of the δ-function with an arbitrary function $f(x)$ integrated over all values of x satisfies the important relation

$$\int_{-\infty}^{\infty} f(x)\delta(x - x_0)\mathrm{d}x = f(x_0). \tag{1.74}$$

In other words, the effect of the δ-function is the following: when the product of $\delta(x - x_0)$ with $f(x)$ is integrated over the entire range of values of x, it picks out one value of the function $f(x)$, the one that corresponds to the value of x for which the argument of the δ-function vanishes.

A useful representation of the δ-function in terms of a smooth continuous function is a Gaussian, studied already in Section 1.5.2

$$\delta(x - x_0) = \lim_{\sigma \to 0} \left[\frac{1}{\sigma\sqrt{2\pi}} e^{-(x - x_0)^2/2\sigma^2} \right]. \tag{1.75}$$

The behavior of this function for different values of the parameter σ is shown in Fig. 1.24.

To prove that this function has the proper behavior, we note that around $x = x_0$ it has width σ which vanishes for $\sigma \to 0$; its value at $x = x_0$ is $1/\sigma\sqrt{2\pi}$, which tends to ∞ for $\sigma \to 0$; and its integral over all values of x is 1 for any σ (see Section 1.5.2).

Another useful representation of the δ-function is the Lorentzian,

$$\delta(x - x_0) = \lim_{\beta \to 0} \left[\frac{1}{\pi} \frac{\beta}{(x - x_0)^2 + \beta^2} \right]. \tag{1.76}$$

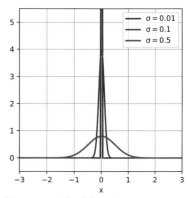

Figure 1.24: The δ-function as represented by a normalized Gaussian defined in Eq. (1.75), for $\sigma = 0.5$ (green), 0.1 (red) and 0.01 (blue) at $x_0 = 0$.

To prove that this function has the proper behavior, we note that around $x = x_0$ it has width β which vanishes for $\beta \to 0$; its value at $x = x_0$ is $1/\pi\beta$ which is infinite for $\beta \to 0$; and its integral over all values of x_0 is 1 for any value of β, which can be shown using the contour integration technique discussed in detail in Chapter 5 (the function has simple poles at $x = x_0 \pm i\beta$).

Yet another useful representation of the δ-function is the following:

$$\delta(x - x_0) = \lim_{\gamma \to 0} \left[\frac{1}{\pi\gamma} \sin\left(\frac{x - x_0}{\gamma} \right) \frac{\gamma}{x - x_0} \right]. \tag{1.77}$$

To prove that this function has the proper behavior, we note that the function $\sin(y)/y \to 1$ for $y \to 0$ and its width around $y = 0$ is determined by half the interval between y values at which it first becomes zero, which occurs at $y = \pm\pi$, that is, its width in the variable y is π. With $y = (x - x_0)/\gamma$, we conclude that around $x = x_0$ the function on the right-hand side of Eq. (1.77) has width $\gamma\pi$ in the variable x, which vanishes for $\gamma \to 0$; its value at $x = x_0$ is $1/\pi\gamma$ which is infinite for $\gamma \to 0$. Finally, its integral over all values of x_0 is 1 for any value of γ. The easiest way to prove this fact is by expressing the $\sin((x - x_0)/\gamma)$ in terms of exponentials of complex arguments and using the technique of contour integration, topics that we discuss in detail in Chapters 4 and 5 (each term in the integrand has a simple pole at $x = x_0$).

We note that all the expressions we used above to define the δ-function are even functions of their argument and we can assume that this property holds in general, namely,

$$\delta(x - x_0) = \delta(x_0 - x). \tag{1.78}$$

Moreover, from the definition of the θ-function, it follows that the derivative of a smooth, continuous representation of the θ-function must be a corresponding representation of a δ-function. Hence, we will assume that the following relation holds:

$$\frac{d}{dx}\theta(x - x_0) = \delta(x - x_0). \tag{1.79}$$

For instance, the derivative of the representation of the θ-function given in Eq. (1.68) is

$$\lim_{\alpha \to 0}\left[\frac{e^{-(x-x_0)/\alpha}}{\alpha(e^{-(x-x_0)/\alpha} + 1)^2}\right], \tag{1.80}$$

and this expression is indeed, as expected from Eq. (1.79), another representation of the δ-function.

An expression often arises in which the δ-function has as its argument another function, $f(x)$, and the integral of its product with yet another function, $g(x)$, must be evaluated:

$$I = \int_{-\infty}^{\infty} g(x)\delta(f(x))dx.$$

We consider the result of this evaluation, assuming that $f(x)$ has only simple roots, that is, $f(x_0) = 0, f'(x_0) \neq 0$, with $f'(x)$ the derivative of $f(x)$. Since the δ-function is zero when its argument is not zero, we first analyze the neighborhood of one simple root x_0 of $f(x)$, that is, $x \in [x_0^-, x_0^+]$, $x_0^- < x_0 < x_0^+$ assuming that no other root exists in $[x_0^-, x_0^+]$. We define the variable u though the relation

$$u = f(x) \Rightarrow x = f^{-1}(u).$$

At the values of $x = x_0, x_0^-, x_0^+$, we denote the values of u as

$$u_0 = f(x_0) = 0, \quad u_1 = f(x_0^-), \quad u_2 = f(x_0^+).$$

From the definition of u we have

$$du = f'(x)dx \Longrightarrow dx = \frac{1}{f'(x)}du.$$

These expressions, when inserted in the integral under consideration with x in the neighborhood of x_0, give

$$I_0 = \int_{u_1}^{u_2} g(f^{-1}(u))\delta(u)\frac{1}{f'(f^{-1}(u))}du.$$

In order to evaluate this integral, we only need to consider the order of the limits: with x_0^-, x_0^+ close to x_0, we will have the results below:

$$f'(x_0) > 0 \Rightarrow f(x_0^-) = u_1 < f(x_0) = u_0 = 0 < f(x_0^+) = u_2,$$

$$f'(x_0) < 0 \Rightarrow f(x_0^-) = u_1 > f(x_0) = u_0 = 0 > f(x_0^+) = u_2.$$

The above considerations imply the following:

$$f'(x_0) > 0 \Rightarrow I_0 = \int_{u_1}^{u_2} g(f^{-1}(u))\delta(u)\frac{1}{f'(f^{-1}(u))}du = \frac{g(f^{-1}(u_0))}{f'(f^{-1}(u_0))},$$

where we have used $f^{-1}(u_0) = x_0$. Similarly

$$f'(x_0) < 0 \Rightarrow I_0 = -\int_{u_2}^{u_1} g(f^{-1}(u))\delta(u)\frac{1}{f'(f^{-1}(u))}du = \frac{g(f^{-1}(u_0))}{-f'(f^{-1}(u_0))}.$$

Thus, in both cases we find that the integral is equal to $g(x)/|f'(x)|$ evaluated at $x = x_0$. If the argument of $\delta(f(x))$ vanishes at all the roots $x_0^{(i)}$ of $f(x)$, we will get similar contributions from each root, which gives the final result

$$\int_{-\infty}^{\infty} g(x)\delta(f(x))dx = \sum_i \frac{g(x_0^{(i)})}{\left|f'(x_0^{(i)})\right|}, \quad \text{where } f(x_0^{(i)}) = 0, \qquad (1.81)$$

with the summation on i running over all the simple roots of $f(x)$. This result implies directly the following relations:

$$\delta(a(x - x_0)) = \frac{1}{|a|}\delta(x - x_0), \qquad (1.82)$$

and

$$\delta(f(x)) = \sum_i \frac{1}{\left|f'(x_0^{(i)})\right|}\delta(x - x_0^{(i)}), \quad \text{where } f(x_0^{(i)}) = 0. \qquad (1.83)$$

1.7   Application: Signal filtering, convolution, correlation

Many signals can be thought of as continuous functions of one or more variables. Examples include the familiar acoustic signals (sounds), electromagnetic signals (optical, radio waves, X-rays). Detecting a signal relies on some physical device, for instance, our ear for acoustic signals, a radio receiver for electromagnetic radio waves, an astronomical antenna for X-rays from outer space, etc. This device can respond to the field or other physical medium that carries the signal, but the response is never perfect, in the sense that the device's own physical properties play an important role of how the signal is detected. This is captured by the device's "response function". Thus, the measured signal is captured by the so-called "convolution" of the actual signal and the detector's response function. A closely related concept is the "correlation" of the signal and the response function. We examine these two concepts next.

We consider the original signal to be the function $s(x)$ and the response function to the function $g(x)$. The **convolution** of the two functions, denoted as $s \star g(x)$, is defined by

$$s \star g(x) \equiv \int_{-\infty}^{\infty} s(y)g(x - y)dy. \qquad (1.84)$$

The **correlation** of the two functions, denoted as $c[s, g](x)$, is defined by

$$c[s, g](x) \equiv \int_{-\infty}^{\infty} s(y)g(y - x)dy = \int_{-\infty}^{\infty} s(y + x)g(y)dy. \qquad (1.85)$$

The two definitions, Eqs. (1.84) and (1.85), have similar appearance but differ in important ways. In the convolution, as the argument of the function s moves *forward* (from $-\infty$ to $+\infty$), the argument of the function g moves backward (from $+\infty$ to $-\infty$), and is offset by x. In the correlation, the arguments of the two functions move in the same direction, but they are simply offset by x.

One extreme example of the application of these concepts is the case that the response function is a δ-function; in this case, the measuring apparatus is sensitive to only a single value of the argument. Because the δ-function is an even function [see Eq. (1.78)] its convolution and correlation with the signal $s(x)$ give the same result:

$$s \star \delta(x) = \int_{-\infty}^{\infty} s(y)\delta(x-y)\mathrm{d}y = s(x), \qquad (1.86)$$

$$c[s,\delta](x) = \int_{-\infty}^{\infty} s(y)\delta(y-x)\mathrm{d}y = s(x). \qquad (1.87)$$

These results show that applying either the convolution or the correlation to a function $s(x)$ with the δ function as the response, gives back the function $s(x)$ at each value of its argument.

In the following discussion, we denote the detected signal as $S(x)$, which is given in terms of the actual signal $s(x)$ by $S(x) = s \star g(x)$. To illustrate the effect of the convolution with the instrumental response function $g(x)$, we consider a simple signal consisting of a square pulse: the signal is constant for a range of the variable y in the range $-1 < y < 1$ and zero everywhere else. Moreover, for reasons of simplicity, we will take the response function to be a normalized Gaussian with $\mu = 0$ and $\sigma = 1$, namely the function $\varphi(x)$ defined in Eq. (1.59), which for the purposes of the present discussion we denote as $g_1(x)$. Then the measuring process involves the signal $s(y)$ passing through the detector with response function $g_1(x-y)$,

$$g_1(x-y) = \frac{1}{\sqrt{2\pi}}e^{-(x-y)^2/2}, \qquad (1.88)$$

producing the measured signal $S_1(x)$; this process involves the convolution integral of Eq. (1.84).

We may think of this process as the signal $s(y)$ being "scanned" by the detector with its response function $g_1(x-y)$, with y fixed while x takes all possible values, $x \in (-\infty, \infty)$; this is shown schematically in Fig. 1.25.

Alternatively, we can view this process as the signal $s(y)$ "sweeping" through the detector's response function $g_1(x-y)$, with x fixed while y takes all possible values, $y \in (-\infty, \infty)$. The resulting measured signal $S(x)$ depends sensitively on the response function. In fact, for our first choice of response function, $g_1(x)$, the measured signal $S_1(x) = s \star g_1(x)$ looks rather different from the original one, as shown in Fig. 1.26. The very broad response function $g_1(x)$ has resulted in a measured signal much smoother than the square pulse. This effect is referred to as "signal filtering". Essentially, as the signal passes through the detector, the convolution mixes different parts of the signal, since for a given value of x the two functions $s(y)$ and $g_1(x-y)$ are multiplied for all values of y and the values of the product are summed to give the convolution integral, leading to the value $S_1(x)$.

The result of the filtering process is improved if the response function becomes more sharp. To illustrate this fact, we choose as response functions

$$g_2(x) = f_g(x; 0, 0.25) = \frac{4}{\sqrt{2\pi}}e^{-8x^2}, \quad \text{and} \quad g_3(x) = f_g(x; 0, 0.1) = \frac{10}{\sqrt{2\pi}}e^{-50x^2},$$

which are shown in Fig. 1.26. As the width of the Gaussian filter becomes smaller [$\sigma_2 = 0.25$ for $g_2(x)$ and $\sigma_3 = 0.1$ for $g_3(x)$], the filtering becomes less severe and the measured signal is closer to the original one. In the limit of $\sigma \to 0$, the Gaussian filter approaches a δ-function and all the details or the

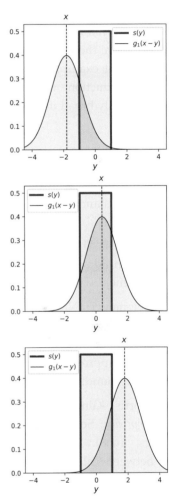

Figure 1.25: Illustration of the convolution process of the signal $s(y)$ (blue square pulse, in the range $-1 < y < 1$) with the response function $g_1(x-y)$, Eq. (1.88) for three relative positions of the two curves, namely $x < -1$ (top panel), $-1 < x < 1$ (middle panel) and $1 < x$ (bottom panel).

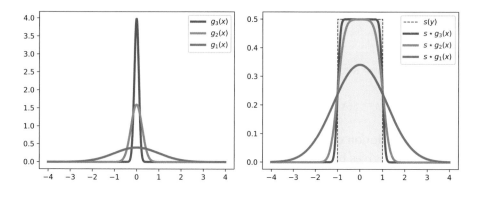

Figure 1.26: **Left**: The three different choices of response functions, $g_1(x)$ (green), $g_2(x)$ (orange), $g_3(x)$ (blue), which are normalized Gaussians with width $\sigma_1 = 1$, $\sigma_2 = 0.25$, $\sigma_3 = 0.1$, respectively. **Right**: The corresponding measured signal $S_1(x)$, $S_2(x)$, $S_3(x)$; in each case $S_i(x) = s \star g_i(x)$, $i = 1, 2, 3$. The original signal $s(x)$ is shown in dashed lighter lines and is shaded, in the background.

original signal (like the sharp rise at its onset and the sharp fall at its end) are fully captured by the detector. Conversely, the broader the response function is, the more the measured signal looks like the response itself (a Gaussian in the present example) than like the original signal. In other words, a broad response function "washes out" most of the signal features and what remains is an equally broad measured signal similar in shape to the response function.

Further reading

Most of the material presented in this chapter can be found in standard textbooks on calculus. Suggested choices are:

1. George B. Thomas, Jr., *Calculus and Analytic Geometry*, Fourth Ed. (Addison-Wesley, 1972). This is a classic treatment with careful and thorough explanations.

2. Zbigniew H. Nitecki, *Calculus Deconstructed: A Second Course in First-Year Calculus* (American Mathematical Society, 2009). This is a modern and concise treatment, appropriate for students with some calculus background.

3. Morris Kline, *Calculus: An Intuitive and Physical Approach*, Second Ed. (Dover Books on Mathematics, 1998). This book makes useful connections between calculus and its applications to physical problems. The treatment is at a level of rigor similar to the present book.

4. Michael Spivak, *Calculus*, Fourth Ed. (Publish or Perish, 2008). This is a useful reference book for those with some familiarity with calculus, with a rigorous treatment of the various topics.

Problems

1. For any two real numbers x_1, x_2 show that:

$$|x_1 + x_2| \leq |x_1| + |x_2|.$$

This is known as the "triangle inequality". Show that this inequality can also be generalized to an arbitrary number of real numbers, that is,

$$|x_1 + \cdots + x_n| \leq |x_1| + \cdots + |x_n|. \tag{1.89}$$

2. Find the condition that the real coefficients of the cubic polynomial must satisfy so that all three of its roots, given by the expressions in Eqs (1.6), are real numbers.

3. The function $f(x) = \cos(x)$ is defined in the domain $x \in (-\infty, +\infty)$; is $f^{-1}(x)$ a properly defined function? Is $f^{-1}(x)$ a properly defined function, if the domain is $x \in [0, 2\pi]$? Is $f^{-1}(x)$ a properly defined function, if the domain is $x \in [0, \pi]$? What about the inverse of the function $g(x) = \sin(x)$ for the same three domains?

4. Using the definition of the exponential function as the limit

$$e^x = \lim_{N \to \infty} \left(1 + \frac{x}{N}\right)^N,$$

prove that the derivative of the exponential function is the same function. [Note: You may assume that it is allowed to interchange the order of the derivative and the limit $N \to \infty$. We will investigate the legitimacy of this interchange in more detail in Chapter 3.]

5. Consider the functions $\sinh(x)$, $\cosh(x)$, and $\tanh(x) = \sinh(x)/\cosh(x)$.

   (a) Which of these functions has an inverses function that is properly defined in the domain $x \in (-\infty, +\infty)$?

   (b) What is the answer if the domain is $x \in [0, +\infty)$?

6. Consider the function defined as

$$f(x) = \begin{cases} (1+x)^3, & x > 0, \\ \cos(x), & x \leq 0. \end{cases}$$

   Show that this function is continuous at $x = 0$. Is the first derivative of the function, $f'(x)$, continuous or discontinuous at $x = 0$? Give a proof of your answer.

7. Show that the derivative of Eq. (1.68) is the expression given in Eq. (1.80), and that this expression has all the properties of the δ-function.

8. Consider the following definite integral:

$$\int_a^b \frac{1}{x}dx, \quad \text{where } a, b \in \mathbb{R}, \ a < 0, \ b > 0, \ -a \neq b.$$

   Explain why this integral is not properly defined if we do not take the principal value. Calculate the principal value of this integral.

9. The function $f(x)$ is differentiable at $f(0) = 0$. The function $g(y)$ is defined as

$$g(y) = \int_0^y \frac{f(x)}{\sqrt{y-x}}dx.$$

   Prove that

$$\frac{dg}{dy} = \int_0^y \left(\frac{df}{dx}\right) \frac{1}{\sqrt{y-x}}dx.$$

   For the case $f(x) = x^n$, prove that

$$\frac{d^n g}{dy^n} = 2(n!)\sqrt{y}.$$

10. Using integration by parts and the fact that the sine and cosine functions are derivatives of each other (with a minus sign) prove the following formulas:

$$\int \cos^n(x)dx = \frac{\cos^{n-1}(x)\sin(x)}{n}$$
$$+ \frac{n-1}{n}\int \cos^{n-2}(x)dx, \quad (1.90a)$$

$$\int \sin^n(x)dx = -\frac{\sin^{n-1}(x)\cos(x)}{n}$$
$$+ \frac{n-1}{n}\int \sin^{n-2}(x)dx, \quad (1.90b)$$

where n is an integer. These are called "reduction formulas" because they lead to integrals that involve a power of the sine or cosine which is smaller by 2 than the original one. By continuing the integration by parts we end up with an integrand with power zero (if n is even) or one (if n is odd), which can be trivially calculated.

11. Using integration by parts prove the following formula:

$$\int e^{ax}\cos(bx)dx = \frac{e^{ax}}{a^2+b^2}\left\{b\sin(bx) + a\cos(bx)\right\} + c,$$

where c is an arbitrary constant. [Note: The use of Euler's formula for complex exponentials makes this integral a trivial one, see Chapter 4, Section 4.1.]

12. Show that the convolution of the normalized Gaussian $f_g(x, 0, \sigma)$ with the square pulse, defined as the function of constant height A for $x \in [-a, a]$ and zero everywhere else, is given by the expression

$$A\left\{\Phi(a - x; 0, \sigma) - \Phi(-a - x; 0, \sigma)\right\}.$$

In the above expression, $\Phi(w; \mu, \sigma)$ is the generalization of the cumulative normal distribution function, $\Phi(w)$, defined in Eq. (1.60a), for the normalized Gaussian $f_g(x; \mu, \sigma)$. Choose values for the parameters A and a and plot the above expression for various values of σ; compare your results to those shown in Section 1.7, which were generated with $A = 0.5$, $a = 1$. [Note: $\Phi(w, \mu, \sigma)$ is available in modern computing environments as a library function; in python it is called "norm.cdf", which stands for "normal cumulative distribution function".]

13. Prove the following result by using the change of variables $x = \sin(\pi t/2)$:

$$\int_{-1}^1 \frac{1}{\sqrt{1-x^2}}dx = \pi. \quad (1.91)$$

14. The "autocorrelation function", $g(x)$, and the "auto-convolution function", $h(x)$, of a function $f(x)$ are defined as

$$g(x) \equiv \int_{-\infty}^\infty f(t)f(t-x)dt,$$

$$h(x) \equiv \int_{-\infty}^\infty f(t)f(x-t)dt.$$

(a) Prove that the autocorrelation is a symmetric function, that is $g(-x) = g(x)$. What is the condition that $f(x)$ must satisfy for this result to hold? [Hint: Consider the integral of $|f(x)|^2$ over all values of x.]

(b) Show that

$$\int_{-\infty}^{\infty} f(t)f(t-x)\,dt \le \int_{-\infty}^{\infty} [f(t)]^2\,dt.$$

[Hint: Consider the integral

$$\int_{-\infty}^{\infty} [f(t) - f(t-x)]^2\,dt,$$

and expand the square in the integrand.] Use this result to show that the autocorrelation satisfies $|g(x)| \le g(0)$.

(c) Apply the methods described in part (b) to the autoconvolution $h(x)$ to get a bound on $|h(x)|$, assuming that $f(x)$ satisfies the same conditions as in part (a). Is it possible to assert that $|h(x)| \le h(0)$?

(d) Consider the function

$$f(x) = \begin{cases} 1-x, & 0 \le x \le 1, \\ 0 & \text{everywhere else.} \end{cases}$$

Find the autocorrelation and the autoconvolution of $f(x)$ and show that they satisfy the general relations derived in parts (a), (b), and (c).

15. Consider the function

$$f(x) = \begin{cases} 1-e^{-bx}, & x \ge 0, \quad \text{where } b > 0, \\ 0, & x < 0. \end{cases}$$

(a) Find the autocorrelation $g(x)$ of $f(x)$, as defined in the previous problem. Is $g(x)$ continuous or discontinuous? Does $g(x)$ satisfy the relation $|g(x)| \le g(0)$ that was derived in the previous problem? Why?

(b) Find the autoconvolution $g(x)$ of $f(x)$, as defined in the previous problem. Is $h(x)$ continuous or discontinuous?

Chapter 2

Functions of Many Variables

As you will find in multi-variable calculus, there is often a number of solutions for any given problem.

John Forbes Nash (American mathematician, 1928–2015)

2.1   General considerations

In many situations, in order to describe the phenomena or systems of interest, we need to handle more than one independent variables. We then have to work with functions that depend simultaneously on all these variables. In this chapter, we describe the tools and techniques necessary to handle problems that involve functions of many variables. We will think of each variable as introducing a new dimension in the problem, that is, as expanding our space. We first discuss coordinate systems in the familiar two and three dimensions and then explore how we can handle differentiation and integration of function in many-dimensional space.

In the following, we use the symbol $\mathbf{r}$ for a point in the n-dimensional space,

$$\mathbf{r} \rightarrow (x_1, \ldots, x_n).$$

In the same way that the variable x takes real values along the x-axis, the variables x_i, with $i = 1, \ldots, n$, take real values along different axes. Our assumption that the variables are independent implies that we can choose the axes to be orthogonal to each other in an n-dimensional space. We associate a unit vector $\hat{x}_i$, $i = 1, \ldots, n$, with each one of these orthogonal axes pointing along the direction of increasing values on this axis; as a result, the point $\mathbf{r}$ is represented by a vector in the n-dimensional space, given by

$$\mathbf{r} = x_1\hat{x}_1 + \cdots + x_n\hat{x}_n.$$

By analogy to the case of functions of a single variable, we are typically interested in two basic aspects of a function's behavior:

- the rate at which a multi-variable function changes value when its variables change values, that is, the *derivatives* of the multi-variable function;

- the sum of all the values of the function over the entire range of each of its variables, that is, the *integral* of the multi-variable function.

To quantify the change in the function's value when its variables change values, we consider what happens when only one variable changes keeping the remaining variables fixed; this introduces the notion of the *partial derivatives*: Suppose that we are dealing with a function of n variables; $f(x_1, \ldots, x_n)$; to take its derivative with respect to x_1, at the value $x_1 = x_1^{(0)}$ holding all the other variables $x_2, \ldots, x_n$ constants at the values $x_i = x_i^{(0)}$ for $i = 2, \ldots, n$. This provides the definition of the "partial derivative with respect to x_1", evaluated at $(x_1, \ldots, x_n) = (x_1^{(0)}, \ldots, x_n^{(0)})$:

$$\frac{\partial f}{\partial x_1}\left(x_1^{(0)}, \ldots, x_n^{(0)}\right) \equiv \lim_{x_1 \to x_1^{(0)}} \frac{f\left(x_1, x_2^{(0)}, \ldots, x_n^{(0)}\right) - f\left(x_1^{(0)}, x_2^{(0)}, \ldots, x_n^{(0)}\right)}{x_1 - x_1^{(0)}},$$

(2.1)

and similarly for all the other partial derivatives with respect to x_i, $i = 2, \ldots, n$. The higher derivatives can also be defined in the same way. To keep the notation simple, we define the following symbol for the partial derivative with respect to the variable x:

$$\partial_x \equiv \frac{\partial}{\partial x}.$$

(2.2)

We will also extend this convention to denote the first- and higher-order derivatives of a function $u(x)$ of a single variable, or the partial derivatives of functions of several variables, namely, we will employ subscripts of the variable to denote the derivative with respect to which the derivative is defined. To be specific, for the function $u(x)$ of the single variable x, the first and second derivatives will be denoted as

$$u_x(x) \equiv \frac{du}{dx}, \quad u_{xx}(x) \equiv \frac{d^2u}{dx^2}.$$

(2.3)

Similarly, for the function $u(x, y)$ of the two independent variables x, y, the two first partial derivatives will be denoted as

$$u_x(x, y) \equiv \frac{\partial u}{\partial x}, \quad u_y(x, y) \equiv \frac{\partial u}{\partial y},$$

(2.4)

and the three second partial derivatives as

$$u_{xx}(x, y) \equiv \frac{\partial^2 u}{\partial x^2}, \quad u_{yy}(x, y) \equiv \frac{\partial^2 u}{\partial y^2}, \quad u_{xy}(x, y) \equiv \frac{\partial^2 u}{\partial x \partial y}.$$

(2.5)

To integrate a function of many independent variables, we follow similar rules as in the case of differentiation, that is, we integrate with respect to one of the independent variables treating the other ones as constants. For instance, if the function of n variables can be written as a sum of single-variable functions

$$f(x_1, x_2, \ldots, x_n) = f_1(x_1) + f_2(x_2) + \cdots + f_n(x_n),$$

we would have for its integral:

$$\int_V f(x_1, x_2, \ldots, x_n) \, dx_1 \, dx_2 \cdots dx_n = \sum_{j=1}^{n} V_j \int_{x_j^{(i)}}^{x_j^{(f)}} f_j(x_j) \, dx_j,$$

where V is the n-dimensional volume spanned by the n variables, V_j is the $(n-1)$-dimensional volume spanned by the $n-1$ variables *other* than x_j, and $x_j^{(i)}, x_j^{(f)}$ are the initial and final values of the variable x_j.

Similarly, if the function of n variables can be written as a product of single-variable functions,

$$f(x_1, x_2, \ldots, x_n) = f_1(x_1) \cdot f_2(x_2) \cdots f_n(x_n),$$

we would have for its integral:

$$\int_V f(x_1, x_2, \ldots, x_n) \, dx_1 \, dx_2 \, \cdots dx_n = \prod_{j=1}^{n} \int_{x_j^{(i)}}^{x_j^{(f)}} f_j(x_j) \, dx_j,$$

with the same meaning of symbols as above.

2.2   Coordinate systems and change of variables

2.2.1   Coordinate systems in two dimensions

In the two-dimensional (2D) space, in the conventional system of two orthogonal independent variables (x, y), we express **r** in terms of the independent variables and the unit vectors $\hat{x}, \hat{y}$ along the two axes of these coordinates, as

$$\mathbf{r} = x\hat{x} + y\hat{y}. \tag{2.6}$$

The x and y coordinates are referred to as "Cartesian". The infinitesimal change in the position vector in Cartesian coordinates is given by

$$d\mathbf{r} = dx\,\hat{x} + dy\,\hat{y},$$

from which we can deduce that the volume element[1] in this coordinate system is given by

$$dV = dx\,dy. \tag{2.7}$$

Depending on the symmetries of the problem, it is often more convenient to use a different coordinate system. A common coordinate system is the so-called "polar coordinate" system. The **polar coordinates**, (ρ, ϕ), are defined in terms of the Cartesian coordinates as shown in Fig. 2.1, through the relations:

$$\rho = \sqrt{x^2 + y^2}, \quad \phi = \tan^{-1}(y/x). \tag{2.8}$$

The inverse transformation, polar to Cartesian, is given by

$$x = \rho\cos(\phi), \quad y = \rho\sin(\phi). \tag{2.9}$$

The unit vectors of the polar coordinates are given in term of the unit vectors of the Cartesian coordinates as

$$\hat{\rho} = \cos(\phi)\,\hat{x} + \sin(\phi)\,\hat{y}, \quad \hat{\phi} = -\sin(\phi)\,\hat{x} + \cos(\phi)\,\hat{y},$$

and the vector **r** in polar coordinates is simply given by

$$\mathbf{r} = \rho\,\hat{\rho}.$$

The infinitesimal change in the position vector in polar coordinates is given by

$$d\mathbf{r} = d\rho\,\hat{\rho} + \rho d\phi\,\hat{\phi},$$

[1] We refer to this quantity as the "volume", although in 2D space it is often called the "area"; the choice of the term "volume" is appropriate for the more general use of the concept, independent of the dimensionality of space. This convention is employed in the rest of the book as well.

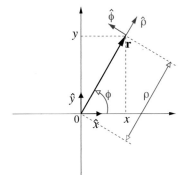

Figure 2.1: Definition of the polar coordinates (ρ, ϕ) in 2D space in terms of the Cartesian coordinates (x, y).

from which we can deduce that the volume element in this coordinate system is given by

$$\mathrm{d}V = (\mathrm{d}\rho)(\rho\mathrm{d}\phi). \qquad (2.10)$$

This result is shown schematically in Fig. 2.2. The infinitesimal volumes in the Cartesian and polar coordinate systems are therefore related by:

$$\mathrm{d}x\,\mathrm{d}y = (\mathrm{d}\rho)\,(\rho\mathrm{d}\phi) = \rho\mathrm{d}\rho\,\mathrm{d}\phi. \qquad (2.11)$$

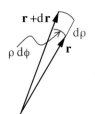

Figure 2.2: The calculation of the infinitesimal volume in polar coordinates, Eq. (2.11).

Derivation: Rotation of axes and vectors in two dimensions

An interesting exercise is to consider a new set of axes, (x', y'), rotated with respect to the original ones (x, y), by an angle ω in the *counterclockwise* direction. Our goal is to express the vector $\mathbf{r}$, which has not changed, in terms of the new coordinates. In polar coordinates, this is a very easy task: the new coordinates are $(\rho', \phi') = (\rho, \phi - \omega)$ and the unit vectors have not changed, that is, $\hat{\rho}' = \hat{\rho}, \hat{\phi}' = \hat{\phi}$. In Cartesian coordinates the situation is a bit more complicated. The situation is illustrated in the diagram below.

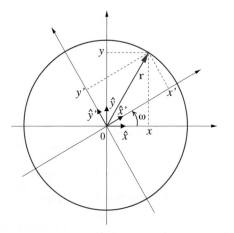

The unit vectors have been rotated by ω, so projecting $\hat{x}'$ and $\hat{y}'$ onto the (x, y) axes we obtain the following expressions for the new unit vectors, $\hat{x}', \hat{y}'$ in terms of the old ones, $\hat{x}, \hat{y}$:

$$\hat{x}' = \cos(\omega)\,\hat{x} + \sin(\omega)\,\hat{y}, \quad \hat{y}' = -\sin(\omega)\,\hat{x} + \cos(\omega)\,\hat{y}.$$

Solving this as a system of two equations in the two unknown quantities $\hat{x}$ and $\hat{y}$ we obtain expressions for the old unit vectors, $\hat{x}, \hat{y}$, in terms of the new ones, $\hat{x}', \hat{y}'$ (the inverse relations):

$$\hat{x} = \cos(\omega)\,\hat{x}' - \sin(\omega)\,\hat{y}', \quad \hat{y} = \sin(\omega)\,\hat{x}' + \cos(\omega)\,\hat{y}'.$$

Using these expressions for the unit vectors we find:

$$\mathbf{r} = x\hat{x} + y\hat{y} = x\left[\cos(\omega)\,\hat{x}' - \sin(\omega)\,\hat{y}'\right] + y\left[\sin(\omega)\,\hat{x}' + \cos(\omega)\,\hat{y}'\right]$$

$$\Rightarrow \quad \mathbf{r} = [x\cos(\omega) + y\sin(\omega)]\,\hat{x}' + [-x\sin(\omega) + y\cos(\omega)]\,\hat{y}' = x'\hat{x}' + y'\hat{y}'.$$

By equating the factors in front of each unit vector in the last expression for **r** we obtain the following expressions for the new variables x', y' in terms of the old ones x, y:

$$x' = x\cos(\omega) + y\sin(\omega), \quad y' = -x\sin(\omega) + y\cos(\omega). \tag{2.12}$$

Solving this as a system of two equations in the two unknown quantities x and y we find expressions for the old variables x, y in terms of the new ones x', y' (the inverse relations):

$$x = x'\cos(\omega) - y'\sin(\omega), \quad y = x'\sin(\omega) + y'\cos(\omega). \tag{2.13}$$

These relations can be expressed in matrix notation as:

$$\begin{bmatrix} x' \\ y' \end{bmatrix} = \begin{bmatrix} \cos(\omega) & \sin(\omega) \\ -\sin(\omega) & \cos(\omega) \end{bmatrix} \begin{bmatrix} x \\ y \end{bmatrix} \tag{2.14a}$$

$$\Rightarrow \begin{bmatrix} x \\ y \end{bmatrix} = \begin{bmatrix} \cos(\omega) & -\sin(\omega) \\ \sin(\omega) & \cos(\omega) \end{bmatrix} \begin{bmatrix} x' \\ y' \end{bmatrix}. \tag{2.14b}$$

Notice that the two matrices that appear in the two transformations, from $(x, y) \to (x', y')$ and from $(x', y') \to (x, y)$ are the inverse of each other, as expected.

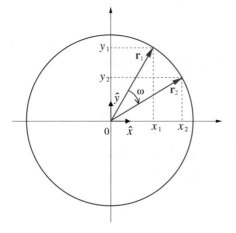

Another very similar transformation is to consider the original vector, $\mathbf{r}_1 = x_1\hat{x} + y_1\hat{y}$, rotated by an angle ω in the *clockwise* direction, to produce a new vector, $\mathbf{r}_2 = x_2\hat{x} + y_2\hat{y}$, as shown in the diagram above. In this case, using the fact that the length ρ of the vector is unaffected by the rotation, and employing the relations

$$x_1 = \rho\cos(\phi_1), \quad y_1 = \rho\sin(\phi_1), \quad x_2 = \rho\cos(\phi_2), \quad y_2 = \rho\sin(\phi_2),$$

where $\phi_2 = \phi_1 - \omega$, we find that the coordinates of the new vector are given in terms of the coordinates of the old vector as

$$\begin{bmatrix} x_2 \\ y_2 \end{bmatrix} = \begin{bmatrix} \cos(\omega) & \sin(\omega) \\ -\sin(\omega) & \cos(\omega) \end{bmatrix} \begin{bmatrix} x_1 \\ y_1 \end{bmatrix}, \tag{2.15}$$

that is, by the same transformation as the rotation of the axes by the angle ω in the *clockwise* direction. This makes sense from a physical perspective, as the two operations are equivalent.

The operations we considered above are referred to as "linear transformations" because they involve linear combinations of the original variables (x, y) to produce the new variables. The coefficients involved in these operations are sines and cosines of the rotation angle.

2.2.2   Coordinate systems in three dimensions

By analogy to the 2D case, in three-dimensional (3D) space we can define the corresponding Cartesian and polar coordinates. In 3D, it is also convenient to introduce another coordinate system, called "spherical coordinates". These three 3D systems are discussed next.

- **Cartesian coordinates:** The position vector $\mathbf{r}$ is expressed in terms of the Cartesian coordinates x, y, z and the unit vectors along the three axes of these independent coordinates, $\hat{x}, \hat{y}, \hat{z}$:

$$\mathbf{r} = x\hat{x} + y\hat{y} + z\hat{z}. \tag{2.16}$$

The infinitesimal change in the position vector in Cartesian coordinates is given by

$$\mathbf{dr} = dx\,\hat{x} + dy\,\hat{y} + dz\,\hat{z},$$

from which we can deduce that the infinitesimal volume element in this coordinate system is given by

$$dV = dx\,dy\,dz. \tag{2.17}$$

- **Polar coordinates:** These coordinates, denoted as (ρ, ϕ, z) in 3D, are defined in terms of the Cartesian coordinates through the following relations:

$$\rho = \sqrt{x^2 + y^2}, \quad \phi = \tan^{-1}(y/x), \quad z. \tag{2.18}$$

The inverse transformation, polar to Cartesian, is given by

$$x = \rho\cos(\phi), \quad y = \rho\sin(\phi), \quad z. \tag{2.19}$$

The infinitesimal change in the position vector in polar coordinates is given by

$$\mathbf{dr} = d\rho\,\hat{\rho} + \rho dy\,\hat{\phi} + dz\,\hat{z},$$

from which we can deduce that the volume element in this coordinate system is given by

$$dV = (d\rho)(\rho d\phi)(dz). \tag{2.20}$$

Therefore, the infinitesimal volumes in the Cartesian and polar coordinate systems are related by the equation

$$dx\,dy\,dz = (d\rho)(\rho d\phi)(dz) = \rho d\rho\,d\phi\,dz. \tag{2.21}$$

- **Spherical coordinates**: These coordinates, denoted as (r, θ, ϕ) in 3D, are defined in terms of the Cartesian coordinates through the following relations:

$$r = \sqrt{x^2 + y^2 + z^2}, \quad \phi = \tan^{-1}(y/x), \quad \theta = \cos^{-1}(z/r), \tag{2.22}$$

shown schematically in Fig. 2.3.

The inverse transformation, spherical to Cartesian, is given by

$$x = r\sin(\theta)\cos(\phi), \quad y = r\sin(\theta)\sin(\phi), \quad z = r\cos(\theta). \tag{2.23}$$

The infinitesimal change in the position vector in spherical coordinates is given by

$$\mathbf{dr} = dr\,\hat{r} + r\sin(\theta)d\phi\,\hat{\phi} + rd\theta\,\hat{\theta},$$

from which we can deduce that the volume element in this coordinate system is given by

$$dV = (dr)(r\sin(\theta)d\phi)(rd\theta). \tag{2.24}$$

This is shown schematically in the inset of Fig. 2.4. Therefore, the infinitesimal volumes in the two coordinate systems are related by the equation

$$dx\,dy\,dz = (dr)(r\sin(\theta)d\phi)(rd\theta) = r^2 dr\,\sin(\theta)d\theta\,d\phi. \tag{2.25}$$

The relations between the infinitesimal volume in the Cartesian coordinate system and the infinitesimal volume in the polar coordinate system, Eq. (2.21), or the spherical coordinate system, Eq. (2.25), were derived by geometric considerations. Namely, we examined how the infinitesimal volume element is created by small changes in the variables of each system along the three orthogonal directions of the respective unit vectors. We also took for granted that the infinitesimal volumes created in this way are the same across different coordinate systems. For a more precise statement of these relations we can use the general rule relating the infinitesimal volume elements in different coordinates that involves the "Jacobian"; this topic is discussed in the next subsection.

To illustrate the importance of using the proper coordinate system for a given problem, we calculate the volume of a few simple, highly symmetric geometric objects. We consider first a symmetric cone whose base is a circle of radius R, and whose height is h. In order to calculate the volume of the cone we simply perform an integral over all space with boundaries the outer surfaces of the cone. This is achieved by slicing the cone in horizontal disks of infinitesimal width dz along the vertical z-axis, as shown in Fig. 2.5, and adding up the volumes of these disks. The volume dV_{disk} of a disk with radius ρ is its area times its thickness,

$$dV_{\text{disk}} = \pi\rho^2 dz.$$

From the similarity of right-angle triangles ABD and ACE we obtain the relations

$$\frac{|CE|}{|BD|} = \frac{|AC|}{|AB|} \Rightarrow \frac{\rho}{R} = \frac{h-z}{h} = 1 - \frac{z}{h} \Rightarrow \rho = R\left(1 - \frac{z}{h}\right),$$

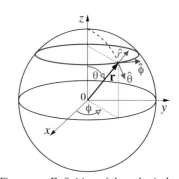

Figure 2.3: Definition of the spherical coordinates (r, θ, ϕ) in 3D space, in terms of the Cartesian coordinates (x, y, z).

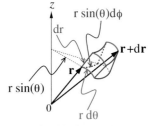

Figure 2.4: The calculation of the infinitesimal volume in spherical coordinates, Eq. (2.25).

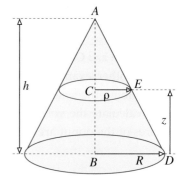

Figure 2.5: A symmetric cone with circular base of radius R and height h, sliced in horizontal disks of infinitesimal width along the vertical axis (shown in red). A is the apex point of the cone, B is the center of its base, C is the center of the horizontal disk at height z and D and E are point on the perimeter of the base and the perimeter of the disk.

which gives for the volume of the disk

$$dV_{\text{disk}} = \pi R^2 \left(1 - \frac{z}{h}\right)^2 dz.$$

To get the volume of the cone, we simply integrate this expression over the range of values that the variable z takes:

$$V_{\text{cone}} = \int dV_{\text{disk}} = \int_0^h \pi R^2 \left(1 - \frac{z}{h}\right)^2 dz = -\frac{h}{3}\pi R^2 \left[\left(1 - \frac{z}{h}\right)^3\right]_0^h = \frac{h}{3}\pi R^2,$$

which is a well-known result, namely, the volume of the cone is equal to one-third its height times the area of its base.

For the sphere, we might be tempted to use the same process, namely slice the sphere into horizontal disks of infinitesimal thickness dz along the vertical axis, as shown in Fig. 2.6. The volume of the disk dV_{disk} is given by the same expression as above, but the radius of the disk now takes the form $\rho^2 = R^2 - z^2$, and integrating again over all values of z we obtain

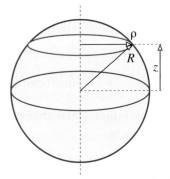

Figure 2.6: A sphere of radius R, sliced in horizontal disks of infinitesimal width dz along the vertical axis (shown in red).

$$V_{\text{sphere}} = \int dV_{\text{disk}} = \int_{-R}^R \pi\rho^2 dz = \pi \int_{-R}^R (R^2 - z^2)dz = \pi \left[R^2 z - \frac{1}{3}z^3\right]_{-R}^R$$

$$\Rightarrow \quad V_{\text{sphere}} = \frac{4\pi}{3}R^3.$$

Alternatively, we can obtain the volume of the sphere by simply integrating the volume element in spherical coordinates over the range of values that each of the variables takes:

$$V_{\text{sphere}} = \int_0^R r^2 dr \int_0^\pi \sin(\theta)d\theta \int_0^{2\pi} d\phi = \frac{1}{3}R^3 \left[-\cos(\theta)\right]_0^\pi 2\pi = \frac{4\pi}{3}R^3.$$

This approach is more direct and simpler because it takes into consideration the full spherical symmetry of the problem by expressing it in terms of the natural coordinates for the symmetry.

Example 2.1: Calculation of the volume of 3D solid objects
As another example of how to choose the right set of coordinates for describing the system, which can significantly simplify the task at hand, we calculate the volume of a donut (a circular tube). The geometric features of the donut are shown in the diagram below.

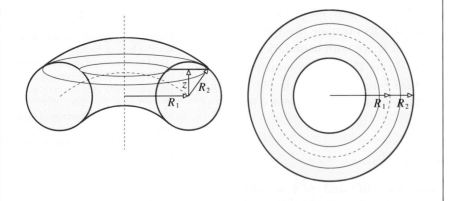

The center of the donut is a circle of radius R_1 on the horizontal plane, and its vertical cross-section at any point is a circle of radius R_2. On the left, is a perspective view of the donut cut in half by a vertical plane; the vertical axis of symmetry is shown by a dashed line and the center of the donut is denoted by a dashed half-circle which lies on the horizontal plane. On the right, is a top view of the donut along its vertical axis. The red circular ring in both cases shows a horizontal section at height $z < R_2$ above the horizontal plane where the donut center lies (dashed circle).

The volume of this object can be calculated using polar coordinates as follows: we consider thin vertical slices which are disks of radius R_2 and width $dw = R_1 d\phi$. The volume of a vertical disk is

$$dV_{\text{disk}} = \pi R_2^2 dw = \pi R_2^2 R_1 d\phi.$$

Integrating over all the values of the polar angle ϕ we obtain the volume of the donut

$$V_{\text{donut}} = \int dV_{\text{disk}} = \int_0^{2\pi} \pi R_2^2 R_1 d\phi = 2\pi^2 R_1 R_2^2.$$

Alternatively, we can consider slicing the donut horizontally, at a height z above the plane where the center of the donut lies, which produces slices that are circular rings of infinitesimal thickness dz along the vertical axis. The inner radius of the circular ring is $\rho_1 = R_1 - \sqrt{R_2^2 - z^2}$ while the outer radius of the ring is $\rho_2 = R_1 + \sqrt{R_2^2 - z^2}$. The volume of the ring is

$$dV_{\text{ring}} = (\pi \rho_2^2 - \pi \rho_1^2)dz = \pi \left[\left(R_1 + \sqrt{R_2^2 - z^2} \right)^2 - \left(R_1 - \sqrt{R_2^2 - z^2} \right)^2 \right] dz$$

$$= 4\pi R_1 \sqrt{R_2^2 - z^2} dz.$$

From this, we can obtain the volume of the donut by integrating over all the values that the variable z takes, namely, $z \in [-R_2, R_2]$:

$$V_{\text{donut}} = \int dV_{\text{ring}} = 4\pi R_1 \int_{-R_2}^{R_2} \sqrt{R_2^2 - z^2} dz.$$

The last integral can be found in tables of integrals [see I. S. Gradshteyn and I.M. Ryzhik, *Table of Integrals, Series and Products* (Academic Press, 1980), c.f. §2.27, p. 86, Eq. (2.271.3)], and leads to the following result for the volume:

$$V_{\text{donut}} = 4\pi R_1 \left[\frac{1}{2}(R_2^2 - z^2)z + \frac{1}{2} R_2^2 \arcsin\left(\frac{z}{R_2} \right) \right]_{-R_2}^{R_2} = 4\pi R_1 \left[\frac{1}{2} \pi R_2^2 \right],$$

which is the same result as before. We note that the second approach based on the horizontal slices of the donut does not take into consideration the natural axial symmetry of the object, which suggests the use of polar coordinates, hence it results in a more laborious calculation.

2.2.3 *Change of variables*

In general, we want to be able to express functions of many independent variables in terms of different sets of such variables, some of which may be more convenient for certain applications than others. We then need to derive the expressions that arise as a result of the transformations from one set to a different set of independent variables. We illustrate this for the case of two independent variables. Suppose that we have two functions of the independent variables x, y, which we call $p(x,y), q(x,y)$, with p, q linearly independent. In principle, we can invert these two functions to obtain the variables x, y as functions of p, q; that is, we can find the relations $x(p,q)$ and $y(p,q)$. Now consider another function, w which depends on x, y, or equivalently, on p, q; since x, y can be expressed in terms of p, q, we can write

$$w(x,y) = w(p,q).$$

We want to be able to express partial derivatives with respect to x, y in terms of partial derivatives with respect to p, q. We first consider the infinitesimal changes in p, q, the quantities dp and dq, which are given by

$$dp = \frac{\partial p}{\partial x}dx + \frac{\partial p}{\partial y}dy, \quad dq = \frac{\partial q}{\partial x}dx + \frac{\partial q}{\partial y}dy. \tag{2.26}$$

Similarly, we can compute the infinitesimal changes in x, y, the quantities dx and dy:

$$dx = \frac{\partial x}{\partial p}dp + \frac{\partial x}{\partial q}dq, \quad dy = \frac{\partial y}{\partial p}dp + \frac{\partial y}{\partial q}dq. \tag{2.27}$$

Next, we calculate the infinitesimal change in the function w, the quantity dw, where w is viewed first as a function of x, y and second as a function of p, q:

$$dw = \frac{\partial w}{\partial x}dx + \frac{\partial w}{\partial y}dy = \frac{\partial w}{\partial p}dp + \frac{\partial w}{\partial q}dq. \tag{2.28}$$

We use the expressions from Eq. (2.26) to substitute dp and dq in the last part of the above equation in terms of dx and dy:

$$dw = \frac{\partial w}{\partial p}\left(\frac{\partial p}{\partial x}dx + \frac{\partial p}{\partial y}dy\right) + \frac{\partial w}{\partial q}\left(\frac{\partial q}{\partial x}dx + \frac{\partial q}{\partial y}dy\right)$$

$$= \left[\left(\frac{\partial p}{\partial x}\partial_p + \frac{\partial q}{\partial x}\partial_q\right)w\right]dx + \left[\left(\frac{\partial p}{\partial y}\partial_p + \frac{\partial q}{\partial y}\partial_q\right)w\right]dy, \tag{2.29}$$

where in the last expression we have simply regrouped term to factor out the dx and dy parts and we employed the notation of Eq. (2.2) for the partial derivatives with respect to the variables p and q. Comparing the last expression for dw to the first part of Eq. (2.28), since x, y are independent variables and therefore dx, dy are independent, we can identify the contents of the square brackets in Eq. (2.29) with $\partial w/\partial x$ and $\partial w/\partial y$; taking into account that w is an arbitrary function, we conclude that:

$$\partial_x = \frac{\partial p}{\partial x}\partial_p + \frac{\partial q}{\partial x}\partial_q, \quad \partial_y = \frac{\partial p}{\partial y}\partial_p + \frac{\partial q}{\partial y}\partial_q. \tag{2.30}$$

Reversing the roles of (x,y) and (p,q), we obtain

$$\partial_p = \frac{\partial x}{\partial p}\partial_x + \frac{\partial y}{\partial p}\partial_y, \quad \partial_q = \frac{\partial x}{\partial q}\partial_x + \frac{\partial y}{\partial q}\partial_y. \tag{2.31}$$

We can put these expressions in matrix notation:

$$\begin{bmatrix} \partial_x \\ \partial_y \end{bmatrix} = \begin{bmatrix} \partial_x p & \partial_x q \\ \partial_y p & \partial_y q \end{bmatrix}\begin{bmatrix} \partial_p \\ \partial_q \end{bmatrix}, \tag{2.32a}$$

$$\begin{bmatrix} \partial_p \\ \partial_q \end{bmatrix} = \begin{bmatrix} \partial_p x & \partial_p y \\ \partial_q x & \partial_q y \end{bmatrix}\begin{bmatrix} \partial_x \\ \partial_y \end{bmatrix}. \tag{2.32b}$$

These two equations show that the 2×2 matrices appearing in them must be the inverse of each other, that is:

$$\begin{bmatrix} \partial_x p & \partial_x q \\ \partial_y p & \partial_y q \end{bmatrix} \begin{bmatrix} \partial_p x & \partial_p y \\ \partial_q x & \partial_q y \end{bmatrix} = \begin{bmatrix} 1 & 0 \\ 0 & 1 \end{bmatrix}.$$

The determinants of the 2×2 transformation matrices are known as the "Jacobians" and are symbolized as

$$\frac{\partial(p,q)}{\partial(x,y)} = \left[(\partial_x p)(\partial_y q) - (\partial_x q)(\partial_y p) \right], \tag{2.33a}$$

$$\frac{\partial(x,y)}{\partial(p,q)} = \left[(\partial_p x)(\partial_q y) - (\partial_p y)(\partial_q x) \right]. \tag{2.33b}$$

As an example, suppose we have a function $\Psi(x,y)$ and we want to use the transformation to polar coordinates to rewrite in terms of (ρ, ϕ). Applying the rules of transformation between the two sets of partial derivatives, Eq. (2.32), we find:

$$\begin{bmatrix} \partial_\rho \\ \partial_\phi \end{bmatrix} = \begin{bmatrix} \cos(\phi) & \sin(\phi) \\ -\rho \sin(\phi) & \rho \cos(\phi) \end{bmatrix} \begin{bmatrix} \partial_x \\ \partial_y \end{bmatrix}, \tag{2.34}$$

$$\begin{bmatrix} \partial_x \\ \partial_y \end{bmatrix} = \begin{bmatrix} \cos(\phi) & -(1/\rho)\sin(\phi) \\ \sin(\phi) & (1/\rho)\cos(\phi) \end{bmatrix} \begin{bmatrix} \partial_\rho \\ \partial_\phi \end{bmatrix}, \tag{2.35}$$

and, as expected, the two 2×2 transformation matrices are the inverse of each other, which can be easily verified by multiplying them.

The Jacobian can be generalized to any number of dimensions and serves to connect the volume elements in different coordinate sets as follows. Suppose we have two sets of coordinates spanning an n-dimensional space, $(x_1, x_2, \ldots, x_n)$ and $(x_1', x_2', \ldots, x_n')$ which are related by

$$x_j = f_j(x_1', x_2', \ldots, x_n'), \quad j = 1, 2, \ldots, n,$$

where f_j are given functions of the n-variables $(x_1', x_2', \ldots, x_n')$. The Jacobian for this transformation is denoted as

$$\frac{\partial(f_1, f_2, \ldots, f_n)}{\partial(x_1', x_2', \ldots, x_n')},$$

and is defined as the determinant of the square matrix of size $n \times n$ whose matrix elements are all the partial derivatives $(\partial f_k / \partial x_j')$, $j, k = 1, \ldots, n$. Then, the infinitesimal volume elements in the two sets of coordinates are related by:

$$dx_1 dx_2 \cdots dx_n = \left| \frac{\partial(f_1, f_2, \ldots, f_n)}{\partial(x_1', x_2', \ldots, x_n')} \right| dx_1' dx_2' \cdots dx_n'. \tag{2.36}$$

For instance, in the case of the transformation from 3D Cartesian coordinates to spherical coordinates we have:

$$(x_1, x_2, x_3) = (x, y, z), \quad \text{and} \quad (x_1', x_2', x_3') = (r, \phi, \theta).$$

Therefore, from Eq. (2.23), we identify the functions $f_j, j = 1, 2, 3$, as

$$f_1(r, \phi, \theta) = r \cos(\phi) \sin(\theta), \quad f_2(r, \phi, \theta) = r \sin(\phi) \sin(\theta), \quad f_3(r, \phi, \theta) = r \cos(\theta).$$

Using these expressions we obtain:

$$
\begin{bmatrix}
\partial_r f_1 & \partial_\phi f_1 & \partial_\theta f_1 \\
\partial_r f_2 & \partial_\phi f_2 & \partial_\theta f_2 \\
\partial_r f_3 & \partial_\phi f_3 & \partial_\theta f_3
\end{bmatrix}
=
\begin{bmatrix}
\cos(\phi)\sin(\theta) & -r\sin(\phi)\sin(\theta) & r\cos(\phi)\cos(\theta) \\
\sin(\phi)\sin(\theta) & r\cos(\phi)\sin(\theta) & r\sin(\phi)\cos(\theta) \\
\cos(\theta) & 0 & -r\sin(\theta)
\end{bmatrix}.
$$

The Jacobian of this matrix can be easily evaluated to obtain the relation between the infinitesimal volume elements in the two sets of coordinates, namely Eq. (2.25), which was derived earlier from geometric considerations.

2.3   Vector operations

Having defined vectors in the various coordinate systems, it is useful to define the operations of addition and multiplication. Before we discuss these operations, we mention that the *length* or *magnitude* $r = |\mathbf{r}|$ of the vector $\mathbf{r}$ in n-dimensions is defined by

$$
\mathbf{r} = x_1 \hat{x}_1 + x_2 \hat{x}_2 + \cdots + x_n \hat{x}_n \Rightarrow r = |\mathbf{r}| = \left(x_1^2 + x_2^2 + \cdots + x_n^2 \right)^{\frac{1}{2}}. \tag{2.37}
$$

The addition of two vectors $\mathbf{r}_1$ and $\mathbf{r}_2$ is a straightforward operation. For instance, in 2D Cartesian coordinates it takes the following form:

$$
\text{if } \mathbf{r}_1 = x_1 \hat{x} + y_1 \hat{y}, \quad \mathbf{r}_2 = x_2 \hat{x} + y_2 \hat{y}, \quad \text{then } \mathbf{r}_1 + \mathbf{r}_2 = (x_1 + x_2)\hat{x} + (y_1 + y_2)\hat{y},
$$

in other words, to add two vectors we simply add the components in each dimension separately, as shown in Fig. 2.7. Similarly, multiplication of a vector $\mathbf{r}$ by a scalar λ is a straightforward operation, namely

$$
\mathbf{r} = x\hat{x} + y\hat{y} + z\hat{z} \Rightarrow \lambda \mathbf{r} = (\lambda x)\hat{x} + (\lambda y)\hat{y} + (\lambda z)\hat{z}.
$$

However, multiplication of a vector with another vector offers the possibility of creating new interesting mathematical objects. The two most common and useful products are referred to as the "dot" product (also called the "scalar" or "inner" product), and the "cross" product (also called the "vector" or "outer" product).

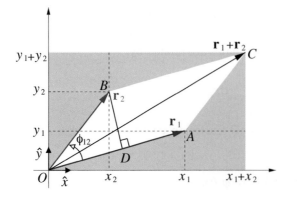

Figure 2.7: Diagram for the definition of products between the 2D vectors $\mathbf{r}_1 = x_1 \hat{x} + y_1 \hat{y}$ (blue arrow, from O to A) and $\mathbf{r}_2 = x_2 \hat{x} + y_2 \hat{y}$ (red arrow, from O to B); the vector sum, $\mathbf{r}_1 + \mathbf{r}_2$, is also shown (black arrow, from O to C); ϕ_{12} is the angle between the two vectors, measured from vector $\mathbf{r}_1$; and the segment OD is the projection of $\mathbf{r}_2$ onto the direction of the $\mathbf{r}_1$, equal to $r_2 \cos(\phi_{12})$.

It is easiest to demonstrate the definition and the geometrical meaning of the dot and cross products using the example of 2D vectors, shown in Fig. 2.7.

2.3.1 The dot or scalar product

For the two vectors with Cartesian coordinates $\mathbf{r}_1 = x_1\hat{x} + y_1\hat{y}$, $\mathbf{r}_2 = x_2\hat{x} + y_2\hat{y}$, the dot product $\mathbf{r}_1 \cdot \mathbf{r}_2$ is defined as

$$\mathbf{r}_1 \cdot \mathbf{r}_2 = x_1 x_2 + y_1 y_2, \tag{2.38}$$

and is a scalar quantity. From geometric considerations, it can be shown that the dot product is equivalent to the expression

$$\mathbf{r}_1 \cdot \mathbf{r}_2 = r_1 r_2 \cos(\phi_{12}), \tag{2.39}$$

where r_1, r_2 are the magnitudes of the two vectors and ϕ_{12} is the angle (in radians) between them; by convention, this angle is measured from the first vector to the second one in an anti-clockwise direction, with $0 \leq \phi_{12} < 2\pi$ (see Problem 1). We note that from the first definition of the dot product, Eq. (2.38), it follows that this product is **symmetric**, namely

$$\mathbf{r}_1 \cdot \mathbf{r}_2 = \mathbf{r}_2 \cdot \mathbf{r}_1.$$

This can also be seen from the second definition of the dot product, Eq. (2.39), because the cosine is an even function of its argument.

From Fig. 2.7 we see that the dot product is equivalent to taking the projection of vector $\mathbf{r}_2$ onto vector $\mathbf{r}_1$, which is equal to $r_2 \cos(\phi_{12})$, and multiplying it with the magnitude of vector $\mathbf{r}_1$. The dot product is a measure of the co-linearity of the two vectors: If they are on the same line and are pointing in the same direction, then $\phi_{12} = 0$ and the dot product assumes its maximum value, namely $r_1 r_2$: the two vectors are collinear and parallel. If they are on the same line but point in opposite directions, then $\phi_{12} = \pi$ and the dot product assumes its minimum value, namely $-(r_1 r_2)$: the two vectors are collinear but anti-parallel. For any other relative orientation of the two vectors, $0 \leq \phi_{12} < 2\pi$, and the dot product takes values between its minimum and maximum values. Finally, when two vectors are orthogonal, $\phi_{12} = \pi/2$ and therefore the dot product vanishes: the two vectors are orthogonal to each other. Indeed, all the coordinate axes we have defined earlier involve orthogonal vectors. This is a convenient choice for handling independent variables.

The dot product is readily generalized to the case of n-dimensional vectors:

$$\mathbf{r}_j = x_1^{(j)}\hat{x}_1 + x_2^{(j)}\hat{x}_2 + \cdots + x_n^{(j)}\hat{x}_n \Rightarrow \mathbf{r}_1 \cdot \mathbf{r}_2 = \sum_{i=1}^{n} x_i^{(1)} x_i^{(2)}, \tag{2.40}$$

where we use the superscript index (j) to identify the different vectors and subscript index i to identify the dimensions in the n-dimensional space.

2.3.2   The cross or vector product

The cross product of the two vectors shown in Fig. 2.7, which lie on the xy-plane, is defined as:

$$\mathbf{r}_1 \times \mathbf{r}_2 = (x_1 y_2 - x_2 y_1)\hat{z}. \tag{2.41}$$

This is different than the dot product in several aspects. First, it is a vector, not a scalar quantity; the direction of this vector, $\hat{z}$, is perpendicular to the plane defined by the two vectors $\mathbf{r}_1$ and $\mathbf{r}_2$. Moreover, this quantity is **antisymmetric**, that is, it changes sign if the order of the two vectors is inverted:

$$\mathbf{r}_2 \times \mathbf{r}_1 = (x_2 y_1 - x_1 y_2)\hat{z} = -\mathbf{r}_1 \times \mathbf{r}_2.$$

We can also deduce from geometric considerations that another equivalent definition of the cross product is (see Problem 2)

$$\mathbf{r}_1 \times \mathbf{r}_2 = r_1 r_2 \sin(\phi_{12})\hat{z}, \tag{2.42}$$

which leads to

$$(x_1 y_2 - x_2 y_1) = r_1 r_2 \sin(\phi_{12}). \tag{2.43}$$

This is an important relation to which we return below.

The generalization of the cross product to three-dimensional vectors is

$$\mathbf{r}_1 \times \mathbf{r}_2 = \begin{vmatrix} \hat{x} & \hat{y} & \hat{z} \\ x_1 & y_1 & z_1 \\ x_2 & y_2 & z_2 \end{vmatrix}$$
$$= (y_1 z_2 - y_2 z_1)\hat{x} + (z_1 x_2 - z_2 x_1)\hat{y} + (x_1 y_2 - x_2 y_1)\hat{z}. \tag{2.44}$$

This expression fulfills several important requirements. First, it reduces to the previous expression, Eq. (2.41), when the two vectors lie on the xy-plane, in which case $z_1 = z_2 = 0$. Second, it produces a vector that is orthogonal to both $\mathbf{r}_1$ and $\mathbf{r}_2$, because its dot products with both $\mathbf{r}_1$ and $\mathbf{r}_2$ vanish,

$$\mathbf{r}_1 \cdot (\mathbf{r}_1 \times \mathbf{r}_2) = 0, \quad \text{and} \quad \mathbf{r}_2 \cdot (\mathbf{r}_1 \times \mathbf{r}_2) = 0,$$

as can be easily checked. This fact implies that the vector represented by the cross product is perpendicular to both $\mathbf{r}_1$ and $\mathbf{r}_2$ and is therefore perpendicular to the entire plane defined by the vectors $\mathbf{r}_1, \mathbf{r}_2$. Finally, this is an antisymmetric product, namely

$$\mathbf{r}_2 \times \mathbf{r}_1 = -(\mathbf{r}_1 \times \mathbf{r}_2),$$

as can easily checked from the definition.

2.3.3   The wedge product

The absolute value of the quantity defined in Eq. (2.43) is the area of the parallelogram $OACB$ shown in Fig. 2.7, but the quantity itself can become negative, as is evident from the expression on the right-hand side, in which r_1 and r_2 are positive (the magnitudes of the vectors $\mathbf{r}_1$ and $\mathbf{r}_2$, respectively), but the sine can take positive or negative values or zero, depending on the value of ϕ_{12}.

Moreover, this expression is *antisymmetric*, because $\phi_{21} = 2\pi - \phi_{12}$ and therefore $\sin(\phi_{21}) = -\sin(\phi_{12})$. Because of these properties, this quantity is useful and appears in other applications beyond the cross product; indeed, it is also denoted by the symbol $\wedge$ and is referred to as the "wedge" product:

$$\mathbf{r}_1 \wedge \mathbf{r}_2 = r_1 r_2 \sin(\phi_{12}), \tag{2.45}$$

which is a scalar quantity.

The definition of the wedge product can be extended to non-vector entities, as long as there is a unique way of defining the angle ϕ_{12} between the two entities whose wedge product we want to calculate. For example, we can define the wedge product of any pair of Cartesian coordinates, say the pair (dx, dy), as follows:

$$dx \wedge dy = dx\ dy\ \sin\left(\phi_{\hat{x}\hat{y}}\right) = dx\ dy\ \sin\left(\frac{\pi}{2}\right) = dx\ dy. \tag{2.46}$$

Since the wedge product is antisymmetric by construction, when applied to the infinitesimals of any pair of Cartesian coordinates it gives

$$dy \wedge dx = -dx \wedge dy. \tag{2.47}$$

Indeed, applying the analogous expression of Eq. (2.46) for the pair (dy, dx) we obtain

$$dy \wedge dx = dy\ dx\ \sin\left(\phi_{\hat{y}\hat{x}}\right) = dy\ dx\ \sin\left(\frac{3\pi}{2}\right) = -dx\ dy.$$

Hence,

$$dw \wedge dw = 0, \tag{2.48}$$

with w being any of the Cartesian coordinates.

It turns out that the wedge product plays an important role in the theory of the so-called *differential forms*, discussed in Section 2.5.

2.4   *Differential operators*

We refer to functions of many variables as "fields", since they are often used to represents fields with physical meaning. These can be scalar quantities, which we will denote as $\Psi(x_1, x_2, \ldots, x_n)$, or vector quantities, which we will denote as

$$\mathbf{F}(x_1, x_2, \ldots, x_n) = F_1(x_1, x_2, \ldots, x_n)\hat{x}_1 + \cdots + F_n(x_1, x_2, \ldots, x_n)\hat{x}_n, \tag{2.49}$$

with each one of the functions $F_j(x_1, x_2, \ldots, x_n), j = 1, \ldots, n$ being a scalar field, and $\hat{x}_j$ the corresponding unit vectors along each independent dimension.

Certain special combinations of the partial derivatives are called **differential operators**. For a function of the n variables $(x_1, x_2, \ldots, x_n)$, which is a *scalar* quantity, the basic differential operator in n-dimensional space is the **gradient**, defined below:

$$\text{gradient}: \nabla_{\mathbf{r}} \equiv \sum_{i=1}^{n} \frac{\partial}{\partial x_i} \hat{x}_i, \tag{2.50}$$

which, when applied to the function $f(x_1, \ldots, x_n)$ takes the form:

$$\nabla_{\mathbf{r}} f = \frac{\partial f}{\partial x_1} \hat{x}_1 + \cdots + \frac{\partial f}{\partial x_n} \hat{x}_n.$$

This has the same meaning as the derivative, only now it has acquired directionality, expressed by the presence of the unit vectors $\hat{x}_i$. The physical meaning of the gradient of an n-dimensional function is that it gives the direction in which the function changes most rapidly and toward higher values, at a given point in the n-dimensional space.

By analogy to the second derivative, we define the **Laplacian** operator in n-dimensional space as:

$$\text{Laplacian}: \ \nabla_{\mathbf{r}}^2 \equiv \sum_{i=1}^{n} \frac{\partial^2}{\partial x_i^2}. \tag{2.51}$$

When the gradient is zero, if all three partial second derivatives are positive we have a *local minimum* of the field, if they are all negative, we have a *local maximum* of the field. If some are positive and some are negative, we have *saddle points*, that is, points in which the function increases along certain directions (the ones along which the partial second derivatives are positive) and decreases along other directions (the ones along which the partial second derivatives are negative). Our definition of the unit vectors along the orthogonal axes leads to

$$\nabla_{\mathbf{r}}^2 = \nabla_{\mathbf{r}} \cdot \nabla_{\mathbf{r}},$$

that is, we can write the Laplacian as the dot product of the gradient with itself.

The gradient and the Laplacian of the 2D scalar function $\Psi(x, y)$ in Cartesian and in polar coordinates, the second obtained from the first by using the transformation of the partial derivatives, Eqs. (2.34), (2.35), give

$$\nabla_{\mathbf{r}} \Psi(\mathbf{r}) = \frac{\partial \Psi}{\partial x} \hat{x} + \frac{\partial \Psi}{\partial y} \hat{y} = \frac{\partial \Psi}{\partial \rho} \hat{\rho} + \frac{1}{\rho} \frac{\partial \Psi}{\partial \phi} \hat{\phi}, \tag{2.52}$$

and

$$\nabla_{\mathbf{r}}^2 \Psi(\mathbf{r}) = \frac{\partial^2 \Psi}{\partial x^2} + \frac{\partial^2 \Psi}{\partial y^2} = \frac{1}{\rho} \frac{\partial}{\partial \rho} \left(\rho \frac{\partial \Psi}{\partial \rho} \right) + \frac{1}{\rho^2} \frac{\partial^2 \Psi}{\partial \phi^2}. \tag{2.53}$$

Example 2.2: Consider the function

$$\Psi(x, y) = A_1 e^{-[(x-x_1)^2 + (y-y_1)^2]/2\sigma_1^2} + A_2 e^{-[(x-x_2)^2 + (y-y_2)^2]/2\sigma_2^2}, \tag{2.54}$$

which is a sum of two two-dimensional Gaussians, the first centered at (x_1, y_1) with standard deviation σ_1 in both the x and y dimensions and strength A_1, and the second centered at (x_2, y_2) with standard deviation σ_2 in both the x and y dimensions and strength A_2. For $A_1 < 0, A_2 < 0$, this function represents two wells as shown in the two diagrams below: the first diagram shows the behavior of the function in terms of contours of equal height as in a topographic map, while the second diagram depicts the function in terms of a perspective three-dimensional (3D) figure with the values of the function displayed on the vertical axis.

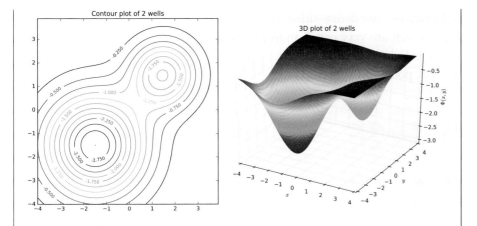

The gradient of this function is given by the following expression:

$$\nabla_{\mathbf{r}}\Psi(x,y) = -\frac{A_1}{\sigma_1^2}e^{-[(x-x_1)^2+(y-y_1)^2]/2\sigma_1^2}\left[(x-x_1)\hat{x} + (y-y_1)\hat{y}\right]$$
$$- \frac{A_2}{\sigma_2^2}e^{-[(x-x_2)^2+(y-y_2)^2]/2\sigma_2^2}\left[(x-x_2)\hat{x} + (y-y_2)\hat{y}\right].$$

The gradient is a vector field and is represented by the magnitude and direction of arrows, calculated on a uniform grid on the (x,y) plane.

The Laplacian of this function is given by the following expression:

$$\nabla_{\mathbf{r}}^2\Psi(x,y) = \frac{A_1}{\sigma_1^2}e^{[(x-x_1)^2+(y-y_1)^2]/2\sigma_1^2}\left[\frac{(x-x_1)^2}{\sigma_1^2} + \frac{(y-y_1)^2}{\sigma_1^2} - 2\right]$$
$$+ \frac{A_2}{\sigma_2^2}e^{[(x-x_2)^2+(y-y_2)^2]/2\sigma_2^2}\left[\frac{(x-x_2)^2}{\sigma_2^2} + \frac{(y-y_2)^2}{\sigma_2^2} - 2\right].$$

The Laplacian is a scalar field, and is represented by contours of constant value.

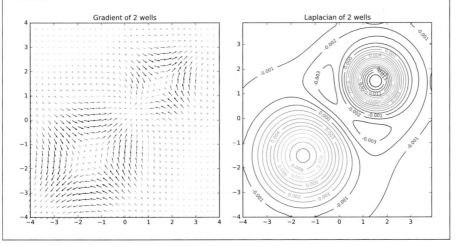

When dealing with vector fields, additional operators can be defined that involve vector products of the gradient with the field $\mathbf{F}(\mathbf{r})$, namely, the

$$\text{divergence}: \nabla_{\mathbf{r}} \cdot \mathbf{F},$$

and the

$$\text{curl}: \nabla_{\mathbf{r}} \times \mathbf{F}.$$

Instead of giving the most general expressions, we define these differential operators for the case of 3D vector fields which are very common in physical applications. We give these expressions for all three coordinate systems in 3D that we have mentioned, namely, Cartesian, polar and spherical.

Cartesian coordinates

$$\nabla_{\mathbf{r}} \Psi(\mathbf{r}) = \frac{\partial \Psi}{\partial x} \hat{x} + \frac{\partial \Psi}{\partial y} \hat{y} + \frac{\partial \Psi}{\partial z} \hat{z}, \tag{2.55}$$

$$\nabla_{\mathbf{r}}^2 \Psi(\mathbf{r}) = \frac{\partial^2 \Psi}{\partial x^2} + \frac{\partial^2 \Psi}{\partial y^2} + \frac{\partial^2 \Psi}{\partial z^2}, \tag{2.56}$$

$$\mathbf{F}(\mathbf{r}) = F_1(x, y, z)\hat{x} + F_2(x, y, z)\hat{y} + F_3(x, y, z)\hat{z}, \tag{2.57}$$

$$\nabla_{\mathbf{r}} \cdot \mathbf{F} \equiv \frac{\partial F_1}{\partial x} + \frac{\partial F_2}{\partial y} + \frac{\partial F_3}{\partial z}, \tag{2.58}$$

$$\nabla_{\mathbf{r}} \times \mathbf{F} \equiv \left(\frac{\partial F_3}{\partial y} - \frac{\partial F_2}{\partial z} \right) \hat{x} + \left(\frac{\partial F_1}{\partial z} - \frac{\partial F_3}{\partial x} \right) \hat{y} + \left(\frac{\partial F_2}{\partial x} - \frac{\partial F_1}{\partial y} \right) \hat{z}. \tag{2.59}$$

Polar coordinates

$$\nabla_{\mathbf{r}} \Psi(\mathbf{r}) = \frac{\partial \Psi}{\partial \rho} \hat{\rho} + \frac{1}{\rho} \frac{\partial \Psi}{\partial \phi} \hat{\phi} + \frac{\partial \Psi}{\partial z} \hat{z}, \tag{2.60}$$

$$\nabla_{\mathbf{r}}^2 \Psi(\mathbf{r}) = \frac{1}{\rho} \frac{\partial}{\partial \rho} \left(\rho \frac{\partial \Psi}{\partial \rho} \right) + \frac{1}{\rho^2} \frac{\partial^2 \Psi}{\partial \phi^2} + \frac{\partial^2 \Psi}{\partial z^2}, \tag{2.61}$$

$$\mathbf{F}(\mathbf{r}) = F_1(\rho, \phi, z)\hat{\rho} + F_2(\rho, \phi, z)\hat{\phi} + F_3(\rho, \phi, z)\hat{z}, \tag{2.62}$$

$$\nabla_{\mathbf{r}} \cdot \mathbf{F}(\mathbf{r}) = \frac{1}{\rho} \frac{\partial}{\partial \rho} (\rho F_1) + \frac{1}{\rho} \frac{\partial F_2}{\partial \phi} + \frac{\partial F_3}{\partial z}, \tag{2.63}$$

$$\nabla_{\mathbf{r}} \times \mathbf{F}(\mathbf{r}) = \left(\frac{1}{\rho} \frac{\partial F_3}{\partial \phi} - \frac{\partial F_2}{\partial z} \right) \hat{\rho} + \left(\frac{\partial F_1}{\partial z} - \frac{\partial F_3}{\partial \rho} \right) \hat{\phi} + \left(\frac{\partial F_2}{\partial \rho} - \frac{1}{\rho} \frac{\partial F_1}{\partial \phi} + \frac{F_2}{\rho} \right) \hat{z}. \tag{2.64}$$

Spherical coordinates

$$\nabla_{\mathbf{r}} \Psi(\mathbf{r}) = \frac{\partial \Psi}{\partial r} \hat{r} + \frac{1}{r} \frac{\partial \Psi}{\partial \theta} \hat{\theta} + \frac{1}{r \sin(\theta)} \frac{\partial \Psi}{\partial \phi} \hat{\phi}, \tag{2.65}$$

$$\nabla_{\mathbf{r}}^2 \Psi(\mathbf{r}) = \frac{1}{r^2} \frac{\partial}{\partial r} \left(r^2 \frac{\partial \Psi}{\partial r} \right) + \frac{1}{r^2 \sin(\theta)} \frac{\partial}{\partial \theta} + \frac{1}{r^2 \sin^2 \theta} \frac{\partial^2 \Psi}{\partial \phi^2}, \tag{2.66}$$

$$\mathbf{F}(\mathbf{r}) = F_1(r, \theta, \phi)\hat{r} + F_2(r, \theta, \phi)\hat{\theta} + F_3(r, \theta, \phi)\hat{\phi}, \tag{2.67}$$

$$\nabla_{\mathbf{r}} \cdot \mathbf{F}(\mathbf{r}) = \frac{1}{r^2} \frac{\partial (r^2 F_1)}{\partial r} + \frac{1}{r \sin(\theta)} \frac{\partial (\sin(\theta) F_2)}{\partial \theta} + \frac{1}{r \sin(\theta)} \frac{\partial F_3}{\partial \phi}, \tag{2.68}$$

$$\nabla_{\mathbf{r}} \times \mathbf{F}(\mathbf{r}) = \frac{1}{r\sin(\theta)}\left[\frac{\partial(\sin(\theta)F_3)}{\partial\theta} - \frac{\partial F_2}{\partial\phi}\right]\hat{r} + \frac{1}{r}\left[\frac{1}{\sin(\theta)}\frac{\partial F_1}{\partial\phi} - \frac{\partial(rF_3)}{\partial r}\right]\hat{\theta}$$
$$+ \frac{1}{r}\left[\frac{\partial(rF_2)}{\partial r} - \frac{\partial F_1}{\partial\theta}\right]\hat{\phi}. \tag{2.69}$$

A useful relation that can be easily proved in any set of coordinates is that curl of the gradient of scalar field is identically zero:

$$\nabla_{\mathbf{r}} \times \nabla_{\mathbf{r}}\Psi(\mathbf{r}) = 0. \tag{2.70}$$

For example, using Cartesian coordinates, we can define $\mathbf{F}(\mathbf{r}) = \nabla_{\mathbf{r}}\Psi(\mathbf{r})$, and then we have

$$F_1 = \frac{\partial\Psi}{\partial x}, \quad F_2 = \frac{\partial\Psi}{\partial y}, \quad F_3 = \frac{\partial\Psi}{\partial z}.$$

Then the x component of the curl becomes

$$\left(\frac{\partial F_3}{\partial y} - \frac{\partial F_2}{\partial z}\right) = \left(\frac{\partial^2\Psi}{\partial z\partial y} - \frac{\partial\Psi}{\partial y\partial z}\right) = 0,$$

and similarly for the y and z components. Another relation of the same type is that the divergence of the curl of a vector field vanishes identically:

$$\nabla_{\mathbf{r}} \cdot \nabla_{\mathbf{r}} \times \mathbf{F} = \partial_x\left(\frac{\partial F_3}{\partial y} - \frac{\partial F_2}{\partial z}\right) + \partial_y\left(\frac{\partial F_1}{\partial z} - \frac{\partial F_3}{\partial x}\right) + \partial_z\left(\frac{\partial F_2}{\partial x} - \frac{\partial F_1}{\partial y}\right)$$
$$\Rightarrow \nabla_{\mathbf{r}} \cdot (\nabla_{\mathbf{r}} \times \mathbf{F}(\mathbf{r})) = 0. \tag{2.71}$$

A third useful relation expresses the curl of the curl of a vector field in terms of its divergence and its Laplacian (see Problem 3):

$$\nabla_{\mathbf{r}} \times (\nabla_{\mathbf{r}} \times \mathbf{F}(\mathbf{r})) = \nabla_{\mathbf{r}}(\nabla_{\mathbf{r}} \cdot \mathbf{F}(\mathbf{r})) - \nabla_{\mathbf{r}}^2\mathbf{F}(\mathbf{r}). \tag{2.72}$$

The last three results can be intuitively understood by thinking of $\nabla_{\mathbf{r}}$ as a common vector, say $\mathbf{A}$, the gradient of a scalar function, $\nabla_{\mathbf{r}}f(\mathbf{r})$, as a vector *parallel* to $\mathbf{A}$, say $\lambda\mathbf{A}$, and the curl of a vector function, $\nabla_{\mathbf{r}} \times \mathbf{F}(\mathbf{r})$, as a vector *perpendicular* to both $\mathbf{A}$ and $\mathbf{F}(\mathbf{r})$, say $\mathbf{A} \times \mathbf{F}$. Then, Eq. (2.70) becomes the equivalent of the vector identity:

$$\mathbf{A} \times (\lambda\mathbf{A}) = 0,$$

Eq. (2.71) becomes the equivalent of the vector identity:

$$\mathbf{A} \cdot (\mathbf{A} \times \mathbf{F}) = 0,$$

and Eq. (2.72) becomes the equivalent of the vector identity:

$$\mathbf{A} \times (\mathbf{A} \times \mathbf{F}) = \mathbf{A}(\mathbf{A} \cdot \mathbf{F}) - (\mathbf{A} \cdot \mathbf{A})\mathbf{F}.$$

These three vector identities are trivial to prove from the definitions of the dot and cross products for common vectors.

2.5 The vector calculus integrals

Certain important rules apply when the integrand involves vector fields and the differential operators applied to such fields. We discuss here the fundamental theorems that apply in the following cases:

1. How the integral of a vector field's divergence over a closed surface S is related to the integral of the field over the volume Ω enclosed by this surface; this is referred to as the "divergence theorem" or "Gauss' theorem".

2. How the integral of a vector field's curl over a finite surface S is related to the integral of the field along the closed curve C that forms the boundary of this surface referred to as the "curl theorem" or "Stokes' theorem".

Another important theorem that underlies the Gauss' and Stokes' theorems, relates the integral of a field along a closed surface to the integral of certain combinations of the partial derivatives of the field over a volume enclosed by the surface, which is known as "Green's theorem". Although these theorems apply to n-dimensional spaces, we will discuss them here in the context of two-dimensional (2D) space.

2.5.1 Simple curves and simply-connected domains

Since we will be working in two dimensions, the equivalent of a "surface" in 2D is a line on the plane, and the equivalent of a "volume" is a region (domain) of the plane, enclosed by a curve. In order to state the theorems precisely and to provide a proof for the Green's and Gauss' theorems in 2D, we need to introduce the notions of the "simple closed curve" and the corresponding "simply-connected domain" enclosed by a simple closed curve.

A simply-connected domain is a region of the plane such that its bounding curve does not intersect itself, as shown in Fig. 2.8. This requirement is necessary to ensure two conditions: first that it is possible to define a unique sense of traversing the boundary of the domain, and second that all interior points of the domain are accessible by other interior points without having to cross the boundary. If the boundary curve intersects itself, then it is not possible to decide unambiguously the contour traversal direction at the intersection points. Moreover, there exist regions that are "cut-off" from other regions, in the sense that they can be accessed only by crossing the boundary, as is shown in the examples of Fig. 2.8. The boundary of the simply-connected domain is a simple curve. If the bounding curve is not simple (it intersects itself at one or more points), the domain is called "multiply connected".

In the proofs of the Green's and Gauss' theorems in two dimensions, we will assume without loss of generality that the simple curve forming the boundary of a simply-connected domain has a unique extreme left point, that is, all points in the neighborhood of this point lie to its right (at higher values of x than the value of x corresponding to the extreme left point). Similarly, we will assume that the contour enclosing the simple domain has unique extreme right, extreme top and extreme bottom points, as illustrated in Fig. 2.9. The definitions of these points are analogous to the one for the extreme left point. If this is not the case, then we can break the integration over the entire domain

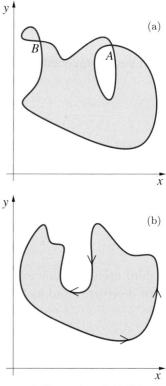

Figure 2.8: Illustration of: (a) "Non-simple" contour enclosing a multiply-connected domain (blue region), which does not have a unique sense of traversing; at the points marked A and B there is an ambiguity on how to continue traversing the contour. (b) "Simple" contour that encloses a simply-connected domain (blue region); the contour can be traversed in a unique sense (in this example counter-clockwise).

into smaller domains, each of which satisfy the above requirement of having extreme left, right, top and bottom points. The volume integral over the entire domain is the sum over these subdomains, and the additional surface integrals introduced by the subdivisions cancel because they are traversed in opposite directions in the two integrals that share a common dividing surface, which leaves the total surface integral unchanged. An example is shown in Fig. 2.9.

2.5.2 Green's theorem

Green's theorem relates the integral of the dot product of vector field $\mathbf{F}(\mathbf{r})$ with the infinitesimal displacement $\mathbf{ds}$ along a closed path to the integral of the partial derivatives of the field components over the area enclosed by the path. We define the vector field as

$$\mathbf{F}(\mathbf{r}) = F_1(x,y)\hat{x} + F_2(x,y)\hat{y},$$

the simple domain as Ω, and the simple closed curve C that forms its boundary as $C = \partial\Omega$; the infinitesimal displacement $\mathbf{ds}$ along the boundary is

$$\mathbf{ds} = \mathrm{d}x\,\hat{x} + \mathrm{d}y\,\hat{y}.$$

With these definitions, Green's theorem states:

$$\text{Green's theorem}: \quad \oint_{\partial\Omega} \mathbf{F}(\mathbf{r})\cdot\mathbf{ds} = \int_{\Omega}\left[\frac{\partial F_2}{\partial x} - \frac{\partial F_1}{\partial y}\right]\mathrm{d}\Omega, \qquad (2.73)$$

where the last integral is a two-dimensional integral over the volume Ω with the infinitesimal volume element $\mathrm{d}\Omega = \mathrm{d}x\,\mathrm{d}y$.

To prove Green's theorem, we start by expanding the dot product in the surface integral:

$$\oint_{\partial\Omega}\mathbf{F}(\mathbf{r})\cdot\mathbf{ds} = \oint_{\partial\Omega} F_1(x,y)\mathrm{d}x + \oint_{\partial\Omega} F_2(x,y)\mathrm{d}y.$$

The first term in the last expression for the surface integral, integrated around the closed contour $\partial\Omega$ in the counter-clockwise sense starting at the point with abscissa x_1 as shown in Fig. 2.10, gives:

$$\oint_{\partial\Omega} F_1(x,y)\mathrm{d}x = \int_{x_1}^{x_2} F_1(x,y_B)\mathrm{d}x + \int_{x_2}^{x_1} F_1(x,y_T)\mathrm{d}x$$

$$= \int_{x_1}^{x_2}[F_1(x,y_B) - F_1(x,y_T)]\mathrm{d}x = \int_{x_1}^{x_2}\left[-\int_{y_B(x)}^{y_T(x)}\frac{\partial F_1}{\partial y}\mathrm{d}y\right]\mathrm{d}x$$

$$= -\iint_{\Omega}\frac{\partial F_1}{\partial y}\mathrm{d}y\mathrm{d}x,$$

where $F_1(x,y_B)$ and $F_2(x,y_T)$ refer to the values that $F_1(x,y)$ takes while the independent variable x spans the values from x_1 to x_2; the dependent variable y takes the values prescribed by the curve $\partial\Omega$ (the contour of integration), where y is defined through $y = y_B(x)$ or $y = y_T(x)$, in the *top* and *bottom* parts of the contour, respectively, as those are identified in Fig. 2.10.

The second term in the last expression for the surface integral, integrated around the closed contour $\partial\Omega$ in the anticlockwise sense starting at the point

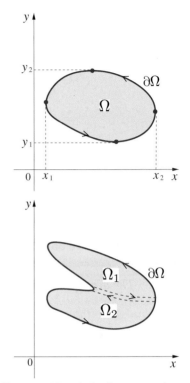

Figure 2.9: **Top:** A simply-connected domain Ω bounded by a simple closed contour $\partial\Omega$ that possesses unique extreme left–right, at x_1 and x_2, respectively, and extreme low–high points, at y_1 and y_2, respectively. **Bottom:** Example of a bounding contour $\partial\Omega$ that does not fulfill the requirements of a shape with identifiable extreme points, as it has two possible extreme left points. The domain can be broken into two parts, Ω_1 and Ω_2, by introducing the dividing surfaces shown as red dashed lines, which are traversed in opposite directions in the surface integral and therefore their contributions cancel.

with ordinate y_1 as shown in Fig. 2.10, gives

$$\oint_{\partial\Omega} F_2(x,y)dy = \int_{y_1}^{y_2} F_2(x_R,y)dy + \int_{y_2}^{y_1} F_2(x_L,y)dy$$

$$= \int_{y_1}^{y_2} [F_2(x_R,y) - F_2(x_L,y)]dy = \int_{y_1}^{y_2}\left[\int_{x_L(y)}^{x_R(y)} \frac{\partial F_2}{\partial x}dx\right]dy$$

$$= \iint_\Omega \frac{\partial F_2}{\partial x}dxdy,$$

where $F_2(x_R,y)$ and $F_2(x_L,y)$ refer to the values that $F_2(x,y)$ takes while the independent variable y spans the values from y_1 to y_2; the dependent variable x takes the values prescribed by the curve $\partial\Omega$ (the contour of integration), where x is defined through $x = x_L(y)$ or $x = x_R(y)$, in the *left* and *right* parts of the contour, respectively, as those identified in Fig. 2.10. Putting together the two results, we obtain the expression for Green's theorem, Eq. (2.73).

2.5.3 The divergence and curl theorems

The divergence theorem: The divergence theorem, also referred to as **Gauss'** **theorem**, relates the integral of the divergence of the vector field $\mathbf{G}(\mathbf{r})$, denoted as $[\nabla_\mathbf{r} \cdot \mathbf{G}(\mathbf{r})]$, over a volume Ω to the integral of the vector field over the surface $S = \partial\Omega$ which encloses the volume of integration:

$$\text{Gauss' theorem}: \quad \int_\Omega [\nabla_\mathbf{r} \cdot \mathbf{G}(\mathbf{r})]\, d\Omega = \oint_{S=\partial\Omega} \mathbf{G}(\mathbf{r}) \cdot \hat{\mathbf{n}}_S\, dS, \quad (2.74)$$

where $\hat{\mathbf{n}}_S$ is the surface-normal unit vector on the surface element dS.

The meaning of Gauss' theorem can be intuitively understood by considering the one-dimensional situation. In this case, we are dealing with a finite domain for the single variable x, namely $x \in [a,b]$, and a scalar function $g(x)$. Then, the role of the divergence is played by the derivative of the function $g(x)$, and the domain boundary consists simply of the endpoints, $x = a$ and $x = b$, while the surface normal vector is the unit vector $\hat{x}$ at the point b and $-\hat{x}$ at the point a, pointing in both cases outward from the domain. Thus, the left-hand side of Eq. (2.74) becomes the integral of the derivative of the function $g(x)$ over the domain $[a,b]$, while the right-hand side becomes the sum of the values $g(b)$ and $-g(a)$. With these considerations, Gauss' theorem for this case takes the following form:

$$\int_a^b \frac{dg}{dx}dx = g(b) - g(a),$$

which is the fundamental theorem of calculus (see Section 1.4).

We provide next a more formal proof of Gauss' theorem in two dimensions. In this case, the "volume" is a domain Ω on the (x,y) plane, and its "surface" $\partial\Omega$ is a closed curve. The proof proceeds as follows: we define the components of the vector field $\mathbf{G}(\mathbf{r})$ to be G_1 and G_2,

$$\mathbf{G}(x,y) = G_1(x,y)\hat{x} + G_2(x,y)\hat{y}.$$

This implies that the volume integral of the divergence satisfies

$$\int_\Omega [\nabla_\mathbf{r} \cdot \mathbf{G}(\mathbf{r})]\, d\Omega = \int_\Omega \left[\frac{\partial G_1}{\partial x} + \frac{\partial G_2}{\partial y}\right] dxdy. \quad (2.75)$$

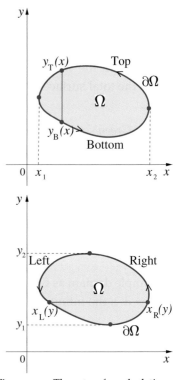

Figure 2.10: The setup for calculating the surface integrals in Green's theorem, Eq. (2.73): the first surface term is obtained with the help of the top diagram and the second surface term with the help of the bottom diagram.

For the surface integral along the curve $\partial\Omega$, we define the unit *tangent* vector $\hat{\mathbf{t}}_S$ at the point (x,y), which is given by

$$\hat{\mathbf{t}}_S = \cos(\phi)\hat{x} + \sin(\phi)\hat{y},$$

where $\tan(\phi) = \mathrm{d}y/\mathrm{d}x$ is the slope of the tangent at this point, as shown in Fig. 2.11.

Using the expression for the unit tangent vector, we deduce that the unit *normal* vector, $\hat{\mathbf{n}}_S$, which must be perpendicular to the tangent vector, satisfies the condition $\hat{\mathbf{t}}_S \cdot \hat{\mathbf{n}}_S = 0$, see Fig. 2.11. Hence, the unit normal vector is given by the following expression:

$$\hat{\mathbf{n}}_S = \sin(\phi)\hat{x} - \cos(\phi)\hat{y} = \frac{1}{\sqrt{1+\tan^2(\phi)}} [\tan(\phi)\hat{x} - \hat{y}]. \tag{2.76}$$

Note that when defining the normal unit vector we have two possible choices, the one given in Eq. (2.76) and the opposite of it, since both are orthogonal to the tangent unit vector. By convention, the normal unit vector is chosen so that it points to the "exterior" of the curve C if it is a closed curve, or, equivalently to the "right" of the curve if it traversed in the anticlockwise direction, as shown in Fig. 2.11. Substituting $\tan(\phi) = \mathrm{d}y/\mathrm{d}x$ in the expression for the unit normal vector we find

$$\hat{\mathbf{n}}_S = \frac{1}{\sqrt{1+(\mathrm{d}y/\mathrm{d}x)^2}} [\hat{x}(\mathrm{d}y/\mathrm{d}x) - \hat{y}] = \frac{1}{\mathrm{d}S} [\hat{x}\mathrm{d}y - \hat{y}\mathrm{d}x],$$

where we have used for the infinitesimal length along the surface the expression

$$\mathrm{d}S = \sqrt{(\mathrm{d}x)^2 + (\mathrm{d}y)^2},$$

as is evident from Fig. 2.11. With these results, the surface integral takes the form:

$$\oint_{\partial\Omega} \mathbf{G}(\mathbf{r}) \cdot \hat{\mathbf{n}}_S \, \mathrm{d}S = \oint_{\partial\Omega} (G_1(x,y)\hat{x} + G_2(x,y)\hat{y}) \cdot \left(\frac{1}{\mathrm{d}S} [\hat{x}\mathrm{d}y - \hat{y}\mathrm{d}x] \right) \mathrm{d}S$$
$$= \oint_{\partial\Omega} G_1(x,y)\mathrm{d}y - \oint_{\partial\Omega} G_2(x,y)\mathrm{d}x. \tag{2.77}$$

This last expression is precisely the type of expression that appears in Green's theorem, and therefore with the correspondence $G_2 \to -F_1, \quad G_1 \to F_2$, by applying Green's theorem we find

$$\oint_{\partial\Omega} G_1(x,y)\mathrm{d}y - \oint_{\partial\Omega} G_2(x,y)\mathrm{d}x = \iint_{\Omega} \left[\frac{\partial G_1}{\partial x} + \frac{\partial G_2}{\partial y} \right] \mathrm{d}x\mathrm{d}y, \tag{2.78}$$

which is the same result as the volume integral of the divergence, Eq. (2.75). This completes the proof of Gauss' theorem.

The curl theorem: The curl theorem, also referred to as **Stokes' theorem**, relates the integral of the curl of the vector field $\mathbf{H}(\mathbf{r})$, denoted as $[\nabla_{\mathbf{r}} \times \mathbf{H}(\mathbf{r})]$, over a surface S to the line integral of the field along the closed contour C which is the boundary of the surface of integration, a fact denoted as $C = \partial S$:

$$\text{Stokes' theorem} : \int_S [\nabla_{\mathbf{r}} \times \mathbf{H}(\mathbf{r})] \cdot \hat{\mathbf{n}}_S \, \mathrm{d}S = \oint_{C=\partial S} \mathbf{H}(\mathbf{r}) \cdot \mathbf{ds}, \tag{2.79}$$

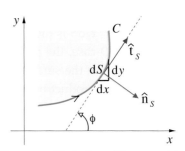

Figure 2.11: Illustration of the geometric relations between the infinitesimals $\mathrm{d}S$, $\mathrm{d}x$ and $\mathrm{d}y$, and the normal unit vector $\hat{\mathbf{n}}_S$ and tangent unit vector $\hat{\mathbf{t}}_S$, at a point of the directed curve C (green line); the shaded region corresponds to the "interior" of the curve. The expression $\tan(\phi)$ is the slope of the tangent, shown by a dashed line, at this point.

where $\hat{\mathbf{n}}_S$ is the surface-normal unit vector associated with the surface element dS and $\mathbf{ds} = \hat{\mathbf{t}}_S dS$ is the differential vector along the contour $C = \partial S$.

In the 2D case, the proof of the curl theorem is straightforward. First, we note that since the surface must lie on the xy-plane, its normal unit vector is $\hat{\mathbf{n}}_S \hat{z}$. The dot product between the curl of the vector field and the normal unit vector is then simply the z-component of the curl, which according to Eq. (2.59) is given by

$$[\nabla_{\mathbf{r}} \times \mathbf{H}(\mathbf{r})] \cdot \hat{\mathbf{n}}_S = [\nabla_{\mathbf{r}} \times \mathbf{H}(\mathbf{r})] \cdot \hat{z} = \frac{\partial H_2}{\partial x} - \frac{\partial H_1}{\partial y}.$$

In the present case, S plays the role of the domain Ω, so from Green's theorem, Eq. (2.73), we find that the integral of the above expression over this 2D domain is related to the integral along the boundary of the domain, the closed curve $C = \partial S$:

$$\int_S \left[\frac{\partial H_2}{\partial x} - \frac{\partial H_1}{\partial y} \right] dS = \oint_{\partial S} \mathbf{H} \cdot \mathbf{ds}.$$

This completes the proof of Stokes' theorem.

Derivation of Green's and Gauss' theorems using Poincaré's lemma

It turns out that Green's, Gauss' and Stokes' theorems are particular cases of a general result known as "Poincaré's lemma". This result states that the integral of a *differential form* $\mathcal{F}$ over an $(n-1)$-dimensional space denoted as $\partial\Omega$, equals the integral of $d\mathcal{F}$ over the associated n-dimensional space denoted as Ω:

$$\int_{\partial\Omega} \mathcal{F} = \int_\Omega d\mathcal{F}. \tag{2.80}$$

The theory of differential forms is based on the wedge operator, which was discussed in Section 2.3. We provide an elementary introduction to differential forms below.

Let F_j, with $j = 1, \ldots, n$, be n functions depended on the variables $x_1, \ldots, x_n$; the usual differential dF_j of the function F_j is given by

$$dF_j = \frac{\partial F_j}{\partial x_1} dx_1 + \cdots + \frac{\partial F_j}{\partial x_n} dx_n. \tag{2.81}$$

The differential form $\mathcal{F}$ is defined by

$$\mathcal{F} = \sum_{j=1}^n F_j dx_j, \tag{2.82}$$

and the differential of $\mathcal{F}$ is given by the expression

$$d\mathcal{F} = \sum_{j=1}^n dF_j \wedge dx_j = \sum_{j=1}^n \left(\frac{\partial F_j}{\partial x_1} dx_1 + \cdots + \frac{\partial F_j}{\partial x_n} dx_n \right) \wedge dx_j. \tag{2.83}$$

Thus, in calculating the differential $d\mathcal{F}$, each infinitesimal change dF_j must be multiplied by the corresponding infinitesimal change dx_j in a wedge sense, which includes weighing by the sine of the angle between the directions of the two entities. dF_j acquires direction through its dependence on all the infinitesimals $dx_i, i = 1, \ldots, n$, each with a coefficient equal to $\partial F_j / \partial x_i$, by analogy to a vector in n-dimensional space that has components along each direction $\hat{x}_i$. In particular, the terms

$$\frac{\partial F_j}{\partial x_j}\mathrm{d}x_j \wedge \mathrm{d}x_j$$

in the summation on the right-hand side of Eq. (2.83) vanish identically because of Eq. (2.48).

As an illustration of the usefulness of Poincaré's lemma, we use Eq. (2.80) to derive Green's theorem. Letting the differential form $\mathcal{F}$ be

$$\mathcal{F} = F_1\mathrm{d}x + F_2\mathrm{d}y,$$

and using Eq. (2.83) to calculate $\mathrm{d}\mathcal{F}$ we find the following:

$$\mathrm{d}\mathcal{F} = \left(\frac{\partial F_1}{\partial x}\mathrm{d}x + \frac{\partial F_1}{\partial y}\mathrm{d}y\right)\wedge \mathrm{d}x + \left(\frac{\partial F_2}{\partial x}\mathrm{d}x + \frac{\partial F_2}{\partial y}\mathrm{d}y\right)\wedge \mathrm{d}y$$

$$= \left(\frac{\partial F_2}{\partial x} - \frac{\partial F_1}{\partial y}\right)\mathrm{d}x\mathrm{d}y,$$

where we have used the properties of the wedge product, Eq. (2.47) and Eq. (2.48) for $w = x$ and $w = y$. Using the above expressions for $\mathcal{F}$ and $\mathrm{d}\mathcal{F}$ in Eq. (2.80) we obtain Green's theorem.

Similarly, letting the differential form $\mathcal{G}$ be

$$\mathcal{G} = \mathbf{G}\cdot \hat{\mathbf{n}}_S\mathrm{d}S = (G_1\hat{x} + G_2\hat{y})\cdot (\hat{x}\mathrm{d}y - \hat{y}\mathrm{d}x) = G_1\mathrm{d}y - G_2\mathrm{d}x,$$

and using Eq. (2.83) to calculate $\mathrm{d}\mathcal{G}$ we find the following:

$$\mathrm{d}\mathcal{G} = \left(\frac{\partial G_1}{\partial x}\mathrm{d}x + \frac{\partial G_1}{\partial y}\mathrm{d}y\right)\wedge \mathrm{d}y - \left(\frac{\partial G_2}{\partial x}\mathrm{d}x + \frac{\partial G_2}{\partial y}\mathrm{d}y\right)\wedge \mathrm{d}x$$

$$= \left(\frac{\partial G_1}{\partial x} + \frac{\partial G_2}{\partial y}\right)\mathrm{d}x\mathrm{d}y = [\nabla_{\mathbf{r}}\cdot \mathbf{G}(\mathbf{r})]\,\mathrm{d}\Omega.$$

Using the above expressions for $\mathcal{G}$ and $\mathrm{d}\mathcal{G}$ in Eq. (2.80) we obtain Gauss' theorem.

Finally, as we mentioned above, in the 2D case Stokes' theorem is equivalent to Green's theorem because the surface normal, $\hat{z}$, projects out the z-component of the curl, which leads to the same expression for the integral of the curl as in Green's theorem. Thus, the proof of Stokes' theorem in 2D using Poincaré's lemma is the same as the proof of Green's theorem.

2.5.4 Geometric interpretation of the divergence and curl theorems

The divergence and curl theorems are useful in many contexts where a physical quantity of interest is described by a vector field. This could be, for example, the velocity field of a fluid, or the components of the electromagnetic field. To gain some insight on how these theorems emerge, we provide a geometric interpretation of these theorems in the 2D case.

The integral of the field over the surface appearing on the right-hand side of Gauss' theorem, Eq. (2.74), defines the net *flux*, which we denote by Ψ, through a surface S of a physical quantity described by the vector field $\mathbf{G}(\mathbf{r})$.

Thus, Gauss' theorem states that this flux is equal to the strength of the sources or sinks of this physical quantity; the sources or sinks are described by the divergence of the field over the volume enclosed by the surface. For example, the net flux of an incompressible fluid through the surface S, whose flow is described by the vector field $\mathbf{G}(\mathbf{r})$, is equal to the net amount of fluid produced by sources (corresponding to positive $\nabla_{\mathbf{r}} \cdot \mathbf{G}(\mathbf{r})$) or consumed by sinks (corresponding to negative $\nabla_{\mathbf{r}} \cdot \mathbf{G}(\mathbf{r})$) within the volume enclosed by the surface.

To see how the divergence arises, we tile the surface within the closed curve C by squares of infinitesimal size $dx = dy = 2\epsilon \to 0$, as shown in Fig. 2.12. The flux through one such square centered at $\mathbf{r} = x\hat{x} + y\hat{y}$ is given by the net flow entering and leaving the square, as shown in Fig. 2.13. The flux through the surface dS (a segment of the curve) is given by the product $\mathbf{G} \cdot \hat{\mathbf{n}}_S dS$; thus, we need to calculate this product for each side of the square and sum all the contributions. From the diagram of Fig. 2.13, with the field given by $\mathbf{G}(\mathbf{r}) = G_1(x,y)\hat{x} + G_2(x,y)\hat{y}$, we find:

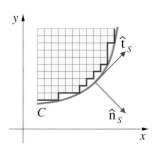

Figure 2.12: Illustration of the tiling of the interior of a curve C (green line) by squares of infinitesimal size $dx = dy = 2\epsilon$; the sides of squares which are missing one or more neighbors (thicker blue line) coincide with the bounding curve C in the limit $\epsilon \to 0$.

$$\mathbf{G}^{(1)} \cdot \hat{\mathbf{n}}_S^{(1)} dS^{(1)} = \mathbf{G}^{(1)} \cdot \hat{x} dy = G_1^{(1)} dy,$$

$$\mathbf{G}^{(2)} \cdot \hat{\mathbf{n}}_S^{(2)} dS^{(2)} = \mathbf{G}^{(2)} \cdot \hat{y} dx = G_2^{(2)} dx,$$

$$\mathbf{G}^{(3)} \cdot \hat{\mathbf{n}}_S^{(3)} dS^{(3)} = \mathbf{G}^{(3)} \cdot (-\hat{x} dy) = -G_1^{(3)} dy,$$

$$\mathbf{G}^{(4)} \cdot \hat{\mathbf{n}}_S^{(4)} dS^{(4)} = \mathbf{G}^{(4)} \cdot (-\hat{y} dx) = -G_2^{(4)} dx.$$

Evidently, in this simple example, flux is flowing into the square at the sides labeled "3" and "4", and is flowing out of the square at the sides labeled "1" and "2". For the net amount of flux $d\Psi$ flowing through the square, we sum over its four sides to obtain:

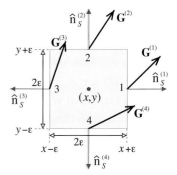

Figure 2.13: A zoomed-in representation of an infinitesimal square centered at the point (x, y) (see Fig. 2.12), with a field $\mathbf{G}^{(i)}$ (thick black arrows) and the normal unit vectors $\hat{\mathbf{n}}_S^{(i)}$ (red arrows) on the sides labeled $i = 1, 2, 3, 4$, for the calculation of the flux.

$$\begin{aligned} d\Psi &= \left[G_1^{(1)} - G_1^{(3)}\right] dy + \left[G_2^{(2)} - G_2^{(4)}\right] dx \\ &= \left[G_1(x+\epsilon, y) - G_1(x-\epsilon, y)\right] dy + \left[G_2(x, y+\epsilon) - G_2(x, y-\epsilon)\right] dx \\ &= \left[\frac{\partial G_1}{\partial x} dx\right] dy + \left[\frac{\partial G_2}{\partial y} dy\right] dx = \left[\frac{\partial G_1}{\partial x} + \frac{\partial G_2}{\partial y}\right] dx dy = \left[\nabla_{\mathbf{r}} \cdot \mathbf{G}(\mathbf{r})\right] dx dy, \end{aligned}$$

where we have used the identities

$$G_j(x+\epsilon, y) - G_j(x-\epsilon, y) = \frac{\partial G_j}{\partial x} dx, \tag{2.84a}$$

$$G_j(x, y+\epsilon) - G_j(x, y-\epsilon) = \frac{\partial G_j}{\partial y} dy, \tag{2.84b}$$

for $j = 1, 2$ and $dx = dy = 2\epsilon$, which are valid in the limit $\epsilon \to 0$. We observe that when we sum over all the infinitesimal squares, in the interior region the contributions coming from the sides of two adjacent square that are in contact are traversed in opposite directions and therefore cancel. Thus, the only contributions arise from the sides of squares which are missing one or more neighbors; but in the limit $\epsilon \to 0$ these sides become the bounding curve, as shown in Fig. 2.12. Therefore, the integral of the flux Ψ over the interior of the curve, which we denote as Ω, must be equal to the contribution along the curve which is the boundary of Ω, denoted as $S = \partial\Omega$:

$$\Psi = \int_{\Omega} d\Psi = \int_{\Omega} \left[\nabla_{\mathbf{r}} \cdot \mathbf{G}(\mathbf{r})\right] dx dy = \int_{S} \mathbf{G} \cdot \hat{\mathbf{n}}_S dS.$$

The last equation is Gauss' theorem, Eq. (2.74).

In an analogous fashion, we may be interested in calculating the *vorticity* $\mathcal{V}$ (or *circulation*) of a field $\mathbf{H}$ around a closed curve which encloses a surface S. We again tile the interior of the curve by infinitesimal squares, as shown in Fig. 2.12. The vorticity around the perimeter of a square is given by the product $\mathbf{H} \cdot \hat{\mathbf{t}}_S dS$; thus, we need to calculate this product for each side of the square and sum all the contributions. From the diagram of Fig. 2.14, with the field given by $\mathbf{H}(\mathbf{r}) = H_1(x,y)\hat{x} + H_2(x,y)\hat{y}$, we find

$$\mathbf{H}^{(1)} \cdot \hat{\mathbf{t}}_S^{(1)} dS^{(1)} = \mathbf{H}^{(1)} \cdot \hat{y} dy = H_2^{(1)} dy,$$
$$\mathbf{H}^{(2)} \cdot \hat{\mathbf{t}}_S^{(2)} dS^{(2)} = \mathbf{H}^{(2)} \cdot (-\hat{x}) dx = -H_1^{(2)} dx,$$
$$\mathbf{H}^{(3)} \cdot \hat{\mathbf{t}}_S^{(3)} dS^{(3)} = \mathbf{H}^{(3)} \cdot (-\hat{y} dy) = -H_2^{(3)} dy,$$
$$\mathbf{H}^{(4)} \cdot \hat{\mathbf{t}}_S^{(4)} dS^{(4)} = \mathbf{H}^{(4)} \cdot \hat{x} dx = H_1^{(4)} dx.$$

Summing over the four sides of this square we obtain

$$
\begin{aligned}
d\mathcal{V} &= \left[H_2^{(1)} - H_2^{(3)} \right] dy - \left[H_1^{(2)} - H_1^{(4)} \right] dx \\
&= \left[H_2(x+\epsilon, y) - H_2(x-\epsilon, y) \right] dy - \left[H_1(x, y+\epsilon) - H_1(x, y-\epsilon) \right] dx \\
&= \left[\frac{\partial H_2}{\partial x} dx \right] dy - \left[\frac{\partial H_1}{\partial y} dy \right] dx = \left[\frac{\partial H_2}{\partial x} - \frac{\partial H_1}{\partial y} \right] dx dy \\
&= \left[\nabla_{\mathbf{r}} \times \mathbf{H}(\mathbf{r}) \right] \cdot \hat{z} \, dx dy,
\end{aligned}
$$

where we have used identities for the $\mathbf{H}$ field analogous to those of Eq. (2.84) for the $\mathbf{G}$ field. The same cancellation as mentioned above takes place in the interior of the region enclosed by the boundary. Therefore, the integral of the vorticity over the interior of the curve, denoted here as S, must be equal to the contribution along the curve which forms the boundary of S, denoted as $C = \partial S$:

$$\mathcal{V} = \int_S d\mathcal{V} = \int_S \left[\nabla_{\mathbf{r}} \times \mathbf{H}(\mathbf{r}) \right] \cdot \hat{z} \, dx dy = \int_C \mathbf{H} \cdot \hat{\mathbf{t}}_S dS.$$

Recognizing that $\hat{z} = \hat{\mathbf{n}}_S$ is the normal unit vector corresponding to the 2D surface element $dx dy = dS$, and that $\hat{\mathbf{t}}_S dS = d\mathbf{s}$ is the infinitesimal vector along the direction of the curve C, we obtain Stokes' theorem, Eq. (2.79).

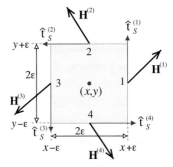

Figure 2.14: A zoomed-in representation of an infinitesimal square centered at the point (x,y) (see Fig. 2.12), with a field $\mathbf{H}^{(i)}$ and the tangent unit vectors $\hat{\mathbf{t}}_S^{(i)}$ (blue arrows) for the calculation of the vorticity.

2.6 Function optimization

2.6.1 Finding the extrema of a function

In the study of functions, especially in high-dimensional spaces, it is often desirable to determine the extrema of the function, namely the points at which the function assumes a local minimum or maximum value. We illustrate how this can be accomplished by using information about the derivative of the function. As discussed in Chapter 1, Section 1.3.3, the extrema of a function of one variable occur at points where the derivative is zero. The actual type of the extremum (a minimum or a maximum) depends on the value of the second derivative at the point where the first derivative vanishes.

The vanishing of the derivative at the extrema suggests a simple strategy (in what follows, we assume the small change Δx in the value of the variable x is always positive):

- If we are searching for a *minimum*, then:

  - if the derivative is *negative* at a point x, we *increase* the value of x by a small amount to $(x + \Delta x)$, since this will bring us to a lower value of the function (see Fig. 2.15);

  - if the derivative is *positive* at a point x, we *decrease* x by a small amount to $(x - \Delta x)$, since this will bring us to a lower value of the function (see Fig. 2.15).

    After changing the value of x we evaluate again the derivative at the new position. This process can be repeated until the derivative becomes vanishingly small, in which case we have reached the desired minimum value. If the derivative changes sign before it becomes zero, this means that we have passed the desired minimum value. In this case, we simply *decrease* the size of the step Δx and repeat the process from the current value of x.

- If we are searching for a *maximum*, the process is similar as in the search for a minimum, but we make each time changes in the position of x that are *opposite* to those for the minimum search, namely, if the derivative at x is *positive* we *increase* x to $(x + \Delta x)$, and if the derivative at x is *negative* we *decrease* x to $(x - \Delta x)$.

In order to generalize the above discussion to the case of functions of many variables, we make the following observation: we can associate the *direction of change* in the value of x with the positive, $+\hat{x}$, or the negative, $-\hat{x}$ direction along the x-axis. With this identification, it is evident from Fig. 2.15 that in order to reach a minimum of the function from either side of it, the *direction* of change in x must be *opposite* to the sign of derivative. This motivates the generalization to the case of a multi-dimensional function: in that case, the role of the derivative is played by the gradient; we recall that the gradient is a vector, and therefore points toward a specific direction in the multi-dimensional space where the function is defined. Accordingly, in order to move from any given point toward a *minimum* of the function we must move in a direction *opposite* to the gradient. This approach is referred to as "gradient descent" or "steepest descent" toward the minimum. Conversely, in order to move from any given point toward a *maximum* of the function we must move along the direction of the gradient. Of course, when the extremum point has been reached, the gradient vanishes, just like the derivative in the single-variable case.

To complete the discussion of finding the extrema of a function, it is important to point out that the processes we described above only guarantee that we can reach iteratively a *local* extremum, namely, the point closest to the starting point where the function assumes a minimum or maximum value. Once this extremum has been reached, the derivative (or gradient) becomes zero and the iterative search has converged and stops. However, there is no guarantee that the value we have reached is the *global* extremum of the function, which may be an important objective. Finding the global extremum, when it exists, is a much more challenging problem when dealing with multi-dimensional spaces and requires sophisticated search procedures.

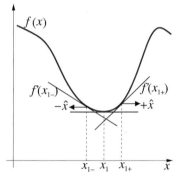

Figure 2.15: Illustration of the process of finding an extremum (minimum or maximum) of a function. The inclined lines labeled $f'(x_{1-})$ and $f'(x_{1+})$ represent the slope (derivative) of the function at points just to the left and just to the right, respectively, of the point x_1. At x_1 the function has a minimum so $f'(x_1) = 0$, as indicated by the horizontal (zero slope) red line. The unit vectors $+\hat{x}$ and $-\hat{x}$, whose sign is that of the slope at x_{1+} and x_{1-} respectively, point in the direction *opposite* which the value of x must change to approach the minimum at x_1.

2.6.2   Constrained optimization: Lagrange multipliers

It is often useful to find the extrema of a function subject to a condition that its arguments must satisfy. This is referred to as "constrained optimization".

To motivate the need for such a process, we use the example of a function $f(x,y)$ in a two-dimensional space, which exhibits a so-called "saddle point" at (x_0, y_0). The saddle-point behavior is shown schematically in Fig. 2.16: $f(x,y)$ assumes a value which is both a *maximum* along a certain direction (marked as AEB in Fig. 2.16) as well as a *minimum* along a *different* direction (marked as CED in Fig. 2.16) in the two-dimensional domain of the function. A constrained minimization would, for instance, correspond to searching for the minimum of the function $f(x,y)$ while moving along the direction of the path CED in Fig. 2.16, which we can assume to be described by the relation $g(x,y) = 0$.

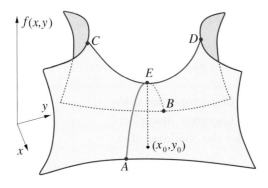

Figure 2.16: Illustration of "saddle point" behavior of a function of two variables, $f(x,y)$: the function's values are represented by the surface outlined in blue lines. The function assumes a maximum value along the green path and a minimum value along the red path, at the same point, denoted as (x_0, y_0).

The task of constrained optimization can be accomplished by employing a very powerful concept, the so-called "Lagrange multipliers". To illustrate the method of Lagrange multipliers, suppose we wish to find the minimum of the function $f(x,y)$ under the constraint that $g(x,y) = 0$, where $g(x,y)$ is a different function than $f(x,y)$. The constraint implies that the two variables x and y are not independent. In fact, we can imagine solving the constraint equation for y in terms of x, introducing this $y(x)$ in the original expression for $f(x,y)$ which then becomes a function of x only, and then minimizing with respect to x, as usual. However, finding y in terms of x from the constraint $g(x,y) = 0$ may not be easy or even feasible. The method of Lagrange multipliers provides an alternative way to solve the problem.

The Lagrange-multiplier method works as follows: We introduce the Lagrange multiplier λ, and construct the new function $F(x,y,\lambda)$:

$$F(x,y,\lambda) = f(x,y) - \lambda g(x,y). \tag{2.85}$$

We then minimize the function $F(x,y,\lambda)$ without imposing any constraint. The function $F(x,y,\lambda)$ has three independent variables, since the value of λ is not known. The conditions that lead to minimization of $F(x,y,\lambda)$ are as follows:

$$\frac{\partial F}{\partial x} = 0 \quad \Rightarrow \quad \frac{\partial f}{\partial x} - \lambda \frac{\partial g}{\partial x} = 0, \tag{2.86a}$$

$$\frac{\partial F}{\partial y} = 0 \quad \Rightarrow \quad \frac{\partial f}{\partial y} - \lambda \frac{\partial g}{\partial y} = 0, \tag{2.86b}$$

$$\frac{\partial F}{\partial \lambda} = 0 \quad \Rightarrow \quad g(x,y) = 0. \tag{2.86c}$$

This system of three equations with three unknowns can in principle be solved for the values that satisfy all three equations. Assuming that such a solution can be found, the last condition ensures that the original constraint is satisfied.

The reason why this approach works is the following: Finding the extremum of f under the condition $g = 0$ implies that the problem *cannot* be reduced to finding the values of the variables for which *all* partial derivatives of f vanish, as in the case of a normal extremum. Indeed, the only requirements are that $df = 0$ and $dg = 0$, that is, in the case of functions of two variables (x, y),

$$df = f_x dx + f_y dy = 0, \quad \text{and} \quad dg = g_x dx + g_y dy = 0,$$

where the first relation holds at the point of the extremum and the second relation holds everywhere because of the constraint. However, the variables x and y are no longer *independent* because of the constraint, so the partial derivatives f_x, f_y do not have to vanish at the extremum; they must simply take values that make the differential df vanish. From Eqs. (2.86a) and (2.86b), we conclude that

$$f_x = \lambda g_x \quad \text{and} \quad f_y = \lambda g_y \Rightarrow df = \lambda(g_x dx + g_y dy) = \lambda dg = 0,$$

because $dg = 0$. Thus, by making the partial derivatives of f proportional to those of g, with the *same constant of proportionality* (the Lagrange multiplier λ), we make sure that df vanishes at the extremum under the condition that $g = 0$, and this is truly independent of the value of λ.

In practice, the equations relating the partial derivatives of f to those of g can be used to eliminate the unknown λ, and this step is often sufficient to lead to a solution. The Lagrange multiplier method is readily extended to more variables and more constraints, with each constraint requiring the introduction of a different Lagrange multiplier.

2.6.3   Calculus of variations

A different type of function optimization is to determine the behavior of a function in a range of values of its arguments, relative to all possible functional behaviors and subject to certain constraints. To handle this situation we need to introduce the concept of "functions of functions", which are called "functionals". We denote a functional as $F(x, y)$, where y is a function of x; the dependence of the functional on the variable x can be implicit, through the dependence of y on x, or explicit, if there exist x-dependent terms in F in addition to the dependence on y. The calculus of functionals is referred to as "calculus of variations".

We start with the general expression for a "functional derivative": The derivative with respect to a variable y that appears either in the limits of integration or in the integrand of a definite integral over a variable x, can be computed as follows:

$$\frac{d}{dy} \int_{f_1(y)}^{f_2(y)} F(x, y) dx = \int_{f_1(y)}^{f_2(y)} \frac{\partial F}{\partial y}(x, y) dx$$
$$+ \frac{d f_2(y)}{dy} F(x = f_2(y), y) - \frac{d f_1(y)}{dy} F(x = f_1(y), y),$$

which is an application of the chain rule for differentiation and the expression of Eq. (1.48).

We next consider the functional $F[y]$ of the function $y = f(x)$. Typically, this functional does not depend on x explicitly; for example, it may involve an integration over all values of x. In this case, the infinitesimal change in $F[y]$ is due to the small change in the function $f(x)$, which is denoted as $\delta y = \delta f(x)$ [in this section, we use the symbol "δ" to denote differentiation with respect to a *function*, in distinction to the symbols "d" and "∂" used to denote differentiation with respect to a *variable*]. To represent this small change, we choose an arbitrary function $\eta(x)$ and multiply it by the small constant ϵ,

$$\delta f(x) = \epsilon\eta(x).$$

By appropriately choosing ϵ, $\delta f(x)$ can be very small everywhere, no mater what the values of $\eta(x)$ are in the interval of interest (provided they are finite). We also impose the condition that the function $\eta(x)$ vanishes identically at the boundaries of the interval spanned by the variable x:

$$\text{for } x \in [x_1, x_2] \ : \ \eta(x_1) = \eta(x_2) = 0,$$

as illustrated in Fig. 2.17. The values x_1, x_2 need not be finite.

The functional derivative of $F[f(x)]$ with respect to variations in $f(x)$ is given by

$$\frac{\partial F}{\partial y}(x, y) = \frac{\delta F[f(x)]}{\delta f(x)}.$$

The change in the functional $F(y)$ due to the change in y by δy is, to first order in δy,

$$F(y + \delta y) \approx F(y) + \int \frac{\partial F}{\partial y}\delta y \mathrm{d}x,$$

where we are assuming that the functional involves an integration over the values of x. Using in this expression the quantities defined earlier we find

$$F[f(x) + \delta f(x)] = F[f(x)] + \epsilon\int_{x_1}^{x_2} \frac{\delta F[f(x)]}{\delta f(x)}\eta(x)\mathrm{d}x. \tag{2.87}$$

For a stationary point in $F[f(x)]$, that is, an extremum (minimum or maximum) with respect to variations in $f(x)$, we must have to first order in $\delta f(x)$ that

$$F[f(x) + \delta f(x)] - F[f(x)] = 0.$$

Thus, the term proportional to ϵ in Eq. (2.87) must vanish. Since $\eta(x)$ is an arbitrary function, this requires that the functional derivative of $F[f(x)]$ with respect to changes in $f(x)$ vanish identically for all x:

$$\text{stationary condition}: \ \frac{\delta F[f(x)]}{\delta f(x)} = 0, \quad \forall x \in [x_1, x_2].$$

This is the equivalent of the first derivative of an ordinary function vanishing at the extrema of the function.

An important class of functionals are those that can be cast in the form below:

$$F[y] = \int_{x_1}^{x_2} G(y, y_x, x)\mathrm{d}x \Rightarrow F[f(x)] = \int_{x_1}^{x_2} G(f(x), f_x(x), x)\mathrm{d}x. \tag{2.88}$$

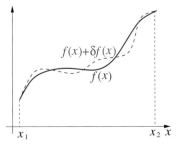

Figure 2.17: Illustration of the idea of variation of a function $f(x)$, given by $\delta f(x) = \epsilon\eta(x)$. The variation vanishes at the points of the interval, where $\eta(x_1) = 0 = \eta(x_2)$.

In this case, the functional derivative of $F[f(x)]$ with respect to small changes $\delta f(x)$ is given by

$$\frac{\delta F[f(x)]}{\delta f(x)} = \int_{x_1}^{x_2} \left[\frac{\partial G}{\partial f} \delta f(x) + \frac{\partial G}{\partial f_x} \delta f_x(x) \right] dx.$$

The small variations in $f_x(x)$ satisfy the following:

$$\delta f_x(x) = \delta \left(\frac{df}{dx} \right) = \frac{d}{dx} (\delta f(x)) = \frac{d}{dx} (\epsilon \eta(x)) = \epsilon \eta_x(x).$$

Using this result in the expression for the functional derivative of $F[f(x)]$ in terms of partial derivatives of $G(f, f_x, x)$, we find

$$\frac{\delta F[f(x)]}{\delta f(x)} = \epsilon \int_{x_1}^{x_2} \left[\frac{\partial G}{\partial f} \eta(x) + \frac{\partial G}{\partial f_x} \eta_x(x) \right] dx.$$

Performing an integration by parts in the second term and using the fact that $\eta(x)$ vanishes identically at the endpoints of the integral, we arrive at

$$\frac{\delta F[f(x)]}{\delta f(x)} = \epsilon \int_{x_1}^{x_2} \left[\frac{\partial G}{\partial f} - \frac{d}{dx} \left(\frac{\partial G}{\partial f_x} \right) \right] \eta(x) dx.$$

For a stationary point, we use again the fact that $\eta(x)$ is arbitrary to find:

$$\text{stationary condition :} \quad \frac{\partial G}{\partial f} - \frac{d}{dx} \left(\frac{\partial G}{\partial f_x} \right) = 0, \quad \forall x \in [x_1, x_2]. \qquad (2.89)$$

This relation is known as the "Euler–Lagrange" equation.

Example 2.3: Curve of shortest distance between two points

As an example of how the Euler–Lagrange equation can be used in practice we solve the following problem: Find the curve that gives the shortest distance between two points (x_1, y_1) and (x_2, y_2) on a plane.

We consider all possible curves passing through these points; these curves are characterized by functions $y = f(x)$ such that $f(x_1) = y_1$ and $f(x_2) = y_2$ [in the following, $f_x(x)$ and $f_{xx}(x)$ denote the first and second derivative of $f(x)$, respectively]. The infinitesimal length dl along the curve $y = f(x)$ is given by

$$dl = \left[(dx)^2 + (dy)^2 \right]^{1/2} = \left[1 + \left(\frac{dy}{dx} \right)^2 \right]^{1/2} dx = \left[1 + (f_x(x))^2 \right]^{1/2} dx.$$

Thus, the total length L is given by

$$L[f(x)] = \int_{x_1}^{x_2} \left[1 + (f_x(x))^2 \right]^{1/2} dx.$$

We want to find the minimum of this functional with respect to variations in $f(x)$, subject to the boundary conditions at the points x_1, x_2, that is, the variations must vanish identically at the endpoints of the interval. This is a functional precisely of the type defined in Eq. (2.88) with

$$G(f, f_x, x) = \left[1 + (f_x(x))^2 \right]^{1/2}.$$

To find its stationary point, we calculate the partial derivatives of the function $G(f, f_x, x)$ with respect to f and f_x:

$$\frac{\partial G}{\partial f} = 0, \quad \frac{\partial G}{\partial f_x} = \left[1 + (f_x(x))^2\right]^{-1/2} f_x(x),$$

and apply the Euler–Lagrange equation, which leads to:

$$\frac{\mathrm{d}}{\mathrm{d}x}\left(\frac{\partial G}{\partial f_x}\right) = \frac{\mathrm{d}}{\mathrm{d}x}\left\{\left[1 + (f_x(x))^2\right]^{-1/2} f_x(x)\right\}$$
$$= \left[1 + (f_x(x))^2\right]^{-3/2} f_{xx}(x) = 0.$$

For a function $f(x)$ that takes real, finite values in the interval $[x_1, x_2]$, the term in square brackets in the last expression cannot be zero, which implies

$$f_{xx}(x) = 0, \quad \forall x \in [x_1, x_2] \Rightarrow f(x) = \alpha x + \beta,$$

where α, β are real constants. Thus, as expected, the curve that minimizes the distance between two points on a plane is a straight line.

2.7 Application: Feedforward neural network

A neural network is a model for representing functions of many variables. This type of model is gaining a lot of attention for its versatility and applicability to many different types of problems. The topic is very rich, and its full treatment is beyond the scope of this book. However, the main idea of the model can be presented in terms of the concepts we have discussed so far. Suppose we have a number of values represented as an L-dimensional vector $\mathbf{x} = (x_1, \ldots, x_L)$, which we will call the **input**; from the input we obtain a number of other values represented as an M-dimensional vector $\tilde{\mathbf{y}} = (\tilde{y}_1, \ldots, \tilde{y}_M)$, which we will call the **output**. The number of values involved in the input vector $\mathbf{x}$ are referred to as **features**. We view this as a functional relation, that is, from the *known* vector $\mathbf{x}$ (the values x_l, $l = 1, \ldots, L$), we obtain the *desired* vector $\tilde{\mathbf{y}}$ (the values $\tilde{y}_m$, $m = 1, \ldots, M$), but the functional form that transforms $\mathbf{x}$ to $\tilde{\mathbf{y}}$ is unknown. In general, the size of the input vector L and the output vector M are unequal, and typically $L > M$.

Since the functional relation between $\mathbf{x}$ and $\tilde{\mathbf{y}}$ is unknown, we construct a **model**, consisting of a set of linear and nonlinear transformations of $\mathbf{x}$ that lead to $\tilde{\mathbf{y}}$; the transformations involve parameters which are called the **weights**. If the model was perfect, the output values would be the same as the desired values $\tilde{y}_m$, $m = 1, \ldots, M$; typically, the output of the model and desired values differ, with this difference being referred to as the "error". The usefulness of neural networks is based on the expectation that the model can be used for other values of x, beyond those in the input set, to predict what the output of the model would be for these new values. This has applications to diverse fields including image and speech recognition, signal analysis, and many fields of science for analyzing large datasets and making useful predictions such as predicting earthquakes and the weather.

In the following, we illustrate this concept in the simple case where the input is provided (is "fed") to the network which then produces the desired output through a sequence of transformations of the information contained in the input. This type of network is referred to as a "feedforward" neural network.

2.7.1 Definition of feedforward neural network

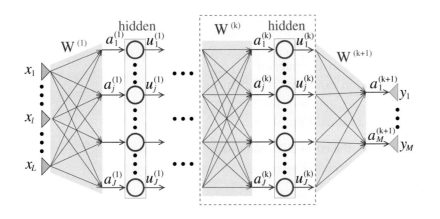

Figure 2.18: Schematic illustration of the neural network scheme consisting of the input, $x_1, \ldots, x_L$, linear transforms by the matrices $\mathbf{W}^{(1)}$ and $\mathbf{W}^{(k)}$, each followed by one hidden layer of nonlinear transforms (red circles), the last linear transform by the matrix $\mathbf{W}^{(k+1)}$ and the output $y_1, \ldots, y_M$. The basic unit, consisting of the linear transform $\mathbf{W}^{(k)}$ followed by the nonlinear transform $u_j^{(k)} = S(a_j^{(k)})$, is shown highlighted in gray.

The transformations that define the network, illustrated schematically in Fig. 2.18, are as follows:

1. A linear transformation of the input data:

$$a_j^{(1)} = w_{j0}^{(1)} + \sum_{l=1}^{L} w_{jl}^{(1)} x_l, \quad j = 1, \ldots, J,$$

where $w_{jl}^{(1)}$ are the elements of the matrix $\mathbf{W}^{(1)}$ of size $J \times L$, and $w_{j0}^{(1)}$ are the elements of a column of length J, which is considered as part of the matrix; this matrix contains the weights that need to be determined. For a more concise notation, we define $x_0 = 1$ and write this relation as

$$a_j^{(1)} = \sum_{l=0}^{L} w_{jl}^{(1)} x_l, \quad j = 1, \ldots, J. \tag{2.90}$$

Each linear combination of the features is referred to as a **node**; thus, there are J nodes. In general, $J > L$.

2. A nonlinear transformation of the new variables $a_j^{(1)}$:

$$u_j^{(1)} = S(a_j^{(1)}), \quad j = 1, \ldots, J. \tag{2.91}$$

3. The combination of a linear transform followed by a nonlinear one can be repeated at will; each such step is called a "hidden layer". A network with more than one hidden layer is called a "deep network". For example, to create more hidden layers we take:

$$a_j^{(k)} = w_{j0}^{(k)} + \sum_{i=1}^{J} w_{ji}^{(k)} u_i^{(k-1)} \longrightarrow u_j^{(k)} = S(a_j^{(k)}), \quad j = 1, \ldots, J, \tag{2.92}$$

with the variables $u_i^{(k-1)}$ being the output of the previous nonlinear transform, and use the values of the new variables $u_j^{(k)}$ as input to the next linear transform. Each such step introduces a new matrix of weights, $\mathbf{W}^{(k)}$.

4. A final linear transformation of the output of the previous step:

$$y_m = a_m^{(k+1)} = w_{m0}^{(k+1)} + \sum_{j=1}^{J} w_{mj}^{(k+1)} u_j^{(1)}, \quad m = 1, \ldots, M. \qquad (2.93)$$

The last result is identified as the model output, y_m, where $w_{mj}^{(k+1)}$ are the elements of a matrix of size $M \times J$ and $w_{m0}^{(k+1)}$ is a column of length M of weights to be determined.

5. A variation on the basic method outlined above is that the last linear transform $\mathbf{W}^{(k+1)}$ can be omitted and the output of the kth nonlinear transform, $u_j^{(k)}$, can be directly identified with the output. This requires choosing the size of the linear transform $\mathbf{W}^{(k)}$ such that the desired number of output values is obtained, that is, we must make sure that for this linear transform $J = M$. Alternatively, we can apply a nonlinear transform to the last set of values $a_m^{(k+1)}$ before we identify them as the output values y_m. These two options are essentially equivalent, differing by just the total number of hidden layers.

6. The error for each value is defined by

$$\varepsilon_m = y_m - \tilde{y}_m, \quad m = 1, \ldots, M. \qquad (2.94)$$

The values of the weight matrices $\mathbf{W}^{(1)}, \ldots, \mathbf{W}^{(k+1)}$ are adjusted so that the error values are minimized; once the error values have been reduced below a certain level, the model produces the desired results. The total number of parameters involved is given by

$$P_{\text{tot}} = J(L+1) + (k-1)J(J+1) + M(J+1) + J. \qquad (2.95)$$

The standard nonlinear transformation of the input data involves the sigmoid function,

$$S(x) = \frac{1}{1+e^{-x}} = \frac{1}{2}\left[1 + \tanh\left(\frac{x}{2}\right)\right], \qquad (2.96)$$

the role of which is similar to the firing of neurons, hence the name of the approach. The derivative of this function is

$$S'(x) = S(x)[1 - S(x)]. \qquad (2.97)$$

In the definition of the error above, we used only one instance of the input and output vectors. In order to properly train the network, we need to supply many examples of the input–output pairs. We label these pairs as

$$x_l^{(n)} \leftrightarrow y_m^{(n)}, \quad n = 1, \ldots, N.$$

Using these pairs, an "error function" can be defined in different ways as a collective measure of the error. A common measure is the "root mean squared error" (RMSE), see Eq. (3.38); in the present case, this error will be denoted as $\mathcal{E}_m^{(2)}$, and is given by

$$\mathcal{E}_m^{(2)} \equiv \left\{ \frac{1}{N} \sum_{n=1}^{N} \left(y_m^{(n)} - \tilde{y}_m^{(n)} \right)^2 \right\}^{1/2}. \qquad (2.98)$$

With this definition, the weights are adjusted to minimize the error, and the error value depends on each output value. For a measure of the overall error, we average over the index m of the output values:

$$\mathcal{E}^{(2)} = \sum_{m=1}^{M} \beta_m \mathcal{E}_m^{(2)}, \quad \sum_{m=1}^{M} \beta_m = 1,$$

where the quantities β_m are additional variables used to take into consideration that some output values may be more important than others; if all of output values are equally important, we can simply set $\beta_m = 1/M$.

To check the validity of the model, the available information (the "data", that is, the N pairs of $x_l^{(n)}, y_m^{(n)}$ values) is split into two sets, the **training** set and the **validation** set. A typical division is $\sim$80% of the data being in the training and the rest in the validation set. Accordingly, the collective measure of the error can refer to the training set, which gives a measure of how well the model fits the data in this set, and to the validation set, which a measure of how well the model performs for data that it has not been exposed to; this is an indication of the ability of the model to make **predictions** for other possible values of the input, not part of the available data set.

Derivation: Weights of a network in an illustrative toy model

We consider a simple toy model: suppose we have a system of three binary variables (they take values 0 or 1), and want to represent a function that gives as output, for every state of the system, the majority rule, that is, 0 if there are two or more 0's and 1 if there two or more 1's. Though seemingly naive, this example is not too far from real problems; for instance, an image can be viewed as a long vector of three variables with values in a given range, say from 1 to 256, representing the color content (RGB, R = red, G = green, B = blue) of each pixel.

In order to construct the desired function, we need to describe each state of the system with a numerical value. This corresponds to passing from the actual system, defined as states $s^{(n)} = [b_1^{(n)}, b_2^{(n)}, b_3^{(n)}]$ for all possible combinations of the three binary values, to a numerical value for each state. In the present case, a sensible choice of descriptor for each state is the variable $x^{(n)}$, defined as

$$s^{(n)} = [b_1^{(n)}, b_2^{(n)}, b_3^{(n)}] \longrightarrow x^{(n)} = b_1^{(n)} + b_2^{(n)} + b_3^{(n)}, \quad n = 1, \dots, 8,$$

with the obvious choices that $\tilde{y}^{(n)} = 0$ for $x^{(n)} = 0, 1$ and $\tilde{y}^{(n)} = 1$ for $x^{(n)} = 2, 3$. Our goal is to devise a neural network that will produce the desired outcome from the input values $x^{(n)}$. Let's also split the dataset to four states that will be used for training, selected "randomly" to be the states $s^{(1)}, s^{(2)}, s^{(6)}, s^{(8)}$ and the rest to be reserved for validation; these are listed in the following table.

n	$s^{(n)} =$	$[b_1^{(n)}$	$b_2^{(n)}$	$b_3^{(n)}]$	$x^{(n)}$	$\tilde{y}^{(n)}$	
1		0	0	0	0	0	Training
2		0	0	1	1	0	Training
3		0	1	0	1	0	Validation
4		1	0	0	1	0	Validation
5		0	1	1	2	1	Validation
6		1	0	1	2	1	Training
7		1	1	0	2	1	Validation
8		1	1	1	3	1	Training

We will also assume that the value of all the weights must be smaller than a maximum allowed value w_{max}, which we arbitrarily choose here to be 10,

$$|w_j^{(k)}| \le w_{max} = 10,$$

because otherwise, from the simplicity of the model, we could assign very large values to the initial weights and get the desired output in a trivial manner.

We next need to find the weights that give an output as close to the desired one as possible. We will consider only one node per layer, that is, $J = 1$ in the notation of the general problem defined above, and try models with different number of layers, $k = 1, 2, 3, 4$. Since we have only one variable, $x^{(n)}$, for each layer's node we can have only two weights, $w_0^{(k)}$ and $w_1^{(k)}$. The model will then give for the first node, $k = 1$:

$$a_n^{(1)} = w_0^{(1)} + w_1^{(1)} x^{(n)} \Rightarrow u_n^{(1)} = S(a_n^{(1)}) = y_1^{(n)},$$

with n taking the values for the states in the training set. Note that here we have taken the output $y_1^{(n)}$ to be the same as the value of the first nonlinear transform, $u_n^{(1)}$, because the simplicity of the model does not justify another linear transform. In order to get the right answer, we see that we must have:

$$w_0^{(1)} \ll 0, \; w_0^{(1)} + w_1^{(1)} < 0, \; w_0^{(1)} + 2w_1^{(1)} > 0, \; w_0^{(1)} + 3w_1^{(1)} \gg 0,$$

so that the output for $x^{(n)} = 0, 1$ is close to 0 and the output for $x^{(n)} = 2, 3$ is close to 1. We can readily see that a reasonable choice of weights involved in the first hidden node, is:

$$w_0^{(1)} = -w_{max} = -10, \; w_1^{(1)} = \frac{2}{3} w_{max} = 6.666\ldots,$$

within the limits of the parameters set above; with these values we get the column of $y_1^{(n)}$ values shown in the table below.

n	$x^{(n)}$	$y_1^{(n)}$	$y_2^{(n)}$	$y_3^{(n)}$	$y_4^{(n)}$	$\tilde{y}^{(n)}$
1	0	0.00005	0.00670	0.00715	0.00719	0
2	1	0.03445	0.00942	0.00735	0.00720	0
6	2	0.96555	0.99058	0.99265	0.99280	1
8	3	0.99995	0.99330	0.99285	0.99281	1
$\mathcal{E}_{\text{trn}}^{(2)}$		0.02436	0.00817	0.00725	0.00719	
3	1	0.03445	0.00942	0.00735	0.00720	0
4	1	0.03445	0.00942	0.00735	0.00720	0
5	2	0.96555	0.99058	0.99265	0.99280	1
7	2	0.96555	0.99058	0.99265	0.99280	1
$\mathcal{E}_{\text{val}}^{(2)}$		0.03445	0.00942	0.00735	0.00720	

The result is encouraging: all output values are quite close to the target values, $\tilde{y}_1^{(n)}$, especially those for $n = 1, 8$. However, the training set RMSE, $\mathcal{E}_{\text{trn}}^{(2)}$, is not very low, because the values for $n = 2, 6$ are rather poor.

Using a second hidden layer, we can improve the results. Now the input values are those of $u_n^{(1)}$, which are either very close to 0 or very close to 1. The output of the second linear transformation should be such that, upon the second nonlinear transform, the values close to 0 should stay close to 0 and those close to 1 should stay close to 1, that is, we want to have:

$$w_0^{(2)} + w_1^{(2)} u_n^{(1)} \ll 0, \text{ for } u_n^{(1)} \approx 0, \quad w_0^{(2)} + w_1^{(2)} u_n^{(1)} \gg 0, \text{ for } u_n^{(1)} \approx 1.$$

The optimal weights for this case are then easily guessed to be $w_0^{(2)} = -0.5\, w_{\text{max}} = -5$ and $w_1^{(2)} = w_{\text{max}} = 10$. These give the set of output values $y_2^{(n)}$ in the table. While these values are worse that the corresponding values $y_1^{(n)}$ for $n = 1, 8$, they are much improved for $n = 2, 6$ and $\mathcal{E}_{\text{trn}}^{(2)}$ is significantly lower. Using more hidden layers with the same weights as for $k = 2$ gives even better results regarding the RMSE, as shown in the table above. Beyond $k = 4$ there is no more improvement: the model has reached its full potential. Regarding the RMSE $\mathcal{E}_{\text{val}}^{(2)}$, it is noted that it starts rather large for the first hidden node and keeps decreasing as we add more hidden nodes. It is interesting that the validation error is always larger than the training set error. In the present case, this is the result of our choice for training and validation sets, which was rather deliberate. It turns out that in actual applications the validation error is always larger than the training error.

2.7.2   Training of the network

The essential idea is to determine the values of the weights so that the final output y_m of the network is as close to the set of desired values $\tilde{y}_m$ as possible, that is, minimize the error, as defined in Eq. (2.98). For concreteness and simplicity, we assume here a neural network with only one hidden layer ($k = 1$ in the illustration of Fig. 2.18), as defined by Eqs. (2.90)–(2.93). Also, in order to retain as simple notation as possible, we drop the index over input–output pairs of vectors keeping in mind that in the end we need to take some type of average over all such pairs, as in the examples of Eq. (2.98).

The output is evidently a function of all the weights, that is, all the matrix elements in the matrices $\mathbf{W}^{(1)}, \ldots, \mathbf{W}^{(2)}$. For some initial guess of the values of the weights, the error will be non-zero. If we assume that we are not too far from the desired value of the output we can use an expansion that involves the first derivatives of the error with respect to the weights, that is, we can write:

$$\varepsilon_m = y_m - \tilde{y}_m = \sum_{l=0}^{L} \sum_{j=0}^{J} \frac{\partial y_m}{\partial w_{jl}^{(1)}} \left(w_{jl}^{(1)} - \tilde{w}_{jl}^{(1)} \right) + \sum_{j=0}^{J} \frac{\partial y_m}{\partial w_{mj}^{(2)}} \left(w_{mj}^{(2)} - \tilde{w}_{mj}^{(2)} \right), \quad (2.99)$$

where the quantities $\tilde{w}_{jl}^{(1)}, \tilde{w}_{mj}^{(2)}$ are the values of the weights which give the desired output, or equivalently minimize the value of the error. This is the Taylor expansion of the function $\varepsilon(w_{jl}^{(1)}, w_{mj}^{(2)})$ to first order in all its independent variables. But minimizing the errors is equivalent to finding a stationary point of the multi-variable function $\varepsilon_m = y_m - \tilde{y}_m$, that is, reaching a minimum at which the errors will vanish. We can accomplish this by following the negative gradient of the multi-variable function ε_m toward the stationary point. This is called the "steepest descent" method. From our definition of the various steps, we can easily obtain the partial derivatives involved in these expressions:

$$\frac{\partial \varepsilon_m}{\partial w_{mj}^{(2)}} = \frac{\partial y_m}{\partial w_{mj}^{(2)}} = \frac{\partial a_m^{(2)}}{\partial w_{mj}^{(2)}} = u_j^{(1)},$$

$$\frac{\partial \varepsilon_m}{\partial w_{jl}^{(1)}} = \frac{\partial y_m}{\partial w_{jl}^{(1)}} = \frac{\partial a_m^{(2)}}{\partial w_{jl}^{(1)}} = w_{mj}^{(2)} \frac{\partial u_j^{(1)}}{\partial w_{jl}^{(1)}},$$

$$\frac{\partial u_j^{(1)}}{\partial w_{jl}^{(1)}} = S'(a_j^{(1)}) \frac{\partial a_j^{(1)}}{\partial w_{jl}^{(1)}} = S'(a_j^{(1)}) x_l = u_j^{(1)} \left(1 - u_j^{(1)} \right) x_l,$$

where we have used the property of the sigmoid function, Eq. (2.97). We have also assumed that the weights can be treated as independent variables, hence

$$\frac{\partial w_{mj}^{(2)}}{\partial w_{jl}^{(1)}} = 0.$$

This makes it possible to use the gradient of the errors to correct the weight choices, so that the values of the errors can be gradually reduced until the optimal point $\tilde{w}$ is reached at which the error takes the minimum value $\varepsilon(\tilde{w}) = 0$. As explained in our discussion of function optimization (see Section 2.6), this process involves following the direction of the *negative* gradient of the error with respect to the weights; in other words, we perform a gradient descent search toward the minimum value of the error.

These consideration lead to the following recipe for updating the weights, given an initial set of values for which the error is non-zero: Starting from the initial choice of weight values, we update them according to the rules:

$$w_{mj}^{(2)} \longrightarrow w_{mj}^{(2)} - \text{sign}(\varepsilon_m) \frac{\partial \varepsilon_m}{\partial w_{mj}^{(2)}} = w_{mj}^{(2)} - \lambda_2 \, \varepsilon_m \, u_j^{(1)},$$

$$w_{jl}^{(1)} \longrightarrow w_{jl}^{(1)} - \text{sign}(\varepsilon_m) \frac{\partial \varepsilon_m}{\partial w_{jl}^{(1)}} = w_{jl}^{(1)} - \lambda_1 \, \varepsilon_m \, w_{mj}^{(2)} \, u_j^{(1)} \left(1 - u_j^{(1)} \right) x_l.$$

In the final expressions we have introduced λ_1, λ_2, two positive constants used to control the size of the updates along the gradients to ensure that the process converges; these constants are often referred to the "learning rates" for training the network. Moreover, since we introduced these constants to adjust the size of the update, we can also replace the $\text{sign}(\varepsilon_m)$ by simply ε_m, which has the added benefit of making the updates smaller in magnitude as we approach closer to the zero value of the error, thus helping the convergence. This is equivalent to reducing the step of updating the value of the variables while iterating toward the minimum. After enough iterations of this scheme, the errors become sufficiently small and the model produces the desired output values with adequate accuracy.

Invoking again our discussion of function optimization (see Section 2.6), we note that the final value of the error will not necessarily be the *global* minimum (which is zero); it is simply a *local* minimum which depends on the initial values of the weights and the features of the error as a function of the weights. Hopefully, the final value of the error is sufficiently small to provide a very good approximation of the target values, that is, $y_m \approx \tilde{y}_m$ with the desired accuracy.

Example 2.4: We give a specific example of how the function,

$$f(x) = \sin(2\pi x) + x^2 - 3x^3,$$

can be represented by a neural network in the interval $x \in [-1, 1]$. In this case, the input and output vectors are both of size one, that is, $L = M = 1$, and will employ a neural network consisting of two hidden layers, that is, $k = 2$ following the scheme of Fig. 2.18. In all cases, we use $J = 6$ nodes in each hidden layer and we will not use the last linear transform, that is, the output is set equal to the result of the second nonlinear transform, $y = u^{(2)}$. Thus, from Eq. (2.95), since the last linear transform is not used, the total number of parameters is $P_{\text{tot}} = 54$. We use three different choices of training values drawn directly from function values: In the first case we use $N = 5$, in the second case $N = 9$ and in the third case $N = 21$ equally spaced points (these are the samples of the input–output pairs), to train the network for 40×10^3 iterations (called "epochs").

The results are compared in the panels below. The points for $N = 21$ include noise. The training set points are shown as black dots, the validation set points as open circles. The logarithms of the Root Mean Square Error for the training set (blue dots) and the validation set (red circles). From this comparison, the importance of having enough training points (large N) is evident. In the first case, even though the fit is better judging from RMSE value, the function is poorly represented at points other than those on which the network has been trained, but still within the range of training values, $x \in [-1, 1]$; this is the difference between the red dashed line (true values of the function) and blue dashed line (output of the neural network). In the second case, the actual fit to the training set is worse as judged from the RMSE, but the value obtained from the network for points other than those on which it was trained are much closer to the actual function values: the red and blue dashed lines are much closer in the range $x \in [-1, 1]$.

To quantify these observations, we introduce a set of $N' = N - 1$ points between successive x values of the training set, which we call the "validation" set, and calculate the RMSE for the values of the function corresponding to the validation set. In both cases, the validation set error is larger than the training set error, as expected. This validation error is much larger in the case of $N = 5$ training points than in the case of $N = 9$ training points, by more that two orders of magnitude.

In both cases, if we extend the range of the variable x to values beyond the training range, that is, $x < -1$ or $x > 1$, the performance of the network is very poor. In other words, extrapolating from the training data is very dangerous whereas interpolating between the training values can be quite accurate.

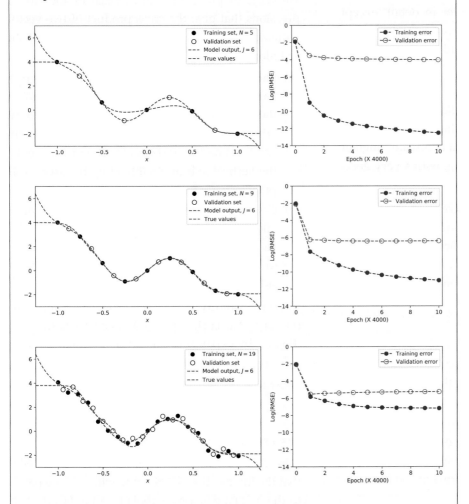

The real power of the method lies in its ability to draw conclusions from large number of points ("big data") even if they include noise. To illustrate this, we show below the case of a training set with $N = 21$ points that include noise. In this case, even though the fit to the target data (the points that include noise) is not very good judging from the RMSE error, the model has actually "learned" the correct behavior, in the sense that it gives a good fit to the true function underlying the data (shown by the red dashed line), which compares well with the output of the model (blue dashed line) in the range $x \in [-1, 1]$. Using again a set of $N' = N - 1$ validation points at x values between successive training points, gives a qualitative measure of this important fact: the training and validation errors are actually quite close for this example. The problem of poor performance when extrapolating beyond the training range is still rather vexing.

Further reading

The mathematical-methods textbooks mentioned under "General Sources" in the Preface cover the material discussed in the present chapter in detail, except for the topic discussed in the Application section (Section 2.7), which has become of broad interest more recently. For additional reading on this and related topics we recommend:

1. Christopher M. Bishop, *Pattern Recognition and Machine Learning* (Springer, 2006). This is a comprehensive and very accessible treatment of a wide range of topics related to machine learning, with a very good introduction to neural networks.

2. Sergios Theodoridis, *Machine Learning, a Bayesian and Optimization Perspective* (Academic Press, 2015). This is a modern overview of the field of machine learning, with useful discussions of neural networks, high-dimensional spaces and optimization methods.

Problems

1. We wish to prove the equivalence of the two expressions that give the dot product of two vectors, namely, Eqs. (2.38) and (2.39).

   (a) Using the result of Eq. (2.15), show that if a 2D vector with components (x_1, y_1) is rotated *counterclockwise* by $\pi/2$, the new values of its components (x_2, y_2) are given in terms of the old ones by $x_2 = -y_1, y_2 = x_1$.

   (b) Based on the result from part (a) and the first dot product definition, Eq. (2.38), show that the dot product between *any* two vectors at a right angle relative to each other is zero.

   (c) From the result (b), show that the second definition of the dot product, Eq. (2.39), is equivalent to the first one. [Hint: Consider the projection of the second vector along two perpendicular directions, one of which is along the direction of the first vector, as indicated in Fig. 2.7.]

2. We wish to prove the equivalence of the two expressions that give the cross product of two vectors, namely, Eqs. (2.41) and (2.42). This can be established as follows:

   (a) Show that the expression $x_1 y_2 - x_2 y_1$, the right-hand side of Eq. (2.41), is equal to the area of the white parallelogram in Fig. 2.7, by considering the entire area of the rectangle with sides $(x_1 + x_2)$ on the horizontal axis and $(y_1 + y_2)$ on the vertical axis and subtracting the areas of the purple shaded triangles and rectangles.

   (b) Calculate the area of the white parallelogram by taking one side as the base (say, the length of the vector $\mathbf{r}_1$), and multiplying by the height relative to this base; this turns out to be the right-hand side of Eq. (2.42).

3. Show the validity of the vector identity that involves the curl of a curl, Eq. (2.72), using Cartesian coordinates, by expanding the various terms on the two sides of the equation.

4. Show that the function

$$f(x, y) = x^2 - 4x - y^2 + 6y + 4$$

   has a saddle point at $(x, y) = (2, 3)$.

5. Consider the symmetric cone shown in Fig. 2.5. Show that the intersection of this cone with a plane parallel to the yz-plane is a hyperbola [see Eq. (1.11) for the definition of the hyperbola].

6. Using the method of Lagrange multipliers, find the point in the hyperbola defined in Problem 5 that is at a maximum distance from the origin. [Hint: The function to be maximized is the distance from the origin,

$$f(x, y, z) = x^2 + y^2 + z^2,$$

under the constraints that the point must lie both on the surface of the cone and on the hyperbola.]

7. Using the method of Lagrange multipliers, show that the shape of a rectangular parallelepiped whose sides add up to a constant value $c > 0$, and has the maximum volume, is the cube. What is the value of the maximum volume in terms of c? [Hint: The volume of a rectangular parallelepiped with sides x, y, z, is

$$f(x,y,z) = xyz, \quad x,y,z > 0,$$

so the problem reduces to finding the maximum of $f(x,y,z)$ under the constraint imposed by the sum of the sides being constant.]

8. A particle experiences a potential energy U in a three-dimensional space which is given by

$$U(\mathbf{r}) = 2(x+y-4)^2 - (y+z-2)^2 - (z+x-2)^2,$$

where $\mathbf{r}$ is the position vector $\mathbf{r} = x\hat{x} + y\hat{y} + z\hat{z}$. The force $\mathbf{F}$ is given in terms of the potential energy by $\mathbf{F} = -\nabla_{\mathbf{r}}U(\mathbf{r})$.

(a) Calculate the force experienced by the particle at each point in space.

(b) Calculate the lagrangian of the potential.

(c) Using information from parts (a) and (b) find the equilibrium positions and characterize them as stable or unstable. An "equilibrium position" corresponds to zero force; a "stable equilibrium" corresponds to a local minimum, and an "unstable equilibrium" to a local maximum, of the potential energy.

9. Prove the equivalence of the first and the second expressions given in the definition of the sigmoid function, $S(x)$, in Eq. (2.96). Prove the relation between the sigmoid function and its derivative, given in Eq. (2.97).

Chapter 3

Series Expansions

The divergent series are the invention of the devil, and it is a shame to base on them any demonstration whatsoever — by using them, one may draw any conclusion he pleases.

Niels Henrik Abel (Norwegian mathematician, 1802–1829)

3.1 Infinite sequences and series

A sequence is a well-defined set of numbers (integers or real numbers). The definition is usually expressed by a formula, like the sequence of numbers denoted by a_n, where the index n takes integer values, and a_n is defined in terms of this index, as in the example

$$a_n = \frac{1}{n}.$$

In order to specify the sequence we need to define the range of values of the index n. In the above example, we could for instance let n be any set of integers that does not contain $n = 0$, since a_0 is meaningless. The most interesting types of sequences are those that consist of an infinite number of terms, for example:

$$a_n = \frac{1}{n}, \quad n = 1, 2, \ldots, \infty, \tag{3.1}$$

or, written out explicitly,

$$1, \frac{1}{2}, \frac{1}{3}, \frac{1}{4}, \cdots$$

For infinite sequences, a key question is whether or not the sequence tends to some constant (independent of n), and *finite* value as the index goes to infinity. This value is called the "limit" of the sequence, and if it exists, then we say that the sequence "converges" to this value. In the above example, it is obvious that as $n \to \infty$ the value of $a_n \to 0$.

The precise definition of the convergence of a sequence is the following: A sequence a_n converges if for any $\epsilon > 0$, we can find a value of the index $n = N$ such that

$$|a_n - a| < \epsilon, \quad \forall n > N,$$

where a is the limit (also called the limiting or asymptotic value) of the sequence. Since this must hold for any $\epsilon > 0$, by taking $\epsilon \to 0$ the above definition ensures that beyond a certain value of the index n all terms of the sequence are within the narrow range $a \pm \epsilon$, that is, arbitrarily close to the limit of the sequence. In many cases, it is not easy to answer the question of convergence; to this end, several "convergence tests" or "criteria" have been developed to provide a definitive answer; we examine some of them in the next section.

Another interesting case is when the sequence is not defined by a simple formula, but by a rule. A famous example of this case is the Fibonacci sequence, defined by the recursive relation (rule):

$$y_{n+2} = y_n + y_{n+1}, \quad n = 0, 1, 2, \ldots, \infty; \quad y_0 = 0, \ y_1 = 1.$$

This rule says that one element of the sequence y_{n+2} is given as the sum of the previous two elements y_{n+1} and y_n. This obviously means that we need to know the first two elements in order to calculate the third one; then, we have all the information needed to keep calculating successive elements all the way to infinity. This is the reason the definition of the sequence includes the values of y_0 and y_1, which are 0 and 1, respectively. Applying the rule, we find the first few elements of the Fibonacci sequence to be:

$$0, 1, 1, 2, 3, 5, 8, 13, 21, 34, 55, 89, 144, 233, 377, 610, 987, 1597, \ldots$$

The Fibonacci sequence does not converge as $n \to \infty$, since its terms keep growing as n increases (recall that the limit must be a *finite* number). However, there is another interesting limit, namely the *ratio* of two successive terms:

$$\lim_{n \to \infty} \left[\frac{y_{n+1}}{y_n} \right] = \frac{1 + \sqrt{5}}{2},$$

which is the value of the golden ratio (see Chapter 1, Section 1.1).

Series is another very useful mathematical concept. A series is the sum of the infinite terms of an infinite sequence. For example, for the sequence defined in Eq. (3.1), the corresponding series (called the "harmonic series") is

$$\sum_{n=0}^{\infty} a_n = \sum_{n=1}^{\infty} \frac{1}{n} = 1 + \frac{1}{2} + \frac{1}{3} + \frac{1}{4} + \frac{1}{5} + \cdots$$

Again, the important question is whether or not the series converges to a *finite* number. Notice that the difficulty arises from the infinite number of terms as well as from the values that these terms take for large values of n. For this reason, typically we are not interested in the sum of the first few terms, or for that matter in the sum of any finite number of terms because this part always gives a finite number, as long as the terms themselves are finite. Accordingly, we usually focus on what happens to the "tail" of the series, namely, starting at some large value of the index n and going all the way to infinity. We will therefore often denote the infinite series as a sum over n of terms a_n, without specifying the starting value of n and with the understanding that the final value of the index is always ∞.

In order to determine the convergence of an infinite series we can turn it into a sequence, by considering the partial sums

$$s_n = \sum_{j=1}^{n} a_j.$$

Then we can ask whether the sequence s_n converges. If it does, so will the series, since the asymptotic term of the sequence s_n for $n \to \infty$ is the same as the infinite series. Thus, a series is said to converge if the sequence of partials sums converges, otherwise it diverges.

It is useful to notice that if the series of absolute values converges, then so does the original series:

$$\sum_n |a_n| \; : \; \text{converges} \Rightarrow \sum_n a_n \; : \; \text{converges}$$

The reason for this fact is simple: the absolute values are all positive, whereas the original terms can be positive or negative. If all terms have the same sign, then it is the same as summing all the absolute values with the overall sign outside, so if the series of absolute values converges so does the original series. If the terms can have either sign, then their total sum in absolute value will necessarily be smaller than the sum of the absolute values, and therefore if the series of absolute values converges so does the original series. If the series $\sum |a_n|$ converges, then we say that the series $\sum a_n$ converges absolutely (and of course it converges in the usual sense).

3.1.1 Series convergence tests

A number of tests have been devised, which by looking at the behavior of the nth term, can determine whether or not a series converges. We give some of these tests here.

1. **Comparison test:** *Consider two series, $\sum a_n, \sum b_n$:*

$$\text{If } |b_n| \le |a_n| \quad and \quad \sum a_n \; : conv. \; abs. \Rightarrow \sum b_n \; : conv. \; abs. \quad (3.2)$$
$$\text{If } |a_n| \le |b_n| \quad and \quad \sum a_n \; : diver. \; abs. \Rightarrow \sum b_n \; : diver. \; abs. \quad (3.3)$$

   If a series diverges absolutely we cannot conclude that it diverges.

2. **Ratio test:** *For the series $\sum a_n$,*

$$if \; \left| \frac{a_{n+1}}{a_n} \right| \le t < 1, \quad \forall \, n > N \Rightarrow \sum a_n \; : converges. \quad (3.4)$$

   Proof: From the inequality satisfied by a_{n+1}/a_n for $n > N$, we obtain the following:

$$|a_{N+2}| \le t|a_{N+1}|,$$
$$|a_{N+3}| \le t|a_{N+2}| \le t^2|a_{N+1}|,$$
$$|a_{N+4}| \le t|a_{N+3}| \le t^3|a_{N+1}|,\dots$$

   By summing the terms on the left side and the terms on the right side of the above inequalities we obtain

$$|a_{N+1}| + |a_{N+2}| + |a_{N+3}| + \cdots \le |a_{N+1}|(1 + t + t^2 + \cdots) = \frac{|a_{N+1}|}{(1-t)},$$

   where we have used the result for the geometric series summation that will be derived later (see Eq. (3.5) below). The last relation for the sum of a_n with $n > N$ shows that the "tail" of the series of absolute values is bounded by a finite number, namely the quantity $|a_{N+1}|/(1-t)$, therefore the series converges absolutely. ∎

3. **Limit tests:** *For the series $\sum a_n$, we define the following limits:*

$$\lim_{n \to \infty} \frac{|a_{n+1}|}{|a_n|} = L_1, \qquad \lim_{n \to \infty} \sqrt[n]{|a_n|} = L_2.$$

If L_1 or $L_2 < 1$ then the series converges absolutely.
If L_1 or $L_2 > 1$ then the series diverges.
If L_1 or $L_2 = 1$ then convergence cannot be determined.

4. **Integral test:** *If $0 < a_{n+1} \le a_n$, and $f(x)$ is a continuous non-increasing function such that $a_n = f(n)$, then the series $\sum_{n=1}^{\infty} a_n$ converges if and only if*

$$\int_1^{\infty} f(x) \, \mathrm{d}x < \infty.$$

5. **Dirichlet test:** *Suppose that the terms c_j defining the series can be expressed as follows:*

$$\sum_{j=1}^{\infty} c_j = \sum_{j=1}^{\infty} a_j b_j.$$

Consider a_j and b_j as separate series: if the partial sums of a_j are bounded and the terms of b_j are positive and tend monotonically to 0, the series c_j converges.

Example 3.1: A series that satisfies the Dirichlet test

We consider the series with terms given by

$$c_j = \cos\left((j-1)\frac{\pi}{4}\right)\frac{1}{j}.$$

Making the obvious choices $a_j = \cos((j-1)\pi/4)$ and $b_j = (1/j)$, and defining the partial sums as

$$s_n = \sum_{j=1}^{n} a_j$$

leads to the following table of values for a_n, b_n and s_n:

n	1	2	3	4	5	6	7	8	9
a_n	1	$\frac{1}{\sqrt{2}}$	0	$-\frac{1}{\sqrt{2}}$	-1	$-\frac{1}{\sqrt{2}}$	0	$\frac{1}{\sqrt{2}}$	1
s_n	1	$1+\frac{1}{\sqrt{2}}$	$1+\frac{1}{\sqrt{2}}$	1	0	$-\frac{1}{\sqrt{2}}$	$-\frac{1}{\sqrt{2}}$	0	1
b_n	1	$\frac{1}{2}$	$\frac{1}{3}$	$\frac{1}{4}$	$\frac{1}{5}$	$\frac{1}{6}$	$\frac{1}{7}$	$\frac{1}{8}$	$\frac{1}{9}$

As is evident from this table, the partial sums of a_j are bounded by the value $(1+\frac{1}{\sqrt{2}})$, and the b_j terms are positive and monotonically decreasing. Hence, the original series c_j converges because it satisfies the Dirichlet test.

3.1.2   Some important number series

The geometric series: An important series is the "geometric sequence", where

$$a_j = t^j, \quad 0 < t < 1, \quad j = 0, 1, \ldots$$

The "geometric series" is given by

$$\sum_{j=0}^{\infty} a_j = \sum_{j=0}^{\infty} t^j = 1 + t + t^2 + t^3 + \cdots + t^n + \cdots$$

The sequence of partials sums is

$$s_n = \sum_{j=0}^{n} t^j = 1 + t + t^2 + t^3 + \cdots + t^n = \frac{1 - t^{n+1}}{1 - t},$$

which for large n converges provided that $t < 1$:

$$t < 1 \Rightarrow \lim_{n \to \infty} t^n = 0 \Rightarrow \lim_{n \to \infty} \frac{1 - t^{n+1}}{1 - t} = \frac{1}{1 - t}. \tag{3.5}$$

Thus, the limit of the geometric series, for $0 < t < 1$, is $1/(1 - t)$.

The harmonic and related series: Some of the simplest and most important series are those involving $a_n = 1/n^p$, where $p > 0$:

$$a_n = \frac{1}{n^p} \to \sum_{n=1}^{\infty} \frac{1}{n^p} = 1 + \frac{1}{2^p} + \frac{1}{3^p} + \frac{1}{4^p} + \frac{1}{5^p} + \cdots \tag{3.6}$$

For $p = 1$ the resulting series is called the "harmonic" series. For this series we cannot use the comparison test for an obvious choice of b_n. The ratio test is also not useful because:

$$\frac{|a_{n+1}|}{|a_n|} = \frac{n^p}{n^p + 1},$$

which cannot be bounded by a number $t < 1$ because it tends to 1 for $n \to \infty$. The limit tests give:

$$L_1 = \lim_{n \to \infty} \frac{|a_{n+1}|}{|a_n|} = \lim_{n \to \infty} \frac{n^p}{n^p + 1} = 1,$$

$$L_2 = \lim_{n \to \infty} \sqrt[n]{|a_n|} = \lim_{n \to \infty} \frac{1}{\sqrt[n]{n^p}} = 1,$$

so they are both inconclusive. However, the integral test works:

$$a_n = \frac{1}{n^p} \Rightarrow f(x) = \frac{1}{x^p},$$

$$\text{for } p = 1 : \quad \int_1^{\infty} \frac{1}{x} \, dx = [\ln(x)]_1^{\infty} \to \infty,$$

which proves that the harmonic series diverges, while

$$\text{for } p > 1 : \quad \int_1^{\infty} \frac{1}{x^p} \, dx = -\frac{1}{p-1} \left[\frac{1}{x^{p-1}} \right]_1^{\infty} = \frac{1}{p-1} \left[1 - \frac{1}{\infty} \right] = \frac{1}{p-1},$$

which proves that the series converges for $p > 1$.

This example is useful in illustrating why the integral test works: as shown in Fig. 3.1, the sum of the terms in the harmonic series is equivalent to calculating the total area covered by rectangles whose width is 1 and height is $1, 1/2, 1/3, 1/4, \ldots$. The function $f(x) = 1/x$ passes through the top left corners of all of these rectangles. The area under the function $f(x)$ is equal to the integral of the function from 1 to ∞, but the rectangles cover a larger area than the one under the curve representing $f(x)$. Therefore, if the integral of

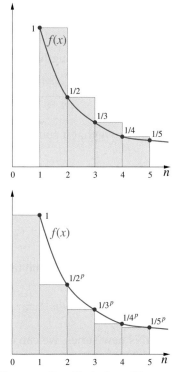

Figure 3.1: **Top:** Illustration of the integral test for the harmonic series, Eq. (3.6) with $p = 1$, using the function $f(x) = 1/x$ (red line) which passes through the individual values $1/n$ (blue dots). **Bottom:** Illustration of the integral test for the convergence of the series of Eq. (3.6) with $p > 1$, using the function $f(x) = 1/x^p$ (red line) which passes through the individual values $1/n^p$ (blue dots).

the function diverges, so will the infinite sum of the rectangles which is larger than the integral. Conversely, as also shown in Fig. 3.1, the sum of the terms in the series with terms $1/n^p$ is equivalent to calculating the total area covered by rectangles whose width is 1 and height is $1, 1/2^p, 1/3^p, 1/4^p, \ldots$. The function $f(x) = 1/x^p$ passes through the top right corners of all of these rectangles. The area under the function $f(x)$ is equal to the integral of the function from 1 to ∞, but the rectangles cover a smaller area than the one under the curve representing $f(x)$. Therefore, if the integral of the function converges, so will the infinite sum of the rectangles which is smaller than the value of the integral.

For example, the integral test for the inverse-square series, $p = 2$, gives

$$a_n = \frac{1}{n^2} \rightarrow \sum_{n=1}^{\infty} \frac{1}{n^2} = 1 + \frac{1}{4} + \frac{1}{9} + \frac{1}{16} + \frac{1}{25} + \cdots , \qquad (3.7)$$

and if we identify the function $f(x)$ as below

$$a_n = \frac{1}{n^2} \Rightarrow f(x) = \frac{1}{x^2},$$

then the infinite series is bounded by

$$\sum_{1}^{\infty} \frac{1}{n^2} \leq \int_{1}^{\infty} \frac{1}{x^2} \, dx = \left[-\frac{1}{x} \right]_{1}^{\infty} = 1,$$

which shows that this series converges.

For the alternating harmonic series

$$c_n = \frac{(-1)^{n+1}}{n} \rightarrow \sum_{n=1}^{\infty} \frac{(-1)^{n+1}}{n} = 1 - \frac{1}{2} + \frac{1}{3} - \frac{1}{4} + \frac{1}{5} - \cdots , \qquad (3.8)$$

we can prove that it converges by grouping its terms in pairs into a new series:

$$d_n = c_{2n-1} + c_{2n}, \ n = 1, 2, 3, \ldots,$$

and for this new series we have

$$\sum_{n=1}^{\infty} d_n = (c_1 + c_2) + (c_3 + c_4) + (c_5 + c_6) + \cdots = \sum_{n=1}^{\infty} c_n.$$

From the definition of the terms of the new series, we obtain

$$d_n = \frac{(-1)^{2n}}{2n - 1} + \frac{(-1)^{2n+1}}{2n} = \frac{1}{2n(2n - 1)}.$$

For this new series, we can establish convergence by the comparison test:

$$d_n = \frac{1}{4n(n - \frac{1}{2})} \leq \frac{1}{4} \frac{1}{(n-1)^2},$$

since the last expression is simply the inverse-square series of Eq. (3.7), multiplied by the constant factor $1/4$ (we can think of the series as starting at $n = 2$, which makes no difference in the convergence of the infinite sum). This is an interesting result, in the sense that the series of the absolute value of the terms of the alternating harmonic series does not converge (it is exactly equal to the harmonic series, which we proved above that it diverges), whereas when the alternating signs are included the series converges.

The binomial series: A very useful power series expansion is the binomial expansion:

$$(a+b)^p = \sum_{n=0}^{\infty} \frac{p(p-1)(p-2)\cdots(p-(n-1))}{n!} a^{p-n} b^n$$

$$= a^p \left[1 + \frac{p}{1!}\left(\frac{b}{a}\right) + \frac{p(p-1)}{2!}\left(\frac{b}{a}\right)^2 + \frac{p(p-1)(p-2)}{3!}\left(\frac{b}{a}\right)^3 + \cdots \right],$$

(3.9)

which is valid for $a > 0$, b, p real and $|b/a| < 1$. In the second line, we have written the first few terms of this expansion in terms of the ratio (a/b), which is a useful expression when $|(b/a)| = \epsilon \ll 1$: in this case, these are the dominant terms in the Taylor expansion in the small quantity ϵ. It is evident from Eq. (3.9) that for $p = N$ an integer, the binomial expansion is finite and has a total number of $(N+1)$ terms; if p is not an integer, it has infinite terms. In the former case, the binomial expansion can be written in the form:

$$(a+b)^N = \sum_{n=0}^{N} \frac{N!}{(N-n)!n!} a^{N-n} b^n$$

$$= a^N \left[1 + \frac{N!}{(N-1)!1!}\left(\frac{b}{a}\right) + \frac{N!}{(N-2)!2!}\left(\frac{b}{a}\right)^2 + \cdots \right].$$

(3.10)

3.2 Series expansions of functions

We are often interested in representing a function $f(x)$ as a series expansion in terms of other functions, $g_n(x)$, with $n = 1, 2, \ldots$, which can be manipulated more easily than the original function:

$$f(x) = c_0 + \sum_{n=1}^{\infty} c_n g_n(x).$$

(3.11)

Usually, it is straightforward to evaluate, differentiate or integrate the functions $g_n(x)$. There are three important issues for constructing such series expansions:

- *What are the functions $g_n(x)$ that should be employed in the series?*

- *What are the corresponding numerical coefficients c_n?*

- *Does the series converge and how fast?*

The series expansion is an exact representation of the function only if we include an infinite number of terms. In practice, this approach is useful if the exact infinite series expansion can be truncated to a few terms which give an adequate approximation of the function. Thus, the choice of $g_n(x)$ and c_n regarding the above questions should be such that:

- *The functions $g_n(x)$ are easy to manipulate (evaluate, differentiate, integrate).*

- *The coefficients c_n can be obtained easily from the definition of $g_n(x)$ and the function $f(x)$.*

- *The series must converges fast to $f(x)$, so that truncation to the first few terms gives a good approximation.*

We give below several examples of series expansions. The most common and useful ones are those corresponding to the powers of $(x - x_0)$, discussed in Section 3.2.1, and those for which $g_n(x)$ is a type of function with convenient properties. Examples of such functions include the trigonometric functions $\sin(n\pi x/L)$ and $\cos(n\pi x/L)$, discussed in Section 3.2.3, or a special type of polynomial, discussed in Section 3.2.4. The advantage of these choices is that each term in the series satisfies a type of differential equation for some value of the parameters that appear in the differential equation. This property of the individual terms can be used to construct the general solution of the differential equation with properly chosen values for the coefficients c_n in the expression of Eq. (3.11). We discuss these issues in more detail in Chapter 8.

3.2.1   Power series expansion: The Taylor series

An important type of series involves the powers of $(x - x_0)$ where x_0 is a specific value of the variable near which we want to know the values of the given function $f(x)$. This series is known as a "power series" expansion:

$$f(x) = c_0 + \sum_{n=1}^{\infty} c_n(x - x_0)^n.$$

For this type of series, there exists an expression, known as the *Taylor series* expansion, which gives a unique representation of the function in terms of powers of $(x - x_0)$ near the point x_0. The Taylor series expresses a function $f(x)$ in terms of its derivatives and powers of $(x - x_0)^n$, for values of x near x_0. This expansion, which applies to continuous and infinitely differentiable functions of x, takes the following form:

$$f(x) = f(x_0) + \frac{1}{1!}f^{(1)}(x_0)(x - x_0) + \frac{1}{2!}f^{(2)}(x_0)(x - x_0)^2$$
$$+ \cdots + \frac{1}{n!}f^{(n)}(x_0)(x - x_0)^n + \cdots ,$$

(3.12)

with $f^{(n)}(x)$ denoting the nth derivative of $f(x)$. The term which involves the power $(x - x_0)^n$ is called the "term of order n" and its coefficient c_n is equal to the value of the nth derivative at $x = x_0$, multiplied by $(1/n!)$. For x close to x_0, the expansion can be truncated to the first few terms, giving a very good approximation for $f(x)$ in terms of the function and its lowest few derivatives evaluated at $x = x_0$. We will derive this expansion later, using the properties of functions of complex variables (see Chapter 4, Section 4.7). For the first few terms, we have already provided an explanation in the discussion of geometric interpretation of the derivative (see Chapter 1, Section 1.3.3).

Using the general expression for the Taylor series, we obtain for the common exponential, logarithmic, trigonometric and hyperbolic functions, at $x_0 = 0$ the following expressions:

$$e^x = 1 + x + \frac{1}{2}x^2 + \frac{1}{6}x^3 + \cdots$$

(3.13)

$$\log(1 + x) = x - \frac{1}{2}x^2 + \frac{1}{3}x^3 + \cdots$$

(3.14)

$$\cos(x) = 1 - \frac{1}{2}x^2 + \frac{1}{24}x^4 + \cdots \qquad (3.15)$$

$$\sin(x) = x - \frac{1}{6}x^3 + \frac{1}{120}x^5 + \cdots \qquad (3.16)$$

$$\tan(x) = x + \frac{1}{3}x^3 + \frac{2}{15}x^5 + \cdots \qquad (3.17)$$

$$\cot(x) = \frac{1}{x} - \frac{1}{3}x - \frac{1}{45}x^3 + \cdots \qquad (3.18)$$

$$\cosh(x) = 1 + \frac{1}{2}x^2 + \frac{1}{24}x^4 + \cdots \qquad (3.19)$$

$$\sinh(x) = x + \frac{1}{6}x^3 + \frac{1}{120}x^5 + \cdots \qquad (3.20)$$

$$\tanh(x) = x - \frac{1}{3}x^3 + \frac{2}{15}x^5 + \cdots \qquad (3.21)$$

$$\coth(x) = \frac{1}{x} + \frac{1}{3}x - \frac{1}{45}x^3 + \cdots \qquad (3.22)$$

Remarks

(i) The Taylor series expansion can be used to generate many useful results. For instance, the binomial expansion, Eq. (3.9), can be obtained as a Taylor series expansion, see Problem 3.

(ii) When performing a Taylor series expansion of the function $f(x)$ around $x = x_0$, it is important to make sure that all terms that contain the same power of $(x - x_0)$ are properly calculated, otherwise the expression may contain errors. For an example of possible pitfalls, see Problem 7.

Derivation: The Taylor expansion of the exponential

With the help of the Taylor expansion of the exponential and the binomial expansion, we can decipher the deeper meaning of the exponential function: e^x is the result of continuous compounding, which mathematically is expressed by the relation

$$\lim_{N \to \infty} \left[1 + \frac{\kappa t}{N} \right]^N = e^{\kappa t}, \qquad (3.23)$$

where we use the variable t multiplied by the constant factor κ, instead of x, to make the analogy to compounding with time more direct. Indeed, suppose that at time $t = 0$ we have a capital C, to which we add interest at a rate κ every time interval Δt; in other words, our original capital increases by a factor $(1 + \kappa \Delta t)$ at every time interval, where κ, has the dimensions of $1/[\text{time}]$, for example, if the increase is 1% per day, in one day the capital would increase by a factor 1.01. In the Nth interval the capital will then have increased by a factor $(1 + \kappa \Delta t)^N$. In order to reach the continuous compounding limit we will let $\Delta t \to 0$ and N large enough, $N \Delta t = t$ with t a finite value. From the binomial expansion, we get

$$(1 + \kappa \Delta t)^N = \sum_{n=0}^{N} \frac{N!}{n!(N-n)!}(\kappa \Delta t)^n.$$

The general term in this series can be written in the form

$$\frac{N!}{n!(N-n)!}(\kappa\Delta t)^n = \frac{1}{n!}(\kappa\Delta t)^n(N-n+1)(N-n+2)\cdots N.$$

Since N is very large, each factor involving N in the parentheses is approximately equal to N, giving

$$\frac{N!}{n!(N-n)!}(\kappa\Delta t)^n \approx \frac{N^n}{n!}(\kappa\Delta t)^n = \frac{1}{n!}(\kappa N\Delta t)^n = \frac{1}{n!}(\kappa t)^n.$$

For $N\to\infty, \Delta t\to 0, N\Delta t = t$:finite, we find

$$\sum_{n=0}^{\infty}\frac{1}{n!}(\kappa t)^n = e^{\kappa t},$$

where the last equality follows from the Taylor expansion of the exponential function.

In the plot below we show the value of $\exp(x)$ for $x = 1$ (horizontal dashed line) and its approximation from Eq. (3.23) for values of $N = 2,\ldots,20$ (red dots), as well as the approximation in terms of the Taylor series expansion with $x = 1$ and keeping terms up to $n = 2,\ldots,10$ (blue dots).

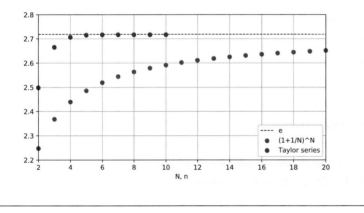

Example 3.2: Taylor expansion of the normal cumulative distribution
The normal cumulative distribution function was introduced in Chapter 1, Eq. (1.60a):

$$\Phi(w) = \frac{1}{\sqrt{2\pi}}\int_{-\infty}^{w}e^{-x^2/2}\,dx.$$

This function does not have an analytical expression but can be computed by using its Taylor expansion. To obtain its first derivative, we use the formula of Eq. (1.48) with

$$f(x) = \frac{1}{\sqrt{2\pi}}e^{-x^2/2},$$

and

$$a(w) = -\infty \Rightarrow \frac{da}{dw} = 0, \quad b(w) = w \Rightarrow \frac{db}{dw} = 1,$$

which yield

$$\frac{d\Phi}{dw} = \frac{1}{\sqrt{2\pi}} e^{-w^2/2}.$$

From this expression, it is straightforward to obtain higher derivatives; the derivatives up to ninth order are:

$$\frac{d^2\Phi}{dw^2} = -\frac{e^{-w^2/2}}{\sqrt{2\pi}} w,$$

$$\frac{d^3\Phi}{dw^3} = -\frac{e^{-w^2/2}}{\sqrt{2\pi}} \left[1 - w^2 \right],$$

$$\frac{d^4\Phi}{dw^4} = \frac{e^{-w^2/2}}{\sqrt{2\pi}} \left[3w - w^3 \right],$$

$$\frac{d^5\Phi}{dw^5} = \frac{e^{-w^2/2}}{\sqrt{2\pi}} \left[3 - 6w^2 + w^4 \right],$$

$$\frac{d^6\Phi}{dw^6} = -\frac{e^{-w^2/2}}{\sqrt{2\pi}} \left[15w - 10w^3 + w^5 \right],$$

$$\frac{d^7\Phi}{dw^7} = -\frac{e^{-w^2/2}}{\sqrt{2\pi}} \left[15 - 45w^2 + 15w^4 - w^6 \right],$$

$$\frac{d^8\Phi}{dw^8} = \frac{e^{-w^2/2}}{\sqrt{2\pi}} \left[105w - 105w^3 + 21w^5 - w^7 \right],$$

$$\frac{d^9\Phi}{dw^9} = \frac{e^{-w^2/2}}{\sqrt{2\pi}} \left[105 - 420w^2 + 210w^4 - 28w^6 + w^8 \right].$$

Evaluating these expressions at $w_0 = 0$, and generalizing the result, we find that the nth derivative evaluated at $w_0 = 0$ vanishes for n even, while for n odd, with $n = 2k + 1, k = 1, 2, \ldots$, it takes the form:

$$\Phi^{(2k+1)}(0) = \frac{(-1)^k (2k-1)!!}{\sqrt{2\pi}}, \quad k = 1, 2, \ldots,$$

where $(2k-1)!!$ is the product of the odd integers from 1 to $(2k-1)$ (the first derivative gives $\Phi^{(1)}(0) = 1/\sqrt{2\pi}$). With these expressions, and taking into consideration that $\Phi(0) = 0.5$, we can write the Taylor expansion of $\Phi(w)$ near $w_0 = 0$ as

$$\Phi(w) = \frac{1}{2} + \frac{1}{\sqrt{2\pi}} w + \sum_{k=1}^{\infty} \frac{(-1)^k}{\sqrt{2\pi}} \frac{(2k-1)!!}{(2k+1)!} w^{2k+1}.$$

Using this expression we can then obtain the values of $\Phi(w)$ in an interval around $w_0 = 0$ with the desired accuracy, by including a sufficient number of terms in the expansion. As an illustration, we show in the plot below the values of $\Phi(w)$ in the interval $w \in [-2.5, 2.5]$ obtained through the Taylor series expansion by keeping terms up to the nth-order derivative (shown as colored lines for $n = 3, \ldots, 13$), and compare those approximations to the exact result (shown as thicker black line) obtained from tabulated values. In this range of w values, and on the scale of the plot, the Taylor approximation matches very well the exact result by keeping derivatives up to order $n = 15$: for $n = 13$ the difference from the exact result is barely visible near the endpoints of the interval, $w = \pm 2.5$, while the approximation for $n = 15$ is indistinguishable from the exact result. The approximation becomes better the closer w is to the value 0 and therefore fewer terms are required for narrower intervals around 0; for instance, in the interval $w \in [-1, 1]$, keeping terms up to $n = 3$ provides an adequate approximation.

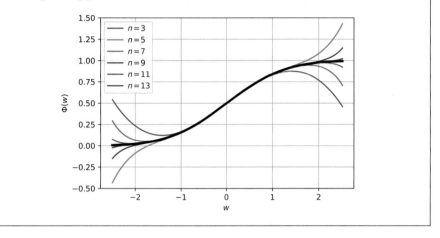

3.2.2 Inversion of power series

In certain applications we encounter the following problem: Suppose the variable y is expressed in a power series in terms of the variable x as:

$$y = x(1 + a_1 x + a_2 x^2 + a_3 x^3 + \cdots), \quad |x| \ll 1, \tag{3.24}$$

with coefficients $|a_n| \sim 1$ for all n, and we wish to express the variable x as a power series in terms of the variable y. This problem is referred to as "series inversion". Note that because of the relation expressed by Eq. (3.24) the variable y also satisfies $|y| \ll 1$, and therefore a power expansion in terms of this variable is meaningful in the sense that higher powers will have smaller contributions. The application of the Taylor series expansion makes it possible to solve this problem in a systematic way.

The zeroth-order approximation consists of neglecting the powers of x beyond the zeroth-order term inside the parenthesis of Eq. (3.24), which leads to

$$\text{zeroth order}: \quad x \approx y. \tag{3.25}$$

For the next order, we keep the first-order power of x inside the parenthesis of Eq. (3.24), which leads to

$$y = x(1 + a_1 x) \Rightarrow x = y(1 + a_1 x)^{-1}.$$

We define the function

$$f_1(x) = (1 + a_1 x)^{-1},$$

and calculate its first derivative, $f_1^{(1)}(x)$, which we use to produce its Taylor expansion at $x = 0$:

$$f_1^{(1)}(x) = -a_1(1 + a_1 x)^{-2} \Rightarrow f_1^{(1)}(0) = -a_1$$
$$\Rightarrow f_1(x) \approx f_1(0) + f_1^{(1)}(0)x = 1 - a_1 x.$$

Then we use the value of x from the zeroth-order approximation, Eq. (3.25), to obtain the first-order approximation for x:

$$\text{first order}: \quad x \approx y(1 - a_1 y). \tag{3.26}$$

For the next order, we keep the second-order power of x inside the parenthesis of Eq. (3.24), which leads to

$$y = x(1 + a_1 x + a_2 x^2) \Rightarrow x = y \left(1 + a_1 x + a_2 x^2\right)^{-1}.$$

We define the function

$$f_2(x) = (1 + a_1 x + a_2 x^2)^{-1},$$

and calculate its first and second derivatives, $f_2^{(1)}(x)$ and $f_2^{(2)}(x)$, respectively, which we evaluate at $x = 0$, to obtain

$$f_2^{(1)}(x) = -(1 + a_1 x + a_2 x^2)^{-2}(a_1 + 2a_2 x)$$

$$\Rightarrow f_2^{(1)}(0) = -a_1,$$

$$f_2^{(2)}(x) = 2(1 + a_1 x + a_2 x^2)^{-3}(a_1 + 2a_2 x)^2 - (1 + a_1 x + a_2 x^2)^{-2} 2a_2$$

$$\Rightarrow f_2^{(1)}(0) = 2a_1^2 - 2a_2.$$

We then use these values to produce the Taylor expansion of $f_2(x)$ at $x = 0$:

$$f_2(x) \approx 1 - a_1 x + (a_1^2 - a_2)x^2,$$

and substitute x in terms of y from the first-order approximation, Eq. (3.26), to arrive at:

$$x \approx y \left[1 - a_1 \left[y(1 - a_1 y)\right] + (a_1^2 - a_2) \left[y(1 - a_1 y)\right]^2\right].$$

In this last expression, we need to keep terms up to second order in y inside the square bracket, which leads to the following result:

$$\text{second order}: \quad x \approx y \left[1 - a_1 y - \left(a_2 - 2a_1^2\right) y^2\right]. \tag{3.27}$$

This process can be continued in the same fashion to generate higher-order terms in the variable y in the inverted series.

3.2.3 Trigonometric series expansions

In many contexts, a very useful series representation of a function $f(x)$ is in terms of the trigonometric functions

$$\sin\left(\frac{n\pi x}{L}\right) \text{ and } \cos\left(\frac{n\pi x}{L}\right), \quad n = 1, 2, \ldots, \tag{3.28}$$

where L is the finite size of the interval where we want to represent the function. These functions are shown in Fig. 3.2 for the lowest four values of n in each case. The trigonometric functions are solutions of the following differential equation for the unknown function $u(x)$:

$$\frac{d^2 u}{dx^2} = -k_n^2 u(x), \quad k_n = \frac{n\pi}{L}, \quad n = 1, 2, \ldots \tag{3.29}$$

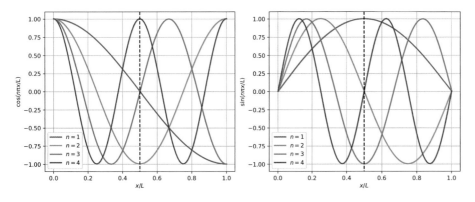

Figure 3.2: The first four cosine functions (left panel) and first four sine functions (right panel) of the type defined in Eq. (3.28), in the interval $(x/L) \in [0,1]$.

The series expansion of a function $f(x)$ in terms of the trigonometric functions defined in Eq. (3.28) takes the following form:

$$f(x) = a_0 + \sum_{n=1}^{\infty} a_n \cos\left(\frac{n\pi x}{L}\right) + \sum_{n=1}^{\infty} b_n \sin\left(\frac{n\pi x}{L}\right), \quad 0 \leq x \leq L. \quad (3.30)$$

The coefficients a_n and b_n involved in this series expansion are given by

$$a_0 = \frac{1}{L}\int_0^L f(x)\mathrm{d}x, \quad a_n = \frac{2}{L}\int_0^L f(x)\cos\left(\frac{n\pi x}{L}\right)\mathrm{d}x, \quad n \neq 0,$$

and

$$b_n = \frac{2}{L}\int_0^L f(x)\sin\left(\frac{n\pi x}{L}\right)\mathrm{d}x, \quad n \neq 0.$$

These expressions can be easily derived by multiplying both sides of Eq. (3.30) by one of the terms in the expansion, integrating both sides over the interval $[0, L]$, and taking advantage of the following identities:

$$\int_0^L \sin\left(\frac{n\pi x}{L}\right)\sin\left(\frac{m\pi x}{L}\right)\mathrm{d}x = \frac{L}{2}\delta_{nm}, \quad (3.31a)$$

$$\int_0^L \cos\left(\frac{n\pi x}{L}\right)\cos\left(\frac{m\pi x}{L}\right)\mathrm{d}x = \frac{L}{2}\delta_{nm}, \quad (3.31b)$$

$$\int_0^L \sin\left(\frac{n\pi x}{L}\right)\cos\left(\frac{m\pi x}{L}\right)\mathrm{d}x = 0. \quad (3.31c)$$

The sine and cosine functions defined in Eq. (3.28) have many useful properties. For example, all sine functions are equal to zero at the endpoints of the interval, $x = 0$ and $x = L$, so if the function $f(x)$ represented by this expansion happens to take values $f(0) = f(L) = 0$ then these are the natural terms to use in the expansion, that is, we can set $a_n = 0$ and calculate only the values of b_n coefficients. Alternatively, if the function $f(x)$ takes finite values at the end points of the interval, the natural terms to include in the expansion are the cosine terms. Moreover, if $f(0) = f(L)$ only cosines with even of n are needed, whereas if $f(0) = -f(L)$ only cosines with odd values of n are needed, since those are the terms that embody the behavior of the function at the endpoints, as shown in Fig. 3.2.

The sine and cosine series expansion of the arbitrary function $f(x)$ are part of a broader type of representation of the function referred to as "Fourier series expansion". We note that both the sine and the cosine functions take finite values in the interval $[-1, 1]$ for any value of their argument, so L can be arbitrarily large. Actually, it is possible to let $L \to \infty$, in which case we obtain a different type of representation of the function, known as the "Fourier transform", which is a *continuous* transform as opposed to the *discrete* expansion in a series. We return to these topics in Chapter 7, where we give a more complete discussion of the sine and cosine expansions and their generalization to the Fourier expansion and Fourier transform.

3.2.4 *Polynomial series expansion*

Another useful type of series expansions is one in which the functions $g_n(x)$ are polynomials, denoted in the following as $p_n(x)$, with special properties (often referred to as "special functions"). There are many such examples of special polynomials, motivated by different considerations of how these polynomials should behave. This type of expansion is designed to be a good representation of the given function $f(x)$ over a finite range of values of x and not just in the neighborhood of some specific value x_0, as was the case of the Taylor series. Because of this requirement, an important consideration is to make the special polynomials as "different" from each other as possible so that they are not redundant, and can perform well in representing the function $f(x)$ accurately in the desired range of x values. The concept of "difference" between functions is captured by the notion of "orthogonality". Two members of the series, $p_n(x)$ and $p_m(x)$, are called "orthogonal" if they satisfy the following relation:

$$\int_a^b p_n(x)p_m(x)w(x)\mathrm{d}x = \lambda_n \delta_{nm}, \tag{3.32}$$

where $x \in [a, b]$ being the interval where the function $f(x)$ needs to be represented, and $w(x)$ is a "weighting function". In other words, the integral of the product, weighted by the weighting function, of any two *different* functions $p_n(x)$ and $p_m(x)$ (the case $m \neq n$) vanishes, whereas the integral of the square, weighted by the weighting function, of any function $p_n(x)$ (the case $m = n$) is equal to a constant, called λ_n. In the more general case of functions of complex variables, one of the two functions in the integral of Eq. (3.32) involves complex conjugation.

Legendre polynomials: An example of special polynomials that appear in many physical contexts are the so-called "Legendre polynomials", denoted by $P_n(x)$. The Legendre polynomials are defined in the interval $x \in [-1, 1]$, through the recursive relation:

$$P_0(x) = 1, \quad P_1(x) = x, \quad P_{n+1}(x) = \frac{1}{n+1}\left[(2n+1)xP_n(x) - nP_{n-1}(x)\right].$$

They are solutions of the following differential equation for the unknown function $u(x)$:

$$(1-x^2)\frac{\mathrm{d}^2 u}{\mathrm{d}x^2} - 2x\frac{\mathrm{d}u}{\mathrm{d}x} = -\kappa_n u(x), \quad \kappa_n = n(n+1), \quad n = 0, 1, 2, \ldots,$$

which is encountered in problems of quantum mechanics.

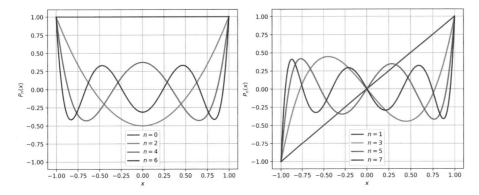

Figure 3.3: The Legendre polynomials, $P_n(x)$, for even $n = 0, 2, 4, 6$ (left panel) and odd $n = 1, 3, 5, 7$ (right panel), in their domain of definition, $x \in [-1, 1]$.

The first eight Legendre polynomials are shown in Fig. 3.3; their analytical forms are as follows:

$$P_0(x) = 1, \quad P_1(x) = x, \quad P_2(x) = \frac{1}{2}(3x^2 - 1), \quad P_3(x) = \frac{1}{2}(5x^3 - 3x),$$

$$P_4(x) = \frac{1}{8}(35x^4 - 30x^2 + 3), \quad P_5(x) = \frac{1}{8}(63x^5 - 70x^3 + 15x),$$

$$P_6(x) = \frac{1}{16}(231x^6 - 315x^4 + 105x^2 - 5),$$

$$P_7(x) = \frac{1}{16}(429x^7 - 693x^5 + 315x^3 - 35x).$$

These polynomials have several interesting properties: As is evident from the figure, they are even functions of x for n even, $P_n(-x) = P_n(x)$ and odd functions of x for n odd, $P_n(-x) = -P_n(x)$. The Legendre polynomials are orthogonal,

$$\int_{-1}^{1} P_n(x) P_m(x) dx = \frac{2}{2n+1} \delta_{nm}, \tag{3.33}$$

with the weighting function for these polynomials being $w(x) = 1$. Furthermore, they form a "complete basis" for expanding any function. We can calculate all the coefficients of the series expansion of $f(x)$ in Legendre polynomials from the formula

$$a_n = \frac{2n+1}{2} \int_{-1}^{1} f(x) P_n(x) dx.$$

This formula for the coefficients is easily proven by inserting the series expansion for the function in the integral and using the orthogonality of the Legendre polynomials, Eq. (3.33).

We note that because of their nature, the Legendre polynomials can only be defined in a finite interval, like $x \in [-1, 1]$; this interval cannot be extended to infinity because in this limit, $P_n(x)$ is unbounded (tends to $\pm\infty$).

Chebyshev polynomials: Another example of special polynomials are the so-called "Chebyshev polynomials". These polynomials, denoted by $T_n(x)$, are defined in the interval $x \in [-1, 1]$ through the recursive relation:

$$T_0(x) = 1, \quad T_1(x) = x, \quad T_{n+1}(x) = 2x T_n(x) - T_{n-1}(x).$$

The Chebyshev polynomials are solutions of the following differential equation for the unknown function $u(x)$:

$$(1-x^2)\frac{d^2u}{dx^2} - x\frac{du}{dx} = -\kappa_n u(x), \quad \kappa_n = n^2, \quad n = 0,1,2,\dots$$

The first eight Chebyshev polynomials are shown in Fig. 3.4; their analytical forms are as follows:

$$T_0(x) = 1, \quad T_1(x) = x, \quad T_2(x) = 2x^2 - 1, \quad T_3(x) = 4x^3 - 3x,$$

$$T_4(x) = 8x^4 - 8x^2 + 1, \quad T_5(x) = 16x^5 - 20x^3 + 5x,$$

$$T_6(x) = 32x^6 - 48x^4 + 18x^2 - 1, \quad T_7(x) = 64x^7 - 112x^5 + 56x^3 - 7x.$$

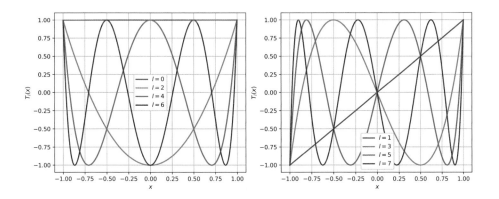

Figure 3.4: The Chebyshev polynomials, $T_n(x)$, for even $n = 0,2,4,6$ (left panel) and odd $n = 1,3,5,7$ (right panel), in their domain of definition, $x \in [-1,1]$.

These polynomials have several interesting properties: As is evident from the figure, they are even functions of x for n even, $T_n(-x) = T_n(x)$ and odd functions of x for n odd, $T_n(-x) = -T_n(x)$. The Chebyshev polynomials are orthogonal,

$$\int_{-1}^{1} T_n(x)T_m(x)\frac{dx}{\sqrt{1-x^2}} = \frac{(1+\delta_{n0})\pi}{2}\delta_{nm}, \qquad (3.34)$$

with the weighting function for these polynomials being $w(x) = 1/\sqrt{1-x^2}$. Furthermore, they form a "complete basis" for expanding any function. We can calculate all the coefficients of the series expansion of $f(x)$ in Chebyshev polynomials from the formula

$$a_n = \frac{2}{(1+\delta_{n0})\pi}\int_{-1}^{1} f(x)T_n(x)\frac{dx}{\sqrt{1-x^2}}.$$

This formula for the coefficients is easily proven by inserting the series expansion for the function in the integral and using the orthogonality relation, Eq. (3.34).

3.3   Convergence of series expansion of functions

A series of functions like the one defined in Eq. (3.11), for a given value of x, is just a series of real numbers, since given the value of x each function $g_n(x)$ takes on a specific real value. But summing an infinite set of numbers has the danger of yielding infinity — a series is useful only if it converges. Establishing whether a series expansion of a given function converges or not for a given value of x, is therefore of crucial importance. The expansion may converge at certain values of the variable x and may diverge at other values. Thus, one needs to establish the convergence of a series expansion for all the values of interest of the variable x.

3.3.1   Uniform convergence

As in the case of series of numbers, we study the convergence of a series of functions by turning them into sequences through the introduction of partial sums:

$$s_n(x) = \sum_{j=1}^{n} c_j g_j(x),$$

where $g_j(x)$ is the set of functions that have been chosen for the expansion. If the sequence $s_n(x)$ converges to the limiting function $s(x)$, the series will also converge:

$$\sum_{j=1}^{\infty} c_j g_j(x) = \lim_{n \to \infty} s_n(x) = s(x).$$

In this case, the condition for convergence of the sequence $s_n(x)$ is that for any $\epsilon > 0$ and a given value of x, we can find N such that for all $n > N$, $|s_n(x) - s(x)| < \epsilon$. The dependence of this condition on x introduces further complexity to the problem. This leads to the notion of *uniform convergence*, which reflects the possibility of convergence of the series for only a range of values of x. The definition of uniform convergence for $x \in [x_1, x_2]$ is the following: for any $\epsilon > 0$ we can find N such that $|s_n(x) - s(x)| < \epsilon$ for *all* $n > N$ and for all x in the interval of interest $x \in [x_1, x_2]$.

The meaning of uniform convergence of the sequence $s_n(x)$ for $x \in [x_1, x_2]$ is that for $\epsilon > 0$, we can find N such that for $n > N$ all terms of $s_n(x)$ lie within $\pm \epsilon$ of $s(x)$ and this applies to all x in the interval $[x_1, x_2]$. The value of N always depends on ϵ (usually, the smaller the value of ϵ the larger the N we have to choose), but as long as we can find one N for all x in the interval $[x_1, x_2]$ that satisfies this condition (that is, N is finite and does not depend on x), the sequence converges uniformly. However, the value of N does depend on the interval $[x_1, x_2]$.

To illustrate these notions, we examine the convergence of the series:

$$e^{-x} = 1 - x + \frac{1}{2!}x^2 - \frac{1}{3!}x^3 + \frac{1}{4!}x^4 - \frac{1}{5!}x^5 + \cdots \tag{3.35}$$

in the range $x \in [0, a]$, and then analyze the sequence produced by its partial sums. The partial sums $s_n(x)$ of this series, and the limit $s(x)$ are given by

$$s_n(x) = 1 + \sum_{k=1}^{n} (-1)^k \frac{1}{k!} x^k \Rightarrow s(x) = \lim_{n \to \infty} s_n(x) = e^{-x}. \tag{3.36}$$

In order to establish uniform convergence of the original series in the given interval, we must take the difference $|s(x) - s_n(x)|$ and ask whether or not it

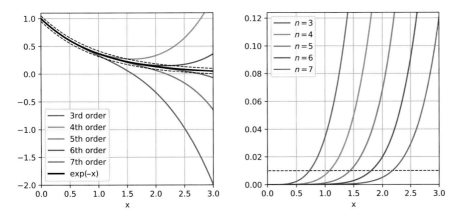

Figure 3.5: **Left**: The exponential function e^{-x} for $0 \leq x \leq 3$ (black line) and its approximation by a Taylor expansion keeping terms up to seventh order (color lines for orders $3, \ldots, 7$). The black dashed lines show the function $e^{-x} \pm \epsilon$. **Right**: The difference of the nth-order partial sum from the limit function, $|s_n(x) - s(x)|$, for $n = 3, \ldots, 7$ (color lines) in the interval $[0,3]$. The black horizontal dashed line corresponds to $\epsilon = 0.01$.

is bounded by some $\epsilon > 0$ for $n > N$ for all $x \in [0,a]$. The differences $|s_n(x) - s(x)|$ for $n = 3, \ldots, 7$ are shown in in Fig. 3.5 for x in the range $[0,3]$. We note that if we restrict our interest in the interval $[0,1]$ and choose $\epsilon = 0.01$, then we have $|s_3(1) - s(1)| > \epsilon$ while $|s_4(1) - s(1)| < \epsilon$; moreover, all differences obey $|s_n(x) - s(x)| < \epsilon$, for all values of $x \in [0,1]$ and for all $n > 4$. This suggest that for $\epsilon = 0.01$ we can choose $N = 4$ to satisfy the condition of uniform convergence. If we had restricted our interest in the interval $[0,2]$, for $\epsilon = 0.01$, we would find $|s_6(2) - s(2)| > \epsilon$ while $|s_7(2) - s(2)| < \epsilon$; therefore for the interval $[0,2]$ we need to choose $N = 7$ to satisfy the condition of uniform convergence. Had we chosen a smaller value of ϵ, the corresponding values of N for the intervals $[0,1]$ and $[0,2]$ would need to be higher. It is easy to convince oneself that as long as the interval is $[0,a]$ with a finite, no matter how small ϵ is, we can always find a *finite* value of N such that $|s_n(x) - s(x)| < \epsilon$ for all $n > N$ and all $x \in [0,a]$. In other words, the sequence of Eq. (3.35) converges uniformly to $s(x) = e^{-x}$ in any finite interval $[0,a]$. This argument does not work though in the *infinite* interval $[0,\infty)$, that is, the series is *not* uniformly convergent in this interval! Evidently, when we include the limit $x \to \infty$ in the interval, it is impossible to find a finite value of N, for any given ϵ, to satisfy the uniform convergence condition because for sufficiently large x all differences $|s_n(x) - s(x)|$ for $n > N$ blow up, no matter how large the value of N is.

Example 3.3: A series with uniform convergence

We consider the sequence of partial sums

$$s_n(x) = \frac{n}{n+1}(1-x), \quad \text{for } x \in [0,1]$$

(we are not concerned here with the question of analyzing the series that produced this sequence). We will show that the corresponding series converges uniformly in the chose interval. We first need to determine the limit of the sequence for $n \to \infty$:

$$s(x) = \lim_{n \to \infty} s_n(x) = \lim_{n \to \infty} \frac{n}{n+1}(1-x) = (1-x).$$

These functions for $n = 1, 2, 3, 4$ are illustrated in the diagram below, with the sequence of functions $s_n(x)$ shown in color and the limiting function $s(x)$ shown as a black thicker line; the functions $s(x) \pm \epsilon$ are shown as black dashed curves.

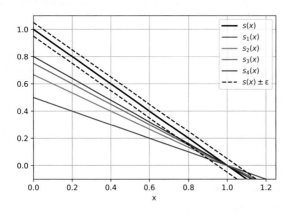

To determine whether the sequence is uniformly convergent, we examine the following differences:

$$|s_n(x) - s(x)| = \left| \frac{n}{n+1}(1-x) - (1-x) \right| = \left| \frac{1}{n+1}(1-x) \right| = \frac{1}{n+1}(1-x).$$

Notice that because $x \in [0, 1]$, we have

$$0 \leq (1-x) \leq 1 \Rightarrow |s_n(x) - s(x)| = \frac{1-x}{n+1} \leq \frac{1}{n+1}.$$

Given an $\epsilon > 0$, we want to find the value of N such that $(1-x)/(n+1) < \epsilon$ for $n > N$. But $(1-x)/(n+1) \leq 1/(n+1)$, so it is sufficient to find N such that $1/(n+1) < \epsilon$. If we choose

$$\frac{1}{N+1} < \epsilon \Rightarrow N > \frac{1}{\epsilon} - 1,$$

then $1/(n+1) < 1/(N+1) < \epsilon$ and

$$|s_n(x) - s(x)| \leq \frac{1}{n+1} < \frac{1}{N+1} < \epsilon, \quad \text{for } n > N.$$

Since our choice of N is independent of x, we conclude that the sequence $s_n(x)$ converges uniformly.

Example 3.4: A series with non-uniform convergence
We consider the sequence of partial-sum functions

$$s_n(x) = \frac{nx}{1 + n^2 x^2}, \quad x \in [0, \infty),$$

and determine the limit $s(x)$: for a fixed value of x, we have

$$s(x) = \lim_{n \to \infty} s_n(x) = \lim_{n \to \infty} \frac{nx}{1 + n^2 x^2} = 0.$$

The functions $s_n(x)$ for $n = 1, 2, 3, 4$ are shown in the diagram below as colored curves; the function $s(x)$ is shown as a black thicker line and the functions $s(x) \pm \epsilon$ are shown as black dashed lines.

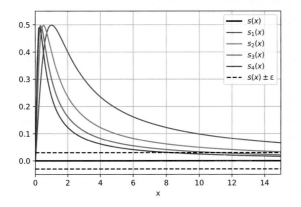

To determine whether the sequence converges uniformly or not, we examine the behavior of the difference

$$|s_n(x) - s(x)| = \frac{nx}{1 + n^2x^2} = s_n(x).$$

Let us study the behavior of $s_n(x)$ for a given n, as a function of x. This function has a maximum at $x = 1/n$, which is equal to $1/2$. For fixed x, and given $\epsilon > 0$, if the sequence converges uniformly we should be able to find N such that $nx/(1 + n^2x^2) < \epsilon$, for $n > N$. How large should N be in order to attain this inequality? We define n_0 to be the index of the function with its maximum at our chosen value of $x \Rightarrow n_0 = 1/x$ (we assume that x is small enough so that $1/x$ is arbitrarily close to an integer). For n larger than n_0, we will have $s_n(x) < s_{n_0}(x)$ and for sufficiently large n we will achieve $s_n(x) < \epsilon$. Therefore, n must be at least larger than n_0. How much larger? Call N the first value of n for which $s_n(x) < \epsilon$, then

$$\frac{Nx}{1 + N^2x^2} = \frac{N\frac{1}{n_0}}{1 + N^2(\frac{1}{n_0})^2} < \epsilon \Rightarrow 1 + \left(\frac{N}{n_0}\right)^2 > \frac{1}{\epsilon}\frac{N}{n_0}.$$

This inequality is satisfied by the choice $(N/n_0) > 1/\epsilon$:

$$\frac{N}{n_0} > \frac{1}{\epsilon} \Rightarrow \left(\frac{N}{n_0}\right)^2 > \frac{N}{n_0}\frac{1}{\epsilon} \Rightarrow 1 + \left(\frac{N}{n_0}\right)^2 > \frac{N}{n_0}\frac{1}{\epsilon}.$$

But $N > n_0/\epsilon \Rightarrow N > 1/(\epsilon x)$, which makes the choice of N depend on x. This contradicts the requirements for uniform convergence. If we concentrate for a moment to values $x \in [x_0, \infty)$ where x_0 is a finite number > 0, the choice

$$N > \frac{1}{x_0\epsilon} \geq \frac{1}{x\epsilon}$$

covers all possibilities in the interval $[x_0, \infty)$, so that $s_n(x)$ converges uniformly for x in that interval. But for $x \in [0, \infty)$ it is *not* possible to find a single N to satisfy $N > 1/(x\epsilon)$ for all x, as required for uniform convergence. This implies that the sequence $s_n(x)$ does *not* converge uniformly for $x \in [0, \infty)$. The problem with this sequence is that there is a spike near $x = 0$, and as n gets larger the spike moves closer to zero and becomes narrower, but it never goes away. This spike cannot be contained by an ϵ within $s(x)$ which is zero everywhere. If we truncate the interval to $[x, \infty)$ with x_0 a finite value, then we can "squeeze" the spike to values below x_0 and the sequence converges uniformly in this range, but if we include $x = 0$ in the interval we can never squeeze the spike away.

It is evident from these two examples that non-uniform convergence implies that the difference between $s_n(x)$ and $s(x)$ can be made arbitrarily small for each x with suitable choice of $n > N$, but cannot be made uniformly small for all x *simultaneously* for the same *finite* value of N.

3.3.2 *Uniform convergence criteria*

By analogy to the tests for convergence of series of numbers, we can devise tests for the uniform convergence of series of functions. Two particularly useful tests are given below.

(i) **Weierstrass test**

Consider the sequence of functions $s_j(x)$ and the sequence of positive numbers a_j. If

$$|s_j(x)| < a_j, \quad x \in [x_1, x_2],$$

and a_j is a convergent sequence, then the sequence $s_j(x)$ converges in the interval $[x_1, x_2]$. Since a_j does not involve x, this implies uniform convergence of the sequence $s_j(x)$.

(ii) **Cauchy test**

A sequence for which

$$|s_n(x) - s_m(x)| < \epsilon, \quad n, m > N$$

is called a "Cauchy sequence". This is a necessary condition for convergence of the sequence.

Proof: If the sequence is convergent we will have the following:

$$|s_n(x) - s(x)| < \frac{\epsilon}{2}, \ |s_m(x) - s(x)| < \frac{\epsilon}{2}, \quad n, m > N,$$

where we have called $\epsilon/2$ the small quantity that restricts $s_n(x)$ and $s_m(x)$ close to $s(x)$ for $n, m > N$. We can rewrite the difference between $s_n(x)$ and $s_m(x)$ in the form

$$|s_n(x) - s_m(x)| = |s_n(x) - s(x) + s(x) - s_m(x)|.$$

Using the triangle inequality and the previous relation, we obtain

$$|s_n(x) - s_m(x)| \leq |s_n(x) - s(x)| + |s_m(x) - s(x)| < \epsilon,$$

which completes the proof that it is a necessary condition. ■

A sequence being a Cauchy sequence is also a sufficient condition for convergence, in other words if a sequence is a Cauchy sequence then it converges uniformly.

Finally, we mention below four useful theorems related to uniform convergence, without delving into their proofs.

- **Theorem 1:** *If there is a convergent series of constants, $\sum c_n$, such that the following relation holds:*

$$|g_n(x)| \le c_n, \quad \forall x \in [a,b],$$

then the series $\sum g_n(x)$ is uniformly and absolutely convergent in $[a,b]$.

- **Theorem 2:** *Let $\sum g_n(x)$ be a series such that each $g_n(x)$ is a continuous function of x in the interval $[a,b]$. If the series is uniformly convergent in $[a,b]$, then the sum of the series is also a continuous function of x in $[a,b]$.*

- **Theorem 3:** *If a series of continuous functions $\sum g_n(x)$ converges uniformly to $g(x)$ in $[a,b]$, then*

$$\int_\alpha^\beta g(x)\mathrm{d}x = \int_\alpha^\beta g_1(x)\mathrm{d}x + \int_\alpha^\beta g_2(x)\mathrm{d}x + \cdots + \int_\alpha^\beta g_n(x)\mathrm{d}x + \dots,$$

where $a \le \alpha \le b$ and $a \le \beta \le b$. Moreover, the convergence is uniform with respect to α and β.

- **Theorem 4:** *Let $\sum g_n(x)$ be a series of differentiable functions that converges to $g(x)$ in $[a,b]$, and let $g_n^{(1)}(x)$ be the first derivative of $g_n(x)$. If the series $\sum g_n^{(1)}(x)$ converges uniformly in $[a,b]$, then it converges to the function $g^{(1)}(x)$ which is the first derivative of $g(x)$.*

The usefulness of uniform convergence, as is evident from the above theorems, is that it allows us to interchange the order of the integration or differentiation operations with the infinite summation. This makes possible the evaluation of integrals and derivatives through term-by-term integration or differentiation of a series. Non-uniform convergence implies that in general the order of these operations *cannot* be interchanged, although in some cases this may be possible.

Since integrals are limits of series, it is not surprising that the notion of uniform convergence can be extended to integrals. It should be noted that the series and integral representations obtained through the usual transform methods for the solution of partial differential equations are *not* uniformly convergent. Surprisingly, this serious disadvantage is *not* mentioned in the literature. One of the consequences of this lack of uniform convergence is that such classical representations are *not* suitable for numerical computations. The main advantage of the unified transform that will be introduced in Chapter 9 is that it always produces representations that *are* uniformly convergent. Furthermore, these representations are particularly suitable for numerical evaluation of the solution.

3.4   Truncation error in series expansions

Using series expansions to represent a function is useful because the basis for the expansion is formed by functions which are very simple, thus can be easily manipulated. However, the representation of the original function in terms of a series expansion is exact only if an infinite number of terms in the expansion is included, which is not possible in practice. Therefore, the hope is that we can truncate the series to a finite number of terms and still obtain a reasonable representation. The truncation introduces an error as the function is now represented by the finite number of terms we have retained in the expansion. It is useful to have a measure of this error, typically as a function of the number of terms retained, denoted by N. Let us assume that we have an infinite series representation of the function $f(x)$

$$f(x) = \sum_{n=0}^{\infty} c_n p_n(x),$$

where $p_n(x)$ is a set of orthogonal functions in the interval $[-L, L]$:

$$\int_{-L}^{L} p_n(x) p_m(x) \mathrm{d}x = \delta_{nm} \lambda_n, \quad \lambda_n = \int_{-L}^{L} [p_n(x)]^2 \, \mathrm{d}x. \tag{3.37}$$

We define

$$\tilde{f}_N(x) = \sum_{n=0}^{N} c_n p_n(x),$$

namely, the approximation of $f(x)$ by the truncated series with the index n running from 0 to N, thus containing a finite number of $(N+1)$ terms. A conventional measure of the error introduced by the truncation is the square root of the average of the absolute value square of the difference $f(x) - \tilde{f}_N(x)$, referred to as "root mean squared error" (RMSE):

$$\mathcal{E}^{(2)}(N) \equiv \left\{ \frac{1}{2L} \int_{-L}^{L} \left| f(x) - \tilde{f}_N(x) \right|^2 \mathrm{d}x \right\}^{1/2}$$

$$= \left\{ \frac{1}{2L} \int_{-L}^{L} \left| \sum_{n=N+1}^{\infty} c_n p_n(x) \right|^2 \mathrm{d}x \right\}^{1/2}$$

$$= \left\{ \sum_{n=N+1}^{\infty} \frac{|c_n|^2}{2L} \int_{-L}^{L} |p_n(x)|^2 \mathrm{d}x \right.$$

$$\left. + \sum_{n,m=N+1; n \neq m}^{\infty} \frac{c_n c_m}{2L} \int_{-L}^{L} p_n(x) p_m(x) \mathrm{d}x \right\}^{1/2} = \left\{ \sum_{n=N+1}^{\infty} \frac{|c_n|^2}{2L} \lambda_n \right\}^{1/2}, \tag{3.38}$$

where the last equality is a consequence of the orthogonality of the functions $p_n(x)$, Eq. (3.37). The reason for taking the absolute value squared is to avoid accidental cancellation of the negative and positive contributions to the error, which would lead to a meaningless estimate; this is referred to as the "L2" measure of the error. Another way to avoid the accidental cancellation of positive and negative contributions to the error is to simply take the absolute value of the difference $|f(x) - \tilde{f}_N(x)|$ in the integrand, which is called the "L1" measure. Other error definitions are also possible, such as the "L4" measure. Each one has its advantages or disadvantages and the choice depends on what aspect of the error is interesting for a particular application.

By considering c_n and λ_n as functions of n, we can find an upper bound for the $\mathcal{E}^{(2)}(N)$:

$$\mathcal{E}^{(2)}(N) = \left\{ \frac{1}{\Delta n} \sum_{n=N+1}^{\infty} \frac{|c_n|^2}{2L} \lambda_n \Delta n \right\}^{1/2} \leq \left\{ \frac{1}{2L\Delta n} \int_{N}^{\infty} |c(n)|^2 \lambda(n) \mathrm{d}n \right\}^{1/2}, \tag{3.39}$$

where Δn is the spacing of the values of the index n. For example, let us assume that c_n and λ_n take the following values:

$$\lambda_n = \pi,$$

$$c_n = \begin{cases} -\dfrac{4}{\pi^2 n^2} & \text{when } n \text{ is odd,} \\ 0 & \text{when } n \text{ is even} \end{cases}$$

(this is actually the set of values involved in the cosine expansion of a triangular wave, see Section 7.1). Substituting these values viewed as functions of n which we treat as a continuous variable in the expression for the bound, Eq. (3.39) gives the following estimate of the truncation error at the value N:

$$\mathcal{E}^{(2)}(N) \leq \left\{ \frac{1}{4\pi} \left(\frac{16}{\pi^4} \right) \pi \int_N^\infty \frac{1}{n^4} \mathrm{d}n \right\}^{1/2} = \left\{ \frac{4}{\pi^4} \left[\frac{-1}{3n^3} \right]_N^\infty \right\}^{1/2} = \frac{2}{\sqrt{3}\pi^2} \frac{1}{N^{3/2}},$$

where we have used $\Delta n = 2$ because n is odd. If we have a certain value for the tolerable error in mind, we can set it equal to the above result, which determines the value of N up to which we have to sum the series in order to have an error less than or equal to the tolerable value.

3.5 Application: Wavelet analysis

Although series expansions based on polynomials or trigonometric functions are a powerful tool for representing any given function, they suffer from certain limitations. In particular, they are not suited for representing function which exhibit variations on different scales. This is illustrated in Fig. 3.6 by a function $f(x)$ which has many sharp variations in the interval $x \in [0, L]$, namely, many local minima and maxima separated by intervals that are very small on the scale of L. This behavior will require a very large number of terms in a series expansion, like the sine series shown in Fig. 3.6, to capture all the features of the function $f(x)$. This is because the first few terms (small values of n) capture the large-scale behavior and the higher-order terms (large values of n) are needed to represent the small scale behavior. Indeed, the example of Fig. 3.6, the first four terms in the sine expansion, $\sin(n\pi x/L)$ with $n = 1, \ldots, 4$, can only capture the broad scale features of the function $f(x)$. This means that we will need to include a very large number of terms in the sine series expansion in order to obtain a reasonable representation of the function.

The so-called "wavelets" provide a far more efficient approach to the above problem. Wavelets are step-like functions, or more elaborate functions of special shape, whose scale is changed by a constant factor while they remain orthogonal to each other, just like the terms in the sine series expansion, Eq. (3.31). Because wavelets are defined locally and are not periodic functions, the total interval of interest L can be divided into smaller subintervals; following convention, we choose to divide L into 2^m subintervals of equal size, $l = L/2^m$, with m integer. For each subinterval, we can create a sequence of wavelets that can capture finer and finer features on this scale. We note that such a subdivision is not possible for the sine series expansion, because the

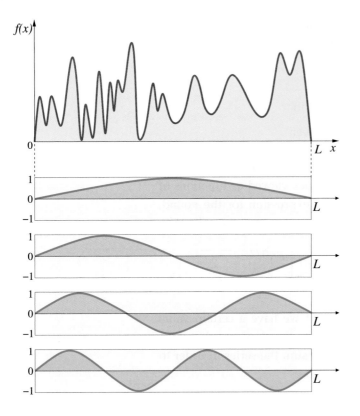

Figure 3.6: Illustration of a function $f(x)$ (in red) defined in the interval $x \in [0, L]$ that has both large-scale features (variations on the scale of L) and small-scale features (variations on a scale much smaller than L; this makes its representation in terms of a trigonometric expansion, Eq. (3.30), challenging. The first four functions of the sine expansion, $\sin(n\pi x/L)$ with $n = 1, 2, 3, 4$, are shown below (in green).

basis functions in that case are periodic, and this periodicity is imposed on a range beyond the fundamental length over which the basis functions are defined.

3.5.1 Haar wavelet expansion

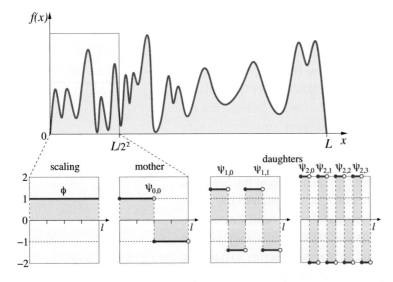

Figure 3.7: Illustration of wavelet analysis of a function $f(x)$ (in red) decomposed into its components in the window $x \in [x_0, x_0 + l]$ outlined in dashed border, with $l = L/2^m$. The scaling function $\phi(x)$, mother function $\psi_{0,0}(x)$ and two additional wavelet sets $\psi_{1,k}(x)$ with $k = 0, 1$ and $\psi_{2,k}(x)$ with $k = 0 - 3$ are shown (in blue).

The wavelet construction is illustrated in Fig. 3.7 for the simplest case, that of step-functions which are known as Haar wavelets. In this example, for the interval $x \in [0, l]$, with $l = L/2^m$ and $m = 2$, we use the function $\phi(x)$, called the "scaling function", which is simply equal to 1 for x in this interval; this interval then is repeated in the other $2^m - 1$ positions so that the entire range of $[0, L]$ is covered. The operations defined in each sub-interval are exactly the same, so for the sake of simplicity in notation we deal with only the first interval. We define the "mother function" as

$$
\psi_{0,0}(x) = \begin{cases} +1, & x \in \left[0, \dfrac{l}{2}\right), \\[2mm] -1, & x \in \left[\dfrac{l}{2}, l\right), \\[2mm] 0 & \text{everywhere else.} \end{cases} \tag{3.40}
$$

Then, we define the "daughter function" of order $n > 0$, denoted by $\psi_{n,k}(x)$, with $k = 0, 1, \ldots, 2^n - 1$, recursively by shifting the mother function by k and scaling the range by $1/2^n$:

$$
\psi_{n,k}(x) = 2^{n/2}\psi_{0,0}(2^n x - k), \quad n > 0, \quad k = 0, 1, \ldots, 2^n - 1, \tag{3.41}
$$

where the factor $2^{n/2}$ is included to ensure proper normalization. The daughter functions can be expressed explicitly as

$$
\psi_{n,k}(x) = \begin{cases} +2^{n/2}, & x \in \left[k\dfrac{l}{2^n}, \dfrac{2k+1}{2}\dfrac{l}{2^n}\right), \\[3mm] -2^{n/2}, & x \in \left[\dfrac{2k+1}{2}\dfrac{l}{2^n}, (k+1)\dfrac{l}{2^n}\right), \\[3mm] 0 & \text{everywhere else.} \end{cases} \tag{3.42}
$$

In the example of Fig. 3.7, the mother and daughter functions up to order $n = 2$ already capture both most of the fine-scale behavior of the function $f(x)$, *as well as* the large scale behavior. It is evident that the use of these wavelets as basis functions will be much more economical regarding the number of terms required for the accurate description of the function $f(x)$.

From the definition of the Haar wavelets, it is easy to see that the average of each one is zero, and that each one is properly normalized:

$$
\int_0^l \psi_{n,k}(x)\mathrm{d}x = 0, \quad \int_0^l |\psi_{n,k}(x)|^2\mathrm{d}x = 1. \tag{3.43}
$$

Furthermore, these wavelets are mutually orthogonal, as well as orthogonal to the scaling function:

$$
\int_0^l \psi_{n,k}(x)\, \psi_{m,j}(x)\mathrm{d}x = \delta_{nm}\, \delta_{kj}, \quad \int_0^l \psi_{n,k}(x)\, \phi(x)\mathrm{d}x = 0. \tag{3.44}
$$

These properties make the Haar wavelets a convenient and complete basis for expanding any function:

$$
f(x) = A\phi(x) + \sum_{n=0}^{\infty} \sum_{k=0}^{2^n-1} a_{n,k}\psi_{n,k}(x), \quad 0 \le x < l. \tag{3.45}
$$

Using the orthonormality relations we find:

$$
A = \int_0^l f(x)\mathrm{d}x, \quad a_{n,k} = \int_0^l f(x)\psi_{n,k}(x)\mathrm{d}x. \tag{3.46}
$$

Another interesting feature of wavelets is that the wavelet transform can be cast in the form of matrix multiplication. This provides an easy way to obtain the coefficients involved in the wavelet expansion. To give an example, suppose we are using the Haar wavelets up to order $n = 2$, as illustrated in Fig. 3.7. In this case, we have a total of eight basis functions, including the scaling function: $\phi, \psi_{0,0}, \psi_{1,0}, \psi_{1,1}, \psi_{2,0}, \psi_{2,1}, \psi_{2,2}, \psi_{2,3}$. Note that in general, from the construction of the Haar wavelets, the number of basis functions for wavelets up to order n is 2^{n+1}. We evaluate the *average* of the function in the eight intervals between the points $x_i, i = 0, \ldots, 8$, with $x_i = i \, 1/2^{n+1}$, and call these values $\tilde{f}_i$:

$$\tilde{f}_i = \int_{x_{i-1}}^{x_i} f(x)\mathrm{d}x, \quad i = 1, \ldots, 8.$$

From the definition of the wavelets and the wavelet expansion, Eq. (3.45), with the summation over n truncated at the value $n = 2$, these eight values will be given by the following matrix equation:

$$\begin{pmatrix} \tilde{f}_1 \\ \tilde{f}_2 \\ \tilde{f}_3 \\ \tilde{f}_4 \\ \tilde{f}_5 \\ \tilde{f}_6 \\ \tilde{f}_7 \\ \tilde{f}_8 \end{pmatrix} = \begin{bmatrix} 1 & 1 & \sqrt{2} & 0 & 2 & 0 & 0 & 0 \\ 1 & 1 & \sqrt{2} & 0 & -2 & 0 & 0 & 0 \\ 1 & 1 & -\sqrt{2} & 0 & 0 & 2 & 0 & 0 \\ 1 & 1 & -\sqrt{2} & 0 & 0 & -2 & 0 & 0 \\ 1 & -1 & 0 & \sqrt{2} & 0 & 0 & 2 & 0 \\ 1 & -1 & 0 & \sqrt{2} & 0 & 0 & -2 & 0 \\ 1 & -1 & 0 & -\sqrt{2} & 0 & 0 & 0 & 2 \\ 1 & -1 & 0 & -\sqrt{2} & 0 & 0 & 0 & -2 \end{bmatrix} \begin{pmatrix} A \\ a_{0,0} \\ a_{1,0} \\ a_{1,1} \\ a_{2,0} \\ a_{2,1} \\ a_{2,2} \\ a_{2,3} \end{pmatrix},$$

where each column of the above 8×8 matrix, which we will call $\mathbf{H}$, represents one of the basis functions, including the scaling function as the first column. A shorthand version of this expression, treating the values of the average function as the column vector $\tilde{\mathbf{f}}$ and the coefficients as the column vector $\mathbf{a}$, both of length 8, is

$$\tilde{\mathbf{f}} = \mathbf{Ha}.$$

$\mathbf{H}$ has the useful property that its inverse is its transpose multiplied by $1/2^{n+1}$, where n is the highest order of wavelets included in the expansion:

$$\mathbf{H}^{-1} = \frac{1}{2^{n+1}}\mathbf{H}^T.$$

This fact, which can be easily checked by direct matrix multiplication, is a consequence of the orthonormality of the wavelets and their structure. It is a very convenient feature, because it allows a direct evaluation of the coefficients in the wavelet expansion through the inverse matrix (obtained as the transpose of $\mathbf{H}$): the solution for the wavelet coefficients takes the simple form

$$\mathbf{a} = \mathbf{H}^{-1}\tilde{\mathbf{f}} = \frac{1}{2^{n+1}}\mathbf{H}^T\tilde{\mathbf{f}}.$$

For the case $n = 2$, we have:

$$\begin{pmatrix} A \\ a_{0,0} \\ a_{1,0} \\ a_{1,1} \\ a_{2,0} \\ a_{2,1} \\ a_{2,2} \\ a_{2,3} \end{pmatrix} = \frac{1}{8} \begin{bmatrix} 1 & 1 & 1 & 1 & 1 & 1 & 1 & 1 \\ 1 & 1 & 1 & 1 & -1 & -1 & -1 & -1 \\ \sqrt{2} & \sqrt{2} & -\sqrt{2} & -\sqrt{2} & 0 & 0 & 0 & 0 \\ 0 & 0 & 0 & 0 & \sqrt{2} & \sqrt{2} & -\sqrt{2} & -\sqrt{2} \\ 2 & -2 & 0 & 0 & 0 & 0 & 0 & 0 \\ 0 & 0 & 2 & -2 & 0 & 0 & 0 & 0 \\ 0 & 0 & 0 & 0 & 2 & -2 & 0 & 0 \\ 0 & 0 & 0 & 0 & 0 & 0 & 2 & -2 \end{bmatrix} \begin{pmatrix} \tilde{f}_1 \\ \tilde{f}_2 \\ \tilde{f}_3 \\ \tilde{f}_4 \\ \tilde{f}_5 \\ \tilde{f}_6 \\ \tilde{f}_7 \\ \tilde{f}_8 \end{pmatrix}.$$

Example 3.5: We consider the function $f(x) = \cos(3\pi x)e^{-x^2}$ expanded in terms of Haar wavelets of different orders in the interval $0 \leq x \leq 1$.

In the diagrams below, we show the original function $f(x)$ and its approximation by Haar wavelets. On the left panels, we show the average values of the function in the intervals $x \in [x_{i-1}, x_i]$, $i = 1, \ldots, 2^{n+1}$, where $x_i = i/2^{n+1}$ (red dots) and the approximation by the wavelet expansion (blue lines) and the actual function (red dashed line); on the right panels we show the values of the wavelet coefficients, $A, a_{n,k}$, labeled from $j = 0$ to $2^{n+1} - 1$, for the cases $n = 3$ (top row), corresponding to a total of 16 basis functions and $n = 4$ (bottom row), corresponding to a total of 32 basis functions. It is evident that the first approximation is not very accurate, whereas the second one is quite close to the values of the function $f(x)$ throughout the range of interest.

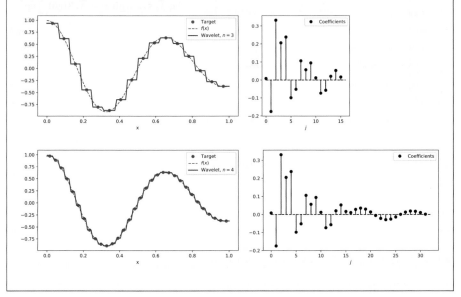

3.5.2   The wavelet transform

In the example shown above, the smooth function $f(x)$ is represented by series of small steps, the center of each step passing through uniformly spaced values of the function. Given the nature of the Haar wavelets, this feature is not surprising, but it is not satisfactory because in this approach the representation of a smooth continuous function is achieved through a set of discontinuous steps. This problem can be bypassed by designing wavelet approximations based on smooth basis functions. We can then employ this so-called "wavelet transform", using these continuous wavelets. In this case, the scaling relation in terms of the mother function $\psi(x)$ is defined by

$$\psi_{\sigma,\tau}(x) = \frac{1}{\sqrt{|\sigma|}}\psi\left(\frac{x-\tau}{\sigma}\right), \quad \sigma,\tau \in \mathbb{R}. \tag{3.47}$$

In other words, the argument of the mother function is scaled by the factor σ and is translated by τ. This transform is meaningful provided that the following condition is satisfied:

$$\int_{-\infty}^{\infty} \frac{1}{|k|^2}\left[\left|\int_{-\infty}^{\infty}\psi(x)\cos(kx)dx\right|^2 + \left|\int_{-\infty}^{\infty}\psi(x)\sin(kx)dx\right|^2\right]dk < \infty. \tag{3.48}$$

The wavelet transform involves two independent parameters, σ and τ. The arbitrary function $f(x)$ is then represented in the form

$$f(x) = \int_{-\infty}^{\infty}\int_{-\infty}^{\infty} W_\psi(\sigma,\tau)\psi_{\sigma,\tau}(x)\frac{d\sigma}{\sigma^2}d\tau, \tag{3.49}$$

where the coefficients $W_\psi(\sigma,\tau)$ are given by

$$W_\psi(\sigma,\tau) = \int_{-\infty}^{\infty} f(x)\psi_{\sigma,\tau}(x)dx. \tag{3.50}$$

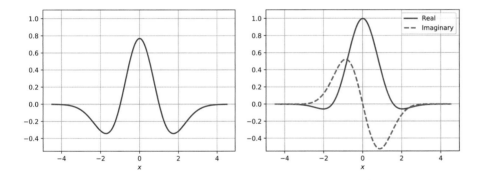

Figure 3.8: **Left**: The Ricker or "Mexican hat" wavelet, $\psi_R(x)$ defined in Eq. (3.51), with $\alpha = 1$. **Right**: The real and imaginary components of the Gabor wavelet, $\psi_G(x)$ defined in Eq. (3.52), with $\beta = 1$, $\mu = 0$, $\nu = 1$.

An example of a continuous wavelet is the Ricker wavelet, also known as the "Mexican hat" wavelet, shown in Fig. 3.8

$$\psi_R(x) = \frac{1}{\sqrt{3\alpha}\pi^{1/4}}\left(1 - \frac{x^2}{\alpha^2}\right)e^{-x^2/2\alpha^2}. \tag{3.51}$$

Another example is the Gabor wavelet, shown in Fig. 3.8, which has both real and imaginary components:

$$\psi_G(x) = e^{-(x-\mu)^2/\beta^2}\left\{\cos\left(\nu(x-\mu)\right) + i\sin\left(\nu(x-\mu)\right)\right\}. \tag{3.52}$$

The Gabor wavelet minimizes the variance $(\Delta x)^2$ in real space (x-space), as well as the variance $(\Delta k)^2$ in reciprocal space (k-space). We elaborate on these topics in later chapters, in particular in Chapter 7.

Further reading

The mathematical-methods textbooks mentioned under "General Sources" in the Preface cover the material discussed in the present chapter in detail, except for the topic discussed in the Application section (Section 3.5), which is more specialized. For this and related topics and their applications see the additional suggestions below.

1. Amir-Homayoon Najmi, *Wavelets, A Concise Guide* (Johns Hopkins University Press, 2012). This is a useful guide to the wavelet transform and its connection to the Fourier transform, at an introductory level.

2. Martin Vetterli, Jelena Kovacevic and Vivek K. Goyal, *Foundations of Signal Processing* (Cambridge University Press, 2014). This is a compilation of various approaches that use series expansions, including wavelets, for representing signals. It also provides a useful discussion of bases, error estimation and convergence issues.

3. Jelena Kovacevic, Vivek K. Goyal and Martin Vetterli, *Fourier and Wavelet Signal Processing* (Cambridge University Press, 2014). This is a companion volume by the same authors as the previous reference; it provides a more detailed discussion of wavelets.

Problems

1. Determine whether the following series converge or diverge. Can you evaluate the convergent ones?

(a) $\displaystyle\sum_{n=1}^{\infty} \frac{n+2}{n^2+n}$,

(b) $\displaystyle\sum_{n=1}^{\infty} \frac{n!}{n^n}$,

(c) $\displaystyle\sum_{n=1}^{\infty} n^n 2^{-n}$,

(d) $\displaystyle\sum_{n=1}^{\infty} \frac{e^{2n}}{n!}$,

(e) $\displaystyle\sum_{n=1}^{\infty} \left(\frac{\ln(n)}{n}\right)^2$,

(f) $\displaystyle\sum_{n=1}^{\infty} \frac{1}{n\ln(n)}$,

(g) $\displaystyle\sum_{n=0}^{\infty} \frac{(-1)^n 2^{2n}}{(n!)^2}$.

2. Determine whether the following series converge or diverge:

(a) $\displaystyle\sum_{n=2}^{\infty} \frac{i^n}{n-1}$,

(b) $\displaystyle\sum_{n=1}^{\infty} \frac{i^n}{n^2}$,

(c) $\displaystyle\sum_{n=0}^{\infty} \frac{\{x\}_n}{n!} 2^{-n}$, $x \in \mathbb{R}$,

where in (a) and (b) i is the imaginary unit, for which it is given that $i^2 = -1$ and $|i| = 1$, and in (c) the meaning of the symbol $\{x\}_n$ is

$$\{x\}_n = x(x+1)(x+2)\cdots(x+n-1) \text{ for } n \geq 1,$$

and we define $\{x\}_0 = 1$.

3. Derive the binomial expansion, Eq. (3.9), from the Taylor series expansion of the function

$$f(x) = a^p(1+x)^p,$$

with $x = (b/a)$, expanded around $x_0 = 0$.

4. Find the Taylor expansion of the function

$$f(x) = \sqrt{1+x}$$

at the origin $x = 0$. Can you find the Taylor expansion of

$$f(x) = \sqrt{x}$$

at the origin? Justify your answer.

5. Using the Taylor expansion, find power series representations of the functions

(a) $\dfrac{1}{\sqrt{1+x}}$,

(b) $\dfrac{1}{\sqrt{1-x}}$,

(c) $\dfrac{1}{\sqrt{1+x^2}}$,

(d) $\dfrac{1}{\sqrt{1-x^2}}$.

For what range of values of x can these power series expansions be used effectively as approximations of the function, that is, they can be truncated at some term, with the remainder of the terms making an insignificant contribution?

6. Find the Taylor expansion of the following functions:

(a) $f(x) = \cos(\pi e^x)$, at $x_0 = 0$,

(b) $f(x) = \cosh(\sin(x))$, at $x_0 = 0$.

7. Find the Taylor expansion of the function

$$f(x) = \left(1 + x + \frac{7}{10}x^2\right)^{2/3},$$

around $x_0 = 0$, up to second order in x. Repeat this exercise by first changing variables to $y = x + (7/10)x^2$, obtaining the Taylor expansion in powers of y around $y_0 = 0$, and then substituting back in terms of x. Do your results for the two expansions in powers of x agree?

8. Calculate the third-order term in the inversion of the series of Eq. (3.24) by using similar steps as those that led to the first- and second-order terms, Eqs. (3.26) and (3.27), respectively.

9. Consider the following two sequences of functions which represent the partial sums of some infinite series:

(i) $s_n(x) = 2nxe^{-nx^2}$, (ii) $s_n(x) = \left(1 - \frac{x}{n}\right)^n$,

and are both defined for $x \in [0,1]$ and $n \in \mathbb{N}$. Our goal is to study the uniform convergence of each series in the given interval through the following steps:

(a) Find the limit $s(x) = \lim_{n \to \infty} s_n(x)$.

(b) Plot the first few $s_n(x)$ as functions of x. Can you establish whether or not the sequence $s_n(x)$ converges uniformly in the given interval, using this graphical representation of its terms?

(c) If the interval had been specified as $x \in [x_0, 1]$, where $0 < x_0 < 1$, how might the index N be chosen for a given bound ϵ, so that $|s(x) - s_n(x)| < \epsilon$ for $n > N$ and $x \in [x_0, 1]$?

(d) Evaluate the integral

$$\int_0^1 s_n(x)dx$$

and take its limit for $n \to \infty$. Compare this limit to the following integral:

$$\int_0^1 s(x)dx.$$

10. Consider the following two sequences of functions which represent the partial sums of some infinite series:

(i) $s_n(x) = n^2xe^{-nx}$, (ii) $s_n(x) = xe^{-n^2x^2/2}$,

and are both defined for $x \in [0, \infty)$, and $n \in \mathbb{N}$. Our goal is to study the uniform convergence of each series in the given interval through the following steps:

(a) Find the limit $s(x) = \lim_{n \to \infty} s_n(x)$.

(b) For fixed n, find the position and the value of the maximum of $s_n(x)$.

(c) Plot the first few $s_n(x)$ as functions of x. Use a graphical argument to show whether $s_n(x)$ converges uniformly or not in the interval $[0, \infty)$.

(d) Find the integral

$$I_n = \int_0^\infty s_n(x)dx$$

and the limit $\lim_{n \to \infty} [I_n]$. Compare this limit with the value of the integral

$$I = \int_0^\infty s(x)dx.$$

Part II

Complex Analysis and the Fourier Transform

Chapter 4

Functions of Complex Variables

The miracle of the appropriateness of the language of mathematics in physical sciences is a wonderful gift which we never understood or deserved.

Eugene Wigner (Hungarian-American theoretical physicist, 1902–1995)

4.1 Complex numbers

We begin with some historical background of complex numbers. Complex numbers were originally introduced to solve the innocent looking equation

$$x^2 + 1 = 0, \tag{4.1}$$

which in the realm of real numbers has no solution because it requires the square of a number to be a negative quantity. This led to the introduction of the imaginary unit "i" defined as

$$i = \sqrt{-1} \Rightarrow i^2 = -1. \tag{4.2}$$

This solved the original problem, seemingly by brute force. This solution however, has far reaching implications. For starters, the above equation is just a special case of the second-order polynomial equation

$$ax^2 + bx + c = 0. \tag{4.3}$$

This equation has real solutions if and only if $\Delta = (b^2 - 4ac) \geq 0$: the solutions are

$$x = \frac{-b \pm \sqrt{\Delta}}{2a}. \tag{4.4}$$

For $\Delta < 0$, using the imaginary unit, the solutions become

$$z = \frac{-b \pm i\sqrt{|\Delta|}}{2a}. \tag{4.5}$$

The above expression defines a *complex number*, z, with real part $-b/2a$ and imaginary part $\pm\sqrt{|\Delta|}/2a$. Thus, the introduction of the imaginary unit implies that the second-order polynomial equation always has exactly two solutions (also called "roots"). In fact, every polynomial equation of degree n

$$a_n x^n + a_{n-1} x^{n-1} + \cdots + a_1 x + a_0 = 0, \tag{4.6}$$

with real coefficients $a_j, j = 0, 1, 2, \ldots, n$, has exactly n solutions, where some of them may be complex numbers.

We denote the real part of a complex number z by $x = \text{Re}[z]$ and the imaginary part by $y = \text{Im}[z]$. An important concept is the complex conjugate $\bar{z}$ of the complex number z:

$$z = x + iy \Rightarrow \bar{z} = x - iy. \tag{4.7}$$

A complex number and its conjugate have equal magnitude, given by

$$|z| = |\bar{z}| = \sqrt{z \cdot \bar{z}} = \sqrt{x^2 + y^2}. \tag{4.8}$$

Significant progress in understanding complex numbers has been made by the representation of the complex number $z = x + iy$ on the so-called "complex plane", with abscissa (x) along the real axis (horizontal), and ordinate (y) along the imaginary axis (vertical).

In terms of the polar coordinates, $r = \sqrt{x^2 + y^2}$ and $\phi = \tan^{-1}(y/x)$, a complex number takes the form:

$$z = x + iy = r[\cos(\phi) + i\sin(\phi)], \tag{4.9}$$

as illustrated in Fig. 4.1. Since $r = |z|$, r is the **magnitude** of z; ϕ is called the **argument** of z and is denoted by $\text{Arg}[z]$. Note that the use of polar coordinates introduces an ambiguity: if the argument is changed by an integral multiple of 2π the complex number stays the same:

$$r[\cos(\phi + 2n\pi) + i\sin(\phi + 2n\pi)] = r[\cos(\phi) + i\sin(\phi)], \tag{4.10}$$

where n is any integer. The ambiguity can actually be turned into an advantage when finding the roots of a given complex equation, as it will be discussed in detail below.

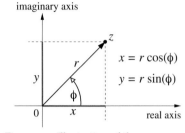

Figure 4.1: Illustration of the representation of a complex number z on the complex plane, in terms of its real x and imaginary y parts, and the polar coordinates r and ϕ.

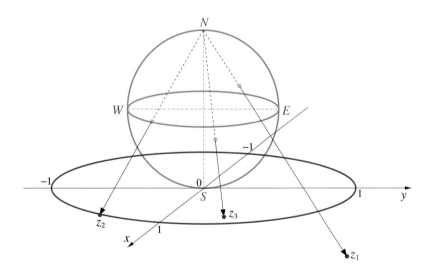

Figure 4.2: Illustration of the representation of a complex number on the Riemannian sphere and its mapping to the complex z-plane. The South Pole (S) corresponds to the origin, $z = 0$, while the North Pole (N) corresponds to the point $z \to \infty$. Points in the northern hemisphere correspond to points outside the unit circle, $|z_1| > 1$, points in the equator correspond to points on the unit circle, $|z_2| = 1$, and points in the southern hemisphere correspond to points inside the unit circle, $|z_3| < 1$.

An extension of the complex plane is the so-called Riemannian sphere, which includes the representation of the point at infinity. It is also called the *extended complex plane*. In the Riemannian sphere, illustrated in Fig. 4.2, the South Pole corresponds to the value $z = 0$, the sphere has diameter equal to unity, and the North Pole corresponds to the point $z \to \infty$. Each point on the z-plane is mapped onto a point on the sphere through the following construction: consider a line that begins at the North Pole and ends on the z-plane;

the point where this line intersect the surface of the sphere is the image of the point where the line meets the z-plane. Through this procedure, the equator of the sphere is mapped to the unit circle of the z-plane, the points in the southern hemisphere are mapped to the interior of this circle, and the points in the northern hemisphere are mapped to the exterior of this circle.

We examine next the form of various mathematical operations when real numbers are replaced by complex numbers. The product of two complex numbers z_1, z_2 in terms of their real and imaginary components, $z_1 = x_1 + iy_1$, $z_2 = x_2 + iy_2$, is easily defined following the associative rule:

$$z_1 z_2 = (x_1 + iy_1)(x_2 + iy_2) = (x_1 x_2 - y_1 y_2) + i(x_1 y_2 + x_2 y_1). \qquad (4.11)$$

The definition of the quotient can be reduced to the definition of the product, using complex conjugation:

$$\frac{z_1}{z_2} = \frac{z_1 \bar{z}_2}{z_2 \bar{z}_2} = \frac{z_1 \bar{z}_2}{|z_2|^2} = \frac{(x_1 + iy_1)(x_2 - iy_2)}{(x_2 + iy_2)(x_2 - iy_2)}$$

$$\Rightarrow \quad \frac{z_1}{z_2} = \left(\frac{x_1 x_2 + y_1 y_2}{x_2^2 + y_2^2} \right) + i \left(\frac{x_2 y_1 - x_1 y_2}{x_2^2 + y_2^2} \right). \qquad (4.12)$$

An extremely useful expression is the so-called **Euler's formula**, that will be derived below:

$$e^{ix} = \cos(x) + i\sin(x), \quad x \in \mathbb{R}. \qquad (4.13)$$

Combining Eq. (4.9) and Eq. (4.13) we obtain the most useful representation

$$z = re^{i\phi}, \qquad (4.14)$$

which is called the "polar coordinate" representation of the complex number z because it employs the polar coordinates r and ϕ. We can prove Euler's formula by using the series expansions of the trigonometric functions (see Chapter 1):

$$\cos(x) = 1 - \frac{1}{2!}x^2 + \frac{1}{4!}x^4 - \cdots = 1 + \frac{1}{2!}(ix)^2 + \frac{1}{4!}(ix)^4 + \cdots,$$

$$i\sin(x) = ix - i\frac{1}{3!}x^3 + i\frac{1}{5!}x^5 - \cdots = (ix) + \frac{1}{3!}(ix)^3 + \frac{1}{5!}(ix)^5 + \cdots,$$

where we have used $i^2 = -1, i^3 = -i, i^4 = 1, i^5 = i$, etc. Adding these two expressions we find

$$\cos(x) + i\sin(x) = 1 + (ix) + \frac{1}{2!}(ix)^2 + \frac{1}{2!}(ix)^3 + \frac{1}{4!}(ix)^4 + \frac{1}{5!}(ix)^5 + \cdots = e^{ix},$$

in which we recognize the Taylor series expansion for the exponential of (ix), which is the desired result. It may seem strange that we identify the above infinite series with an exponential, since the Taylor series was earlier defined for real numbers. However, the laws of addition and multiplication are the same for real and imaginary numbers, therefore the term-by-term operations involved in the above series make it identical to the exponential of ix power series expansion *à la* Taylor. This will be discussed further in Section 4.7.

Applying Euler's formula for $x = \pi$ we obtain:

$$e^{i\pi} = \cos(\pi) + i\sin(\pi) = -1, \qquad (4.15)$$

which relates the real and imaginary units, 1 and i, with the transcendental numbers e and π whose meaning and values were given in Chapter 1 (see Section 1.1). A consequence of this result is

$$e^{i2n\pi} = \cos(2n\pi) + i\sin(2n\pi) = \left(e^{i\pi}\right)^{2n} = 1, \quad n \text{ integer.} \qquad (4.16)$$

Equations (4.15) and (4.16) are consistent with the geometrical implementation of Eq. (4.14): the complex number $e^{i\pi}$ has length 1 and argument π, thus it is a distance 1 from the origin and angle $\phi = \pi$ with the positive real axis; hence, it represents the number -1. Similarly, the number $e^{i2n\pi}$ is at a distance 1 from the origin and makes an angle $\phi = 2n\pi$ with the positive real axis which is equivalent to the angle 0; hence, this number is 1.

The use of Euler's formula and the polar coordinate representation simplifies significantly operations with complex numbers:

$$\text{product} : z_1 z_2 = r_1 r_2 e^{i(\phi_1 + \phi_2)}$$
$$= r_1 r_2 [\cos(\phi_1 + \phi_2) + i\sin(\phi_1 + \phi_2)], \qquad (4.17a)$$
$$\text{inverse} : \frac{1}{z} = \frac{1}{r}e^{-i\phi} = r^{-1}[\cos(\phi) - i\sin(\phi)], \qquad (4.17b)$$
$$\text{quotient} : \frac{z_1}{z_2} = \frac{r_1}{r_2}e^{i(\phi_1 - \phi_2)}$$
$$= \frac{r_1}{r_2}[\cos(\phi_1 - \phi_2) + i\sin(\phi_1 - \phi_2)], \qquad (4.17c)$$
$$n\text{th power} : z^n = r^n e^{in\phi} = r^n[\cos(n\phi) + i\sin(n\phi)]. \qquad (4.17d)$$

In the last equation, raising the complex number z to a positive integer power n does *not* have any adverse effects. However, raising z to a power p different than a positive integer has severe implications, that are related to the ambiguity of using the polar coordinate representation. We deal with this important issue in detail in the next section.

In general

$$|e^{i\phi}| = \sqrt{\cos^2(\phi) + \sin^2(\phi)} = 1, \quad \phi \in \mathbb{R}. \qquad (4.18)$$

Applying the expression for the nth power to a complex number of magnitude 1, we obtain

$$\left[e^{i\phi}\right]^n = [\cos(\phi) + i\sin(\phi)]^n = \cos(n\phi) + i\sin(n\phi). \qquad (4.19)$$

This last equation is known as "De Moivre's formula". As an application of this powerful formula, we can prove the trigonometric relations discussed in Chapter 1, Eqs. (1.16b) and (1.16d):

$$[\cos(\phi) + i\sin(\phi)]^2 = \cos^2(\phi) + i2\cos(\phi)\sin(\phi) - \sin^2(\phi) = \cos(2\phi) + i\sin(2\phi),$$

and from this expression, equating the real and imaginary parts on the two sides of the last equation we find:

$$\cos(2\phi) = \cos^2(\phi) - \sin^2(\phi), \quad \sin(2\phi) = 2\cos(\phi)\sin(\phi).$$

4.2   Complex variables

The next natural step is to consider functions of the complex variable z. In Chapter 1, we introduced basic functions of the real variable x: $\sin(x)$, $\cos(x)$,

e^x, $\ln(x)$. Since $\ln(x)$ is the inverse function of e^x, it follows that we have three basic functions. Now, with the help of complex numbers we will see that we only have one basic function, namely the exponential.

4.2.1   The complex exponential

From Euler's formula we can derive the following expressions for the trigonometric functions $\sin(x)$ and $\cos(x)$ in terms of $\exp(\pm ix)$:

$$\cos(x) = \frac{1}{2}\left[e^{ix} + e^{-ix}\right], \quad \sin(x) = \frac{1}{2i}\left[e^{ix} - e^{-ix}\right]. \tag{4.20}$$

The exponential of (ix) involved in the above expressions is a new type of function, because it contains the imaginary unit i. Generalizing this idea, we can consider functions of complex numbers; for example,

$$e^z = e^{x+iy} = e^x e^{iy}.$$

Since we know how to handle both the exponential of a real number $\exp(x)$ and the exponential of an imaginary number $\exp(iy)$, this expression is well-defined. It turns out that the exponential of z can be expressed as

$$e^z = 1 + z + \frac{1}{2!}z^2 + \frac{1}{3!}z^3 + \cdots, \tag{4.21}$$

which is consistent with the Taylor expansion of the exponential of a real variable (see Chapter 1) or the exponential of a purely imaginary variable (see above, proof of Euler's formula). We can also define the hyperbolic and trigonometric functions of the complex variable z using *analytic continuation*, namely replacing x with z:

$$\cosh(z) = \frac{1}{2}\left[e^z + e^{-z}\right], \quad \sinh(z) = \frac{1}{2}\left[e^z - e^{-z}\right], \tag{4.22a}$$

$$\cos(z) = \frac{1}{2}\left[e^{iz} + e^{-iz}\right], \quad \sin(z) = \frac{1}{2i}\left[e^{iz} - e^{-iz}\right]. \tag{4.22b}$$

These definitions lead to the following relations:

$$\cosh(iz) = \cos(z), \quad \sinh(iz) = i\sin(z), \tag{4.23a}$$

$$\cos(iz) = \cosh(z), \quad \sin(iz) = i\sinh(z). \tag{4.23b}$$

4.2.2   Multi-valuedness in polar representation

In the polar coordinate representation of the complex variable $z = re^{i\phi}$, there is an important ambiguity, namely, changing the value of the argument ϕ by an integral multiple of 2π leaves z unchanged:

$$e^{i(\phi + k2\pi)} = \cos(\phi + k2\pi) + i\sin(\phi + k2\pi) = \cos(\phi) + i\sin(\phi) = e^{i\phi}, \quad k \text{ integer.}$$

This ambiguity has important consequences: it has the advantage of providing a simple algorithmic way of finding the roots of a given algebraic equation but at the same time it gives rise to multi-valuedness that must be avoided in order to be able to define a function of the variable z, which must have a unique value for any value of z.

As a case in point, multi-valuedness is actually helpful in the evaluation of the nth root of z, as it produces n different values for $z^{1/n}$:

$$z^{1/n} = r^{1/n} \left[\cos\left(\frac{\phi + 2k\pi}{n} \right) + i\sin\left(\frac{\phi + 2k\pi}{n} \right) \right], \quad k = 0, \ldots, n-1. \quad (4.24)$$

The existence of n roots is consistent with representing the unknown root by w; then

$$w = z^{1/n} \Rightarrow w^n - z = 0,$$

which implies that there should be n values of w satisfying the equation. A value of k outside the range $[0, n-1]$ gives the same result as the value of k within this range. For example, consider the equation

$$w^4 - 16 = 0 \Rightarrow w^4 = 2^4 e^{i(0 + 2k\pi)} \Rightarrow w = 2e^{i(2n\pi/4)}, \quad n = 0, 1, 2, 3$$

$$\Rightarrow w = 2, \ 2e^{i\pi/2}, \ 2e^{i\pi}, \ 2e^{i3\pi/2}, \quad \text{or} \quad w = 2, \ 2i, \ -2, \ -2i.$$

It can be easily verified that by raising each of these four solutions to the fourth power one finds the number 16.

Multi-valuedness can be eliminated by what is called the *principal value*, namely, limiting the values of the argument to one specific interval of 2π, usually chosen as the interval $0 \le \phi < 2\pi$ (which corresponds to $k = 0$ in the above expressions). This restriction is referred to as considering one "branch" of the multi-valued expression, thus turning it into a proper function.

4.2.3   The complex logarithm

Another use of the Euler formula is to provide a definition for the logarithm of a complex number:

$$\ln(z) = \ln\left(re^{i(\phi + 2k\pi)} \right) = \ln(r) + i(\phi + 2k\pi), \quad 0 \le \phi < 2\pi, \ k \text{ integer.} \quad (4.25)$$

In this case, the ambiguity in the argument of z produces an ambiguity in both the argument and the magnitude of $\ln(z)$ which both depend on the value of k:

$$|\ln(z)| = \sqrt{[\ln(r)]^2 + (\phi + 2k\pi)^2}, \quad \text{Arg}[\ln(z)] = \tan^{-1}\left(\frac{\phi + 2k\pi}{\ln(r)} \right).$$

This ambiguity can be eliminated by considering the principal value, namely limiting ϕ to the range $0 \le \phi < \pi$.

Example 4.1: We are interested in finding the magnitude and argument of w defined by the equation

$$w = (1 + i)^{(1+i)},$$

assuming that we working with principal values, that is, $0 \le \phi < 2\pi$ for the arguments of complex numbers. We first express $(1 + i)$ in polar coordinates:

$$1 + i = \sqrt{2}\left(\frac{1}{\sqrt{2}} + i\frac{1}{\sqrt{2}} \right) = \sqrt{2}e^{i(\pi/4)} = e^{\ln(\sqrt{2})}e^{i(\pi/4)}.$$

Substituting the above expression in the representation of w, we find

$$w = \left[e^{\ln(\sqrt{2})+i(\pi/4)} \right]^{(1+i)} = e^{\ln(\sqrt{2})+i\ln(\sqrt{2})} e^{i\pi/4-\pi/4} = e^{\ln(\sqrt{2})-\pi/4} e^{i(\pi/4+\ln(\sqrt{2}))}.$$

Introducing the polar representation of w in terms of the magnitude r_w and the argument ϕ_w, we find

$$w = r_w e^{i\phi_w} = \sqrt{2} e^{-\pi/4} e^{i(\ln(\sqrt{2})+\pi/4)} \Rightarrow r_w = \sqrt{2} e^{-\pi/4}, \quad \phi_w = \ln(\sqrt{2}) + \frac{\pi}{4}.$$

4.3 Continuity and derivatives

As in the case of a complex number $z = x + iy$, a function of a complex variable $f(z)$ can always be separated into its real and imaginary parts:

$$f(z) = w = u(x,y) + iv(x,y), \tag{4.26}$$

where $u(x,y)$ and $v(x,y)$ are real functions of the real variables x, y. The function $w = f(z)$ implies a mapping of the z-plane to the w-plane, a topic that will be discussed in Chapter 6.

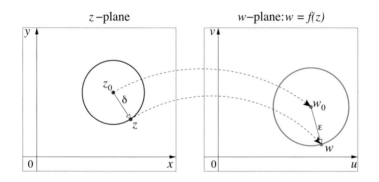

Figure 4.3: Illustration of continuity of the function $f(z) = w$ near the point $f(z_0) = w_0$: when z is closer than δ to z_0, then w is closer than ϵ to w_0.

We first consider the issue of continuity for functions of complex variables. We define continuity of a function of the complex variable z in a similar way as in the real case: the function $f(z)$ is continuous if for any $\epsilon > 0$, there exists $\delta > 0$ such that

$$|z - z_0| < \delta \Rightarrow |f(z) - f(z_0)| < \epsilon. \tag{4.27}$$

The essence of this condition is illustrated in Fig. 4.3, with $\delta \to 0$ when $\epsilon \to 0$.

Example 4.2: Consider the function

$$w = f(z) = z^2.$$

In order to show that it is continuous, we need to establish that for any given $\epsilon > 0$, assuming that $|z - z_0| < \delta$, then $|w - w_0| < \epsilon$. This involves finding a way of relating ϵ and δ. Since $|z - z_0| < \delta$, we have

$$|w - w_0| = |z^2 - z_0^2| = |(z - z_0)(z + z_0)| < \delta|z + z_0| = \delta|(z - z_0) + 2z_0|,$$

where in the last step we changed the content of the absolute value by adding and subtracting z_0, so that the quantities $(z - z_0)$ and z_0 appear inside the absolute value. Using the triangle inequality, the last relation implies

$$|w - w_0| \leq \delta[|z - z_0| + |2z_0|] < \delta[\delta + 2|z_0|], \tag{4.28}$$

where in the last step we have again taken advantage of the fact that $|z - z_0| < \delta$. If we express ϵ in terms of δ through the relations below, we find

$$\epsilon = \delta[\delta + 2|z_0|] \Rightarrow \delta^2 + 2|z_0|\delta - \epsilon = 0 \Rightarrow \delta = \sqrt{|z_0|^2 + \epsilon} - |z_0| > 0$$

(where we have chosen the positive root of the second-order polynomial equation in δ). The last expression shows that no matter how small ϵ is, we can always find a value for $\delta > 0$ that satisfies the general continuity condition, Eq. (4.27), which in the present case takes the form of Eq. (4.28). From the above relation between ϵ and δ, it is also evident that the smaller the ϵ the smaller the corresponding δ, a situation that is common but not always necessary.

We next wish to define derivatives of a complex function. This involves several complications: first, we can treat the complex function as a function that depends on the two independent variables x, y; then, the derivatives will involve partial derivatives with respect to x or y of the two functions $u(x, y)$ and $v(x, y)$. It turns out that the most efficient way to treat a function of a complex variable $z = x + iy$ is to introduce the complex conjugate of z, namely the complex variable $\bar{z} = x - iy$ as the second independent variable. The crucial realization is that we may choose to treat a function $f(z)$ as depending only on one, namely z, and not on the other, $\bar{z}$, of these two independent variables. This point of view has important consequences. Indeed, first, since the original variables x, y and the new variables $z, \bar{z}$ are related by a linear transformation, their derivatives can be expressed in terms of each other. Specifically, the partial derivatives with respect to x and y can be expressed in terms of the partial derivatives with respect to z and $\bar{z}$:

$$z = x + iy, \quad \bar{z} = x - iy \quad \longleftrightarrow \quad x = \frac{z + \bar{z}}{2}, \quad y = \frac{z - \bar{z}}{2i} = \frac{-iz + i\bar{z}}{2}. \tag{4.29}$$

The chain rule gives the following relations:

$$\frac{\partial}{\partial x} = \frac{\partial z}{\partial x}\frac{\partial}{\partial z} + \frac{\partial \bar{z}}{\partial x}\frac{\partial}{\partial \bar{z}} = \frac{\partial}{\partial z} + \frac{\partial}{\partial \bar{z}}, \tag{4.30a}$$

$$\frac{\partial}{\partial y} = \frac{\partial z}{\partial y}\frac{\partial}{\partial z} + \frac{\partial \bar{z}}{\partial y}\frac{\partial}{\partial \bar{z}} = i\left(\frac{\partial}{\partial z} - \frac{\partial}{\partial \bar{z}}\right). \tag{4.30b}$$

Similarly,

$$\frac{\partial}{\partial z} = \frac{\partial x}{\partial z}\frac{\partial}{\partial x} + \frac{\partial y}{\partial z}\frac{\partial}{\partial y} = \frac{1}{2}\left(\frac{\partial}{\partial x} - i\frac{\partial}{\partial y}\right), \tag{4.31a}$$

$$\frac{\partial}{\partial \bar{z}} = \frac{\partial x}{\partial \bar{z}}\frac{\partial}{\partial x} + \frac{\partial y}{\partial \bar{z}}\frac{\partial}{\partial y} = \frac{1}{2}\left(\frac{\partial}{\partial x} + i\frac{\partial}{\partial y}\right). \tag{4.31b}$$

Combining the expressions for the partial derivatives with respect to z and $\bar{z}$, we find the interesting result

$$\frac{\partial^2}{\partial z \partial \bar{z}} = \frac{1}{4}\left(\frac{\partial^2}{\partial x^2} + \frac{\partial^2}{\partial y^2}\right). \tag{4.32}$$

By analogy to the case of functions of real variables, we define the derivative of a function $f(z)$ of the complex variable z, at some point z_0, through the limit:

$$f'(z_0) = \lim_{\Delta z \to 0} \frac{f(z_0 + \Delta z) - f(z_0)}{\Delta z} = \lim_{z \to z_0} \frac{f(z) - f(z_0)}{z - z_0}. \qquad (4.33)$$

In order for the derivative $f'(z)$ to make sense, the limit must exist and must be independent of how $\Delta z \to 0$, that is, independent of the direction in which the point z_0 on the complex plane is being approached. This aspect did not exist in the case of the derivative of a function of a real variable, since the point x_0 on the real axis can only be approached along values on the real axis. The possible dependence of the ratio $\Delta f(z)/\Delta z$ on the approach to z_0 imposes important restrictions on $f(z)$ in order for $f'(z_0)$ to exist. As will be shown in the next section, these restrictions are consistent with the requirement that $f(z)$ depends only on z as opposed to depending on both z and $\bar{z}$.

4.4 Analyticity

A function $f(z)$ is called *analytic* in the domain $\mathcal{D}$ of the xy-plane if:

- $f(z)$ is defined in $\mathcal{D}$, that is, it is single-valued; and

- $f(z)$ is differentiable in $\mathcal{D}$, that is, the derivative $f'(z)$ exists and is finite.

If these conditions are not met, the function is called *non-analytic*. There are two important consequences of the analyticity of a function $f(z)$:

   (i) $f'(z)$ is continuous (this is known as "Coursat's theorem").

   (ii) $f(z)$ has continuous derivatives of all orders.

In the following, when employing the notion of analyticity we will assume that it holds at least in a finite *region* around some point z_0. By region we mean an area that is part of the complex plane and has the same dimensionality as the plane itself. For instance, the set of all points in a straight line do not constitute a region, because the line is one-dimensional whereas the plane is two-dimensional. In other words, we adopt the convention that the notion of analyticity applies strictly in a finite or infinite, two-dimensional portion of the complex plane.

By this convention, it does not make sense to talk about analyticity at a *single* point of the complex plane. On the other hand, it *does* makes sense to talk about a function being non-analytic at certain *points* inside a region of analyticity of the function. For example, the function

$$f(z) = \frac{1}{z}$$

is analytic everywhere on the complex plane except at $z = 0$, where it is non-analytic because it does not have a finite value. For any other point, even arbitrarily close to zero, the function is properly defined and so are all its derivatives. Isolated points where an otherwise analytic function is non-analytic are called **singularities**. By extension, we note that the function $f(z) = z^n$, where n is a positive integer, is analytic and so are all simple functions containing it, except if z^n is in the denominator, because then it can vanish for $z = 0$; in fact $f(z) = 1/z^n$ is also analytic everywhere except for $z = 0$, which is a singularity. There exist different types of singularities, depending on how a function becomes non-analytic. We will explore the types and implications of singularities in detail in Chapter 5.

Derivation: The analyticity of the natural logarithm of z

We wish to establish the analyticity of the natural logarithm function with complex argument z,

$$f(z) = \ln(z), \quad \mathcal{D} : \{z \in \mathbb{C}, \quad z \neq 0, \infty, \quad 0 \leq \text{Arg}[z] < 2\pi\},$$

where we have chosen as the domain $\mathcal{D}$ where this function is defined to be the entire complex plane, excluding the points $z = 0$ and $z = \infty$. As we will see in Chapter 5, these two points are singularities of a particular type (branch points) for the logarithm. By excluding those two points and restricting the argument ϕ of z to the range $0 \leq \phi < 2\pi$, we ensure that the logarithm is analytic in $\mathcal{D}$.

To show the analyticity of $\ln(z)$ in $\mathcal{D}$, we express it in terms of polar coordinates:

$$z = re^{i\phi} \Rightarrow \ln(z) = \ln(r) + i\phi, \quad r = (x^2 + y^2)^{1/2}, \quad \phi = \tan^{-1}\left(\frac{y}{x}\right).$$

From this expression, it is evident that both the real and the imaginary part of $\ln(z)$ are single-valued functions in the domain $\mathcal{D}$. Moreover, the derivative with respect to z can be calculated using the expression of Eq. (4.31a):

$$\frac{d}{dz} \ln(z) = \frac{1}{2}\left(\frac{\partial}{\partial x} - i\frac{\partial}{\partial y}\right)[\ln(r) + i\phi].$$

For the partial derivatives with respect to x and y of the real part, we find

$$\frac{\partial}{\partial x}\left[\ln\left((x^2 + y^2)^{1/2}\right)\right] = \frac{x}{x^2 + y^2}, \quad \frac{\partial}{\partial y}\left[\ln\left((x^2 + y^2)^{1/2}\right)\right] = \frac{y}{x^2 + y^2},$$

and combining the two contributions we obtain

$$\frac{d\ln(r)}{dz} = \frac{1}{2}\left(\frac{\partial r}{\partial x} - i\frac{\partial r}{\partial y}\right) = \frac{1}{2}\left(\frac{x}{x^2 + y^2} - i\frac{y}{x^2 + y^2}\right) = \frac{1}{2z}.$$

For the partial derivative with respect to x of the imaginary part, we use the chain rule:

$$\frac{\partial}{\partial x}[\tan(\phi)] = \frac{\partial}{\partial \phi}[\tan(\phi)]\frac{\partial \phi}{\partial x}$$

$$\Rightarrow \quad \frac{\partial \phi}{\partial x} = \frac{\partial}{\partial x}[\tan(\phi)]\left[\frac{\partial}{\partial \phi}[\tan(\phi)]\right]^{-1}. \tag{4.34}$$

To calculate the partial derivative with respect to x on the right-hand side of Eq. (4.34), we observe:

$$\phi = \tan^{-1}\left(\frac{y}{x}\right) \Rightarrow \frac{y}{x} = \tan(\phi). \Rightarrow \frac{\partial}{\partial x}[\tan(\phi)] = \frac{\partial}{\partial x}\left(\frac{y}{x}\right) = -\frac{y}{x^2},$$

and for the derivative of $\tan(\phi)$ with respect to ϕ which appears on the right-hand side of Eq. (4.34) we have

$$\frac{\partial}{\partial \phi}\left[\tan(\phi)\right] = \frac{\partial}{\partial \phi}\left[\frac{\sin(\phi)}{\cos(\phi)}\right] = \frac{1}{\cos^2(\phi)} = 1 + \tan^2(\phi) = \frac{x^2 + y^2}{x^2}.$$

Combining the results of the two calculations we arrive at

$$\frac{\partial \phi}{\partial x} = \left(-\frac{y}{x^2}\right)\left(\frac{x^2}{x^2 + y^2}\right) = -\frac{y}{x^2 + y^2}.$$

By analogous steps we also obtain the partial derivative of ϕ with respect to y:

$$\frac{\partial \phi}{\partial y} = \frac{x}{x^2 + y^2}.$$

From these expressions we can then calculate

$$\frac{d\phi}{dz} = \frac{1}{2}\left(\frac{\partial \phi}{\partial x} - i\frac{\partial \phi}{\partial y}\right) = \frac{1}{2}\left(\frac{-y}{x^2 + y^2} - i\frac{x}{x^2 + y^2}\right) = -i\frac{1}{2z}.$$

With the derivatives of the real and imaginary parts of the logarithm at hand, we can use them to obtain

$$\frac{d}{dz}\ln(z) = \frac{d}{dz}\left[\ln(r) + i\phi\right] = \frac{d\ln(r)}{dz} + i\frac{d\phi}{dz} = \frac{1}{z}.$$

Thus, through steps based entirely on derivatives of real functions of the real variables x and y, we were able to derive a familiar expression for the derivative of the logarithm with complex argument z, which exists and is finite in the domain $\mathcal{D}$. From this derivation, we conclude that the function $\ln(z)$ is an analytic function in $\mathcal{D}$; its derivative, $1/z$, is evidently continuous in this domain, as Coursat's theorem requires. Furthermore, we can calculate higher derivatives using the same process and we find that they exist and are continuous to all orders in the domain $\mathcal{D}$.

In contrast, any function that contains $\bar{z}$ is *non-analytic*. For instance, consider the function $f(z, \bar{z}) = \text{Re}[z] = x$ which is non-analytic everywhere on the complex plane, because:

$$f(z, \bar{z}) = \text{Re}[z] = x = \frac{z + \bar{z}}{2}$$

and this last expression contains $\bar{z}$. The key point is that any function $f(x, y)$, through the use of the relations (4.29) can be written in terms of $z, \bar{z}$. Analytic functions belong to the special set where $\bar{z}$ is canceled out. Consider, for example, the function

$$f(x, y) = x^2 - y^2 + 2ixy.$$

Using the last two equations of (4.29), it follows that $\bar{z}$ cancels out and $f(x, y) = z^2$; thus, this function is an analytic function and we can write it as $f(z)$.

4.4.1 The d-bar derivative

An important consequence of the idea of analyticity, is that an analytic function $f(z)$ cannot contain $\bar{z}$, thus

$$f(z) \text{ analytic} \Rightarrow \frac{\partial f}{\partial \bar{z}} = 0, \qquad (4.35)$$

which is known as the "d-bar derivative". As an example, the function $f(z) = \bar{z}$ has a non-vanishing d-bar derivative

$$f(z, \bar{z}) = \bar{z} \Rightarrow \frac{\partial f}{\partial \bar{z}} = \frac{\partial \bar{z}}{\partial \bar{z}} = 1,$$

which implies that $f(z, \bar{z}) = \bar{z}$ is non-analytic everywhere on the complex plane.

The vanishing of the d-bar derivative has important consequences on the behavior of the real and imaginary parts of the function. Using the expressions for the partial derivatives derived earlier, Eq. (4.31b), we obtain

$$\frac{\partial f}{\partial \bar{z}} = \frac{\partial}{\partial \bar{z}} (u + iv) = \frac{1}{2} \left(\frac{\partial}{\partial x} + i \frac{\partial}{\partial y} \right) (u + iv) = 0.$$

Separating out the real and imaginary parts of the last expression, which must vanish independently, we arrive at

$$\left(u_x - v_y \right) + i \left(u_y + v_x \right) = 0 \Rightarrow u_x = v_y, \ \ u_y = -v_x. \qquad (4.36)$$

The last two equations are known as the "Cauchy–Riemann relations".[1] They play an important role in many applications, including partial differential equations, as we discuss below and in later chapters.

4.4.2 Cauchy–Riemann relations and harmonic functions

We take another look at the Cauchy–Riemann relations, Eq. (4.36), from a different perspective. We will prove these relations by relying only on the definition of the derivative $f'(z)$ of an analytic function $f(z)$ at some point z_0, which must exist and must be independent of the approach $z \to z_0$ or $\Delta z \to 0$:

$$f'(z) = \lim_{\Delta z \to 0} \frac{\Delta f}{\Delta z}. \qquad (4.37)$$

Expressing $f(z)$ in terms of its real and imaginary parts, $u(x, y)$ and $v(x, y)$, which are both functions of x and y, we can write the differential of $f(z)$ in terms of the partial derivatives with respect to x and y:

$$\Delta f = \frac{\partial f}{\partial x} \Delta x + \frac{\partial f}{\partial y} \Delta y = (u_x + iv_x) \Delta x + \left(u_y + iv_y \right) \Delta y = f_x \Delta x + f_y \Delta y.$$

The last equation defines the definition of the partial derivatives f_x and f_y in terms of partial derivatives of u and v. Let us assume that f_x and f_y are not both zero; without loss of generality, we take $f_x \neq 0$. Then, the derivative can be put in the form

$$f'(z_0) = \lim_{\Delta x, \Delta y \to 0} \left[\frac{1 + (f_y / f_x)(\Delta y / \Delta x)}{1 + i(\Delta y / \Delta x)} f_x \right].$$

[1] In these expressions and in the following discussion, we use the notation introduced in Chapter 2 for partial derivatives, Eqs. (2.4)–(2.5), namely that partial derivatives are denoted by the subscript of the variable with respect to which we are differentiating.

The derivative must be independent of the direction of approach of $\Delta x, \Delta y \to 0$, and this can be achieved only for $(f_y/f_x) = i$, because otherwise the expression for the derivative would depend on the direction of approach, namely, the ratio $(\Delta y/\Delta x)$. Consequently,

$$f_x = -if_y \Rightarrow u_x + iv_x = -i\left(u_y + +iv_y\right) = v_y - iu_y.$$

Equating the real and imaginary parts of f_x and f_y we obtain the Cauchy–Riemann relations, Eq. (4.36).

Taking cross-derivatives of the Cauchy–Riemann relations and adding them leads to:

$$u_{xy} = v_{yy} \quad \text{and} \quad u_{yx} = -v_{xx} \quad \Rightarrow \quad -u_{yx} + u_{xy} = v_{xx} + v_{yy} = 0,$$
$$u_{xx} = v_{xy} \quad \text{and} \quad u_{yy} = -v_{yx} \quad \Rightarrow \quad v_{xy} - v_{yx} = u_{xx} + u_{yy} = 0.$$

Thus, both the real and imaginary parts of an analytic function satisfy the so-called Laplace equation

$$w_{xx} + w_{yy} = 0. \tag{4.38}$$

Functions that satisfy Laplace's equation are called **harmonic**.

Example 4.3: In the derivation of the analyticity of the logarithm, we established that the real and imaginary parts of this function are given by:

$$\ln(z) = \ln(r) + i\phi = \ln\left(\left(x^2 + y^2\right)^{1/2}\right) + i\tan^{-1}\left(\frac{y}{x}\right)$$

$$\Rightarrow u(x,y) = \ln\left(\left(x^2 + y^2\right)^{1/2}\right), \quad v(x,y) = \tan^{-1}\left(\frac{y}{x}\right).$$

In the same derivation, we also established that

$$\frac{\partial}{\partial x}\left[\ln\left((x^2 + y^2)^{1/2}\right)\right] = \frac{x}{x^2 + y^2} = \frac{\partial u}{\partial x},$$

$$\frac{\partial}{\partial y}\left[\ln\left((x^2 + y^2)^{1/2}\right)\right] = \frac{y}{x^2 + y^2} = \frac{\partial u}{\partial y},$$

$$\frac{\partial \phi}{\partial x} = -\frac{y}{x^2 + y^2} = \frac{\partial v}{\partial x}, \quad \frac{\partial \phi}{\partial y} = \frac{x}{x^2 + y^2} = \frac{\partial v}{\partial y}.$$

Comparing the partial derivatives of u and v we see that they obey the Cauchy–Riemann equations, Eq. (4.36),

$$u_x = v_y, \quad u_y = -v_x,$$

as expected, since $\ln(z)$ is an analytic function in the domain $\mathcal{D}$ defined in the derivation of its analyticity. Moreover, taking one more partial derivative for each component we find:

$$\frac{\partial^2 u}{\partial x^2} = \frac{-x^2 + y^2}{(x^2 + y^2)^2} \quad \text{and} \quad \frac{\partial^2 u}{\partial y^2} = \frac{x^2 - y^2}{(x^2 + y^2)^2} \Rightarrow u_{xx} + u_{yy} = 0,$$

$$\frac{\partial^2 v}{\partial x^2} = \frac{2xy}{(x^2 + y^2)^2} \quad \text{and} \quad \frac{\partial^2 v}{\partial y^2} = \frac{-2xy}{(x^2 + y^2)^2} \Rightarrow v_{xx} + v_{yy} = 0,$$

in other words, both $u(x,y)$ and $v(x,y)$ are harmonic functions as they satisfy Laplace's equation, Eq. (4.38).

Harmonic functions appear frequently in applications in connection with several physical phenomena, such as electrostatic fields produced by electrical charges, the temperature field in an isotropic thermal conductor under steady heat flow, and the pressure field for incompressible liquid flow in a porous medium. Eq. (4.38) will be discussed in Chapters 8 and 12.

4.4.3   Consequences of analyticity

It was shown earlier that the condition $\partial f / \partial \bar{z} = 0$ implies the Cauchy–Riemann relations. We next show that the inverse is also true: starting with the Cauchy–Riemann relations, we will prove that the d-bar derivative must vanish, Eq. (4.35). From the Cauchy–Riemann relations and using the expressions for the partial derivatives that we found earlier, Eqs. (4.30a), (4.30b), we obtain

$$\frac{\partial u}{\partial x} = \frac{\partial v}{\partial y} \Rightarrow \frac{\partial u}{\partial z} + \frac{\partial u}{\partial \bar{z}} = i\frac{\partial v}{\partial z} - i\frac{\partial v}{\partial \bar{z}},$$

$$\frac{\partial u}{\partial y} = -\frac{\partial v}{\partial x} \Rightarrow i\frac{\partial u}{\partial z} - i\frac{\partial u}{\partial \bar{z}} = -\frac{\partial v}{\partial z} - \frac{\partial v}{\partial \bar{z}} \Rightarrow i\frac{\partial v}{\partial z} + i\frac{\partial v}{\partial \bar{z}} = \frac{\partial u}{\partial z} - \frac{\partial u}{\partial \bar{z}}.$$

Adding the two equations on the far right of the above lines, we find

$$\frac{\partial}{\partial z}(u + iv) + \frac{\partial}{\partial \bar{z}}(u + iv) = \frac{\partial}{\partial z}(u + iv) - \frac{\partial}{\partial \bar{z}}(u + iv),$$

which simplifies to the desired result:

$$\frac{\partial f}{\partial \bar{z}} = -\frac{\partial f}{\partial \bar{z}} \Rightarrow \frac{\partial f}{\partial \bar{z}} = 0.$$

The CR relations are necessarily obeyed by an analytic function, as proven above, but they are not sufficient to ensure that a function is analytic. This can be shown by the following example [invented by S. Pollard (1928)]:

$$f(x,y) = \frac{(x^3 - y^3) + i(x^3 + y^3)}{(x^2 + y^2)}, \quad z \neq 0; \quad f(0) = 0,$$

which obeys the Cauchy–Riemann relations but is not differentiable at $z = 0$. However, the lack of analyticity of the above function becomes obvious by rewriting the expression of $f(z)$ for $z \neq 0$ in terms of z and $\bar{z}$:

$$f(x,y) = \frac{1}{2}\frac{(z+\bar{z})^3 - i(z-\bar{z})^3 + i(z+\bar{z})^3 - (z-\bar{z})^3}{(z+\bar{z})^2 - (z-\bar{z})^2} = \left(\frac{1+i}{4}\right)\frac{3z^2 + \bar{z}^2}{z}.$$

Since $f(x,y)$ depends on *both* z and $\bar{z}$, clearly this function is *not* analytic. This example emphasizes further the importance of treating an analytic function as a "reduction" of the general function $f(z,\bar{z})$ to one which is *independent* of $\bar{z}$. In addition, the CR equations necessitate the introduction of $u(x,y)$ and $v(x,y)$, namely of the real and imaginary parts of $f(z)$, whereas the condition

$$f(z) \text{ analytic} \iff \frac{\partial f}{\partial \bar{z}} = 0$$

expresses the necessary and sufficient condition of analyticity directly in terms of $f(z)$.

The Cauchy–Riemann relations imply the following important statements:

(i) since $u(x,y), v(x,y)$ (often called "fields") satisfy Laplace's equation, they can be determined by knowledge of their boundary values, namely, by the values they take on the boundary of the given domain;

(ii) if one field is known, the conjugate field (u from v or v from u) is determined up to a constant of integration;

(iii) since u and v are determined from the boundary values, a complex analytic function is determined by its boundary values as well;

(iv) any linear combination of the two fields $u(x,y), v(x,y)$ of an analytic function is a harmonic function.

Example 4.4: To illustrate the above facts, we consider the field

$$u(x,y) = \cos(x)e^{-y}.$$

We will first verify that it is harmonic, and then we will determine its conjugate field $v(x,y)$ and the corresponding analytic function $f(z) = u(x,y) + iv(x,y)$. To verify harmonicity, we calculate the second partial derivatives of the given field and check that their sum vanishes:

$$u_{xx} = -\cos(x)e^{-y}, \quad u_{yy} = +\cos(x)e^{-y};$$

thus, since the sum of the above terms vanishes, $u(x,y)$ is a harmonic function. We will next determine the conjugate field $v(x,y)$: from the Cauchy–Riemann relations, we find the following equations:

$$v_x = -u_y = \cos(x)e^{-y}$$

$$\Rightarrow \quad v(x,y) = \int [\cos(x)e^{-y}]dx = e^{-y}\sin(x) + c_1(y),$$

$$v_y = u_x = -\sin(x)e^{-y}$$

$$\Rightarrow \quad v(x,y) = \int [-\sin(x)e^{-y}]dy = \sin(x)e^{-y} + c_2(x),$$

where $c_1(y)$ and $c_2(x)$ are constants of integration, produced from the indefinite integral over dx and dy, respectively. The two different expressions for $v(x,y)$ obtained from the two integrations must be equal, so

$$c_1(y) = c_2(x),$$

which can only be satisfied if they are both equal to the same constant c (since the first is a function of y and the second is a function of y). Thus,

$$v(x,y) = \sin(x)e^{-y} + c.$$

Then, the complex function $f(z)$ is given by

$$f(z) = u(x,y) + iv(x,y) = [\cos(x) + i\sin(x)]e^{-y} + ic = e^{ix-y} + ic = e^{iz} + ic,$$

an expression that indeed involves only z, and is therefore an analytic function.

From the properties of analyticity, we have the following conditions:

- If $f(z)$ is an analytic function in a domain $\mathcal{D}$, then so is the function $1/f(z)$, except at the roots of $f(z)$, namely at the points in $\mathcal{D}$ for which $f(z) = 0$.

- If $g(z), h(z)$ are two analytic functions in a domain $\mathcal{D}$, then so are the functions defined by the following expressions in terms of $g(z), h(z)$:

$$f_1(z) = g(z) + h(z), \quad f_2(z) = g(z)g(z), \quad f_3(z) = g(h(z))$$

in the same domain $\mathcal{D}$.

By analogy to the case of functions of real variables, the following rules also apply to the differentiation of analytic functions:

$$f(z) = g(z)h(z) \quad \Rightarrow \quad \frac{\partial}{\partial z}f(z) = \left[\frac{\partial}{\partial z}g(z)\right]h(z) + g(z)\left[\frac{\partial}{\partial z}h(z)\right],$$

$$f(z) = g(h(z)) \quad \Rightarrow \quad \frac{\partial}{\partial z}f(z) = \left(\frac{\partial}{\partial h}g(h)\right)\left(\frac{\partial}{\partial z}h(z)\right).$$

4.4.4 Explicit formulas for computing an analytic function in terms of its real or imaginary parts

In Example 4.4, the conjugate field $v(x, y)$ was determined from $u(x, y)$ through the solutions of the Cauchy–Riemann relations. This approach involves the "inverse" operations of differentiation and integration. This suggests that maybe it is possible to avoid these operations altogether and to obtain $v(x, y)$ through an *algebraic* procedure. It turns out that this is indeed the case. This result is a consequence of the understanding that an analytic function is independent of $\bar{z}$. We present this approach that relies on algebraic procedures next.

Let the complex-valued function $f(z)$ be analytic in the neighborhood of the point z_0 and let $u(x, y), v(x, y)$ be the real and imaginary parts of $f(z)$. The function $f(z)$ can be written explicitly in terms of u and v and the value $f(z_0)$:

$$f(z) = 2u\left(\frac{z + \bar{z}_0}{2}, \frac{z - \bar{z}_0}{2i}\right) - \overline{f(z_0)}, \tag{4.39a}$$

$$f(z) = 2iv\left(\frac{z + \bar{z}_0}{2}, \frac{z - \bar{z}_0}{2i}\right) + \overline{f(z_0)}. \tag{4.39b}$$

To prove these expressions, we will use repeatedly the important property of the analytic function $f(z)$ that it is independent of $\bar{z}$, Eq. (4.35). Expressing x and y in terms of z and $\bar{z}$ through Eq. (4.29), we can rewrite $f(z)$ in the form

$$f(z) = u\left(\frac{z + \bar{z}}{2}, \frac{z - \bar{z}}{2i}\right) + iv\left(\frac{z + \bar{z}}{2}, \frac{z - \bar{z}}{2i}\right). \tag{4.40}$$

The functions $u(x, y)$ and $v(x, y)$ are harmonic, thus they admit analytic continuation, namely, $u(x, y)$ and $v(x, y)$ make sense if x and y are replaced by *complex* variables defined near the point z_0 in the analytic domain. Thus, we can evaluate the general expression of Eq. (4.40) at $z = z_0$:

$$f(z_0) = u\left(\frac{z_0 + \bar{z}}{2}, \frac{z_0 - \bar{z}}{2i}\right) + iv\left(\frac{z_0 + \bar{z}}{2}, \frac{z_0 - \bar{z}}{2i}\right).$$

In the above expression, we have *not* set $\bar{z} = \bar{z}_0$ as might have been suggested from our choice of $z = z_0$, because in the expression of $f(z)$ that involves both z and $\bar{z}$, we treat z and $\bar{z}$ as *independent* variables. Computing the complex conjugate of the last equation, we find

$$\overline{f(z_0)} = u\left(\frac{\bar{z}_0 + z}{2}, \frac{z - \bar{z}_0}{2i}\right) - iv\left(\frac{\bar{z}_0 + z}{2}, \frac{z - \bar{z}_0}{2i}\right). \qquad (4.41)$$

Using again the fact that the analytic function $f(z)$ is independent of $\bar{z}$, we can now replace on the right-hand side of Eq. (4.40) $\bar{z}$ by $\bar{z}_0$, which leads to

$$f(z) = u\left(\frac{z + \bar{z}_0}{2}, \frac{z - \bar{z}_0}{2i}\right) + iv\left(\frac{z + \bar{z}_0}{2}, \frac{z - \bar{z}_0}{2i}\right).$$

Adding and subtracting this last equation to Eq. (4.41), we find the desired results, that is, Eq. (4.39). It is remarkable that the above expressions allow one to determine the analytic function $f(z)$ when only its real part, or only its imaginary part, is known, *without the need of differentiation or integration*. We illustrate the application of this approach with a couple of examples.

Example 4.5: We provide two examples of how Eqs. (4.39a) and (4.39b) can be used to compute an analytic function from its real and imaginary parts.

(a) Find all analytic functions whose real part is given by

$$u(x, y) = \frac{x}{x^2 + y^2}.$$

We use Eq. (4.39a) with $x = (z + \bar{z}_0)/2, y = (z - \bar{z}_0)/2i$:

$$f(z) = 2\frac{(z + \bar{z}_0)/2}{(z+\bar{z}_0)^2/4 - (z - \bar{z}_0)^2/4} - \overline{f(z_0)} \Rightarrow f(z) + \overline{f(z_0)} = \frac{z + \bar{z}_0}{z\bar{z}_0} = \frac{1}{z} + \frac{1}{\bar{z}_0}.$$

It is obvious from this last equation that if we choose

$$f(z) = \frac{1}{z} + ia, \quad a \in \mathbb{R},$$

the equation is satisfied and this provides the general solution of this problem.

(b) Find all analytic functions whose complex part is given by

$$v(x, y) = \sin(x)e^{-y}.$$

We use Eq. (4.39b) with $x = (z + \bar{z}_0)/2, y = (z - \bar{z}_0)/2i$:

$$f(z) = 2i \sin\left(\frac{z + \bar{z}_0}{2}\right) \exp\left(-\frac{z - \bar{z}_0}{2i}\right) + \overline{f(z_0)}$$

$$\Rightarrow \quad f(z) - \overline{f(z_0)} = \left[\exp\left(i\frac{z + \bar{z}_0}{2}\right) - \exp\left(-i\frac{z + \bar{z}_0}{2}\right)\right] \exp\left(i\frac{z - \bar{z}_0}{2}\right)$$

$$= e^{iz} - e^{-i\bar{z}_0}.$$

From the last expression, it is obvious that this equation is satisfied by the choice

$$f(z) = e^{iz} + c, \quad c \in \mathbb{R}.$$

4.5 Complex integration: Cauchy's theorem

4.5.1 Integration over a contour

By analogy to the case of integrals of real functions, Eqs. (1.40) and (1.47), we define the integral of a complex function as

$$\int_{z_i}^{z_f} f(z)\mathrm{d}z = \lim_{\Delta z \to 0} \left[\sum_{z=z_i}^{z_f} f(z)\Delta z \right]. \tag{4.42}$$

There is a qualitative difference between definite integrals of functions of real variables and functions of complex variables. In the case of real variables, a definite integral is the integral between two points on the real axis x_i and x_f and there is only *one* way to go from x_i to x_f. In the case of complex variables $z_i = x_i + \mathrm{i}y_i$ and $z_f = x_f + \mathrm{i}y_f$, there are infinitely many ways of going from z_i to z_f. Thus, in general the definite integral between z_i and z_f depends on the path followed on the complex plane. Thus, the integral

$$\int_{z_i}^{z_f} f(z)\mathrm{d}z$$

depends on the contour along which the integration is performed. A useful relation that bounds the value of an integral of the function $f(z)$ over a contour is the so-called *ML* inequality:

$$\left| \int_{z_1}^{z_2} f(z)\mathrm{d}z \right| \le ML, \quad M = \max(|f(z)|), \quad L = \left| \int_{(x_1,y_1)}^{(x_2,y_2)} \mathrm{d}s \right|, \tag{4.43}$$

where M is the maximum value of $|f(z)|$ along the contour and L is the length of the contour, where $(x_1,y_1), (x_2,y_2)$ correspond to the end points of the path from z_1 to z_2 and

$$\mathrm{d}s = \sqrt{(\mathrm{d}x)^2 + (\mathrm{d}y)^2}$$

is the elemental length along the path.

The simplest way to perform an integration of a function of a complex variable along a contour is to parameterize this contour: suppose that the contour C can be described by $z = z(t)$, with t a real variable. Then

$$\mathrm{d}z = \frac{\mathrm{d}z}{\mathrm{d}t}\mathrm{d}t = \dot{z}(t)\mathrm{d}t = [\dot{x}(t) + \mathrm{i}\dot{y}(t)]\mathrm{d}t$$

$$\Rightarrow \quad \int_C f(z)\mathrm{d}z = \int_{t_{\min}}^{t_{\max}} [u(t) + \mathrm{i}v(t)]\dot{z}(t)\mathrm{d}t = \int_{t_{\min}}^{t_{\max}} [u(t) + \mathrm{i}v(t)][\dot{x}(t) + \mathrm{i}\dot{y}(t)]\mathrm{d}t$$

$$= \int_{t_{\min}}^{t_{\max}} [u(t)\dot{x}(t) - v(t)\dot{y}(t)]\mathrm{d}t + \mathrm{i}\int_{t_{\min}}^{t_{\max}} [u(t)\dot{y}(t) + v(t)\dot{x}(t)]\mathrm{d}t.$$

The last expression involves integrals of functions of real variables. Although this method is straightforward, it is not always applicable because there is no guarantee that the variable z can be put in a suitable parametric form along the path of integration. Furthermore, as it will be made clear later there exists powerful tools for complex integration, like the Residue theorem, which are expressed directly in terms of z.

Related to the integral of $f(z)$ over a contour are the notions of a contour enclosing a simply-connected region, which is called a "simple contour" and has a unique sense of traversing, and of a multiply-connected region which we will call a "non-simple contour" and has *no* unique sense of traversing. The term non-simple refers to the fact that such a contour intersects itself; the points of intersection create ambiguity with the sense of traversing since at each intersection there are more than one choices for the direction of the path. These notions were discussed in Section 2.5.

4.5.2   Cauchy's theorem

There exist contour integrals which are *independent* of the contour and depend only on the initial and final points. Recall that the derivative evaluated at $z = z_0$,

$$\frac{\partial f}{\partial z}(z_0),$$

in general depends on the direction of approach to z_0, but for analytic functions it does not depend on the direction of approach. An analogous statement applies to the integral of $f(z)$: the integral of $f(z)$ from z_i to z_f along two different contours is the *same*, provided that the function $f(z)$ is analytic in the domain enclosed by these two contours. This is equivalent to the statement that the integral of a function $f(z)$ along a *closed contour*, that is, one for which the starting and ending points are the same, vanishes, provided that $f(z)$ is analytic in the domain enclosed by the contour. This is known as Cauchy's theorem. This statement makes sense intuitively, because just like the analyticity of the function guarantees the existence of its derivative, it should also guarantee the existence of its antiderivative, as was defined in Chapter 1. If the antiderivative exists, then the integral around a closed contour can be computed by subtracting the values of the antiderivative at the same points, which is the starting and ending point of the closed contour; for a closed contour $z_i = z_f$, and this can be any point on the closed contour. This argument breaks down if the contour encloses singularities. We examine these notions in more detail in the following sections. But first, we prove the validity of Cauchy's theorem.

Cauchy's Theorem: *If $f(z)$ is analytic in the domain $\mathcal{D}$ interior to a contour C (this domain includes C), its integral over C is zero*:

$$f(z): \text{ analytic in } C \Longrightarrow \oint_C f(z)\mathrm{d}z = 0. \tag{4.44}$$

Proof: The basic setup of the proof is illustrated in Fig. 4.4. We will assume that the shape of the contour is simple, so that we can define two points as the leftmost and right-most points in the horizontal direction and two points as the lowest and highest ones in the vertical direction. If these assumptions are not valid, we can break the path into smaller closed paths, with the sum over all such paths being equal to the original path, since the additional segments are traversed in two opposite directions and therefore their contributions cancel.

This argument was explained in detail in Section 2.5. Then, we can define between these pairs of points the bottom (B) and top (T) part of the path, and the left (L) and right (R) part of the path, respectively, as shown in Fig. 4.4. For these parts, we assume that the contour is described by proper functions $y_T(x), y_B(x)$ or $x_L(y), x_R(y)$.

Writing out separately the real and imaginary parts of the integral, we find

$$\int f(z)dz = \int [u(x,y) + iv(x,y)][dx + idy] \tag{4.45}$$

$$= \int u(x,y)dx - \int v(x,y)dy + i \int u(x,y)dy + i \int v(x,y)dx.$$

In the following, we denote by $\mathcal{D}$ the domain enclosed by the contour C. The first term in the above expression, integrated around the closed contour C, gives

$$\oint_C u(x,y)dx = \int_{x_1}^{x_2} u(x,y_B)dx + \int_{x_2}^{x_1} u(x,y_T)dx$$

$$= \int_{x_1}^{x_2} [u(x,y_B) - u(x,y_T)]dx = \int_{x_1}^{x_2} \left[-\int_{y_B(x)}^{y_T(x)} u_y dy \right] dx$$

$$= -\iint_{\mathcal{D}} u_y dy dx, \tag{4.46}$$

where $u(x,y_B)$ and $u(x,y_T)$ refer to the values that $u(x,y)$ takes while the independent variable x spans the values from x_1 to x_2; the dependent variable y takes the values prescribed by the curve (the contour of integration), where y is defined through $y = y_B(x)$ or $y = y_T(x)$, in the *top* and *bottom* parts of the contour, respectively, as identified in Fig. 4.4. The second term gives

$$\oint_C v(x,y)dy = \int_{y_1}^{y_2} v(x_R,y)dy + \int_{y_2}^{y_1} v(x_L,y)dy$$

$$= \int_{y_1}^{y_2} [v(x_R,y) - v(x_L,y)]dy = \int_{y_1}^{y_2} \left[\int_{x_L(y)}^{x_R(y)} v_x dx \right] dy$$

$$= \iint_{\mathcal{D}} v_x dx dy, \tag{4.47}$$

where $v(x_R,y)$ and $v(x_L,y)$ refer to the values that $v(x,y)$ takes while the independent variable y spans the values from y_1 to y_2; the dependent variable x takes the values prescribed by the curve (the contour of integration), where x is defined through $x = x_L(y)$ or $x = x_R(y)$, in the *left* and *right* parts of the contour, respectively, as identified in Fig. 4.4. Combining these two terms we obtain for the real part of the last line in Eqs. (4.46) the following:

$$\oint_C [u(x,y)dx - v(x,y)dy] = \iint_{\mathcal{D}} (-u_y - v_x) \, dx dy. \tag{4.48}$$

Similar expressions can be derived for the imaginary part of the last line in Eq. (4.46):

$$\oint_C [u(x,y)dy + v(x,y)dx] = \iint_{\mathcal{D}} (-v_y + u_x) \, dx dy. \tag{4.49}$$

For a function that obeys the Cauchy–Riemann relations, Eq. (4.36), which can be expressed as $(u_y + v_x) = 0$ and $(u_x - v_y) = 0$, we conclude that the integral over C vanishes identically. ∎

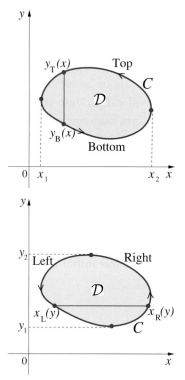

Figure 4.4: The paths for calculating the different integral contributions in the proof of Cauchy's integral theorem. $\mathcal{D}$ is the domain interior to the closed contour C over which we perform the integral of the function $f(z)$. The pair of points x_1, x_2 correspond to the left-most and right-most points of the contour C; the points y_1, y_2 correspond to the lowest and highest points of the contour C.

We emphasize that the expressions in Eqs. (4.48) and (4.49) are valid for any function $f(z,\bar{z}) = u(x,y) + iv(x,y)$, since we made no assumptions in their derivation. In fact, both of these expressions can be thought of as an application of Green's theorem, Eq. (2.73).

Corollary: *A consequence of Cauchy's theorem is that the integral of an analytic function between two points A and B on the complex plane is independent of the path connecting A and B, as long as A, B and the entire path lie within the domain of analyticity of the function.*

Proof: Consider two different paths going from A to B. The combination of these paths forms a closed contour, if one of the paths is traversed in the opposite sense (from B to A, rather than from A to B); the result of the integration over the entire closed contour is zero. Therefore, the integration over either path gives the same result. ∎

As a counter example, we consider the integral of the non-analytic function $f(z) = \bar{z}$ over a simple closed contour C. Its integral around C is

$$\oint_C \bar{z}\,\mathrm{d}z = \iint_{\mathcal{D}} \left(-u_y - v_x\right) \mathrm{d}x\mathrm{d}y + \mathrm{i} \iint_{\mathcal{D}} \left(-v_y + u_x\right) \mathrm{d}x\mathrm{d}y,$$

where we have used Eqs. (4.48) and (4.49), which are valid for any function $f(z,\bar{z})$. We next take into account the fact that

$$\bar{z} = x - \mathrm{i}y \Rightarrow u(x,y) = x \text{ and } v(x,y) = -y,$$

from which we readily obtain

$$u_y = 0, \ v_x = 0, \ u_x = 1, \ v_y = -1.$$

Using these results, the evaluation the integral yields

$$\oint_C \bar{z}\,\mathrm{d}z = 2\mathrm{i}A,$$

where A is the total area of the domain $\mathcal{D}$ enclosed by the contour C.

4.5.3   Contour deformation

A most important notion relevant to integration is the concept of contour deformation: it is often convenient to change the original contour of integration into a simpler one without changing the value of the integral. For example, consider an arbitrarily shaped simple contour C such as the one shown in Fig. 4.5. Let us denote the integral of the function around the contour as I.

Now assume that we deform the contour as shown in Fig. 4.5: specifically, we cut the original contour labeled C at some point and then join it with two straight segments labeled L_1 and L_2 to another circular contour labeled C_0. Furthermore, assume that the value of the integral over the entire new contour, comprising of the union of paths labeled C, L_1, L_2, C_0, vanishes, that is

$$I + I_0 + I_1 + I_2 = 0,$$

where the indices in each integral denote the part over the contour where the integral is evaluated. In the limit where the inside and outside contours are fully closed, we will have $I_1 = -I_2$ because along the two straight segments

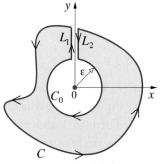

Figure 4.5: Illustration of contour deformation. The simple contour C which contains the point $z = 0$ where the function $f(z) = 1/z$ is not analytic, is deformed by adding the circular contour C_0 centered at this point, through straight segments labeled L_1 and L_2 that are traversed in opposite directions.

$f(z)$ takes the same values while $dz_1 = -dz_2$. This means that in order to find the value of I_0, we only have to evaluate the integral I since in this limit $I = -I_0$. But the path for this new integral is a circle, and on a circle it is easy to perform a parametric integration (reducing to the problem to integrating real functions as described earlier).

To illustrate the usefulness of this concept, we consider the integral

$$\oint_C \frac{1}{z} dz,$$

where C is an *arbitrary* simple closed contour around the point $z = 0$ which is clearly the only singularity of the integrand. We perform contour deformation as indicated in Fig. 4.5. For the deformed contour, which *does not* enclose the singularity $z = 0$, we conclude by Cauchy's theorem that the integral is zero, since the function $1/z$ is analytic everywhere inside and on this *deformed* contour. Since the integrals along the straight line segments cancel each other (by the argument mentioned above), we conclude that $I = -I_0$. Thus, we can evaluate the original integral by calculating the integral along the inner circular contour: assuming that this circle is of radius ϵ we have

$$z = \epsilon e^{i\phi}, \ \ 0 \leq \phi < 2\pi \Rightarrow dz = i\epsilon e^{i\phi} d\phi,$$

and the contour is traversed in the clockwise sense, which means ϕ goes from 2π to 0. Substituting these relations for the integral along the circular path we obtain

$$\oint_{C[z,\epsilon]} \frac{1}{z} dz = \int_{2\pi}^{0} \frac{1}{\epsilon e^{i\phi}} i\epsilon e^{i\phi} d\phi = -2\pi i.$$

Hence, the original integral is

$$I = \oint_C \frac{1}{z} dz = 2\pi i. \tag{4.50}$$

This is a very useful result, because it holds for an *arbitrary* simple closed contour which encloses the singularity of the integrand at $z = 0$. Moreover, for n a positive integer, and for an *arbitrary* simple closed contour C enclosing $z = 0$, a similar computation yields

$$\oint_C \frac{1}{z^n} dz = 2\pi i \delta_{1n}, \tag{4.51}$$

where δ_{ij} is the Kronecker delta, equal to one for $i = j$ and zero otherwise. To derive Eq. (4.50), we consider $n \neq 1$, since we have already dealt with the case $n = 1$. For $n \neq 1$, deforming the arbitrary contour C that encloses the value $z = 0$ to a circle of radius ϵ centered at the origin, for which $z = \epsilon e^{i\phi}$, we find the following:

$$\oint_C \frac{1}{z^n} dz = \int_0^{2\pi} \frac{1}{\epsilon^n e^{in\phi}} \epsilon i e^{i\phi} d\phi = \frac{i}{\epsilon^{n-1}} \int_0^{2\pi} e^{-i(n-1)\phi} d\phi$$
$$= \frac{-1}{\epsilon^{n-1}(n-1)} \left[e^{-i(n-1)\phi} \right]_0^{2\pi}.$$

This result is equal to 0 for $n \neq 1$ because in this case we have

$$e^{-i(n-1)2\pi} = 1,$$

independent of the value of ϵ.

4.5.4 Cauchy's Integral Formula

Corollary: *Cauchy's Integral Formula states that for a function $f(z)$ which is analytic everywhere inside the domain $\mathcal{D}$ enclosed by a simple contour C and containing the point z, the following relation holds:*

$$f(z) = \frac{1}{2\pi i} \oint_C \frac{f(z')}{z' - z} dz'. \tag{4.52}$$

Proof: Considering a circular contour of constant radius ϵ around z, and letting $\epsilon \to 0$ we find

$$C[z, \epsilon] \; : \; z' = z + \epsilon e^{i\phi} \Rightarrow dz' = i\epsilon e^{i\phi} d\phi, \;\; 0 \leq \phi < 2\pi.$$

Hence, the integral of $f(z')/(z' - z)$ around it is given by

$$\lim_{\epsilon \to 0} \oint_{C[z,\epsilon]} \frac{f(z')}{z' - z} dz' = \lim_{\epsilon \to 0} \int_0^{2\pi} \frac{f(z + \epsilon e^{i\phi})}{\epsilon e^{i\phi}} i\epsilon e^{i\phi} d\phi = 2\pi i f(z).$$

By using contour deformation we can easily prove that this result holds for any simple contour C that encloses z, as illustrated in Fig. 4.6. ∎

The deeper meaning of Cauchy's Integral Formula (CIF) is the following: if an analytic function is known at the boundary of a region (the contour of integration C), then it can be calculated everywhere within that region by performing an appropriate contour integral over the boundary. Indeed, if z is in the interior of the contour, CIF gives the value of $f(z)$ by performing an integral of $f(z')/(z' - z)$ over the contour, that is, by using only values of the function on the contour C. This is consistent with the statements made earlier that harmonic functions are fully determined by their boundary values and that an analytic function has real and imaginary parts which are harmonic functions. The CIF provides a recipe for obtaining all the values of an analytic function $f(z)$ (and hence the harmonic functions $u(x, y), v(x, y)$ which are its real and imaginary parts) when the boundary values (the values of $f(z)$ on the contour C) are known.

Taking the CIF one step further, we can evaluate the derivatives of an analytic function, by differentiating the expression in Eq. (4.52) with respect to z:

$$f'(z) = \frac{d}{dz} \left[\frac{1}{2\pi i} \oint_C \frac{f(z')}{(z' - z)} dz' \right] = \frac{1}{2\pi i} \oint_C \frac{f(z')}{(z' - z)^2} dz',$$

and similarly for higher orders:

$$f^{(n)}(z) = \frac{n!}{2\pi i} \oint_C \frac{f(z')}{(z' - z)^{n+1}} dz'. \tag{4.53}$$

This is also consistent with a statement made earlier that all derivatives of an analytic function exist. We should caution the reader that in the case of derivatives, we cannot use a simple contour of constant radius ϵ and let $\epsilon \to 0$ to obtain explicit expressions, but we must perform the contour integration over C.

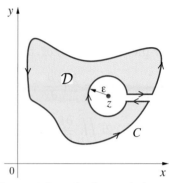

Figure 4.6: Setup for the proof of Cauchy's Integral Formula around a contour C: the variable of integration around C is z' and the contour encloses the point z, which is avoided by the circle of radius ϵ centered at z, through contour deformation.

4.6 Extensions of Cauchy's theorem

We can take advantage of the results obtained in the proof of Cauchy's theorem to give a more general form of this important theorem. This has interesting implications, including the derivation of a useful expression for the δ-function with complex argument.

4.6.1   General form of Cauchy's theorem

Theorem: *Suppose that the complex-valued function $f(z, \bar{z})$ is continuously differentiable in the domain $\mathcal{D}$ enclosed by a regular closed contour C, that is, the first derivatives of the real and imaginary parts of $f(z, \bar{z})$ exist on $\mathcal{D}$ and on its boundary C and are continuous; then*

$$\oint_C f(z, \bar{z}) dz = 2i \iint_{\mathcal{D}} \frac{\partial f}{\partial \bar{z}} dx dy. \tag{4.54}$$

Proof: Starting with the right-hand side, we use the expression from Eq. (4.31b) for the derivative $\partial/\partial\bar{z}$, and set $f = u + iv$, to obtain

$$\frac{\partial f}{\partial \bar{z}} = \frac{1}{2}\left(\frac{\partial}{\partial x} + i\frac{\partial}{\partial y}\right)(u + iv) = \frac{1}{2}\left[(u_x - v_y) + i(u_y + v_x)\right]$$

$$\Rightarrow \quad 2i \iint_{\mathcal{D}} \frac{\partial f}{\partial \bar{z}} dx dy = \iint_{\mathcal{D}} (-u_y - v_x) dx dy + i \iint_{\mathcal{D}} (u_x - v_y) dx dy.$$

The two integrals on the right-hand side of the last equation are the same as the integrals on the right-hand sides of Eq. (4.48) and (4.49), respectively. Thus, adding them, with the second multiplied by i, gives the initial integral of Eq. (4.46) performed over the closed contour C, which is the desired result. ∎

If $f(z, \bar{z})$ is an analytic function of z everywhere in the domain $\mathcal{D}$, we will have

$$\frac{\partial f}{\partial \bar{z}} = 0 \text{ in } \mathcal{D}$$

by virtue of Eq. (4.35), therefore the integral over the closed contour on the left-hand side of Eq. (4.54) vanishes identically, which is Cauchy's theorem. In other words, the expression of Eq. (4.54) contains Cauchy's theorem as a special case.

4.6.2   Pompeiu's formula

Theorem: *Suppose that the complex-valued function $f(z, \bar{z})$ is continuously differentiable in the domain $\mathcal{D}$ enclosed by a regular closed contour C, that is, the first derivatives of the real and imaginary parts of $f(z, \bar{z})$ exist on $\mathcal{D}$ and on its boundary C and are continuous; then*

$$f(z, \bar{z}) = \frac{1}{2i\pi} \oint_C \frac{f(z', \bar{z}')}{z' - z} dz' - \frac{1}{\pi} \iint_{\mathcal{D}} \frac{\partial f}{\partial \bar{z}'}(z', \bar{z}') \frac{1}{z' - z} dx' dy', \quad z \in \mathcal{D}. \tag{4.55}$$

Proof: The proof consists of three steps. First, we apply the expression of Eq. (4.54) for the complex-valued function $f(z', \bar{z}')/(z' - z)$, by creating the deformed contour shown in Fig. 4.6; with this choice, we avoid the point $z' = z$ through the circular contour $C[z, \epsilon]$ of radius ϵ centered at z, thus making sure that $1/(z' - z)$ is analytic everywhere in the remaining domain. We call the remaining domain $\mathcal{D}'$. Then, Eq. (4.54) gives

$$\oint_C \frac{f(z', \bar{z}')}{z' - z} dz' - \oint_{C[z, \epsilon]} \frac{f(z', \bar{z}')}{z' - z} dz' = 2i \iint_{\mathcal{D}'} \frac{\partial f}{\partial \bar{z}'}(z', \bar{z}') \frac{1}{z' - z} dx' dy', \quad z \in \mathcal{D}',$$

where we have taken advantage of the fact that $1/(z'-z)$ is analytic everywhere in the domain $\mathcal{D}'$, therefore

$$\frac{\partial}{\partial \bar{z}'}\left(\frac{1}{z'-z}\right) = 0.$$

Second, in the limit $\epsilon \to 0$ the circular contour evidently shrinks to zero and $\mathcal{D}'$ becomes the same as $\mathcal{D}$, thus we expect the double integral over $\mathcal{D}'$ on the right-hand side of the above equation to become equal to the double integral on the right-hand side of Eq. (4.55) over the domain $\mathcal{D}$. To show this explicitly, we note that the difference between the two integrals is equal to a double integral over the circular domain bounded by $C[z, \epsilon]$. Using $z' = z + r\exp(i\phi)$ with $0 \le r \le \epsilon$ for this circular domain, it follows that the difference between the two integrals is bounded by the following quantity:

$$\int_0^\epsilon \int_0^{2\pi} \left|\frac{\partial f}{\partial \bar{z}'}\right| \frac{r\,dr\,d\phi}{r} = 2\pi M\epsilon,$$

where we have used the fact that the continuity of $\partial f/\partial \bar{z}'$ in a bounded domain implies that there exists a constant M such that $|\partial f/\partial \bar{z}'| \le M$. Thus, in the limit $\epsilon \to 0$, the difference between the two double integrals vanishes.

Third, we use $z' = z + \epsilon\exp(i\phi)$ for the integral over the circular contour $C[z, \epsilon]$ to obtain

$$\oint_{C[z,\epsilon]} \frac{f(z', \bar{z}')}{z'-z}\,dz' = \int_0^{2\pi} \frac{f(z+\epsilon e^{i\phi}, \bar{z}+\epsilon e^{-i\phi})}{\epsilon e^{i\phi}}\epsilon i e^{i\phi}\,d\phi$$

$$\Rightarrow \quad \lim_{\epsilon \to 0}\left[\oint_{C[z,\epsilon]} \frac{f(z', \bar{z}')}{z'-z}\,dz'\right] = 2\pi i f(z, \bar{z}).$$

Putting the results of the three steps together we arrive at Eq. (4.55).   ∎

If $f(z, \bar{z})$ is an analytic function of z everywhere in the domain $\mathcal{D}$, we have

$$\frac{\partial f}{\partial \bar{z}} = 0 \text{ in } \mathcal{D}$$

by virtue of Eq. (4.35), and the second integral of the right-hand side of Eq. (4.55) vanishes identically. The remaining expression is Cauchy's Integral Formula, Eq. (4.52). In other words, Pompeiu's formula, Eq. (4.55), contains Cauchy's Integral Formula as a special case.

4.6.3   The δ-function of complex argument

We apply Pompeiu's formula to the case of the function $f(z) = 1/(z-z_0)$, with the domain $\mathcal{D}$ being the entire complex plane, $\mathbb{C}$. In this case, the boundary of the domain C is at infinity, and the integrand of the first integral on the right-hand side of Eq. (4.55) vanishes for any finite value z_0. Therefore, Pompeiu's formula gives

$$\frac{1}{z-z_0} = \iint_{\mathbb{C}} \left[\frac{1}{\pi}\frac{\partial}{\partial \bar{z}'}\left(\frac{1}{z'-z_0}\right)\right]\frac{1}{z-z'}\,dx'dy'.$$

This last expression implies that

$$\frac{1}{\pi}\frac{\partial}{\partial \bar{z}}\left(\frac{1}{z-z_0}\right) = \delta(z-z_0), \tag{4.56}$$

by virtue of the basic property that the δ-function must obey from its definition, Eq. (1.74). The expression of Eq. (4.56) is a very useful representation of the δ-function with a complex variable argument.

4.7   Complex power series expansions

We next consider the power series expansion of $f(z)$ at a reference point z_0 that involves only non-negative powers of $(z - z_0)$ (these terms are all analytic everywhere on the complex plane):

$$f(z) = \sum_{n=0}^{\infty} a_n (z - z_0)^n. \tag{4.57}$$

The analysis of this equation will establish an important result, whose particular cases were encountered earlier, namely the Taylor series expansion. By analogy to what was discussed for power series expansions of functions of real variables, and using our knowledge of convergence criteria, we expect that the following statements must hold for the power series on the right-hand side of Eq. (4.57):

- if it converges for z_1 and $|z_2 - z_0| < |z_1 - z_0|$ then it converges for z_2;

- if it diverges for z_3 and $|z_4 - z_0| > |z_3 - z_0|$ then it diverges for z_4.

These statements can be justified by considering the convergence of the series in absolute terms and since $|z - z_0|$ is the real quantity

$$|z - z_0| = \sqrt{(x - x_0)^2 + (y - y_0)^2},$$

we can apply the tests valid for the convergence of real series. In particular, the comparison tests, Eqs. (3.2), (3.3), are useful in establishing the two statements mentioned above. These lead us to the definition of the *radius of convergence*: it is the largest circle of radius $R = |z - z_0|$ that encloses all points around z_0 for which the series converges. If $R \to \infty$ then the series converges everywhere on the complex plane; if $R = 0$ the series diverges everywhere on the complex plane.

Example 4.6: To illustrate these points, we consider the familiar binomial expansion, Eq. (3.9), involving a complex number z:

$$(a + z)^p = a^p + p a^{p-1} z + \frac{p(p - 1)}{2!} a^{p-2} z^2 + \frac{p(p - 1)(p - 2)}{3!} a^{p-3} z^3 + \cdots$$

In this expression, we have put the function $f(z) = (a + z)^p$ in an infinite series form of the type shown in Eq. (4.57), with the well-defined coefficients a_n around the point $z_0 = 0$ given by

$$a_n = \frac{p(p - 1) \cdots [p - (n - 1)]}{n!} a^{p-n}.$$

Indeed, to compute the radius of convergence of this series, we let $\zeta = z/a$:

$$(a + z)^p = a^p \left[1 + p\zeta + \frac{p(p - 1)}{2!} \zeta^2 + \frac{p(p - 1)(p - 2)}{3!} \zeta^3 + \cdots \right]. \tag{4.58}$$

We then apply the ratio test, Eq. (3.4), to determine the radius of convergence of this series. The ratio of the $(n + 1)$th to the nth term is given by

$$\left|\left[\frac{p(p-1)\cdots(p-n)\zeta^{n+1}}{(n+1)!}\right]\left[\frac{p(p-1)\cdots(p-(n-1))\zeta^{n}}{n!}\right]^{-1}\right| = \left|\frac{p-n}{n+1}\zeta\right|.$$

The ratio test requires that this last expression be bounded by a positive number smaller than unity for all $n > N$, but

$$n \to \infty \Rightarrow \left|\frac{p-n}{n+1}\right| \to 1.$$

Hence, the series converges if

$$|\zeta| < 1 \Rightarrow |z| < |a|,$$

which gives for the series around $z_0 = 0$ the radius of convergence $R = |a|$.

For the power series representation of a function $f(z)$, as given in Eq. (4.57), assuming that it convergences absolutely and uniformly inside the disk $|z - z_0| < R$, the following statements are true (some for obvious reasons):

1. Since $(z - z_0)^n$ is continuous, $f(z)$ is also continuous.

2. We can differentiate and integrate the series term by term inside the disk of convergence, to obtain a new convergent series, since the resulting series has the same ratio of the $(n+1)$th to the nth term as the original series.

3. The power series representation about the point z_0 is unique inside the disk.

4. For $R > 0$, the power series represents an *analytic* function, which is infinitely differentiable within the disk of its convergence.

5. Conversely, every function has a unique power series representation about z_0 when it is analytic in a disk of radius R around it, and hence all correct methods for obtaining this series yield the same result.

The last two statements suggest that given a function $f(z)$ it is possible to immediately determine its radius of convergence by identifying a point where analyticity is lost. This fact emphasizes the importance of using complex functions. As an illustration, we attempt to determine the radius of convergence of the expansion $1/(1 + x^2)$ with x real. In this connection, recall the geometric series expansion

$$\frac{1}{1-x} = 1 + x + x^2 + \cdots,$$

which converges for $|x| < 1$. This makes sense since for $x = 1$ the function $1/(1-x)$ is not defined. However, although the function $1/(1+x^2)$ is defined for *all* x, the series

$$\frac{1}{1+x^2} = 1 - x^2 + x^4 + \cdots$$

also converges *only* for $|x| < 1$. This can be understood by noting that the function $1/(1+x^2)$ is the restriction of the complex function $1/(1+z^2)$ to the real axis, $z = x$, and the latter function has singularities at $z = \pm i$, thus its radius of convergence is 1.

4.7.1   Taylor series

Theorem: *The series expansion for a function $f(z)$ which is analytic everywhere in the domain enclosed by a regular contour C, including the contour C, is given by the expression (the Taylor series)*

$$f(z) = f(z_0) + f'(z_0)(z - z_0) + \frac{1}{2!}f''(z_0)(z - z_0)^2$$
$$+ \cdots + \frac{1}{n!}f^{(n)}(z_0)(z - z_0)^n + \cdots, \tag{4.59}$$

where $f^{(n)}(z_0)$ is the nth-order derivative of the function evaluated at z_0.

Proof: By contour deformation, we can always consider the contour of interest C' to be a circle of fixed radius centered around z_0. First, we note that for every point z inside the contour C', with the variable z' taking values on the contour C', we have

$$|z - z_0| < |z' - z_0| \Rightarrow \frac{|z - z_0|}{|z' - z_0|} < 1,$$

since the point z inside the circular contour is closer to the center z_0 than the point z' which lies on the contour. Applying CIF, Eq. (4.52), for the analytic function $f(z)$ on a contour C that encloses the point z_0 we find:

$$f(z) = \frac{1}{2\pi i}\oint_C \frac{f(z')}{(z' - z)}dz' = \frac{1}{2\pi i}\oint_{C'} \frac{f(z')}{[1 - (z - z_0)/(z' - z_0)](z' - z_0)}dz'$$
$$= \frac{1}{2\pi i}\oint_{C'} \frac{f(z')}{(z' - z_0)}\left[1 + \left(\frac{z - z_0}{z' - z_0}\right) + \left(\frac{z - z_0}{z' - z_0}\right)^2 + \cdots\right]dz'$$
$$= f(z_0) + f'(z_0)(z - z_0) + \frac{1}{2!}f''(z_0)(z - z_0)^2$$
$$+ \cdots + \frac{1}{n!}f^{(n)}(z_0)(z - z_0)^n + \cdots$$

In the next-to-last step above we used the geometric series summation

$$\frac{1}{1 - t} = 1 + t + t^2 + t^3 + \cdots,$$

see Eq. (3.5), which is valid for $t = (z - z_0)/(z' - z_0)$ since $|t| < 1$, and for the last step we used the CIF for the derivatives of $f(z)$, Eq. (4.53). ■

A Taylor series around $z_0 = 0$ is referred to as a "Maclaurin series".

Example 4.7: Suppose we want to calculate the power series of

$$f(z) = \sin^{-1}(z),$$

about the point $z_0 = 0$. We first note that this function is analytic in a disk of non-zero radius around the point of interest. Therefore, it should have a Taylor series expansion around this point. As it is difficult to take derivatives of the inverse sine function, we use an indirect approach: since for an analytic function any way of obtaining the power series expansion is acceptable, we call the original function w and examine its inverse, which is simply the sine function:

$$w = \sin^{-1}(z) \Rightarrow z = \sin(w)$$

$$\Rightarrow \quad \frac{dz}{dw} = \cos(w) = \sqrt{1 - \sin^2(w)} = \sqrt{1 - z^2}$$

$$\Rightarrow \quad \frac{dw}{dz} = \frac{1}{\sqrt{1 - z^2}} = [1 + (-z^2)]^{-1/2}.$$

In the last expression we can use the familiar binomial expression, Eq. (4.58) with $a = 1, \zeta = -z^2, p = -1/2$, which, when expanded in powers of $(-z^2)$ and integrated term by term leads to

$$w = z + \frac{1}{6}z^3 + \frac{3}{40}z^5 + \frac{5}{112}z^7 + \cdots$$

Note that we *can* apply integration term by term to the series obtained for dw/dz because the binomial series converges uniformly for $|\zeta| = |-z^2| < 1$, a condition that is certainly satisfied for z in the neighborhood of zero. Moreover, the constant of integration, which appears when the series for dw/dz is integrated in order to obtain the series for w, can be determined by observing that the definition of w implies $z = 0 \Rightarrow w = 0$, and hence the constant of integration must be zero.

4.7.2   Laurent series

Theorem: *Consider a function $f(z)$ which is analytic everywhere in a circular annulus with outer radius ρ_1 on the circular contour C_1 and inner radius ρ_2 on the circular contour C_2, both centered at z_0, including the contours C_1, C_2. Then for a closed contour C that lies entirely within the circular annulus bounded by the circular contours C_1 and C_2, the function can be expressed as*

$$f(z) = \sum_{n=-\infty}^{\infty} \left[\frac{1}{2\pi i} \oint_C \frac{f(z')}{(z' - z_0)^{n+1}} dz' \right] (z - z_0)^n. \tag{4.60}$$

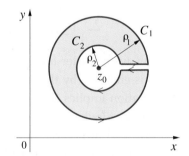

Figure 4.7: Circular annulus around the point z_0 on the complex plane used for the derivation of the Laurent series expansion.

Proof: We notice that if we join the two circular contours by two line segments traversed in the opposite direction, as illustrated in Fig. 4.7, the function $f(z)$ is analytic everywhere in and on the resulting closed contour, and therefore we can apply the CIF. In the limit when the two line segments lie on top of each other, and hence their contributions cancel out, we obtain:

$$f(z) = \frac{1}{2\pi i} \oint_{C_1} \frac{f(z')}{(z' - z)} dz' - \frac{1}{2\pi i} \oint_{C_2} \frac{f(z')}{(z' - z)} dz'. \tag{4.61}$$

In the last expression, each contour is traversed in the counter-clockwise sense and the integration over C_2 has an overall minus sign because that part was traversed in the clockwise direction in the joined contour. The first integral can be written in the form

$$I_1 = \frac{1}{2\pi i} \oint_{C_1} \frac{f(z')}{(z' - z)} dz' = \frac{1}{2\pi i} \oint_{C_1} \frac{f(z')}{[1 - (z - z_0)/(z' - z_0)](z' - z_0)} dz'. \tag{4.62}$$

For z in the annulus and z' on the C_1 contour we have

$$|z - z_0| < |z' - z_0| \Rightarrow \left| \frac{z - z_0}{z' - z_0} \right| < 1.$$

Hence, we can apply the geometric series expansion to obtain:

$$I_1 = \frac{1}{2\pi i} \oint_{C_1} \frac{f(z')}{(z'-z_0)} \left[1 + \left(\frac{z-z_0}{z'-z_0}\right) + \left(\frac{z-z_0}{z'-z_0}\right)^2 + \cdots \right] dz'$$

$$= \sum_{n=0}^{\infty} \left[\frac{1}{2\pi i} \oint_{C_1} \frac{f(z')}{(z'-z_0)^{n+1}} dz' \right] (z-z_0)^n. \qquad (4.63)$$

For the second integral, we have

$$I_2 = -\frac{1}{2\pi i} \oint_{C_2} \frac{f(z')}{(z'-z)} dz' = \frac{1}{2\pi i} \oint_{C_2} \frac{f(z')}{[1-(z'-z_0)/(z-z_0)](z-z_0)} dz'.$$

For z in the annulus and z' on the C_2 contour we have

$$|z'-z_0| < |z-z_0| \Rightarrow \left| \frac{z'-z_0}{z-z_0} \right| < 1.$$

Hence, we can apply the geometric series expansion to obtain:

$$I_2 = \frac{1}{2\pi i} \oint_{C_2} \frac{f(z')}{(z-z_0)} \left[1 + \left(\frac{z'-z_0}{z-z_0}\right) + \left(\frac{z'-z_0}{z-z_0}\right)^2 + \cdots \right] dz'$$

$$= \sum_{n=1}^{\infty} \left[\frac{1}{2\pi i} \oint_{C_2} f(z')(z'-z_0)^{n-1} dz' \right] \frac{1}{(z-z_0)^n}$$

$$= \sum_{n=1}^{\infty} \left[\frac{1}{2\pi i} \oint_{C_2} \frac{f(z')}{(z'-z_0)^{-n+1}} dz' \right] (z-z_0)^{-n}$$

$$= \sum_{n=-1}^{-\infty} \left[\frac{1}{2\pi i} \oint_{C_2} \frac{f(z')}{(z'-z_0)^{n+1}} dz' \right] (z-z_0)^n. \qquad (4.64)$$

Now, we consider a closed contour C that lies entirely within the circular annulus enclosed by the circular contours C_1 (outer) and C_2 (inner). Contour deformation implies that the integral over C_1 or over C_2 of a function that is analytic everywhere in the circular annulus is equal to the integral over C, as illustrated in Fig. 4.8. Specifically, we can rewrite the integrals that appear in the square brackets under the summations of Eq. (4.63) and Eq. (4.64) as follows:

$$\oint_{C_i} \frac{f(z')}{(z'-z_0)^{n+1}} dz' = \oint_C \frac{f(z')}{(z'-z_0)^{n+1}} dz', \quad i = 1, 2,$$

since the integrands in each case are analytic everywhere in the region enclosed by C and C_1 or C_2, respectively. We can then express both integrals I_1 and I_2 evaluated earlier in terms of the common contour C for all the quantities that appear inside the square brackets in each sum, and this yields Eq. (4.60). ∎

The expression in Eq. (4.60) is known as the Laurent power series expansion and contains *both negative and positive powers of* $(z-z_0)^n$. In the following, we denote this expansion by

$$f(z) = \sum_{n=0}^{\infty} a_n (z-z_0)^n + \sum_{n=1}^{\infty} \frac{b_{-n}}{(z-z_0)^n} \qquad (4.65a)$$

$$a_n = \frac{1}{n!} f^{(n)}(z_0), \quad b_{-n} = \frac{1}{2\pi i} \oint_{C[z_0]} \frac{f(z')}{(z'-z_0)^{-n+1}} dz'. \qquad (4.65b)$$

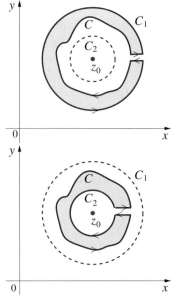

Figure 4.8: A path C that lies entirely within the circular annulus defined by the circular paths C_1 (outer) and C_2 (inner): For a function that is analytic everywhere on the circular annulus, the integrals over C_1 and C_2 can be related to the one over C using contour deformation; the contributions along the cut joining C to C_1 or to C_2 cancel.

Example 4.8: To illustrate the Taylor and Laurent series expansions we consider the function

$$f(z) = \frac{1}{(z-1)(z-2)} = \frac{1}{1-z} + \frac{1}{z-2}$$

and ask what is its power series expansion around $z = 0$. To answer this question, we need to separate the complex plane in three regions in which we can easily determine the analyticity of the function. These are as follows:

- Region I: $|z| < 1$, where both $1/(1-z)$ and $1/(z-2)$ are analytic;

- Region II: $1 < |z| < 2$, which encloses a singularity of $1/(1-z)$;

- Region III: $2 < |z|$, which encloses a singularity of both $1/(1-z)$ and of $1/(z-2)$.

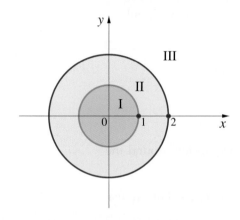

In each region, we can determine separately the expansion of the two fractions that appear in $f(z)$, that is, $1/(1-z)$ and $1/(z-2)$. The first fraction, $1/(1-z)$, is analytic in region I, so we expect a Taylor expansion for this function, which is easily obtained from the geometric series, since in this region $|z| < 1$. Regions II and III can be considered as circular annuli which contain a singularity at $z = 1$, so we expect a Laurent expansion; in both regions $|z| > 1 \Rightarrow 1/|z| < 1$, hence if we rewrite the denominator so as to contain the factor $(1 - 1/z)$ we can use the geometric series again to obtain:

$$\text{I:} \quad \frac{1}{1-z} = 1 + z + z^2 + \cdots$$

$$\text{II, III:} \quad \frac{1}{1-z} = \frac{1}{(1 - \frac{1}{z})(-z)} = \frac{-1}{z}\left[1 + \frac{1}{z} + \frac{1}{z^2} + \cdots \right].$$

The second fraction, $1/(z-2)$, is analytic in regions I and II, so we expect a Taylor expansion for this function, which is easily obtained from the geometric series, since in these regions $|z| < 2 \Rightarrow |z/2| < 1$. Region III can be considered as a circular annulus which contains a singularity at $z = 2$, so we expect a Laurent expansion; in this region $|z| > 2 \Rightarrow 2/|z| < 1$, hence if we rewrite the denominator so as to contain the factor $(1 - 2/z)$

we can use the geometric series again to obtain

$$\text{I, II:} \quad \frac{1}{z-2} = \frac{-1}{2} \frac{1}{(1-\frac{z}{2})} = \frac{-1}{2} \left[1 + \frac{z}{2} + \left(\frac{z}{2}\right)^2 + \cdots \right],$$

$$\text{III:} \quad \frac{1}{z-2} = \frac{1}{(1-\frac{2}{z})z} = \frac{1}{z} \left[1 + \frac{2}{z} + \left(\frac{2}{z}\right)^2 + \cdots \right].$$

Combining the results in the different regions, we obtain

$$\text{I:} \quad f(z) = \frac{1}{2} + z\left(1 - \frac{1}{4}\right) + z^2\left(1 - \frac{1}{8}\right) + \cdots$$

$$\text{II:} \quad f(z) = \cdots - \frac{1}{z^2} - \frac{1}{z} - \frac{1}{2} - \frac{1}{4}z - \frac{1}{8}z^2 + \cdots$$

$$\text{III:} \quad f(z) = (-1+2)\frac{1}{z^2} + (-1+2^2)\frac{1}{z^3} + \cdots + (-1+2^{n-1})\frac{1}{z^n} + \cdots$$

We note that $f(z)$ has a Taylor expansion in region I, a Laurent expansion with all powers of z in region II, and a Laurent expansion with only negative powers of z starting at $n = -2$ in region III.

Generalizing the above results for a function of the form

$$f(z) = \sum_{k=1}^{N} \frac{A_n}{z - z_k},$$

we note that in order to express $f(z)$ as a power series expansion around the point z_0, we must consider two possibilities:

(i) If the domain of interest around $z = z_0$ *includes* the point z_k, then we use the identity

$$\frac{1}{z - z_k} = \frac{1}{z - z_0}\left[1 - \left(\frac{z_k - z_0}{z - z_0}\right)\right]^{-1}, \quad \left|\frac{z_k - z_0}{z - z_0}\right| < 1, \qquad (4.66)$$

where the last inequality is valid on a circular contour around z_0 with radius *larger* than the distance $|z_k - z_0|$; in this case, we expand the factor in the square brackets as a geometric series which produces a Laurent series in powers of $(z - z_0)$.

(ii) If the domain of interest around $z = z_0$ does *not* include the point z_k, then we use the identity

$$\frac{1}{z - z_k} = \frac{1}{z_0 - z_k}\left[1 - \left(\frac{z - z_0}{z_k - z_0}\right)\right]^{-1}, \quad \left|\frac{z - z_0}{z_k - z_0}\right| < 1, \qquad (4.67)$$

where the last inequality is valid on a circular contour around z_0 with radius *smaller* than the distance $|z_k - z_0|$; in this case, we expand the factor in the square brackets as a geometric series which produces a Taylor series in powers of $(z - z_0)$.

4.8   Application: The 2D Fourier transform

A very useful tool for exploring the behavior of functions of one or more variables is their transformation to the space of so-called reciprocal variables.

This is the dual space used to analyze the function's properties. Perhaps the most widely used such transform is the Fourier transform (FT), a subject we discuss in detail in Chapter 7. An elegant application of a key result introduced in the present chapter, Pompeiu's formula, Eq. (4.55), is the derivation of the FT in two dimensions (2D), namely, for functions of two independent variables. We describe this derivation here.

The starting point is a given function $f(x,y)$ of the independent real variables x,y. The domain $\mathcal{D}$ where the function is defined is the entire plane, $-\infty < x, y < \infty$. We define the FT of the function, a new function called $\hat{f}(p,q)$, as:

$$\hat{f}(p,q) = \int_{-\infty}^{\infty} \int_{-\infty}^{\infty} e^{-ipx} e^{-iqy} f(x,y) \mathrm{d}x \mathrm{d}y, \quad -\infty < p, q < \infty, \qquad (4.68)$$

where p,q are the two "reciprocal" (or "inverse") space variables. The reason for the terminology "reciprocal" or "inverse" space is that the products px and qy must be dimensionless quantities, since they appear in the argument of the complex exponentials. Thus, if x,y have a dimension (for instance, length or time), then p,q must have the inverse or reciprocal dimension (inverse length or inverse time). The inverse length has a physical interpretation, namely it is the wavenumber that characterizes a wave. Similarly, the inverse time has the physical meaning of a frequency, which is also a characteristic of waves.

The next goal is to derive an expression which makes it possible to reconstruct the original function using $\hat{f}(p,q)$. This expression turns out to be the following formula:

$$f(x,y) = \int_{-\infty}^{\infty} \int_{-\infty}^{\infty} e^{ipx} e^{iqy} \hat{f}(p,q) \frac{\mathrm{d}p}{2\pi} \frac{\mathrm{d}q}{2\pi}, \quad -\infty < x, y < \infty, \qquad (4.69)$$

which is known as the "inverse Fourier transform" (IFT). The function $\hat{f}(p,q)$ contains exactly the same amount of information as the original function, $f(x,y)$, but in the reciprocal space. The FT and the IFT constitute a pair of extremely useful transforms in many applications, for instance in analyzing signals to determine their spatial or temporal information, as we discuss in more detail in Chapter 7.

To derive the formula of Eq. (4.69) we consider a so-called "auxiliary function", $F(x,y;p,q)$. Instead of using the variables x,y and p,q, we will use the complex variables

$$z = x + iy, \qquad \bar{z} = x - iy, \qquad (4.70a)$$

$$k = k_R + ik_I, \quad \bar{k} = k_R - ik_I, \quad \text{where } k_R = -\frac{q}{2}, \quad k_I = \frac{p}{2}. \qquad (4.70b)$$

We require that the function $F(z,\bar{z};k,\bar{k})$ satisfies the equation:

$$\frac{\partial F}{\partial \bar{z}} - kF = f(x,y). \qquad (4.71)$$

For the purposes of the following derivation, the only condition that the functions $f(x,y)$ and $\hat{f}(p,q)$ must satisfy is that they vanish at the boundary of their domain, namely $f(x,y) = 0$ for $x,y \to \pm\infty$, and $\hat{f}(p,q) = 0$ for $p,q \to \pm\infty$. Consistent with these constraints on $f(x,y)$ and $\hat{f}(p,q)$, $F(z,\bar{z};k,\bar{k})$ must also vanish in the limits $|z| \to \infty$, $|k| \to \infty$. Additional constraints on

$f(x, y)$ must be imposed in order to make the FT finite, which have to do with the integrability of $f(x, y)$; this is addressed in Chapter 7 for the FT of a function of a single variable.

As the first step of the derivation, we multiply both sides of Eq. (4.71) by $\exp[-k\bar{z}]$ to obtain

$$\frac{\partial F}{\partial \bar{z}}e^{-k\bar{z}} - kFe^{-k\bar{z}} = \frac{\partial}{\partial \bar{z}}\left(Fe^{-k\bar{z}}\right) = f(x, y)e^{-k\bar{z}}. \tag{4.72}$$

We next multiply both sides in the last equality of the above equation by $\exp[\bar{k}z]$ to obtain

$$\frac{\partial}{\partial \bar{z}}\left(Fe^{\bar{k}z - k\bar{z}}\right) = f(x, y)e^{\bar{k}z - k\bar{z}}. \tag{4.73}$$

Regarding the above step, we note that $\exp[\bar{k}z]$ is independent of $\bar{z}$, thus it commutes with the derivative with respect to $\bar{z}$ (in other words, we can interchange the order of the multiplication and the derivative). The introduction of the two exponential factors gives rise to the term $\exp[\bar{k}z - k\bar{z}]$, which, as we show below, is the exponential of a purely imaginary number, and thus is bounded for all complex values of k and z. We can now apply Pompeiu's formula with the expression of Eq. (4.73) for the derivative of F with respect to $\bar{z}$; using also the fact that F must vanish at the domain boundary, we find

$$Fe^{\bar{k}z - k\bar{z}} = -\frac{1}{\pi}\int_{-\infty}^{\infty}\int_{-\infty}^{\infty}\frac{f(x', y')e^{\bar{k}z' - k\bar{z}'}}{z' - z}dx'dy'. \tag{4.74}$$

This result can easily be rewritten as

$$F(z, \bar{z}; k, \bar{k}) = -\frac{1}{\pi}\int_{-\infty}^{\infty}\int_{-\infty}^{\infty}\frac{f(x', y')}{z' - z}e^{k(\bar{z} - \bar{z}')}e^{-\bar{k}(z - z')}dx'dy'. \tag{4.75}$$

Equation (4.75) expresses F in terms of $f(x, y)$. It turns out that, by using Pompeiu's formula in reciprocal space, it is possible to obtain an expression for F in terms of $\hat{f}(p, q)$. For this purpose, we differentiate the formula of Eq. (4.75) with respect to $\bar{k}$ to obtain

$$\frac{\partial F}{\partial \bar{k}} = -\frac{1}{\pi}\int_{-\infty}^{\infty}\int_{-\infty}^{\infty}f(x', y')e^{k(\bar{z} - \bar{z}')}e^{-\bar{k}(z - z')}dx'dy' = e^{k\bar{z} - \bar{k}z}\check{f}(k, \bar{k}), \tag{4.76}$$

where we have introduced the function $\check{f}(k, \bar{k})$, defined as

$$\check{f}(k, \bar{k}) = \check{f}(k_R, k_I) = -\frac{1}{\pi}\int_{-\infty}^{\infty}\int_{-\infty}^{\infty}f(x, y)e^{\bar{k}z - k\bar{z}}dxdy, \quad -\infty < k_R, k_I < \infty. \tag{4.77}$$

With the result of Eq. (4.76) we can apply Pompeiu's formula again, but now in the $k, \bar{k}$ variables to obtain

$$F(z, \bar{z}; k, \bar{k}) = -\frac{1}{\pi}\int_{-\infty}^{\infty}\int_{-\infty}^{\infty}\frac{\check{f}(k'_R, k'_I)e^{k'\bar{z} - \bar{k}'z}}{k' - k}dk'_R dk'_I. \tag{4.78}$$

Finally, we can use the above expression for $F(z, \bar{z}; k, \bar{k})$ and the original definition of F through Eq. (4.71) to obtain

$$f(x, y) = \frac{\partial F}{\partial \bar{z}} - kF = -\frac{1}{\pi}\int_{-\infty}^{\infty}\int_{-\infty}^{\infty}\check{f}(k_R, k_I)e^{k\bar{z} - \bar{k}z}dk_R dk_I, \quad -\infty < x, y < \infty. \tag{4.79}$$

From the definition of the variables $z, \bar{z}$ and $k, \bar{k}$, Eqs. (4.70a) and (4.70b), it is a trivial exercise to obtain

$$k\bar{z} - \bar{k}z = \mathrm{i}(px + qy), \qquad (4.80)$$

which shows that $(k\bar{z} - \bar{k}z)$ is a purely imaginary number, as mentioned above. Inserting this result in Eq. (4.77) and comparing the expression for $\check{f}(k_R, k_I)$ to the definition of the FT, Eq. (4.68), we find

$$\check{f}(k_R, k_I) = -\frac{1}{\pi}\hat{f}(p, q). \qquad (4.81)$$

When we substitute the expressions from Eqs. (4.80) and (4.81) in the right-hand side of Eq. (4.79) we obtain the desired result for the inverse FT, namely Eq. (4.69).

The FT and the IFT are very useful in analyzing multi-dimensional functions, like images. A two-dimensional image is equivalent to a set of values of the function $f(x, y)$ on the xy-plane that represents the brightness at each position (x, y). A brightness function will produce a single-color or a black-and-white image. Mixing three different pure colors (like red, green, blue, or some other carefully chosen combination), each with brightness represented by a different function of the variables x, y, produces full color images.

By considering a function $f(x, y)$ that vanishes for x and y outside the rectangle of sides $2L_1, 2L_2$, it follows that if we define its FT, $\hat{f}(p, q)$, as

$$\hat{f}(p, q) = \int_{-L_2}^{L_2} \int_{-L_1}^{L_1} f(x, y)\mathrm{e}^{-\mathrm{i}(px+qy)}\mathrm{d}x\mathrm{d}y, \qquad (4.82)$$

the IFT, according to Eq. (4.69), is given by

$$f(x, y) = \int_{-M_2}^{M_2} \int_{-M_1}^{M_1} \hat{f}(p, q)\mathrm{e}^{\mathrm{i}(px+qy)}\frac{\mathrm{d}p}{2\pi}\frac{\mathrm{d}q}{2\pi}, \qquad (4.83)$$

where $M_1, M_2 \to \infty$. However, if $\hat{f}(p, q)$ decays sufficiently fast for large values of p and q, we can approximate this integral with M_1, M_2 finite and sufficiently large. This turns out to be the case in applications like the FT of images, and can lead to big gains in image storage requirements (image compression).

The features of an image can be altered globally by manipulating its FT, namely by changing the values of the function $\hat{f}(p, q)$ for certain values (or ranges of values) of the variables p, q. For example, assuming that $\hat{f}(p, q)$ decays sufficiently fast for large values of p and q, a reasonable representation of an image can be obtained by keeping only a small set of values of $\hat{f}(p, q)$ for small p, q values, and taking the IFT of the truncated $\hat{f}(p, q)$ function. The small values of p, q are referred to as "low frequencies", and the operation of keeping only values of $\hat{f}(p, q)$ for the low frequencies is referred to as a "low-pass filter". The image resulting from taking the IFT of a low-pass filtered image is a blurry version of the original one, because the low-frequency components of the FT cannot reproduce the details (fine-scale features) of the image. Conversely, we can choose to keep only values of $\hat{f}(p, q)$ for the large values of p, q (the "high frequencies"), an operation referred to as a "high-pass filter". The image resulting from taking the IFT of a high-pass filtered image is a faint image with only fine-scale details. By combining the low-pass and high-pass filtered images, we can reconstruct the original image in all its detail. This effect is illustrated in the example below. We note that in real applications an ingenious algorithm invented by J.W. Cooley and J.W. Tukey[2] makes the computation very efficient; this algorithm is known as the "fast Fourier transform" (FFT) and is in very wide use in signal analysis.

[2] J.W. Cooley, J.W. Tukey, An algorithm for the machine calculation of complex Fourier series, *Mathematics of Computation* **19**, pp. 297–301 (1965).

Example 4.9: Image compression using the Fourier transform
Digital images are represented on a discrete grid of points on the (x, y)-plane, that is, the values of the variables x and y are discrete, with fixed spacing $\Delta x = \Delta y = \mu$; a square of size μ^2 is called a "pixel" (a term coined from the words "picture" and "element"). It turns out that for this discrete version of the FT, the size of the rectangle $2L_1 \times 2L_2$ in (x, y) space is equal to the size of the rectangle $2M_1 \times 2M_2$ in reciprocal space; in other words, the number of values where the FT $\hat{f}(p, q)$ is computed in the discrete grid of points on the pq-plane is the same as the number of values (pixels) for which the original function $f(x, y)$ is known on the xy-plane. Consequently, the function $\hat{f}(p, q)$ contains the same amount of information as the original picture, $f(x, y)$, and the two spaces have the same resolution. Thus, the IFT can reproduce the original picture in every detail. The FFT and its inverse (IFFT) are provided as standard library routines in modern programming environments. Thus, the fast Fourier transform of an image (also referred to as "forward FFT") and its inverse can be obtained by in a single line of computer code.

We show below a digital image consisting of $770 \times 770 = 592,900$ pixels, in grey scale (the picture is a drawing by M.C. Escher, titled *Relativity*).

Original

The following two panels are the forward FFT and the inverse FFT, each consisting of 770×770 pixels. The forward FFT is actually shown in absolute value and on a logarithmic scale because only a minuscule fraction of points near the center of the FFT have large positive or large negative values. To produce these plots we employed a simple code in python, starting with the two-dimensional array called "originalimage" and using the routines of the numerical python (numpy) library to perform the FFT and the IFFT with the commands

```
FFTimage=numpy.fft.fft2(originalimage)
IFFTimage=numpy.fft.ifft2(FFTimage).
```

We then used standard image plotting commands to display the three 770×770 two-dimensional arrays, namely `originalimage`, `FFTimage` and `IFFTimage`. The inverse FFT reproduces the original figure in every detail, as expected, and is indistinguishable from the original image.

| Forward FFT | Inverse FFT |

An obvious question is: What will happen if we use a low-pass or a high-pass filter to alter the image? In the plot below we show on the left panel the result of using a low-pass filter on $\hat{f}$ and then taking the inverse FFT. In this example, we retain elements of the function $\hat{f}(p,q)$ for values of p and q within a radius of 50 pixels from the origin of the FFT, and set all other values of $\hat{f}$ to zero. This amounts to keeping only 7850 non-zero values for $\hat{f}(p,q)$, a mere 1.3% of the total number of values. The resulting image is a somewhat blurry version of the original, but all the important features of are clearly recognizable, despite the reduction of the information contained in the full FFT function by 98.7%.

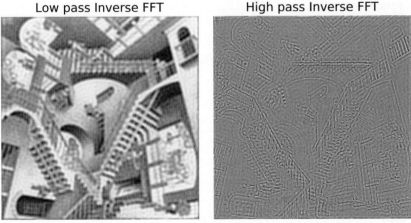

If, on the other hand, we set the values of $\hat{f}(p,q) = 0$ for the same range of p,q values around the origin, that is, we apply a high-pass filter, and then take the inverse FFT of this modified function, we obtain an image shown in the right panel. This image is rather different from the original one: it contains only the fine details of the original image which are captured by the high-frequency values of the FT. In other words, most of the useful information about the image is contained in the few low-frequency components of the FT, while the high-frequency components contain information about the details (fine features) of the image. We can reconstruct the full image by adding the two contributions from the low-pass and the high-pass filters; the reconstructed image is indistinguishable from the original one. This result is due to the linearity of the Fourier transform, a property discussed in more detail in Chapter 7. This example naturally leads to the idea of image compression, that is, the elimination of information that is not essential.

Further reading

The mathematical-method textbooks mentioned under "General Sources" in the Preface cover the material discussed in the present chapter. For more extensive and in-depth discussion of functions of complex variables see:

1. Mark J. Ablowitz and Athanassios S. Fokas, *Complex Variables, Introduction and Applications* (Cambridge University Press, 1997). This is a valuable resource that contains extensive discussions of the topics covered in this chapter as well as in the next two chapters. The coverage is at a higher level and with more mathematical detail than in the present book.

2. Tristan Needham, *Visual Complex Analysis* (Oxford University Press, 1997). This book offers a very interesting perspective of complex functions with many geometric arguments that lead to useful visualizations.

3. E.B. Saff and A.D. Snider, *Fundamentals of Complex Analysis*, Third Ed. (Prentice Hall, Pearson Education Inc., 2003). This is a useful book on complex analysis, at the same level as the present book, but with more extensive discussions of many topics. It also contains useful numerical examples.

4. John H. Mathews and Russell W. Howell, *Complex Analysis for Mathematics and Engineering*, Fifth Ed. (Jones and Bartlett Publishing, 2006). This book covers the topics of complex analysis in the same order as in the present book and offers many useful examples.

5. Jerrold E. Marsden and Michael J. Hoffman, *Basic Complex Analysis*, Third Ed. (W. H. Freeman, 1998). This book offers many formal proofs as well as a large number of examples and problems, in a style and level similar to the present book.

6. M. Ya. Antimirov, Andrei A. Kolyshkin, and Rémi Vaillancourt, *Complex Variables* (Academic Press, 1998). This book offers a formal treatment of many topics.

Problems

1. Write the following in the form $a + ib$, where $a, b \in \mathbb{R}$:

   (a) $(1+i)^3$,

   (b) $2i/(3+i)$,

   (c) $[(2+i)(1-i)]^{-1}$,

   (d) i^{197},

   (e) $e^{2+5\pi i}$,

   (f) $\cos(2-i)$,

   (g) $\sqrt{1-i\sqrt{3}}$,

   (h) $\tan(-i)$,

   (i) $(1+i)^n + (1-i)^n$.

2. Write the following in the form $re^{i\phi}$, where $r, \phi \in \mathbb{R}$:

   (a) $(\bar{2}i)^3$,

   (b) $1-i$,

   (c) $1-2i$,

   (d) $1/(2+i)$,

   (e) $i^{-\ln(i)}$,

   (f) $i^{(i^i)}$,

   (g) $(i^i)^i$.

3. Evaluate the following expressions:

   (a) $|(2+3i)/(1-i)|$,

   (b) $\text{Im}\left[(1+i)^3\right]$,

   (c) $\text{Re}\left[\sin(-i)\right]$,

   (d) $\text{Arg}\left[1/(1-i)\right]$.

4. Prove algebraically the following equalities or inequalities, where z, z_1, z_2 are complex numbers:

   (a) $\text{Re}\left[z\right] \leq |z|$,

(b) $|z_1 + z_2| \leq |z_1| + |z_2|$,

(c) $|z_1 + z_2| \geq ||z_1| - |z_2||$,

(d) $|z_1 + z_2|^2 + |z_1 - z_2|^2 = 2\left(|z_1|^2 + |z_2|^2\right)$.

5. Prove that for any n complex numbers $z_1, \ldots, z_n$ the following inequality holds:

$$|z_1 + \cdots + z_n| \leq |z_1| + |z_2| + \cdots + |z_n|.$$

Let $z_1, \ldots, z_n$ be complex numbers with the property $|z_i| < 1$ for $i = 1, \ldots, n$. Let λ_i be non-negative real number, such that $\lambda_1 + \lambda_2 + \cdots + \lambda_n = 1$. Show that

$$|\lambda_1 z_1 + \lambda_2 z_2 + \cdots + \lambda_n z_n| < 1.$$

6. Find all the values of z that satisfy the following equations or inequalities in the complex plane:

(a) $|1/z| \geq 1$,

(b) $|z + i| = 2$,

(c) $|z - 1| \leq 2$ and $|z - i| \leq 1$,

(d) $|\bar{z} - i + 1| = 1$,

(e) $|1/(z - 3i)| \geq 1$,

(f) $|z + 2 - i| < 2$,

(g) $|z - 2| \leq 3$, and $|z - 2i| \leq 2$,

(h) $e^{iz} = i$,

(i) $|\bar{z} - 2i - 1| = 1$,

(j) $\text{Re}[z - i] > 3$,

(k) $z^{2\pi i} + 1 = 0$,

(l) $\cos(z) = \cosh(z)$.

7. Find all the solutions of the following equations:

(a) $z^3 - 8 = 0$,

(b) $z^4 - 3z^3 - z + 3 = 0$,

(c) $z^5 + 3125 = 0$,

(d) $z^5 - iz^4 - z + i = 0$.

8. Using De Moivre's formula, Eq. (4.19), prove the following:

(a) $\cos(3\theta) = \cos^3(\theta) - 3\cos(\theta)\sin^2(\theta)$,

(b) $\sin(3\theta) = 3\cos^2(\theta)\sin(\theta) - \sin^3(\theta)$.

9. Consider the following relations. Are they true for all $z \in \mathbb{C}$? Provide a justification of your answer.

(a) $|e^z| > 0$,

(b) $|\cos(z)| \geq 1$,

(c) $\overline{e^z} = e^{\bar{z}}$,

(d) $|e^z| = e^{|z|}$.

10. Consider the following proposed derivation:

$$\left|e^{iz}\right| = |\cos(z) + i\sin(z)| = \sqrt{\cos^2(z) + \sin^2(z)} = 1.$$

Find a counter example for this relation, that is, a value of z for which it is obviously not true, and explain the flaws in the derivation.

11. Determine where in the complex plane the following functions are analytic:

(a) $z^3 + z^2 - i$,

(b) $2z/(z^2 + 1)$,

(c) $|z - a|$,

(d) $\tanh(z)$,

(e) $\text{Re}[z]$,

(f) $(1 - 2z^3)^5$,

(g) $(x + iy)/(x^2 + y^2)$,

(h) $\bar{z}/z$,

(i) $\ln\left[(z^2 + a^2)/(z^2 - a^2)\right]$,

(j) $\sin(z)/z$ for $z \neq 0$, 1 for $z = 0$.

12. Determine whether or not the given function $u(x, y)$ is harmonic. If it is, find the conjugate function $v(x, y)$ and the corresponding analytic function $f(z)$ and express f in terms of z:

(a) $u(x, y) = e^x \cos(y)$,

(b) $u(x, y) = \cos(x)\cosh(y)$,

(c) $u(x, y) = r^2 \cos(2\phi)$,

where $r = \sqrt{x^2 + y^2}$, $\phi = \tan^{-1}(y/x)$.

13. Find the real positive value of λ so that the real function

$$v(x, y) = y - \frac{2y}{\lambda^2 x^2 + y^2}$$

is harmonic. For this value of λ, determine the conjugate function $u(x, y)$ and the corresponding analytic function $f(z)$ as a function of z.

14. Show that the function $\ln(|f(z)|)$ is analytic in the domain $\mathcal{D}$ of the z-complex plane if $f(z) = u(x, y) + iv(x, y)$ is analytic and non-zero in $\mathcal{D}$. Then show that in $\mathcal{D}$ the following inequality holds:

$$\left(\frac{\partial^2}{\partial x^2} + \frac{\partial^2}{\partial y^2}\right)\sqrt{[u(x, y)]^2 + [v(x, y)]^2} \geq 0.$$

Does the equality sign ever hold? Explain your answer. What happens if $\mathcal{D}$ is extended to include zeros of $f(z)$? [Hint: Calculate explicitly $\nabla^2 \ln(|f|)$, set this equal to zero, and solve with respect to $\nabla^2(\sqrt{u^2 + v^2})$.]

15. Prove that:
    (a) $f(z)$ and $\overline{f(z)}$ can both be analytic in a given region if and only if $f(z)$ is a constant.
    (b) If $f(z)$ is analytic in a region $\mathcal{D}$ and $f'(z) = 0$ in $\mathcal{D}$, then $f(z)$ is a constant.
    (c) If a harmonic function $u(x, y)$ has two harmonic conjugates $v_1(x, y)$ and $v_2(x, y)$, then their difference $v_1(x, y) - v_2(x, y)$ is a constant function.

16. Simplify the following expression:
$$\sin(\phi) + \sin(2\phi) + \cdots + \sin(n\phi).$$
    [Hint: This is the imaginary part of some complex-valued function of z.]

17. Let $f(z)$ be a function which is analytic everywhere in $\mathbb{C}$. Is the function $\overline{f(\bar{z})}$ analytic in $\mathbb{C}$?

18. For each of the functions below find the Taylor or the Laurent series expansion in powers of $(z - z_0)$, as appropriate in different regions around z_0:
    (a) $\dfrac{1}{(z + i)(z - 2i)}$, $z_0 = 1$,
    (b) $\dfrac{1}{z^2 + 1} + \dfrac{1}{3 - z}$, $z_0 = 0$,
    (c) $f(z) = \dfrac{1}{z}$, $z_0 = 1 + i$,
    (d) $\dfrac{1}{z^3 - 3z^2 + 2z}$, $z_0 = 0$,
    (e) $\sqrt{z} \sin\left(\dfrac{1}{\sqrt{z}}\right)$, $z_0 = 0$,
    (f) $\dfrac{\cosh(z) - 1}{z^2}$, $z_0 = 0$,
    (g) $\dfrac{1}{z^3 + 8}$, $z_0 = 0$,
    (h) $\dfrac{1}{z^2 + 4}$, $z_0 = 0$,
    (i) $\dfrac{1}{(iz - 2)(iz + 3)}$, $z_0 = \dfrac{i}{2}$,
    (j) $\dfrac{1}{(e^z - 1)^2}$, $z_0 = 0$ and $|z| < 2\pi$,
    (k) $\dfrac{1}{\sin(z)}$, $z_0 = \pi$ and $|z - \pi| < \pi$.

19. Derive the Maclaurin series of $f(z) = \tanh^{-1}(z)$. What is the radius of convergence of the series? Is it possible to derive a Laurent series expansion for this function in powers of z for $z \to \infty$? If so, derive the series, if not explain why it is not possible. [Hint: Try deriving a series for $z = 1/\zeta$ with $\zeta \to 0$.]

20. Compare the Maclaurin series of the functions $f_1(z)$ and $f_2(z)$:
$$f_1(z) = \ln(1 + z), \quad f_2(z) = \frac{1}{z + 1}.$$

How are the two series related? Find the radius of convergence of each series. Write the second function as a Laurent series in powers of z for $z \to \infty$. Can this series be integrated term by term? What is the problem with performing such an integration?

21. Use Euler's formula, Eq. (4.13), to evaluate the real, indefinite integral I
$$I = \int e^{ax} \cos(bx) \mathrm{d}x,$$
in two different ways:
    (a) Write the integral as the real part of an expression involving a complex exponential, perform the integration, and take the real part of the result.
    (b) Express $\cos(bx)$ in terms of complex exponentials, then perform the integrations involving these complex exponentials separately, and combine the results.

Do you get the same answer in both cases? Compare your answer to that obtained using integration by parts, see Problem 11 of Chapter 1.

22. Evaluate the integral
$$\int_0^{1+i\pi} e^z \mathrm{d}z$$
using two different contours: (i) The line segment from 0 to $1 + i\pi$. (ii) The line segment from 0 to $i\pi$ and then another segment from $i\pi$ to $1 + i\pi$. Do you get the same answer? If so, explain why.

23. Let $\Gamma(t)$ be a path that connects the origin with a point b on the z-complex plane, defined by $\Gamma(t) = z(t)$, where $z(\alpha) = 0$, $z(\beta) = b$, with $\alpha, \beta, t \in \mathbb{R}$, and $t \in [\alpha, \beta]$. Prove that
$$\int_\Gamma \mathrm{d}z = b, \quad \int_\Gamma z \mathrm{d}z = \frac{b^2}{2}.$$

24. If the function $f(z)$ is analytic on and within a closed contour C, and z_0 lies within C, show that:
$$\oint_C \left[f'(z) - \frac{f(z)}{z - z_0} \right] \frac{\mathrm{d}z}{z - z_0} = 0.$$

25. Show that
$$\oint_C \frac{\sin nz}{z - \pi} \mathrm{d}z = 0, \quad \text{and} \quad \oint_C \frac{\cos nz}{z - \pi} \mathrm{d}z = (-1)^n 2\pi i,$$
with $n \in \mathbb{N}$,
where C is a circle centered at the origin with the radius 2π.

Chapter 5

Singularities, Residues, Contour Integration

The Universe must have begun with a singularity, if Einstein's general theory of relativity is correct. That appeared to indicate that science could not predict how the Universe would begin.

Stephen William Hawking (English theoretical physicist, 1942–2018)

According to Liouville's theorem, if a function is analytic everywhere in the complex plane including infinity, then this function must be constant. Thus, in order to construct interesting functions, it is necessary to "break" analyticity. The points where analyticity is broken are referred to as **singularities**.

An interesting question is how to quantify the "departure" from analyticity. Actually, the d-bar derivative, Eq. (4.35), measures this departure of a function from analyticity. The simplest type of singularity is a pole, that is, the function "departs" from analyticity only at a single point on the complex plane. For instance, the function $1/(z - z_0)$ has a pole at $z = z_0$ and its d-bar derivative tells us that its departure from analyticity is expressed by the δ-function, Eq. (4.56). Indeed, since the function is not analytic at this point we expect that its derivative does not exist; this is captured by the behavior of $\delta(z - z_0)$ for $z \to z_0$. This chapter is devoted to exploring the nature of poles and other types of singularities, and how they can be used to calculate integrals on the complex plane by employing Cauchy's theorem.

5.1 Types of singularities

With the help of the Laurent series expansion, it is possible to classify the isolated singularities of a function into three categories:

1. z_0 is a **removable** singularity if the coefficients of the negative exponents in the Laurent series, $b_{-1}, b_{-2}, \ldots$, are zero. In this case, the function has a Taylor series expansion around z_0.

2. z_0 is a **pole of order m**, where m is a positive integer, if the coefficient $b_{-m} \neq 0$ and all coefficients with lower indices ($-n < -m < 0$ or $0 < m < n$) are zero. A pole of order $m = 1$ is also called a **simple pole**.

3. z_0 is an **essential** singularity if an infinite number of the coefficients of the negative exponents in the Laurent series are non-zero.

A removable singularity is one for which the function itself is not properly defined at $z = z_0$ (hence a singularity), but the Taylor series expansion that approximates it near z_0 is well-defined for all values of z, including z_0. If we define the value of the function to be the same as the value of the series expansion for $z = z_0$, that is, the function at z_0 is assigned the value of the constant term in the series expansion, then the function becomes continuous and analytic. For an example, consider the function

$$f(z) = \frac{\sin(z)}{z},$$

which for $z = 0$ is not properly defined: both numerator and denominator are zero, so strictly speaking any value is possible for this function. However, if we consider the power series expansion of the numerator for finite z and divide by the denominator, we have:

$$\sin(z) = z - \frac{1}{3!}z^3 + \frac{1}{5!}z^5 - \cdots \Rightarrow \frac{\sin(z)}{z} = 1 - \frac{1}{3!}z^2 + \frac{1}{5!}z^4 - \cdots, \quad z \text{ finite.}$$

The value of the last expression for $z = 0$ is unity; if we assign this value to the function for $z = 0$, that is, $f(0) \equiv 1$, then the function becomes analytic in the neighborhood of $z = 0$, including $z = 0$ itself.

As far as poles are concerned, the standard example of a pole of order m is the function

$$f(z) = \frac{1}{(z - z_0)^m}.$$

Occasionally, the pole is a little harder to identify, as for example with the function

$$f(z) = \frac{1}{\sin^m(z)},$$

which has a pole of order m at $z = 0$. To establish this fact, consider the expansion of $\sin(z)$ near $z = 0$:

$$\sin(z) = z - \frac{1}{3!}z^3 + \cdots \Rightarrow \sin^m(z) = z^m \left(1 - \frac{1}{3!}z^2 + \cdots\right)^m$$

$$\Rightarrow \quad f(z) = \frac{1}{z^m}\left(1 - \frac{1}{3!}z^2 + \cdots\right)^{-m} \to \frac{1}{z^m},$$

which behaves like $1/z^m$ as $z \to 0$.

An example of an essential singularity is the point $z = 0$ for the function $e^{1/z}$:

$$f(z) = e^{1/z} = 1 + \frac{1}{z} + \frac{1}{2!}\frac{1}{z^2} + \frac{1}{3!}\frac{1}{z^3} + \cdots,$$

where we have used the familiar power series expansion of the exponential. In the above expansion, all negative powers of z appear, hence we have an infinite number of coefficients of the negative powers which are not zero (as $z \to 0$, each term z^{-m} gives a larger and larger contribution as m increases).

There exists an interesting theorem that describes the behavior of a function with an essential singularity near this point, known as "Picard's theorem":

Picard's Theorem: *If an analytic function has an essential singularity at some point z_0, then in any neighborhood of z_0, excluding this point, the function takes all possible complex values infinitely often, with at most one possible exception.*

For the example, we considered above, the function $e^{1/z}$ in any neighborhood of $z_0 = 0$, excluding z_0 itself, takes all possible complex values except 0.

5.2   Residue theorem

In the following, we adopt the notation $C[z_0]$ to denote that the contour C encloses the point z_0 at which the integrand has a singularity. On occasion, we will also use a second argument which will specify the radius of a circular contour, as for example in $C[z_0, \epsilon]$, with ϵ being the radius of the circular contour centered at z_0.

An important consequence of the Laurent series expansion is that the integral of a function $f(z)$ around a closed contour $C[z_0]$ which encloses a singularity of the function at z_0, can be evaluated using the coefficients of the Laurent series expansion.

Residue Theorem: *Assume that for a function $f(z)$ the Laurent series*

$$f(z) = \sum_{n=0}^{\infty} a_n(z - z_0)^n + \sum_{n=1}^{\infty} \frac{b_{-n}}{(z - z_0)^n},$$

converges uniformly for z inside a circular annulus around the point z_0, then

$$\oint_{C[z_0]} f(z)\mathrm{d}z = 2\pi\mathrm{i}\, b_{-1} = 2\pi\mathrm{i}\, \text{Residue}\,[f(z)]_{z=z_0}, \qquad (5.1)$$

where the "Residue" is defined as the value of the coefficient b_{-1} in the Laurent expansion of the function $f(z)$ around the point z_0.

Proof: From the uniform convergence of the Laurent expansion, we have

$$\oint_{C[z_0]} f(z)\mathrm{d}z = \sum_{n=0}^{\infty} a_n \oint_{C[z_0]} (z - z_0)^n \mathrm{d}z + \sum_{n=1}^{\infty} b_{-n} \oint_{C[z_0]} \frac{1}{(z - z_0)^n}\mathrm{d}z.$$

Equation (4.51) implies

$$\oint_{C[z_0]} \frac{1}{(z - z_0)^n}\mathrm{d}z = 2\pi\mathrm{i}\delta_{1n}.$$

Furthermore, all the terms $(z - z_0)^n$ with $n \geq 0$ give vanishing integrals because they are analytic everywhere on the complex plane. Substituting the above expression in the summation that involves the b_{-n} terms, we obtain the desired result, Eq. (5.1). ∎

The usefulness of the theorem lies in the fact that the coefficients in the Laurent expansion can be determined by much simpler methods than having to evaluate the integrals which appear in the general expression of Eq. (4.60). This was illustrated in the example of a Laurent series expansion that was presented in Example 4.8 of Chapter 4.

A generalization of this result is the calculation of the contribution from a pole of order m at $z = z_0$, that is, a root of order m in the denominator.

Theorem: *For a function $f(z)$ which is analytic on an annulus around z_0 and has a pole of order m at $z = z_0$, the residue at z_0 is given by*

$$\text{Residue}\,[f(z)]_{z=z_0} = \frac{1}{(m - 1)!} \left[\frac{\mathrm{d}^{m-1}}{\mathrm{d}z^{m-1}} \{(z - z_0)^m f(z)\}\right]_{z=z_0}. \qquad (5.2)$$

Proof: For a pole of order m, the function of interest can be written in general form as:

$$f(z) = \frac{b_{-m}}{(z - z_0)^m} + \frac{b_{-m+1}}{(z - z_0)^{m-1}} + \cdots + \frac{b_{-1}}{(z - z_0)} + a_0 + a_1(z - z_0) + \cdots$$

Thus, it has a Laurent expansion starting with the term of order $-m$. We define a new function $F(z)$ by multiplying $f(z)$ with the factor $(z - z_0)^m$:

$$F(z) = (z - z_0)^m f(z)$$
$$= b_{-m} + b_{-m+1}(z - z_0) + \cdots + b_{-1}(z - z_0)^{m-1} + a_0(z - z_0)^m + \cdots,$$

which is a Taylor series expansion for the function $F(z)$, since it contains no negative powers of $(z - z_0)$. Therefore, the coefficient of the term of order $m - 1$ is the derivative of the same order evaluated at $z = z_0$, with a factor of $1/(m-1)!$ in front:

$$b_{-1} = \frac{1}{(m-1)!} \left[\frac{d^{m-1}}{dz^{m-1}} F(z) \right]_{z=z_0},$$

which is the desired result since $(z - z_0)^m f(z) = F(z)$. ∎

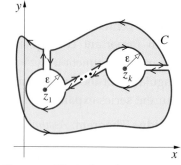

Figure 5.1: Illustration of contour deformation for the integration of a function with several simple residues at $z = z_1, \ldots, z_k$ in the closed contour C: the circles centered at each pole have radius $\epsilon \to 0$ and the contributions from the straight segments joining these circles to the original contour C and to other circles, cancel out.

We can also generalize the Residue theorem to any *finite* number N of singularities at $z = z_k$ $(k = 1, \ldots, N)$ within the contour C, each singularity being a pole of arbitrary order using the construction indicated in Fig. 5.1. We can then use the expressions derived above to calculate the residue for each singularity. In this way we obtain for the integral around this contour the expression

$$\oint_{C[z_k(k=1,\ldots,n)]} f(z)\, dz = 2\pi i \sum_{k=1}^{n} \text{Residue}\, [f(z)]_{z=z_k}, \quad z_k \text{ inside } C. \tag{5.3}$$

Note that in the case of an essential singularity the approach fails because in this case we cannot evaluate the residue by using the above expressions, since the negative powers in the Laurent expansion for an essential singularity go all the way to infinity. Also, in this case we cannot even use the general expression for the Laurent coefficients, Eq. (4.65b), because for $n = -1$ it gives

$$b_{-1} = \frac{1}{2\pi i} \oint_{C[z_0]} f(z')dz',$$

which is an identity.

5.3 Integration by residues

In this section, we provide a number of representative cases in which the Residue theorem proves a powerful tool for calculating definite integrals.

5.3.1 Integrands with simple poles

The most straightforward application of the Residue theorem involves integrands which have simple roots in the denominator.

Theorem: *Consider a function that can be written as a ratio of two other functions, $f(z) = h(z)/g(z)$, such that $h(z)$ is analytic inside the region enclosed by a simple contour C and $g(z)$ has simple roots at $z = z_k, k = 1, \ldots, n$ inside this region. The integral of $f(z)$ over the contour C is given by*

$$\oint_{C[z_k(k=1,\ldots,n)]} f(z)dz = 2\pi i \sum_{k=1}^{n} \left[\frac{h(z_k)}{g'(z_k)} \right]. \tag{5.4}$$

Proof: A simple root of $g(z)$ at $z = z_k$ means that the function vanishes at this value of z but its first derivative does not. Near the roots, where $g(z_k) = 0$, we can expand the function $g(z)$ in Taylor series in the form

$$g(z) = g'(z_k)(z - z_k) + \frac{1}{2!}g''(z_k)(z - z_k)^2 + \cdots,$$

where the dots represent higher-order terms in $(z - z_k)$ which vanish much faster than the first-order term as $z \to z_k$. We also know that because we are dealing with simple roots of $g(z)$, $g'(z_k) \neq 0$ (otherwise the root would be of higher order). We can then rewrite $g(z)$ in the following form:

$$g(z) = g'(z_k)(z - z_k)\left[1 + \frac{1}{2!}\frac{g''(z_k)}{g'(z_k)}(z - z_k) + \cdots\right]$$

$$\Rightarrow \quad g(z) = g'(z_k)(z - z_k) \quad \text{for} \quad z \to z_k.$$

The integral of the ratio $h(z)/g(z)$ over the contour C is then given by the formula

$$\sum_{k=1}^{n}\oint_{C[z_k]}\frac{h(z)}{g'(z_k)(z - z_k)}dz = 2\pi i\sum_{k=1}^{n}\left[\frac{h(z_k)}{g'(z_k)}\right].$$

The above equality follows from performing contour deformation as indicated in Fig. 5.1, and using the small circular paths with radii tending to zero around each simple root of the denominator, for which the result of Eq. (4.51) applies. This is the result we want to prove, Eq. (5.4). Evidently, in this case the residues at $z = z_k$ ($k = 1, \ldots, n$) are given by $h(z_k)/g'(z_k)$. ∎

These results are extremely useful in evaluating both complex integrals as well as integrals of real functions. In fact, they can be used to evaluate many real integrals by cleverly manipulating the expressions that appear in the integrand and constructing appropriate contour integrals. We illustrate these facts by several representative examples.

Example 5.1: We consider the real integral

$$I = \int_0^\infty \frac{1}{x^2 + a^2}dx, \quad a > 0. \tag{5.5}$$

We will evaluate this integral as one part of the integral of the function

$$f(z) = \frac{1}{z^2 + a^2}$$

over a closed contour C which consists of

$$C = C_R \cup L' \cup L,$$

namely, the semicircle of radius R centered at the origin, the line segment L from 0 to R and the line segment L' from R to 0, with the implied limit $R \to \infty$; this contour is shown in the diagram below. In the limit $R \to \infty$, the integral along the line segment L gives the desired integral I, while the contribution of the line segment L' is another integral I' on the real axis and the contribution along the semicircle is an integral I_R. The complex integral along the contour C is given by

$$I_C = I + I' + I_R = \oint_C \frac{1}{z^2 + a^2}dz.$$

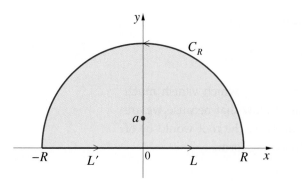

From the Residue theorem, Eq. (5.1), we have for the total contribution to the contour C

$$I_C = 2\pi i \, \text{Residue} \, [f(z)]_{z=z_k}, \ z_k \text{ in } C.$$

To calculate the Residue at $z = ia$, we use the identity

$$\frac{1}{z^2 + a^2} = \frac{1/(z + ia)}{z - ia}.$$

Thus, the integrand has the simple pole, $z = ia$ inside the contour C. Therefore,

$$\text{Residue} \, [f(z)]_{z=ia} = \left(\frac{1}{z + ia} \right)_{z=ia} = \frac{1}{2ia} \Rightarrow 2\pi i \, \text{Residue} \, [f(z)]_{z=ia} = \frac{\pi}{a}.$$

The integrand along the real axis is an even function of x, that is, it is the same for $x \to -x$, which leads to

$$I' = \int_{-\infty}^{0} \frac{1}{x^2 + a^2} dx = \int_{0}^{\infty} \frac{1}{(-x)^2 + a^2} d(-x) = I.$$

It turns out that the integral I_R over the large semicircle with radius $R \to \infty$ vanishes: we can write the variable of integration z and the differential dz over the semicircle in the form

$$z = Re^{i\phi}, \quad 0 \le \phi \le \pi \Rightarrow dz = Rie^{i\phi} d\phi.$$

Thus, the integral I_R in the limit $R \to \infty$ takes the form

$$\lim_{R \to \infty} (I_R) = \lim_{R \to \infty} \left[\int_{0}^{\pi} \frac{1}{R^2 e^{2i\phi} + a^2} iRe^{i\phi} d\phi \right]$$

$$= \lim_{R \to \infty} \frac{i}{R} \left[\int_{0}^{\pi} \frac{e^{i\phi}}{e^{2i\phi} + a^2/R^2} d\phi \right] = 0,$$

since the integral over ϕ gives a finite value. Having established that $I' = I$ and $I_R \to 0$ for $R \to \infty$, we conclude that the value of the original integral is

$$I = \int_{0}^{\infty} \frac{1}{x^2 + a^2} dx = \frac{\pi}{2a}. \tag{5.6}$$

This result was obtained in Chapter 1, Section 1.4.2, using a change of variables.

5.3.2 Integrands that are ratios of polynomials

We present a generalization of the approach used for proving that I_R vanishes for $R \to \infty$ in the preceding example.

Theorem: *Suppose that the contour of integration is a circular arc of radius R, centered at the origin. Assume that the integrand is a ratio of two polynomials, where the order of the polynomial in the denominator is larger by at least two than the order of the polynomial in the numerator. Then $I_R \to 0$ as $R \to \infty$.*

Proof: Suppose we want to evaluate a real integral involving two polynomials, of degree p in the numerator and q in the denominator:

$$I = \int_{x_i}^{x_f} \frac{a_0 + a_1 x + a_2 x^2 + \cdots + a_p x^p}{b_0 + b_1 x + b_2 x^2 + \cdots + b_q x^q} \, dx.$$

We can try to evaluate this integral by turning it into a complex integral over a contour C one part of which can be identified as the real integral I:

$$\int_C \frac{a_0 + a_1 z + a_2 z^2 + \cdots + a_p z^p}{b_0 + b_1 z + b_2 z^2 + \cdots + b_q z^q} \, dz.$$

Let us assume that the contour C contains an arc of radius R centered at the origin, which contributes I_R to the value of the contour integral. Along this arc, the integration variable z and the differential dz take the form

$$z = R e^{i\phi}, \quad dz = i R e^{i\phi} d\phi, \quad \phi_1 \le \phi \le \phi_2.$$

Using these substitutions in the complex integral, we find

$$I_R = \int_{\phi_1}^{\phi_2} \frac{a_0 + a_1 R e^{i\phi} + a_2 R^2 e^{i2\phi} + \cdots + a_p R^p e^{ip\phi}}{b_0 + b_1 R e^{i\phi} + b_2 R^2 e^{i2\phi} + \cdots + b_q R^q e^{iq\phi}} \, i R e^{i\phi} d\phi.$$

Taking the common factors $R^p e^{ip\phi}$ and $R^q e^{iq\phi}$ from the numerator and the denominator and keeping only the dominant terms (because eventually we will let $R \to \infty$) we arrive at the expression

$$I_R = \frac{a_p}{b_q} R^{p-q+1} \int_{\phi_1}^{\phi_2} e^{i(p-q+1)\phi} i d\phi = \frac{a_p}{b_q} R^{p-q+1} \frac{1}{p-q+1} \left[e^{i\phi_2} - e^{i\phi_1} \right]. \quad (5.7)$$

This integral gives a vanishing contribution when $R \to \infty$, provided that

$$p - q + 1 \le -1 \Rightarrow q \ge p + 2.$$

In this case, the factor R^{p-q+1} makes the value of I_R tend to zero. ∎

Example 5.2: We now consider a similar real integral, for which a simple factorization of the denominator of the integral does not work, as was the case for the previous example. The Residue theorem, however, can still be applied. We will evaluate the real integral

$$I = \int_0^\infty \frac{1}{x^8 + 1} dx. \quad (5.8)$$

As in the previous example, we turn this integral into a complex integral over the closed contour C_A consisting of

$$C_A = C_R \cup L' \cup L,$$

with C_R a semicircle of radius R centered at the origin, the line segment L from 0 to R on the real axis, and the line segment L' from $-R$ to 0 on the real axis, with the implied limit $R \to \infty$; this contour is shown in the diagram on the left below, is labeled as "Contour A".

The corresponding complex integral is

$$I_{C_A} = I + I' + I_R = \oint_{C_A} f(z)\mathrm{d}z, \quad \text{where } f(z) = \frac{1}{z^8 + 1},$$

with the integral over the line segment L being equal to the desired integral I, the integral over the line segment L' giving rise to the value I' and the integral over the semicircle of radius R giving a contribution I_R.

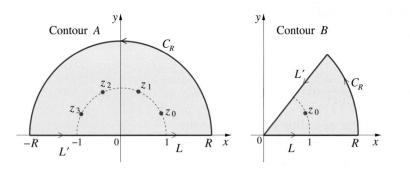

We notice that the integrand is an even function under $x \to -x$, so the integral on the negative real axis (on the line segment L') is the same as I, hence, $I' = I$. We can also argue, for the same reasons as in the previous example, that $I_R \to 0$ for $R \to \infty$, therefore

$$I_{C_A} = I + I' + \lim_{R \to \infty}(I_R) = 2I = 2\pi i \sum_k \left\{ \text{Residue}\left[f(z)\right]_{z=z_k}, \ z_k \text{ in } C_A \right\}.$$

For the residues, we can use the expression of Eq. (5.4) since the denominator $g(z) = z^8 + 1$ has only simple roots inside the contour C_A and the numerator is simply unity; for each residue this yields

$$\frac{1}{8z_k^7}, \quad \text{for } z_k^8 = -1 \Rightarrow z_k = \mathrm{e}^{\mathrm{i}(2k+1)\pi/8}, \ k = 0,\ldots,7.$$

Among the above poles the ones with $k = 0,1,2,3$ are within C_A. Moreover, we note from the above relations that

$$z_k^8 = -1 \Rightarrow z_k^7 = -\frac{1}{z_k} \Rightarrow \frac{1}{8z_k^7} = -\frac{z_k}{8}.$$

Hence, we obtain the following result for the original integral:

$$2I = 2\pi i \frac{-1}{8} \left[\mathrm{e}^{\mathrm{i}\pi/8} + \mathrm{e}^{\mathrm{i}3\pi/8} + \mathrm{e}^{\mathrm{i}5\pi/8} + \mathrm{e}^{\mathrm{i}7\pi/8} \right].$$

Using the trigonometric relations

$$\cos\left(\frac{7\pi}{8}\right) = -\cos\left(\frac{\pi}{8}\right), \quad \sin\left(\frac{7\pi}{8}\right) = \sin\left(\frac{\pi}{8}\right),$$

$$\cos\left(\frac{5\pi}{8}\right) = -\cos\left(\frac{3\pi}{8}\right), \quad \sin\left(\frac{5\pi}{8}\right) = \sin\left(\frac{3\pi}{8}\right),$$

we arrive at the final result

$$I = \frac{\pi}{4}\left[\sin\left(\frac{\pi}{8}\right) + \sin\left(\frac{3\pi}{8}\right)\right] = \frac{\pi}{4}\left[\sin\left(\frac{\pi}{8}\right) + \cos\left(\frac{\pi}{8}\right)\right].$$

We next evaluate the same integral with the use of a different contour C_B consisting of

$$C_B = C_R \cup L' \cup L,$$

shown on the diagram of the previous page labeled "Contour B". This contour contains an eighth of a circle of radius R centered at the origin and labeled C_R, the line segment on the real axis L from 0 to R, and the line segment L' at angle of $\pi/4$ relative to the real axis, with the implied limit $R \to \infty$. As for the previous path, the part on the real axis (line segment L) gives the desired integral I, while the path along the circular arc gives a vanishing contribution by the same arguments as before, $I_R \to 0$ for $R \to \infty$. For the last part of this contour along the line segment L', we can write the variable z and the differential dz in the form

$$z = re^{i\pi/4}, \quad dz = e^{i\pi/4}dr.$$

This yields the following contribution:

$$I' = \int_{r=\infty}^{0} \frac{1}{(re^{i\pi/4})^8 + 1}d(re^{i\pi/4}) = -e^{i\pi/4}\int_0^\infty \frac{1}{r^8+1}dr = -e^{i\pi/4}I.$$

Therefore, for the entire contour we find

$$I_{C_B} = I + I' + \lim_{R\to\infty}(I_R) = I(1 - e^{i\pi/4})$$

$$= 2\pi i \sum_k \left\{\text{Residue}\,[f(z)]_{z=z_k},\ z_k \text{ in } C_B\right\}.$$

Since the only singularity within C_B is at $z_0 = \exp(i\pi/8)$, using the relations we established earlier for the value of the residues, we find the final answer:

$$I(1 - e^{i\pi/4}) = 2\pi i \frac{-1}{8}e^{i\pi/8} \Rightarrow I = -\frac{\pi i}{4}\frac{e^{i\pi/8}}{1 - e^{i\pi/4}}$$

or, after using Euler's formula for the complex exponentials,

$$I = -\frac{\pi i}{4}\frac{1}{e^{-i\pi/8} - e^{i\pi/8}} = \frac{\pi}{8\sin(\pi/8)} = \frac{\pi}{4}\left[\sin\left(\frac{\pi}{8}\right) + \cos\left(\frac{\pi}{8}\right)\right].$$

This is identical to the answer obtained through the contour C_A.

The advantage of the second approach in Example 5.2 (Contour B) is that it does not use the fact that x^8 is an even function. Hence, this approach can also be used for the cases that this property is not valid, such as the case of $1/(x^3+1)$.

Example 5.3: A slightly more complicated situation arises when the integrand has poles on the real axis, as for example in the case of the integral

$$I = \fint_0^\infty \frac{1}{x^8 - 1} dx. \tag{5.9}$$

This integral makes sense only as a principal value integral discussed in Section 1.4.4; the principal value is indicated by the symbol $\fint$. This notation is vague because it does not specify the points of singularity, which in the above integral is at $x = 1$. In order to compute the above principal value integral, we will follow the approach used earlier and will turn it into a complex integral over a closed contour, a part of which can be identified as the desired real integral. Two possible contour choices, labeled Contour A (C_A) and Contour B (C_B), are shown in the diagram below. The first consist of

$$C_A = C_R \cup L' \cup C_2 \cup L \cup C_1,$$

with C_R a semicircle of radius R centered at the origin, the line segment L from 0 to R on the real axis, the line segment L' from $-R$ to 0 on the real axis, with the implied limit $R \to \infty$, and the two semicircles labeled C_1 and C_2 centered at ± 1 and of radius ϵ, with the implied limit $\epsilon \to 0$. The second contour consists of

$$C_B = C_R \cup L' \cup C_2 \cup L \cup C_1,$$

with C_R an eighth of a circle of radius R centered at the origin, the line segment L from 0 to R on the real axis, the line segment L' at an angle $\pi/4$ relative to the real axis, with the implied limit $R \to \infty$, and the two semicircles labeled C_1 and C_2 centered at 1 and at $z_1 = \exp(i\pi/4)$ and of radius ϵ, with the implied limit $\epsilon \to 0$. Both contours are shown on the diagram below.

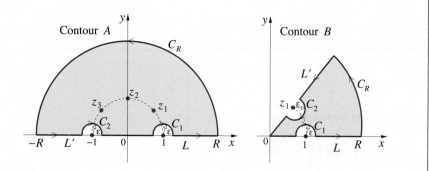

We concentrate on the first choice, which we will call contour C_A:

$$I_{C_A} = I + I' + I_R + I_1 + I_2 = \oint_{C_A} f(z) dz, \quad \text{where } f(z) = \frac{1}{z^8 - 1}.$$

This contour includes the usual contributions from the positive real axis (line segment L, which corresponds to the desired integral I), and from the negative real axis (line segment L', which corresponds to I'). As in the previous two examples, we can easily show that $I' = I$. The contour also contains a contribution from the semicircle centered at the origin with radius R, which, for the same reasons as in the previous examples vanishes for $R \to \infty$. Finally, we also have two more contributions from the semicircles with radius ϵ, centered at $x = \pm 1$, labeled I_1, I_2, respectively. These contributions are required in order to avoid the singularities of the integrand at $z = \pm 1$. This situation is analogous to the example we had discussed in Chapter 1, for the integral

$$\oint_0^\infty \frac{1}{x^2 - 1} dx,$$

where we had argued that approaching the root of the denominator at $x = 1$ symmetrically from the left and the right of ± 1 is required in order to produce a well-defined integral, called the principal value integral. This is precisely what the two semicircles with vanishing radii lead to: In the limit $\epsilon \to 0$, the integrals I and I' along the line segments L and L' are avoiding the poles at $x = \pm 1$ and are evaluated in the principal value sense:

$$I = \lim_{\epsilon \to 0} \left[\int_0^{1-\epsilon} \frac{1}{x^8 - 1} dx + \int_{1+\epsilon}^\infty \frac{1}{x^8 - 1} dx \right],$$

$$I' = \lim_{\epsilon \to 0} \left[\int_{-\infty}^{-1-\epsilon} \frac{1}{x^8 - 1} dx + \int_{-1+\epsilon}^0 \frac{1}{x^8 - 1} dx \right].$$

For the semicircle centered at $x = +1$ we have

$$C_1 \; : \; z = 1 + \epsilon e^{i\phi}, \; \pi \le \phi \le 0$$

$$\Rightarrow \quad I_1 = \int_{C_1} \frac{1}{z^8 - 1} \, dz = -\int_0^\pi \frac{1}{(1 + \epsilon e^{i\phi})^8 - 1} \epsilon i e^{i\phi} \, d\phi.$$

Similarly, for the semicircle centered at $x = -1$ we have

$$C_2 \; : \; z = -1 + \epsilon e^{i\phi}, \; \pi \le \phi \le 0$$

$$\Rightarrow \quad I_2 = \int_{C_2} \frac{1}{z^8 - 1} \, dz = -\int_0^\pi \frac{1}{(-1 + \epsilon e^{i\phi})^8 - 1} \epsilon i e^{i\phi} \, d\phi.$$

We can expand the denominators in the integrands of I_1 and I_2 with the use of the binomial expansion: for the pole at $z = +1$ we have

$$z^8 = (1 + \epsilon e^{i\phi})^8 = 1 + 8\epsilon e^{i\phi} + \cdots \Rightarrow z^8 - 1 = 8\epsilon e^{i\phi},$$

while for the pole at $z = -1$ we have

$$z^8 = (-1 + \epsilon e^{i\phi})^8 = (1 - \epsilon e^{i\phi})^8 = 1 - 8\epsilon e^{i\phi} + \cdots \Rightarrow z^8 - 1 = -8\epsilon e^{i\phi},$$

where we have neglected higher-order terms, since we are interested in the limit $\epsilon \to 0$. When these expansions are substituted in the integrals I_1 and I_2, we find

$$\lim_{\epsilon \to 0}(I_1) = -i\pi \frac{1}{8}, \quad \lim_{\epsilon \to 0}(I_2) = -i\pi \frac{-1}{8}.$$

Thus, the two contributions from the infinitesimal semicircles centered at $x = \pm 1$ cancel each other. Putting all these results together, we arrive at the final result:

$$I_{C_A} = I + I' + \lim_{\epsilon \to 0}(I_1) + \lim_{\epsilon \to 0}(I_2) + \lim_{R \to 0}(I_R) = 2I$$

$$= 2\pi i \sum_k \left\{ \text{Residue}\,[f(z)]_{z=z_k}\,,\ z_k \text{ in } C_A \right\}$$

$$= 2\pi i \frac{1}{8} \left[e^{i\pi/4} + e^{i\pi/2} + e^{i3\pi/4} \right].$$

This is a consequence of the fact that the only poles enclosed by the entire contour C_A are the simple poles at

$$z_1 = e^{i\pi/4},\ z_2 = e^{i2\pi/4},\ z_3 = e^{i3\pi/4},$$

all of which are solutions of the equation

$$z_k^8 = 1 \Rightarrow \frac{1}{z_k^7} = z_k,\quad k = 1, 2, 3.$$

Using the identities:

$$\cos\left(\frac{\pi}{4}\right) = -\cos\left(\frac{3\pi}{4}\right),\quad \sin\left(\frac{\pi}{4}\right) = \sin\left(\frac{3\pi}{4}\right),\quad e^{i\pi/2} = i,$$

we find

$$I = -\frac{\pi}{8}(1 + \sqrt{2}). \tag{5.10}$$

5.3.3   Integrands with trigonometric or hyperbolic functions

Another interesting case involves integration requiring a rectangular rather than a semicircular contour. The reason for this choice is that for the associated integrand a circular contour with radius $R \to \infty$ would enclose an infinite number of residues. Integrals of that type appear when the integrand involves trigonometric or hyperbolic functions. We work out such an example next.

Example 5.4: Our example consists of the following real integral:

$$I_{\text{real}} = \int_{-\infty}^{\infty} \frac{\sin(\lambda x)}{\sinh(x)}\,dx = \text{Im}[I], \tag{5.11}$$

where we have defined

$$I = \int_{-\infty}^{\infty} \frac{e^{i\lambda x}}{\sinh(x)}\,dx,\quad \lambda > 0. \tag{5.12}$$

We note that the definition of the hyperbolic functions imply the following:

$$\sinh(x) = \frac{1}{2}\left[e^x - e^{-x}\right] \Rightarrow \sinh(x + 2\pi i) = \frac{1}{2}\left[e^{x+2\pi i} - e^{-x-2\pi i}\right] = \sinh(x).$$

Hence, for the evaluation of the integral I in Eq. (5.12) we can use the function $f(z)$

$$f(z) = \frac{e^{i\lambda z}}{\sinh(z)},$$

and the contour shown in the diagram below.

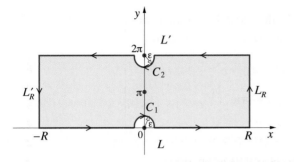

This contour consists of

$$C = L \cup C_1 \cup L_R \cup L' \cup C_2 \cup L'_R,$$

the line segment L on the real axis from $-R$ to R, the vertical line segment L_R at $x = R$ going from $y = 0$ to $y = 2\pi$, the horizontal line segment L' at $x = 2\pi$ from R to $-R$, the vertical line segment L'_R at $x = -R$ going from $y = 2\pi$ to $y = 0$, in the implied limit $R \to \infty$, and the semicircles of radius ϵ centered at 0 and at $x = 0, y = 2\pi$, labeled C_1, C_2, respectively, in the implied limit $\epsilon \to 0$.

The contribution of the line segment L gives the desired integral I. Along the two horizontal parts of this contour, the denominators in the integrands are the same. For the line segment L' of this contour along $z = x + 2\pi i$, we have

$$I' = -\int_{-\infty}^{\infty} \frac{e^{i\lambda(x+2\pi i)}}{\sinh(x + 2\pi i)} d(x + 2\pi i) = -e^{-2\lambda\pi} I.$$

For the contribution of the line segment L_R along the vertical path $z = R + iy$, $0 \le y < 2\pi$, the integrand becomes

$$\frac{2e^{i\lambda(R+iy)}}{\left(e^{R+iy} - e^{-R-iy}\right)} = \frac{2e^{-\lambda y}e^{i\lambda R}}{e^R\left(e^{iy} - e^{-2R-iy}\right)}$$

$$\Rightarrow \quad \lim_{R\to\infty}\left(I_{L_R}\right) = \lim_{R\to\infty}\left(\frac{1}{e^R}\int_0^{2\pi} \frac{2ie^{-\lambda y}e^{i\lambda R}}{e^{iy} - e^{-2R-iy}}\,dy\right) = 0,$$

because the integration over y gives a finite value (the integrand is finite in magnitude). Similarly for $I_{L'_R}$ along the vertical path L'_R $z = -R + iy$, $2\pi \ge y \ge 0$, we find

$$\frac{2e^{i\lambda(-R+iy)}}{\left(e^{-R+iy} - e^{R-iy}\right)} = \frac{2e^{-\lambda y}e^{-i\lambda R}}{e^R\left(e^{-2R+iy} - e^{-iy}\right)}$$

$$\Rightarrow \quad \lim_{R\to\infty}\left(I_{L'_R}\right) = \lim_{R\to\infty}\left(-\frac{1}{e^R}\int_0^{2\pi} \frac{2ie^{-\lambda y}e^{-i\lambda R}}{e^{-2R+iy} - e^{-iy}}\,dy\right) = 0,$$

because again the integration over y gives a finite value. From the Residue theorem, we have

$$I_C = I + I' + \lim_{R \to \infty}(I_{L_R}) + \lim_{R \to \infty}(I_{L'_R}) + \lim_{\epsilon \to 0}(I_{C_1}) + \lim_{\epsilon \to 0}(I_{C_2})$$

$$= I - e^{-2\pi\lambda}I + \lim_{\epsilon \to 0}(I_{C_1}) + \lim_{\epsilon \to 0}(I_{C_2}) = 2\pi i \text{ Residue}\,[f(z)]_{z=\pi i}\,.$$

The singularity at $z = \pi i$ is the only one inside the closed contour is at $z = i\pi$, and since the poles at $z = 0, \pi i, 2\pi i$ are simple poles, we obtain

$$\lim_{\epsilon \to 0}(I_{C_1}) = (-1)\pi i \text{ Residue}\,[f(z)]_{z=0} = -\pi i,$$

$$\lim_{\epsilon \to 0}(I_{C_2}) = (-1)\pi i \text{ Residue}\,[f(z)]_{z=2\pi i} = -\pi i e^{-2\lambda\pi},$$

$$\text{Residue}\,[f(z)]_{z=\pi i} = -e^{-\lambda\pi}.$$

Substituting the above expressions in the previous equation we find

$$I_{\text{real}} = \text{Im}[I] = \pi\left(\coth(\lambda\pi) - \frac{1}{\sinh(\lambda\pi)}\right).$$

5.3.4   Skipping a simple pole

Finally, we work out a generalization of the contribution incurred when skipping a simple pole by going around it in an arc of radius ϵ and angular width ψ.

Theorem: *Consider a function $f(z)$ analytic in an annulus centered at z_0, with a simple pole at $z = z_0$. Then*

$$\lim_{\epsilon \to 0} \int_{C_{\epsilon,\psi}[z_0]} f(z)\mathrm{d}z = i\psi \text{ Residue}\,[f(z)]_{z=z_0}, \tag{5.13}$$

where $C_{\epsilon,\psi}[z_0]$ is a circular arc of radius ϵ and angular width ψ, which lies entirely within the annulus centered at z_0 : $z = z_0 + \epsilon e^{i\phi}$, with $\phi_0 \le \phi \le \phi_0 + \psi$.

Proof: The setup is illustrated in Fig. 5.2. We note first that since $f(z)$ is analytic on an annulus around z_0 we can use its Laurent series expansion in the neighborhood of z_0 assuming that it converges uniformly in the annulus:

$$f(z) = \sum_{n=-\infty}^{\infty} b_n(z - z_0)^n.$$

Since the function has a simple pole at z_0, $b_n = 0$ for $n \le -2$ which leads to

$$\int_{C_{\epsilon,\psi}[z_0]} f(z)\mathrm{d}z = \sum_{n=-\infty}^{\infty} b_n \int_{\phi_0}^{\phi_0+\psi} i\epsilon e^{i\phi}\epsilon^n e^{in\phi}\mathrm{d}\phi = \sum_{n=-1}^{\infty} b_n I_n,$$

where the integral I_n is defined as

$$I_n = \epsilon^{n+1}i \int_{\phi_0}^{\phi_0+\psi} e^{i(n+1)\phi}\mathrm{d}\phi.$$

Evaluating this integral we find the following conditions:

$$\text{for } n = -1 \quad \rightarrow \quad I_{-1} = i\psi,$$

$$\text{for } n \ge 0 \quad \rightarrow \quad I_n = \epsilon^{n+1}\frac{1}{n+1}e^{i\phi_0(n+1)}\left(e^{i\psi(n+1)} - 1\right). \tag{5.14}$$

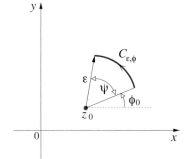

Figure 5.2: Illustration of the circular arc $C_{\epsilon,\psi}[z_0]$ of radius ϵ and angular width ψ, centered at z_0, used to skip a simple pole at z_0.

In the limit $\epsilon \to 0$ the terms with $n \neq -1$ vanish, and only the term $n = -1$ survives, leading to

$$\int_{C_{\epsilon,\psi}[z_0]} f(z) \, dz = i\psi \, b_{-1} = i\psi \, \text{Residue} \left[f(z)\right]_{z=z_0},$$

which proves the desired formula, Eq. (5.13). ∎

Note that if $\psi = 2\pi$ this formula simply reduces to the Residue theorem. We emphasize, however, that the formula of Eq. (5.13) *applies to simple poles only*, whereas the Residue theorem is valid for a pole of any order. Indeed, we observe that for an arbitrary value of $\psi \neq 2\pi$ the expression on the right-hand side of Eq. (5.14) vanishes only for $n \geq 0$, because for these values of n the power of ϵ is positive and the limit $\epsilon \to 0$ produces a zero value for I_n. In contrast to this, when $\psi = 2\pi$ the expression on the right-hand side of Eq. (5.14) vanishes identically for any value of n, because of the identity

$$e^{i2\pi(n+1)} = 1.$$

Thus, for $\psi = 2\pi$ we can afford to have terms with $n < -1$ in the Laurent expansion, that is, a pole of order higher than one, since the integrals I_n produced by these terms will vanish and we will be left with the value of the residue.

We also note that if the circular arc is traversed in the counter-clockwise sense, the angle ϕ increases, therefore $\psi > 0$, whereas if it is traversed in the clockwise sense, $\psi < 0$ which produces the (-1) factor mentioned earlier.

The formula of Eq. (5.13) allows us to avoid simple poles that happen to lie on the path of integration, which is equivalent to evaluating integrals by taking their principal value, as already implemented in Examples 5.3 and 5.4. The above discussion also indicates that a principal value integral can be defined only if the integrand vanishes in the form $c(x - x_0)$ as x approaches the singularity x_0, with c a constant.

5.4   Jordan's lemma

We saw in the previous section the usefulness of introducing circular arcs as part of a closed contour over which a complex integral can be evaluated by using the Residue theorem. In fact, in many situations it is convenient to choose a circular path such that its contribution to the integral vanishes in the limit of the radius going to infinity. Accordingly, we discuss next a very useful general criterion for when an integral over a circular arc of infinite radius vanishes; this criterion is known as "Jordan's lemma".

Jordan's Lemma: *Consider the integral*

$$I_R = \int_{C_R} e^{i\lambda z} f(z) \, dz, \tag{5.15}$$

where λ is a real, positive constant, C_R is a circular arc of radius R centered at the origin and lying on the upper-half complex plane defined by $z = Re^{i\phi}$ with $0 \leq \phi_1 \leq \phi \leq \phi_2 \leq \pi$. Assume that the function $f(z)$ is bounded on C_R, $|f(z)| \leq F_R, z \in \mathbb{C}$ where F_R is finite. Then, the integral I_R over the circular arc is bounded:

$$|I_R| = \left| \int_{C_R} e^{i\lambda z} f(z) dz \right| \leq \int_{C_R} |e^{i\lambda z} f(z)| dz \leq \frac{\pi}{\lambda} F_R,$$

and, as $R \to \infty$, if $F_R \to 0 \Rightarrow I_R \to 0$.

Proof: On the circular arc extending from ϕ_1 to ϕ_2 we have

$$z = Re^{i\phi}, \quad dz = iRe^{i\phi}d\phi.$$

Then, letting $z = x + iy$, the absolute value of the integral over this arc takes the form

$$|I_R| = \left| \int_{\phi_1}^{\phi_2} e^{i\lambda(x+iy)} f(z) iRe^{i\phi}\, d\phi \right| \le \int_{\phi_1}^{\phi_2} \left| e^{i\lambda(x+iy)} f(z) iRe^{i\phi} \right|\, d\phi,$$

where the last relation is derived from the triangle inequality. Using $|i| = 1$, $|\exp(i\lambda x)| = 1$, $|\exp(i\phi)| = 1$ and $y = R\sin(\phi)$, we find

$$|I_R| \le F_R R \int_{\phi_1}^{\phi_2} e^{-\lambda R \sin(\phi)}\, d\phi.$$

Assume $\phi_1 = 0, \phi_2 = \pi$, which covers the entire range of allowed values for the situation we are considering, and using the integrand in the last expression is a symmetric function of ϕ with respect to $\pi/2$, it follows that it is sufficient to evaluate the above integral only in the range $0 \le \phi \le \pi/2$ and multiply the resulting expression by a factor of 2:

$$\sin(\pi - \phi) = \sin(\phi) \Rightarrow \int_0^{\pi} e^{-\lambda R \sin(\phi)}\, d\phi = 2 \int_0^{\pi/2} e^{-\lambda R \sin(\phi)}\, d\phi.$$

In this range, we have

$$0 \le \phi \le \frac{\pi}{2} \ : \ \sin(\phi) \ge \frac{2\phi}{\pi}.$$

This inequality is illustrated by the graph in Fig. 5.3.

Replacing $\sin(\phi)$ in the integrand by $2\phi/\pi$ we find the following:

$$0 \le \phi \le \frac{\pi}{2} \ : \ -\sin(\phi) \le -\frac{2\phi}{\pi} \Rightarrow \int_0^{\pi/2} e^{-\lambda R \sin(\phi)}\, d\phi \le \int_0^{\pi/2} e^{-\lambda R 2\phi/\pi}\, d\phi.$$

Hence,

$$|I_R| \le 2F_R R \int_0^{\pi/2} e^{-\lambda R 2\phi/\pi}\, d\phi.$$

Using the change of variables $\tau = 2\lambda R\phi/\pi$, the above inequality becomes

$$|I_R| \le \frac{F_R \pi}{\lambda} \int_0^{\lambda R} e^{-\tau}\, d\tau = \frac{F_R \pi}{\lambda}(1 - e^{-\lambda R}).$$

As $R \to \infty$ we find

$$R \to \infty \Rightarrow |I_R| \le \frac{F_R \pi}{\lambda},$$

which shows that the integral I_R is bounded since F_R is finite and in the limit $F_R \to 0$ it vanishes. ∎

If λ is real and *negative*, then Jordan's lemma is still valid provided that the path of integration lies on the *lower* half of the complex plane ($y \le 0$).

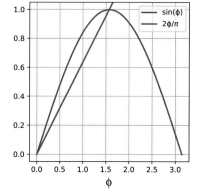

Figure 5.3: Justification of the inequality $\sin(\phi) \ge 2\phi/\pi$ for $0 \le \phi \le \pi/2 = 1.570795$: the red line represents the values of $y = \sin(\phi)$ while the green line represents the values of $y = 2\phi/\pi$.

Example 5.5: To illustrate the usefulness of Jordan's lemma, we consider the following example of a real integral which can be evaluated by contour integration:

$$I = \int_{-\infty}^{\infty} \frac{\cos(\lambda x)}{x - a}\, dx, \tag{5.16}$$

with a, λ real positive constants, where f denotes principal value integral at $x = a$. First, note that we can rewrite this integral in terms of two other integrals:

$$I = \frac{1}{2}\fint_{-\infty}^{\infty} \frac{e^{i\lambda x} + e^{-i\lambda x}}{x - a}\, dx = \frac{1}{2}I_A + \frac{1}{2}I_B,$$

$$I_A = \fint_{-\infty}^{\infty} f_1(x)dx, \quad \text{where } f_1(x) = \frac{e^{i\lambda x}}{x - a},$$

$$I_B = \fint_{-\infty}^{\infty} f_2(x)dx, \quad \text{where } f_2(x) = \frac{e^{-i\lambda x}}{x - a}.$$

I_A and I_B can be considered as parts of contour integrals on the complex plane which satisfy the conditions of Jordan's lemma for the contours C_A and C_B shown in the diagram below:

$$C_A = C_R \cup L \cup C_a, \quad C_B = C_R' \cup L \cup C_a'.$$

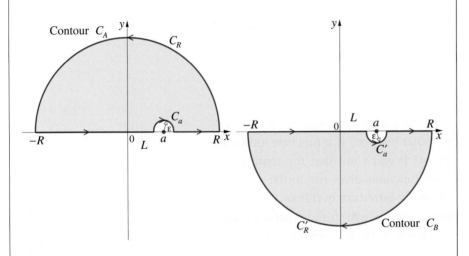

C_A consists of the line segment L from $-R$ to R on the real axis and the semicircle C_R of radius R centered at the origin, with the implied limit $R \to \infty$, as well as the semicircle C_a centered at $x = a$, of radius ϵ, with the implied limit $\epsilon \to 0$, with all semicircles lying on the *upper* half complex plane ($y > 0$); C_B consists of the corresponding components C_R', C_a', which lie on the *lower* half complex plane ($y < 0$). These choices make sure that Jordan's lemma is satisfied for the C_R portion of each contour, since for C_A the complex exponential involves $\lambda > 0$ and for C_B the complex exponential involves $-\lambda < 0$. The contour integration over C_A, which contains no singularities, yields:

$$I_A + \lim_{\epsilon \to 0}(I_{C_a}) + \lim_{R \to \infty}(I_{C_R}) = 0 \Rightarrow I_A = -\lim_{\epsilon \to 0}(I_{C_a}),$$

where the integral I_{C_R} for the semicircle C_R on the upper half-plane is eliminated by Jordan's lemma. The other integral at the singularity on the x-axis gives

$$\lim_{\epsilon \to 0}(I_{C_a}) = -\pi i\, \text{Residue}\,[f_1(z)]_{z=a} = -\pi i e^{i\lambda a} \Rightarrow I_A = \pi i e^{i\lambda a}.$$

Similarly, the contour integration over C_B, which contains no singularities, yields:

$$I_B + \lim_{\epsilon \to 0}(I_{C_a'}) + \lim_{R \to \infty}(I_{C_R'}) = 0 \Rightarrow I_B = -\lim_{\epsilon \to 0}(I_{C_a'}),$$

where the integral $I_{C_R'}$ on the lower half-plane is eliminated by Jordan's lemma. The other integral at the singularity on the x-axis gives

$$\lim_{\epsilon \to 0}(I_{C_a'}) = \pi\mathrm{i}\,\text{Residue}\,[f_2(z)]_{z=a} = \pi\mathrm{i}\mathrm{e}^{-\mathrm{i}\lambda a} \Rightarrow I_B = -\pi\mathrm{i}\mathrm{e}^{-\mathrm{i}\lambda a}.$$

In both cases, the integrals around the singularity at $x = a$ were evaluated in the usual way, using the results discussed above for avoiding simple poles. This is equivalent to evaluating the integrals I_A and I_B in the principal value sense. Note that for I_{C_a} the semicircle of radius ϵ is traversed in the *clockwise* sense in contour C_A but for $I_{C_a'}$ the semicircle of radius ϵ is traversed in the *anti-clockwise* sense in contour C_B. Combining the above results, we find

$$I = \frac{1}{2}[I_A + I_B] = \frac{1}{2}\left[\pi\mathrm{i}\mathrm{e}^{\mathrm{i}\lambda a} - \pi\mathrm{i}\mathrm{e}^{-\mathrm{i}\lambda a}\right] = -\pi\sin(\lambda a).$$

5.5 Scalar Riemann–Hilbert problems

Up to this point, we have considered singularities at single points on the complex plane. An interesting question is: What happens if a function looses its analyticity on a curve instead of a point? It turns out that the mathematical machinery needed for analyzing such functions gives rise to the so-called Riemann–Hilbert formalism, which we briefly introduce in this section.

A prototypical function that is analytic everywhere in the complex plane, including infinity, except on a curve L is given by the following integral:

$$F(z) = \int_L \frac{f(s)}{s-z}\mathrm{d}s, \quad z \in \mathbb{C}/L, \tag{5.17}$$

where the notation $\mathbb{C}/L$ refers to all points on the complex plane except those on the curve L. We first assume that the function $f(s)$ is analytic in the neighborhood of L. The curve L can be a sufficiently smooth arc in the complex plane or a sufficiently smooth closed curve. We also define the domains labeled '+' and '−' on either side of the curve L, as shown in Fig. 5.4: the '+' domain is on the *left* of the increasing direction of L, and the '−' domain is on the *right* of L. With this choice, if the curve is closed and is traversed in the anti-clockwise direction, the '+' domain is *inside* and the '−' domain is *outside* the curve. Clearly, as long as z is not on L, the integral in Eq. (5.17) is well-defined and it gives rise to an analytic function vanishing at infinity. Indeed, using the Laurent expansion

$$\frac{1}{s-z} = -\frac{1}{z}\left[1 + \frac{s}{z} + \mathcal{O}\left(\frac{1}{z^2}\right)\right], \quad z \to \infty,$$

it follows that

$$F(z) = -\frac{1}{z}\int_L f(s)\mathrm{d}s + \mathcal{O}\left(\frac{1}{z^2}\right). \tag{5.18}$$

Thus, $F(z)$ is of the form $c/z + \mathcal{O}(1/z^2)$ as $z \to \infty$.

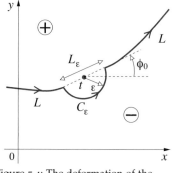

Figure 5.4: The deformation of the curve L in the neighborhood of the point $z = t$ that lies on L, into $(L - L_\epsilon) \cup C_\epsilon$ in order to skip this point. The domains '+' and '−' are also shown.

Regarding the function $F(z)$, it is natural to pose the following two questions:

(i) Does the function $F(z)$ have a limit as z approaches L along a non-tangential curve in the '+' domain? Similarly, does $F(z)$ have a limit as z approaches L along a non-tangential curve in the '−' domain?

(ii) If these limits exist, can they be computed?

The answers to both questions are affirmative. In what follows we will denote these limits by $F^{(+)}(z)$ and $F^{(-)}(z)$, respectively. Furthermore, we will compute them in the very special case that $f(s)$ is analytic in the neighborhood of the curve L.

The main difficulty with the above limit is that if z becomes t, where t is on the curve L, then the integrand is meaningless. However, if $f(s)$ is locally analytic we can solve this problem by using contour deformation. We deal with the case of the function $F^{(+)}(z)$ first, which is defined in the '+' domain: We can replace L by the contour $(L - L_\epsilon) \cup C_\epsilon$, where L_ϵ is a small segment of the curve with length 2ϵ centered at t, and C_ϵ is a semicircle of radius ϵ centered at t, as shown in Fig. 5.4. Since the point t is not on the new contour because it has been "skipped over" by the semicircle C_ϵ of radius ϵ, the limit of the integrand as z approaches t is non-singular. Notice that our choice of the semicircle makes sure that it lies in the *complementary* domain, in this case the '−' domain. This is a crucial choice: since we are working with a function defined on the '+' domain, the value $z = t$ *can* be approached in the '+' domain but *cannot* be approached in the '−' domain. Thus, our choice to place the semicircle C_ϵ in the complementary domain ensures that in the limit $\epsilon \to 0$ the singularity at $z = t$ is avoided by using the deformed path. Along this semicircle, we can write the complex variable as $\zeta = \epsilon \exp[i\phi]$, where ζ denotes a complex extension of s, namely in the neighborhood of t we take

$$s = t + \zeta \Rightarrow ds = d\zeta = i\epsilon e^{i\phi} d\phi, \quad \phi_0 - \pi \le \phi \le \phi_0.$$

Using the method of skipping a simple pole described in Section 5.3.4, we find

$$F^{(+)}(t) = \lim_{\epsilon \to 0} \int_{\phi_0 - \pi}^{\phi_0} \frac{f(t + \epsilon e^{i\phi})}{\epsilon e^{i\phi}} i\epsilon e^{i\phi} d\phi + \lim_{\epsilon \to 0} \int_{L_\epsilon} \frac{f(\tau)}{\tau - t} d\tau,$$

where the second integral is the usual principal value integral. Taking the limit $\epsilon \to 0$, we obtain

$$F^{(+)}(t) = i\pi f(t) + \fint_L \frac{f(\tau)}{\tau - t} d\tau. \tag{5.19}$$

For the case of $F^{(-)}(z)$, skipping the point t by a semicircle that lies entirely in the '+' domain,

$$s = t + \zeta \Rightarrow ds = d\zeta = i\epsilon e^{i\phi} d\phi, \quad \phi_0 + \pi \ge \phi \ge \phi_0,$$

we obtain

$$F^{(-)}(t) = -i\pi f(t) + \fint_L \frac{f(\tau)}{\tau - t} d\tau, \tag{5.20}$$

because the semicircle is now traversed in the *clockwise* direction.

Remarkably, the above expressions for $F^{(\pm)}(t)$ are valid even if the function $f(s)$ is *not* locally analytic. Actually, it does not even need to be differentiable; it suffices to belong to the so-called class of Hölder functions which are "better" than continuous but not "as good" as differentiable functions. If $F(z)$ were analytic, then $F^{(+)}(z) = F^{(-)}(z)$. Hence, the difference

$$\Delta F(z) \equiv F^{(+)}(z) - F^{(-)}(z)$$

measures the departure of the function from analyticity.

The problem of constructing a sectionally analytic function $F(z)$, vanishing at infinity, from the knowledge of the "jump" $\Delta F(z)$ is known as a scalar Riemann–Hilbert (RH) problem.

Subtracting Eq. (5.20) from Eq. (5.19) we find

$$F^{(+)}(t) - F^{(-)}(t) = 2\pi i f(t) \Rightarrow F(z) = \frac{1}{2\pi i} \int_L \frac{F^{(+)}(\tau) - F^{(-)}(\tau)}{\tau - t} d\tau. \quad (5.21)$$

It should be noted that if L is the real axis, then the functions $F^{(\pm)}(t)$ take the form:

$$F^{(\pm)}(x) = \lim_{\epsilon \to 0} \left[\int_{-\infty}^{\infty} \frac{f(\xi)}{\xi - (x \pm i\epsilon)} d\xi \right]. \quad (5.22)$$

Example 5.6: Find the sectionally analytic function vanishing at infinity, whose jump across the real axis is given by $\sin(x)/x$.

Let $F^{(+)}(z)$ and $F^{-)}(z)$ denote this function in the upper and lower complex plane, respectively. Using the familiar expression for the $\sin(s)$

$$\sin(s) = \frac{e^{is} - e^{-is}}{2i},$$

using the equation $\Delta F(x) = \sin(x)/x$, Eq. (5.21) implies that the function $f(s)$ must be

$$f(s) = \frac{\Delta F(s)}{2\pi i} = \frac{1}{2\pi i} \frac{e^{is} - e^{-is}}{2is}.$$

Thus, we can write the part of the function $F(z)$ that is analytic on either side of the real axis as

$$F^{(\pm)}(z) = \frac{1}{2\pi i} \int_{-\infty}^{\infty} \frac{e^{is} - e^{-is}}{2is} \frac{ds}{s - z},$$

where the '+' sign refers to $\text{Im}[z] > 0$ and the '−' sign to $\text{Im}[z] < 0$. In order to compute this integral by contour integration, we "complexify" s and split it into two terms, the first containing the exponential $\exp[is]$ and the second containing the exponential $\exp[-is]$, so that we can close the contour either in the upper or in the lower half of the complex plane. Before splitting, we add and subtract 1 so that we can retain the removable nature of the singularity at the origin:

$$F^{(\pm)}(z) = \frac{1}{2\pi i} \int_{-\infty}^{\infty} \frac{e^{i\zeta} - 1}{2i\zeta} \frac{d\zeta}{\zeta - z} - \frac{1}{2\pi i} \int_{-\infty}^{\infty} \frac{e^{-i\zeta} - 1}{2i\zeta} \frac{d\zeta}{\zeta - z}.$$

Thus, we have to compute the following two integrals:

$$I_1 = \int_{-\infty}^{\infty} f_1(\zeta) d\zeta, \quad \text{where } f_1(\zeta) = \frac{e^{i\zeta} - 1}{2i\zeta(\zeta - z)},$$

$$I_2 = \int_{-\infty}^{\infty} f_2(\zeta)\mathrm{d}\zeta, \quad \text{where } f_2(\zeta) = \frac{\mathrm{e}^{-\mathrm{i}\zeta} - 1}{2\mathrm{i}\zeta(\zeta - z)}.$$

We discuss first the case $\mathrm{Im}[z] > 0$. The integral I_1 can be performed by using contour integration, with the contour C_1 shown in the diagram below, with the semicircle C_R closing in the upper half-plane, so that the exponential $\exp[\mathrm{i}\zeta]$ gives a vanishing contribution in the limit $R \to \infty$. We note that the value $\zeta = 0$ is a removable singularity, so it does not need to be excluded from the path L along the real axis. The contour C_1 then encloses only the singularity at $\zeta = z$, and the Residue theorem gives for this integral

$$I_1 = 2\pi\mathrm{i}\, \mathrm{Residue}\,[f_1(\zeta)]_{\zeta=z} = 2\pi\mathrm{i}\left(\frac{\mathrm{e}^{\mathrm{i}z} - 1}{2\mathrm{i}z}\right).$$

The integral I_2 must be computed by the contour labeled C_2 which closes in the lower half-plane, so that the exponential $\exp[-\mathrm{i}\zeta]$ gives a vanishing contribution in the limit $R \to \infty$. The contour C_2 encloses no singularities, since $\mathrm{Im}[z] > 0$ and therefore for this case $I_2 = 0$. Combining results we arrive at

$$F^{(+)}(z) = \frac{\mathrm{e}^{\mathrm{i}z} - 1}{2\mathrm{i}z}.$$

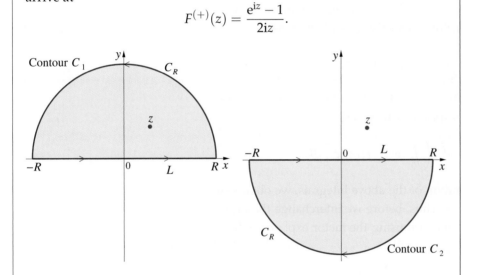

By similar steps we find that in the case when the point z lies in the lower half-plane, $\mathrm{Im}[z] < 0$, the first integral vanishes and the second integral gives

$$I_2 = -2\pi\mathrm{i}\, \mathrm{Residue}\,[f_2(\zeta)]_{\zeta=z} = -2\pi\mathrm{i}\left(\frac{\mathrm{e}^{-\mathrm{i}z} - 1}{2\mathrm{i}z}\right),$$

where the extra minus sign comes from the fact that contour C_2 is traversed in the *clockwise* direction, that is, in the opposite direction than the normal. Combining these results we arrive at

$$F^{(-)}(z) = \frac{\mathrm{e}^{-\mathrm{i}z} - 1}{2\mathrm{i}z}.$$

Clearly, both $F^{(+)}(z)$ and $F^{(-)}(z)$ are analytic functions which vanish for large $|z|$ in the upper and lower half complex plane, respectively.

Moreover, the jump as the real axis is approached from above and from below is

$$\Delta F(x) = \lim_{\mathrm{Im}[z] \to 0} [F^{(+)}(z) - F^{(-)}(z)]$$

$$= \frac{e^{ix} - 1}{2ix} - \frac{e^{-ix} - 1}{2ix} = \frac{e^{ix} - e^{-ix}}{2ix} = \frac{\sin(x)}{x}.$$

5.5.1 Derivation of the inverse Fourier transform

The formalism discussed above is useful in deriving the inverse of an important transform, namely the inverse Fourier transform.

Let $f(x)$ be a function of the Hölder type for $x \in [0, X]$ where $X > 0$. Let $\hat{f}(k)$ denote the following function:

$$\hat{f}(k) = \int_0^X e^{-ik\xi} f(\xi) d\xi, \quad k \in \mathbb{C}. \tag{5.23}$$

This expression is a generalization of the Fourier transform which will be discussed in detail in Chapter 7; the usual Fourier transform involves the lower limit of the integral being $-\infty$, $X \to \infty$ and $k \in \mathbb{R}$. In what follows, we will derive the inverse of the Fourier transform, namely we will show that

$$f(x) = \frac{1}{2\pi} \int_{-\infty}^{\infty} e^{ikx} \hat{f}(k) dk, \quad x \in \mathbb{R}. \tag{5.24}$$

To prove this result, we multiply both sides of Eq. (5.23) by $\exp[ikx]$ and integrate the resulting expression with respect to k from 0 to ∞:

$$\int_0^{\infty} e^{ikx} \hat{f}(k) dk = \int_0^{\infty} e^{ikx} \left(\int_0^X e^{-ik\xi} f(\xi) d\xi \right) dk.$$

If we interchange the order of integration of the above integrals, we obtain an integral in k which does not converge. Thus, before we interchange the order of integration we first regulate the k-integral using the factor $\exp[-\epsilon k]$ where $\epsilon > 0$:

$$\int_0^{\infty} e^{ikx} \hat{f}(k) dk = \lim_{\epsilon \to 0} \left[\int_0^{\infty} e^{ikx - \epsilon k} \left(\int_0^X e^{-ik\xi} f(\xi) d\xi \right) dk \right]$$

$$= \lim_{\epsilon \to 0} \left[\int_0^X f(\xi) \left(\int_0^{\infty} e^{ik(x - \xi + i\epsilon)} dk \right) d\xi \right].$$

The above k-integral can be computed explicitly. Hence,

$$\int_0^{\infty} e^{ikx} \hat{f}(k) dk = \lim_{\epsilon \to 0} \left[(-i) \int_0^X \frac{f(\xi)}{\xi - (x + i\epsilon)} d\xi \right].$$

By similar steps we obtain

$$\int_{-\infty}^0 e^{ikx} \hat{f}(k) dk = \lim_{\epsilon \to 0} \left[i \int_0^X \frac{f(\xi)}{\xi - (x - i\epsilon)} d\xi \right].$$

Taking the limit $\epsilon \to 0$ for the first integral, with $0 < k < \infty$, and using the formula of Eq. (5.19), we obtain

$$\int_0^{\infty} e^{ikx} \hat{f}(k) dk = -i \left[i\pi f(x) + \int_0^X \frac{f\xi)}{\xi - x} d\xi \right],$$

while for the second integral, with $-\infty < k < 0$, taking the limit $\epsilon \to 0$ and using the formula of Eq. (5.20), we obtain

$$\int_{-\infty}^{0} e^{ikx} \hat{f}(k) dk = i \left[-i\pi f(x) + \int_{0}^{X} \frac{f\xi)}{\xi - x} d\xi \right].$$

Adding the above expressions, we get the desired result, namely Eq. (5.24).

The above considerations suggest that the RH problems are based on a very natural way of "loosing analyticity", thus they should arise often in applications. Surprisingly, for many years the only applications of RH problems occurred in connection with the so-called Wiener–Hopf technique which is a particular case of a RH problem. However, for the last 50 years they have appeared in a great variety of applications, including: the analysis of nonlinear integrable ordinary differential equations like the classical Painlevé equations; the solution of nonlinear integrable partial differential equations; random matrices and orthogonal polynomials; and in deep problems arising in statistics and in medical imaging (see the Section 5.8).

5.6   Branch points and branch cuts

There exists an important type of singularities of functions of complex variables which gives rise to multi-valuedness. These singularities are called **branch points**. To illustrate the nature of a branch point we consider the function $z^{1/n}$, n integer, discussed earlier:

$$w = z^{1/n} = \left(re^{i(\phi + 2k\pi)} \right)^{1/n} = r^{1/n} e^{i(\phi + 2k\pi)/n}, \quad k = 0, 1, \ldots, n - 1.$$

For any point other than zero, this function takes n different values depending on the choice of k. Only for $z = 0$ it does not matter what the value of k is. In this sense, $z = 0$ is a special point of the function w. We can also look at $z = 0$ from a different perspective: suppose that we are at some point $z = r \neq 0$ and we start increasing the argument from $\phi = 0$ keeping r constant. All the points we encounter while doing so are different points on the complex plane for $0 \leq \phi < 2\pi$, but when ϕ hits the value 2π we get back to the same point where we had started. If we continue increasing the value of ϕ we will hit this point again every time ϕ is an integral multiple of 2π. When z is raised to $1/n$ power, these values of z will in general be different from each other, since their argument will be divided by n. Thus, for the same value of the complex variable z we get multiple values for the function w. *This only happens if we go around the point $z = 0$ more than once.* If we go around a path that encircles any other point but not zero, the argument of the variable will not keep increasing, and therefore this problem will not arise.

The same problematic behavior was encountered for the natural logarithm: if the argument of the variable z is allowed to increase without limit, then both the argument and the magnitude of the function $w = \ln(z)$ change when we go through points on the complex plane whose argument differs by a multiple of 2π, even though these are *the same* points on the complex plane. It is useful to notice that in the case of the logarithm because of the relation

$$\ln \left(\frac{1}{z} \right) = -\ln(z),$$

the behavior of the function in the neighborhood of the points z and $1/z$ is the same, with an overall minus sign. Thus, if $z = 0$ is a special point in the sense described above, the same is valid for the point $z \to \infty$. We conclude that the logarithm has two branch points, $z = 0$ and $z \to \infty$. In fact, since as shown earlier all non-integer powers of complex numbers can be expressed with the help of the logarithm, we conclude that the functions involving non-integer powers of z have the same two branch points, namely 0 and ∞. These special points are values of the variable where the quantity whose logarithm or the base of the of the non-integer power we want to evaluate, either vanishes or becomes infinite. For example, the function

$$f(z) = \ln\left(\frac{\sqrt{z^2 - 1}}{\sqrt{z^2 + 1}}\right)$$

has branch points at $z = \pm 1$, where the content of the parenthesis becomes zero, and at $z = \pm i$ where the content of the parenthesis becomes infinite. As other examples the function

$$f(z) = \sqrt{(z - a)}, \quad a \text{ finite constant}$$

has branch points $z = a$ and $z \to \infty$, while the function

$$f(z) = \left(\frac{1}{z} - b\right)^{\pi}, \quad b \neq 0$$

has branch points $z = 1/b$ and $z = 0$.

Recall that one of the basic requirements of a function is single-valuedness. Thus, in order to properly define functions involving branch points we must eliminate the multi-valuedness associated with these points. This can be achieved through the introduction of the so-called **branch cuts**. These are cuts the complex plane which join two branch points: we imagine that with the help of a pair of "mathematical scissors" we literally cut the complex plane along a line so that it is not allowed to cross from one side of the line to the other. In the case of the logarithm one possible cut would be along the x-axis form $x = 0$ to $x \to \infty$. This choice restricts the argument ϕ in the range of values $[0, 2\pi)$, where we imagine that we can approach the cut from one side but not from the other. That is, we can reach the values of z that lie on the x-axis approaching from above the axis, but we cannot reach them from below the axis. This is a matter of convention, and the reverse would have been equally suitable, namely we could approach the values of z that lie on the x-axis from below the axis but not from above the axis; this would be equivalent to restricting the argument ϕ to the range $(0, 2\pi]$. The insertion of a branch cut eliminates any ambiguity in the values that the function $\ln(z)$ is allowed to take!

There are usually many possible choices for a branch cut. In fact, for the case of the logarithm and of a non-integer power, any line joining the points o and ∞ is a legitimate choice, as long as it does not cross itself; in the latter case a portion of the complex plane would be completely cut off (it could not be reached from the rest of the plane), which is not allowed to happen. Each such choice of branch cut restricts the range of the argument of z to one set of values spanning an interval of 2π. For example, if we had chosen the positive y-axis as the branch cut, the allowed values of ϕ would be in the interval $[-3\pi/2, \pi/2)$, if we had chosen the negative x-axis as the branch cut, the values of ϕ would be in the interval $[-\pi, \pi)$, etc. Notice that the introduction of a branch cut that restricts the values of the argument to $0 \leq \phi < 2\pi$ is equivalent to taking the principal value of the argument. There are also more complicated possible choices, for which the range of ϕ may be different for each value of the radius r, but it always spans an interval of 2π for a given value of r.

Example 5.7: We give an example of a type of contour integral which requires a branch cut:

$$I = \int_0^\infty \frac{x^{-\lambda}}{x+1} dx, \quad 0 < \lambda < 1, \tag{5.25}$$

which can be expressed as part of a contour integral of a complex function. Note that raising the variable x to the non-integer power λ implies that we will need to introduce a branch cut in the complex plane from $z = 0$ to $z \to \infty$. We therefore consider the function

$$f(z) = \frac{z^{-\lambda}}{z+1},$$

and the contour shown in the diagram below, with the branch cut along the real axis, from $z = x = 0$ to $z = x \to \infty$. The contour C consists of

$$C = C_R \cup L' \cup C_\epsilon \cup L,$$

namely, the union of the circle C_R of radius R centered at the origin, the circle C_ϵ of radius ϵ centered at the origin, and the two horizontal line segments infinitesimally above (L) and below (L') the real axis, extending from ϵ to R; this contour is shown in the diagram below, with the implied limits $\epsilon \to 0$ and $R \to \infty$.

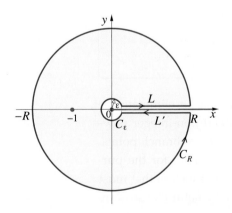

The point $z = 0$ is a singularity of the integrand (a branch point), so it must be avoided and this is achieved through the small circle of radius ϵ around it. The integral along the line of the contour which lies just above the real axis, L, approaches the value of the desired integral I. For the line of the contour which lies just below the real axis, L', where the argument of the complex variable z approaches the value 2π, the corresponding integral I' is obtained as follows:

$$\phi \to 2\pi : z = xe^{i2\pi} \Rightarrow \frac{z^{-\lambda}}{z+1} = \frac{x^{-\lambda}}{x+1}e^{-i2\pi\lambda}, \quad dz = dxe^{i2\pi} = dx$$

$$\Rightarrow \quad I' = \int_{\infty}^{0} \frac{z^{-\lambda}}{z+1}\,dz = -\int_{0}^{\infty} \frac{x^{-\lambda}e^{-i2\pi\lambda}}{x+1}\,dx = -e^{-i2\pi\lambda}I.$$

The circle of radius ϵ around the origin yields a zero contribution, with $z = \epsilon\exp[i\phi]$:

$$I_{\epsilon} = -\int_{0}^{2\pi} \frac{(\epsilon e^{i\phi})^{-\lambda}}{\epsilon e^{i\phi}+1}\epsilon i e^{i\phi}\,d\phi = -i\epsilon^{1-\lambda}\int_{0}^{2\pi} \frac{e^{i\phi(1-\lambda)}}{\epsilon e^{i\phi}+1}\,d\phi \to 0 \text{ for } \epsilon \to 0,$$

since the contribution of the integral over ϕ for any ϵ is a finite quantity and $1 - \lambda > 0$. By analogous considerations, the contribution from the circle of radius R around the origin vanishes:

$$z = Re^{i\phi}, \quad dz = Rie^{i\phi}d\phi$$

$$\Rightarrow \quad I_R = \int_{0}^{2\pi} \frac{(Re^{i\phi})^{-\lambda}}{Re^{i\phi}+1}Rie^{i\phi}\,d\phi = iR^{1-\lambda}\int_{0}^{2\pi} \frac{e^{i\phi(1-\lambda)}}{Re^{i\phi}+1}\,d\phi$$

$$\Rightarrow \quad |I_R| \le R^{1-\lambda}\int_{0}^{2\pi} |Re^{i\phi}+1|^{-1}\,d\phi \le \frac{R^{1-\lambda}}{R-1}2\pi$$

$$\Rightarrow \quad \lim_{R\to\infty}|I_R| \le \lim_{R\to\infty}(2\pi R^{-\lambda}) = 0,$$

where we have used the ML inequality, Eq. (4.43), to find an upper bound for the integral $|I_R|$ before taking the limit $R \to \infty$. Finally, the entire contour encloses only one singularity, namely the point $z = -1$, which evidently is a simple pole with a residue given by $(e^{i\pi})^{-\lambda} = e^{-i\lambda\pi}$. Combining these results we find:

$$I + I' + \lim_{\epsilon\to0}(I_{\epsilon}) + \lim_{R\to\infty}(I_R) = 2\pi i \text{ Residue }[f(z)]_{z=-1}$$

$$\Rightarrow \quad I(1 - e^{-i2\pi\lambda}) = 2\pi i e^{-i\lambda\pi} \Rightarrow I = \frac{\pi}{\sin(\lambda\pi)}.$$

With the introduction of branch cuts we have been able to eliminate the problem of multi-valuedness for the natural logarithm and non-integer powers. This approach can be applied to any function that involves branch points. Since the logarithm and non-integer powers are in fact related, for the purposes of the present discussion the only truly multi-valued function that must be fixed by the introduction of branch cuts, essentially, is the natural logarithm.

5.7 The source–sink function and related problems

In several types of problems, we are interested in describing a source or a sink of physical quantity. This may involve, for instance, the circumferential fluid flow generated around a source or sink, or the behavior of fields, for example the electromagnetic field arising from charges or currents, or the elastic-energy field arising from dislocations in crystals. As a case in point, the function $\pm\ln(z - z_0)$ on the complex plane is often used to simplify the description of a radial or circumferential flow from a source (with the + sign) or a sink (with the − sign) situated at $z = z_0$. A combination that includes this function with the plus and the minus sign, each component centered at a different point, namely at $z = a$ and $z = -a$, is called the source–sink pair:

$$f(z) = \ln(z + a) - \ln(z - a) = \ln\left(\frac{z + a}{z - a}\right) = \ln\left(\frac{\rho_1}{\rho_2}\right) + i(\phi_1 - \phi_2), \quad (5.26)$$

where $\rho_1 = |z + a|, \phi_1 = \text{Arg}[z + a]$ and $\rho_2 = |z - a|, \phi_2 = \text{Arg}[z - a]$, as shown in Fig. 5.5. To handle this function, we need to introduce branch cuts because there are at least two branch points, $z = \pm a$, where the function is non-analytic. We notice that for a closed path that encircles both or neither of $z = +a$ and $z = -a$ the quantity $(\phi_1 - \phi_2)$ takes the same value at the end of the path as at the beginning, that is, it does not increase by a factor of 2π even though each of the angles ϕ_1 and ϕ_2 increases by 2π. Thus, such a path would not lead to multi-valuedness. For a closed path that encircles only one of $z = a$ or $z = -a$ the quantity $(\phi_1 - \phi_2)$ increases by 2π at the end of the path relative to its value at the beginning. These facts are established by inspecting Fig. 5.5.

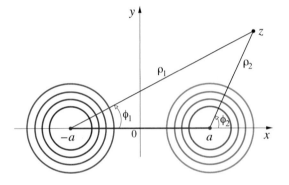

Figure 5.5: Branch cut for the function $f(z) = \ln(z + a) - \ln(z - a)$, along the real axis $-a \leq x \leq +a$, represented by the thick red line, joining the two branch point $z = \pm a$. The concentric circles centered at $z = \pm a$ represent equipotential curves of the source (magenta shades) and the sink (green shades).

The function $f(z)$ defined in Eq. (5.26) has no other branch points beyond $z = \pm a$. Indeed, it is analytic for any finite $z \neq \pm a$, and in the limit $z \to \infty$ we find

$$\lim_{z \to \infty}[f(z)] = \lim_{z \to \infty}\left[\ln\left(\frac{z + a}{z - a}\right)\right] = \ln(1) = 0.$$

Another way to establish that the only branch points are at $z = \pm a$ is to evaluate the derivative $f'(z)$ of the function:

$$f'(z) = \frac{1}{z + a} - \frac{1}{z - a},$$

which is well-defined everywhere except for $z = \pm a$. Consequently, we need a branch cut joining the points $z = \pm a$. Any such path will do, and it does not have to pass through $z \to \infty$. A possible choice for a branch cut is shown in Fig. 5.5.

Integrals that involve functions with branch points similar to those of the source-sink function, containing the logarithm or roots of polynomials, arise often in the context of continuum elasticity theory. For instance, the calculation of the elastic strain energy associated with a dislocation gives rise to the following integral[1]:

$$\int_{-a}^{a} \frac{\ln(1 + x^2)}{1 + x^2}\,dx, \quad a > 1.$$

[1] See J.P. Hirth and J. Lothe, *Theory of Dislocations*, 2nd Ed., p. 227 (Krieger Publishing Co., 1992).

Recasting this problem as a contour integral in the complex plane implies that the integrand will contain the function $\ln(1 + z^2)$ which has branch points at $z = \pm i$ (in this particular case the branch points are also roots of the denominator).

As a representative example of how these types of integrals can be computed we will prove the following result[2]:

$$I_{\text{real}} \equiv \int_{-1}^{1} \frac{\sqrt{1 - x^2}}{1 + x^2} \, dx = \pi(\sqrt{2} - 1). \tag{5.27}$$

In order to derive Eq. (5.27), we consider the contour shown in Fig. 5.6, consisting of

$$C = C_R \cup L_1 \cup L \cup C_2 \cup L' \cup C_1 \cup L_2,$$

namely, the union of the circle C_R of radius R centered at the origin, with $R \to \infty$, the two circles C_1, C_2 of radius ϵ centered at $x = -1$ and $x = +1$ respectively, with $\epsilon \to 0$, the two vertical line segments labeled L_1, L_2 and the two horizontal line segments infinitesimally above (L) and below (L') the real axis, extending from $-1 + \epsilon$ to $1 - \epsilon$. To compute the desired integral I, we consider the integral along the closed contour C,

$$I_C = \int_C f(z) \, dz, \quad \text{where } f(z) = \frac{\sqrt{z^2 - 1}}{z^2 + 1}.$$

The contributions of the vertical line segments L_1, L_2 cancel each other so we neglect them. Hence, the result of integration over the entire contour is given by

$$I_C = I + I' + I_{C_1} + I_{C_2} + I_{C_R},$$

with the part corresponding to the line segment L giving the integral I and the part corresponding to the line segment L' giving the integral I'. It is clear that to evaluate I_C we need to apply Cauchy's theorem in the shaded region of Fig. 5.6, thus the function $\sqrt{z^2 - 1} = \sqrt{z - 1}\sqrt{z + 1}$ must be analytic in this region. Accordingly, due to the presence of the square-root function in the integrand, we have chosen the branch consisting of a cut along the real axis connecting the points -1 and $+1$.

We deal first with the contributions from the horizontal line segments above and below the real axis, labeled I, I'. The function $\sqrt{z^2 - 1}$ is the product of the functions $\sqrt{z + 1}$ and $\sqrt{z - 1}$. Thus, if we choose the branch cuts $(-1, \infty)$ and $(1, \infty)$ respectively for these two functions, these branch cuts cancel from $(1, \infty)$ and the function $\sqrt{z^2 - 1}$ has a branch cut in $(-1, 1)$. This can be checked explicitly: let

$$(z + 1) = \rho_1 e^{i\phi_1}, \quad (z - 1) = \rho_2 e^{i\phi_2}, \quad 0 < \phi_1, \phi_2 < 2\pi, \tag{5.28}$$

as shown in Fig. 5.6. Then:

$$\sqrt{z^2 - 1} = \sqrt{\rho_1 \rho_2} \, e^{i(\phi_1 + \phi_2)/2}, \tag{5.29}$$

and the values of $\phi = (\phi_1 + \phi_2)/2$ can easily be computed from the definition of ϕ_1, ϕ_2. Specifically, approaching the real axis from above ($y > 0$) we have,

$$\text{for } x < -1: \quad \phi_1 = \phi_2 = \pi;$$
$$\text{for } -1 < x < 1: \quad \phi_1 = 0, \phi_2 = \pi;$$
$$\text{for } x > 1: \quad \phi_1 = \phi_2 = 0.$$

[2] This integral can also be computed by numerical techniques, specifically by using Monte Carlo integration (see Section 14.9).

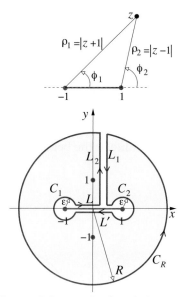

Figure 5.6: Contour for the calculation of the integral of Eq. (5.27).

Similarly, approaching the real axis from below ($y < 0$) we have

$$\text{for } x < -1: \quad \phi_1 = \phi_2 = \pi;$$
$$\text{for } -1 < x < 1: \quad \phi_1 = 2\pi, \phi_2 = \pi;$$
$$\text{for } x > 1: \quad \phi_1 = \phi_2 = 2\pi.$$

For $|x| < 1$, $|z^2 - 1| = 1 - x^2$, hence for z above and below the branch cut, $\sqrt{z^2 - 1}$ equals $i\sqrt{1 - x^2}$ and $-i\sqrt{1 - x^2}$, respectively. The factor $x^2 + 1$ in the denominator is even in x and has no roots along these paths. Therefore, integration along the horizontal line segment that lies above the real axis gives the desired integral I defined in Eq. (5.27) times a factor of i, that is, $I = iI_{\text{real}}$. Similarly, integration along the horizontal line segment that lies below the real axis, taking into account that it is performed with x running from $+1 - \epsilon$ to $-1 + \epsilon$, gives the same result, that is, $I' = I = iI_{\text{real}}$.

The contributions around I_{C_1}, I_{C_2} vanish. For example, for C_1 we have: $z = 1 + \epsilon \exp(i\phi)$, and

$$\left| \int_{C_1} \frac{\sqrt{z^2 - 1}}{z^2 + 1} dz \right| \leq \int_0^{2\pi} \frac{\sqrt{|z - 1|}\sqrt{|z + 1|}}{|2 + \epsilon^2 e^{2i\phi} + 2\epsilon e^{i\phi}|} \epsilon d\phi \leq \int_0^{2\pi} \frac{\sqrt{\epsilon^2 + 2\epsilon}}{2 - \epsilon^2 - 2\epsilon} \epsilon d\phi \to 0,$$

as $\epsilon \to 0$. The contribution around I_{C_R} can be computed explicitly: For large R,

$$z = Re^{i\phi}, \quad \phi_1 \approx \phi_2 \approx \phi, \quad \sqrt{z^2 - 1} = R \exp[i(\phi_1 + \phi_2)/2] \approx Re^{i\phi}.$$

We also have, using the appropriate Taylor expansions for functions of z in the limit $|z| = R \to \infty$, and therefore $1/|z| = 1/R \to 0$:

$$\frac{\sqrt{z^2 - 1}}{z^2 + 1} = \left[z^2 \left(1 - \frac{1}{z^2} \right) \right]^{1/2} \left[z^2 \left(1 + \frac{1}{z^2} \right) \right]^{-1}$$

$$= \frac{1}{z} \left(1 - \frac{1}{2z^2} + \cdots \right) \left(1 - \frac{1}{z^2} + \cdots \right) = \frac{1}{z} \left(1 - \frac{3}{2z^2} + \cdots \right).$$

Thus,

$$\lim_{R \to \infty} \left[\int_{C_R} \frac{\sqrt{z^2 - 1}}{z^2 + 1} dz \right] = \lim_{R \to \infty} \left[\int_0^{2\pi} \frac{1}{Re^{i\phi}} iRe^{i\phi} d\phi \right] = 2\pi i.$$

Combining the above results, the integral over the whole contour C gives

$$2i \int_{-1}^{1} \frac{\sqrt{1 - x^2}}{1 + x^2} dx + 2\pi i = 2\pi i \left\{ \text{Residue} \left[f(z) \right]_{z=i} + \text{Residue} \left[f(z) \right]_{z=-i} \right\}. \tag{5.30}$$

The final step involves evaluating the residues at $z = \pm i$:

$$\text{at } z = i : \rho_1 = \rho_2 = \sqrt{2}, \phi_1 = \pi/4, \phi_2 = 3\pi/4,$$

which, from the relation of Eq. (5.29) yields the following:

$$z = i : \sqrt{z^2 - 1} = \sqrt{\rho_1 \rho_2} e^{i(\phi_1 + \phi_2)/2} = \sqrt{2} e^{i\pi/2} = \sqrt{2}i$$

$$\Rightarrow \left(\frac{\sqrt{z^2 - 1}}{2z} \right)_{z=i} = \frac{\sqrt{2}}{2}.$$

Similarly,

$$\text{at } z = -i : \rho_1 = \rho_2 = \sqrt{2}, \phi_1 = 7\pi/4, \phi_2 = 5\pi/4,$$

which, from the relation of Eq. (5.29) yields the following:

$$z = -i \; : \; \sqrt{z^2 - 1} = \sqrt{\rho_1 \rho_2} e^{i(\phi_1 + \phi_2)/2} = \sqrt{2} e^{i3\pi/2} = -\sqrt{2}i$$

$$\Rightarrow \left(\frac{\sqrt{z^2 - 1}}{2z} \right)_{z=-i} = \frac{\sqrt{2}}{2}.$$

Inserting the above two results in Eq. (5.30) yields Eq. (5.27).

5.8   Application: Medical imaging

Modern medical imaging techniques have had a truly transformative impact on medicine. In particular, the advent of *Computerized Tomography* (CT) made possible for the first-time direct images of brain tissue. Furthermore, the subsequent development of *Magnetic Resonance Imaging* (MRI) allowed striking discrimination between gray and white matter. These techniques have had a tremendous impact on neurology and neurosurgery. Although the first applications of CT and MRI were in the visualization of the brain, these techniques were later applied to many other areas of medicine, including oncology and cardiology. Indeed, it is impossible to think of medicine today without CT and MRI.

These important medical imaging techniques are based on the solution of certain mathematical problems, known as "inverse problems". For the case of CT, the associative inverse problem involves the determination of a function from the knowledge of its Radon transform,[3] namely of the integral of this function along all possible rays. The function under consideration is the so-called called *x-ray attenuation coefficient* of the tissue; this coefficient, which is related to the density of the medium, characterizes the amount of the energy absorbed when x-rays transverse this tissue. The measurement by the CT camera of the absorbed energy provides the value of the Radon transform.

Following further astonishing development, it is now possible to observe in real life the activation of specific parts of the brain. The associated medical imaging techniques are referred to as *functional imaging techniques*; they include *Positron Emission Tomography* (PET), *Single Photon Emission Computerized Tomography* (SPECT), and *functional Magnetic Resonance Imaging* (fMRI). Again, just like their predecessors, namely CT and MRI, the new techniques have a variety of important medical applications beyond neurology.

It turns out that the mathematical transform associated with PET is again the Radon transform. However, SPECT involves a certain complicated generalization of the latter transform known as the *Attenuated Radon Transform*. The inversion of this transform was constructed only in 2002[4] through the use of the new method for inverting integral transforms introduced in the Application section of Chapter 4, Section 4.8, which makes crucial use of Pompeiu's formula.[5]

In what follows, for simplicity, instead of deriving the inverse Attenuated Radon Transform we will use the new method to indicate how the inverse Radon Transform can be derived. Although the latter transform can be derived in a simpler way via the use of the two-dimensional Fourier transform, the derivation presented here has the advantage that can be easily extended to the case of the attenuated Radon transform.

[3] This transform and inverse were invented by the Austrian mathematician Johann Radon in 1917.

[4] R. G. Novikov, An inversion formula for the attenuated X-ray transformation, *Arkiv för Matematik* **40**, pp. 145–167 (2002).

[5] A. S. Fokas, I. M. Gelfand, Integrability of linear and nonlinear evolution equations, and the associated nonlinear Fourier transforms, *Letters of Mathematical Physics* **32**, pp. 189–210 (1994).

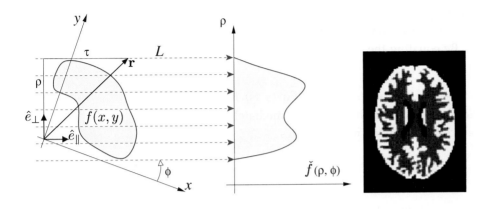

Figure 5.7: **Left**: The definition of the local coordinates ρ, τ, ϕ and unit vectors $\hat{e}_{\parallel}$, $\hat{e}_{\perp}$, parallel and perpendicular to the line L. **Right**: The Hoffman brain phantom (see text for details).

We first present the mathematical formulation of the Radon transform: A line L on the plane can be specified by the signed distance from the origin ρ, with $-\infty < \rho < \infty$, and the angle with the x-axis ϕ, with $0 \leq \phi < 2\pi$, see Fig. 5.7. We denote the corresponding unit vectors parallel and perpendicular to L by $\hat{e}_{\parallel}$ and $\hat{e}_{\perp}$, respectively. These vectors are given by

$$\hat{e}_{\parallel} = (\cos\phi, \sin\phi), \quad \hat{e}_{\perp} = (-\sin\phi, \cos\phi). \tag{5.31}$$

Every point $\mathbf{r} = (x, y)$ on L can be expressed in terms of the local coordinates (ρ, τ) via

$$\mathbf{r} = \rho\hat{e}_{\perp} + \tau\hat{e}_{\parallel}, \tag{5.32}$$

where τ denotes the arc length. The linear transformations that describe the rotation of coordinate axes in two dimensions by an angle ω were worked out in Section 2.2.1. Adopting the results of Eq. (2.12) and (2.13) with the identification $x' \to \tau$, $y' \to \rho$, $\omega \to \phi$, we obtain

$$\tau = y\sin\phi + x\cos\phi, \quad \rho = y\cos\phi - x\sin\phi, \tag{5.33}$$

and for the inverse transformation

$$x = \tau\cos\phi - \rho\sin\phi, \quad y = \tau\sin\phi + \rho\cos\phi. \tag{5.34}$$

We define the Radon transform $\check{f}(\rho, \phi)$ of a two-dimensional function $f(x, y)$ as the following line integral:

$$\check{f}(\rho, \phi) = \int_L f(x, y)\mathrm{d}s = \int_{-\infty}^{\infty} f(\tau\cos\phi - \rho\sin\phi, \tau\sin\phi + \rho\cos\phi)\mathrm{d}\tau, \tag{5.35}$$

for $0 \leq \phi < 2\pi$, $-\infty < \rho < \infty$.

The inversion of the Radon transform can be achieved by analyzing the following equation:

$$\left[\frac{1}{2}\left(k + \frac{1}{k}\right)\partial_x + \frac{1}{2i}\left(k - \frac{1}{k}\right)\partial_y\right]\mu(x, y, k) = f(x, y), \quad k \in \mathbf{C}. \tag{5.36}$$

This equation is the "auxiliary" equation needed for the inversion of the Radon transform; it is the analogue of Eq. (4.71) used in Section 4.8. As explained in Section 4.8, the analysis of this equation consists of two parts:

(i) Solving the direct problem, namely, assuming that f is a given function which decays for large values of x and y, construct a solution μ of Eq. (5.36) which is bounded for all complex values of k.

(ii) Solving the inverse problem, namely, by utilizing the fact that μ is bounded for all complex k, obtain a different expression for μ, which instead of depending on f it depends on $\check{f}$. By equating the above two expression it is possible to express f in terms of $\check{f}$.

The implementation of the step (i) is based on Pompeiu's formula, Eq. (4.55). In this connection, in order to rewrite Eq. (5.36) in a form that we can immediately use this powerful formula, we introduce the following new variables:

$$z = \frac{1}{2i}\left(k - \frac{1}{k}\right)x - \frac{1}{2}\left(k + \frac{1}{k}\right)y, \tag{5.37a}$$

$$\bar{z} = -\frac{1}{2i}\left(\bar{k} - \frac{1}{\bar{k}}\right)x + \frac{1}{2}\left(\bar{k} + \frac{1}{\bar{k}}\right)y. \tag{5.37b}$$

Employing the chain rule, we find

$$\partial_x = \frac{1}{2i}\left(k - \frac{1}{k}\right)\partial_z - \frac{1}{2i}\left(\bar{k} - \frac{1}{\bar{k}}\right)\partial_{\bar{z}}, \tag{5.38a}$$

$$\partial_y = -\frac{1}{2}\left(k + \frac{1}{k}\right)\partial_z - \frac{1}{2}\left(\bar{k} + \frac{1}{\bar{k}}\right)\partial_{\bar{z}}. \tag{5.38b}$$

Using Eq. (5.38b) we can rewire Eq. (5.36) in the form

$$\mu_{\bar{z}} = \frac{f}{v}, \quad |k| \neq 1, \tag{5.39}$$

where v, which depends only on $|k|$, is defined by

$$v(|k|) = \frac{1}{2i}\left(\frac{1}{|k|^2} - |k|^2\right). \tag{5.40}$$

We supplement Eq. (5.36) with the boundary condition

$$\mu = \mathcal{O}\left(\frac{1}{z}\right), \quad z \to \infty. \tag{5.41}$$

This condition implies that the boundary term in Pompeiu's formula, Eq. (4.55), vanishes, and hence Eq. (4.55) yields

$$\mu(x,y,k) = \frac{1}{2\pi i}\text{sgn}\left(\frac{1}{|k|^2} - |k|^2\right)\iint \frac{f(x',y')}{z' - z}dx'dy'. \tag{5.42}$$

The implementation of step (ii) involves the formulation of a Riemann–Hilbert (RH) problem. In this connection we note that Eq. (5.42) implies that if k is either inside or outside the unit circle, the only dependence of μ on k is through z and z'; thus, μ is a sectionally analytic function of k with a "jump" across the unit circle of the k-complex plane. Using straightforward but tedious calculations it can be shown that the limits as k approaches the unit circle from inside or outside can be computed; their difference is given by

$$\mu^{(+)} - \mu^{(-)} = -\frac{1}{i\pi}(\mathcal{H}\check{f})(\rho,\phi), \quad -\infty < \rho < \infty, \quad 0 \leq \phi < 2\pi,$$

where $\mathcal{H}$ denotes the Hilbert transform, namely

$$(\mathcal{H}\check{f})(\rho,\phi) \equiv \frac{1}{\pi}\int_{-\infty}^{\infty}\frac{\check{f}(r,\phi)}{r - \rho}dr, \quad -\infty < \rho < \infty, \quad 0 \leq \phi < 2\pi, \tag{5.43}$$

and $\fint$ denotes the Cauchy principal value integral. Also, Eq. (5.42) implies that μ vanishes as $k \to \infty$. Hence, using the solution of the RH problem obtained in Section 5.5 we find

$$\mu = -\frac{1}{2i\pi^2} \int_0^{2\pi} \frac{e^{i\phi}(\mathcal{H}\check{f})(\rho,\phi)}{e^{i\phi} - k} d\phi, \quad k \in \mathbb{C}, \ |k| \neq 1, \ \rho \in \mathbb{R}.$$

The large $|k|$ limit of this equation implies[6]

$$\mu = \left[\frac{1}{2\pi i} \int_0^{2\pi} e^{i\phi}(\mathcal{H}\check{f})(\rho,\phi) d\phi \right] \frac{1}{k} + \mathcal{O}\left(\frac{1}{k^2}\right), \quad k \to \infty. \qquad (5.44)$$

Substituting the above expression in Eq. (5.35) we find a formula for the inverse Radon transform:

$$f(x,y) = \frac{1}{4\pi i} (\partial_x - i\partial_y) \int_0^{2\pi} e^{i\phi}(\mathcal{H}\check{f})(\rho,\phi) \Bigg|_{\rho = y\cos\phi - x\sin\phi} d\phi. \qquad (5.45)$$

Equation (5.45) can be simplified by using the identity

$$(\partial_x - i\partial_y) = e^{-i\phi}(\partial_\tau - i\partial_\rho). \qquad (5.46)$$

In this way we obtain

$$f(x,y) = -\frac{1}{4\pi} \int_0^{2\pi} \left[\frac{\partial(\mathcal{H}\check{f})(\rho,\phi)}{\partial\rho} \right]_{\rho = y\cos\phi - x\sin\phi} d\phi, \qquad (5.47)$$

with $-\infty < x, y < \infty$.

The mathematical problem arising in clinical practice is the computation of the right-hand side of Eq. (5.47) from the knowledge of the function $\check{f}$, which is given at discrete values of its arguments. This is a problem that can be solved by numerical techniques.

An example of the numerical computation of the Radon inversion is provided by the reconstruction of so-called Hoffman phantom,[7] depicted in Fig. 5.7, which is a three-dimensional anatomically accurate representation of the radioisotope distribution found in the normal brain.

[6] Details of the derivation of this result can be found in A. S. Fokas, R. G. Novikov, Discrete analogues of $\bar{\partial}$-equation and of Radon transform, *Comptes rendus de l' Académie des Sciences, (Paris), Série I Mathématiques* **313**, pp. 75–80 (1991).

[7] G. A. Kastis, D. Kyriakopoulou, A. Gaitanis, Y. FernÃąndez, B. F. Hutton, A. S. Fokas, Evaluation of the spline reconstruction technique for PET, *Medical Physics* **41**, 042501 (2014).

Further reading

The textbooks mentioned under Further Reading at the end of Chapter 4 contain much useful material on the subject of singularities and the use of the Residue theorem to calculate definite integrals of real and complex functions.

Problems

1. Consider two different functions of the complex variable z, $f_1(z)$ and $f_2(z)$:

   (a) Is the residue of the function $g(z) = f_1(z) + f_2(z)$ equal to the residue of $f_1(z)$ plus the residue of $f_2(z)$ at some value z_0?

   (b) Is the residue of the function $h(z) = f_1(z)f_2(z)$ equal to the product of the residues of $f_1(z)$ and $f_2(z)$ at some point z_0?

2. Compute the residues of the following functions:

   (a) $f(z) = z\tan(z)$, at $z_0 = \pi/2$,

   (b) $f(z) = \sin z / (1 - 2\cos z)$, at $z_0 = \pi/3$,

   (c) $f(z) = z/[(2 - 3z)(4z + 3)]$, at $z_1 = 2/3$, and $z_2 = -3/4$.

3. The integral of the function $f(z) = 1/z^2$ gives zero for a closed contour C around the origin, even though the function $f(z)$ has a singularity (it is not analytic) at the origin. Does this violate Cauchy's integral theorem? Explain why.

4. Find the residues of the following functions of the complex variable z at all singular points:

   (a) $\dfrac{1}{z} + \dfrac{i}{(z-1)^2} + \pi z$,

   (b) $\dfrac{z}{z^3 + 8}$,

   (c) $z \csc(z)$,

   (d) $\dfrac{z^2}{\sin^2 z}$,

   (e) $\dfrac{z+i}{z^2+i}$,

   (f) $\dfrac{1}{z} + \dfrac{i}{z^2}$,

   (g) $\tanh(z)$,

   (h) $\cot(z^2)$.

5. Show that

$$\oint_C \frac{1}{(z-z_1)(z-z_2)} \, dz = 0,$$

   where C is a simple closed contour that encloses both points z_1 and z_2, and $z_1 \neq z_2$. Can this result be generalized to the integral

$$\oint_C \frac{1}{(z-z_1)(z-z_2)\cdots(z-z_m)} \, dz,$$

   where $m \geq 3$ and C encloses all points $z_1, z_2, \ldots, z_m$? We assume that each point is different from all the other points.

6. Evaluate the following integrals where C is the unit circle centered at the origin:

   (a) $\oint_C z^m \bar{z}^n \, dz$, where $n, m \in \mathbb{N}$,

   (b) $\oint_C \dfrac{|z|^\alpha}{z^\beta} \, dz$, where $\alpha, \beta \in \mathbb{R}$.

7. Evaluate the following integrals where C is a circle of radius 2 centered at origin:

   (a) $\oint_C \sinh^2(z) \, dz$,

   (b) $\oint_C \dfrac{1}{z^2+1} \, dz$,

   (c) $\oint_C \dfrac{z^2}{z^2 - 2z - 3} \, dz$,

   (d) $\oint_C \ln(z) \, dz$,

   (e) $\oint_C \dfrac{\cos(z)}{z^2 - 1 - 2iz} \, dz$.

8. Evaluate the following integrals where C is the unit circle centered at the origin:

   (a) $\oint e^{-1/z} \, dz$,

   (b) $\oint \dfrac{e^z - 1}{\sinh(z)} \, dz$,

   (c) $\oint \dfrac{\sinh z}{\sin z} \, dz$,

   (d) $\oint \sin\left(\dfrac{1}{z}\right) \, dz$,

   (e) $\oint_C \dfrac{1}{\cos(z) - \cosh(z)} \, dz$,

   (f) $\oint_C \dfrac{\sinh^2(z)}{\cosh(z/5)} \, dz$,

   (g) $\oint_C \dfrac{z^2}{z^2 - 8z^2 + 16} \, dz$.

9. Let $f(z) = \ln z$ with the value of the argument $\alpha \leq \text{Arg}[z] < \beta$. Evaluate the integral

$$\int_C f(z) \, dz,$$

   if $C = \{z : |z| = 1\}$ and $\alpha = 2k\pi, \beta = 2(k+1)\pi$, for $k = 0, 1, \ldots$

10. Consider the function

$$F(z) = z - \frac{z^2}{2} + \frac{z^3}{3} - \frac{z^4}{4} + \cdots$$

   (a) What function is represented by the series $F(z)$ and in which region? If necessary, specify a branch of the function.

   (b) Differentiate the series $F(z)$ term by term. We denote this series as $G(z)$. What function is represented by $G(z)$? What is its radius of convergence?

   (c) Where does $G(1/z)$ converge? What function is represented by the term by term integral of $G(1/z)$? Why it does not coincide with $F(1/z)$?

11. Evaluate the following integrals by using contour integration:

   (a) $\displaystyle\int_0^\infty \dfrac{x^2}{x^6+1} \, dx$,

   (b) $\displaystyle\int_{-\infty}^\infty \dfrac{1}{x^4+1} \, dx$,

   (c) $\displaystyle\int_0^{2\pi} \dfrac{1}{a + \sin\theta} \, dx$, $a > 1$,

   (d) $\displaystyle\int_0^\infty \dfrac{x\sin(kx)}{x^2+a^2} \, dx$, $a, k \in \mathbb{R}$, $a > 0$,

   (e) $\displaystyle\int_0^\infty \dfrac{x^\alpha}{(x+1)^2} \, dx$, $0 < \alpha < 1$.

   Introduce appropriate branch cuts where necessary.

12. Evaluate the integral

$$I_n(y) = \int_0^\infty \frac{1}{(x^n + y^n)}dx,$$

where $n \in \mathbb{N}$ and $n \geq 2$, $y \in \mathbb{R}$ and $y > 0$. What is the limit $\lim_{n\to\infty}[I_n(y)]$?

13. Evaluate the following integral:

$$\int_{-\infty}^\infty e^{-x^2}\cos(2kx)dx,$$

where $k > 0$, by contour integration. Use a contour C consisting of the rectangle $-R$ to R to $R + ik$ to $-R + ik$ to $-R$, where R is real, and take the limit $R \to \infty$.

14. Evaluate the Cauchy principal value of the following integral:

$$\int_{-\infty}^\infty \frac{e^{itx}}{x}dx, \quad t \in \mathbb{R}.$$

Use the result to show that

$$\int_{-\infty}^\infty \frac{\sin(tx)}{x}dx = \begin{cases} \pi, & \text{for } t > 0, \\ 0, & \text{for } t = 0, \\ -\pi, & \text{for } t < 0. \end{cases}$$

15. Evaluate the following integrals by contour integration:

(a) $\int_{-\infty}^\infty \frac{\sin x}{x(\pi^2 - x^2)}dx,$

(b) $\int_0^\infty \frac{x\sin x}{16x^2 + 9}dx,$

(c) $\int_{-\infty}^\infty \frac{1}{x^3 + 4x + 5}dx,$

(d) $\int_0^\infty \frac{x^3}{1 + x^6}dx,$

(e) $\int_{-\infty}^\infty \frac{1}{(2 - x)(x^2 + 4)}dx,$

(f) $\int_0^\infty \frac{x^\alpha}{a^2 + x^2}dx, \ |\alpha| < 1,$

where the parameters a and α in case (f) are real. Introduce a branch cut where necessary and use the principal value sense of the integral where appropriate.

16. (a) Determine the value of the parameter a so that the function $f(z)$ is analytic:

$$f(z) = \begin{cases} az^{-4}\left[2\cosh(z) - 2 - z^2\right], & \text{for } z \neq 0, \\ 1/6, & \text{for } z = 0. \end{cases}$$

(b) For the value of a determined in part (a), consider the function

$$g(z) = f(1/\sqrt{z}).$$

Is $g(z)$ a single-valued function or not? Explain your answer.

(c) Evaluate the integral

$$I = \oint_C (z^2 + 1)^2 g(z)dz,$$

where C is the unit circle centered at the origin. Does the value of this integral depend on the choice of the contour C? Explain your answer.

17. Evaluate the real integral

$$\int_{-\infty}^\infty \frac{e^{-y/2}}{\sinh y}dy,$$

using contour integration with an appropriately designed contour. [Hint: Consider a contour similar in nature to that of Example 5.4, with the variable of integration running along the imaginary axis.]

Chapter 6

Mappings Produced by Complex Functions

The shortest and best way between two truths of the real domain often passes through the imaginary one.

Jaques Hadamard (French mathematician, 1865–1963)

In this chapter, we study mappings of the z-plane onto the w-plane through complex functions $w = f(z)$. These mappings take us from a value of z in the z-plane onto a value of w in the w-plane through $f(z)$. A very simple, almost trivial case, is the mapping produced by the function

$$w = az + b, \quad a,\ b \text{ constants.}$$

Expressing w and the constants a and b in terms of their real and imaginary parts, $w = u + iv$, $a = a_r + ia_i$, $b = b_r + ib_i$, we find:

$$w = u + iv = (a_r + ia_i)(x + iy) + (b_r + ib_i)$$
$$\Rightarrow u(x,y) = (a_r x - a_i y) + b_r, \quad v(x,y) = (a_r y + a_i x) + b_i.$$

This represents a shift of the origin by $-b$ and a rotation and scaling of the original set of values of x and y. Specifically, if we define the angle ϕ through

$$\phi = \tan^{-1}\left(\frac{a_i}{a_r}\right) \Rightarrow a_r = |a|\cos(\phi), \quad a_i = |a|\sin(\phi),$$

the original values of x, y are rotated by the angle ϕ and scaled by the factor $|a|$. Thus, if the values of x, y describe a curve in the z-plane, this curve will be shifted, rotated and scaled through the mapping defined by the function $f(z)$, but *its shape will not change.*

It is often useful, from a practical point of view, to view the mapping from the opposite perspective: the central idea is to consider a set of complicated curves in the z-plane which through $f(z)$ gets mapped onto a set of very simple curves in the w-plane. Thus, we will pose the following question: Given the function $f(z)$, what kind of curves in the z-plane produce some simple curves in the w-plane? In this sense, the usefulness of the mapping concept lies in simplifying the curves and shapes of interest. An even more difficult question is: Given a set of complicated curves in the z-plane, what function $f(z)$ can map them onto a set of simple curves in the w-plane? We will not attempt to address this question in general. However, we hope that the familiarity gained through the analysis of mappings of several common functions, will allow the reader to develop an intuition which will be helpful for answering the above question in many situations.

The simplest set of curves in the w-plane correspond to $u = $ constant (these are straight vertical lines, with v assuming any value between $-\infty$ and $+\infty$), and to $v = $ constant (these are straight horizontal lines, with u assuming any value between $-\infty$ and $+\infty$). Other simple sets of curves in the w-plane are those that correspond to a fixed argument ϕ_w and variable magnitude r_w of w, which are rays emanating from the origin at fixed angle ϕ_w with the u-axis, or those that correspond to a fixed magnitude r_w and variable argument ϕ_w of w, which are circular arcs centered at the origin at fixed radius r_w (the arcs are full circles for $0 \leq \phi_w < 2\pi$).

6.1   Mappings produced by positive powers

We consider first a simple power of z:

$$w = z^2 \Rightarrow u(x,y) + iv(x,y) = (x^2 - y^2) + i(2xy)$$
$$\Rightarrow u(x,y) = x^2 - y^2, \quad v(x,y) = 2xy. \tag{6.1}$$

From this expression, we can identify the curves in the z-plane that are mapped onto vertical lines in the w-plane (constant u) as being described by the relations

$$x^2 - y^2 = u_0 \Rightarrow y = \pm\sqrt{x^2 - u_0},$$

with u_0 a constant, which are hyperbolas. These are shown in Fig. 6.1. Similarly, the curves in the z-plane that are mapped onto horizontal lines (constant v) in the w-plane are described by the equation

$$2xy = v_0 \Rightarrow y = \frac{v_0}{2x},$$

with v_0 a constant, which are shown in Fig. 6.2.

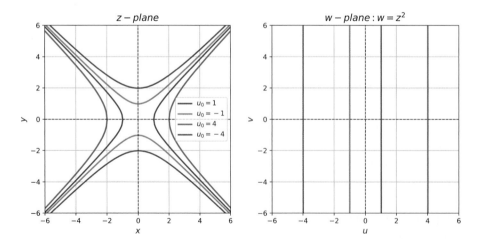

Figure 6.1: Mapping of the z-plane onto the w-plane through $w = f(z) = z^2$ of curves that correspond to constant u values.

From this mapping we see that half of the z-plane maps onto the entire w-plane, because if we restrict the values of z to $x \geq 0$ or to $y \geq 0$, we still can obtain all possible values of u and v. The simplest way to see this is to use polar coordinates:

$$w = z^2 \Rightarrow r_w e^{i\phi_w} = r^2 e^{i2\phi} \Rightarrow r_w = r^2, \quad \phi_w = 2\phi.$$

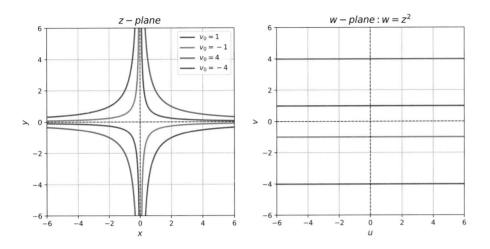

Figure 6.2: Mapping of the z-plane onto the w-plane through $w = f(z) = z^2$ of curves that correspond to constant v values.

Thus, $0 \le \phi < \pi \mapsto 0 \le \phi_w < 2\pi$. Hence, the range of values from zero to π in the argument of z is mapped to a range of values from zero to 2π in the argument of w.

Similar arguments apply to subregions of the z-plane, which through $w = f(z) = z^2$ are "inflated" (or "shrunk") to cover much larger (or smaller) regions on the w-plane, depending on the shape of the original region. For instance, the region enclosed by the rectangle

$$z : \; -l \le x \le l, \;\; 0 \le y \le l,$$

identified by the points $z : (x, y)$ in the z-plane,

$$a : (l, 0), \;\; b : (l, l), \;\; c : (0, l), \;\; d : (-l, -l), \;\; e : (-l, 0),$$

which lies entirely in the upper half of the plane, gets mapped onto the region identified by the points $w : (u, v)$ in the w-plane

$$A : (l^2, 0), \;\; B : (0, 2l^2), \;\; C : (-l^2, 0), \;\; D : (0, -2l^2), \;\; E : (l^2, 0),$$

with the correspondence $a \mapsto A, b \mapsto B, c \mapsto C, d \mapsto D, e \mapsto E$. The domain enclosed by the lines connecting these points in the w-plane extends over both the upper and the lower half, as shown in Fig. 6.3. It is evident that any point

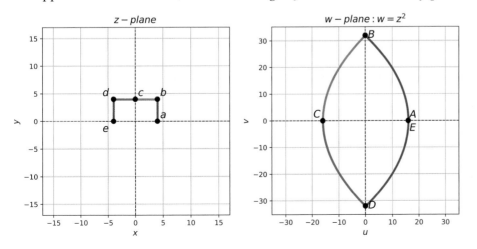

Figure 6.3: Mapping of a rectangular region in the z-plane to the w-plane through $w = f(z) = z^2$: the points a, b, c, d, e in the z-plane get mapped onto the points A, B, C, D, E, respectively, in the w-plane (the segments between successive points are color-coded); the illustration corresponds to $l = 2$ of the general case discussed in the text.

inside the rectangle *abcde* in the z-plane will get mapped to a point inside the domain bounded by the lines connecting the points $ABCDE$ in the w-plane. Thus, we have identified the image of the whole domain in the z-plane under the mapping by the function $w = z^2$. If we consider the limit of the rectangle in the z-plane covering the entire upper half of the plane, that is, we take the limit $l \to \infty$, then its image will cover the entire w-plane. We return to this case in Example 6.5 below, after we have introduced the notion of conformal mapping, which gives a fuller picture of what is involved in the mapping of regions of the z-plane onto regions in the w-plane.

A related mapping involves the square root:

$$w = z^{\frac{1}{2}} \Rightarrow w^2 = z \Rightarrow (u^2 - v^2) + \mathrm{i}2uv = x + \mathrm{i}y$$
$$\Rightarrow x = (u^2 - v^2), \ \ y = 2uv. \tag{6.2}$$

For curves in the z-plane that map onto vertical lines $(u(x,y) = u_0)$, we will have

$$u = u_0 \to v = \frac{y}{2u_0} \to u_0^2 - \frac{y^2}{4u_0^2} = x,$$

which are parabolas, as shown in Fig. 6.4. For curves in the z-plane that map onto horizontal lines $(v(x,y) = v_0)$ we will have

$$v = v_0 \to u = \frac{y}{2v_0} \to \frac{y^2}{4v_0^2} - v_0^2 = x,$$

which are also parabolas, as shown in Fig. 6.4.

In this case, the entire z-plane maps onto half of the w-plane, since the values of u_0 and v_0 appear squared in the final equations, so that two values of opposite sign correspond to the same curve. This is consistent with the fact that the function $z^{1/2}$ is the inverse of the function z^2. This is a troubling fact, however, because we are not sure at which of the two halves of the w-plane we will end up. This is of course a consequence of the fact that the function $z^{1/2}$ is multi-valued. To see this more clearly, we use polar coordinates:

$$w = z^{1/2} \Rightarrow r_w \mathrm{e}^{\mathrm{i}\phi_w} = r^{1/2}\mathrm{e}^{\mathrm{i}(\phi + 2k\pi)/2} \Rightarrow r_w = r^{1/2}, \ \phi_w = \frac{\phi}{2} + k\pi, k = 0, 1,$$

so there are two possible choices for the values of k, giving rise to two possible mappings. In order to remove the ambiguity, we need again to introduce branch cuts. Evidently, the branch points in this case are again $z = 0$ and $z \to \infty$, because of the involvement of the logarithm in the definition of z^a with a non-integer. Any proper branch cut joining these two points removes the ambiguity and renders the function single-valued. For example, the standard choice of the positive x-axis as the branch cut with $0 \le \phi < 2\pi$ (equivalent to the choice $k = 0$), leads to the entire z-plane mapping onto the upper half $(0 \le \phi_w < \pi$ or $v \ge 0)$ of the w-plane; the same choice for a branch cut with $2\pi \le \phi < 4\pi$ (equivalent to the choice $k = 1$), leads to the entire z-plane mapping onto the lower half $(\pi \le \phi_w < 2\pi$ or $v \le 0)$ of the w-plane.

The examples used above can be generalized. For instance, the mapping $w = z^n$ with n a positive integer maps a fraction $1/n$ of the z-plane onto

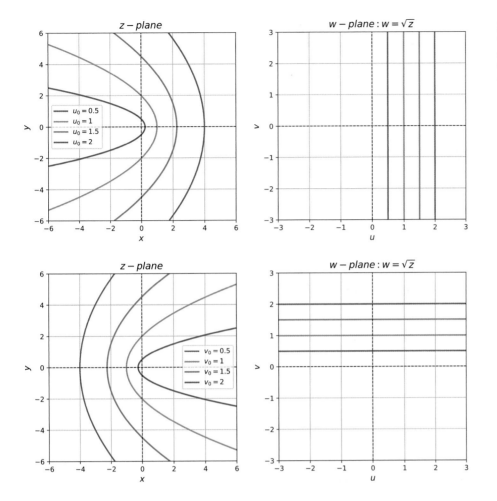

Figure 6.4: Mapping of the z-plane onto the w-plane through $w = f(z) = z^{1/2}$ of curves that correspond to constant u values (top two panels) and to constant v values (bottom two panels).

the entire w-plane. Conversely, the mapping $w = z^{1/n}$ with n a positive integer maps the entire z-plane onto a fraction $1/n$ of the w-plane. In the latter case, appropriate branch cuts need to be introduced to make the mapping single-valued.

6.2   Mappings produced by $1/z$

An interesting set of mappings are produced by the function

$$w = f(z) = \frac{1}{z}.$$

We consider first the image of a circle in the z-plane obtained through this mapping onto the w-plane. The equation of a circle in the z-plane with center at (x_0, y_0) and radius ρ is given by

$$(x - x_0)^2 + (y - y_0)^2 = \rho^2 \Rightarrow a(x^2 + y^2) + bx + cy + d = 0, \qquad (6.3)$$

where the parameters a, b, c, d are related to x_0, y_0, ρ as follows:

$$a = \frac{1}{\rho^2}, \quad b = -\frac{2x_0}{\rho^2}, \quad c = -\frac{2y_0}{\rho^2}, \quad d = \frac{x_0^2 + y_0^2}{\rho^2} - 1.$$

The fact that we pass from a set of three *independent* parameters x_0, y_0, ρ that describe the original curve, to a set of four parameters a, b, c, d, is not a contradiction because the latter set of parameters are not independent:

$$b^2 + c^2 = 4a(d+1).$$

Using the relations

$$x = \frac{z + \bar{z}}{2}, \quad y = \frac{z - \bar{z}}{2i}, \quad x^2 + y^2 = z\bar{z},$$

and the fact that $w = u + iv = 1/z \Rightarrow \bar{w} = u - iv = 1/\bar{z}$, we find

$$u = \frac{w + \bar{w}}{2} = \frac{z + \bar{z}}{2z\bar{z}}, \quad v = \frac{w - \bar{w}}{2i} = \frac{\bar{z} - z}{2iz\bar{z}}, \quad u^2 + v^2 = w\bar{w} = \frac{1}{z\bar{z}}.$$

Thus,

$$x = \frac{1}{w\bar{w}}u = \frac{u}{u^2 + v^2}, \quad y = -\frac{1}{w\bar{w}}v = -\frac{v}{u^2 + v^2}, \quad x^2 + y^2 = \frac{1}{w\bar{w}} = \frac{1}{u^2 + v^2}.$$

Substituting these expressions for x, y, in Eq. (6.3) we find that the equation for a circle in the z-plane is transformed to

$$a + bu - cv + d(u^2 + v^2) = 0. \tag{6.4}$$

This equation has exactly the same functional form in the variables u and v as the equation for the circle in the z-plane, Eq. (6.3). From this fact we conclude that the original circle on the z-plane is mapped onto a circle in the w-plane.

An interesting question is: What happens when $d = 0$? In this case, $x_0^2 + y_0^2 = \rho^2$, which means that one point of the circle coincides with the origin of the axes in the z-plane, because the distance of the origin $(x = 0, y = 0)$ from the center of the circle situated at (x_0, y_0), is

$$\sqrt{(x_0 - 0)^2 + (y_0 - 0)^2} = \rho,$$

that is, exactly equal to the radius of the circle. When $d = 0$, the equation for the curve in the w-plane becomes $a + bu - cv = 0$, which is the equation for a line. Thus, we conclude that circles in the z-plane which pass from the origin, under the mapping $f(z) = 1/z$ become straight lines in the w-plane.

A related mapping is given by the function

$$w = f(z) = \frac{(z - a)}{(z - b)},$$

which also maps a circle onto a circle. To see this, we perform the following changes of variables:

$$z_1 = z - b \; : \; \text{shift the origin by } b,$$

$$z_2 = \frac{1}{z_1} \; : \; \text{maps circles to circles,}$$

$$w = \lambda z_2 + 1 \; : \; \text{linear scaling by } \lambda, \text{ shift the origin by } -1,$$

which proves that the mapping $w = (z - a)/(z - b)$ is also a mapping of a circle onto a circle, since the only relation that involves the mapping is the second change of variables (the other two merely shift or scale the curve without changing its shape). All that remains to do is to find the value of the scaling parameter λ. By requiring that the final expression is equal to the original one, we find

$$w = \lambda z_2 + 1 = \lambda \frac{1}{z_1} + 1 = \lambda \frac{1}{z-b} + 1 = \frac{z-a}{z-b} \Rightarrow \lambda = b - a,$$

which completes the argument. A generalization of the above is the mapping produced by the so-called Möbius function (see Problem 6.5)

$$f(z) = \frac{az + b}{cz + d}, \quad a, b, c, d \in \mathbb{C}. \tag{6.5}$$

Example 6.1: Having studied both the mapping of the function $1/z$ and the function z^2, we can consider the mapping produced by the composition of these two functions, namely, the function

$$w = f(z) = \frac{1}{z^2}.$$

We consider the region in the z-plane consisting of the entire upper half-plane except for a semicircle of radius l centered at the origin. In what follows, we will determine its image in the w-plane under the above mapping. One way to figure out the relevant image is to write out the function in terms of the real and imaginary components and find the image of a few key points, like the ones denoted as a, b, c, as well as the image of $z = 0$ and $z \to \infty$. Specifically, we have

$$w = \frac{1}{z^2} = \frac{\bar{z}^2}{|z|^4} = \frac{x^2 - y^2}{(x^2 + y^2)^2} + i \frac{-2xy}{(x^2 + y^2)^2}.$$

From this we find that the points

$$a : (l, 0), \quad b : (0, l), \quad c : (-l, 0),$$

in the z-plane are mapped onto the points

$$A : (1/l^2, 0), \quad B : (-1/l^2, 0), \quad C : (1/l^2, 0),$$

in the w-plane, as shown in the diagram below, while the point $z \to \infty$ is mapped onto $w = 0$. Using polar coordinates, we find the following mapping for points in a circular path around the origin of radius l in the z-plane:

$$z = le^{i\phi} \mapsto w = (1/l^2)e^{-2i\phi},$$

so that the semicircle of radius l in the upper half z-plane, for which $0 \leq \phi \leq \pi$, is mapped onto a full circle of radius $1/l^2$ in the w-plane. Putting this information together, we find that the region defined above in the z-plane is mapped onto the circle of radius $1/l^2$ and its interior in the w-plane.

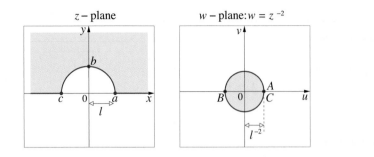

Alternatively, we can use our knowledge of previously studied mappings to obtain the above result without any additional calculation. Specifically, the mapping produced by $w = z^{-2}$ can be broken into two parts: we first define the variable $z' = z^2$ and then consider the mapping $w = 1/z'$. The first step, $z' = z^2$, takes the semicircle with radius l in the z-plane into a full circle of radius l^2 in the z'-plane. The second step, $w = 1/z'$, maps the points *outside* the full circle of radius l^2 in the z'-plane to the points *inside* the full circle of radius $1/l^2$ in the w-plane. Thus, using our knowledge of the earlier mappings we arrive at the final answer in a simpler way rather than doing the analysis for the direct mapping implied by $w = 1/z^2$. The net result from applying the two mappings is shown in the diagram below.

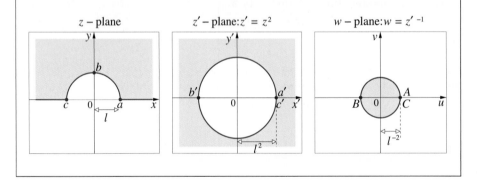

Example 6.2: As another example, we study the mapping of a function which involves both a power of z and an inverse power of z:

$$f(z) = z^2 + z^{-2}.$$

Expressing the complex variable z in terms of its real and imaginary parts $z = x + iy$ leads to

$$f(z) = (x^2 - y^2)\left(1 + \frac{1}{(x^2 + y^2)^2}\right) + i2xy\left(1 - \frac{1}{(x^2 + y^2)^2}\right).$$

Setting as usually the function $f(z)$ equal to $w = u(x, y) + iv(x, y)$, we find expressions for the real and imaginary parts of w:

$$u(x, y) = (x^2 - y^2)\left(1 + \frac{1}{(x^2 + y^2)^2}\right), \quad v(x, y) = 2xy\left(1 - \frac{1}{(x^2 + y^2)^2}\right).$$

As an illustration of the mapping produced by $f(z)$, we consider a curve in the z-plane described by the parametric form

$$x(t) = re^{t^2/a^2} \cos(t), \quad y(t) = re^{t^2/a^2} \sin(t), \quad -\pi < t < \pi,$$

where r, a are real positive constants; this curve is illustrated in the diagram below, on the left, with the choice of parameters $r = 4$ and $a = 2.5$. Its map through $f(z)$ in the w-plane is given by

$$u(t) = r^2 e^{2t^2/a^2} \cos(2t) \left(1 + \frac{1}{r^4 e^{4t^2/a^2}}\right),$$

$$v(t) = r^2 e^{2t^2/a^2} \sin(2t) \left(1 - \frac{1}{r^4 e^{4t^2/a^2}}\right),$$

which is illustrated in the diagram below, on the right.

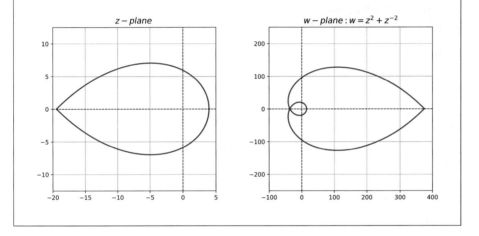

6.3   Mappings produced by the exponential function

We consider the mapping of the z-plane onto the w-plane through the exponential function:

$$w = e^z = e^x e^{iy} = e^x \cos(y) + ie^x \sin(y)$$
$$\Rightarrow \quad u(x,y) = e^x \cos(y), \quad v(x,y) = e^x \sin(y). \tag{6.6}$$

In the spirit discussed above, we consider what curves in the z-plane are mapped onto vertical lines, $u(x,y) = u_0$, or horizontal lines, $v(x,y) = v_0$, in the w-plane. For *vertical* lines in the w-plane (constant u) the curves in the z-plane are characterized as follows:

$$u(x,y) = e^x \cos(y) = u_0 \Rightarrow x = \ln(u_0) - \ln(\cos(y))$$
$$\text{for } \left[u_0 > 0, \cos(y) > 0 \Rightarrow -\frac{\pi}{2} < y < \frac{\pi}{2}\right],$$
$$\Rightarrow x = \ln(-u_0) - \ln(-\cos(y))$$
$$\text{for } \left[u_0 < 0, \cos(y) < 0 \Rightarrow \frac{\pi}{2} < y < \frac{3\pi}{2}\right].$$

These mappings are illustrated in the diagram below for positive values of u_0. Similarly, for *horizontal* lines in the w-plane (constant v) the curves in the

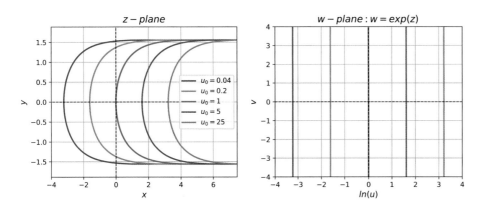

Figure 6.5: Mapping of the z-plane onto the w-plane through $w = f(z) = \exp(z)$ using Cartesian coordinates in the w-plane (note that the scale of the variable u is logarithmic for clarity).

z-plane are characterized as follows:

$$v(x,y) = e^x \sin(y) = v_0 \Rightarrow x = \ln(v_0) - \ln(\sin(y))$$
$$\text{for } [v_0 > 0, \sin(y) > 0 \Rightarrow 0 < y < \pi],$$
$$\Rightarrow x = \ln(-v_0) - \ln(-\sin(y))$$
$$\text{for } [v_0 < 0, \sin(y) < 0 \Rightarrow \pi < y < 2\pi],$$

with the corresponding curves being very similar to the ones shown in Fig. 6.5 but shifted to the appropriate range of y values.

We can examine the mapping produced by the same function using polar coordinates in the w-plane, $w = r_e \exp(i\phi_w)$:

$$w = e^z \Rightarrow \ln(w) = z \Rightarrow \ln(r_w) + i\phi_w = x + iy \Rightarrow \ln(r_w) = x, \quad \phi_w = y.$$

In this case, curves described by a constant value of $x = x_0$ with y in a range of values spanning 2π, map onto circles of constant radius $r_w = e^{x_0}$:

$$[x = x_0, \ 0 \leq y < 2\pi] \rightarrow [r_w = e^{x_0}, \ 0 \leq \phi_w < 2\pi].$$

Similarly, curves described by constant value of $y = y_0$ and x taking all real values $-\infty < x < \infty$, map onto straight lines (rays) emanating from the origin at constant angle angle ϕ_w to the u-axis:

$$[-\infty < x < \infty, \ y = y_0] \rightarrow [0 \leq r_w < \infty, \ \phi_w = y_0].$$

This analysis of the mappings produced by the function $w = \exp(z)$ indicates that every strip of width 2π in y and $-\infty < x < \infty$ on the z-plane maps onto the entire w-plane. This is true for the mapping using Cartesian or polar coordinates in the w-plane.

We next consider the mapping of the inverse function of the exponential, namely the natural logarithm; this mapping is more subtle. We employ polar coordinates for z, which, through the use of Euler's formula make things simpler.

$$w = \ln(z) = \ln(re^{i\phi}) = \ln(r) + i\phi$$
$$\Rightarrow u(x,y) = \ln(r) = \ln(\sqrt{x^2 + y^2}), \quad v(x,y) = \phi = \tan^{-1}\left(\frac{y}{x}\right).$$

This is the inverse mapping of the exponential, thus we can make the correspondence $(x \leftrightarrow u, y \leftrightarrow v; r_w \leftrightarrow r, \phi_w \leftrightarrow \phi)$, and then we can obtain the mapping by reversing what we found for the mapping $w = e^z$. This simple argument gives the following mappings:

$$[r = r_0, \ 0 \le \phi < 2\pi] \mapsto [u(x,y) = u_0 = \ln(r_0), \ 0 \le v(x,y) < 2\pi],$$

$$[\phi = \phi_0, \ 0 \le r < \infty] \mapsto [-\infty < u(x,y) < \infty, \ v(x,y) = v_0 = \phi_0].$$

Thus, circles of radius r_0 on the z-plane map onto straight vertical lines $u(x,y) = u_0 = \ln(r_0)$ on the w-plane of height 2π, while rays at an angle ϕ_0 to the x-axis in the z-plane map to straight horizontal lines $v(x,y) = v_0 = \phi_0$ in the w-plane. This is all fine when ϕ lies within an interval of 2π, which was assumed so far, but leads to difficulties if the range of values of ϕ is not specified. This feature is of course related to the problem of multi-valuedness of the logarithm, which can be eliminated by identifying the branch points and introducing a branch cut. We have already established that the function $\ln(z)$ has two branch points, zero and ∞, and any line between them which does not intersect itself is a legitimate branch cut.

Example 6.3: As an example we will consider the mappings implied by the hyperbolic and trigonometric functions, which are closely related. We begin by examining the mapping implied by the hyperbolic cosine:

$$w = \cosh(z) = \frac{e^z + e^{-z}}{2} = \frac{e^x(\cos(y) + i\sin(y))}{2} + \frac{e^{-x}(\cos(y) - i\sin(y))}{2}$$
$$\Rightarrow u(x,y) = \cosh(x)\cos(y), \ v(x,y) = \sinh(x)\sin(y).$$

We investigate first the types of curves in the z-plane that produce vertical ($u = u_0$: constant) or horizontal ($v = v_0$: constant) lines in the w-plane. For vertical lines we have

$$u_0 = \frac{e^x + e^{-x}}{2}\cos(y) \Rightarrow e^x - \frac{2u_0}{\cos(y)} + e^{-x} = 0.$$

These expressions can be turned into a second-order polynomial in $\exp(x)$ by multiplying through by $\exp(x)$, from which we obtain the solutions:

$$e^x = \frac{u_0}{\cos(y)} \pm \left(\frac{u_0^2}{\cos^2(y)} - 1\right)^{1/2}$$

$$\Rightarrow x = \ln\left[\frac{u_0}{\cos(y)} \pm \left(\frac{u_0^2}{\cos^2(y)} - 1\right)^{1/2}\right].$$

The last expression is meaningful only for values of y which make the quantity under the square root positive; for these values of y, the quantity under the logarithm is positive for either choice of sign in front of the square root. Similarly, for horizontal lines we have

$$v_0 = \frac{e^x - e^{-x}}{2}\sin(y) \Rightarrow e^x - \frac{2v_0}{\sin(y)} - e^{-x} = 0$$

$$\Rightarrow \quad e^x = \frac{v_0}{\sin(y)} \pm \left(\frac{v_0^2}{\sin^2(y)} + 1 \right)^{1/2}$$

$$\Rightarrow \quad x = \ln \left[\frac{v_0}{\sin(y)} \pm \left(\frac{v_0^2}{\sin^2(y)} + 1 \right)^{1/2} \right].$$

Here again we have used the second-order polynomial solutions. In this case, the quantity under the square root is always positive and the argument of the natural logarithm is positive only for the solution with the plus sign in front of the square root. From these equations, the sets of curves in the z-plane that correspond to vertical or horizontal lines in the w-plane can be drawn. These sets of curves are quite similar to the curves that get mapped onto vertical or horizontal lines through the function $\exp(z)$, which is no surprise since the hyperbolic cosine is the sum of $\exp(z)$ and $\exp(-z)$. From this statement, we conclude that a strip of values in the z-plane, extending over the entire x-axis and covering a width of 2π on the y-axis, is mapped onto the entire w-plane, just as it occurs in the case of the mapping produced by $\exp(z)$. In the present case, the values z and $-z$ are mapped onto the same value of w, in other words, the strip of width 2π in y is mapped *twice* onto the w-plane.

From the above analysis, we can easily obtain the mappings due to the other hyperbolic and trigonometric functions. For instance, from the relations

$$z_1 = z + i\frac{\pi}{2} \Rightarrow \cosh(z_1) = \frac{e^{z+i\pi/2} + e^{-z-i\pi/2}}{2} = \frac{ie^z - ie^{-z}}{2} = i\sinh(z),$$

we conclude that the function $\sinh(z)$ can be viewed as the combination of three functions: the first consists of a shift of the origin by $-i\pi/2$, the second is the hyperbolic cosine, and the third is a multiplication by $-i$:

$$z \mapsto z_1 = z + i\frac{\pi}{2}, \quad z_1 \mapsto z_2 = \cosh(z_1), \quad z_2 \mapsto w = -iz_2 = \sinh(z).$$

Since we know the mapping produced by each of these functions, it is straightforward to derive the mapping of the hyperbolic sine. In particular, since multiplication by $-i$ involves a rotation by $\pi/2$, we conclude that under $w = \sinh(z)$ a strip of the z-plane of width 2π on the x-axis and extending over the entire y-axis is mapped twice onto the entire w-plane. Similarly, because of the relations

$$\cos(iz) = \cosh(z), \quad \sin(iz) = i\sinh(z),$$

we can derive the mappings due to the trigonometric sine and cosine functions in terms of the mappings due to the hyperbolic sine and cosine functions.

Example 6.4: As another example we consider the mapping due to an inverse trigonometric function,

$$w = \sin^{-1}(z).$$

Taking the sine of both sides of this equation, we obtain

$$z = \sin(w) = \sin(u + iv) = \frac{(\cos(u) + i\sin(u))e^{-v} - (\cos(u) - i\sin(u))e^{v}}{2i}$$
$$= \sin(u)\cosh(v) + i\cos(u)\sinh(v)$$
$$\Rightarrow x = \sin(u)\cosh(v), \quad y = \cos(u)\sinh(v).$$

From the last two relations, using the properties of the trigonometric and hyperbolic sine and cosine functions, we find

$$\frac{x^2}{\sin^2(u)} - \frac{y^2}{\cos^2(u)} = \cosh^2(v) - \sinh^2(v) = 1.$$

This is the equation of a hyperbola with length scales $2a = 2|\sin(u)|$ on the x-axis and $2b = 2|\cos(u)|$ on the y-axis. Setting $u = u_0$ constant, we see that confocal hyperbolas in the z-plane with foci at $z = \pm 1$ are mapped onto straight vertical lines in the w-plane, as indicated in the diagram below. The thicker red lines indicate the branch cuts connecting the branch points ± 1 and $\pm\infty$, and their images in the w-plane, at $u = \pm\pi/2$, as explained in more detail below.

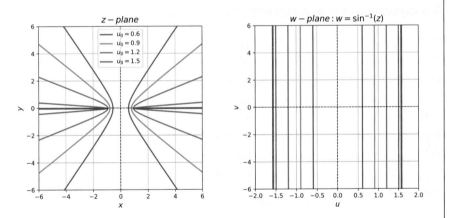

Similarly,

$$\frac{x^2}{\cosh^2(v)} + \frac{y^2}{\sinh^2(v)} = \sin^2(u) + \cos^2(u) = 1,$$

which is the equation for an ellipse with extent $2a = 2\cosh(v)$ on the x-axis and $2b = 2|\sinh(v)|$ on the y-axis. Setting $v = v_0$ constant, we see that confocal ellipses in the z-plane with foci at $z = \pm 1$ are mapped onto straight horizontal lines in the w-plane, as indicated in the diagram below. The thicker red lines indicate the branch cuts connecting the branch points ± 1 and $\pm\infty$ in the z-plane and their images in the w-plane, at $u = \pm\pi/2$, as explained in more detail below.

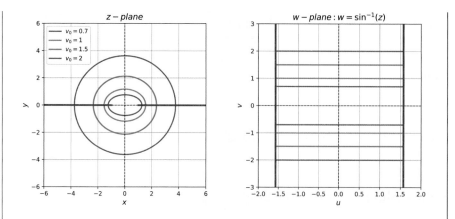

This example deserves closer scrutiny: although we are interested in obtaining the values of w from z, we derived the relations between x and y by going in the inverse direction, that is, by expressing z as a function of w. This may have hidden problems with multi-valuedness, so we examine the expression of w in terms of z in more detail:

$$z = \sin(w) = \frac{e^{iw} - e^{-iw}}{2i} \Rightarrow e^{iw} - 2iz - e^{-iw} = 0$$

$$\Rightarrow \quad e^{iw} = iz \pm \sqrt{1 - z^2} = i\left(z \pm \sqrt{z^2 - 1}\right),$$

where for the last exponential we have used the solution of the above second-order polynomial equation. Taking the logarithm of the above expression, we can obtain the explicit expression of w as a function of z:

$$w = \frac{1}{i} \ln\left[i\left(z \pm \sqrt{z^2 - 1}\right)\right] = \frac{\pi}{2} + \frac{1}{i} \ln\left[z \pm \left(\sqrt{z - 1}\right)\left(\sqrt{z + 1}\right)\right].$$

This expression clearly shows that we *do* need to worry about branch points and branch cuts for this situation. Specifically, there are two square root functions involved as well as the logarithm function, all of which introduce branch points. The branch points of the two square roots are $z = \pm 1, \infty$. The branch points of the logarithm correspond to the values zero and ∞ for its argument. By inspection, we see that the expression under the logarithm above is never zero for finite z:

$$z \pm \sqrt{z^2 - 1} = 0 \Rightarrow z^2 = z^2 - 1,$$

which cannot be satisfied for any finite value of z. On the other hand, for $z \to \infty$ this expression can be 0 and ∞, because in the limit $z \to \infty$ we have $\sqrt{z^2 - 1} \approx \sqrt{z^2} = \pm z$ and $z \pm (\pm z) = 0$ or ∞. Thus, as $z \to \infty$ the argument of the logarithm becomes infinity and hence the only branch point for the logarithm is $z \to \infty$. From this analysis we conclude that the branch cut we need for this function must go through the points $z = \pm 1$ as well as through the point $z \to \infty$. A possible choice for a branch cut is shown in it consists of two parts, $-\infty < x \le -1$ and $+1 \le x < +\infty$. The images of these lines onto the w-plane are the vertical lines $u = \pm\pi/2$. The entire z-plane is mapped onto the strip of values extending over the entire range of v and having width 2π on the u-axis in the w-plane. This is expected from the preceding analysis of the mapping of the function $\sin(z)$!

6.4 Conformal mapping

An important concept in the mapping of functions in the complex plane is that of "conformal mapping" which is characterized by the property that it preserves angles. More specifically, consider two curves in the z-plane with a given direction, as illustrated in Fig. 6.6 with the arrows; the conformal mapping of these curves onto curves in the w-plane preserves the angle at the intersection, namely, the angle ϕ on the z-plane is equal in magnitude and orientation to the angle ψ in the w-plane.

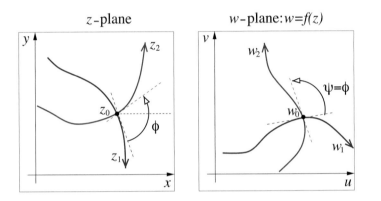

Figure 6.6: Illustration of a conformal mapping: the angle ϕ between the tangents of oriented curves z_1 and z_2 at their intersection z_0 in the z-plane is equal to the angle ψ between the tangents of the corresponding oriented curves w_1 and w_2, which are the images of z_1 and z_2 under $w = f(z)$, at their intersection point w_0, the image of z_0.

Theorem: *The mapping $w = f(z)$ defined by an analytic function $f(z)$ from the domain $\mathcal{D}$ in the z-plane to a domain $\mathcal{G}$ in the w-plane is conformal at the point z_0 in $\mathcal{D}$, provided that $f'(z_0) \neq 0$.*

Proof: If $f(z)$ is analytic at z_0, then we can use its Taylor expansion near $z - z_0$:

$$f(z) \approx f(z_0) + f'(z_0)(z - z_0). \tag{6.7}$$

Now consider two curves in the z-plane, described by the relations $z_1(t)$ and $z_2(t)$, where the real variable t parameterizes the variables x, y, $x = x(t)$, $y = y(t)$. We define $t = t_0$ the value for which $z_1(t_0) = z_2(t_0) = z_0$, where the two curves intersect. Since z_1 and z_2 are functions of t, we can take their derivatives with respect to t in a neighborhood of $t \in [t_0 - \epsilon, t_0 + \epsilon]$, with $\epsilon > 0$, and evaluate them at $t = t_0$:

$$\dot{z}_1(t_0) = \lim_{t \to t_0} \frac{z_1(t) - z_1(t_0)}{t - t_0}, \quad \dot{z}_2(t_0) = \lim_{t \to t_0} \frac{z_2(t) - z_2(t_0)}{t - t_0},$$

where the notation $\dot{z}(t)$ denotes the derivative of the complex variable z with respect to the parameter t. In terms of these derivatives, for t close to t_0 we will have

$$z_1(t) = z_0 + \dot{z}_1(t_0)(t - t_0), \quad z_2(t) = z_0 + \dot{z}_2(t_0)(t - t_0).$$

The slope of $z_1(t)$ near t_0 is given by its derivative at t_0, namely $\dot{z}_1(t_0)$, and similarly for $z_2(t)$. Therefore, the angle between the curves at $z = z_0$ is the angle between these two slopes, $\dot{z}_1(t_0)$ and $\dot{z}_2(t_0)$. Thus, using the definitions

$$\dot{z}_1(t_0) \equiv r_1 e^{i\phi_1}, \quad \dot{z}_2(t_0) \equiv r_2 e^{i\phi_2},$$

we find that the angle ϕ between the two curves in the z-plane at z_0 is

$$\phi = \phi_1 - \phi_2,$$

as shown in Fig. 6.6. From the mapping $w = f(z)$, we obtain two curves in the w-plane, described by $w_1(t) = f(z_1(t)), w_2(t) = f(z_2(t))$, which from the continuity of $f(z)$, in the neighborhood of $w_0 = f(z_0)$ can be written in the following form:

$$w_1(t) = w_0 + f'(z_0)(z_1(t) - z_0) = w_0 + f'(z_0)\dot{z}_1(t_0)(t - t_0),$$

$$w_2(t) = w_0 + f'(z_0)(z_2(t) - z_0) = w_0 + f'(z_0)\dot{z}_2(t_0)(t - t_0).$$

In these equations, we have used the relations derived above for expressing $z_1(t) - z_0$ and $z_2(t) - z_0$ in terms of their derivatives at $t = t_0$. This shows that the slopes of the two curves $w_1(t), w_2(t)$ at w_0, which corresponds to $t = t_0$, are given by their tangents:

$$\dot{w}_1(t_0) = f'(z_0)\dot{z}_1(t_0), \quad \dot{w}_2(t_0) = f'(z_0)\dot{z}_2(t_0).$$

We define

$$f'(z_0) \equiv r_0 e^{i\phi_0}$$

and use the earlier definitions of $\dot{z}_1(t_0), \dot{z}_2(t_0)$ to obtain the following relations for the slopes of the two curves on the w-plane at w_0:

$$\dot{w}_1(t_0) = (r_0 r_1)e^{i(\phi_0 + \phi_1)}, \quad \dot{w}_2(t_0) = (r_0 r_2)e^{i(\phi_0 + \phi_2)}.$$

Hence, the angle between these two slopes is ϕ:

$$\psi = (\phi_0 + \phi_1) - (\phi_0 + \phi_2) = \phi,$$

as required by the definition of conformal mapping.   ∎

The proof given above holds for $f'(z_0) \neq 0$. If some of the lowest-order derivatives of $f(z)$ are zero at z_0, and the first derivative that does not vanish is the derivative of order k, namely, $f^{(k)}(z_0) \neq 0$, then the mapping $w = f(z)$ at z_0 multiplies the angles between oriented curves at z_0 by a factor of k. This can be seen by the same argument as above, noticing that in this case we will have

$$f(z) \approx f(z_0) + \frac{1}{k!}f^{(k)}(z_0)(z - z_0)^k. \tag{6.8}$$

Thus, in the neighborhood of z_0, we will have for a curve $w(t)$ parameterized by t and produced by the curve $z(t)$ the following:

$$w(t) = w_0 + \frac{1}{k!}f^{(k)}(z_0)(z(t) - z_0)^k = w_0 + \frac{1}{k!}f^{(k)}(z_0)(\dot{z}(t_0))^k(t - t_0)^k.$$

The slope of the function $w(t)$ at t_0 now involves a factor $(\dot{z}(t_0))^k$, which means that angles between oriented curves in the z-plane will be multiplied by a factor of k in the w-plane.

Example 6.5: To illustrate how these ideas apply to the mapping of a specific function, we consider again the case of $f(z) = z^2$ and the square of side $l > 1$, with one corner at $z = 0$ and one side each on the x- and y-axes, as shown in the diagram below, with the points a, b, c, d mapped onto A, B, C, D, respectively. First, we note that $f'(z) = 2z$, which is zero only at $z = 0$. Hence, the mapping of this function is conformal everywhere except at $z = 0$.

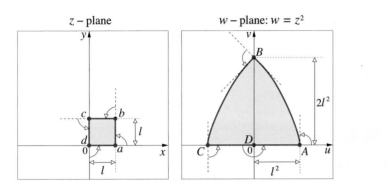

The conformal character of the mapping means that if we consider the perimeter of the square to consist of four oriented curves (in this case lines), which are traversed in the counterclockwise direction, then the angles between these curves are preserved in magnitude and orientation. Specifically, the angles at the corners a, b, c of the square, which are equal to $\pi/2$ in the z-plane, are the same in the corresponding corners A, B, C of the shape obtained from the square through the mapping. Notice that the angles in the shape in the w-plane are those formed by the *tangents* to the curves into which the sides of the square are mapped. The only point where the angle is not preserved is d, as we expect because at this point $f'(0) = 0$. We also notice that at this point the first non-zero derivative of $f(z)$ is the second derivative $f''(z) = 2$. Accordingly, we expect that at this point the angles between oriented curves should be multiplied by a factor of 2 and indeed at this point the angle, which is $\pi/2$ in the original shape, becomes π in the image of the shape in the w-plane.

6.5 Application: Fluid flow around an obstacle

From the physics of fluids, it can be demonstrated that the flow of an ideal fluid, namely, of an incompressible fluid without energy dissipation, can be described by a complex potential $\Psi(z)$ which is an analytic function:

$$\Psi(z) = V_1(x, y) + iV_2(x, y), \tag{6.9}$$

where $V_1(x, y)$ is called the "velocity potential". The curves where this function takes constant values, $V_1(x, y) = c_1$, are called "equipotentials". The function $V_2(x, y)$ has the property that the curves $V_2(x, y) = c_2$ represent the "streamlines" of the flow, namely the curves along which the fluid particles move. It is straightforward to show that the two-dimensional vector $\tau(x, y)$ at the point

(x, y) along a streamline, with components

$$\tau(x, y) = \left(\frac{\partial V_1}{\partial x}, -\frac{\partial V_2}{\partial x} \right), \tag{6.10}$$

represents the velocity of the particles: assuming that the streamlines are given by the parameteric equations $x = x(t), y = y(t)$, we have

$$\frac{dV_2}{dt} = \frac{\partial V_2}{\partial x}\frac{dx}{dt} + \frac{\partial V_2}{\partial y}\frac{dy}{dt} = 0 \Rightarrow \frac{dy}{dt} = \frac{-\partial V_2/\partial x}{\partial V_2/\partial y}\frac{dx}{dt}.$$

However, since $\Psi(z)$ is an analytic function, the Cauchy–Riemann equations satisfy

$$\frac{\partial V_2}{\partial y} = \frac{\partial V_1}{\partial x}.$$

Thus, the above expression relating dy/dt and dx/dt gives

$$\frac{dy}{dx} = \frac{-\partial V_2/\partial x}{\partial V_1/\partial x}.$$

The slope of the function $V_2(x, y) = c_2$, viewed as an implicit relation between y and x, is exactly equal to dy/dx, which means that the tangent vector to the curve at the point (x, y) has components $\partial V_1/\partial x$ in the x direction and $-\partial V_2/\partial x$ in the y direction, as required by Eq. (6.10).

The complex potential $\Psi(x, y)$, and therefore its components $V_1(x, y)$, $V_2(x, y)$, can be determined by the boundary conditions. For an ideal fluid flow, these boundary conditions are that the streamlines are locally parallel to the walls of the container in which the fluid flows, or to the boundary of the obstacle the flow encounters. For an incompressible fluid, the streamlines are also locally parallel to each other.

Furthermore, we can make use of the so-called "Theorem of Invariance of Ideal Fluid Flow" (TIIFF).

Theorem: *Let a function $f(z) = u(x, y) + iv(x, y) = w$ be a one-to-one conformal mapping of a domain $\mathcal{D}$ in the complex z-plane onto a domain $\mathcal{G}$ in the w-plane. Then, if the complex potential $\Psi(u, v) = V_1(u, v) + iV_2(u, v)$ in the complex w-plane describes the ideal fluid flow in the domain $\mathcal{G}$ in w, the corresponding potential obtained through the mapping $w = f(z)$, namely the potential*

$$\Psi(x, y) = V_1(u(x, y), v(x, y)) + iV_2(u(x, y), v(x, y))$$

describes the ideal fluid flow in the domain $\mathcal{D}$ in z.

The above result provides a useful approach for determining two- dimensional ideal fluid flow patterns in vessels of different shapes or around obstacles of different shapes, as will be illustrated below with several examples. The idea is to solve the problem in the w-plane where the shape of the vessel or the obstacle is simple and the corresponding streamlines are easy to determine, and then to use a mapping to obtain the corresponding streamlines in the z-plane. Recall that the streamlines are determined by the relation $V_2(u, v) = c_2$ in the w-plane, which becomes $V_2(u(x, y), v(x, y)) = c_2$ in the z-plane, giving the shape of the original curves in the z-plane from which the streamlines in the w-plane are produced through the mapping $w = f(z)$.

The above argument can be inverted, provided we interchange the roles of the $w(u, v)$ and $z(x, y)$ variables: if we know the streamlines in the z-plane,

$V_2(x,y) = c_2$, then we can obtain the streamlines in the w-plane, assuming that we find a mapping $z = g(w)$ which maps the shape in the w-plane that we are interested in, onto the corresponding shape in the z-plane for which we know the streamlines. In order for this to be consistent with the previous mapping $w = f(z)$, we must have

$$z = g(w) = g(f(z)) \Rightarrow g(z) = f^{-1}(z).$$

This is a very useful corollary of the TIIFF: it states that if we already know the streamlines around a shape in the z-plane, given by $V_2(x,y) = c_2$, and we can find a function $w = f(z)$ that describes another desired shape in the w-plane, then the streamlines around this shape will be given by $V_2(u,v) = c_2$, where we obtain u, v through $w = u + iv = f(z)$. The only requirement for this to be valid is that we can find the inverse function $f^{-1}(z)$, since the application of the TIIFF in going from the z-plane streamlines onto the w-plane streamlines was based on using the function $g(w) = z$ with $g(z) = f^{-1}(z)$.

Example 6.6: Ideal fluid flow around a circle

Consider the circle of radius 1 centered at the origin in the z-plane. In what follows, we will determine the streamlines for an ideal fluid flow around it, originating on the far left ($x \to -\infty$). In this connection, we will use the mapping

$$w = f(z) = z + \frac{1}{z}. \tag{6.11}$$

This function produces a conformal mapping since it is analytic everywhere except at $z = 0$, which is not part of the domain as we are interested in fluid flow around the unit circle centered at the origin. Equation (6.11) implies

$$w = (x + iy) + \frac{1}{x + iy} = x\left(1 + \frac{1}{x^2 + y^2}\right) + iy\left(1 - \frac{1}{x^2 + y^2}\right),$$

thus

$$u(x,y) = x\left(1 + \frac{1}{x^2 + y^2}\right), \quad v(x,y) = y\left(1 - \frac{1}{x^2 + y^2}\right).$$

Using these relations, it is easy to find the image of the unit circle in the w-plane, which turns out to be the line on the u-axis from $u = -2$ to $u = 2$. The streamlines from the far left that are parallel to this line are given by $v(x,y) = c_0$, where c_0 can take any real value. Taking $v = c_0$, we solve for x in terms of y using the above expression for $v(x,y)$:

$$x = \pm\sqrt{\frac{y}{y - c_0} - y^2}.$$

This is the relation that describes the curves in the z-plane which under the above mapping are mapped onto lines of $v = c_0$ constant in the w-plane. From the Invariance of Ideal Fluid Flow theorem, we conclude that this expression gives the streamlines in the z-plane for the flow around the unit circle centered at the origin.

It is useful to determine the limiting values of x and y: from the above relation we find that for $y = c_0 \Rightarrow x \to \pm\infty$, while for $y = (c_0/2) \pm \sqrt{(c_0/2)^2 + 1} \Rightarrow x = 0$, with the $(+)$ sign used for $c_0 > 0$ and the $(-)$ sign used for $c_0 < 0$. For $c_0 = 0$, the only possible solutions to the above relationship are for $y = 0$ and $|x| > 1$. In the diagram below, we show the streamlines for an ideal fluid flow around a circular obstacle (represented by the thicker red line), for various values of the constant c_0.

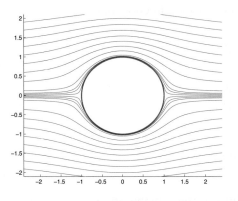

Having solved the problem of the flow in the z-plane around the circular obstacle, we can take advantage of this solution to obtain the flow in more complicated situations. For example, what is the flow around a slit extending from -2 to $+2$ on the horizontal axis, with the boundary condition that at a distance far from the slit the streamlines are at given angle ψ relative to the horizontal axis? For the flow around the circular obstacle, if the entire pattern is rotated by ψ, the streamlines are still the same, and asymptotically the flow is at the desired angle relative to the horizontal axis. This corresponds to the intermediate mapping $z_1 = e^{i\psi}z$. The image of the circle of radius 1 under the mapping

$$w_1 = z_1 + \frac{1}{z_1}, \quad z_1 = e^{i\psi}z, \tag{6.12}$$

is still the same as before, that is, a line segment from $u_1 = -2$ to $u_1 = +2$ (because the circle has just been rotated by ψ, which leaves the circle invariant). Therefore, using the TIIFF as discussed above, we can directly calculate the streamlines around the horizontal slit which are asymptotically at an angle ψ relative to the horizontal axis by simply taking the same mapping as before, that is, $f(z) = z + 1/z$, but with the new variable z_1. The only issue is to establish that the inverse of $f(z)$ exists. This can easily be shown by starting with $w = z + 1/z$ and solving for z in terms of w:

$$z^2 - wz + 1 = 0 \Rightarrow z = \frac{1}{2}\left[w + \left(w^2 - 4\right)^{1/2}\right] = g(w)$$

$$\Rightarrow \quad f^{-1}(z) = \frac{1}{2}\left[z + \left(z^2 - 4\right)^{1/2}\right],$$

where we have chosen the $+$ sign in front of the root in the solution of the second-order equation for z in terms of w. In the diagram below, we show the streamlines for ideal fluid flow around a horizontal slit (represented by the thicker red line), for a flow that far from the slit flows in a direction at $\psi = \pi/6 = 30°$ to the horizontal axis.

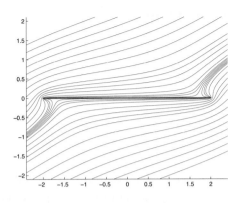

We can use these results to analyze an even more interesting case known as the "Joukowsky airfoil". Consider the circle in the z-plane with center at the point $(x_0, y_0) = (-d, c(1+d))$, and radius $\rho = (1+d)\sqrt{1+c^2}$, where d, c are real, positive constants. This circle can be thought of as produced by the linear mapping $z_2 = az + b$, with $a = (1+d)\sqrt{1+c^2}$, $b = -d + ic(1+d)$, where the values of the variable z now are given by $z = x + iy = \cos(\phi) + i\sin(\phi), 0 \le \phi < 2\pi$, that is, they describe the circle of radius 1, centered at the origin. We note that the circle described by the variable z_2 passes through the point $z_2 = 1$ and encloses the origin.

The mapping

$$w = f(z_2) = z_2 + \frac{1}{z_2}, \quad z_2 = az_1 + b, \quad z_1 = e^{i\psi}z, \qquad (6.13)$$

for $z_1 = z$ or $\psi = 0$, produces a shape that looks like the cross-section of the wing of an airplane (we refer to this shape as the "airfoil"). For this shape, the streamlines for ideal fluid flow originating far to the left will be the same as the streamlines around the original circle described by the variable z_2 because the mapping is a one-to-one conformal mapping: $f(z)$ is analytic everywhere except at $z = 0$, which is not part of the domain. But we have already figured out in the previous example the streamlines for the unit circle. Therefore, all we have to do is to use the equations that define the streamlines of the unit circle example, use the linear mapping of the unit circle onto the circle described by the variable z_2, and then apply the $w = f(z_2)$ mapping, which will give the streamlines of the airfoil. Moreover, we can combine the results of the previous example, and obtain the mapping of the streamlines around the airfoil which asymptotically flows in a direction at an angle ψ with respect to the horizontal axis: we simply rotate the flow in the z-plane before shifting and scaling the circular obstacle, namely, we use the variable $z_1 = e^{i\psi}z$ in the linear transformation $z_2 = az_1 + b$. The streamlines for several values of the constant s_0 are shown in the diagram below, for $\psi = 0$ (left) and $\psi = 30°$ (right).

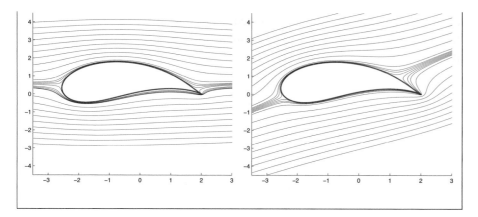

Further reading

The textbooks mentioned under Further Reading at the end of Chapter 4 contain much useful material on the subject of mappings by functions of complex variables. More specifically, the book by T. Needham contains many interesting and beautifully illustrated examples of mappings and an extended discussion of the Möbius transformation. The book by E.B. Saff and A.D. Snider contains a useful table of conformal mappings, as well as a very nice appendix on the numerical construction of conformal maps (by L.N. Trefethen and T. Driscoll).

Problems

1. In each of the following cases, graph the set of points in the z-plane and its image in the w-plane produced by the mapping $w = f(z)$:

   (a) $\left\{ z : \text{Re}[z] \le 0, \quad -\dfrac{\pi}{2} \le \text{Im}[z] \le \dfrac{\pi}{2} \right\}, \quad w = e^z,$

   (b) $\left\{ z : 0 < |z| \le 2, \quad 0 \le \text{Arg}[z] \le \dfrac{\pi}{3} \right\}, \quad w = z^3,$

   (c) $\left\{ z : 0 \le \text{Re}[z] \le a, \quad -\dfrac{\pi}{2} \le \text{Im}[z] \le 0 \right\},$
   $$w = i \cosh(z + 1),$$

   (d) $\{ z : |z| \ge 1, \quad 0 \le \text{Arg}[z] \le \pi \}, \quad w = z + \dfrac{1}{z},$

   (e) $\{ z : 0 \le \text{Re}[z] \le \pi, \quad 0 \le \text{Im}[z] \le 2 \}, \quad w = \cos(z).$

2. Find the image of the region $\{ z : 1 < |z| < 2, 0 < \text{Arg}[z] < \pi/6 \}$, under the following mappings:

   (a) $w = e^{i\pi/4} z,$

   (b) $w = -2iz + 1 + i,$

   (c) $w = z^{13},$

   (d) $w = \dfrac{1}{z},$

   (e) $w = \ln(z), \quad$ where $0 \le \text{Im}[z] < 2\pi.$

3. Let t be a real number with $0 < t < 2\pi$ and a be a complex number with $|a| > 1$. Show that the function
   $$w = f(z) = e^{it} \frac{z - a}{1 - \bar{a}z}$$
   maps $\{ z : |z| < 1 \}$ onto $\{ w : |w| > 1 \}$. Is the inverse of this map defined for the given regions?

4. Consider the mappings
   $$w_1 = \frac{z}{1 + z}, \quad w_2 = \frac{z}{1 - z}.$$

   (a) Show that w_1 maps the half-plane $\{ z : \text{Re}[z] > -1/2 \}$ onto the interior of the unit circle $\{ w : |w| < 1 \}$.

   (b) Show that w_2 maps the interior of the unit circle $\{ z : |z| < 1 \}$ onto the half-plane $\{ w : \text{Re}[w] > -1/2 \}$.

5. Consider the mapping from the z-plane onto the w-plane by the function
   $$f(z) = z^2 = w = u + iv.$$

   (a) Find a region in the z-plane whose image is the square domain in the w-plane bounded by the lines $u = 1, u = 2, v = 1, v = 2$.

   (b) Show that this mapping is a one-to-one mapping of the lines $\text{Re}[z] = a > 0$ onto the parabolas defined by the relation $v^2 = -4a^2(u - a^2)$.

   (c) Show that this mapping is a one-to-one mapping of the lines $\text{Im}[z] = b > 0$ onto the parabolas defined by the relation $v^2 = -4b^2(u + b^2)$.

6. Consider the mapping from the z-plane onto the w-plane by the function
   $$f(z) = e^z = w = \rho e^{i\phi}.$$

   Show that under this mapping the lines in the z-plane defined by $y = x/a$, where $a \ne 0$, are mapped onto the spirals $\rho = e^{a\phi}$ in the w-plane.

7. Consider the mapping from the z-plane onto the w-plane by the function

$$f(z) = \sin(z) = w.$$

Show that under this mapping the rectangular region in the z-plane defined by $-\pi \leq \text{Re}[z] \leq \pi$, $0 < a \leq \text{Im}[z] \leq b$ is mapped onto a region bounded by two ellipses.

8. Consider the mapping from the z-plane onto the w-plane by the function

$$f(z) = \ln(z) = w.$$

Find the image of the z-plane in the w-plane using the following branch cuts:

   (a) A cut consisting of the line segments $(-\infty, -1]$ and $[0, 1]$ on the real axis and the unit semicircle in the upper half-plane, with the convention $0 \leq \theta < 2\pi$ for $|r| < 1$ and $\pi \leq \theta < 3\pi$ for $|r| \geq 1$.

   In addition, find the image of a line through the origin at an angle of $\pi/4$ with the x-axis.

   (b) A spiral cut defined by the curve $r_c(t) = ae^{bt}$, $\phi_c(t) = t$, for all real t, where a, b are real, positive constants, and the restriction that for any point in the z-plane with $z = re^{i\phi}$, $t \leq \phi < t + 2\pi$ for the value of t that gives $r = r_c(t)$.

   In addition, find the image of the circle of radius R centered at the origin.

   (c) A cut consisting of the line segments $[1, \infty)$ and $[-1, 0]$ on the real axis and the unit semicircle in the upper half-plane, with the convention $-\pi \leq \phi < \pi$ for $r < 1$ and $2\pi \leq \phi < 4\pi$ for $r \geq 1$.

   In addition, find the image of the following curve: $r_c(t) = ae^{bt}$, $\phi_c(t) = t$, where t is a real variable that takes all possible values and a, b are real, positive constants.

9. Consider the mapping produced by the function of Eq. (6.5), which is known as the "Möbius transformation".

   (a) Show that the inverse of any Möbius transformation is a Möbius transformation.

   (b) Show that a Möbius transformation with non-zero parameter a, b, c, d can be written as a composition of the following transformations: translations by complex numbers, multiplication by complex numbers, and inversion $w = 1/z$.

   (c) Show that a Möbius transformation maps lines and circles onto lines and circles.

   (d) Find a Möbius transformation that maps the triplet of points $\{0, 1, i\}$ into the triplet of points $\{-1, \infty, 1\}$ (in the same order).

   (e) A point z_0 in the complex plane is called a fixed point of a map $f(z)$, if $f(z_0) = z_0$. How many fixed points may a Möbius transformation have?

   (f) Show that the combination of two Möbius transformations is another Möbius transformation. Find the explicit rule of transformation for the composition, that is, put the composition w in the Möbius form in terms of the coefficients of the respective transformations.

   (g) Show that the Möbius transformation of Eq. (6.5) maps the upper half-plane onto itself if a, b, c, $d \in \mathbb{R}$ and $ad - bc > 0$. What happens if $ad - bc < 0$?

   (h) Plot the images of the sets of curves in the z-plane described by

   (i) $|z - k| = k$, $k \in \mathbb{N}$,   (ii) $\left| z - \dfrac{i}{k} \right| = \dfrac{1}{k}$, $k \in \mathbb{N}$,

   under a Möbius transformation with $a = -1$, $b = c = d = 1$.

Chapter 7

The Fourier Transform

Fourier's theorem is not only one of the most beautiful results of modern analysis, but it may be said to furnish an indispensable instrument in the treatment of nearly every recondite question in modern physics.

William Thomson — Baron Kelvin (British physicist, 1824–1907)

7.1 Fourier series expansions

7.1.1 Real Fourier expansion

We consider a function $f(x)$ of the real variable x and its representation as a series expansion in terms of the functions $\cos(nx)$ and $\sin(nx)$ with n integer.

$$f(x) = a_0 + \sum_{n=1}^{\infty} a_n \cos(nx) + \sum_{n=1}^{\infty} b_n \sin(nx), \qquad (7.1)$$

where a_n and b_n are real constants. We have explicitly separated out the cosine term with $n = 0$, which is a constant a_0 (there is no analogous sine term because $\sin(0) = 0$). The expression of Eq. (7.1) is referred to as the "real Fourier series representation" or "real Fourier expansion". Note that we only need positive values for the indices n. Indeed, since $\cos(-nx) = \cos(nx)$ and $\sin(-nx) = -\sin(nx)$, the sine and cosine functions with negative indices give terms that can be absorbed in a_n and b_n; thus, terms with negative n do not provide any additional flexibility in representing the general function $f(x)$. Since each term in the expansion is periodic the expression on the right-hand side of Eq. (7.1) implies that the function $f(x)$ is periodic,

$$f(x + 2\pi) = f(x).$$

The fact that a periodic function $f(x)$, when it satisfies certain conditions of continuity and boundedness to be discussed later, can be represented by a series expansion that involves sines and cosines is referred to as "Fourier's theorem".

Our first task is to determine the values of the coefficients a_n and b_n that appear in the real Fourier expansion in terms of $f(x)$. Toward this goal, we multiply both sides of Eq. (7.1) by $\cos(mx)$ or by $\sin(mx)$ and integrate over x from $-\pi$ to π; this produces the following terms on the right-hand side:

$$\cos(nx)\cos(mx) = \frac{1}{2}\cos(nx - mx) + \frac{1}{2}\cos(nx + mx),$$

$$\sin(nx)\sin(mx) = \frac{1}{2}\cos(nx - mx) - \frac{1}{2}\cos(nx + mx),$$

$$\sin(nx)\cos(mx) = \frac{1}{2}\sin(nx - mx) + \frac{1}{2}\sin(nx + mx),$$

where we have used the trigonometric identities

$$\cos(a \pm b) = \cos(a)\cos(b) \mp \sin(a)\sin(b),$$

$$\sin(a \pm b) = \sin(a)\cos(b) \pm \cos(a)\sin(b),$$

that allow to rewrite the products of two cosines, two sines and a sine with a cosine in terms of sums. For $n \neq m$, we can easily integrate these terms over $-\pi \leq x \leq \pi$ and conclude that these integrals vanish because of the periodicity of the sine and cosine functions:

$$\int_{-\pi}^{\pi} \cos(nx)\cos(mx)\mathrm{d}x = \frac{1}{2}\left[\frac{\sin(nx - mx)}{n - m} + \frac{\sin(nx + mx)}{n + m}\right]_{-\pi}^{\pi} = 0,$$

$$\int_{-\pi}^{\pi} \sin(nx)\sin(mx)\mathrm{d}x = \frac{1}{2}\left[\frac{\sin(nx - mx)}{n - m} - \frac{\sin(nx + mx)}{n + m}\right]_{-\pi}^{\pi} = 0,$$

$$\int_{-\pi}^{\pi} \sin(nx)\cos(mx)\mathrm{d}x = -\frac{1}{2}\left[\frac{\cos(nx - mx)}{n - m} + \frac{\cos(nx + mx)}{n + m}\right]_{-\pi}^{\pi} = 0.$$

The only non-vanishing terms are those for $n = m$:

$$\int_{-\pi}^{\pi} \cos^2(nx)\mathrm{d}x = \int_{-\pi}^{\pi} \sin^2(nx)\mathrm{d}x = \frac{1}{2}\int_{-\pi}^{\pi}\left[\sin^2(nx) + \cos^2(nx)\right]\mathrm{d}x = \pi.$$

These relations show that the sine and cosine functions are "orthogonal", in the sense discussed in Chapter 3, Section 3.2, see Eq. (3.31). They also form a complete set, and thus can be used to expand any periodic function that satisfies the conditions of continuity and boundedness.

For $n = 0$ the integral is trivial and concerns only the cosine term, producing 2π. Collecting the above results, we arrive at the following expressions for the coefficients a_n, b_n:

$$a_0 = \frac{1}{2\pi}\int_{-\pi}^{\pi} f(x)\mathrm{d}x, \tag{7.2a}$$

$$a_n = \frac{1}{\pi}\int_{-\pi}^{\pi} f(x)\cos(nx)\mathrm{d}x, \tag{7.2b}$$

$$b_n = \frac{1}{\pi}\int_{-\pi}^{\pi} f(x)\sin(nx)\mathrm{d}x. \tag{7.2c}$$

These expressions are known as the "Euler formulas". We next give some examples of real Fourier series expansions.

Example 7.1: Consider the function that describes a so-called square wave:

$$f(x) = \begin{cases} -1, & -\pi < x < 0, \\ +1, & 0 < x < \pi, \end{cases} \tag{7.3}$$

which we want to express as a real Fourier expansion. This function has discontinuities at $x = 0, \pm\pi$ and at the periodic images of these points differing by 2π. We have purposefully avoided to assign a value to the function at these points, and this will be discussed below. However, the values of the function in the limit approaching these points from above or below are well-defined:

$$f(0^{(+)}) = +1, \quad f(0^{(-)}) = -1, \quad f(\pi^{(-)}) = +1, \quad f(-\pi^{(+)}) = -1.$$

Using the Euler formulas we can determine the coefficients of this expansion, which are the following:

$$a_n = 0, \quad \text{for all } n; \quad b_n = \frac{4}{\pi n}, \quad n = \text{odd}; \quad b_n = 0, \quad n = \text{even}.$$

With these coefficients, we can then express the original function as a Fourier expansion:

$$f(x) = \frac{4}{\pi}\left[\sin(x) + \frac{1}{3}\sin(3x) + \frac{1}{5}\sin(5x) + \frac{1}{7}\sin(7x) + \cdots\right].$$

In the diagrams below we show the Fourier expansion (FE) graph using terms up to $n = 10$ (in blue line) to the original function (shown in red-dashed line, superimposed to the fit). We also show the corresponding coefficients b_n.

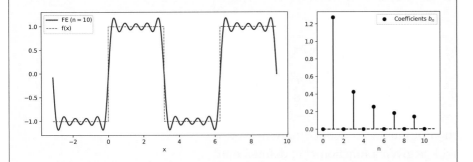

As an application, we evaluate both sides at $x = \pi/2$, where $f(\pi/2) = 1$ to obtain

$$1 = \frac{4}{\pi}\left[\sin\left(\frac{\pi}{2}\right) + \frac{1}{3}\sin\left(\frac{3\pi}{2}\right) + \frac{1}{5}\sin\left(\frac{5\pi}{2}\right) + \cdots\right] = \frac{4}{\pi}\left[1 - \frac{1}{3} + \frac{1}{5} + \cdots\right]$$

$$\Rightarrow \sum_{n=0}^{\infty} \frac{(-1)^n}{2n+1} = \frac{\pi}{4}.$$

Example 7.2: As a second example, consider the function which describes the so-called triangular wave:

$$f(x) = \begin{cases} -\frac{x}{\pi}, & -\pi \leq x \leq 0, \\ +\frac{x}{\pi}, & 0 \leq x \leq \pi. \end{cases} \tag{7.4}$$

Using the Euler formulas, we can determine the coefficients for the Fourier expansion of this function:

$$a_0 = \frac{1}{2}; \quad a_n = -\frac{4}{\pi^2 n^2}, \quad n = \text{odd}; \quad a_n = 0, \quad n = \text{even}; \quad b_n = 0, \quad \text{for all } n.$$

Employing these constants to express $f(x)$ as a Fourier expansion we find

$$f(x) = \frac{1}{2} - \frac{4}{\pi^2} \left[\cos(x) + \frac{1}{9} \cos(3x) + \frac{1}{25} \cos(5x) + \frac{1}{49} \cos(7x) + \cdots \right].$$

In the diagrams below, we show the Fourier expansion (FE) graph using terms up to $n = 10$ (in blue line) to the original function (shown in red-dashed line superimposed to the fit), as well as the corresponding coefficients a_n.

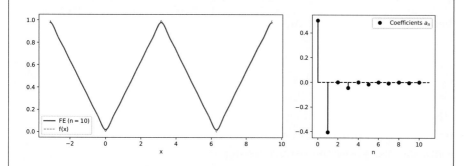

As an application, we evaluate both sides at $x = \pi$, where $f(\pi) = 1$, to obtain

$$1 = \frac{1}{2} - \frac{4}{\pi^2} \left[\cos(\pi) + \frac{1}{9} \cos(3\pi) + \frac{1}{25} \cos(5\pi) + \frac{1}{49} \cos(7\pi) + \cdots \right]$$

$$= \frac{1}{2} + \frac{4}{\pi^2} \left[1 + \frac{1}{9} + \frac{1}{25} + + \frac{1}{49} \cdots \right] \Rightarrow \sum_{n=0}^{\infty} \frac{1}{(2n+1)^2} = \frac{\pi^2}{8}.$$

Half-range expansion: Often, the function of interest is defined only in half of the usual range. For example, we might be given a function $f(x)$ defined only in the interval $[0, \pi]$. In this case, we cannot readily apply the expressions of the Euler formulas to obtain the coefficients of the Fourier expansion. What is usually done, is to assume either a cosine series expansion for the function in the interval $[-\pi, \pi]$, or a sine series expansion in the same interval. The first choice produces an even function in the interval $[-\pi, \pi]$, while the second choice produces an odd function. Since both expansions approximate the original function accurately in the interval $[0, \pi]$ (assuming enough terms in the expansion have been retained), it does not matter which choice we make. The behavior of the function, including its convergence properties, may make it clear which is the preferred expansion choice.

7.1.2   Odd and even function expansions

It is instructive to consider the expansions in the examples considered above (Examples 7.1 and 7.2) in more detail. Both expressions are *approximations* to the original functions which become better the more terms are included in the expansion. The terms that appear in the expansions have the same symmetry as the functions themselves. Specifically, the square wave is an odd function of x, namely, it changes sign when its argument changes sign

$$\text{odd function}: \quad f_{\text{odd}}(-x) = -f_{\text{odd}}(x), \tag{7.5}$$

thus, only the sine terms appear in the expansion, which are also odd functions of x, $\sin(-nx) = -\sin(nx)$. Similarly, the triangular wave is an even function of x, namely, it does not change sign when its argument changes sign,

$$\text{even function}: \quad f_{\text{evn}}(-x) = f_{\text{evn}}(x), \qquad (7.6)$$

thus, only the cosine terms appear in the expansion, which are also even functions of x, since $\cos(-nx) = \cos(nx)$. If the function $f(x)$ has no definite symmetry for $x \to -x$, then the Fourier expansion will contain both types of terms (sines and cosines).

7.1.3   Handling of discontinuities in Fourier expansions

The square wave has discontinuities at $x = 0, \pm\pi$, whereas the triangular wave is a continuous function. The sines and cosines are *continuous* everywhere. In this sense, approximating the original function by a *truncated* sine or cosine expansion, because after all we can only handle a finite number of terms in the expansion, is not accurate, as it misses the discontinuity. This is evident in the example of the square wave, where the value of the truncated expansion at the points $x = 0, \pm\pi$, as well as at the periodic images that differ from these points by 2π, is always equal to zero. The Fourier series at points of discontinuities converges to the average value. Thus, to reconcile this difference between the original function and the truncated expansion, we assign to the original function at the discontinuity at the point x_0 the value

$$f(x_0) = \frac{1}{2}\left[f(x_0^+) + f(x_0^-)\right], \quad f(x_0^\pm) = \lim_{\epsilon \to 0} f(x_0 \pm \epsilon), \quad \epsilon > 0, \qquad (7.7)$$

that is, we assign to the function the average of the two limiting values as the discontinuity is approached from the left or from the right (this is known as a "Dirichlet boundary condition"). In formal terms, we make the function "point-wise" continuous, just like the value the truncated expansion.

7.1.4   Fourier expansion of arbitrary range

In the preceding discussion, we assumed that the functions under consideration were defined in the interval $[-\pi, \pi]$ and were periodic with a period of 2π (as the Fourier expansion requires). The restriction of the values of x in this interval is not crucial. We can consider a periodic function $f(x)$ defined in an interval of arbitrary length $2L$, with $x \in [-L, L]$ and $f(x + 2L) = f(x)$. Then, using the change of variables

$$x' = \frac{\pi}{L}x \Rightarrow x = \frac{L}{\pi}x'$$

and rewriting the function as $f(x) = g(x')$, we see that in the variable x' which takes values in the interval $[-\pi, \pi]$, $g(x')$ has period 2π. Therefore, we can apply the Euler formulas to find the Fourier expansion of $g(x')$, and then the corresponding expansion of $f(x)$. The result, after we change variables back to x, is given by the following formulas:

$$f(x) = a_0 + \sum_{n=1}^{\infty} a_n \cos\left(\frac{\pi n}{L}x\right) + \sum_{n=1}^{\infty} b_n \sin\left(\frac{\pi n}{L}x\right), \tag{7.8a}$$

$$a_0 = \frac{1}{2\pi}\int_{-\pi}^{\pi} g(x')\mathrm{d}x' = \frac{1}{2L}\int_{-L}^{L} f(x)\mathrm{d}x, \tag{7.8b}$$

$$a_n = \frac{1}{\pi}\int_{-\pi}^{\pi} g(x')\cos(nx')\mathrm{d}x' = \frac{1}{L}\int_{-L}^{L} f(x)\cos\left(\frac{\pi n}{L}x\right)\mathrm{d}x, \tag{7.8c}$$

$$b_n = \frac{1}{\pi}\int_{-\pi}^{\pi} g(x')\sin(nx')\mathrm{d}x' = \frac{1}{L}\int_{-L}^{L} f(x)\sin\left(\frac{\pi n}{L}x\right)\mathrm{d}x. \tag{7.8d}$$

For functions that have definite symmetry for $x \to -x$, we can again reduce the computation to finding only the coefficients of terms with the same symmetry. Specifically, in the case of odd functions $f_{\mathrm{odd}}(x)$ only the sine terms survive, whereas in the case of even functions $f_{\mathrm{evn}}(x)$, only the cosine terms survive:

$$f_{\mathrm{evn}}(-x) = f_{\mathrm{evn}}(x) \Rightarrow a_n = \frac{2}{L}\int_0^L f_{\mathrm{evn}}(x)\cos\left(\frac{\pi n}{L}x\right)\mathrm{d}x, \quad b_n = 0 \; \forall n;$$

$$\tag{7.9a}$$

$$f_{\mathrm{odd}}(-x) = -f_{\mathrm{odd}}(x) \Rightarrow b_n = \frac{2}{L}\int_0^L f_{\mathrm{odd}}(x)\sin\left(\frac{\pi n}{L}x\right)\mathrm{d}x, \quad a_n = 0 \; \forall n.$$

$$\tag{7.9b}$$

7.1.5 Complex Fourier expansion

It was shown in Chapter 4 that $\sin(nx)$ and $\cos(nx)$ can be rewritten in terms of e^{inx} and e^{-inx}. Thus, we can generalize the notion of the Fourier expansion to complex numbers by using the complex exponentials $\exp(inx)$ as the basis set, instead of the sines and cosines. With this complex notation we obtain the complex Fourier series expansion

$$f(x) = \sum_{n=-\infty}^{\infty} c_n e^{inx}, \tag{7.10}$$

where c_n are complex constants. The values of these coefficients can be obtained from the Euler's formula using the complex exponential

$$e^{ix} = \cos(x) + i\sin(x),$$

and the Euler formulas for the coefficients that appear in the real Fourier series expansion. Alternatively, we can multiply both sides of Eq. (7.10) by e^{-imx} and integrate over x in $[-\pi, \pi]$, using the identity

$$\int_{-\pi}^{\pi} e^{i(n-m)x}\mathrm{d}x = 2\pi\delta_{nm}.$$

This formula shows that complex exponentials with different indices n and m are orthogonal. The coefficient c_n of $\exp(inx)$ is given by

$$c_n = \frac{1}{2\pi}\int_{-\pi}^{\pi} f(x)e^{-inx}\mathrm{d}x. \tag{7.11}$$

7.1.6   Fourier expansion of convolution and correlation

We discuss next a useful relation that applies to the Fourier expansion of the convolution, as well as the correlation of two periodic functions $f(x)$ and $g(x)$. Let us assume that the Fourier expansions of these functions are known,

$$f(x) = \sum_{n=-\infty}^{\infty} c_n e^{inx}, \quad g(x) = \sum_{m=-\infty}^{\infty} d_m e^{imx},$$

that is, the coefficients c_n and d_m have been determined.

Theorem: *The Fourier expansion of the convolution of two periodic functions is given by the term-by-term product of the Fourier expansions of the two functions:*

$$f \star g(x) = 2\pi \sum_{n=-\infty}^{\infty} (c_n d_n) e^{inx}. \tag{7.12}$$

Proof: The convolution of the two periodic functions is defined by

$$
\begin{aligned}
f \star g(x) &\equiv \int_{-\pi}^{\pi} f(y) g(x-y) dy = \sum_{n=-\infty}^{\infty} \sum_{m=-\infty}^{\infty} c_n d_m \int_{-\pi}^{\pi} e^{iny} e^{im(x-y)} dy \\
&= \sum_{n=-\infty}^{\infty} \sum_{m=-\infty}^{\infty} c_n d_m e^{imx} \int_{-\pi}^{\pi} e^{i(n-m)y} dy \\
&= \sum_{n=-\infty}^{\infty} \sum_{m=-\infty}^{\infty} c_n d_m e^{imx} (2\pi \delta_{nm}),
\end{aligned}
$$

and using the properties of the Kroenecker-δ we arrive at the desired expression, Eq. (7.12). This expresses the convolution in terms of a series in complex exponentials with coefficients equal, term-by-term, to the product of the coefficients of the corresponding Fourier expansions of the two functions (the factor of 2π arises from the definition of the Fourier expansion). ∎

Theorem: *The Fourier expansion of the correlation of two periodic functions is given by the term-by-term product of the Fourier expansions of the functions, with the indices of one set of coefficients running in the opposite direction from the other:*

$$c[f,g](x) = 2\pi \sum_{n=-\infty}^{\infty} (c_{-n} d_n) e^{inx}. \tag{7.13}$$

Proof: The correlation of the two periodic functions is defined by

$$
\begin{aligned}
c[f,g](x) &\equiv \int_{-\pi}^{\pi} f(y) g(x+y) dy = \sum_{n=-\infty}^{\infty} \sum_{m=-\infty}^{\infty} c_n d_m \int_{-\pi}^{\pi} e^{iny} e^{im(x+y)} dy \\
&= \sum_{n=-\infty}^{\infty} \sum_{m=-\infty}^{\infty} c_{-n} d_m e^{imx} \int_{-\pi}^{\pi} e^{-i(n-m)y} dy \\
&= \sum_{n=-\infty}^{\infty} \sum_{m=-\infty}^{\infty} c_{-n} d_m e^{imx} (2\pi \delta_{nm}),
\end{aligned}
$$

and using the properties of the Kroenecker-δ we arrive at the desired expression, Eq. (7.13). This expresses the correlation in terms of a series in complex exponentials with coefficients equal, term-by-term, to the product of the coefficients of the corresponding Fourier expansions of the two functions, with one index, namely that of coefficients "c" running in the opposite direction of the other, namely that of coefficients "d" (the factor of 2π arises from the definition of the Fourier expansion). ∎

7.1.7   Conditions for the validity of a Fourier expansion

For the Fourier expansion to be valid, that is, to converge point-wise to the value of the function $[f(x^+) + f(x^-)]/2$, it is sufficient (but not necessary) that the function $f(x)$ satisfy the following conditions:

- The function must be single-valued in the range $x \in [-L, L]$.

- The function does not need to be continuous everywhere in the range $[-L, L]$, but it can only have at most a *finite* number of *finite* discontinuities.

- The function must have a *finite* number of extrema (maxima or minima) in the range $[-L, L]$.

- The integral of the absolute value of the function must converge (bounded by a finite number):

$$\int_{-L}^{L} |f(x)| \mathrm{d}x \leq F, \quad F : \text{ finite.}$$

It turns out that any function which is piece-wise continuously differentiable (like most functions considered here) satisfies these conditions. The Fourier series expansion may not converge at every point; however, the set of points where it diverges must be small in a certain technical sense (it has a "measure zero").

7.2   The Fourier transform

7.2.1   Definition of the Fourier transform

The Fourier transform is the limit of the Fourier expansion for a non-periodic function $f(x)$ defined on the real line $-\infty < x < \infty$. An important assumption, necessary to make sure that the Fourier transform is a mathematically meaningful expression, is that the integral of the absolute value of the function is bounded

$$\int_{-\infty}^{\infty} |f(x)| \mathrm{d}x \leq F, \quad F \text{ finite.} \tag{7.14}$$

We refer to this condition as the function being "integrable" in absolute value. Moreover, if this relation is valid, it implies that

$$\lim_{x \to \pm\infty} [f(x)] = 0, \tag{7.15}$$

because otherwise, the tail of the integral (as $x \to \pm\infty$) would produce unbounded contributions to the integral.

To provide a heuristic derivation of the Fourier transform, we start with the Fourier expansion

$$f(x) = \sum_{n=-\infty}^{\infty} c_n \mathrm{e}^{in\pi x/L}, \quad c_n = \frac{1}{2L} \int_{-L}^{L} f(x) \mathrm{e}^{-in\pi x/L} \mathrm{d}x. \tag{7.16}$$

We define a new variable $k = n\pi/L$, which takes different values when the value of n changes. Note that the spacing of values of k is given by

$$\Delta k = \frac{\Delta n\,\pi}{L} = \frac{\pi}{L} \Rightarrow \frac{\Delta k}{2\pi} = \frac{1}{2L}, \tag{7.17}$$

since the spacing of values of n is $\Delta n = 1$. When $L \to \infty$, the quantity Δk becomes infinitesimal, which we will denote as dk. From the Fourier expansion coefficients, we obtain in this limit the expression

$$\lim_{L \to \infty} (c_n 2L) = \lim_{L \to \infty} \left[\int_{-L}^{L} f(x) e^{-ikx} dx \right]. \tag{7.18}$$

We define the right-hand side of this equation as the Fourier transform (FT), denoted as $\hat{f}(k)$,

$$\hat{f}(k) \equiv \int_{-\infty}^{\infty} f(x) e^{-ikx} dx, \quad k \in \mathbb{R}. \tag{7.19}$$

$\hat{f}(k)$ is a function of the continuous variable k (because the spacing of its values is now the infinitesimal quantity dk). $\hat{f}(k)$ is a finite quantity:

$$|\hat{f}(k)| = \left| \int_{-\infty}^{\infty} f(x) e^{-ikx} dx \right| \leq \int_{-\infty}^{\infty} |f(x)| dx \leq F,$$

as a consequence of Eq. (7.14). From the expression for $f(x)$ in terms of the Fourier expansion, Eq. (7.16), we obtain

$$f(x) = \sum_{n=-\infty}^{\infty} c_n e^{in\pi x/L} = \sum_{n=-\infty}^{\infty} (2Lc_n) e^{in\pi x/L} \frac{1}{2L} = \sum_{n=-\infty}^{\infty} (2Lc_n) e^{in\pi x/L} \frac{\Delta k}{2\pi},$$

where we have used Eq. (7.17) to obtain the last expression. Taking the limit $L \to \infty$ in the previous expression, we can substitute $\hat{f}(k)$ for the quantity $(2Lc_n)$ using Eq. (7.18), and we can replace Δk by its infinitesimal limit dk using Eq. (7.17). Finally, turning the infinite sum to an integral over the continuous variable k, we obtain

$$f(x) = \int_{-\infty}^{\infty} \hat{f}(k) e^{ikx} \frac{dk}{2\pi}, \quad x \in \mathbb{R}. \tag{7.20}$$

This relation is known as the "Inverse Fourier Transform", which allows us to calculate $f(x)$ if its FT $\hat{f}(k)$ is known.

The function $f(x)$ is well-defined provided that the function $\hat{f}(k)$ defined by Eq. (7.19) is integrable. For this to happen, it is *not* sufficient to assume the absolute integrability of $f(x)$; it is necessary for $f(x)$ to satisfy an additional condition. This precise condition is technically complicated; in practice one assumes that $f(x)$, in addition to satisfying Eq. (7.14), it is also square integrable, namely

$$\int_{-\infty}^{\infty} |f(x)|^2 dx < M, \; M \text{ finite}.$$

7.2.2 *Properties of the Fourier transform*

We discuss several properties of the FT, which can be easily derived from its definition:

1. **Linearity of the FT:** If $\hat{f}(k)$ and $\hat{g}(k)$ are the FTs of the functions $f(x)$ and $g(x)$, then the FT of the linear combination of $f(x)$ and $g(x)$ with two arbitrary constants a, b, is given by

$$af(x) + bg(x) \longrightarrow a\hat{f}(k) + b\hat{g}(k). \tag{7.21}$$

2. **Discontinuities in $f(x)$:** Consider the function $f(x)$ which has a discontinuity Δf at x_0:

$$\Delta f = \lim_{\epsilon \to 0}[f(x_0 + \epsilon) - f(x_0 - \epsilon)] \text{ is finite.}$$

We want to calculate the FT of this function:

$$
\begin{aligned}
\hat{f}(k) &= \int_{-\infty}^{\infty} e^{-ikx} f(x) dx \\
&= \lim_{\epsilon \to 0}\left[\int_{-\infty}^{x_0-\epsilon} e^{-ikx} f(x) dx + \int_{x_0+\epsilon}^{\infty} e^{-ikx} f(x) dx\right] \\
&= \lim_{\epsilon \to 0}\left\{\left[\frac{e^{-ikx}}{-ik} f(x)\right]_{-\infty}^{x_0-\epsilon} + \left[\frac{e^{-ikx}}{-ik} f(x)\right]_{x_0+\epsilon}^{\infty}\right\} \\
&\quad - \int_{-\infty}^{x_0-\epsilon} \frac{e^{-ikx}}{-ik} f'(x) dx - \int_{x_0+\epsilon}^{\infty} \frac{e^{-ikx}}{-ik} f'(x) dx \\
&= \frac{e^{-ikx_0}}{ik}\Delta f + \frac{1}{ik}\left[\int_{-\infty}^{x_0-\epsilon} e^{-ikx} f'(x) dx + \int_{x_0+\epsilon}^{\infty} e^{-ikx} f'(x) dx\right], \quad (7.22)
\end{aligned}
$$

where we have assumed that $f(x) \to 0$ for $x \to \pm\infty$, which is required for the FT to be meaningful. Similarly, a discontinuity $\Delta f^{(n)}$ in the nth derivative produces the FT

$$\frac{e^{-ikx_0}}{(ik)^{n+1}}\Delta f^{(n)}. \tag{7.23}$$

3. **FT of the derivative of $f(x)$:** Given that $\hat{f}(k)$ is the FT of $f(x)$, we can obtain the FT of $f'(x)$, assuming that $f(x) \to 0$ for $x \to \pm\infty$:

$$\int_{-\infty}^{\infty} f'(x) e^{-ikx} dx = \left[f(x)e^{-ikx}\right]_{-\infty}^{\infty} - \int_{-\infty}^{\infty} f(x)(-ik)e^{-ikx} dx = ik\hat{f}(k)$$

$$= ik\hat{f}(k). \tag{7.24}$$

Similarly, the FT of $f^{(n)}(x)$ is given by $(ik)^n \hat{f}(k)$.

4. **FT of the moments of $f(x)$:** The nth moment of the function $f(x)$ is defined by

$$\int_{-\infty}^{\infty} x^n f(x) dx.$$

From the definition of the FT we see that the zeroth moment equals $\hat{f}(0)$. We take the derivative of the FT with respect to k:

$$\frac{d\hat{f}}{dk}(k) = \int_{-\infty}^{\infty} f(x)(-ix)e^{-ikx} dx \Rightarrow i\frac{d\hat{f}}{dk}(0) = \int_{-\infty}^{\infty} xf(x) dx. \tag{7.25}$$

This is easily generalized to an expression for the nth moment:

$$\int_{-\infty}^{\infty} x^n f(x) dx = i^n \frac{d^n \hat{f}}{dk^n}(0). \tag{7.26}$$

5. **FT of the convolution:** Consider two functions $f(x), g(x)$ and their FTs $\hat{f}(k), \hat{g}(k)$. The convolution $f \star g(x)$ is defined by

$$f \star g(x) \equiv \int_{-\infty}^{\infty} f(y)g(x-y) dy = h(x). \tag{7.27}$$

The Fourier transform of the convolution is the product of the Fourier transforms of the two corresponding functions:

$$
\begin{aligned}
\hat{h}(k) &= \int_{-\infty}^{\infty} e^{-ikx} \left[\int_{-\infty}^{\infty} g(x-y)f(y)dy \right] dx \\
&= \int_{-\infty}^{\infty} \left[\int_{-\infty}^{\infty} e^{-ik(x-y)} g(x-y)dx \right] e^{-iky} f(y)dy \\
&= \hat{f}(k)\hat{g}(k).
\end{aligned}
\tag{7.28}
$$

6. **FT of the correlation:** Consider two functions $f(x), g(x)$ and their FTs $\hat{f}(k), \hat{g}(k)$. The correlation $c[f,g](x)$ is defined by

$$
c[f,g](x) \equiv \int_{-\infty}^{\infty} f(y+x)g(y)dy.
\tag{7.29}
$$

The Fourier transform of the correlation is computed as follows:

$$
\begin{aligned}
\hat{c}(k) &= \int_{-\infty}^{\infty} e^{-ikx} \left[\int_{-\infty}^{\infty} f(y+x)g(y)dy \right] dx \\
&= \int_{-\infty}^{\infty} \left[\int_{-\infty}^{\infty} e^{-ik(y+x)} f(y+x)dx \right] e^{-i(-k)y} g(y)dy \\
&= \hat{f}(k)\hat{g}(-k).
\end{aligned}
\tag{7.30}
$$

Example 7.3: Consider the function $f(x)$ defined by

$$
f(x) = \begin{cases} e^{-ax}, & x > 0, \\ 0, & x < 0, \end{cases}
\tag{7.31}
$$

with $a > 0$. The FT of this function is given by

$$
\hat{f}(k) = \int_0^{\infty} e^{-ikx-ax}dx = \left[\frac{1}{ik+a} e^{-(ik+a)x} \right]_0^{\infty} = \frac{1}{ik+a}.
$$

Using the identity

$$
\frac{1}{ik+a} = \frac{-ik+a}{k^2+a^2},
$$

we find

$$
\hat{f}(k) = \left(\frac{a}{k^2+a^2} \right) + i \left(\frac{-k}{k^2+a^2} \right).
\tag{7.32}
$$

In the diagrams below we show the original function $f(x)$ (left panel, in red) and its Fourier transform (FT) $\hat{f}(k)$ (right panel, real and imaginary parts in blue and green respectively), for $a = 1$.

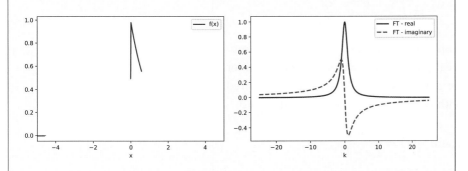

We first explore the implications of this result. At first sight, it seems to contradict our conclusion about the behavior of the FT in the presence of discontinuities, Eq. (7.22). However, this is not the case:

$$\frac{1}{ik+a} = \frac{1}{ik[1-(-a/ik)]} = \frac{1}{ik}\left[1+\frac{-a}{ik}+\left(\frac{-a}{ik}\right)^2+\cdots\right] = \sum_{n=0}^{\infty}\frac{(-a)^n}{(ik)^{n+1}},$$

which converges for $|a/ik| < 1 \Rightarrow |k| > a$. We also note that the nth derivative of the function has a discontinuity

$$\Delta f^{(n)} = (-a)^n,$$

at $x = 0$, and therefore this last expression is exactly what is predicted by our analysis, Eq. (7.23). Finally, since the function $f(x)$ is integrable in absolute value (this can be easily checked by explicitly evaluating the integral of $|f(x)|$), the FT is finite for all values of k, even for $k \to 0$.

An interesting limit of the function $f(x)$ is to take $a \to 0$. In this limit, the function becomes the Heaviside step function, $\theta(x)$, discussed in Chapter 1, Section 1.6. We can then use the above results to find the FT of the $\theta(x)$ function: from Eq. (7.32), the imaginary part of the FT is:

$$\lim_{a\to 0}\left(\mathrm{Im}\left[\hat{f}(k)\right]\right) = \lim_{a\to 0}\left(\frac{-k}{k^2+a^2}\right) = -\frac{1}{k},$$

while the real part is

$$\lim_{a\to 0}\left(\mathrm{Re}\left[\hat{f}(k)\right]\right) = \lim_{a\to 0}\left(\frac{a}{k^2+a^2}\right).$$

The limit $a \to 0$ for the real part gives a function that has an infinite peak and is infinitely narrow, as is evident from the plot of the real and imaginary parts of $\hat{f}(k)$ shown above. This is the behavior of a δ-function in the variable k. Moreover, the integral of this function over all real values of k is equal to π, as shown explicitly in Example 5.1 of Chapter 5, see Eq. (5.6). Putting these results together, we arrive at the conclusion that the FT of the function $\theta(x)$ is given by the expression:

$$\hat{\theta}(k) = \pi\hat{\delta}(k) - i\frac{1}{k} = \frac{1}{ik} + \pi\hat{\delta}(k).$$

We will derive this result again below, in Section 7.3.

Finally, we compute the inverse FT, using Eq. (7.20):

$$f(x) = \int_{-\infty}^{\infty}\hat{f}(k)e^{ikx}\frac{dk}{2\pi} = \int_{-\infty}^{\infty}\frac{e^{ikx}}{i(k-ia)}\frac{dk}{2\pi}.$$

To perform this integral we use contour integration: we define the complex variable $z = k + iq$, whose real part is the variable k, and consider closed contours in the (k,q) plane that allow us to calculate the desired integral by contour integration, with the function $g(z)$ as the integrand:

$$\int_C g(z)\frac{dz}{2\pi}, \quad \text{where} \quad g(z) = \frac{e^{izx}}{i(z-ia)}.$$

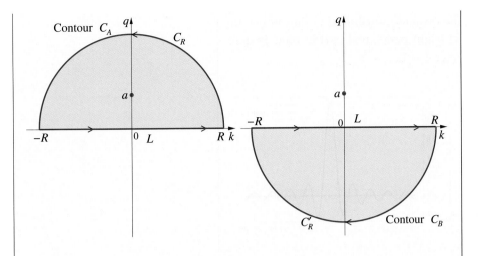

For $x > 0$, in order to satisfy Jordan's lemma, we choose the contour C_A shown in the left diagram, which lies on the upper half-plane. C_A consists of the line segment L extending from $-R$ to R on the real axis, with $R \to \infty$, and the semicircle C_R centered at the origin of radius R closing counter-clockwise on the upper half-plane. The integral along the line segment L gives the desired integral for $f(x)$. Applying the Residue theorem for this contour and noticing that the contour C_A contains only the singularity at $w = ia$ which is a simple pole, we find

$$f(x) = 2\pi i \, \text{Residue} \left[\frac{1}{2\pi} g(z) \right]_{z=ia} = i \frac{e^{i(ia)x}}{i} = e^{-ax}, \quad \text{for } x > 0.$$

For $x < 0$, in order to satisfy Jordan's lemma, we choose the contour C_B shown in the right diagram, which lies on the lower half-plane. C_B consists of the line segment L extending from $-R$ to R on the real axis, with $R \to \infty$, and the semicircle C_R' centered at the origin of radius R closing clockwise on the lower half-plane. Applying the Residue theorem for this contour, and noticing that it contains no singularities, we find

$$f(x) = 0 \quad \text{for } x < 0.$$

Combining these results we recover the original function $f(x)$ for all values of x.

Example 7.4: Consider the function $f(x)$ defined by (assuming $a > 0$)

$$f(x) = \begin{cases} \dfrac{1}{2a} & \text{for } -a \leq x \leq a, \\ 0 & \text{for } x < -a \text{ and } x > a. \end{cases} \tag{7.33}$$

The FT of this function is given by

$$\hat{f}(k) = \frac{1}{2a} \int_{-a}^{a} e^{-ikx} dx = \frac{1}{2ika} e^{ika} - \frac{1}{2ika} e^{-ika} = \frac{\sin(ka)}{ka}. \tag{7.34}$$

In the diagrams below we show the original function $f(x)$ (left panel, in red) and its Fourier transform $\hat{f}(k)$ (right panel, real part in blue, imaginary part is zero), with the parameter $a = 1$.

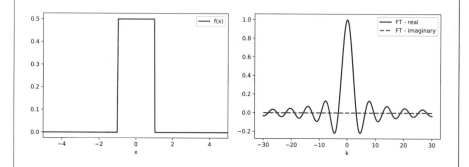

This is precisely the result we would have expected from the properties of the FT, namely, the presence of the two discontinuities at $x = \pm a$, of size $\Delta f(x = -a) = +1/2a$ and $\Delta f(x = +a) = -1/2a$ produces the two terms involving the factors $1/(ik)$. Interestingly, this function becomes a δ-function in the limit $a \to 0$. In this limit, the FT becomes equal to 1 for all finite values of k.

The inverse FT yields

$$f(x) = \int_{-\infty}^{\infty} \left[\frac{e^{ika}}{2ika} - \frac{e^{-ika}}{2ika} \right] e^{ikx} \frac{dk}{2\pi}$$

$$= \frac{1}{2ia} \left[\int_{-\infty}^{\infty} \frac{e^{ik(x+a)}}{k} \frac{dk}{2\pi} - \int_{-\infty}^{\infty} \frac{e^{ik(x-a)}}{k} \frac{dk}{2\pi} \right].$$

We define the last two integrals inside the square bracket as I_1 and I_2. To evaluate these integrals we will employ contour integration on the complex plane with the variable $z = k + iq$:

$$I_1 = \int_{-\infty}^{\infty} g_1(k) \frac{dk}{2\pi}, \quad \text{where} \quad g_1(k) = \frac{e^{ik(x-a)}}{k},$$

$$I_2 = \int_{-\infty}^{\infty} g_2(k) \frac{dk}{2\pi}, \quad \text{where} \quad g_2(k) = \frac{e^{ik(x-a)}}{k}.$$

To calculate I_1, we first need to consider whether $(x + a)$ is greater or smaller than zero. For $(x + a) > 0$, we consider the contour C_A shown in the left diagram below

$$C_A = C_R \cup L \cup C_\epsilon.$$

C_A consists of the semicircle C_R centered at the origin of radius R with $R \to \infty$, the semicircle C_ϵ centered at the origin of radius ϵ with $\epsilon \to 0$, and the line segment L extending from $-R$ to R and avoiding the pole at $z = 0$.

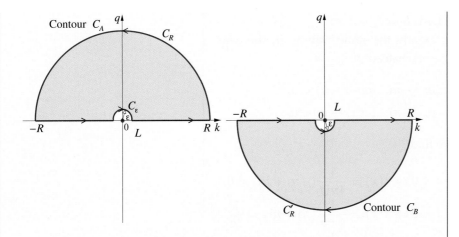

The contour C_A is closed on the upper half of the complex plane to satisfy Jordan's lemma. Calculating the corresponding contour integral through the Residue theorem, since it does not include any singularities, we find

$$I_R + I_L + I_\epsilon = 0.$$

From Jordan's lemma,

$$\lim_{R\to\infty} I_R = 0 \Rightarrow I_L = -I_\epsilon.$$

The contribution of the integral involving the part of the contour that skips the singularity at $z = 0$ can be computed as follows: for this part,

$$z = \epsilon e^{i\theta}, \ \ dz = \epsilon i e^{i\theta} d\theta, \ \ 0 \le \theta \le \pi,$$

and the integral becomes

$$(x+a) > 0 : I_\epsilon = \int_\pi^0 \frac{e^{i\epsilon e^{i\theta}(x+a)}}{\epsilon e^{i\theta}} \epsilon i e^{i\theta} d\theta = i \int_\pi^0 e^{i\epsilon e^{i\theta}(x+a)} d\theta = -\pi i \Rightarrow I_L = \pi i.$$

In the limit $R \to \infty$ and $\epsilon \to 0$ we have $I_1 = I_L = \pi i$.

Similarly, for $(x + a) < 0$, we consider the contour C_B shown in the right diagram

$$C_B = C_R' \cup L \cup C_\epsilon'.$$

C_B consists of the semicircle C_R' centered at the origin of radius R with $R \to \infty$, the semicircle C_ϵ' centered at the origin of radius ϵ with $\epsilon \to 0$, and the line segment L extending from $-R$ to R and skipping the pole at $z = 0$. The contour C_B closes on the lower half-plane to satisfy Jordan's lemma; calculating the corresponding contour integral through the Residue theorem, we find

$$I_R + I_\epsilon + I_L = 0, \quad \lim_{R\to\infty} I_R = 0 \Rightarrow I_L = -I_\epsilon.$$

For the part that skips the singularity at $z = 0$ we have

$$(x+a) < 0 : I_0 = \int_\pi^{2\pi} \frac{e^{i\epsilon e^{i\theta}(x+a)}}{\epsilon e^{i\theta}} \epsilon i e^{i\theta} d\theta = i \int_\pi^{2\pi} e^{i\epsilon e^{i\theta}(x+a)} d\theta = \pi i$$

$$\Rightarrow \quad I_L = -\pi i.$$

In the limit $R \to \infty$ and $\epsilon \to 0$ we have $I_1 = I_L = -\pi i$.

The integral I_2 is evaluated in exactly the same way as I_1, the only difference being that it involves $(x - a)$ instead of $(x + a)$:

$$I_1 = \pi i, \ \text{for } (x + a) > 0; \quad -\pi i, \ \text{for } (x + a) < 0,$$
$$I_2 = \pi i, \ \text{for } (x - a) > 0; \quad -\pi i, \ \text{for } (x - a) < 0.$$

Combining the results for I_1, I_2, we find

$$-\infty < x < -a : \ x + a < 0 , \ x - a < 0 \Rightarrow f(x) = \frac{1}{4\pi i a} (-\pi i + \pi i) = 0,$$
$$-a < x < a : \ x + a > 0 , \ x - a < 0 \Rightarrow f(x) = \frac{1}{4\pi i a} (\pi i + \pi i) = \frac{1}{2a},$$
$$a < x < +\infty : \ x + a > 0 , \ x - a > 0 \Rightarrow f(x) = \frac{1}{4\pi i a} (\pi i - \pi i) = 0.$$

Thus, through the inverse FT we have recovered $f(x)$.

Having obtained the FT of the function, we can easily calculate the moments of $f(x)$ using its FT:

$$\hat{f}(k) = \frac{\sin(ka)}{ka} = \frac{1}{ka} \left[ka - \frac{(ka)^3}{6} + \frac{(ka)^5}{120} - \cdots \right]$$
$$= 1 - \frac{1}{6}(ka)^2 + \frac{1}{120}(ka)^4 - \cdots$$

Thus,

$$\hat{f}(0) = 1, \quad i\frac{d\hat{f}}{dk}(0) = 0, \quad i^2\frac{d^2\hat{f}}{dk^2}(0) = \frac{a^2}{3},$$
$$i^3\frac{d^3\hat{f}}{dk^3}(0) = 0, \quad i^4\frac{d^4\hat{f}}{dk^4}(0) = \frac{a^4}{5}, \ldots,$$

which are the first n moments of $f(x)$, for $n = 0, 1, 2, 3, 4, \ldots$. In this case, the results can be easily verified by evaluating the corresponding integrals:

$$\int_{-\infty}^{\infty} f(x)x^n dx = \frac{1}{2a} \int_{-a}^{a} x^n dx = \frac{1}{2(n+1)a} [a^{n+1} - (-a)^{n+1}] = \frac{a^n}{n+1},$$

for n = even, and 0 for n = odd.

7.2.3 Symmetries of the Fourier transform

We next explore the symmetries of the Fourier transform, namely we investigate what happens when its argument changes sign, or is scaled by a constant factor, or is shifted by a constant amount. All these properties can be easily proven from the definition of the Fourier transform. We assume that in each case $\hat{f}(k)$ is the FT of $f(x)$.

1. The FT of the function $g(x) = f(ax)$ is given by

$$\hat{g}(k) = \int_{-\infty}^{\infty} g(x)e^{-ikx} dx = \frac{1}{a} \int_{-\infty}^{\infty} f(ax)e^{-i(k/a)ax} d(ax) = \frac{1}{|a|}\hat{f}(k/a). \quad (7.35)$$

2. The FT of the function $f(x - x_0)$ is given by

$$\int_{-\infty}^{\infty} f(x - x_0)e^{-ikx}dx = e^{-ikx_0}\int_{-\infty}^{\infty} f(x - x_0)e^{-ik(x-x_0)}dx = e^{-ikx_0}\hat{f}(k).$$

(7.36)

3. The FT of the function $g(x) = e^{ik_0 x}f(x)$ is given by

$$\hat{g}(k) = \int_{-\infty}^{\infty} f(x)e^{ik_0 x}e^{-ikx}dx = \int_{-\infty}^{\infty} f(x)e^{-i(k-k_0)x}dx = \hat{f}(k - k_0).$$

(7.37)

4. The FT of the function $g(x) = f(x)\cos(k_0 x)$ is given by

$$\hat{g}(k) = \int_{-\infty}^{\infty} f(x)\left(\frac{e^{ik_0 x} + e^{-ik_0 x}}{2}\right)e^{-ikx}dx = \frac{1}{2}\left[\hat{f}(k + k_0) + \hat{f}(k - k_0)\right].$$

(7.38)

5. The FT of the function $g(x) = f(x)\sin(k_0 x)$ is given by

$$\hat{g}(k) = \int_{-\infty}^{\infty} f(x)\left(\frac{e^{ik_0 x} - e^{-ik_0 x}}{2i}\right)e^{-ikx}dx = \frac{1}{2i}\left[\hat{f}(k - k_0) - \hat{f}(k + k_0)\right].$$

(7.39)

6. If $f(x)$ is real, $\overline{f(x)} = f(x)$, then

$$\overline{\hat{f}(k)} = \int_{-\infty}^{\infty} \overline{f(x)}\,\overline{e^{-ikx}}dx = \int_{-\infty}^{\infty} f(x)e^{ikx}dx = \hat{f}(-k).$$

7. If $f(x)$ is imaginary, $\overline{f(x)} = -f(x)$, then

$$\overline{\hat{f}(k)} = \int_{-\infty}^{\infty} \overline{f(x)}\,\overline{e^{-ikx}}dx = \int_{-\infty}^{\infty}[-f(x)]e^{ikx}dx = -\hat{f}(-k).$$

8. For an even function $f(-x) = f(x)$, the FT obeys the following symmetry:

$$\hat{f}(k) = \int_{-\infty}^{\infty} f(x)e^{-ikx}dx = -\int_{\infty}^{-\infty} f(-x)e^{-i(-k)(-x)}d(-x)$$
$$= -\int_{\infty}^{-\infty} f(y)e^{-i(-k)y}dy = \hat{f}(-k).$$

9. For an odd function $f(-x) = -f(x)$, the FT obeys the following symmetry:

$$\hat{f}(k) = \int_{-\infty}^{\infty} f(x)e^{-ikx}dx = -\int_{\infty}^{-\infty}[-f(-x)]e^{-i(-k)(-x)}d(-x)$$
$$= \int_{\infty}^{-\infty} f(y)e^{-i(-k)y}dy = -\hat{f}(-k).$$

7.2.4   The sine- and cosine-transforms

The properties of the FT for even and odd functions lead to the definition of two more related transforms, the "sine" and "cosine" transforms.

We define the "sine-transform", $\hat{f}_s(k)$ for $0 < k < \infty$, for a function $f(x)$ defined in $0 < x < \infty$, by

$$\hat{f}_s(k) = \int_0^{\infty} f(x)\sin(kx)dx, \quad 0 < k < \infty.$$

(7.40)

We will show that the inverse sine-transform satisfies

$$f(x) = 4 \int_0^\infty \hat{f}_s(k) \sin(kx) \frac{dk}{2\pi}, \quad 0 < x < \infty. \tag{7.41}$$

Since the sine is an odd function, $\sin(-kx) = -\sin(kx)$, it follows that the sine-transform is an odd function, $\hat{f}_s(-k) = -\hat{f}_s(k)$.

The sine-transform can be derived from the Fourier transform by considering an *odd* function $f(x)$, namely, a function for which $f(-x) = -f(x)$. In this case, the Fourier transform, Eq. (7.19), gives

$$\hat{f}(k) = \int_{-\infty}^0 f(x) e^{-ikx} dx + \int_0^\infty f(x) e^{-ikx} dx = -2i \int_0^\infty f(x) e^{-ikx} dx = -2i \hat{f}_s(k).$$

Using this expression in the inverse Fourier transform, Eq. (7.20), we find:

$$f(x) = -2i \left[\int_{-\infty}^0 e^{ikx} \hat{f}_s(k) \frac{dk}{2\pi} + \int_0^\infty e^{ikx} \hat{f}_s(k) \frac{dk}{2\pi} \right] = 4 \int_0^\infty \sin(kx) \hat{f}_s(k) \frac{dk}{2\pi},$$

where in the last step we have taken advantage of the fact that $\hat{f}_s(k)$ is an odd function to change variables $k \to -k$, and we employed the expression of the sine in terms of the complex exponentials, Eq. (4.20), namely

$$\sin(kx) = \frac{e^{ikx} - e^{-ikx}}{2i} = -\frac{i}{2} \left[e^{ikx} - e^{-ikx} \right].$$

Through similar considerations it can be shown that if the "cosine-transform" is defined by

$$\hat{f}_c(k) = \int_0^\infty f(x) \cos(kx) dx, \quad 0 < k < \infty, \tag{7.42}$$

then its inverse is given by

$$f(x) = 4 \int_0^\infty \hat{f}_c(k) \cos(kx) \frac{dk}{2\pi}, \quad 0 < x < \infty. \tag{7.43}$$

This can be derived from the Fourier transform pair by considering an *even* function $f(x)$, namely a function for which $f(-x) = f(x)$.

7.3 Fourier transforms of special functions

7.3.1 The FT of the normalized Gaussian function

The FT of the normalized Gaussian function is given by

$$g(x) = \frac{1}{\sigma\sqrt{2\pi}} e^{-x^2/2\sigma^2} \xrightarrow{\text{FT}} \hat{g}(k) = \frac{1}{\sigma\sqrt{2\pi}} \int_{-\infty}^\infty e^{-x^2/2\sigma^2 - ikx} dx. \tag{7.44}$$

Without loss of generality, we will assume that $\sigma > 0$. The first step is to complete the square in the exponential:

$$\frac{x^2}{2\sigma^2} + ikx = \left(\frac{x}{\sigma\sqrt{2}} \right)^2 + 2 \left(\frac{x}{\sigma\sqrt{2}} \right) \left(\frac{ik\sigma}{\sqrt{2}} \right) - \frac{k^2\sigma^2}{2} + \frac{k^2\sigma^2}{2}$$

$$= \left(\frac{x}{\sigma\sqrt{2}} + \frac{ik\sigma}{\sqrt{2}} \right)^2 + \frac{k^2\sigma^2}{2}.$$

Inserting the above expression in the integral and introducing the change of variables

$$z = \frac{x}{\sigma\sqrt{2}} + i\frac{k\sigma}{\sqrt{2}} \Rightarrow dz = d\left(\frac{x}{\sigma\sqrt{2}}\right),$$

we find

$$\hat{g}(k) = \frac{1}{\sqrt{\pi}} \left[\int_{x=-\infty}^{x=\infty} e^{-z^2} dz\right]_{y=k\sigma/\sqrt{2}} e^{-k^2\sigma^2/2}.$$

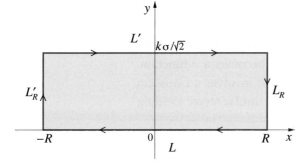

To perform the integral in the square brackets we use the following contour

$$C = L' \cup L_R \cup L \cup L'_R,$$

which is shown in Fig. 7.1. C consists of the horizontal line segments L' parallel to the real axis at $y = k\sigma/\sqrt{2}$ and extending from $x = -R$ to $x = R$, the vertical line segment L_R at $x = R$ and extending from $y = k\sigma/\sqrt{2}$ to $y = 0$, the horizontal line segment L on the real axis from R to $-R$, and the vertical line segment L'_R at $x = -R$ and extending from $y = 0$ to $y = k\sigma/\sqrt{2}$, with $R \to \infty$. The desired integral is the one corresponding to the line segment L' labeled I', where we have assumed that $k > 0$ (for the case $k < 0$ we use a similar contour with the part corresponding to I_1 being on the lower half-plane). The integral from the line segment L along the x-axis in the limit $R \to \infty$ was computed in Chapter 1:

$$I = \int_R^{-R} e^{-x^2} dx = -\int_{-R}^{R} e^{-x^2} dx \Rightarrow \lim_{R\to\infty}(I) = -\sqrt{\pi}.$$

For the integrals I_{L_R} and $I_{L'_R}$ we find that they give vanishing contributions in the limit $R \to \infty$. We show this fact explicitly for the integral $I_{L'_R}$: using

$$z = -R + iy, \quad 0 \le y \le \frac{k\sigma}{\sqrt{2}},$$

the integral $I_{L'_R}$ becomes

$$I_{L'_R} = \int_{-R}^{-R+i(k\sigma)/\sqrt{2}} e^{-(-R+iy)^2} d(-R+iy) = ie^{-R^2} \int_0^{(k\sigma)/\sqrt{2}} e^{2iRy+y^2} dy.$$

Since the integrand of the above integral is finite for any value of y in the range of integration, we conclude that

$$\lim_{R\to\infty}(I_{L'_R}) = \lim_{R\to\infty}\left[ie^{-R^2} \int_0^{(k\sigma)/\sqrt{2}} e^{2iRy+y^2} dy\right] = 0,$$

Figure 7.1: Contour C for obtaining the Fourier transform of a Gaussian, $g(x)$, Eq. (7.44).

due to the exponentially vanishing term e^{-R^2}. By similar considerations, $I_{L_R} = 0$ for $R \to \infty$. Hence,

$$\lim_{R \to \infty} \left[I + I_{L_R} + I_{L'_R} + I' \right] = 0,$$

because the function e^{-z^2} is analytic everywhere on the complex plane. Therefore,

$$I' = -I = \sqrt{\pi} \Rightarrow \hat{g}(k) = \frac{1}{\sqrt{\pi}} I' e^{-k^2 \sigma^2 / 2} = e^{-k^2 \sigma^2 / 2}. \tag{7.45}$$

Hence, the FT of a Gaussian in the x variable is a Gaussian in the k variable. Notice that for $\sigma \to 0$ the normalized Gaussian $g(x)$ becomes a δ-function, and its FT, $\hat{g}(k)$ becomes 1 for all k. The inverse FT also involves a Gaussian integral over the variable k, which can be performed by similar steps, yielding the original function $g(x)$.

7.3.2 FT of the θ-function and of the δ-function

Calculating the FT of the θ-function and the δ-function, is a more demanding task. We begin with a discussion of the FT of the Heaviside step function, or θ-function, defined in Chapter 1, Eq. (1.67). We will choose for simplicity the point where this function has a discontinuity to be $x' = 0$; the derivative of this function is zero everywhere, except at the point of discontinuity where it is not defined. Since the θ-function is not decaying as $x \to \infty$ its FT is not defined in the usual sense. Indeed, it turns out that it involves a δ-function in the k variable. However, since $x = 0$ is a point of discontinuity we expect that whatever is the final form of the FT it must involve the term $1/ik$. If we start with this term,

$$\hat{f}_1(k) = \frac{1}{ik}, \tag{7.46}$$

we can ask what is the function of x corresponding to this FT? Taking the inverse FT of the function $\hat{f}_1(k)$, we find:

$$f_1(x) = \int_{-\infty}^{\infty} \frac{e^{ikx}}{ik} \frac{dk}{2\pi}. \tag{7.47}$$

This can be evaluated by contour integration using the complex variable $z = k + iq$, and utilizing the contours C_A, for $x > 0$, and C_B, for $x < 0$, shown in Fig. 7.2.

For $x > 0$, the contour $C_A = C_R \cup L \cup C_\epsilon$ consists of the semicircle C_R centered at the origin and of radius $R \to \infty$ which lies on the upper half-plane, the line segment L that lies on the real axis and skips the simple pole at $z = 0$, and the semicircle C_ϵ centered at the origin and of radius $\epsilon \to 0$ traversed in the *clockwise* direction. The integral I along the line segment L, after it is divided by the factor 2π, gives the desired value. Because the contour closes on the upper half-plane, it satisfies Jordan's lemma, so $I_R \to 0$. The contour contains no singularities, therefore:

$$x > 0: \quad I + I_R + I_\epsilon = 0 \Rightarrow I = -I_\epsilon = -\left(-i\pi \frac{1}{i} \right) = \pi \Rightarrow f_1(x) = \frac{I}{2\pi} = \frac{1}{2}.$$

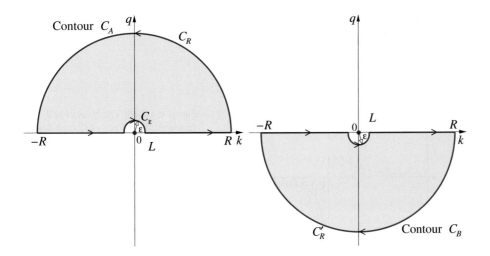

Figure 7.2: Contours C_A and C_B on the complex plane $z = k + iq$ used to evaluate the integral of Eq. (7.47).

For $x < 0$, the contour $C_B = C'_R \cup L \cup C'_\epsilon$ consists of the semicircle C'_R which lies on the lower half-plane, centered at the origin and of radius $R \to \infty$, the line segment L that lies on the real axis and skips the simple pole at $z = 0$, and the semicircle C'_ϵ centered at the origin and of radius $\epsilon \to 0$, traversed in the *counter-clockwise* direction. The integral I along the line segment L, after it is divided by the factor 2π, gives the desired value. Because the contour closes on the lower half-plane, it satisfies Jordan's lemma, so $I'_R \to 0$. The contour contains no singularities, therefore:

$$x < 0: \quad I + I'_R + I'_\epsilon = 0 \Rightarrow I = -I'_\epsilon = -\left(i\pi\frac{1}{i}\right) = -\pi \Rightarrow f_1(x) = \frac{I}{2\pi} = -\frac{1}{2}.$$

Summarizing the results, the function $f_1(x)$ corresponding to $\hat{f}_1(k) = 1/ik$ satisfies

$$f_1(x) = \begin{cases} \dfrac{1}{2} & \text{for } x > 0, \\ -\dfrac{1}{2} & \text{for } x < 0, \end{cases}$$

which is the Heaviside step function or θ-function, $\theta(x)$, shifted down by $1/2$: $f_1(x) = \theta(x) - 1/2$, as shown in Fig. 7.3.

The $\hat{f}_1(k) = 1/ik$ behavior of the FT comes from the fact that $f_1(x)$ is discontinuous:

$$f_1(0^+) = \frac{1}{2}, \quad f_1(0^-) = -\frac{1}{2} \Rightarrow \Delta f(x = 0) = 1,$$

as expected. If we were to calculate the FT from $f_1(x)$ we would take:

$$\hat{f}_1(k) = \int_{-\infty}^{\infty} e^{-ikx} f(x) dx = -\frac{1}{2} \int_{-\infty}^{0^-} e^{-ikx} dx + \frac{1}{2} \int_{0^+}^{\infty} e^{-ikx} dx.$$

The limits at $\pm\infty$ can be evaluated in the principal value sense, that is, by replacing them with R and taking the limit $R \to \infty$:

$$\hat{f}_1(k) = -\frac{1}{2} \int_{-R}^{0^-} e^{-ikx} dx + \frac{1}{2} \int_{0^+}^{R} e^{-ikx} dx = \frac{1}{ik}[1 - \cos(kR)]. \qquad (7.48)$$

The function $\cos(kR)$ is not defined for $R \to \infty$ which is consistent with the fact that $f(x)$ does *not* decay as $x \to \pm\infty$. However, in the limit $R \to \infty$ the argument of the cosine goes through 2π for infinitesimal changes in k, and the *average* of the cosine function over a 2π range vanishes, which leaves as a net result *effectively* $\hat{f}_1(k) = 1/ik$. This is the reason why if we *postulate* that $\hat{f}_1(k) = 1/ik$, we find the correct result. The behavior of the right-hand side of Eq. (7.48) is illustrated in Fig. 7.3.

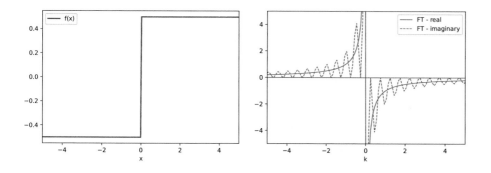

Figure 7.3: **Left**: The Heaviside θ-function, shifted by $-1/2$ (red line). **Right**: Its Fourier transform, $\hat{f}_1(k)$ given by Eq. (7.48): the real part is zero (blue line) while the imaginary part (green dashed line for $R = 50$) approaches $-1/k$ for $R \to \infty$ (green solid line).

The function $f_1(x)$ satisfies $f_1(x) = \theta(x) - 1/2$. From the linearity of the FT, we then deduce that in order to obtain the FT of $\theta(x)$ we must find the FT of the constant function $f_2(x) = 1/2$ for all $x \in (-\infty, \infty)$. Since $f_2(x)$ does not decay as $x \to \pm\infty$, $\hat{f}_2(k)$ is not defined. However, actually we can compute $\hat{f}_2(k)$ in terms of $\delta(k)$ through a limiting process: we take the limits of the integral to be $\pm R$, so that the integral is finite for any finite value of R, and then take the limit of $R \to \infty$. Then, we can calculate $\hat{f}_2(k)$ from $f_2(x)$ in this limit:

$$\hat{f}_2(k) = \lim_{R \to \infty} \left[\int_{-R}^{R} \frac{1}{2} e^{-ikx} dx \right] = \frac{1}{-2ik} \lim_{R \to \infty} \left[e^{-ikx} \right]_{-R}^{R} = \lim_{R \to \infty} R \left(\frac{\sin(kR)}{kR} \right).$$

The function $[\sin(w)/w]$, with $w = kR$, takes the value 1 at $k = 0$, falls off for $k \to \infty$, and has width $2\pi/R$, obtained from the first zero of $\sin(w)$ which occurs at $w = kR = \pm\pi$. This behavior is the same as that of the Fourier transform in Eq. (7.34), studied in Example 7.4. Therefore, the function $R[\sin(w)/w]$ has height $\sim R$ and width $\sim 1/R$; its width goes to 0 and its height goes to ∞ when $R \to \infty$, while its integral is equal to π:

$$\int_{-\infty}^{\infty} \hat{f}_2(k) dk = 2\pi \left[\int_{-\infty}^{\infty} \hat{f}_2(k) e^{ik0} \frac{dk}{2\pi} \right] = 2\pi f_2(0) = \pi.$$

Accordingly, in the limit $R \to \infty$ the function $R[\sin(kR)/(kR)]$ indeed behaves just like a δ-function in the variable k, but is normalized so that it gives π rather than 1 when integrated over all values of k. Hence,

$$\hat{f}_2(k) = \lim_{R \to \infty} \left(R \frac{\sin(kR)}{kR} \right) = \pi \hat{\delta}(k). \tag{7.49}$$

From the linearity of the Fourier transform we have:

$$\theta(x) = f_1(x) + f_2(x) \Rightarrow \hat{\theta}(k) = \hat{f}_1(k) + \hat{f}_2(k),$$

and using Eqs. (7.46), (7.49) we conclude that

$$\hat{\theta}(k) = \frac{1}{ik} + \pi \hat{\delta}(k). \tag{7.50}$$

From this expression, by applying the inverse Fourier transform we obtain:

$$\theta(x - x') = \int_{-\infty}^{\infty} \hat{\theta}(k)e^{ik(x-x')}\frac{dk}{2\pi} = \int_{-\infty}^{\infty} \frac{1}{ik}e^{ik(x-x')}\frac{dk}{2\pi} + \frac{1}{2}, \qquad (7.51)$$

which is a useful representation of the θ-function. This is referred to as the "integral representation" of the θ-function since it involves an integral in the variable k.

We next compute the FT of the δ-function.

$$\hat{\delta}_{x'}(k) = \int_{-\infty}^{\infty} \delta(x - x')e^{-ikx}dx = e^{-ikx'}, \qquad (7.52)$$

where we have kept the value x' as a subscript for the FT of the δ-function because its value depends on it, just like the δ-function itself depends on the value of x' in its argument; in the case $x' = 0$ we have $\hat{\delta}_0(k) = 1$. The Inverse Fourier transform of the δ-function gives

$$\delta(x - x') = \int_{-\infty}^{\infty} \hat{\delta}_{x'}(k)e^{ikx}\frac{dk}{2\pi} = \int_{-\infty}^{\infty} e^{ik(x-x')}\frac{dk}{2\pi}, \qquad (7.53)$$

which is an interesting way of representing the δ-function, referred to as its "integral representation": this expression involves only an integral over the variable k as opposed to taking the limit of a proper function with respect to some parameter, as was done in Chapter 1, Section 1.6. The integral representation can be advantageous in many applications (see, for example, Chapter 9, Section 9.4). Moreover, we can interchange the names of the variables x and k in this expression, while the mathematical formula remains the same, to obtain:

$$\hat{\delta}(k - k') = \int_{-\infty}^{\infty} e^{ix(k-k')}\frac{dx}{2\pi}. \qquad (7.54)$$

The interchange of variable names allowed us to obtain the corresponding expression for the δ-function in the k space, [for consistency with previous notation, we wrote the δ-function with argument k as $\hat{\delta}(k - k')$]. In this way, we have obtained useful representations of the δ-function in the real space, $\delta(x - x')$, and in the reciprocal space, $\hat{\delta}(k - k')$, both of which involve only complex exponentials.

Derivation of the δ-function in k-space
We can use the result of the FT of the θ-function and the shifting properties of the FT to derive the expression of Eq. (7.54) in a different way. We begin with the functions $h_1(x)$ and $h_2(x)$ defined by

$$h_1(x) = \frac{1}{2}e^{ik'x} + \frac{1}{2}e^{-ik'x}, \quad h_2(x) = \frac{1}{2}e^{ik'x} - \frac{1}{2}e^{-ik'x}.$$

Taking the FTs of the above functions, using the result for the function $1/2$,

$$f_2(x) = \frac{1}{2} \Rightarrow \hat{f}_2(k) = \pi\hat{\delta}(k)$$

and employing the shifting properties of the FT, Eq. (7.37), applied first for $k_0 = k'$ and then for $k_0 = -k'$, we find

$$\hat{h}_1(k) = \pi\hat{\delta}(k - k') + \pi\hat{\delta}(k + k'), \quad \hat{h}_2(k) = \pi\hat{\delta}(k - k') - \pi\hat{\delta}(k + k').$$

Adding the functions $h_1(x)$ and $h_2(x)$ we obtain a new function $h(x)$:

$$h(x) = h_1(x) + h_2(x) = e^{ik'x}.$$

The corresponding FTs $\hat{h}_1(k)$ and $\hat{h}_2(k)$, by the linearity of the Fourier Transform, give the FT $\hat{h}(k)$ of this new function

$$\hat{h}(k) = \hat{h}_1(k) + \hat{h}_2(k) = 2\pi\hat{\delta}(k - k').$$

The functions $h(x)$ and $\hat{h}(k)$ are, by construction, a Fourier transform pair:

$$\begin{aligned} \hat{h}(k) &= \int_{-\infty}^{\infty} h(x)\, e^{-ikx} dx \Rightarrow 2\pi\hat{\delta}(k - k') \\ &= \int_{-\infty}^{\infty} e^{ik'x}\, e^{-ikx} dx = \int_{-\infty}^{\infty} e^{-i(k-k')x} dx, \end{aligned}$$

which is the same result derived in Eq. (7.54). The fact that the signs of $(k - k')$ are opposite in the two expressions can be explained by the fact that the δ-function is an even function of its argument, $\hat{\delta}(k - k') = \hat{\delta}(k' - k)$. This restores the exact equivalence of the two expressions.

Note that when the real space variable represents length (usually denoted by x), the reciprocal space variable is the wave-vector $k = 2\pi/\lambda$ with λ the wavelength. The wavelength is the spatial period over which the complex exponentials take the same values. Similarly, when the real space variable represents time (usually denoted by t), the reciprocal space variable is the "angular frequency", denoted by $\omega = 2\pi/\tau = 2\pi\nu$ with τ being the time period and ν being the "frequency", defined as $\nu = 1/\tau$. In both cases, the products kx or ωt that appear in the complex exponential are dimensionless quantities.

7.4   Application: Signal analysis

In what follows, we will consider functions of a real variable and their FTs. We will assume that the real variable represents time and denote it by t, while the variable that appears in the FT denoted by ω represents angular frequency, $\omega = 2\pi/\tau = 2\pi\nu$, where τ is the period and $\nu = 1/\tau$ is the frequency. Very often, when we are trying to measure a time signal $f(t)$, it is convenient to measure its frequency content, namely, its Fourier transform $\hat{f}(\omega)$. This is motivated by the fact that we can construct devices that respond accurately to specific frequencies. Then, we can reconstruct the original signal $f(t)$ by performing the inverse Fourier transform on the measured signal $\hat{f}(\omega)$ in the frequency domain:

$$f(t) \xrightarrow{\text{FT}} \hat{f}(\omega) \xrightarrow{\text{IFT}} f(t).$$

We first recall the relations we found earlier for the representation of the δ-function, translated to the new variables, t and ω: the equivalent of Eq. (7.52) in the time domain is

$$\delta(t - t') = \int_{-\infty}^{\infty} e^{i(t-t')\omega} \frac{d\omega}{2\pi}, \tag{7.55}$$

and the equivalent of Eq. (7.53) in the frequency domain is

$$\hat{\delta}(\omega - \omega') = \int_{-\infty}^{\infty} e^{-i(\omega-\omega')t} \frac{dt}{2\pi}. \tag{7.56}$$

Two useful notions in signal analysis are the "total power" and the "spectral density". The total power P of the signal is defined by

$$P = \int_{-\infty}^{\infty} |f(t)|^2 dt.$$

The spectral density is defined simply as $|\hat{f}(\omega)|^2$. An important relation linking these two notions is the following theorem:

Parseval's Theorem: *The total power P of a signal $f(t)$ is given in terms of its Fourier transform $\hat{f}(\omega)$ by the relation*

$$\int_{-\infty}^{\infty} |\hat{f}(\omega)|^2 d\omega = 2\pi P. \tag{7.57}$$

Proof: To prove this theorem we use the FT of the δ-function, Eq. (7.53); from the definition of the FT of $f(t)$ we have:

$$\hat{f}(\omega) = \int_{-\infty}^{\infty} e^{-i\omega t} f(t) dt \Rightarrow |\hat{f}(\omega)|^2 = \int_{-\infty}^{\infty} e^{-i\omega t} f(t) dt \int_{-\infty}^{\infty} e^{i\omega t'} \bar{f}(t') dt',$$

where $\bar{f}$ is the complex conjugate of f. Integrating the absolute value squared of the above expression over all values of ω yields the following:

$$
\begin{aligned}
\int_{-\infty}^{\infty} |\hat{f}(\omega)|^2 d\omega &= \int_{-\infty}^{\infty}\int_{-\infty}^{\infty} f(t)\bar{f}(t') \left[\int_{-\infty}^{\infty} e^{i\omega(t'-t)} d\omega \right] dt dt' \\
&= \int_{-\infty}^{\infty}\int_{-\infty}^{\infty} f(t)\bar{f}(t') 2\pi\delta(t-t') dt dt' = 2\pi \int_{-\infty}^{\infty} |f(t)|^2 dt.
\end{aligned}
$$

In the above equations, we have used Eq. (7.55) to perform the integration over ω in the square brackets which produced $2\pi\delta(t-t')$, and then performed an integration over one of the two time variables. ∎

7.4.1 Aliasing: The Nyquist frequency

An essential feature of time-signal measurements is their sampling at finite time intervals. This leads to important complications in determining the true signal from the measured signal. To illustrate this problem, suppose that we sample a time signal $f(t)$ at the N discrete time moments separated by a time interval[1] which we call Δt

$$t_n = n\Delta t, \quad n = 1, \ldots, N.$$

Thus, the N values of $f(t_n)$ are the data we can obtain from an experiment. Usually, we are interested in the limit of $N \to \infty$ and $\Delta t \to 0$ with $N\Delta t$, the total duration of the measurements. To make the connection with the earlier discussion easier, we define the total time interval as $2T$ (since we had denoted the total space interval where the function $f(x)$ was defined as $2L$), and we shift the origin of time so that the signal sampling takes place between the initial time $t_i = -T$ and final time $t_f = T$; the corresponding moments of measurement are given by

$$t_n = n\Delta t, \quad n = -\frac{N}{2}, \ldots, 0, \ldots, \frac{N}{2} - 1, \quad \Delta t = \frac{2T}{N}, \tag{7.58}$$

where, since N is large, we can assume N to be an even integer.

[1] This time interval is often called the "sampling rate", but this is a misnomer because a rate has the dimensions of inverse time.

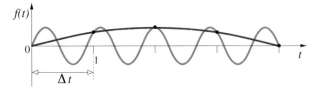

Figure 7.4: Illustration of the sampling problem at finite time intervals of two signals with frequencies ω (green) and ω' (red), measured at $t_n = n\Delta t$, $n = 0, 1, \ldots, 4$. In this example, $\omega\Delta t = 9\pi/4$, so that $\phi = \pi/4$ and $\omega'\Delta t = \pi/4$; for $t_n = n\Delta t$, with $n = 0, 1, \ldots, 4$ the two signals $\sin(\omega t_n)$ and $\sin(\omega' t_n)$ have exactly the same values.

The sampling at finite time intervals has the following limitation: we can only sample unambiguously signals of frequency up to a certain maximum value because signals of higher value give the same values when sampled at the specific moments t_n. If a signal has frequency ω such that $\omega\Delta t > 2\pi$, then the sampled values are the same as the values of a signal whose frequency is $\omega' = \omega - 2\pi/\Delta t$. This can be seen by assuming that $\omega\Delta t = 2\pi + \phi$ with $0 < \phi < 2\pi$, which gives $\omega'\Delta t = \phi$. Therefore, the values of $\omega t_n = \omega\Delta tn = 2\pi n + \phi n$ and $\omega't_n = \phi n$ differ by an integer multiple of 2π; hence, all trigonometric functions (and complex exponentials) with arguments ωt_n and $\omega't_n$ take exactly the same values at these time moments. This is illustrated in Fig. 7.4. Accordingly, any expression involving the Fourier components of the signal and its values at the time moments t_n will produce identical results, making it impossible to tell the two frequencies apart. The maximum frequency where this type of problem is encountered is given by

$$\omega_{\max} = \frac{2\pi}{\Delta t}.$$

To handle these situations, we define the characteristic frequency ω_c (also called "Nyquist frequency") by

$$\omega_c \equiv \frac{\omega_{\max}}{2} = \frac{\pi}{\Delta t}, \tag{7.59}$$

and shift the origin of the frequency axis so that the range is $\omega \in [-\omega_c, \omega_c]$. Let's assume first that we are dealing with "band-width limited" signal, that is, a signal whose FT vanishes for frequencies outside the range $|\omega| > \omega_c$:

$$\hat{f}(\omega) \neq 0 \quad \text{for } \omega \in [-\omega_c, \omega_c], \quad \text{and} \quad \hat{f}(\omega) = 0 \quad \text{for } \omega < -\omega_c, \ \omega > \omega_c.$$

For such a signal, the so-called "sampling theorem" asserts that we can reconstruct the entire signal using the expression

$$f(t) = \sum_{n=-N/2}^{N/2-1} f(t_n) \frac{\sin(\omega_c(t - t_n))}{\omega_c(t - t_n)}, \tag{7.60}$$

where t_n are the moments in time defined in Eq. (7.58). This is a very useful expression, because it gives the time signal $f(t)$ for all values of t (which is a continuous variable), even though it is measured only at the discrete time moments t_n. In other words, the information content of the signal can be determined entirely by sampling it at the time moments t_n. The reconstruction of the signal $f(t)$ from the measurements $f(t_n)$ essentially amounts to the computation of an inverse Fourier transform. Notice that for $t = t_m = m\Delta t$ we have

$$n \neq m : \ t_m - t_n = (m - n)\Delta t \Rightarrow \omega_c(t_m - t_n) = (m - n)\pi$$

$$\Rightarrow \frac{\sin(\omega_c(t_m - t_n))}{\omega_c(t_m - t_n)} = 0,$$

since the argument of the sine is an integer multiple of π; for $n = m$ we recover $f(t_m)$ on the right-hand side of Eq. (7.60), since $\sin(w)/w = 1$ for $w \to 0$.

Typically a time signal is not bandwidth limited, namely, it has Fourier components for all values of ω. If we are still forced to sample such a signal at the time moments t_n to obtain information about the frequency spectrum of the signal, then performing an inverse Fourier transform will create difficulties in this case. To show this explicitly, consider two functions $f_1(t)$ and $f_2(t)$ which are described by the following expressions:

$$f_1(t) = e^{i\omega_1 t}, \quad f_2(t) = e^{i\omega_2 t}, \quad \omega_1 - \omega_2 = 2\omega_c l, \, l \text{ integer.}$$

Each of these functions has a unique frequency content and the two frequencies differ by an integer multiple of $2\omega_c$. Then, if these two functions are sampled at the time moments t_n, we will have

$$f_1(t_n) = e^{i\omega_1 t_n} = e^{i(\omega_2 + 2\omega_c l)t_n} = e^{i\omega_2 t_n} = f_2(t_n);$$

this follows from the definitions of ω_c and t_n:

$$e^{i2\omega_c l t_n} = e^{i2\pi l n} = 1.$$

The result is that the two functions will appear identical when sampled at the time moments t_n.

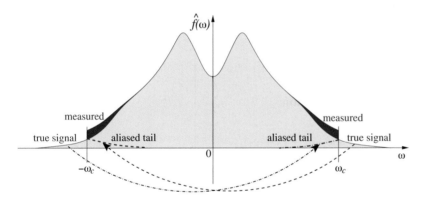

Figure 7.5: Illustration of aliasing: the tails of the FT $\hat{f}(\omega)$ for $\omega < -\omega_c$ and $\omega > \omega_c$ are mapped to values within the interval $[-\omega_c, \omega_c]$ and are added to the true signal, giving a measured signal different from the true signal in the regions close to $\pm\omega_c$.

We can apply the above analysis to the sampling of a signal $f(t)$. If the signal contains frequencies that differ by integer multiples of $2\omega_c$, then the contributions of these frequencies cannot be differentiated. This effect is called "aliasing". When the FT of the signal is measured by sampling the signal at intervals Δt apart, the components that fall outside the interval $-\omega_c \leq \omega \leq \omega_c$ will appear as corresponding to a frequency within that interval, from which they differ by $2\omega_c$, as shown schematically in Fig. 7.5. For instance, the contribution of the frequency $\omega_c + \delta\omega$ (where $\delta\omega > 0$) will appear at the frequency $-\omega_c + \delta\omega$ because these two frequencies differ by $2\omega_c$. Similarly, the contribution of the frequency $-\omega_c - \delta\omega$ will appear at the frequency $\omega_c - \delta\omega$ because these two frequencies also differ by $2\omega_c$. The net result is that the frequency spectrum outside the interval $-\omega_c \leq \omega \leq \omega_c$ is mapped to the spectrum within this interval, which significantly alters the measured spectrum from the true one, at least in the regions near $\pm\omega_c$. Fortunately, typical signals have spectra that fall off with the magnitude of the frequency, so that the aliasing problem is not severe. Moreover, increasing the value of ω_c, that is, reducing the value of the time interval Δt at which the signal is measured, tends to eliminate the aliasing problem for most signals by simply pushing ω_c to values where the signal has essentially died off. Typically, only the positive portion of the frequency spectrum is measured, in which case the negative portion appears in the upper half of the spectrum by aliasing: this is equivalent to shifting all frequencies in the range $[-\omega_c, 0]$ by $2\omega_c$, so that their values appear in the range $[\omega_c, 2\omega_c]$.

Example 7.5: As a practical demonstration, we consider the following time signal:

$$f(t) = \sin(2\pi 25 \cdot t)\cos(2\pi 60 \cdot t)\sin(2\pi 155 \cdot t). \tag{7.61}$$

We will first analyze the signal assuming that we know its form $f(t)$ for all time moments. In the analysis, we will eventually use frequencies rather than angular frequencies to display the FT of the signal, so we have explicitly included the factors of 2π relating the values of frequency ν to angular frequency ω. We recognize that there are three angular frequencies involved in the expression of the signal, namely:

$$\omega_1 = 2\pi\nu_1, \quad \omega_2 = 2\pi\nu_2, \quad \omega_3 = 2\pi\nu_3,$$

$$\text{where}: \quad \nu_1 = 25, \quad \nu_2 = 60, \quad \nu_3 = 155.$$

Expressing the signal in terms of complex exponentials, we obtain:

$$f(t) = -\frac{1}{8}\left[(e^{i\omega_1 t} - e^{-i\omega_1 t})(e^{i\omega_2 t} + e^{-i\omega_2 t})(e^{i\omega_3 t} - e^{-i\omega_3 t})\right].$$

Expanding the products, we obtain the following expression:

$$f(t) = \sum_{j=1}^{8} A_j e^{i\omega_j' t}, \quad \text{where}: \quad \omega_j' = 2\pi\nu_j', \quad \nu_j' = (\pm\nu_1 \pm \nu_2 \pm \nu_3),$$

while the coefficients A_j take the values $\pm 1/8$. If we now calculate the FT of the signal, we obtain

$$\begin{aligned}
\hat{f}(\omega) &= \int_{-\infty}^{\infty} f(t)e^{-i\omega t}\mathrm{d}t = \sum_{j=1}^{8} 2\pi A_j \int_{-\infty}^{\infty} e^{-i(\omega-\omega_j')t}\frac{\mathrm{d}t}{2\pi} \\
&= \sum_{j=1}^{8} 2\pi A_j \hat{\delta}(\omega - \omega_j'),
\end{aligned}$$

where we have taken advantage of Eq. (7.56) for the δ-function in the ω-space. Finally, expressing everything in terms of the frequencies ν_j', and using the property of the δ-function derived in Chapter 1, Eq. (1.82), we arrive at:

$$\hat{f}(\nu) = \sum_{j=1}^{8} A_j \hat{\delta}(\nu - \nu_j'), \quad \nu_j' = \pm 70, \quad \pm 120, \quad \pm 190, \quad \pm 240. \tag{7.62}$$

We next explore how we could reveal the frequency content of the signal (its frequency spectrum), if we did not know the explicit form for all moments in time as defined in Eq. (7.61), but only sampled the signal at specific discrete time moments. We assume that the signal is sampled at $N = 512$ time moments

$$t_n = n\Delta t, \quad n = -256, \ldots, 0, \ldots, 255, \quad \text{with} \quad \Delta t = \frac{1}{512},$$

for a total duration $2T = 1$ (this last choice simply defines the unit of time). In this case, the Nyquist frequency is given by

$$\omega_c = \frac{\pi}{\Delta t} = 2\pi \times 256 \Rightarrow \nu_c \equiv \frac{\omega_c}{2\pi} = 256.$$

For the Fourier transform of the signal we will use the fast Fourier transform (FFT) algorithm and its inverse (IFFT), introduced in Chapter 4 (see Section 4.8). For this example, we employed a simple code in python, starting with the one-dimensional array called "originalsignal", and used the routines of the numerical python (numpy) library to perform the FFT and the IFFT with the commands

```
FFTsignal=numpy.fft.fft(originalsignal)
IFFTsignal=numpy.fft.ifft(FFTsignal).
```

The FFT consists of eight δ-functions: four of them in the interval $[0, 255]$, at the values

$$\nu_1' = 70, \quad \nu_2' = 120, \quad \nu_3' = 190, \quad \nu_4' = 240,$$

and four more in the interval $[256, 511]$:

$$\nu_5' = 272, \quad \nu_6' = 322, \quad \nu_7' = 392, \quad \nu_8' = 442.$$

The diagrams below display the signal sampled at $N = 512$ time moments (left panel) and the absolute value of its FFT (right panel).

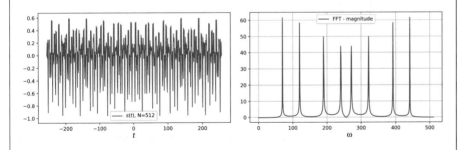

We only show here the absolute value of the FFT because it is easier to recognize in it the main features. However, it should be noted that both real and imaginary parts of the FFT are needed to reconstruct the true signal through the IFFT. We also note that the δ-functions have finite height and finite width, as opposed to being infinitely sharp, because their numerical representation involves approximations of the type discussed in Chapter 1 (see Section 1.6).

We recognize that the last four frequencies can be shifted by twice the Nyquist frequency, $-2\nu_c = -512$, to the equivalent frequencies:

$$\nu_5' = 272 - 512 = -240, \quad \nu_6' = 322 - 512 = -190,$$

$$\nu_7' = 392 - 512 = -120, \quad \nu_8' = 442 - 512 = -70.$$

In this way, we have established that the FFT consists of δ-functions in frequency at the eight values

$$\nu_j' = \pm 70, \quad \pm 120, \quad \pm 190, \quad \pm 240,$$

exactly as we had anticipated from our analytical expression, Eq. (7.62).

If we sample the signal at $N = 300$ time moments

$$t_n = n\Delta t, \quad n = -150, \ldots, 0, \ldots, 149, \quad \text{with} \quad \Delta t = \frac{1}{300}$$

$$\Rightarrow \quad \omega_c = \frac{\pi}{\Delta t} = 150 \times 2\pi \Rightarrow \nu_c = \frac{\omega_c}{2\pi} = 150,$$

the signal looks different; the values of the characteristic frequencies in the FFT are $\nu_1' = 70$, $\nu_2' = 120$, $\nu_3' = 190$ and $\nu_4' = 240$ as before, as well as $\nu_5' = -240 + 300 = 60$, $\nu_6' = -190 + 300 = 110$, $\nu_7' = -120 + 300 = 180$, and $\nu_8' = -70 + 300 = 230$, that is, the negative values shifted by $2\nu_c = 300$, due to aliasing. In the two diagrams below we show the signal sampled at $N = 300$ time moments (left panel) and the absolute value of its FFT (right panel). In this case separating the values due to aliasing from those due to the true signal (called "de-aliasing") is not immediately obvious, because the midpoint of the total interval is *smaller* than some of the combinations.

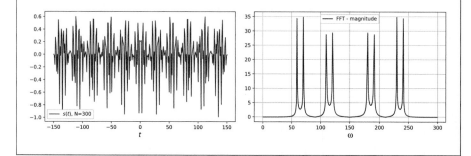

Further reading

The subject of the Fourier expansion and transform is treated in many textbooks, due to its importance in practical applications like signal analysis. All the mathematical-methods textbooks mentioned under "General Sources" in the Preface cover this material in detail. More advanced and technical discussions, with emphasis on applications to signal analysis, can be found in the books by Vetterli, Kovacevic and Goyal mentioned under Further Reading in Chapter 3. For additional works, see the following books:

1. Kenneth B, Howell, *Principles of Fourier Analysis* (Chapman and Hall/CRC Press, 2001). This book offers a detailed discussion of the mathematical foundations of the Fourier expansion and transform, and many examples of their applications.

2. Elias M. Stein and Rami Shakarchi, *Fourier Analysis, an Introduction* (Princeton University Press, 2007). This book offers a useful motivation of the Fourier expansion and transform, with formal proofs of their properties, as well as many examples and problems.

Problems

1. Consider the Fourier series expansion for the unitary square wave discussed in Example 7.1.

   (a) Find the values of x which make the series not uniformly convergent, when included in an interval. Justify your answer through a graphical argument. Since the series is not uniformly convergent, we cannot construct a "Weierstrass M-series" valid for all x in the interval $[-\pi, \pi]$. However, the series does converge when the interval does not include certain values of x. Convergence can be shown by the Dirichlet test in which two series called "a" (with bounded partial sums) and "b" (with positive, monotonically decreasing terms) are multiplied term by term to give the original series, called "c":

   $$\sum_{j=1}^{\infty} c_j = \sum_{j=1}^{\infty} a_j b_j.$$

   [See Chapter 3, Section 3.1.1 for details.]

   (b) Choose $x = \pi/2$ and construct the series "a" and "b" which meet the specified criteria for the Dirichlet test. To what value does the series converge? If the series is truncated at N, construct arguments to show that the neglected remainder satisfies the relation

   $$\left| \sum_{n=N+1}^{\infty} c_n \right| < \frac{4}{\pi} \frac{1}{N+1}.$$

   [Hint: Pair up terms in the remainder, other than the first one.]

   (c) Choose $x = \pi/4$ and construct the series "a" and "b" which satisfy the Dirichlet test.

   (d) Set $x = \epsilon$ where $0 < \epsilon \ll 1$. Again construct the two series and demonstrate that the "a" series does indeed have bounded partial sums. [Hint: Use a planar representation of unit vectors inclined by an angle $(2n+1)\epsilon$ to the x-axis.]

2. Consider the triangular wave:

   $$f(x) = \begin{cases} -\frac{kx}{\pi} + A, & -\pi \leq x \leq 0, \\ \frac{kx}{\pi} + A, & 0 \leq x \leq \pi. \end{cases} \tag{7.63}$$

   Determine the value of the constants A and k so that this wave has zero mean and it is the integral of the unitary square wave.

   (a) Find the Fourier series expansion of this wave and compare it term by term with the Fourier series expansion of the unitary square wave discussed in Example 7.1. What is your conclusion?

   (b) Let the partial sum of the first N terms of this series be denoted by S_N. We wish to determine how many terms must be included so that the root-mean-square error

   $$\mathcal{E}^{(2)}(N) = \left\{ \int_{-\pi}^{\pi} [f(x) - S_N(x)]^2 \, dx \right\}^{1/2},$$

   is smaller than some prescribed ϵ. Find an integral that provides an upper bound to this error, and find N for $\epsilon = \pi/100$ (see Chapter 3, Section 3.4 for details).

   (c) Find the function $g(x)$ which is the integral of $f(x)$, choosing the integration constant to be such that the integral of $f(x)$ in the given range vanishes.

   (d) Find the Fourier series expansion for $g(x)$. Does this series converge uniformly in $[-\pi, \pi]$? Compare the result to that obtained by integrating the Fourier expansion of $f(x)$ term by term.

3. Let $f(x)$ be a periodic function with period 2, defined in the interval $x \in [-1, 1]$ as $f(x) = x$. Find the complex Fourier series expansion of $f(x)$. Compare to the Fourier series expansion of the triangular wave discussed in Example 7.2.

4. Consider the function

   $$f(x) = \pi^2 - x^2, \quad \text{for } x \in (0, \pi).$$

   Find a sine series expansion and a cosine series expansion for $f(x)$. Which series converges faster? Can you quantify your answer by calculating the error of the truncated series in each case?

5. Consider the function $f(x)$ defined by

   $$f(x) = \begin{cases} 1, & |x| < \delta, \\ 0, & \delta < |x| \leq \pi, \end{cases}$$

   where $0 < \delta < \pi$ and $f(x + 2\pi) = f(x)$ for $x \in \mathbb{R}$.

   (a) Find a suitable Fourier series expansion for $f(x)$.

   (b) Use the Fourier series expansion to show that

   $$\sum_{n=1}^{\infty} \frac{\sin(n\delta)}{n} = \frac{\pi - \delta}{2}, \quad \text{and} \quad \sum_{n=1}^{\infty} \frac{\sin^2(n^2\delta)}{n} = \frac{\pi - \delta}{2}.$$

[Hint: For the second relation, express the integral of $[f(x)]^2$ in terms of the coefficients of the Fourier expansion of $f(x)$.]

(c) Take the limit $\delta \to 0$ for the two series above. What do you obtain?

6. Find the Fourier expansion coefficients of the periodic function with period 2π, defined as

$$f(x) = x^2, \quad x \in [0, 2\pi].$$

Using this Fourier series representation, evaluate the sum of the series

$$\sum_{n=1}^{\infty} \frac{1}{n^2}.$$

7. Consider the function $f(x)$ defined by

$$f(x) = x(\pi - x), \quad x \in [0, \pi].$$

(a) Extend $f(x)$ as an *odd* periodic function with period 2π and find its Fourier series.

(b) Extend $f(x)$ as an *even* periodic function with period 2π and find its Fourier series.

8. Find the Fourier series expansion of the following functions, defined for $x \in \mathbb{R}$:

(a) $f(x) = \sin^2(x)$,

(b) $f(x) = 4\cos^3(x) - 2\cos(x)$,

(c) $f(x) = T_n(\cos x)$,

where $T_n(x)$ is the nth Chebyshev polynomial (see Chapter 3, Section 3.2.4).

9. We wish to express the real part $u(x, y)$ of an analytic function $f(z)$ in terms of the polar coordinates (r, ϕ). This is convenient in cases where the function $u(r, \phi)$, $0 \le \phi < 2\pi$, is known at the boundary consisting of a circle of radius $r = R$. To this end, we express u in terms a Fourier series expansion as:

$$u(r, \phi) = a_0 + \sum_{n=1}^{\infty} a_n \left(\frac{r}{R}\right)^n \cos(n\phi)$$

$$+ \sum_{n=1}^{\infty} b_n \left(\frac{r}{R}\right)^n \sin(n\theta),$$

and proceed to determine the coefficients a_0, a_n, b_n, $n = 1, \ldots$, in terms of integrals involving $u(R, \phi)$. Show that this yields the general form

$$u(r, \phi) = \frac{1}{2\pi} \int_0^{2\pi} K(r, \phi; R, \phi') u(R, \phi') d\phi',$$

where the so-called "kernel" $K(r, \phi; R, \phi')$ is given by:

$$K(r, \phi; R, \phi') = 1 + 2 \sum_{n=1}^{\infty} \left(\frac{r}{R}\right)^n \cos\left(n(\phi - \phi')\right).$$

This expression is referred to as the "Poisson integral formula".

10. Show that if the Fourier transforms of the functions $f(x)$ and $g(x)$ are $\hat{f}(k)$ and $\hat{g}(k)$, respectively, then the following relation holds:

$$\int_{-\infty}^{\infty} f(ax)g(bx)dx = \frac{1}{2\pi|ab|} \int_{-\infty}^{\infty} \hat{f}\left(\frac{k}{a}\right) \hat{g}\left(-\frac{k}{b}\right) dk,$$

where $a, b \in \mathbb{R}$.

11. If $\hat{f}(k)$ is the Fourier transform of the function $f(x)$ and $a, b \in \mathbb{R}, a > 0$, show that the following functions produce the Fourier transforms given:

(a) $f(ax)\cos(bx) \xrightarrow{\text{FT}} \frac{1}{2a}\left[\hat{f}\left(\frac{k-b}{a}\right) + \hat{f}\left(\frac{k+b}{a}\right)\right]$,

(b) $f(ax+b) \xrightarrow{\text{FT}} \frac{e^{ibk/a}}{a}\hat{f}\left(\frac{k}{a}\right)$.

Calculate the FT of the function $f(ax)\sin(bx)$ in terms of $\hat{f}(k)$.

12. (a) Find the Fourier transform of the function

$$f(x) = e^{-a|x|}, \quad \text{where } a \in \mathbb{R}, \ a > 0.$$

(b) Show that if the FT of $f(x)$ is $\hat{f}(k)$ then the FT of $\hat{f}(x)$ is $2\pi f(-k)$.

(c) Show that the inverse FT of $f(-k)$ is $(1/2\pi)\hat{f}(x)$.

(d) Use the results from parts (a), (b), (c) to show the following FT:

$$g(x) = \frac{1}{(x^2 + a^2)} \xrightarrow{\text{FT}} \hat{g}(k) = \frac{\pi}{a}e^{-a|k|}.$$

13. Find the Fourier transform of the function $f(x)$ defined as

$$f(x) = \frac{2\sin(x)}{x},$$

by taking advantage of the results of Example 7.4 and making use of the inverse FT.

14. Evaluate the real integral, with $k \in \mathbb{R}$,

$$\int_{-\infty}^{\infty} e^{-x^2} \cos(2kx)dx,$$

by taking advantage of the Fourier transform of the normalized Gaussian function, Eq. (7.45).

15. Consider the function:
$$f(x) = \begin{cases} 0, & x < -1, \\ 1, & -1 \le x \le x, \\ 0, & x > 1. \end{cases}$$

(a) Calculate the Fourier transform $\hat{f}(k)$ of $f(x)$.

(b) Define $g(x)$ to be the autoconvolution of $f(x)$ and calculate its Fourier transform, $\hat{g}(k)$. Discuss the origin of the various terms that appear in $\hat{g}(k)$.

(c) Find the Fourier transform of the derivative of the autoconvolution, $g'(x)$, using the properties of the Fourier transform.

(d) Find the derivative of the autoconvolution $g'(x)$ by performing an inverse Fourier transform and evaluating the integrals using contour integration.

16. Consider the functions:
$$f(x) = \begin{cases} 0, & \text{for } x < 0, \\ e^{-ax}, & \text{for } x \ge 0, \end{cases}$$

and
$$g(x) = \begin{cases} -0.5, & \text{for } x < 0, \\ 0.5, & \text{for } x \ge 0, \end{cases}$$

where $a > 0$. We want to calculate the convolution of these two functions using the Fourier transform approach.

(a) Calculate the Fourier Transforms $\hat{f}(k)$ of $f(x)$ and $\hat{g}(k)$ of $g(x)$.

(b) Calculate the Fourier Transform $\hat{h}(k)$ of the convolution $h(x)$ of the two functions.

(c) Find $h(x)$ by performing an Inverse Fourier Transform on $\hat{h}(k)$ and by evaluating the integrals using contour integration.

17. Consider the following three functions of the real variable x:
$$f_0(x) = e^{-x^2/2}, \quad f_1(x) = xe^{-x^2/2},$$
$$f_2(x) = (x^2 + b)e^{-x^2/2},$$

where b is a constant, $b \in \mathbb{R}$.

(a) Show that the Fourier transform of $f_0(x)$, denoted by $\hat{f}_0(k)$, satisfies the relation $\hat{f}_0(k) = c_0 f_0(k)$, where c_0 is a constant. What is the value of c_0?

(b) Show that the FT of $f_1(x)$, denoted by $\hat{f}_1(k)$, satisfies the relation $\hat{f}_1(k) = c_1 f_1(k)$. Find the constant c_1.

(c) Determine the value of b so that the FT of $f_2(x)$, denoted by $\hat{f}_2(k)$, obeys the relation $\hat{f}_2(k) = c_2 f_2(k)$. For this value of b, find the constant c_2.

[Hint: Use the properties of the FT that relate the moments of a function $f(x)$ to the derivatives of its FT, $\hat{f}(k)$, and take advantage of the known FT of the normalized Gaussian function, Eq. (7.45).]

18. The spectral distribution $\hat{f}(\omega)$ of a time signal $f(t)$ is given by the expression:
$$\hat{f}(\omega) = \begin{cases} 1, & \text{if } 1 \le |\omega| \le 3, \\ 0, & \text{otherwise.} \end{cases}$$

Obtain the signal $f(t)$ that produces this spectral distribution.

19. A signal in time t is given by the following function:
$$f(t) = \cos(\omega_0 t)\frac{1}{\tau\sqrt{2\pi}}e^{-t^2/2\tau^2}.$$

Find the Fourier transform $\hat{f}(\omega)$ of the signal. In the following, assume $\tau = 2/\omega_0$ and $\Delta = \pi\tau/2$. Plot $\hat{f}(\omega)$ as a function of ω. The signal is sampled at the rate $1/\Delta$. What is the shape of the measured $\hat{f}(\omega)$? Plot the measured FT when the signal is measured at the rate $2/\Delta$.

Part III

Applications to Partial Differential Equations

Chapter 8

Partial Differential Equations: Introduction

The differential equations of the propagation of heat express the most general conditions, and reduce the physical questions to problems of pure analysis, and this is the proper object of theory.

Jean-Baptiste Joseph Fourier (French mathematician, 1768–1830)

8.1 General remarks

Consider a function $u(x)$ of one variable, x. A relation coupling this function and its derivatives with respect to the variable x is called an "ordinary differential equation" (ODE). An equation involving the nth-order derivative of the function is referred to as "nth-order" equation; an ODE that involves a function and its first derivative only is a first-order ODE, and so on. If a function depends on more than one variable, then a relation coupling this function and its partial derivatives with respect to the independent variables is called a "partial differential equation" (PDE). Differential equations are called "linear" if the function and its derivatives appear only in the first power, and "nonlinear" if there are terms involving higher powers of the function or its derivatives, or products of the function and the derivatives.

Using the notation for derivatives introduced in Eqs. (2.3)–(2.5), some examples follow of linear PDEs that we will study in this part of the book.

1. The heat or diffusion equation,

$$u_t(x,t) = \alpha u_{xx}(x,t), \quad \alpha > 0,$$

and the related evolution equation,

$$u_t(x,t) = \beta u_x(x,t) + \gamma u_{xxx}(x,t), \quad \beta, \gamma \in \mathbb{R}.$$

2. The wave equation,

$$u_{tt}(x,t) = c^2 u_{xx}(x,t), \quad c \in \mathbb{R}.$$

3. The Laplace equation,

$$u_{xx}(x,y) + u_{yy}(x,y) = 0,$$

and the related Poisson equation,

$$u_{xx}(x,y) + u_{yy}(x,y) = f(x,y).$$

4. The Helmholtz equation,

$$u_{xx}(x,y) + u_{yy}(x,y) + \kappa^2 u(x,y) = 0, \quad \kappa > 0,$$

and the related modified Helmholtz equation,

$$u_{xx}(x,y) + u_{yy}(x,y) - \kappa^2 u(x,y) = 0, \quad \kappa > 0.$$

The branch of mathematics that deals with the solution of PDEs is vast, commensurate with the plethora of their applications. The PDEs mentioned above capture some of this breadth, that includes: oscillatory motion with a regular period (harmonic oscillation) and its variations like damped or forced oscillations; diffusion of chemical or biological species, the spread of heat, and the evolution of the price of financial derivatives; the propagation of waves of various forms in matter or in vacuum; the behavior of the potential and the field due to electric charges and currents; and many more.

In order to solve an ODE or a PDE we need more information than just the equation itself. This is because the general solution of such equations typically involves several constants or functions, the values of which cannot be determined by the equation alone. The determination of these values requires the knowledge of the solution for specific values of its variables. The prescribed relations are known as "boundary conditions" or "initial conditions", depending on the nature of the equation. The term "boundary conditions" is used to describe the relations that give the value of the function on the boundary of the domain of the independent variables, where the equation is defined; the term "initial conditions" is usually reserved for cases when one of the variables is the time, t, and denotes relations for the initial value of this variable, with a common choice for the initial value to be prescribed at $t = 0$.

In this book, we cannot give but a minuscule sample of this important area, introducing only the basics and placing emphasis on applications. In particular, we will provide a rudimentary discussion of traditional methods for solving differential equations. Importantly, we will also present a new method, which incorporates many of the concepts introduced in previous chapters, and leads to an elegant and powerful way of solving linear partial differential equations. This method,[1] which is referred to as the "Unified Transform", is the subject of the next four chapters, where its application will be illustrated with the aid of several examples.

[1] The method was developed by one of the authors and is also known as the "Fokas method".

8.2 Overview of traditional approaches

A general strategy for obtaining solutions of PDEs is to reduce them to ODEs, and then to attempt to reduce these ODEs to simple cases for which the solutions are known. This reduction is achieved through the powerful technique of "separation of variables".

Accordingly, we devote this section to a discussion of simple ODEs with known solutions and to traditional approaches for finding the solution of a linear ODE.

8.2.1   Examples of ODEs

We first present examples that involve expressions in the unknown function $u(x)$ and its first and second derivatives, $u_x(x)$, $u_{xx}(x)$, on the left-hand side and zero on the right-hand side of the equation; this type of ODE is called "homogeneous". We will encounter the solutions of these equations repeatedly later in the present chapter and in Chapters 9–12.

In the more general case, when the right-hand side of the equation is not zero but a known function $f(x)$, the corresponding ODE is called "inhomogeneous". After the examples of homogeneous ODEs with known solutions, we discuss some general methods for finding the solution of inhomogeneous linear ODEs.

Exponential growth or decay: The most common first-order linear ODE and its solution are

$$u_x(x) + ku(x) = 0, \quad k \in \mathbb{C}, \tag{8.1a}$$

$$u(x) = Ae^{-kx}, \tag{8.1b}$$

where A is a constant to be determined by the boundary or initial conditions. Specifically, if the value of the function, u_0, is known at the value of the variable $x = x_0$, then the value of A can be easily obtained:

$$u(x_0) = u_0 = Ae^{-kx_0} \Rightarrow A = u_0 e^{kx_0}.$$

For k real, this equation describes exponential growth ($k > 0$) or exponential decay ($k < 0$) as the variable x increases from an initial value x_0. The exponential function was discussed in Chapter 1 (see Section 1.2.5).

Harmonic oscillations: The most common second-order linear ODE and its solution are as follows:

$$u_{xx}(x) + k^2 u(x) = 0, \quad k \in \mathbb{C}, \tag{8.2a}$$

$$u(x) = Be^{ikx} + Ce^{-ikx}, \tag{8.2b}$$

with the constants B, C to be determined by the boundary conditions. For this equation, we need to know the value of the function at two values of the variable x (typically the "boundaries" of the domain of x), or the value of the function and the value of its first derivative at one value of the variable x. For instance, from Eq. (8.2b), the first derivative takes the form

$$u_x(x) = ik \left[Be^{ikx} - Ce^{-ikx} \right].$$

If at $x = x_0$ the value of the function, $u_0 = u(x_0)$, and the value of the first derivative, $u_1 = u_x(x_0)$, are known, we obtain the following system of linear equations for the constants B, C:

$$Be^{ikx_0} + Ce^{-ikx_0} = u_0, \quad ik \left[Be^{ikx_0} - Ce^{-ikx_0} \right] = u_1.$$

This system of equations can be easily solved, yielding

$$B = \frac{1}{2} e^{-ikx_0} \left(u_0 + \frac{u_1}{ik} \right), \quad C = \frac{1}{2} e^{ikx_0} \left(u_0 - \frac{u_1}{ik} \right).$$

For k real, the solution $u(x)$ given in Eq. (8.2b) describes harmonic oscillations of frequency $|k|$ and is a linear combination of sine and cosine functions (these were discussed in Chapter 1, see Section 1.2.4).

Damped oscillations: The general linear second-order ODE and its solution are as follows:

$$u_{xx}(x) + bu_x(x) + cu(x) = 0, \quad b, c \in \mathbb{C}, \tag{8.3a}$$

$$u(x) = e^{-bx/2} \left[De^{ikx} + Ee^{-ikx} \right], \quad k = \pm \left(c - \frac{b^2}{4} \right)^{1/2}, \tag{8.3b}$$

as can be easily verified by substituting the solution in the ODE; the constants D, E are determined by the boundary conditions. For instance, from Eq. (8.3b), the first derivative takes the form

$$u_x(x) = -\frac{b}{2}u(x) + ike^{-bx/2} \left[De^{ikx} - Ee^{-ikx} \right].$$

If at $x = x_0$ the value of the function, $u_0 = u(x_0)$, and the value of the first derivative, $u_1 = u_x(x_0)$, are known we obtain the following system of linear equations for the constants D, E:

$$De^{ikx_0} + Ee^{-ikx_0} = u_0 e^{bx_0/2},$$

$$De^{ikx_0} \left(ik - \frac{b}{2} \right) - Ee^{-ikx_0} \left(ik + \frac{b}{2} \right) = u_1 e^{bx_0/2}.$$

This system of equations can be easily solved, yielding

$$D = \frac{1}{2ik} \left[\left(ik + \frac{b}{2} \right) u_0 + u_1 \right] e^{(-ik+b/2)x_0},$$

$$E = \frac{1}{2ik} \left[\left(ik - \frac{b}{2} \right) u_0 - u_1 \right] e^{(ik+b/2)x_0}.$$

For real coefficients b, c, that satisfy the condition $(c - b^2/4) > 0$, k is real and the solution of Eq. (8.3b) describes oscillatory motion with frequency $|k|$ and with decaying amplitude (for $b > 0$) or increasing amplitude (for $b < 0$) when x increases from the initial value x_0. An example is shown in Fig. 8.1, for $b = 0.5$, $c = 2$, $u_0 = 1$, $u_1 = 0$.

A nonlinear ODE: The Korteweg–de Vries soliton: As an example of a nonlinear ODE we consider the equation

$$u_{yyy} - 6u_y u - cu_y = 0, \quad c \in \mathbb{R}. \tag{8.4a}$$

The second term, the product of the unknown function $u(y)$ and its derivative u_y, makes this equation a nonlinear ODE. This ODE is actually derived from a PDE, namely the so-called Korteweg–de Vries (KdV) equation

$$u_t + u_{xxx} - 6uu_x = 0. \tag{8.4b}$$

This PDE describes surface water waves of small amplitude $u(x, t)$, in a shallow channel, under the assumption that the flow is inviscid and irrotational.

The amplitude is a function of the spatial and temporal variables x and t, respectively. Under the assumption that the waves are traveling with constant speed c,

$$u(x,t) = u(y), \quad y = x - ct, \quad -\infty < y < \infty,$$

and $\partial_x = \partial_y$, $\partial_t = -c\partial_y$. Thus, Eq. (8.4b) becomes Eq. (8.4a). By integrating Eq. (8.4a) once we obtain

$$u_{yy} - 3u^2 - cu = A,$$

where A is a constant of integration. We assume that u and its derivatives vanish as $|y| \to \infty$ and hence $A = 0$. In this case, the solution to the above equation is given by

$$u(y) = -\frac{c}{2}\left[\cosh\left(\frac{\sqrt{c}}{2}(y - y_0)\right)\right]^{-2}, \tag{8.4c}$$

where y_0 is an arbitrary constant. In terms of the original spatial and temporal variables, this solution corresponds to a "hump" traveling with speed c, which is known as a "solitary wave" or "soliton". An example is shown in Fig. 8.1 for $c = 1$ and $y_0 = 0$. We shall encounter this solution again in the Application section of Chapter 12 (see Section 12.5).

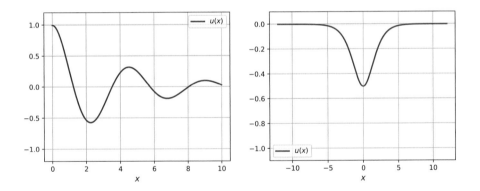

Figure 8.1: Examples of solutions of ODEs: **Left**: The damped harmonic oscillation function $u(x)$ of Eq. (8.3b). **Right**: The soliton function $u(x)$ of Eq. (8.4c).

8.2.2 Eigenvalue equations

Several of the examples discussed above are representative cases of a class of differential equations referred to as "eigenvalue equations"; these equations can be written as an equation with a left-hand side that contains derivatives of the unknown function $u(x)$ and other terms that involve the function itself, while the right-hand side is equal to $\lambda u(x)$, where λ is a constant known as the "eigenvalue". For instance, the equation describing harmonic oscillations, Eq. (8.2a), can be expressed as an eigenvalue equation in the following form:

$$u_{xx} = \lambda u(x), \quad \lambda = -k^2. \tag{8.5}$$

At this point it is also convenient to generalize the notion of the "differential operator", first introduced in Chapter 2, beyond that of specific combinations of partial derivatives of multi-variable functions (like the gradient, the Laplacian, etc., see Section 2.4). Specifically, we call a differential operator, denoted as $\hat{\mathcal{L}}$, a mathematical object that can be applied to a function to produce the left-hand side of a differential equation, including derivatives of the unknown function and other terms. For instance, for the case of Eq. (8.5) the differential operator $\hat{\mathcal{L}}$ is defined as

$$\hat{\mathcal{L}} = \frac{d^2}{dx^2}, \tag{8.6}$$

which, when applied to the function $u(x)$, produces the left-hand side of this equation.

When the eigenvalue λ takes special values, and depending on the relevant boundary conditions, it may be possible to readily obtain the solution of the differential equation. For instance, in the example of Eq. (8.5), if we use $x \in [0, L]$ as the domain of x, with $0 < L < \infty$, and let

$$k = \frac{n\pi}{L}, \quad n \text{ integer}, \tag{8.7}$$

with boundary conditions $u(0) = 0$, $u(L) = 0$, it is straightforward to see that the solution is

$$u(x) = A \sin\left(\frac{n\pi x}{L}\right),$$

where A is a constant. We then say that for these boundary conditions, the set of functions $\sin(n\pi x/L)$ are the "eigenfunctions" of the differential operator of Eq. (8.6), with eigenvalues given by Eq. (8.7). For different boundary conditions we may need to use the other set of functions that satisfy the differential equation with the same eigenvalues, namely $\cos(n\pi x/L)$, or a linear combination of the sines and cosines with argument $(n\pi x/L)$.

Other interesting cases of eigenvalue equations appear in many contexts in the physical and engineering sciences. Some examples were mentioned in Chapter 3, whose eigenfunctions are special functions, namely, the Legendre or the Chebyshev polynomials (see Section 3.2.4). These special functions satisfy a specific type of boundary conditions. They can be used as the "basis" for a series expansion of the general solution that satisfies other types of boundary conditions, by properly choosing the coefficients in the series expansion. In this sense, it is useful for the basis functions to be orthogonal, as the Legendre and Chebyshev polynomials are. Thus, the special functions play a role equivalent to the unit vectors in a multi-dimensional space, which are chosen to be orthogonal, and "span" the entire space. The coefficients in the series expansion then play the role of the projection of a vector onto the different directions in the multi-dimensional space. While this approach can be useful in solving many interesting problems, in this book we concentrate in an alternative way of solving problems that involve differential equations, namely by deriving transforms of the unknown function that satisfy certain conditions and through which we can incorporate the boundary conditions.

8.2.3   The Green's function method

A powerful approach for solving a differential equation is to find the corresponding Green's function. We illustrate this procedure with an example. Suppose we want to solve the second-order linear ODE

$$u_{xx}(x) + k^2 u(x) = f(x) \Rightarrow \left(\frac{d^2}{dx^2} + k^2\right) u(x) = f(x), \tag{8.8}$$

for the unknown function $u(x)$, with $k \in \mathbb{C}$ and $f(x)$ a given (known) function. We observe that for $f(x) = 0$ the problem simplifies to the case of the homogeneous equation, Eq. (8.2a), whose solution is known, Eq. (8.2b). We define the differential operator $\hat{\mathcal{L}}$ for this case,

$$\hat{\mathcal{L}}u(x) = f(x) \Rightarrow \hat{\mathcal{L}} \equiv \frac{d^2}{dx^2} + k^2.$$

In order to solve the general case, $f(x) \neq 0$, we introduce the Green's function, $G(x, \xi, k)$, which satisfies the following equation:

$$\left(\frac{d^2}{dx^2} + k^2\right) G(x, \xi, k) = \delta(x - \xi) \Rightarrow \hat{\mathcal{L}}G(x, \xi, k) = \delta(x - \xi). \qquad (8.9)$$

We multiply both sides of this equation by $f(\xi)$ and then integrate over the variable ξ to obtain:

$$\int \hat{\mathcal{L}}G(x, \xi, k)f(\xi)d\xi = \int \delta(x - \xi)f(\xi)d\xi = f(x).$$

We can move the differential operator $\hat{\mathcal{L}}$ outside of the integral over the variable ξ on the left-hand side of this equation, because $\hat{\mathcal{L}}$ applies only to functions of x:

$$\hat{\mathcal{L}} \int G(x, \xi, k)f(\xi)d\xi = f(x).$$

Comparing this equation with Eq. (8.8) we deduce that

$$\int G(x, \xi, k)f(\xi)d\xi = u(x). \qquad (8.10)$$

In other words, we can compute the solution $u(x)$ of Eq. (8.8) provided we can construct the function $G(x, \xi, k)$. This function is given formally by the application of the inverse of the differential operator $\hat{\mathcal{L}}$ which we will denote as $\hat{\mathcal{L}}^{-1}$, to the δ-function, that is,

$$G(x, \xi, k) = \hat{\mathcal{L}}^{-1}\delta(x - \xi), \quad \hat{\mathcal{L}}^{-1} \equiv \left(\frac{d^2}{dx^2} + k^2\right)^{-1}. \qquad (8.11)$$

However, this last equation is at the moment a fancy notation since we don't know how to handle a differential operator that involves a derivative in the denominator. It turns out that this expression can actually be computed, provided some additional information about the problem is known. Indeed, suppose that we are also given boundary conditions, which, in any case, are necessary to obtain the full solution. We consider periodic boundary conditions:

$$u(0) = u(L), \quad 0 < L < \infty.$$

The solutions of the homogeneous version of Eq. (8.8) that satisfy this boundary condition must be of the form

$$B(e^{ik_n x} \pm e^{-ik_n x}), \quad k_n = \frac{2\pi n}{L}, \quad n \text{ integer},$$

that is, the constants in the solution of Eq. (8.2a) satisfy $C = \pm B$, and only certain complex exponential functions are allowed, namely, those with $k = k_n$.

We next note that, as established in the derivation of the Fourier expansion, we can expand an arbitrary function $g(x)$ in terms of the above complex exponentials, with the expansion coefficients given by the expression below:

$$g(x) = \sum_n g_n e^{ik_n x}, \quad g_n = \int_0^L g(x) e^{-ik_n x} dx.$$

We also note that it is straightforward to compute the differential operator and its inverse applied to these complex exponentials:

$$\hat{\mathcal{L}} e^{ik_n x} = (k^2 - k_n^2) e^{ik_n x} \Rightarrow \hat{\mathcal{L}}^{-1} e^{ik_n x} = \frac{1}{k^2 - k_n^2} e^{ik_n x}. \tag{8.12}$$

According to Eq. (8.11), the calculation of the Green's function requires the application of $\hat{\mathcal{L}}^{-1}$ to the specific function $g(x) = \delta(x - \xi)$; the Fourier expansion coefficients of $\delta(x - \xi)$ are given by

$$g_n = \int_0^L \delta(x - \xi) e^{-ik_n x} dx = e^{-ik_n \xi}.$$

Equation (8.12) implies that in this case the Green's function is

$$G(x, \xi, k) = \hat{\mathcal{L}}^{-1} \delta(x - \xi) = \hat{\mathcal{L}}^{-1} \sum_n e^{ik_n(x-\xi)} = \sum_n \frac{1}{k^2 - k_n^2} e^{ik_n(x-\xi)}.$$

This is already a useful result: it shows that if the value of k is the same as (or very close to) one of the values k_n, then this coefficient in the expansion becomes infinitely large and dominates all other coefficients that have finite values, therefore we can neglect all other terms and keep only the term with $k = k_n$.

To complete the calculation of the solution $u(x)$ for the differential equation we need to integrate the product of the Green's function and the given function $f(\xi)$. We can expand this function in terms of the same complex exponentials as above, with the coefficients in the expansion f_n obtained from the appropriate integrals of the known function $f(\xi)$:

$$f(\xi) = \sum_m f_m e^{ik_m \xi}, \quad f_m = \int_0^L f(\xi) e^{-ik_m \xi} d\xi. \tag{8.13}$$

The integral of Eq. (8.10) takes the following form:

$$u(x) = \int G(x, \xi, k) f(\xi) d\xi = \int_0^L \left(\sum_n \frac{1}{k^2 - k_n^2} e^{ik_n(x-\xi)} \right) \left(\sum_m f_m e^{ik_m \xi} \right) d\xi$$

$$= \sum_{n,m} \frac{f_m}{k^2 - k_n^2} e^{ik_n x} \int_0^L e^{i(k_m - k_n)\xi} d\xi \Rightarrow u(x) = \sum_n \frac{f_n}{k^2 - k_n^2} e^{ik_n x},$$

where we have used the relation

$$\int_0^L e^{i(k_m - k_n)\xi} d\xi = \int_0^L e^{i2\pi(m-n)\xi/L} d\xi = \delta_{nm}.$$

Since the coefficients f_n are known for the given function $f(\xi)$, the expression

$$u(x) = \sum_n \frac{f_n}{k^2 - k_n^2} e^{ik_n x}, \tag{8.14}$$

provides the full solution to the differential equation.

8.2.4 The Fourier series method

An alternative general approach is to assume at the outset a Fourier expansion for the function $u(x)$, as well as for the function $f(x)$ that appears on the right-hand side of the differential equation. This is only feasible under the conditions imposed by the Fourier expansion (see Chapter 7). In particular, if the interval of definition of the variable x is $[0, L]$, then the Fourier expansion implies that both $u(x)$ and $f(x)$ are periodic with period L. We then assume the following expression for $u(x)$:

$$u(x) = \sum_n u_n e^{ik_n x}, \quad k_n = \frac{2\pi n}{L}, \quad n \text{ integer.}$$

Similarly, for $f(x)$ we have the expansion given in Eq. (8.13), with the coefficients f_n known, since $f(x)$ is given. Inserting the Fourier expansions of $u(x)$ and $f(x)$ in the differential Eq. (8.8) we obtain:

$$\sum_n u_n (k^2 - k_n^2) e^{ik_n x} = \sum_n f_n e^{ik_n x}, \quad f_n = \int_0^L f(x) e^{-ik_n x} dx.$$

In order for this to hold for every value of $x \in [0, L]$ the coefficients must be equal, term-by-term:

$$u_n (k^2 - k_n^2) = f_n \Rightarrow u_n = \frac{f_n}{k^2 - k_n^2},$$

which gives the same expression for $u(x)$ as that obtained in Eq. (8.14) through the Green's function method.

8.3 Evolution equations

A variety of physical phenomena can be modeled by linear evolution equations in one spatial dimension, namely by PDEs involving a first-order time-derivative and space-derivatives of at least second order. We first give some examples of such equations and motivate how they arise from physical situations. We then discuss how they can be solved following traditional approaches.

8.3.1 Motivation of evolution equations

The most well-known evolution equation is the "heat equation" (also referred to as the "diffusion equation"). With x denoting the spatial variable and t time, the function of interest is $u(x, t)$ and the heat equation reads

$$u_t = \alpha u_{xx}, \tag{8.15}$$

where α is a constant with dimensions [length]^2/[time]. This equation arises in diverse scientific fields, including the study of Brownian motion through its relation with the Fokker–Planck equation, and in the study of chemical diffusion and other related processes.

The heat equation was first derived in connection with the description of the temperature, represented by the function $u(x, t)$, of a thin rod of cross sectional area A: consider a segment of length Δx of this rod and assume that the temperature at $x + \Delta x$ is higher than the temperature at x, see Fig. 8.2.

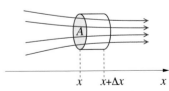

Figure 8.2: A segment of length Δx of a thin rod of cross-sectional area A through which a physical quantity flows, as shown schematically by the blue arrows. In the example discussed here the physical quantity is heat.

According to Fourier's law, the rate of flow of heat energy per unit area through a surface is proportional to the negative temperature gradient across this surface. Thus, the net accumulation of heat in the above segment in time Δt is given by

$$\kappa A \frac{\partial u}{\partial x}(x + \Delta x, t)\Delta t - \kappa A \frac{\partial u}{\partial x}(x, t)\Delta t,$$

where κ denotes thermal conductivity. The change in the internal energy per unit volume in a material of density ρ and specific heat capacity c_p is given by $c_p \rho \Delta u$. Thus, since the volume of the thin rod is $A \Delta x$, we obtain

$$(A\Delta x)c_p \rho \Delta u = \kappa A \left[\frac{\partial u}{\partial x}(x + \Delta x, t) - \frac{\partial u}{\partial x}(x, t) \right] \Delta t$$

$$\Rightarrow \quad \frac{\Delta u}{\Delta t} = \frac{\alpha}{\Delta x} \left[\frac{\partial u}{\partial x}(x + \Delta x, t) - \frac{\partial u}{\partial x}(x, t) \right],$$

where we have defined a new constant, α, the thermal diffusivity,

$$\alpha \equiv \frac{\kappa}{c_p \rho}.$$

In the limit of Δx and Δt tending to zero, the above equation becomes

$$\frac{\partial u}{\partial t} = \alpha \frac{\partial^2 u}{\partial^2 x},$$

which is the heat equation, Eq. (8.15).

Another important evolution equation involving a second-order space derivative is the equation

$$u_t = \alpha u_{xx} + \beta u_x, \tag{8.16}$$

where β is a constant with dimensions [length]/[time]. This equation arises in a variety of different circumstances including financial mathematics and pharmacokinetics. We treat the former case in the Application section at the end of the present chapter (see Section 8.8). Regarding the occurrence of Eq. (8.16) in pharmacokinetics we note that in the convection–dispersion models of oral drug absorption, the small intestine is considered as a one-dimensional tube: Let x be the spatial coordinate denoting the distance from the stomach. It is assumed that the tube contents have constant axial speed v and constant dispersion coefficient D that arises from molecular diffusion stirring, due to the mobility of the intestines. Then, the concentration $c(x, t)$ of a non-absorbable, non-degradable compound satisfies the equation

$$\frac{\partial c}{\partial t} = D\frac{\partial^2 c}{\partial x^2} - v\frac{\partial c}{\partial x},$$

which is the same as Eq. (8.16), with $u(x, t) = c(x, t)$, $\alpha = D$ and $\beta = -v$. A typical boundary value problem for the above equation is formulated in $0 \le x < \infty$ and involves the conditions

$$c(x, 0) = c_0(x), \quad c(0, t) = g_0(t), \quad c \to 0 \text{ as } x \to \infty.$$

The above equation also characterizes a convection-dispersion model describing the concentration $c(x, t)$ for a solute in a tree-like structure modeling the arterial tree.

Another important type of evolution equations involves higher spatial derivatives of the function $u(x,t)$. An example of a linear evolution equation of physical significance evolving a third-order derivative is the so-called "Stokes equation",

$$u_t = \beta u_x + \gamma u_{xxx}, \tag{8.17}$$

where γ is a constant with dimensions $[\text{length}]^3/[\text{time}]$. This equation describes the elevation $u(x,t)$ of surface water waves under certain assumptions and approximations. Indeed, it can be shown that the Navier–Stokes equations, which are the basic equations describing the motion of a fluid, under the assumption that the flow is inviscid (that is, the effect of viscosity can be neglected) and irrotational (its vorticity vanishes), and under the approximations of small amplitude and of long wavelength, give rise to the equation

$$u_t = -\sqrt{\frac{g}{h}} \left(\alpha u_x + \frac{h^3}{6} u_{xxx} \right),$$

where $u(x,t)$ is the elevation of the water above the equilibrium height h, g is the gravitational acceleration, and α is a constant related to the uniform motion of the liquid with dimensions of [length].

Boundary value problems for linear evolution PDEs involving a third-order spatial derivative have received little attention in the literature until recently. Taking into consideration that this type of PDEs *are* important in applications, perhaps the lack of attention is due to the lack of an efficient method for solving such PDEs, until the recent development of the unified transform method.

8.3.2 *Separation of variables and initial value problems*

It is natural to try to reduce a new problem to a problem that has already been solved. Indeed, as soon as d'Alembert wrote the first PDE in the history of mathematics (which was the wave equation), Euler, Bernoulli and others attempted to find solutions of PDEs using the method of *separation of variables*. This technique reduces a PDE in two independent variables to two ODEs.

Consider for example the heat equation, Eq. (8.15). Seeking a separable solution in the form

$$u(x,t) = X(x)T(t), \tag{8.18}$$

where $X(x)$ and $T(t)$ are unknown functions of a single variable each, we find

$$XT' = \alpha X''T,$$

where prime denotes differentiation with respect to the intrinsic variable of each function. Thus,

$$\frac{X''}{X} = \frac{1}{\alpha}\frac{T'}{T}. \tag{8.19}$$

Since the left-hand side of this equation is a function of x, whereas the right-hand side is a function of t, it follows that each of the above ratios is a constant, which for convenience will be denoted by $-k^2$, where k is some complex constant. Thus, Eq. (8.19) implies the two ODEs

$$X''(x) + k^2 X(x) = 0, \tag{8.20}$$

and

$$T'(t) + \alpha k^2 T(t) = 0. \tag{8.21}$$

Clearly the representation of the solution given by the expression of Eq. (8.18) is very limited. However, the intuitive idea is that if we can solve the ODEs of Eqs. (8.20), (8.21), and if we can "sum up" appropriate solutions over k of these ODEs, then perhaps we can obtain the general solution of the heat equation.

Taking into consideration that the exponentials e^{ikx} and $e^{-\alpha k^2 t}$ are particular solutions of Eqs. (8.20) and (8.21), respectively, Eq. (8.18) implies that a particular solution of the heat equation is given by

$$U(k)e^{ikx - \alpha k^2 t},$$

where k is an arbitrary complex constant, and $U(k)$ is an arbitrary function of k. Clearly, the following expression is also a solution of the heat equation:

$$u(x,t) = \int_C U(k)e^{ikx - \alpha k^2 t} dk, \tag{8.22}$$

provided that $U(k)$ and the contour C are chosen in such way that the above integral makes sense.

It turns out that there exists a general, deep result in analysis known as the "Ehrenpreis Principle", which when applied to the particular case of the heat equation states that for a well-posed problem formulated in a bounded, smooth, convex, domain the solution can always be written in the form of Eq. (8.22). Actually, this result can be generalized to unbounded and non-smooth domains (only *convexity* is vital). However, this result does not provide a systematic way for determining the contour C and the function $U(k)$. It turns out that the unified transform always yields representations in the Ehrenpreis form, and moreover it has the advantage that it specifies the contour C and expresses $U(k)$ in terms of the values of the solution u on the boundary.

In order to obtain a well-posed problem for a given PDE, one has to specify the domain of validity of this PDE, as well as appropriate initial and boundary conditions. For evolution PDEs, the simplest such problem is the *initial value problem*. In this case, the domain is given by

$$\Omega = \{-\infty < x < \infty, \ t > 0\}, \tag{8.23}$$

and appropriate initial and boundary conditions are the following:

$$u(x,0) = u_0(x), \quad -\infty < x < \infty, \tag{8.24}$$

where $u_0(x)$ is a given function, and furthermore u as well as a sufficient number of spatial-derivatives of u vanish as $|x| \to \infty$.

For the initial value problem of the heat equation, using the Fourier transform pair, it is straightforward to obtain both the relevant contour C and the function $U(k)$ occurring in Eq. (8.22): C is the infinite line $-\infty < k < \infty$, and $U(k)$ is $\hat{u}_0(k)$ multiplied by $1/2\pi$, where $\hat{u}_0(k)$ is the Fourier transform of $u_0(x)$, namely

$$\hat{u}_0(k) = \int_{-\infty}^{\infty} e^{-ikx} u_0(x) dx, \quad -\infty < k < \infty. \tag{8.25}$$

Indeed, $u(x,t)$ turns out to be:

$$u(x,t) = \int_{-\infty}^{\infty} e^{ikx - \alpha k^2 t} \hat{u}_0(k) \frac{dk}{2\pi}, \quad -\infty < x < \infty, \quad t > 0. \qquad (8.26)$$

Instead of deriving the expression of Eq. (8.26), we choose to verify that the representation defined by the right-hand side of this expression satisfies the heat equation, as well as the initial condition Eq. (8.24): the (x,t) dependence of the right-hand side of Eq. (8.26) is in the form of $\exp[ikx - \alpha k^2 t]$ and this expression is a particular solution of the heat equation for a constant k, thus clearly Eq. (8.26) solves the heat equation. Evaluating the expression of Eq. (8.26) at $t = 0$ we find

$$u(x,0) = \int_{-\infty}^{\infty} e^{ikx} \hat{u}_0(k) \frac{dk}{2\pi}.$$

Since $\hat{u}_0(k)$ is defined by Eq. (8.25), the inverse Fourier transform formula, Eq. (7.20), yields

$$\int_{-\infty}^{\infty} e^{ikx} \hat{u}_0(k) \frac{dk}{2\pi} = u_0(x).$$

The above two equations imply $u(x,0) = u_0(x)$.

There exists an important result in the asymptotic analysis of integrals known as the Riemann–Lebesgue Lemma, which states that if $f(k)$ satisfies appropriate conditions, then,

$$\lim_{|x| \to \infty} \int_{-\infty}^{\infty} e^{ikx} f(k) dk = 0.$$

This result can be used to show that $u(x,t)$ vanishes at $|x| \to \infty$. Actually, for functions $u_0(x)$ such that $\hat{u}_0(k)$ decays fast enough as $|k| \to \infty$, the decay of $u(x,t)$ as $|x| \to \infty$ can be verified directly. Indeed, since the integral of e^{ikx} with respect to k is e^{ikx}/ix, using integration by parts, Eq. (8.26) implies the following:

$$u(x,t) = \frac{1}{2\pi ix} \left[e^{ikx} e^{-\alpha k^2 t} \hat{u}_0(k) \right]_{k=-\infty}^{\infty} - \frac{1}{ix} \int_{-\infty}^{\infty} e^{ikx} \frac{\partial}{\partial k} \left(e^{-\alpha k^2 t} \hat{u}_0(k) \right) \frac{dk}{2\pi}. \qquad (8.27)$$

This is a useful expression since it gives the solution to the heat equation in terms of the initial value only, $u_0(x)$, which appears in the solution through its Fourier transform, $\hat{u}_0(k)$, with the only other condition (boundary value) that the function $u(x,t)$ vanishes for $|x| \to \infty$.

The representation of Eq. (8.26) is very useful because it can be readily generalized to *any* evolution PDE. Using the above procedure of verifying the validity of Eq. (8.26) for the heat equation, it follows that if the expression

$$e^{ikx + \omega(k)t},$$

is a particular solution of a given linear PDE where $\omega(k)$ for k real satisfies $\text{Re}[\omega(k)] \leq 0$; then the solution of the initial value problem of this PDE is given by

$$u(x,t) = \int_{-\infty}^{\infty} e^{ikx + \omega(k)t} \hat{u}_0(k) \frac{dk}{2\pi}. \qquad (8.28)$$

The above restriction on $\omega(k)$, namely $\text{Re}[\omega(k)] \leq 0$, ensures that the term $e^{\omega(k)t}$ is bounded as $|k| \to \infty$.

Example 8.1: Heat equation solution for an initial δ-function pulse

To illustrate the usefulness of Eq. (8.27), we consider the example of the heat equation, Eq. (8.15), with the initial condition:

$$u(x,0) = u_0(x) = \delta(x - x_0),$$

namely, a δ-function pulse of heat at $t = 0$ which occurs at position $x = x_0$. The domain for this case is $x \in (-\infty, \infty)$, and the boundary condition is that $u(x) = 0$ at the boundary, $x \to \pm\infty$, for all $t > 0$.

The Fourier transform of $u_0(x)$ gives

$$\hat{u}_0(k) = \int_{-\infty}^{\infty} e^{-ikx} u_0(x) dx = \int_{-\infty}^{\infty} e^{-ikx} \delta(x - x_0) dx = e^{-ikx_0}.$$

Inserting this expression in the first term on the right-hand side of Eq. (8.27), for $t > 0$ and $\alpha > 0$, we find that this term vanishes in the limits $k \to \pm\infty$. We next deal with the second term of right-hand side of Eq. (8.27), which takes the form:

$$-\frac{1}{ix} \int_{-\infty}^{\infty} e^{ikx} \frac{\partial}{\partial k} \left(e^{-\alpha k^2 t} e^{-ikx_0} \right) \frac{dk}{2\pi} = \frac{1}{ix} \int_{-\infty}^{\infty} e^{-\alpha k^2 t + ik(x - x_0)} \left[2\alpha kt + ix_0 \right] \frac{dk}{2\pi}.$$

We complete the square in the exponential of the integrand by adding and subtracting the appropriate terms, which produces the following expression for the exponent:

$$-\left[\alpha k^2 t - ik(x - x_0) + \left(\frac{i(x - x_0)}{2\sqrt{\alpha t}} \right)^2 \right] + \left(\frac{i(x - x_0)}{2\sqrt{\alpha t}} \right)^2$$

$$= -\left(k\sqrt{\alpha t} - \frac{i(x - x_0)}{2\sqrt{\alpha t}} \right)^2 - \frac{(x - x_0)^2}{4\alpha t}.$$

With the change of variables

$$y = k\sqrt{\alpha t} - \frac{i(x - x_0)}{2\sqrt{\alpha t}},$$

the above integral takes the form:

$$\frac{1}{2\pi ix} e^{-(x-x_0)^2/4\alpha t} \left[\int_{-\infty}^{\infty} e^{-y^2} y \, dy + ix \int_{-\infty}^{\infty} e^{-y^2} dy \right] \frac{1}{\sqrt{\alpha t}}.$$

The first of the integrals over the variable y in the square brackets vanishes because the integrand is an odd function of y, whereas the second integral is equal to $\sqrt{\pi}$, a familiar result from the normalized Gaussian function (see Chapter 1, Eq. (1.57), with $\sigma = 1/\sqrt{2}$. Combining these results, we arrive at

$$u(x,t) = \frac{1}{\sqrt{4\pi\alpha t}} e^{-(x-x_0)^2/4\alpha t}, \tag{8.29}$$

which is a normalized Gaussian function with variance $\sigma^2 = 2\alpha t$.

The solution obtained above for the heat equation is a well-known result, and its validity for $t > 0$ can be easily checked by inserting it in the heat equation and calculating the partial derivatives u_t and u_{xx}. A plot of this solution for various values of t is shown below (in this example we have set $\alpha = 1$ and $x_0 = 0$). For $t \to 0$ the solution takes the form of a δ-function, as was discussed in Chapter 1, see Eq. (1.75). This behavior is familiar from many physical situations involving diffusion of heat in an isotropic medium, or diffusion of other physical quantities, like particles undergoing Brownian motion in a fluid. As is seen from the plots at different time moments, the initial concentration of heat spreads out and becomes more uniform as time passes.

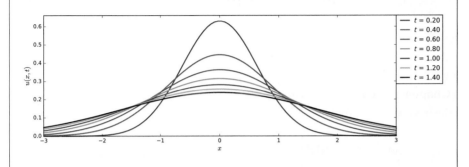

An interesting feature of the solution is that its value at any point x and at time t is larger than the minimum value of the solution and simultaneously smaller than the maximum value of the solution at any previous time $t' < t$, that is:

$$\min_{\{x'\}} \left\{ u(x',t') \right\} \leq u(x,t) \leq \max_{\{x'\}} \left\{ u(x',t') \right\}, \quad \forall x \in (-\infty, \infty), \text{ for } t > t',$$

(8.30)

where the range of x' for finding the minimum or maximum value of $u(x',t')$ is the entire interval $-\infty < x' < \infty$. This remarkable property is referred to as the "maximum principle" of the heat equation and applies when the boundary conditions are such that $u(x,t)$ vanishes at the domain boundary for all $t > 0$.

Derivation: The general solution of the heat equation
The solution found in the above example offers yet another convenient way of expressing the solution of the heat equation in the domain of Eq. (8.23) with the initial and boundary conditions of Eq. (8.24). Consider the function

$$g(x,t) = \frac{1}{\sqrt{4\pi\alpha t}} e^{-(x-x')^2/4\alpha t},$$

which for any value of $t > 0$ is a normalized Gaussian in the variable x centered at x'. This function automatically satisfies the heat equation because its first partial derivative with respect to t is

$$\frac{\partial g}{\partial t} = g(x,t) \left(\frac{(x-x')^2}{4\alpha t^2} - \frac{1}{2t} \right)$$

and its second partial derivative with respect to x is

$$\frac{\partial^2 g}{\partial x^2} = g(x,t)\frac{1}{\alpha}\left(\frac{(x-x')^2}{4\alpha t^2} - \frac{1}{2t}\right).$$

Therefore, the relation $g_t = \alpha g_{xx}$ holds for all x and $t > 0$. In order to ensure that the solution also satisfies the given boundary and initial conditions, we multiply $g(x,t)$ with the function $u_0(x')$ and integrate over x', namely we choose the solution to be

$$u(x,t) = \int_{-\infty}^{\infty} u_0(x')\frac{1}{\sqrt{4\pi\alpha t}}e^{-(x-x')^2/4\alpha t}dx'. \tag{8.31}$$

Indeed, the function $u(x,t)$ satisfies the heat equation because its dependence on the variables x and t is identical to that of the function $g(x,t)$, and in the limit $t \to 0$ the function $g(x,t)$ becomes a δ-function with argument $(x-x')$, as was discussed in Chapter 1, see Eq. (1.75) with $\sigma^2 = 2\alpha t$; therefore, for $t \to 0$ we obtain the desired result, namely

$$\begin{aligned}
u(x,0) &= \lim_{t\to 0}\left[\int_{-\infty}^{\infty} u_0(x')\frac{1}{\sqrt{4\pi\alpha t}}e^{-(x-x')^2/4\alpha t}dx'\right]\\
&= \int_{-\infty}^{\infty} u_0(x')\lim_{t\to 0}\left[\frac{1}{\sqrt{4\pi\alpha t}}e^{-(x-x')^2/4\alpha t}\right]dx'\\
&= \int_{-\infty}^{\infty} u_0(x')\delta(x-x')dx' = u_0(x).
\end{aligned}$$

We observe that the general solution of Eq. (8.31) is the convolution of the initial value $u_0(x)$ with a normalized Gaussian which has mean value x and variance $2\alpha t$.

8.3.3   Traditional integral transforms for problems on the half-line

Initial–boundary value problems for simple evolution PDEs on the half-line are traditionally solved through an integral transform in the space variable x. For example, the heat equation in the domain

$$\Omega = \{0 < x < \infty, \ t > 0\}, \tag{8.32}$$

supplemented with the initial and boundary conditions

$$u(x,0) = u_0(x), \ 0 < x < \infty; \ u(0,t) = g_0(t), \ t > 0, \tag{8.33}$$

where u and u_x vanish as $x \to \infty$, can be solved through the sine-transform pair, Eqs. (7.40) and (7.41). Similarly, the heat equation in the domain Ω defined in (8.32) supplemented with the initial and boundary conditions

$$u(x,0) = u_0(x), \ 0 < x < \infty; \ u_x(0,t) = g_1(t), \ t > 0, \tag{8.34}$$

can be solved by the cosine-transform pair, Eqs. (7.42) and (7.43). In what follows, we will employ the sine-transform pair to solve the heat equation satisfying the conditions of Eq. (8.33).

We use the sine-transform, Eq. (7.40), to define $\hat{u}_s(k,t)$ by the expression

$$\hat{u}_s(k,t) = \int_0^{\infty} \sin(kx)u(x,t)dx, \quad k \in \mathbb{R}. \tag{8.35}$$

Differentiating both sides of Eq. (8.35) with respect to time we find

$$\frac{\partial \hat{u}_s}{\partial t} = \int_0^\infty \sin(kx)\frac{\partial u}{\partial t}dx = \int_0^\infty \sin(kx)\frac{\partial^2 u}{\partial x^2}dx,$$

where we have used the heat equation to replace u_t with u_{xx} (we take here $\alpha = 1$ for simplicity). Employing integration by parts for the integral over x, and recalling that u and u_x vanish as $x \to \infty$, we find

$$\frac{\partial \hat{u}_s}{\partial t} + k^2\hat{u}_s = kg_0(t). \tag{8.36}$$

In the process of integrating by parts, the unknown boundary value $u_x(0, t)$ occurs in the form

$$\lim_{x \to 0}\left[u_x(x, t)\sin(kx)\right],$$

thus its contribution vanishes. This is precisely the reason that the sine-transform is the *correct* transform for solving this particular boundary value problem: its implementation eliminates the unknown boundary value $u_x(0, t)$.

The ODE of Eq. (8.36) can be solved using the so-called "integrating factor": if we multiply the unknown function $\hat{u}_s(k, t)$ by the factor $\exp[k^2 t]$, the combined function obeys the following equation:

$$\frac{d}{dt}\left(e^{k^2 t}\hat{u}_s\right) = \frac{\partial \hat{u}_s}{\partial t}e^{k^2 t} + k^2\hat{u}_s e^{k^2 t} = \left[\frac{\partial \hat{u}_s}{\partial t} + k^2\hat{u}_s\right]e^{k^2 t}$$

$$\Rightarrow \quad \frac{d}{dt}\left(e^{k^2 t}\hat{u}_s\right) = kg_0(t)e^{k^2 t},$$

where the last step is obtained by virtue of Eq. (8.36). This last expression can also been written as:

$$d\left(e^{k^2 t}\hat{u}_s\right) = \left(kg_0(t)e^{k^2 t}\right)dt \Rightarrow dw = f(t)dt,$$

where we have introduced the definitions:

$$w(t) \equiv e^{k^2 t}\hat{u}_s(t), \quad f(t) \equiv kg_0(t)e^{k^2 t}.$$

Integrating both sides of $dw = f(t)dt$ from 0 to t we obtain:

$$w(t) - w(0) = e^{k^2 t}\hat{u}_s(k, t) - \hat{u}_s(k, 0) = k\int_0^t e^{k^2 \tau}g_0(\tau)d\tau. \tag{8.37}$$

Then, using the identity

$$\hat{u}_s(k, 0) = \int_0^\infty \sin(kx)u(x, 0)dx = \int_0^\infty \sin(kx)u_0(x)dx = \hat{u}_s^{(0)}(k),$$

we find

$$e^{k^2 t}\hat{u}_s(k, t) = \hat{u}_s^{(0)}(k) + k\int_0^t e^{k^2 \tau}g_0(\tau)d\tau, \quad 0 < k < \infty. \tag{8.38}$$

Replacing $\hat{u}_s(k, t)$ in the inverse sine-transform formula, Eq. (7.41), by the expression obtained from Eq. (8.38), we find

$$u(x, t) = 4\int_0^\infty \sin(kx)\hat{u}_s(k, t)\frac{dk}{2\pi}$$

$$= \int_0^\infty \sin(kx)e^{-k^2 t}\left[\int_0^\infty \sin(ky)u_0(y)dy + k\int_0^t e^{k^2 \tau}g_0(\tau)d\tau\right]\frac{2dk}{\pi}. \tag{8.39}$$

An alternative approach to analyze a general evolution PDE is through the Laplace transform in the variable t. However, this procedure is usually problematic because it requires the solution of nonlinear algebraic equations and also it involves $0 < t < \infty$. We briefly illustrate this issue by an example: consider Eq. (8.17) supplemented with Eqs. (8.32) and (8.33). Let $\hat{u}_L(x, s)$ denote the Laplace transform of $u(x, t)$, namely,

$$\hat{u}_L(x, s) = \int_0^\infty e^{-st} u(x, t) dt, \quad \mathrm{Re}[s] \geq 0.$$

Equation (8.17) with $\beta = 1$ implies

$$\frac{\partial^3 \hat{u}_L}{\partial x^3} + \frac{\partial \hat{u}_L}{\partial x} + \gamma\, s \hat{u}_L = u_0(x), \quad 0 < x < \infty. \tag{8.40}$$

We now have to solve this third-order ODE supplemented with the boundary condition

$$\hat{u}_L(s, 0) = \int_0^\infty e^{-st} g_0(t) dt. \tag{8.41}$$

Letting $\hat{u}_L(s, x) = \exp[\lambda(s)x]$, we find for $\lambda(s)$ the following cubic equation:

$$[\lambda(s)]^3 + \lambda(s) + \gamma\, s = 0, \tag{8.42}$$

which in general gives rise to complicated expressions involving several roots.

8.3.4 Traditional infinite series for problems on a finite interval

Simple evolution PDEs with the space coordinate in a finite interval are traditionally solved through infinite series expansions. For example, the heat equation (8.15) in the domain

$$\Omega = \{0 < x < L, 0 < L < \infty, t > 0\}, \tag{8.43}$$

supplemented with the initial and boundary conditions

$$u(x, 0) = u_0(x), \;\; 0 < x < L; \;\; u(0, t) = g_0(t), \;\; u(L, t) = h_0(t), \;\; t > 0, \tag{8.44}$$

can be solved through the sine series, which is defined as follows: let

$$f_n = \int_0^L \sin\left(\frac{n\pi x}{L}\right) f(x) dx, \;\; n = 1, 2, \ldots \tag{8.45}$$

Then, $f(x)$ can be expressed in terms of f_n through the formula

$$f(x) = \frac{2}{L} \sum_{n=1}^\infty \sin\left(\frac{n\pi x}{L}\right) f_n, \;\; 0 < x < L. \tag{8.46}$$

In order to solve the heat Eq. (8.15) in the domain of Eq. (8.43) with the initial and boundary condition of Eq. (8.44), we define u_n by

$$u_n(t) = \int_0^L \sin\left(\frac{n\pi x}{L}\right) u(x, t) dx. \tag{8.47}$$

Differentiating the above equation with respect to t, we find

$$\frac{du_n}{dt} = \int_0^L \sin\left(\frac{n\pi x}{L}\right) \frac{\partial u}{\partial t} dx = \int_0^L \sin\left(\frac{n\pi x}{L}\right) \frac{\partial^2 u}{\partial x^2} dx, \tag{8.48}$$

where we have used the heat equation to replace u_t with u_{xx} (again we set $\alpha = 1$ for simplicity). Employing integration by parts, Eq. (8.48) yields

$$\frac{du_n}{dt} + \left(\frac{n\pi}{L}\right)^2 u_n = \frac{n\pi}{L}[g_0 - \cos(n\pi)h_0]. \tag{8.49}$$

As with the case of the sine-transform, the fact that the sine series is the correct transform technique for this problem follows from the fact that the unknown derivatives $u_x(0, t)$ and $u_x(L, t)$ occur in the form

$$\lim_{x \to 0}\left[\sin\left(\frac{n\pi x}{L}\right) u_x(x, t)\right], \text{ and } \lim_{x \to L}\left[\sin\left(\frac{n\pi x}{L}\right) u_x(x, t)\right], \tag{8.50}$$

and therefore are eliminated. Following the same steps as for the derivation of Eq. (8.37), we can solve the above ODE for $u_n(t)$ using the integrating factor $\exp[(n\pi/L)^2 t]$, to find:

$$e^{(n\pi/L)^2 t} u_n(t) = u_n^{(0)} + \frac{n\pi}{L}\int_0^t e^{(n\pi/L)^2 \tau}[g_0(\tau) - \cos(n\pi)h_0(\tau)]\,d\tau, \tag{8.51}$$

where

$$u_n^{(0)} = \int_0^L \sin\left(\frac{n\pi x}{L}\right) u_0(x)\,dx. \tag{8.52}$$

Inserting the above expression for $u_n(t)$ in the equation

$$u(x, t) = \frac{2}{L}\sum_{n=1}^\infty \sin\left(\frac{n\pi x}{L}\right) u_n(t),$$

we find

$$u(x, t) = \frac{2}{L}\sum_{n=1}^\infty \sin\left(\frac{n\pi x}{L}\right) e^{-(n\pi x/L)^2 t}$$
$$\times \left[u_n^{(0)} + \frac{n\pi}{L}\int_0^t e^{(n\pi/L)^2 \tau}[g_0(\tau) - \cos(n\pi)h_0(\tau)]\,d\tau\right]. \tag{8.53}$$

8.4 The wave equation

The wave equation is the first PDE derived in the history of mathematics. It was formulated by d'Alembert in the 1746. Euler, Bernoulli and d'Alembert solved this equation in connection with a vibrating string using separation of variables, as well as employing the d'Alembert formula, that will be derived in Section 8.4.2.

8.4.1 Derivation of the wave equation

The problem of a vibrating string can be formulated as follows: consider a segment of a string of length Δx between the x values x_1 and $x_2 = x_1 + \Delta x$, see Fig. 8.3. We assume that the only forces acting on the string are gravity which acts in the vertical direction, and the pull or tension, whose magnitude is denoted by $T(x, t)$, that acts in a direction tangential to the center line of the string. Furthermore, we assume that the string is perfectly flexible, offering no resistance to bending. Finally, we assume that each point on the string moves only in the vertical direction, thus, the horizontal component of the acceleration vanishes.

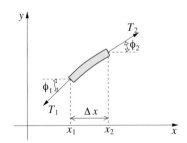

Figure 8.3: Segment of a string with tension T acting at two points labeled x_1 and x_2 of distance $\Delta x = x_2 - x_1$ apart along the x-axis, with $T(x_1) = T_1, T(x_2) = T_2$. The angles between the tension vectors and the x-axis at these two points are $\phi(x_1) = \phi_1, \phi(x_2) = \phi_2$.

We denote the displacement of the string in the vertical direction by $u(x, t)$, so with the assumptions we have made we have

$$\tan(\phi) = \frac{\partial u}{\partial x}, \tag{8.54}$$

and the second derivative of u with respect to t is the vertical acceleration. Applying Newton's second law $\mathbf{F} = m\mathbf{a}$ in the horizontal direction to the segment of Fig. 8.3 yields the following equation:

$$-T(x, t)\cos(\phi(x, t)) + T(x + \Delta x, t)\cos(\phi(x + \Delta x, t)) = 0. \tag{8.55}$$

Applying Newton's second law in the vertical direction yields the equation

$$-T(x, t)\sin(\phi(x, t)) + T(x + \Delta x, t)\sin(\phi(x + \Delta x, t)) - mg = m\frac{\partial^2 u(x, t)}{\partial t^2}. \tag{8.56}$$

In Eq. (8.56) we can express the mass of the segment as $m = \rho\Delta x$, where ρ is the linear density of the string measured in units of mass per unit length.

We can use the following approximations:

$$T(x + \Delta x, t) \approx T(x, t) + \Delta x\frac{\partial T}{\partial x}, \tag{8.57}$$

and

$$\phi(x + \Delta x, t) \approx \phi(x, t) + \Delta x\frac{\partial \phi(x, t)}{\partial x}.$$

Taking the cosine of both sides of the last equation, and using the Taylor expansions for the sine and cosine in which we keep the lowest-order terms in Δx, we arrive at

$$\cos(\phi(x + \Delta x, t)) \approx \cos(\phi(x, t)) - \sin(\phi(x, t))\Delta x\frac{\partial \phi(x, t)}{\partial x}.$$

Then, Eq. (8.55) implies

$$-T\cos(\phi) + \left(T + \Delta x\frac{\partial T}{\partial x}\right)\left[\cos(\phi) - \sin(\phi)\,\Delta x\frac{\partial \phi}{\partial x}\right] \approx 0,$$

or, after removing the two terms $\pm T\cos(\phi)$ and neglecting the term of order $(\Delta x)^2$,

$$\frac{\partial T}{\partial x} \approx T\tan(\phi)\frac{\partial \phi}{\partial x}. \tag{8.58}$$

Similarly, using $m = \rho\Delta x$ in Eq. (8.56) and employing Eq. (8.57) as well as the approximation

$$\sin(\phi(x + \Delta x, t)) \approx \sin(\phi(x, t)) + \cos(\phi(x, t))\,\Delta x\frac{\partial \phi(x, t)}{\partial x},$$

which is obtained by the same steps as the expression for $\cos(\phi(x + \Delta x, t))$, Eq. (8.56) yields

$$\frac{\partial T}{\partial x}\sin(\phi) + T\cos(\phi)\frac{\partial \phi}{\partial x} = \rho\frac{\partial^2 u}{\partial t^2} + g\rho.$$

Replacing in the above equation $\partial T/\partial x$ from Eq. (8.58) and dividing by T we find

$$\cos(\phi)\frac{\partial \phi}{\partial x} + \sin(\phi)\tan(\phi)\frac{\partial \phi}{\partial x} = \frac{\rho}{T}\frac{\partial^2 u}{\partial t^2} + \frac{g\rho}{T}.$$

Differentiating both sides of Eq. (8.54) with respect to x gives

$$\tan(\phi) = \frac{\partial u}{\partial x} \Rightarrow \frac{\partial \tan(\phi)}{\partial x} = \frac{1}{\cos^2(\phi)}\frac{\partial \phi}{\partial x} = \frac{\partial^2 u}{\partial x^2} \Rightarrow \frac{\partial \phi}{\partial x} = \cos^2(\phi)\frac{\partial^2 u}{\partial x^2}.$$

Assuming that ϕ is small, $\sin(\phi) \approx \phi$ and $\cos(\phi) \approx 1$, the previous two equations yield

$$\frac{\partial^2 u}{\partial x^2} = \frac{1}{c^2}\frac{\partial^2 u}{\partial t^2} + \frac{g}{c^2}, \quad c^2 = \frac{T}{\rho}. \tag{8.59}$$

For small ϕ, $\tan(\phi) \approx \phi$, thus Eq. (8.58) becomes

$$\frac{1}{T}\frac{\partial T}{\partial x} \approx \phi\frac{\partial \phi}{\partial x}.$$

Integrating this equation, we find

$$\ln T \approx \frac{\phi^2}{2} + \text{constant}.$$

Thus, since ϕ is small, ϕ^2 can be neglected, and hence T can be approximated by a constant. Furthermore, usually c^2 is quite large, thus the term g/c^2 can be neglected. Hence, Eq. (8.59) becomes the celebrated wave equation

$$u_{tt} - c^2 u_{xx} = 0. \tag{8.60}$$

For a vibrating string, one needs to prescribe the initial position, that is, $u(x,0)$, as well as the initial vertical speed, that is, $u_t(x,0)$. In addition, appropriate boundary conditions must be prescribed. For example, if the two ends of a finite string of length L are fixed, we have the following initial–boundary value problem: the wave equation is formulated in $0 < x < L$, $t > 0$ with the initial conditions

$$u(x,0) = u_0(x), \quad u_t(x,0) = u_1(x), \quad 0 < x < L, \tag{8.61}$$

and the boundary conditions

$$u(0,t) = u(L,t) = 0, \quad t > 0. \tag{8.62}$$

8.4.2 The solution of d'Alembert

The wave equation belongs to a general class of PDEs known as "hyperbolic".

These PDEs can be analyzed through the "method of characteristics", which involves the introduction of a new, more convenient, set of independent variables. For the wave equation, this set is given by

$$\xi = x + ct, \quad \zeta = x - ct. \tag{8.63}$$

The chain rule implies the equations

$$\frac{\partial}{\partial x} = \frac{\partial}{\partial \xi} + \frac{\partial}{\partial \zeta}, \quad \frac{\partial}{\partial t} = c\frac{\partial}{\partial \xi} - c\frac{\partial}{\partial \zeta}. \tag{8.64}$$

Thus, in the new coordinates the wave equation becomes

$$c^2 \left(\frac{\partial}{\partial \xi} - \frac{\partial}{\partial \zeta} \right)^2 u = c^2 \left(\frac{\partial}{\partial \xi} + \frac{\partial}{\partial \zeta} \right)^2 u,$$

or

$$\frac{\partial^2 u}{\partial \xi \partial \zeta} = 0. \tag{8.65}$$

The general solution of Eq. (8.65) is given by

$$u = f_1(\xi) + f_2(\zeta), \tag{8.66}$$

where $f_1(\xi)$ and $f_2(\zeta)$ are two arbitrary functions. Thus, using the definitions of ξ and ζ, we find

$$u(x,t) = f_1(x + ct) + f_2(x - ct). \tag{8.67}$$

This representation is known as "d'Alembert's solution".

In what follows, we will determine the functions f_1 and f_2 in the case that the wave equation is defined in the domain $0 < x < L, t > 0$, and it satisfies the initial and boundary conditions specified by Eqs. (8.61) and (8.62).

The initial conditions of Eqs. (8.61) imply,

$$f_1(x) + f_2(x) = u_0(x), \ \ 0 < x < L, \tag{8.68}$$
$$c f_1'(x) - c f_2'(x) = u_1(x), \ \ 0 < x < L, \tag{8.69}$$

where prime denotes differentiation. Integrating Eq. (8.69) we find

$$f_1(x) - f_2(x) = \frac{1}{c} \int_0^x u_1(y) \mathrm{d}y + A,$$

where A is an arbitrary constant. Solving the above equation together with Eq. (8.68), we find

$$f_1(x) = \frac{1}{2} \left[u_0(x) + U_1(x) + A \right], \ \ 0 < x < L, \tag{8.70}$$

$$f_2(x) = \frac{1}{2} \left[u_0(x) - U_1(x) - A \right] \ \ 0 < x < L, \tag{8.71}$$

where $U_1(x)$ is defined by

$$U_1(x) = \int_0^x u_1(y) \mathrm{d}y, \ \ 0 < x < L. \tag{8.72}$$

The right-hand side of Eqs. (8.70) and (8.71) define $f_1(x)$ and $f_2(x)$ only for $0 < x < L$. However, since $x \pm ct$ can take any value we need $f_1(x)$ and $f_2(x)$ for all values of their argument. Thus, we seek functions $\tilde{u}_0$ and $\tilde{U}_1$ such that

$$f_1(x) = \frac{1}{2} \left[\tilde{u}_0(x) + \tilde{U}_1(x) + A \right], \tag{8.73}$$

$$f_2(x) = \frac{1}{2} \left[\tilde{u}_0(x) - \tilde{U}_1(x) - A \right], \tag{8.74}$$

for all x, where

$$\tilde{u}_0(x) = u_0(x), \ \ \tilde{U}_1(x) = U_1(x), \ \ 0 < x < L. \tag{8.75}$$

The boundary conditions of Eq. (8.62) imply

$$f_1(ct) + f_2(-ct) = 0, \quad t > 0,$$
$$f_1(L + ct) + f_2(L - ct) = 0, \quad t > 0.$$

Replacing in these equations f_1 and f_2 by Eqs. (8.73) and (8.74), we find

$$\tilde{u}_0(ct) + \tilde{U}_1(ct) + \tilde{u}_0(-ct) - \tilde{U}_1(-ct) = 0, \quad t > 0, \qquad (8.76)$$
$$\tilde{u}_0(L + ct) + \tilde{U}_1(L - ct) + \tilde{u}_0(L - ct) - \tilde{U}_1(L - ct) = 0, \quad t > 0. \quad (8.77)$$

Equation (8.76) is satisfied by choosing

$$\tilde{u}_0(ct) = -\tilde{u}_0(-ct), \quad \tilde{U}_1(ct) = \tilde{U}_1(-ct), \quad t > 0.$$

Hence, $\tilde{u}_0$ is an odd function whereas $\tilde{U}_1$ is an even function. Similarly, Eq. (8.77) is satisfied by choosing

$$\tilde{u}_0(L - ct) = -\tilde{u}_0(L + ct), \quad \tilde{U}_1(L - ct) = \tilde{U}_1(L + ct).$$

Since $\tilde{u}_0$ and $\tilde{U}_1$ are odd and even functions, respectively, we find that they satisfy the same property:

$$\tilde{u}_0(-L + ct) = -\tilde{u}_0(L - ct) = \tilde{u}_0(L + ct),$$

$$\tilde{U}_1(-L + ct) = \tilde{U}_1(L - ct) = \tilde{U}_1(L + ct).$$

Hence, $\tilde{u}_0$ and $\tilde{U}_1$ are periodic functions of period $2L$. Thus,

$$u(x,t) = \frac{1}{2}\left[\tilde{u}_0(x + ct) + \tilde{U}_1(x + ct) + \tilde{u}_0(x - ct) - \tilde{U}_1(x - ct)\right], \qquad (8.78)$$

where $\tilde{u}_0$ is the odd, $2L$-periodic extension of u, and $\tilde{U}_1$ is the even, $2L$-periodic extension of u.

The remarkable feature of d'Alembert's solution is that it is explicit in terms of $u_0(x)$ and the integral of $u_1(x)$. However, this expression is only valid for the particular case of homogeneous boundary conditions. For the general case of non-vanishing boundary conditions, the wave equation is traditionally solved via transform methods.

8.4.3 Traditional transforms

Let us consider the wave equation in the half-line, $0 < x < \infty$, $t > 0$, with the initial conditions of Eq. (8.61) and the boundary condition

$$u(0,t) = g_0(t), \quad t > 0, \qquad (8.79)$$

where we assume that u and its derivatives vanish as $x \to \infty$. Using the sine-transform pair for the function $u(x,t)$, namely,

$$\hat{u}(k,t) = \int_0^\infty \sin(kx)u(x,t)dx, \quad 0 < x < \infty, \qquad (8.80a)$$

$$u(x,t) = 4\int_0^\infty \sin(kx)\hat{u}(k,t)\frac{dk}{2\pi}, \quad 0 < k < \infty, \qquad (8.80b)$$

we find

$$\hat{u}_{tt} = \int_0^\infty \sin(kx)u_{tt}dx = c^2 \int_0^\infty \sin(kx)u_{xx}dx.$$

Integrating by parts the integral over x we obtain

$$\hat{u}_{tt}(k,t) + c^2 k^2 \hat{u}(k,t) = c^2 k g_0(t). \qquad (8.81)$$

This ODE is supplemented with the following initial conditions:

$$\hat{u}(k,0) = \int_0^\infty \sin(kx)u_0(x)dx = \hat{u}_0(k), \qquad (8.82)$$

$$\hat{u}_t(k,0) = \int_0^\infty \sin(kx)u_t(x,0)dx = \int_0^\infty \sin(kx)u_1(x)dx = \hat{u}_1(k). \qquad (8.83)$$

In summary, $u(x,t)$ is given by Eq. (8.80b), where $\hat{u}(k,t)$ is the solution of the ODE of Eq. (8.81) with the initial conditions

$$\hat{u}(k,0) = \hat{u}_0(k), \quad \hat{u}_t(k,0) = \hat{u}_1(k), \qquad (8.84)$$

where $\hat{u}_0(k)$ and $\hat{u}_1(k)$ are the sine-transforms of the given functions $u_0(x)$ and $u_1(x)$, respectively.

In the particular case that $g_0 = 0$, Eq. (8.81) yields

$$\hat{u}(k,t) = a(k)\cos(ckt) + b(k)\sin(ckt).$$

This equation together with Eq. (8.84) imply

$$\hat{u}_0(k) = a(k), \quad \hat{u}_1(k) = ckb(k).$$

Substituting the above expressions in Eqs. (8.80a), (8.80b) we conclude that if $g_0 = 0$, then

$$u(x,t) = \int_0^\infty \sin(kx)\left[\hat{u}_0(k)\cos(ckt) + \frac{\hat{u}_1(k)}{ck}\sin(kt)\right]\frac{2dk}{\pi}, \qquad (8.85)$$

where $\hat{u}_0(k)$ and $\hat{u}_1(k)$ are the sine-transforms of $u_0(x)$ and $u_1(x)$ defined by Eqs. (8.82) and (8.83), respectively.

Example 8.2: A vibrating finite string

Let us consider the wave equation in the domain $0 < x < L,\ t > 0$, with the initial conditions of Eq. (8.61) and the boundary conditions

$$u(0,t) = g_0(t), \quad u(L,t) = h_0(t), \quad t > 0. \qquad (8.86)$$

Using the sine series,

$$u^{(n)}(t) = \int_0^L \sin\left(\frac{n\pi x}{L}\right)u(x,t)dx, \quad n = 1,2,\ldots, \qquad (8.87)$$

$$u(x,t) = \frac{2}{L}\sum_{n=1}^\infty \sin\left(\frac{n\pi x}{L}\right)u^{(n)}(t), \quad 0 < x < L, \qquad (8.88)$$

we find

$$u_{tt}^{(n)} = \int_0^L \sin\left(\frac{n\pi x}{L}\right)u_{tt}dx = c^2 \int_0^L \sin\left(\frac{n\pi x}{L}\right)u_{xx}dx.$$

Integrating by parts the last integral over dx, we obtain

$$u_{tt}^{(n)}(t) + c^2 \left(\frac{n\pi}{L}\right)^2 u^{(n)}(t) = \frac{c^2 n\pi}{L} g_0(t) - \frac{c^2 n\pi}{L} \cos(n\pi) h_0(t), \quad t > 0.$$
(8.89)

In summary, $u(x,t)$ is given by Eq. (8.88), where $u^{(n)}(t)$ solves the ODE of Eq. (8.89) supplemented with the following initial conditions:

$$u^{(n)}(0) = \int_0^L \sin\left(\frac{n\pi x}{L}\right) u_0(x) dx,$$
(8.90a)

$$u_t^{(n)}(0) = \int_0^L \sin\left(\frac{n\pi x}{L}\right) u_1(x) dx.$$
(8.90b)

In the particular case of fixed boundaries, that is, $g_0(t) = h_0(t) = 0$ for all $t > 0$, Eq. (8.89) yields

$$u^{(n)}(t) = a_n \cos\left(\frac{n\pi c}{L} t\right) + b_n \sin\left(\frac{n\pi c}{L} t\right),$$

with $u^{(n)}(0) = a_n$, $u_t^{(n)}(0) = b_n \frac{n\pi c}{L}$.

Thus, the full solution for $u(x,t)$ takes the form

$$u(x,t) = \frac{2}{L} \sum_{n=1}^{\infty} \sin\left(\frac{n\pi x}{L}\right) \left[u^{(n)}(0) \cos\left(\frac{n\pi c}{L} t\right) + \frac{L}{n\pi c} u_t^{(n)}(0) \sin\left(\frac{n\pi c}{L} t\right)\right],$$
(8.91)

where $u^{(n)}(0)$ and $u_t^{(n)}(0)$ can be computed in terms of $u_0(x)$ and $u_1(x)$ through Eqs. (8.90a) and (8.90b).

For a concrete example, consider the situation where the string is plucked in the middle, so at $t = 0$ its shape is an isosceles triangle of height equal to 1:

$$u(x,0) = u_0(x) = \begin{cases} \frac{2x}{L}, & 0 \le x \le \frac{L}{2}, \\ 2 - \frac{2x}{L}, & \frac{L}{2} \le x \le L. \end{cases}$$

We first obtain the sine expansion of the function $u_0(x)$, which corresponds to the triangular wave with a half-period L (see Chapter 7, Examples 2):

$$u^{(n)}(0) = \left(\frac{L}{2}\right) \frac{8}{n^2 \pi^2} (-1)^l, \quad \text{for } n = 2l+1, \ l = 0,1,2,\dots,$$

with $u^{(n)}(0) = 0$ for n even. If we take the initial velocity $u_t(x,0) = u_1(x) = 0$, then the coefficients are $u_t^{(n)}(0) = 0$ and the solution has a shape oscillating between the original triangular shape and its mirror image with respect to the horizontal axis, and has a symmetric trapezoid shape at moments between the two extremes. This is shown in the figure below for different moments during one full period of oscillation, which is given by $\tau_0 = c/2L$: the solid colored lines correspond to moments of time $0 \le t < \tau_0/2$ while the dashed color lines correspond to $\tau_0/2 \le t < \tau_0$ (t is given on the right in units of τ_0). At $t = \tau_0/4$ and $t = 3\tau_0/4$ the shape is a straight line on the horizontal axis.

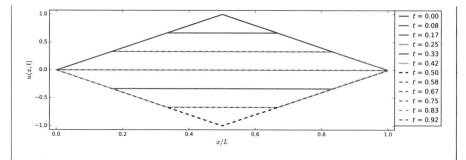

For a slightly more complicated case we take the initial velocity to be given by

$$u_t(x,0) = u_1(x) = \begin{cases} \dfrac{1}{\tau}, & 0 < x < \dfrac{L}{2}, \\ -\dfrac{1}{\tau}, & \dfrac{L}{2} < x < L, \end{cases}$$

where τ is a constant with units of time. In this case, $u_1(x)$ is the square wave of half-period L (see Chapter 7, Example 7.1), and the coefficients $u_t^{(n)}(0)$ of its sine expansion take the form

$$u_t^{(n)}(0) = \left(\frac{L}{2}\right)\frac{8}{n\pi\tau}, \quad \text{for } n = 4l + 2, \; l = 0, 1, 2, \ldots,$$

with $u_t^{(n)}(0) = 0$ for n odd, or a multiple of 4. Using these and the coefficients $u^{(n)}(0)$ calculated above for the sine expansion of $u_0(x)$, we can obtain the shape of the string at any moment in time. This is shown in the figure below, for the same moments in time as in the above example and with the same conventions.

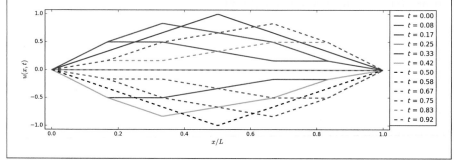

8.5   The Laplace and Poisson equations

The Laplace equation, which in two space dimensions takes the form

$$u_{xx} + u_{yy} = 0, \tag{8.92}$$

is the most important equation among the class of PDEs called "elliptic". Other physically significant elliptic PDEs are the inhomogeneous version of the Laplace equation, which is called the Poisson equation, as well as the Helmholtz and modified Helmholtz equations that will be considered in later sections. If the function u satisfies the Laplace equation, then u is called a "harmonic function" (see also Chapter 4, Section 4.4.2). The theory associated with the analysis of the Laplace equation is called "potential theory".

As a brief reminder of the discussion in Chapter 4, harmonic functions are associated with the real and imaginary parts of an analytic function, $f(z)$, which are denoted by $u(x,y)$ and $v(x,y)$, respectively:

$$f(z) = u(x,y) + iv(x,y).$$

These functions satisfy the Cauchy–Riemann relations

$$u_x = v_y, \quad u_y = -v_x.$$

Differentiating these equations with respect to x and y respectively, and then adding the resulting equations we find that $u(x,y)$ is harmonic. Similarly for $v(x,y)$.

Actually, harmonic functions can be associated directly with analytic functions: for a function $u(x,y)$, not necessarily real, *u is harmonic if and only if u_z is an analytic function.* Indeed, the transformations

$$z = x + iy, \quad \bar{z} = x - iy, \tag{8.93}$$

together with the chain rule, imply the identities

$$\frac{\partial}{\partial z} = \frac{1}{2}\left(\frac{\partial}{\partial x} - i\frac{\partial}{\partial y}\right), \quad \frac{\partial}{\partial \bar{z}} = \frac{1}{2}\left(\frac{\partial}{\partial x} + i\frac{\partial}{\partial y}\right). \tag{8.94}$$

Thus,

$$\frac{\partial^2}{\partial z \partial \bar{z}} = \frac{1}{4}\left(\frac{\partial^2}{\partial x^2} + \frac{\partial^2}{\partial y^2}\right). \tag{8.95}$$

Hence, the Laplace equation, Eq. (8.92), can be written in the form

$$u_{z\bar{z}} = 0. \tag{8.96}$$

If u_z is an analytic function, then its *d*-bar derivative vanishes, that is, $u_{z\bar{z}} = \partial u_z/\partial \bar{z} = 0$, thus u is harmonic. Conversely, if u is harmonic, then $\partial u_z/\partial \bar{z} = 0$, thus u_z is an analytic function. These notions were discussed in more detail in Section 4.4.

Let Ω be a bounded two-dimensional domain and let $\partial\Omega$ denote the boundary of Ω. Suppose that the Laplace equation is valid in the interior of Ω. In order to construct a unique solution $u(x,y)$ it is necessary to prescribe one boundary condition along $\partial\Omega$. The most well-known boundary value problems are the so-called "Dirichlet problem" where u is given on the boundary, and the "Neumann problem" where the derivative in the direction normal to the curve $\partial\Omega$ is given on the boundary.

It turns out that using the so-called fundamental solution of the Laplace equation, which will be discussed in Section 8.5.2, it is straightforward to construct in the physical plane (x,y) an integral representation for the solution $u(x,y)$. However, this representation involves both the Dirichlet and Neumann boundary values, and since only one of these values is prescribed as a boundary condition, this representation is *not* effective, unless one can determine the so-called *generalized Dirichlet to Neumann map*: for the Dirichlet problem, one must determine the Neumann boundary value in terms of the Dirichlet datum, and for the Neumann problem one must determine the Dirichlet boundary value in terms of the Neumann datum.

8.5.1 Motivation of the Laplace and Poisson equations

The Laplace and Poisson equations appear in a wide range of applications, which include astronomy, electromagnetism, steady-state heat conduction, and fluid mechanics.

Regarding the occurrence of the Laplace equation in fluid mechanics, let us denote by u and v the horizontal and vertical components of the velocity field vector, $\mathbf{v} = u\hat{x} + v\hat{y}$, of a steady-state, incompressible, irrotational fluid in two dimensions. The condition that the fluid is incompressible is expressed by the continuity equation

$$u_x + v_y = 0. \tag{8.97}$$

This equation can be solved by introducing the so-called "stream function" w (which is constant along fluid lines) through the equations

$$w_x = v, \quad w_y = -u. \tag{8.98}$$

The condition that the flow is irrotational is expressed by the equation $\nabla \times \mathbf{v} = 0$, which in two dimensions becomes

$$v_x - u_y = 0.$$

Replacing in this equation v and u by the expressions of Eq. (8.98), it follows that w satisfies the Laplace equation.

Regarding the occurrence of the Poisson equation in electromagnetism, the quasistatic Maxwell equations are given by

$$\nabla \times \mathbf{E} = 0, \quad \nabla \times \mathbf{B} = \mu(\mathbf{J} + \sigma\mathbf{E}), \quad \nabla \cdot \mathbf{B} = 0, \tag{8.99}$$

where $\mathbf{E}$ and $\mathbf{B}$ denote the electric and magnetic fields, respectively, σ and μ denote electric conductivity and magnetic permeability, respectively, and $\mathbf{J}$ denotes the electric current. The first of Eq. (8.99) implies the existence of a function u, called the electric potential, such that

$$\mathbf{E} = -\nabla u. \tag{8.100}$$

Using the identity $\nabla \cdot (\nabla \times \mathbf{B}) = 0$, the second of Eq. (8.99) implies the compatibility condition

$$\nabla \cdot \mathbf{J} + \sigma \nabla \cdot \mathbf{E} = 0 \Rightarrow -\nabla \cdot \mathbf{E} = \frac{1}{\sigma}\nabla \cdot \mathbf{J}.$$

Employing in this equation the representation of Eq. (8.100) we find in two dimensions the Poisson equation

$$u_{xx} + u_{yy} = \frac{1}{\sigma}\left[\frac{\partial J_1}{\partial x} + \frac{\partial J_2}{\partial y}\right], \tag{8.101}$$

where J_1 and J_2 are the components of the vector $\mathbf{J}$ in the x and y directions.

In electrostatics, Gauss' law reads:

$$\nabla \cdot \mathbf{E} = 4\pi\rho, \tag{8.102}$$

where $\mathbf{E}$ is the electric field and ρ is the charge density. Since the first of Eq. (8.99) still holds, we can again invoke the definition of the electric potential

u through Eq. (8.100), in terms of which the electric field $\mathbf{E}$ can be obtained. In this case, inserting Eq. (8.100) in Eq. (8.102) we obtain

$$u_{xx} + u_{yy} = -4\pi\rho. \tag{8.103}$$

This is the Poisson equation for an electrostatic charge distribution.

8.5.2 Integral representations through the fundamental solution

In order to derive an integral representation for the Poisson equation

$$u_{xx} + u_{yy} = -f(x, y), \quad (x, y) \in \Omega, \tag{8.104}$$

where $f(x, y)$ is a given function, we introduce the concept of a "fundamental solution".

Let $u(x, y)$ satisfy a PDE with a "forcing term" $f(x, y)$, namely the term on the right-hand side of the PDE, with the left-hand side containing all the terms that involve $u(x, y)$ and its derivatives. A fundamental solution of this PDE is a particular solution $G(x, y; \xi, \zeta)$ obtained by replacing $f(x, y)$ with $\delta(x - \xi)\delta(y - \zeta)$, where $\delta(x)$ denotes the Dirac δ-function. This concept is similar with the notion of the Green's function introduced in Section 8.2.3, but a fundamental solution is defined with respect to the given inhomogeneous equation without the need to satisfy the associated boundary conditions. Thus, a fundamental solution associated with the Laplace equation satisfies

$$G_{xx} + G_{yy} = -\delta(x - \xi)\delta(y - \zeta). \tag{8.105}$$

Multiplying Eq. (8.104) by G and Eq. (8.105) by u and then subtracting the resulting equations we find

$$u(x, y)\delta(x - \xi)\delta(y - \zeta) - G(x, y; \xi, \zeta)f(x, y)$$
$$= [Gu_x - uG_x]_x + [Gu_y - uG_y]_y. \tag{8.106}$$

This equation yields an integral representation of $u(x, y)$ through the use of Green's theorem, Eq. (2.73), with $F_1 = uG_y - Gu_y$, $F_2 = Gu_x - uG_x$: integrating Eq. (8.106) over Ω we find

$$u(\xi, \zeta) = \iint_\Omega G(x, y; \xi, \zeta)f(x, y)\mathrm{d}x\mathrm{d}y$$
$$+ \int_{\partial\Omega} [G(u_x\mathrm{d}y - u_y\mathrm{d}x) - u(G_x\mathrm{d}y - G_y\mathrm{d}x)]. \tag{8.107}$$

Let s denote the arc-length parameterization of the curve $\partial\Omega$ and let $\partial/\partial\mathcal{N}$ denote the derivative normal to $\partial\Omega$ in the outward direction. The following identities are valid for the function $f(x, y)$ evaluated on $\partial\Omega$:

$$\frac{\partial f}{\partial x}\mathrm{d}x + \frac{\partial f}{\partial y}\mathrm{d}y = f_T\mathrm{d}s, \tag{8.108}$$

where f_T denotes the derivative of f along the tangent of the curve $\partial\Omega$; and

$$\frac{\partial f}{\partial x}\mathrm{d}y - \frac{\partial f}{\partial y}\mathrm{d}x = \frac{\partial f}{\partial\mathcal{N}}\mathrm{d}s, \tag{8.109}$$

where $\partial f/\partial\mathcal{N}$ denotes the derivative of f along the outward normal of the curve $\partial\Omega$. Indeed, differentiating the function $f(x(s), y(s)) = f(s)$ with respect to s we find Eq. (8.108), where $f_T = df(s)/ds$. Since the infinitesimal vector $(dy, -dx)$ is normal to the infinitesimal vector (dx, dy), Eq. (8.109) follows. Employing in Eq. (8.107), Eqs. (8.108) and (8.109), and interchanging x, y with ξ, ζ, we find

$$u(x,y) = \int_{\partial\Omega}\left[G\frac{\partial u}{\partial\mathcal{N}} - u\frac{\partial G}{\partial\mathcal{N}}\right]ds + \iint_{\Omega}G(\xi,\zeta;x,y)f(\xi,\zeta)d\xi d\zeta. \quad (8.110)$$

A fundamental solution of the Laplace equation is given by the expression

$$G(\xi,\zeta;x,y) = -\frac{1}{4\pi}\ln[(\xi-x)^2 + (\zeta-y)^2]. \quad (8.111)$$

To prove this result, we will rely on the fact that the δ-function with complex argument can be expressed in terms of the $\partial/\partial\bar{z}$ derivative: namely, we will use the equation

$$\frac{1}{\pi}\frac{\partial}{\partial\bar{\zeta}}\left(\frac{1}{\zeta-z}\right) = \delta(\zeta-z), \quad (8.112)$$

which was derived in Section 4.6, with $\zeta = \xi + i\zeta$, $z = x + iy$. Equation (8.95) implies that Eq. (8.105) defining G can be written in the form

$$\frac{\partial}{\partial\bar{\zeta}}(4G_{\zeta}) = -\delta(\zeta-z). \quad (8.113)$$

Hence, Eq. (8.112) leads to

$$G_{\zeta} = -\frac{1}{4\pi}\frac{1}{\zeta-z}, \quad \text{and} \quad G_{\bar{\zeta}} = -\frac{1}{4\pi}\frac{1}{\bar{\zeta}-\bar{z}},$$

from which, integrating with respect to ζ and $\bar{\zeta}$, we obtain

$$G(\xi,\zeta;x,y) = -\frac{1}{4\pi}[\ln(\zeta-z) + \ln(\bar{\zeta}-\bar{z})]. \quad (8.114)$$

Using the identities

$$\ln(\zeta-z) + \ln(\bar{\zeta}-\bar{z}) = \ln[(\zeta-z)(\bar{\zeta}-\bar{z})]$$
$$= \ln|\zeta-z|^2 = \ln|(\xi-x) + i(\zeta-y)|^2,$$

we find Eq. (8.111).

In order to simplify Eq. (8.110) we note that the identities of Eq. (8.94) imply the following:

$$\frac{\partial G}{\partial\mathcal{N}}ds = G_{\xi}d\zeta - G_{\zeta}d\xi = [G_{\zeta} + G_{\bar{\zeta}}]d\zeta - i[G_{\zeta} - G_{\bar{\zeta}}]d\xi$$

$$= -iG_{\zeta}d\zeta + iG_{\bar{\zeta}}d\bar{\zeta} = \frac{i}{4\pi}\left[\frac{d\zeta}{\zeta-z} - \frac{d\bar{\zeta}}{\bar{\zeta}-\bar{z}}\right].$$

Using this result, Eq. (8.110) becomes

$$u(x,y) = -\frac{1}{4\pi}\int_{\partial\Omega}\ln[(x-\xi)^2 + (y-\zeta)^2]\frac{\partial u}{\partial\mathcal{N}}ds - \frac{i}{4\pi}\int_{\partial\Omega}u\left[\frac{d\zeta}{\zeta-z} - \frac{d\bar{\zeta}}{\bar{\zeta}-\bar{z}}\right]$$
$$- \frac{1}{4\pi}\iint_{\Omega}\ln[(x-\xi)^2 + (y-\zeta)^2]f(\xi,\zeta)d\xi d\zeta, \quad (8.115)$$

which provides the complete solution for $u(x,y)$, assuming that the function $f(\xi,\zeta)$ is known everywhere inside the domain Ω and we are given the appropriate boundary values for u and $\partial u/\partial\mathcal{N}$ on the boundary $\partial\Omega$ that make it possible to calculate the two integrals involving the boundary.

Example 8.3: The electrostatic potential in two dimensions

We can apply the expression of Eq. (8.115) to solve for the electrostatic potential of a distribution of point charges. We consider two point charges q_1 and q_2 at the positions (x_1, y_1) and (x_2, y_2). The charge density in this case is given by

$$\rho(x, y) = \sum_{i=1,2} q_i \delta(x - x_i)\delta(y - y_i).$$

The domain Ω is the entire (x, y) plane, and the boundary $\partial\Omega$ is at $|x| \to \infty$, $|y| \to \infty$. The proper boundary condition is the vanishing of the potential at infinity, namely,

$$\lim_{|x|,|y|\to\infty} u(x, y) = 0.$$

Accordingly, the integrals on the boundary $\partial\Omega$ in Eq. (8.115) vanish.

Inserting the expression for $\rho(x, y)$ in the Poisson equation for the electrostatic potential in two dimensions (2D), we find

$$f(x, y) = 4\pi\rho(x, y) \Rightarrow u(x, y) = -\sum_{i=1,2} q_i \ln[(x - x_i)^2 + (y - y_i)^2].$$

Equipotential contours of this solution are shown in the plot below, labeled "2D Electrostatic potential"; for this example we have used for the charges the values $q_1 = q_2 = 1$ and for their positions $(x_1, y_1) = (-0.25, -0.25)$, $(x_2, y_2) = (0.25, 0.25)$. In this plot, we also compare this solution with the equipotential contours from the expression

$$V(x, y, z = 0) = \sum_{i=1,2} \frac{1}{|\mathbf{r} - \mathbf{r}_i|} = \sum_{i=1,2} \frac{1}{\sqrt{(x - x_i)^2 + (y - y_i)^2}},$$

which is the familiar electrostatic potential for the two point charges at the same positions, in the case of a three-dimensional (3D) system, evaluated on the plane $z = 0$ (this is labeled "3D Electrostatic potential").

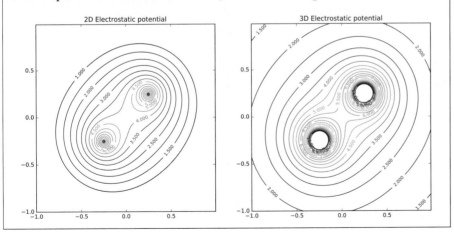

8.6   The Helmholtz and modified Helmholtz equations

To complete our discussion of elliptic PDEs we introduce briefly two other important equations, known as the Helmholtz and modified Helmholtz equations.

The Helmholtz equation is given by

$$u_{xx} + u_{yy} + \kappa^2 u = 0, \quad \kappa > 0. \tag{8.116}$$

This equation is intimately related with both the two-dimensional wave equation

$$U_{tt} - U_{xx} - U_{yy} = 0, \tag{8.117}$$

and with the two-dimensional heat equation

$$U_t - U_{xx} - U_{yy} = 0. \tag{8.118}$$

Indeed, letting

$$U(x, y, t) = e^{ikt} u(x, y),$$

in Eq. (8.117), and

$$U(x, y, t) = e^{-k^2 t} u(x, y),$$

in Eq. (8.118), we find the Helmholtz equation, Eq. (8.116).

A related, important elliptic PDE is the modified Helmholtz equation

$$u_{xx} + u_{yy} - \kappa^2 u = 0, \quad \kappa > 0. \tag{8.119}$$

This equation appears in several physical applications including diffusion coalescence and pattern formation. In particular, the modified Helmholtz equation arises in reaction kinetics in connection with the evolution of interacting "particles". In its general form, reaction kinetics is important to a broad range of disciplines, including: chemistry, where "particles" are chemical reactants undergoing some chemical process; sociology, where "particles" are people interacting in various social ways; biology, where examples of "particles" are predator–prey systems, or the swapping of genetic material between viruses; economy, where "particles" are corporations merging or dissolving. In physics, reaction kinetics is a branch of Statistical Mechanics.[2] A widely applicable paradigm is that of particles moving through the process of diffusion and interacting with some special rate upon encounter, or interacting when approaching each another within some critical distance. Particular types of such models, called coalescence models, can be analyzed through the so-called method of "empty intervals": let $E(x, y, t)$ represent the probability that the interval (x, y) at time t is empty, that is, devoid of particles. $E(x, y, t)$ obeys the simple diffusion equation

$$E_{xx} + E_{yy} = E_t,$$

where the associated diffusion constant has been set equal to unity; the coalescence reaction is expressed in the boundary condition:

$$\lim_{x \to y} E(x, y, t) = 1.$$

The density of the particles can then be obtained from the empty interval probability, through the equation

$$\rho(x, t) = \left[\frac{\partial}{\partial y} E(x, y, t) \right]_{y=x}.$$

[2] See for example, D. ben-Avraham and S. Havlin, *Diffusion and Reactions in Fractals and Disordered Systems* (Cambridge University Press, 2000).

If in addition to coalescence, each particle can also give birth spontaneously to an additional identical particle with rate $\gamma > 0$, the process can yield an equilibrium steady state, represented by the PDE

$$E_{xx} + E_{yy} + \frac{\gamma}{2}(-E_x + E_y) = 0. \tag{8.120}$$

Using the substitution

$$E(x,y,t) = 1 - \exp\left[-\frac{\gamma}{2}(y-x)\right] u(x,y),$$

Eq. (8.120) becomes the modified Helmholtz equation, Eq. (8.119):

$$u_{xx} + u_{yy} - \kappa^2 u = 0, \quad \kappa = \frac{\gamma}{\sqrt{2}} > 0.$$

A fundamental solution, $G(x,y;\xi,\zeta)$, associated with the Helmholtz equation, Eq. (8.116), satisfies

$$G_{xx} + G_{yy} + \kappa^2 G = -\delta(x-\xi)\delta(y-\zeta). \tag{8.121}$$

Then, following identical steps with those used for the derivation of the fundamental solutions of Laplace's equation, it follows that if the Helmholtz equation is valid in the domain Ω, then

$$u(x,y) = \int_{\partial\Omega} \left[G(\xi,\zeta;x,y)\frac{\partial u(\xi,\zeta)}{\partial \mathcal{N}} - u(\xi,\zeta)\frac{\partial G(\xi,\zeta;x,y)}{\partial \mathcal{N}} \right] ds, \tag{8.122}$$

where ξ and ζ depend on s which is the arc-length of the curve $\partial\Omega$ and $\partial f(\xi,\zeta)/\partial\mathcal{N}$ denotes the derivative of f along the outward normal of the curve $\partial\Omega$. Introducing the complex variable

$$\zeta = \xi + i\zeta, \tag{8.123}$$

and using the identities

$$\frac{\partial G}{\partial \mathcal{N}} ds = \frac{\partial G}{\partial \xi} d\zeta - \frac{\partial G}{\partial \zeta} d\xi = \left(\frac{\partial G}{\partial \zeta} + \frac{\partial G}{\partial \bar{\zeta}} \right) d\zeta - i\left(\frac{\partial G}{\partial \zeta} - \frac{\partial G}{\partial \bar{\zeta}} \right) d\xi, \tag{8.124}$$

or

$$\frac{\partial G}{\partial \mathcal{N}} ds = -i\frac{\partial G}{\partial \zeta} d\zeta + i\frac{\partial G}{\partial \bar{\zeta}} d\bar{\zeta}, \tag{8.125}$$

Eq. (8.122) becomes

$$u(x,y) = \int_{\partial\Omega} \left[G\frac{\partial u}{\partial \mathcal{N}} ds + iu\left(\frac{\partial G}{\partial \zeta} d\zeta - \frac{\partial G}{\partial \bar{\zeta}} d\bar{\zeta} \right) \right]. \tag{8.126}$$

Similar considerations are valid for the modified Helmholtz equation, Eq. (8.119). It turns out that the following formulas are valid:

$$\text{Helmholtz}: \ G(\xi,\zeta;x,y) = \frac{i}{4}H_0^{(1)}(k|\zeta - z|), \tag{8.127}$$

and

$$\text{Modified Helmholtz}: \ G(\xi,\zeta;x,y) = \frac{1}{2\pi}K_0(k|\zeta - z|), \tag{8.128}$$

with $z = x + iy$, $\zeta = \xi + i\zeta$, where $H_n^{(1)}(z)$ and $K_n(z)$ denote the Hankel function and the modified Bessel function of the second kind, respectively. The integral representations of these functions are as follows:

$$K_\alpha(x) = \int_0^\infty e^{-x\cosh(t)} \cosh(\alpha t)dt, \qquad (8.129)$$

$$H_\alpha^{(1)}(x) = \frac{1}{i\pi} \int_{-\infty}^{\infty+i\pi} e^{x\sinh(t)-\alpha t}dt, \qquad (8.130)$$

where the path of integration for the Hankel function goes from $-\infty$ to 0 on the real axis, from 0 to $i\pi$ on the imaginary axis, and from $i\pi$ to $\infty + i\pi$ on a line parallel to the real axis. For $\alpha = n$ integer, these functions are evaluated as the limit of the above expressions for $\alpha \to n$.

8.7   Disadvantages of traditional integral transforms

1. *Traditional transforms yield representations which are* not *uniformly convergent at the boundaries.*

   For example, for $g_0(t) \neq 0$, the right-hand side of Eq. (8.39) cannot be uniformly convergent at $x = 0$, otherwise one could compute $u(0, t)$ by inserting the limit $x \to 0$ inside the integral of the right-hand side of Eq. (8.39) and this would yield zero instead of $g_0(t)$.

2. *It is not straightforward to verify that the relevant representation provides a solution of the given initial–boundary value problem.*

   For example, the (x, t) dependence of the right-hand side of Eq. (8.39) is *not* of the simple exponential form e^{ikx-k^2t}, thus it is not obvious by inspection that $u(x, t)$ satisfies the heat equation. Nevertheless, a simple computation shows that this is indeed the case. Moreover, using the inverse sine-transform formula it is immediately obvious that $u(x, 0) = u_0(x)$. However, due to the lack of uniform convergence at $x = 0$, it is *not* straightforward to prove that $u(0, t) = g_0(t)$.

   It is important to emphasize that the derivation of Eq. (8.39) is based on the *assumption* that $u(x, t)$ exists and that it has nice properties, otherwise the starting equation, Eq. (8.35), is *not* well-defined. Thus, unless one can eliminate this assumption using *a priori* PDE estimates, one must verify *a posteriori* that the expression of Eq. (8.39) satisfies the heat equation and the given conditions, Eq. (8.33). It is interesting that this verification procedure is *not* discussed in text books on initial–boundary value problems. Actually, although the lack of uniform convergence at the boundaries is a serious *generic* disadvantage of *any* expression obtained through a traditional transform pair, this disadvantage has *not* been emphasized in the literature.

3. *Representations obtained through traditional transforms are not suitable for numerical computations.*

   This fact is also a direct consequence of the lack of uniform convergence at the boundaries.

4. *The implementation of traditional transforms requires separability of the given domain and of the given boundary conditions.*

   For example, for the heat equation the sine-transform approach fails if the boundary condition in Eq. (8.33) is replaced with the non-local condition

   $$\int_0^\infty K(x)u(x,t)dx = g_0(t), \quad t > 0, \qquad (8.131)$$

   where $K(x)$ is a given function. Such a boundary condition occurs, for example, in modeling the concentration of a dispersed substance in a colloidal suspension, which is discussed in the Application section of Chapter 10 (see Section 10.5).

5. *In general, it is difficult to construct a transform for the particular cases where such a transform exists.*

   For example, in the case of the heat equation on the half-line supplemented with the conditions of Eq. (8.33), we were able to guess the correct transform. However, if the condition $u(0,t) = g_0(t)$ is replaced by the so-called Robin condition $u_x(0,t) - \gamma u(0,t) = g_R(t)$, where γ is a constant, then it is not possible to guess the correct transform. There does exist a systematic approach for deriving the appropriate transform pair for a given boundary value problem, but this approach is quite complicated. For example, in the case of the sine-transform, the first step of this approach is to compute the associated Green's function, namely, to solve the following ODE:

   $$\frac{\partial^2}{\partial x^2}G(x,\xi,k) + k^2 G(x,\xi,k) = \delta(x - \xi), \quad 0 < x < \infty, \ 0 < \xi < \infty,$$

   $$G(0,\xi,k) = 0, \quad \lim_{x \to \infty} G(x,\xi,k) = 0.$$

   Then, one computes the integral of $G(x,\xi,k)$ around an appropriate contour in the k-complex plane, and this yields the sine-transform pair.

   The above approach in addition to being complicated, it also *assumes* certain "completeness" which in general is difficult to justify.

6. *There exist traditional transform pairs only for a very limited case of boundary value problems.*

   For example, consider the Stokes equation, Eq. (8.17), defined on the half-line and supplemented with the conditions of Eq. (8.33). It can be shown that for this problem there does *not* exist a traditional transform pair in the variable x. Actually, in general there does *not* exist the analogue of the sine- and cosine-transforms for solving initial–boundary value problems involving a third-order spatial derivative.

   These disadvantages for the traditional integral transforms are also relevant for the traditional infinite series. For example, as in point 1 above for the integral transforms, if $g_0(t) \neq 0$ and $h_0(t) \neq 0$, then clearly the representation of Eq. (8.53) is *not* uniformly convergent both at $x = 0$ and $x = L$, otherwise one could insert the limits $x \to 0$ and $x \to L$ inside the sum and this would imply that $u(0,t)$ and $u(L,t)$ vanish. Also, it can be shown that it is *not* possible to express the solution of the Stokes equation with generic boundary conditions in terms of an infinite series (the unified transform yields a solution involving appropriate integrals). This implies that the usual "philosophy" that problems defined in a bounded domain can be expressed in terms of an infinite series (due to the existence of the so-called discrete spectrum) is *not* correct — this philosophy is valid only for the limited class of so-called "self-adjoint" problems.

8.8   Application: The Black–Scholes equation

An important concept in financial markets is the ability to trade a security (like a stock) at a pre-determined price within a specified time period, called an "option". There are two basic types of options, the right to buy which is referred to as a "call option", and the right to sell which is referred to as a "put option". The price of the option, denoted as $C(S,t)$ for the call option and $P(S,t)$ for the put option, is a function of the price S of the underlying security at any moment in time t between the time when the option is issued, corresponding to $t = 0$, and its expiration, corresponding to $t = T$.

To provide some intuition, we mention that the call option is the right, but not the obligation, to buy a security with "spot price" $S(t)$ at the price K (also referred to as the "strike" price) up to the maturity of the option which occurs at $t = T$. Near maturity, the call option has value when $S > K$ (this is referred to as the option being *"in the money"*) but it is worthless for $S < K$ (the option is *"out of the money"*). For these two ranges, the variable $\xi(\tau) = \ln(S/K)$, with $\tau = T - t$, takes positive and negative values, respectively, as illustrated in Fig. 8.4. The variables ξ and τ are essential in transformations of the differential equation that describes the value of the option, as discussed in detail below. For the put option, namely the right, but not the obligation, to sell a security at a predetermined value K up to the maturity time $t = T$, the opposite condition holds, that is, near maturity the put option is valuable for $S < K$ and worthless for $S > K$.

The *terminal* value of the option at $t = T$, and its limiting values for $S \to 0$ and $S \to \infty$, are given in terms of K, the discounted future value, and ρ, the risk-free interest rate. For the call option these quantities are

$$\text{Call option}: \quad C(S,T) = \max(S - K, 0), \ S > 0, \tag{8.132a}$$

$$\lim_{S \to 0} C(S,t) = 0, \ t \in [0, T], \tag{8.132b}$$

$$\lim_{S \to \infty} C(S,t) = S, \ t \in [0, T], \tag{8.132c}$$

where $\max(S - K, 0)$ is the payoff. For the put option they are

$$\text{Put option}: \quad P(S,T) = \max(K - S, 0), \ S > 0, \tag{8.133a}$$

$$\lim_{S \to 0} P(S,t) = Ke^{-\rho(T-t)}, \ t \in [0, T], \tag{8.133b}$$

$$\lim_{S \to \infty} P(S,t) = 0, \ t \in [0, T], \tag{8.133c}$$

where $\max(K - S, 0)$ is the payoff.

The two terminal conditions, with the use of the Heaviside θ-function, can be expressed in the following forms:

$$P(S,T) = (K - S)\, \theta(K - S), \tag{8.134}$$

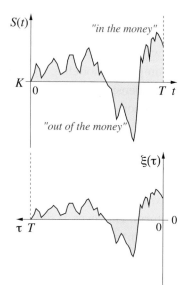

Figure 8.4: **Top**: Schematic illustration of the behavior of the value $S(t)$ of the underlying financial asset as a function of time relative to the discount price K of the call option: when $S(t) > K$ the call option is *"in the money"* and when $S(t) < K$ it is *"out of the money"*. **Bottom**: The corresponding values of the variable $\xi(\tau) = \ln(S/K)$ as a function of the variable $\tau = T - t$.

for the put option, and

$$C(S,T) = (S - K)\,\theta(S - K), \tag{8.135}$$

for the call option; these relations are shown schematically in Fig. 8.5.

The price V of the option for any time $t \in [0, T]$ is described by the celebrated Black–Scholes equation (BSE), which in general form (for either P or C) reads:

$$\frac{\partial V}{\partial t} = \rho V - \frac{1}{2}\sigma^2 S^2 \frac{\partial^2 V}{\partial S^2} - \rho S \frac{\partial V}{\partial S}, \tag{8.136}$$

with the boundary conditions

$$V(S,T) = h(S), \quad S > 0, \tag{8.137}$$

$$\lim_{S \to 0} V(S,t) = R^{(-)}(t), \quad \lim_{S \to \infty} V(S,t) = R^{(+)}(t), \quad t \in [0,T], \tag{8.138}$$

where $R^{(\pm)}(t)$ are two functions appropriately chosen to match the *terminal condition*, $V(S,T) = h(S)$, and σ is a parameter that describes the volatility of the stock price.[3]

In the following, we will concentrate on solving the BSE, Eq. (8.136), subject to the terminal condition, namely the condition obeyed for $t = T$, Eq. (8.137). To proceed, we introduce the change of variables

$$\tau = T - t, \quad V = u e^{-\rho \tau}, \tag{8.139}$$

with which Eq. (8.136) becomes

$$\frac{\partial u}{\partial \tau} = \frac{1}{2}\sigma^2 S^2 \frac{\partial^2 u}{\partial S^2} + \rho S \frac{\partial u}{\partial S}.$$

As the next step, we introduce a new variable

$$\xi = \ln\left(\frac{S}{K}\right), \tag{8.140}$$

which gives for the derivatives with respect to S the expressions

$$\frac{\partial}{\partial S} = \frac{1}{S}\frac{\partial}{\partial \xi}, \quad \frac{\partial^2}{\partial S^2} = -\frac{1}{S^2}\frac{\partial}{\partial \xi} + \frac{1}{S^2}\frac{\partial^2}{\partial \xi^2}.$$

Hence, the BSE becomes

$$\frac{\partial u}{\partial \tau} = \frac{1}{2}\sigma^2 \frac{\partial^2 u}{\partial \xi^2} + \left(\rho - \frac{1}{2}\sigma^2\right)\frac{\partial u}{\partial \xi}.$$

We next introduce the variable x defined by

$$x = \frac{\sqrt{2}}{\sigma}\xi = \frac{\sqrt{2}}{\sigma}\ln\left(\frac{S}{K}\right). \tag{8.141}$$

Using x, the BSE takes the form

$$\frac{\partial u}{\partial \tau} = \frac{\partial^2 u}{\partial x^2} + \left(\frac{\sqrt{2}}{\sigma}\rho - \frac{\sigma}{\sqrt{2}}\right)\frac{\partial u}{\partial x}.$$

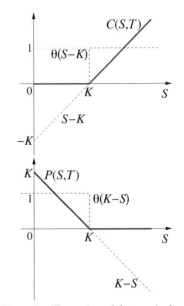

Figure 8.5: Illustration of the terminal conditions for the call option $C(S,T) = \max(S - K, 0)$ (top) and the put option $P(S,T) = \max(K - S, 0)$ (bottom). The diagrams also include in dashed lines the functions $S - K$, $\theta(S - K)$, and $K - S$, $\theta(K - S)$ in the corresponding plots.

[3] For details, see, for example, J.C. Hull, *Options, Futures and Other Derivatives*, Pearson (2014).

Finally, we introduce the parameter β defined as

$$\beta = \left(\frac{\sqrt{2}}{\sigma} \rho - \frac{\sigma}{\sqrt{2}} \right) = \frac{\sqrt{2}}{\sigma} \left(\rho - \frac{1}{2}\sigma^2 \right). \tag{8.142}$$

This leads to the final form of the BSE:

$$\frac{\partial u}{\partial \tau} = \frac{\partial^2 u}{\partial x^2} + \beta \frac{\partial u}{\partial x}, \tag{8.143}$$

which is the same as Eq. (8.16) with $\alpha = 1$.

Considering the changes of variables we have introduced, Eqs. (8.139)–(8.142), and taking into account that the original variables are $t \in [0, T]$ and $S \in [0, \infty)$, we have the following new variables and their ranges:

- time $t = 0$ corresponds to $\tau = T$, and time $t = T$ corresponds to $\tau = 0$;

- the range of the variable x is from $-\infty$ to ∞;

- the *terminal condition* $P(S, T) = h(S)$ has been changed to an *initial condition* for $u(x, \tau)$ which is valid for $\tau = 0$.

The final equation, Eq. (8.143), is therefore supplemented by the following initial–boundary conditions:

$$u_0(x) = u(x, 0) = h(K e^{\sigma x / \sqrt{2}}), \ x \in \mathbb{R},$$

$$\lim_{x \to \pm\infty} u(x, \tau) = e^{\rho \tau} R^{(\pm)}(T - \tau), \ \tau \in [0, T].$$

There is one additional step we can take to simplify further the equation: we introduce the transformation

$$u(x, \tau) = v(x, \tau) \exp\left(-\frac{\beta^2}{4}\tau - \frac{\beta}{2}x \right). \tag{8.144}$$

In this way we find that Eq. (8.143) reduces to the following equation in the function $v(x, \tau)$:

$$\frac{\partial v}{\partial \tau} = \frac{\partial^2 v}{\partial x^2}, \tag{8.145}$$

which is the same as the heat equation, Eq. (8.15), with $\alpha = 1$. However, under the transformation of Eq. (8.144), the initial condition $u_0(x)$ for Eq. (8.143) yields the initial condition

$$v_0(x) = u_0(x) e^{\beta x / 2},$$

for the heat equation. Hence, if $\beta > 0$, generally, $v_0(x)$, will grow exponentially as $x \to \infty$, whereas if $\beta < 0$, $v_0(x)$ will grow exponentially as $x \to -\infty$, unless $u_0(x)$ decays faster than the exponential factors that contain β in the limits $x \to \pm\infty$. Thus, depending on the initial condition, it may be more convenient to solve Eq. (8.143) directly instead of mapping it to the heat equation. Despite this caveat, the form of the BSE given in Eq. (8.145) offers the advantage that a solution can be obtained which is valid for all values of x except possibly for $|x| \to \infty$; and in any case this limit may be of no interest. In what follows, we explore this solution for the case of the call and put options.

For the call option, using Eq. (8.135) for the terminal condition of $C(S, T)$ we find that the initial condition for the function $u_c(x, \tau)$ reads:

$$h_c(S) = (S - K)\theta(S - K)$$

$$\Rightarrow \quad u_{c,0}(x) = h_c(Ke^{\sigma x/\sqrt{2}}) = K\left(e^{\sigma x/\sqrt{2}} - 1\right)\theta(e^{\sigma x/\sqrt{2}} - 1);$$

consequently the initial condition for the function $v_c(x, \tau)$ reads:

$$v_{c,0}(x) = K\left(e^{\sigma x/\sqrt{2}} - 1\right)\theta(e^{\sigma x/\sqrt{2}} - 1)e^{\beta x/2}.$$

Since $v_c(x, \tau)$ obeys the heat equation with $\alpha = 1$, and its initial condition is $v_{c,0}(x)$, we can take advantage of a result derived earlier, namely Eq. (8.31), to write the general solution in the form

$$v_c(x, \tau) = \int_{-\infty}^{\infty} v_{c,0}(x') \frac{1}{\sqrt{4\pi\tau}} e^{-(x-x')^2/4\tau} dx'.$$

From this expression, we can then construct the function $u_c(x, \tau)$ using the transformation of Eq. (8.144). The result of this calculation[4] yields

$$u_c(x, \tau) = K\left[e^{\xi + \tau\rho}u(\tilde{w}_1) - u(\tilde{w}_2)\right], \tag{8.146}$$

[4] This integrations involved are straight-forward if somewhat lengthy, thus we relegate the derivation to Appendix A.

where $u(w)$ is the normal cumulative distribution function introduced in Chapter 1, Eq. (1.60a):

$$u(w) = \frac{1}{\sqrt{2\pi}} \int_{-w}^{\infty} e^{-y^2/2} dy = \frac{1}{\sqrt{2\pi}} \int_{-\infty}^{w} e^{-y^2/2} dy,$$

and the values of $\tilde{w}_1, \tilde{w}_2$ are as follows:

$$\tilde{w}_1 = \frac{1}{\sigma\sqrt{\tau}}\left[\xi + \tau\left(\rho + \frac{1}{2}\sigma^2\right)\right], \quad \tilde{w}_2 = \frac{1}{\sigma\sqrt{\tau}}\left[\xi + \tau\left(\rho - \frac{1}{2}\sigma^2\right)\right]. \tag{8.147}$$

The function $u_c(x, \tau)$ is depicted in Fig. 8.6 for the values of the parameters $\rho = 1, \sigma = 1$, which give for the value of $\beta = 1/\sqrt{2}$ [in the above expressions ξ is given in terms of x through Eq. (8.141)].

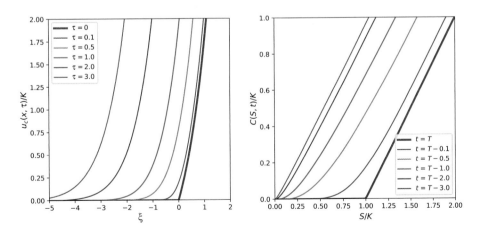

Figure 8.6: **Left**: The function $u_c(x, \tau)$ (in units of K) for the call option plotted as a function of $\xi = x\sigma/\sqrt{2}$, for various values of the time variable τ. **Right**: The function $C(S, t)$ (in units of K) for various values of the time variable t. The values of the parameters are $\rho = 1$, $\sigma = 1$, which give $\beta = 1/\sqrt{2}$.

Having found the solution for the function $u_c(x, \tau)$, we can then construct the function $C(S, t)$ using the relation of Eq. (8.139):

$$C(S, t) = Su(\tilde{w}_1) - Ke^{\rho(t-T)}u(\tilde{w}_2), \tag{8.148}$$

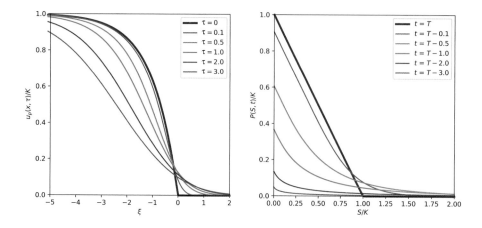

Figure 8.7: **Left:** The function $u_p(x, \tau)$ (in units of K) for the call option plotted as a function of $\xi = x\sigma/\sqrt{2}$, for various values of the time variable τ. **Right:** The function $P(S, t)$ (in units of K) for various values of the time variable t. The values of the parameters are $\rho = 1$, $\sigma = 1$, which give $\beta = 1/\sqrt{2}$.

where the values of $\tilde{w}_1, \tilde{w}_2$ are defined in Eq. (8.147). This function is shown in Fig. 8.6 for various values of the time t, for the parameter values $\rho = 1$, $\sigma = 1$.

For the put option, the terminal condition for $P(S, T)$, Eq. (8.134), translates to the following initial condition for the function $u_p(x, \tau)$:

$$h_p(S) = (K - S)\theta(K - S)$$
$$\Rightarrow u_{p,0}(x) = h_p(Ke^{\sigma x/\sqrt{2}}) = K\left(1 - e^{\sigma x/\sqrt{2}}\right)\theta(1 - e^{\sigma x/\sqrt{2}}).$$

Following exactly the same steps as for the call option we arrive at the result

$$u_p(x, \tau) = K\left[u(-\tilde{w}_2) - e^{\xi + \tau\rho}u(-\tilde{w}_1)\right], \qquad (8.149)$$

which, with the help of Eq. (8.139) leads to the following expression for the function $P(S, t)$:

$$P(S, t) = S\left[u(\tilde{w}_1) - 1\right] - Ke^{\rho(t-T)}\left[u(\tilde{w}_2) - 1\right]. \qquad (8.150)$$

The functions $u_p(x, \tau)$ and $P(S, t)$ are shown in Fig. 8.7 for various values of the time variables, for the parameter values $\rho = 1$, $\sigma = 1$, which give $\beta = 1/\sqrt{2}$.

The solutions for $C(S, t)$ and $P(S, t)$ satisfy the terminal conditions for $t = T$ (compare with the behavior shown in Fig. 8.5), as well as the limiting behavior for $S \to 0$ and $S \to \infty$, as stated in Eq. (8.132) for the call option and Eq. (8.133) for the put option.

Further reading

Classic textbooks covering ODEs and PDEs arising in mathematical physics are the following:

1. Richard Courant and David Hilbert, *Methods of Mathematical Physics, Vol. 2* Second Ed. (Wiley-VCH, 1989).

2. Garrett Birkhoff and Gian-Carlo Rota, *Ordinary Differential Equations*, Fourth Ed. (Wiley, 1989).

Standard textbooks covering PDEs include:

3. Paul R. Garabedian, *Partial Differential Equations*, Revised Ed. (Chelsea Publishing Co., 1998).

4. Fritz John, *Partial Differential Equations*, Fourth Ed. (Springer-Verlag, 1982).

5. Peter J. Olver, *Introduction to Partial Differential Equations* (Springer-Verlag, 2014).

The standard methods for solving initial–boundary values problems, with many nice examples, can be found in:

6. Ivar Stakgold, *Green's Functions and Boundary Value Problems*, Third Ed. (Wiley, 2011).

7. Lawrence C. Evans, *Partial Differential Equations*, Second Ed. (American Mathematical Society, 2010).

8. Richard Haberman, *Elementary Applied Partial Differential Equations with Fourier Series and Boundary Value Problems*, Third Ed. (Prentice Hall, 1997).

9. David L. Powers, *Boundary Value Problems*, Fourth Ed. (Academic Press, 1999).

A rigorous treatment of PDEs is presented in:

10. S. L. Sobolev, *Partial Differential Equations of Mathematical Physics* (Dover, 2011).

11. Rolf Leis, *Initial Boundary Value Problems in Mathematical Physics* (Dover, 2013).

12. Haim Brezis, *Functional Analysis, Sobolev Spaces and Partial Differential Equations* (Springer-Verlag, 2011).

Problems

1. Let $u(x)$ be a solution of the first-order ODE:

$$u_x + au = f(x),$$

where a is a real constant and $f(x)$ is a given function. Prove that the Fourier transforms of $u(x)$ and $f(x)$, denoted by $\hat{u}(k)$ and $\hat{f}(k)$, respectively, satisfy the relation:

$$\hat{u}(k) = \frac{\hat{f}(k)}{ik + a}.$$

Using this result, find the solution $u(x)$, if $f(x) = \delta(x)$ and $a > 0$.

2. Consider the conduction of heat in a metal ring, where the temperature distribution $u(x,t)$ is governed by the heat equation:

$$u_t = \alpha u_{xx},$$

with x measured along the ring and t the time. The constant α is the heat conduction coefficient. The ring has length L, and the initial temperature distribution is given by $u(x,0) = f(x)$, where $f(x)$ is a periodic function of period L.

   (a) Show that for $t \to \infty$, the steady-state solution is a constant temperature that is the average of the value of f.

   (b) Find the temperature distribution for all times if $f(x) = \cos(2\pi x/L)$.

3. We want to solve the following second-order ODE for $u(x)$:

$$u_{xx} + \frac{\pi^2}{b^2}u = i\left(e^{ix/b} - ie^{-ix/b}\right)^2 - 2,$$

using the Fourier expansion technique, in the interval $x \in [0, \pi b]$.

   (a) Choose a Fourier series expansion that automatically satisfies the boundary conditions $u(0) = u(\pi b) = 0$.

   (b) Substitute the Fourier expansion in the differential equation and solve for the coefficients, thus obtaining the solution for $u(x)$.

4. We want to solve the following second-order ODE for $u(x)$:

$$u_{xx} + a^2 u = \frac{\lambda a^2}{\lambda - e^{i\pi ax}},$$

where $|\lambda| > 1$, using the Fourier expansion technique with complex exponentials.

   (a) Find the coefficients that enter in the expansion of $u(x)$.

   (b) Do you expect that this expansion converges fast, and that it could be truncated with a small error? Explain your answer.

5. We want to solve the following second-order ODE for $u(x)$:

$$u_{xx} + \frac{1}{b^2}u = \frac{2}{b^2}\sin\left(\frac{x}{b}\right)\cos\left(\frac{x}{b}\right),$$

using the Fourier expansion technique.

   (a) Choose a Fourier series expansion that automatically satisfies the boundary conditions $u(0) = u(b) = 0$.

   (b) Substitute the Fourier expansion in the differential equation and solve for the coefficients, thus obtaining the solution for $u(x)$.

6. Consider the one-dimensional wave equation, $u_{tt} = c^2 u_{xx}$.

   (a) Show that the function $f(x \pm ct)$ is a solution of the wave equation, where $f(w)$ is an arbitrary function, doubly differentiable with respect to its argument w.

   (b) Which of the following functions are solutions of this equation? Which ones represent traveling waves? Find the speed and the direction of propagation of the wave in each case (A, B, C, D are constants).

$$\text{(i)} \ u(x,t) = A\tanh(x + 10t),$$

$$\text{(ii)} \ u(x,t) = B\sin(x)\cos(10t),$$

$$\text{(iii)} \ u(x,t) = Ce^{-(x^2 - t^2)},$$

$$\text{(iv)} \ u(x,t) = De^{-(x+3t)}\cos(x + 3t).$$

7. Solve the one-dimensional wave equation, $u_{tt} = c^2 u_{xx}$, in the infinite spatial domain, $x \in (-\infty, \infty)$, for the following initial conditions:

$$u(x,0) = \frac{1}{\sqrt{2\pi}\sigma} e^{-x^2/2\sigma^2},$$

$$u_t(x,0) = -\frac{x}{\sqrt{2\pi}\sigma^3} e^{-x^2/2\sigma^2},$$

using the Fourier transform method. What happens in the limit $\sigma \to 0$?

8. Consider the wave equation in two spatial dimensions for the function $u(x,y,t)$ that obeys:

$$u_{tt} = c^2 (u_{xx} + u_{yy}), \quad x \in [0, L], \quad y \in [0, W], \quad t > 0,$$

which describes waves in a surface of orthogonal parallelepiped shape of length L and width W. We will assume the boundary conditions:

$$u(x, 0, t) = u(x, W, t) = 0, \quad t > 0,$$

$$u(0, y, t) = u(L, y, t) = 0, \quad t > 0,$$

which correspond to the boundary of the surface being held fixed, as in a musical drum of this shape. We will use the Fourier series expansion method to obtain the solution u, x, y, t.

(a) Show that the relation between the frequency ω, which appears in the Fourier transform of the time variable, and the wavenumber (k_x, k_y), which appears in the Fourier transform of the spatial variables (x, y), for this case is

$$\omega_{nm} = c \left(\frac{n^2 \pi^2}{L^2} + \frac{m^2 \pi^2}{W^2} \right)^{1/2}, \quad n, m \in \mathbb{Z}.$$

(b) Draw the patterns of the wave for some moment in time t for the lowest few frequencies, that is, the lowest few values of the integers n, m.

9. A simple version of Schrödinger's equation in quantum mechanics for the "wavefunction" $\psi(x, t)$ reads:

$$i\frac{\partial \psi}{\partial t} = -\frac{\partial^2 \psi}{\partial x^2}, \quad -\infty < x < \infty, \quad t \geq 0.$$

This equation describes the motion of a free particle with wave-like properties. Solve this equation by

using the Fourier transform of $\psi(x, t)$ with respect to x, with the initial condition

$$\psi(x, 0) = \frac{1}{\sqrt{2\pi}a} e^{-x^2/2a^2},$$

that is, a normalized Gaussian function of width $a > 0$. Is $\psi(x, t)$ also a Gaussian for $t > 0$? Explain.

10. Consider the two-dimensional Poisson equation:

$$\nabla^2 u(x,y) = \left(\frac{\partial^2}{\partial x^2} + \frac{\partial^2}{\partial y^2} \right) u(x,y) = -4\pi\rho(x,y),$$

with the following charge distribution:

$$\rho(x,y) = 3\cos^2\left(\frac{\pi x}{a}\right)\sin\left(\frac{\pi x}{a}\right)\cos^2\left(\frac{\pi y}{b}\right)$$
$$- \sin^3\left(\frac{\pi x}{a}\right)\cos^2\left(\frac{\pi y}{b}\right)$$
$$+ \sin^3\left(\frac{\pi x}{a}\right)\sin^2\left(\frac{\pi y}{b}\right)$$
$$- 3\cos^2\left(\frac{\pi x}{a}\right)\sin\left(\frac{\pi x}{a}\right)\sin^2\left(\frac{\pi y}{b}\right).$$

Use Fourier expansions for $u(x,y)$ and $\rho(x,y)$ to determine the potential $u(x,y)$.

11. We consider the damped and driven oscillator equation for the function $u(t)$:

$$u_{tt} + bu_t + \omega_0^2 u = f_1 \cos(\omega_1 t) + f_2 \sin(\omega_2 t).$$

We will find a solution of this equation through the following steps:

(a) Consider the corresponding equation in the function $v(t)$:

$$v_{tt} + bv_t + \omega_0^2 v = f_1 \sin(\omega_1 t) + f_2 \cos(\omega_2 t).$$

Multiply the second equation by i, add the two equations and produce the equation for the function $w(t) = u(t) + iv(t)$.

(b) Assume a solution of the form $w(t) = A\exp[i\omega_1 t] + B\exp[-i\omega_2 t]$, and find the values of the constants A and B.

(c) For the case $f_1 = f_2 = f_0$ and $\omega_1 = \omega$, $\omega_2 = 2\omega$, find the solution for $u(t)$ in terms of ω_0, ω and f_0.

Chapter 9

Unified Transform I: Evolution PDEs on the Half-line

The hallmark of an educated mind is that it is satisfied with the level of precision afforded by the nature of the problem, and does not seek exactness when only an approximation of the truth is possible.

Aristotle (Greek philosopher, 384–322 BCE)

In this chapter, we introduce a new method for solving partial differential equations, which employs many of the wonderful properties of functions of complex variables and of the Fourier transform. We refer to this approach as the "unified transform". In what follows, we implement this method to initial–boundary value problems formulated on the half-line and illustrate its effectiveness through several examples of important differential equations. We also point out the advantages of the unified transform over the traditional approaches.

9.1 *The unified transform is based on analyticity*

We describe first some general features of the unified transform for solving PDEs. To make the discussion more transparent, we will use as an example the heat equation, Eq. (8.15), or $u_t = u_{xx}$ (for simplicity, in the following discussion we set the parameter $\alpha = 1$). Suppose that this equation is defined on the infinite line, namely

$$\Omega = \{-\infty < x < \infty, \ t > 0\},$$

with the solution denoted by $u(x,t)$ and the initial condition

$$u(x,0) = u_0(x), \quad -\infty < x < \infty.$$

Using the Fourier transforms of the solution, $\hat{u}(k,t)$, and of the initial condition, $\hat{u}_0(k)$, we can express $u(x,t)$ and $u_0(x)$ in terms of the following integrals:

$$u(x,t) = \int_{-\infty}^{\infty} \hat{u}(k,t)e^{ikx}\frac{dk}{2\pi}, \quad u_0(x) = \int_{-\infty}^{\infty} \hat{u}_0(k)e^{ikx}\frac{dk}{2\pi}.$$

Inserting these expressions in the heat equation, we obtain

$$\frac{\partial \hat{u}(k,t)}{\partial t} = -k^2\hat{u}(k,t) \Rightarrow \hat{u}(k,t) = e^{-k^2t}\hat{u}_0(k), \quad k \in \mathbb{R}. \tag{9.1}$$

The last expression in Eq. (9.1) is an algebraic relation between the Fourier transforms of $u(x,t)$ and $u_0(x)$; these functions are defined on the *boundary* of the domain Ω.

In the case of the heat equation on the half-line,

$$\Omega = \{0 < x < \infty, \ t > 0 \},$$

with the initial condition

$$u(x,0) = u_0(x), \ 0 < x < \infty,$$

and the boundary condition

$$u(0,t) = g_0(t), \ t > 0, \tag{9.2}$$

the sine-transform yields the following analogue of the second of Eq. (9.1), namely Eq. (8.38) derived in Chapter 8:

$$e^{k^2 t}\hat{u}_s(k,t) = \hat{u}_s^{(0)}(k) + k \int_0^t e^{k^2\tau} g_0(\tau)d\tau, \ 0 < k < \infty. \tag{9.3}$$

In this equation $\hat{u}_s(k,t)$ and $\hat{u}_s^{(0)}(k)$ are the sine-transforms of $u(x,t)$ and $u_0(x)$, respectively, and the last term on the right-hand side involves a t-transform of $g_0(t)$. Equation (9.3) is again an algebraic relation between the transforms of functions appearing on the *boundary* of Ω. It should be noted that as a result of the use of the sine-transform, which is the "correct" transform for this boundary value problem, the t-transform of the unknown boundary value $u_x(0,t)$ does not appear in Eq. (9.3).

The starting point of the unified transform is deriving an equation which, like Eqs. (9.1) and (9.3), is valid on the boundary of the domain. This equation will be referred to as the global relation (GR). In contrast to the standard transform methods, the unified transform does not use specialized transforms. Actually, for all evolution PDEs and for the wave equation it uses the same transform, namely, the Fourier transform. This has the major advantage that it eliminates the need to derive a specific transform for the particular problem (for those cases that such a transform exist). However, it has the disadvantage that it gives rise to a global relation containing certain transforms of all boundary values, and some of the boundary values are unknown functions (that is, are not prescribed as boundary conditions). For example, in the case of the heat equation on the half-line, the GR contains a specific t-transform of $u(0,t)$ and $u_x(0,t)$. Remarkably, there is a simple procedure which allows one to eliminate the transforms of the unknown boundary values, using only algebraic manipulations.

In what follows, we, first, describe the specific restrictions of the Fourier transform used for the cases of problems defined on the half-line or the finite interval. Then, we discuss briefly the main idea used for eliminating the transforms of the unknown functions.

(i) For problems on the half-line, the unified transform uses the restriction of the Fourier transform in the domain $0 < x < \infty$:

$$\hat{f}(k) = \int_0^\infty e^{-ikx} f(x)dx, \ \text{Im}[k] \leq 0, \tag{9.4a}$$

$$f(x) = \int_{-\infty}^{\infty} \hat{f}(k)e^{ikx}\frac{dk}{2\pi}, \quad 0 \le x < \infty. \tag{9.4b}$$

Equations (9.4a) and (9.4b) follow from Eqs. (7.19) and (7.20), defining the direct and inverse Fourier transform, respectively, when applied to functions that vanish for $x < 0$. Indeed let

$$F(x) = \begin{cases} f(x), & 0 < x < \infty, \\ 0, & -\infty < x < 0. \end{cases}$$

Then Eqs. (7.19) and (7.20) imply Eqs. (9.4a) and (9.4b), as well as the equation

$$\int_{-\infty}^{\infty} \hat{f}(k)e^{ikx}\frac{dk}{2\pi} = 0, \quad -\infty < x < 0.$$

Replacing in this equation x with $-x$ and k with $-k$, we find

$$\int_{-\infty}^{\infty} \hat{f}(-k)e^{ikx}dk = 0, \quad 0 < x < \infty. \tag{9.5}$$

This equation plays an important role in the implementation of the unified transform.

Similarly, for problems on a finite interval the unified transform uses the restriction of the Fourier transform in the domain $0 < x < L$, where $L > 0$:

$$\hat{f}(k) = \int_{0}^{L} e^{-ikx} f(x)dx, \quad k \in \mathbb{C}, \tag{9.6a}$$

$$f(x) = \int_{-\infty}^{\infty} \hat{f}(k)e^{ikx}\frac{dk}{2\pi}, \quad 0 < x < L. \tag{9.6b}$$

Equations (9.6a) and (9.6b) follow from Eqs. (7.19) and (7.20), when $F(x) = f(x)$ for $0 < x < L$ and $F(x) = 0$ outside this domain. In analogy with Eq. (9.5), we now have

$$\int_{-\infty}^{\infty} \hat{f}(-k)e^{ikx}dk = 0, \quad 0 < x < L. \tag{9.7}$$

(ii) Suppose that the given evolution PDE admits the particular solution $\exp[ikx - \omega(k)t]$, where k is a constant parameter and $\omega(k)$ is a given polynomial of k. It turns out that the t-transforms of the boundary values depend on k only through $\omega(k)$. This implies that these transforms remain invariant if k is replaced by another function of k provided that $\omega(k)$ remains the same. For example, in the case of the heat equation, $\omega(k) = k^2$, hence this function remains the same if k is replaced by $-k$. On the other hand, the GR changes under the above transformations. It turns out that by using the equations obtained through the above transformations, it is possible to eliminate the transforms of the unknown boundary value using only algebra.

The above idea is sufficient for those problems that can be solved by the usual transforms. For problems for which standard transforms do not exist, it is important to supplement the above idea with the powerful techniques of complex variables. In this connection, it is important to notice that the Fourier transform defined by Eq. (9.4a), which sometimes is refereed to as the "half-Fourier transform", is analytic in the lower half of the k-complex plane. Indeed,

$$k = k_R + ik_I \Rightarrow e^{-ikx} = e^{-ik_R x}e^{k_I x},$$

and since $0 < x < \infty$, the exponential appearing in the half-Fourier transform decays for $k_I < 0$ and is bounded for $k_I = 0$. Similarly, the "finite Fourier transform" defined by Eq. (9.6a) is an entire function, that is, it is analytic in the entire finite k-complex plane. These observations are of crucial importance for the applicability of the new method. Indeed, the above analyticity considerations provide the foundation of the unified transform.

In summary, the standard transforms, like the sine- and cosine-transforms, have the advantage that they can eliminate the unknown boundary values directly. However, these specialized transforms can only be applied to a very limited class of boundary value problems. Incidentally, these transforms are only defined for real values of k. On the other hand, the unified transform is based on the restriction of the classical Fourier transform on the domain of interest, and this restriction gives rise to transforms which are analytic functions. By exploring this analyticity, and using the fact that the t-transforms of the boundary values remain invariant under appropriate transformations, it is possible to eliminate indirectly the transforms of the unknown boundary values using only algebraic manipulations. This procedure is *generic*, namely it is not based on any specialized transform or on a specific boundary value problem, thus can be applied to a wide class of initial–boundary value problems.

We illustrate below how the unified transform works by implementing it to evolution equations, starting with the heat equation.

9.2   The heat equation

For the heat equation, if one does not use a specialized transform, the analogue of Eqs. (9.1) and (9.3) will involve the following transforms of the functions $u(x,t)$, $u_0(x)$, $g_0(t) = u(0,t)$, and $g_1(t) = u_x(0,t)$, appearing on the boundary of the domain of Ω: $\hat{u}(k,t)$, $\hat{u}_0(k)$, $\tilde{g}_0(k^2,t)$, $\tilde{g}_1(k^2,t)$, where

$$\hat{u}(k,t) = \int_0^\infty e^{-ikx} u(x,t)\mathrm{d}x, \;\; \mathrm{Im}[k] \le 0, \tag{9.8a}$$

$$\hat{u}_0(k) = \int_0^\infty e^{-ikx} u_0(x)\mathrm{d}x, \;\; \mathrm{Im}[k] \le 0, \tag{9.8b}$$

$$\tilde{g}_0(k,t) = \int_0^t e^{k\tau} g_0(\tau)\mathrm{d}\tau, \;\; k \in \mathbb{C}, \tag{9.8c}$$

$$\tilde{g}_1(k,t) = \int_0^t e^{k\tau} g_1(\tau)\mathrm{d}\tau, \;\; k \in \mathbb{C}. \tag{9.8d}$$

In general, consider a PDF whose highest spatial derivative is of order n. Suppose that this equation admits the particular solution

$$e^{ikx - \omega(k)t}, \;\; k \in \mathbb{C}. \tag{9.9}$$

Then the analogue of Eqs. (9.1) and (9.3) will involve $\hat{u}(k,t)$, $\hat{u}_0(k)$, and $\tilde{g}_m(\omega,t), m = 0,1,\ldots,n-1$, where $g_m(\omega,t)$ is the mth derivative of $u(x,t)$ with respect to x, evaluated at $x = 0$.

For evolution equations the unified transform involves three steps. The second of these steps is the only one that involves complex variables. It turns out that for those problems that can be solved by the usual transform methods this step can be avoided. However, in this way we obtain a solution which has some of the disadvantages of the usual transforms. In particular, the solution is not uniformly convergent at $x = 0$; also, it cannot be easily evaluated by the standard numerical techniques. Despite these serious limitations, we have decided to present this "poor man's" version of the method, in order to help those readers who are not familiar with complex variables to grasp two of the three steps of the method.

Step 1: For the heat equation, the GR is given by

$$e^{k^2 t}\hat{u}(k,t) = \hat{u}_0(k) - \tilde{g}_1(k^2,t) - ik\tilde{g}_0(k^2,t), \ \text{Im}[k] \leq 0. \qquad (9.10)$$

This is consistent with the fact that $\exp[ikx - \omega(k)t] = \exp[ikx - k^2 t]$ solves the heat equation, with $\omega(k) = k^2$. There are several ways of deriving Eq. (9.10). One of them uses the same steps as those used for the derivation of Eq. (9.3): We start by taking a derivative with respect to t of the half-line Fourier transform $\hat{u}(k,t)$ and then use the relation $u_t = u_{xx}$:

$$\frac{\partial \hat{u}}{\partial t} = \int_0^\infty e^{-ikx}u_t(x,t)\mathrm{d}x = \int_0^\infty e^{-ikx}u_{xx}(x,t)\mathrm{d}x.$$

We then perform an integration by parts to obtain

$$\int_0^\infty e^{-ikx}u_{xx}(x,t)\mathrm{d}x = -u_x(0,t) + ik\int_0^\infty e^{-ikx}u_{xx}(x,t)\mathrm{d}x.$$

Another integration by parts yields

$$\int_0^\infty e^{-ikx}u_{xx}(x,t)\mathrm{d}x = -u_x(0,t) + ik\left[-u(0,t) + ik\int_0^\infty e^{-ikx}u(x,t)\mathrm{d}x\right].$$

We have omitted the terms $u_x(x,t)e^{ikx}$ in the first step and $u(x,t)e^{ikx}$ in the second step evaluated at the upper limit of the integral because both expressions contain the factor $e^{k_I x}$ which vanishes when $x \to \infty$ for $k_I < 0$; in the case $k_I = 0$ we appeal to the fact that $u(x,t)$ and $u_x(x,t)$ must also vanish for $x \to \infty$ in order to have physically meaningful solutions. From the last result above, we find

$$\frac{\partial \hat{u}}{\partial t} + k^2\hat{u} = -u_x(0,t) - iku(0,t),$$

which is simply an ODE in the time variable. We can solve this ODE for $\hat{u}(k,t)$ using precisely the same steps used for the solution of Eq. (8.36) which yielded the solution of Eq. (8.38); in particular, we employ the integrating factor $\exp[k^2 t]$. This leads to the desired result, namely Eq. (9.10), where the functions $\hat{u}$, $\hat{u}_0$, $\tilde{g}_1$ and $\tilde{g}_0$ are the integral transforms defined in Eqs. (9.8a)–(9.8d).

Step 2: After obtaining the GR, the second step of the unified transform consists of deriving an integral representation of the solution involving a contour in the complex plane as opposed to one on the real axis. However, as noted earlier we will first avoid the complex plane. In particular, for the case of the heat equation, after inserting in the inverse Fourier transform formula,

$$u(x,t) = \int_{-\infty}^\infty e^{ikx}\hat{u}(k,t)\frac{\mathrm{d}k}{2\pi}, \ \ 0 \leq x < \infty,$$

the expression for $\hat{u}(k,t)$ obtained from the GR, Eq. (9.10), we find

$$u(x,t) = \int_{-\infty}^{\infty} e^{ikx-k^2t}\hat{u}_0(k)\frac{dk}{2\pi}$$
$$- \int_{-\infty}^{\infty} e^{ikx-k^2t}\left[\tilde{g}_1(k^2,t) + ik\tilde{g}_0(k^2,t)\right]\frac{dk}{2\pi}. \tag{9.11}$$

Step 3: The representation Eq. (9.11) involves transforms of both $u(0,t)$ and $u_x(0,t)$, whereas only one of these functions (or their combination) can be prescribed as a boundary condition. Remarkably, it turns out that by using the global relation, the unknown boundary values can be eliminated. For this purpose, we observe that the transforms of the boundary values depend on k^2; thus, these transforms remain invariant if k is replaced by $-k$. The equation obtained from the GR by replacing k with $-k$ takes the form

$$e^{k^2t}\hat{u}(-k,t) = \hat{u}_0(-k) - \tilde{g}_1(k^2,t) + ik\tilde{g}_0(k^2,t), \quad \text{Im}[k] \geq 0. \tag{9.12}$$

Despite the occurrence of the unknown function $\hat{u}(-k,t)$, this equation *can* be used for the elimination of one of the two transforms of the boundary values; indeed, the term $\hat{u}(-k,t)$ does not contribute to the solution representation. For example, in the case that $g_0(t)$ is given, solving Eq. (9.12) for $\tilde{g}_1$ and then substituting the resulting expression in Eq. (9.11) we find the expression

$$u(x,t) = \int_{-\infty}^{\infty} e^{ikx-k^2t}\hat{u}_0(k)\frac{dk}{2\pi} + \int_{-\infty}^{\infty} e^{ikx}\hat{u}(-k,t)\frac{dk}{2\pi}$$
$$- \int_{-\infty}^{\infty} e^{ikx-k^2t}[\hat{u}_0(-k) + 2ik\tilde{g}_0(k^2,t)]\frac{dk}{2\pi}.$$

The second integral on the right-hand side vanishes in view of Eq. (9.5). The resulting equation

$$u(x,t) = \int_{-\infty}^{\infty} e^{ikx-k^2t}\hat{u}_0(k)\frac{dk}{2\pi}$$
$$- \int_{-\infty}^{\infty} e^{ikx-k^2t}\left[\hat{u}_0(-k) + 2ik\tilde{g}_0(k^2,t)\right]\frac{dk}{2\pi}, \tag{9.13}$$

expresses $u(x,t)$ in terms of the half-Fourier transform of $u_0(x)$ and of the t-transform of $g_0(t)$. It is straightforward to rewrite this equation in the form obtained through the sine-transform (see Problem 1).

For problems which cannot be solved by the usual transforms, it is necessary to express the solution along a contour in the complex plane. Furthermore, even for problems which *can* be solved in the standard way, it is advantageous, both for analytical and numerical considerations, to deform part of the real axis to the complex plane.

Step 2 revisited: Throughout this chapter, we will often appeal to the notion of contour deformation, whereby the contour of a given integral is deformed to a different, more convenient one. For instance, the contour $\partial\mathcal{D}^+$ can be deformed to the contour C^+ shown in Fig. 9.1, which lies entirely in the region between the real axis and $\partial\mathcal{D}^+$.

We next show that the second integral in Eq. (9.11) can be deformed from the real axis to the integral along the contour in the k-complex plane consisting of the union of the rays at angles $\pi/4$ and $3\pi/4$, with the orientation shown

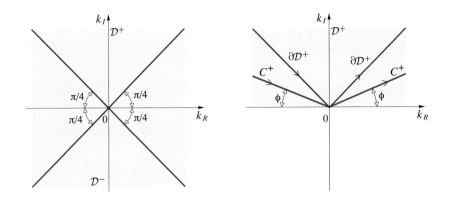

Figure 9.1: **Left:** The domains $\mathcal{D}^+$ and $\mathcal{D}^-$ (light-blue shaded regions) which lie in the upper half and the lower half of the k-complex plane, respectively. **Right:** The boundary $\partial\mathcal{D}^+$ (bold blue lines) of the domain $\mathcal{D}^+$ and the path C^+ (bold purple lines) that lies in the light-red shaded region. C^+ is the deformations of the $\partial\mathcal{D}^+$ boundary toward the real axis from above; for simplicity, it can be chosen as the union of two rays at an angle ϕ with the real axis.

in the contour of Fig. 9.1. For reasons to be explained later, we will denote this contour by $\partial\mathcal{D}^+$.

The contour $\partial\mathcal{D}^+$ is defined by $\mathrm{Re}[k^2] = 0, \mathrm{Im}[k] \geq 0$. Using in this definition, $k = |k|\exp[i\phi]$, we find $\{\cos(2\phi) = 0, 0 \leq \phi \leq \pi\}$, or $\phi = \pi/4$ and $3\pi/4$, as shown in Fig. 9.1.

For problems on the half-line, we only need to consider contours in the $\mathcal{D}^+$ domain. However, for problems on the finite interval we also need to consider contours in its symmetric domain $\mathcal{D}^-$, which lies in the lower half k-complex plane, shown in Fig. 9.1.

It turns out that Eq. (9.11) can be replaced by the equation

$$u(x,t) = \int_{-\infty}^{\infty} e^{ikx - k^2 t} \hat{u}_0(k) \frac{\mathrm{d}k}{2\pi}$$
$$- \int_{\partial\mathcal{D}^+} e^{ikx - k^2 t} [\tilde{g}_1(k^2, t) + ik\tilde{g}_0(k^2, t)] \frac{\mathrm{d}k}{2\pi}. \tag{9.14}$$

Proving this result is the only difficult step of the new method; it requires the use of Jordan's lemma. For this purpose, we write the second integral of the right-hand side of Eq. (9.13) in the form

$$\int_{-\infty}^{\infty} e^{ikx} G(k;t) \frac{\mathrm{d}k}{2\pi}, \quad G(k;t) = \int_0^t e^{-k^2(t-\tau)} g_1(\tau)\mathrm{d}\tau + ik \int_0^t e^{-k^2(t-\tau)} g_0(\tau)\mathrm{d}\tau.$$

The notation $G(k;t)$ emphasizes that in what follows, we will consider G a function of k treating t as a parameter.

According to Jordan's lemma, it is possible to deform the contour from the real axis to $\partial\mathcal{D}^+$ provided that G vanishes as $|k| \to \infty$ in the light-red shaded domain in Fig. 9.1. This is indeed the case: since $t - \tau > 0$, it follows that $\exp[-k^2(t-\tau)]$ decays provided that $\mathrm{Re}[k^2] > 0$, or $\cos(2\theta) > 0$. Hence, $-\pi/2 < 2\theta < \pi/2$ and $3\pi/2 < 2\theta < 5\pi/2$. This implies that G decays in the light-red shaded domain in Fig. 9.1. If k is on $\partial\mathcal{D}^+$ then the relevant exponential oscillates instead of decaying. However, G still decays. This follows directly from a general mathematical result known as the Riemann-Lebesgue lemma. It can also be verified by using integration by parts: noting that the integral of $\exp(k^2\tau)$ with respect to τ is $\exp(k^2\tau)/k^2$, it follows that the leading term of the second integral in G is given by

$$ik\left[\frac{g_0(t)}{k^2} - \frac{g_0(0)e^{-k^2 t}}{k^2}\right];$$

for k^2 purely imaginary the above decays linearly in $1/k$. Similarly, the leading term of the first integral in G decays like $1/k^2$. Using Eq. (9.12) in Eq. (9.14), we now find the expression

$$u(x,t) = \int_{-\infty}^{\infty} e^{ikx-k^2t} \hat{u}_0(k) \frac{dk}{2\pi} + \int_{\partial D^+} e^{ikx} \hat{u}(-k,t) \frac{dk}{2\pi}$$
$$- \int_{\partial D^+} e^{ikx-k^2t} [\hat{u}_0(-k) + 2ik\tilde{g}_0(k^2,t)] \frac{dk}{2\pi}.$$

The second integral on the right-hand side still vanishes. Indeed, both $\exp[ikx]$ and $\hat{u}(-k,t)$ decay exponentially in the domain D^+; thus, Cauchy's theorem in D^+ implies that this integral vanishes. The final expression for $u(x,t)$ is given by the formula

$$u(x,t) = \int_{-\infty}^{\infty} e^{ikx-k^2t} \hat{u}_0(k) \frac{dk}{2\pi}$$
$$- \int_{\partial D^+} e^{ikx-k^2t} [\hat{u}_0(-k) + 2ik\tilde{g}_0(k^2,t)] \frac{dk}{2\pi}. \qquad (9.15)$$

Remark: Suppose that the heat equation is defined for $0 < t < T$. Equation (9.15) can be rewritten in a form which is consistent with the Ehrenpreis fundamental principle:

$$u(x,t) = \int_{-\infty}^{\infty} e^{ikx-k^2t} \hat{u}_0(k) \frac{dk}{2\pi} - \int_{\partial D^+} e^{ikx-k^2t} [\hat{u}_0(-k) + 2ik\tilde{G}_0(k^2)] \frac{dk}{2\pi}, \qquad (9.16)$$

where $\tilde{G}_0(k^2)$ is defined by

$$\tilde{G}_0(k^2) = \int_0^T e^{k^2t} g_0(t) dt. \qquad (9.17)$$

This is a consequence of the following identity:

$$\int_{\partial D^+} e^{ikx-k^2t} k \tilde{g}_0(k^2,t) dk = \int_{\partial D^+} e^{ikx-k^2t} k \tilde{G}_0(k^2) dk$$
$$- \int_{\partial D^+} e^{ikx} F(k,t) dk, \qquad (9.18)$$

where $F(k,t)$ is defined as

$$F(k,t) = k \int_t^T e^{k^2(\tau-t)} g_0(\tau) d\tau.$$

The second integral on the right-hand side of Eq. (9.18) vanishes. This follows from the application of Jordan's lemma in the domain D^+ and from the fact that $F(k,t)$ vanishes as $k \to \infty$.

9.3 The general methodology of the unified transform

The three steps used above for the solution of the heat equation are applicable to a general evolution PDE.

Step 1. *Derive the global relation, namely the equation satisfied by the Fourier transform of the solution, $\hat{u}(k,t)$, on the half-line, $0 \le x < \infty$, for $\mathrm{Im}[k] \le 0$.*

Step 2. *By using the expression for $\hat{u}(k,t)$ obtained through the global relation, together with the inverse Fourier transform, derive an integral representation for the solution involving a contour in the complex plane as opposed to on the real axis. The definition of this contour, denoted by $\partial\mathcal{D}^+$ is*

$$\partial\mathcal{D}^+ = \{k \in \mathbf{C}, \ \text{Im}[k] \geq 0, \ \text{Re}[\omega(k)] = 0\}, \tag{9.19}$$

where $\omega(k)$ appears in Eq. (9.9). Actually, $\partial\mathcal{D}^+$ is the boundary of $\mathcal{D}^+$, which is the part of $\mathcal{D}$ in the upper k-complex plane with $\mathcal{D}$ defined by

$$\mathcal{D} = \{k \in \mathbf{C}, \ \text{Re}[\omega(k)] < 0\}. \tag{9.20}$$

Step 3. *By using the global relation and by employing the transformations in the k-complex plane which leave $\omega(k)$ invariant, it is possible to eliminate from the integral representation of $u(x,t)$ the transforms of the unknown boundary values.*

It is straightforward to show that if an evolution equation consists only of u_t and of a single spatial derivative of order n, then $\mathcal{D}$ consists of m subdomains in the lower half of the k-complex plane and $n - m$ subdomains in the upper half of the k-complex plane, where $m = n/2$ for n even and m is either $(n+1)/2$ or $(n-1)/2$ if n is odd. There exist $n - 1$ functions $\nu(k)$ which leave $\omega(k)$ invariant. These functions can be used to map the m subdomains of $\mathcal{D}$ in the lower half of the k-complex plane (where the GR is valid) to the $n - m$ subdomains of $\mathcal{D}$ in the upper half of the k-complex plane where $\partial\mathcal{D}^+$ is defined. In this way, in each subdomain in $\mathcal{D}^+$ the GR gives rise to m equations. Hence, since the solution representation involves n unknown boundary values, a well-posed problem requires the imposition of $n - m$ boundary conditions at $x = 0$. Similar considerations are valid even if the PDE involves spatial derivatives of lower order: in this case the above decomposition is still valid for large k.

The above discussion implies that the unified transform, in addition to providing effective integral representations for evolution PDEs with spatial derivatives of arbitrary order, it also provides a very simple geometrical approach for determining the number of boundary conditions needed at $x = 0$. Until the development of this new method, the determination of this number, even for a PDE with a third-order derivative, was considered a difficult problem requiring the machinery of rigorous PDE techniques.

Using the above three steps it is possible to obtain the analogue of Eqs. (9.15) and (9.16) for any initial–boundary value problem. It is important to note that the associated derivations involve the manipulation of the *unknown* function $u(x,t)$ and of the boundary values $g_m, m = 0, 1, \ldots n - 1$, some of which are *unknown*. Thus, this derivation is strictly formal, unless it is supplemented with a priori estimates of these unknown functions through rigorous PDE techniques. However, the final representation involves only given functions; furthermore, this representation is uniformly convergent on the boundaries and has explicit (x,t) dependence which occurs only in the form of $\exp[ikx + \omega(k)t]$. Thus, it is straightforward to verify that the relevant integral representation satisfies the given PDE as well as the prescribed initial and boundary conditions. This important feature is further discussed in the next section.

9.3.1 Advantages of the unified transform

1. *The unified transform constructs representations which are always uniformly convergent at the boundaries.*

2. *It is straightforward to verify that the relevant representation provides a solution of the given initial–boundary value problem.*

   For example, for the heat equation:

   (i) Since, the only (x, t) dependence of Eq. (9.16) is in the form of $e^{ikx - k^2 t}$, it is obvious that $u(x, t)$ satisfies the heat equation.

   (ii) Evaluating Eq. (9.16) at $t = 0$ we find

   $$u(x, 0) = \int_{-\infty}^{\infty} e^{ikx} \hat{u}_0(k) \frac{dk}{2\pi} - \int_{\partial \mathcal{D}^+} e^{ikx} \left[\hat{u}_0(-k) + 2ik\tilde{G}_0(k^2) \right] \frac{dk}{2\pi}.$$

   The function in the square bracket in the above integral is an analytic function of k which decays as $k \to \infty$ in $\mathcal{D}^+$. Thus, the application of Cauchy's theorem and Jordan's Lemma in $\mathcal{D}^+$ implies that the associated integral vanishes. Indeed, the function $\hat{u}_0(-k)$ decays exponentially for large $|k|$ in $\mathcal{D}^+$, whereas $\tilde{G}_0(k^2)$ decays exponentially in the interior of $\mathcal{D}^+$; furthermore, integrating by parts implies that $k\tilde{G}_0(k^2)$ decays linearly in $1/k$ on $\partial \mathcal{D}^+$. Hence,

   $$u(x, 0) = \int_{-\infty}^{\infty} e^{ikx} \hat{u}_0(k) \frac{dk}{2\pi} = u_0(x),$$

   in view of the inverse Fourier transform identity.

   (iii) Evaluating Eq. (9.16) at $x = 0$ we find

   $$u(0, t) = \int_{-\infty}^{\infty} e^{-k^2 t} \hat{u}_0(k) \frac{dk}{2\pi} - \int_{\partial \mathcal{D}^+} e^{-k^2 t} \hat{u}_0(-k) \frac{dk}{2\pi}$$
   $$- 2i \int_{\partial \mathcal{D}^+} k e^{-k^2 t} \tilde{G}_0(k^2) \frac{dk}{2\pi}. \qquad (9.21)$$

   By deforming the second integral on the right-hand side of Eq. (9.21) to the real axis and then replacing k with $-k$ we find that this integral cancels the first integral. Furthermore, letting $ik^2 = l$ in the third integral and using the inverse Fourier transform formula, we find

   $$u(0, t) = \int_{-\infty}^{\infty} e^{ilt} \left(\int_0^T e^{-il\tau} g_0(\tau) d\tau \right) \frac{dl}{2\pi} = g_0(t).$$

3. *The implementation of the unified transform does not require separation of the given domain and boundary conditions.*

   An example is discussed in the application of Chapter 10.

4. *The unified transform uses only the Fourier transform, thus there is no need for the complicated derivation of specialized transforms. Furthermore, for the particular cases that the traditional transforms exist, the unified transform provides a much simpler way for deriving these transforms.*

   In deriving (9.15), the real line was deformed to $\partial \mathcal{D}^+$. This deformation is always possible *before* using the global relation. However, *after* using the global relation we introduce $\hat{u}_0$ and then it is *not* always possible to return to the real axis. Actually, the cases where there *do* exist traditional transforms, are precisely the cases where this "return" is possible.

In the particular case of (9.15), we note that $\hat{u}_0(-k)$ is bounded and analytic in the upper half of the complex k plane, thus it *is* possible to return to the line axis:

$$u(x,t) = \int_{-\infty}^{\infty} e^{ikx-k^2t}[\hat{u}_0(k) - \hat{u}_0(-k)]\frac{dk}{2\pi} - 2i\int_{-\infty}^{\infty} ke^{ikx-k^2t}\tilde{g}_0(k^2,t)\frac{dk}{2\pi}.$$

Splitting the integral along $\mathbb{R}$ to an integral from $-\infty$ to 0 plus an integral from 0 to ∞, and letting $k \to -k$ in the former integral we obtain the representation obtained in Eq. (8.39) through the sine-transform.

If a traditional integral transform exists, there is a simple way to relate the unified transform to it. For example, in the particular case of the heat equation, the global relation together with the equation obtained from the global relation by replacing k with $-k$ are the following equations:

$$e^{k^2t}\hat{u}(k,t) = \hat{u}_0(k) - \tilde{g}_1 - ik\tilde{g}_0, \quad \text{Im}[k] \leq 0,$$

$$e^{k^2t}\hat{u}(-k,t) = \hat{u}_0(-k) - \tilde{g}_1 + ik\tilde{g}_0, \quad \text{Im}[k] \geq 0.$$

If k is real, then both these equations are valid. Hence if g_0 is given, we subtract the above equations and we obtain the equation for the sine-transform of $u(x,t)$. Similarly, if $u_x(0,t)$ is given we add the above equations and we obtain

$$e^{k^2t}\hat{u}_c(k,t) = \hat{u}_0(k) - \tilde{g}_1(k^2,t), \quad k \in \mathbb{R},$$

where $\hat{u}_c$ and $\hat{u}_0$ denote the cosine-transform of $u(x,t)$ and $u_0(x)$, respectively, namely:

$$\hat{u}_c(k,t) = \int_0^{\infty} \cos(kx)u(x,t)dx, \quad \hat{u}_0(k) = \int_0^{\infty} \cos(kx)u_0(x)dx.$$

The sine-transform pair can be obtained by evaluating Eq. (8.39) at $t = 0$. Thus, since Eq. (8.39) *can* be derived by the unified transform, it follows that the latter method provides a much easier approach for constructing transform pairs that the usual approach involving the construction of Green's function. This is further discussed in Section 9.5 and in the Application section.

5. *The unified transform constructs a representation that can be employed for a variety of boundary conditions.*

For example, in the case of the heat equation with $u(0,t)$ given, traditionally one employs the sine-transform, whereas if $u_x(0,t)$ is given one employs the cosine-transform. On the other hand, the integral representation (9.14) together with the global relation (9.12), immediately yield the solution if either $u(0,t)$ or $u_x(0,t)$ are given. Indeed, regarding the latter case, solving Eq. (9.12) for $ik\tilde{g}_0(k^2,t)$, substituting the resulting expression in Eq. (9.14) and noting that the contribution of the unknown term $\hat{u}(-k,t)$ vanishes, Eq. (9.14) becomes

$$u(x,t) = \int_{-\infty}^{\infty} e^{ikx-k^2t}\hat{u}_0(k)\frac{dk}{2\pi} - \int_{\partial D^+} e^{ikx-k^2t}[-\hat{u}_0(-k) + 2\tilde{g}_1(k^2,t)]\frac{dk}{2\pi}.$$

The case of the Robin boundary condition is discussed in Section 9.5.

6. *The unified transform constructs representations in the Ehrenpreis form.*

   It has already been noted that for convex domains the unified transform yields integral representations in the Ehrenpreis form. In this sense, the unified transform constructs the proper analogues of the beautiful representation for the solution of the initial–value problem obtained via the Fourier transform.

7. *The unified transform yields in a simple way the number of required boundary conditions at $x = 0$.*

   As illustrated in the derivation of the domain $\mathcal{D}$ for evolution equations with a third-order spatial derivative, this is based on very simple geometrical considerations.

8. *The unified transform can be applied to evolution equations with spatial derivatives of arbitrary order.*

   For example, it is straightforward to solve an initial–boundary value problem for the Stokes equation, discussed in Section 9.4, for which as stated earlier there does not exist a standard x-transform. Remarkably, it can be shown that there exist physically significant evolution PDEs involving only second-order derivatives, which *cannot* be solved by a standard x-transform. An example is provided by the simple PDE

   $$u_t = u_{xx} + \beta u_x, \quad \beta > 0, \quad 0 < x < \infty, \quad t > 0,$$

   which is discussed in the derivation below.

9. *It is straightforward to compute the solution numerically.*

   For simplicity, we concentrate on the case that the relevant transforms can be computed analytically. As an example, we consider the following choices for $u_0(x)$ and $g_0(t)$:

   $$u_0(x) = e^{-ax}, \ 0 < x < \infty; \quad g_0(t) = \cos(bt), \ t > 0; \ a, b > 0, \quad (9.22a)$$

   which give the corresponding transforms $\hat{u}_0(k)$, $\tilde{g}_0(k, t)$ given below:

   $$\hat{u}_0(k) = \frac{1}{ik + a}, \quad \tilde{g}_0(k, t) = \frac{1}{2} \left[\frac{e^{(k+ib)t} - 1}{k + ib} + \frac{e^{(k-ib)t} - 1}{k - ib} \right]. \quad (9.22b)$$

   Then, Eq. (9.15) becomes

   $$u(x, t) = \int_{-\infty}^{\infty} \frac{e^{ikx - k^2 t}}{ik + a} \frac{dk}{2\pi}$$
   $$- \int_{\partial \mathcal{D}^+} e^{ikx - k^2 t} \left[\frac{1}{-ik + a} + ik \frac{e^{(k^2 + ib)t} - 1}{k^2 + ib} + ik \frac{e^{(k^2 - ib)t} - 1}{k^2 - ib} \right] \frac{dk}{2\pi}.$$
   $$(9.23)$$

   The term $\exp[ikx - k^2 t]$ is analytic in $\mathcal{D}^+$; thus, the first integral in the right-hand side of Eq. (9.23) can be replaced with an integral involving $\partial \mathcal{D}^+$ instead of the real axis. For k on $\partial \mathcal{D}^+$, the term e^{ikx} decays exponentially as $|k| \to \infty$. Also by deforming $\partial \mathcal{D}^+$ to C^+, where C^+ is a contour between the real axis and $\partial \mathcal{D}^+$, it follows that for k on C^+ the term $e^{-k^2 t}$ also decays exponentially as $|k| \to \infty$. After deforming the real axis to the contour C^+ for the first integral as well , Eq. (9.23) becomes

$$u(x,t) = \int_{C^+} e^{ikx - k^2 t} \left[\frac{1}{ik+a} + \frac{1}{ik-a} + ik\left(\frac{1}{k^2+ib} + \frac{1}{k^2-ib} \right) \right] \frac{dk}{2\pi}$$
$$- \int_{C^+} e^{ikx} ik \left(\frac{e^{ibt}}{k^2+ib} + \frac{e^{-ibt}}{k^2-ib} \right) \frac{dk}{2\pi}. \tag{9.24}$$

We denote the first integral in the right-hand side of (9.24) by $u_1(x,t)$ and the second by $u_2(x,t)$. The above integrals can be computed efficiently using elementary numerical techniques, as shown in several examples in this chapter. However, if x is very small, the integral in u_2 loses its exponential decay for large k of the term $\exp[ikx]$, thus its numerical integration becomes problematic. In contrast, the integral u_1 can be computed efficiently even for small values of x due to the exponential decay of the term $\exp[-k^2 t]$.

The above difficulty can be bypassed in an elementary way: the integral which is harder to compute numerically can be computed analytically for all x; whereas the integral which cannot be computed analytically, is easily computed numerically. Indeed, using the Residue theorem in the domain above C^+, and noting that the terms $k^2 + ib$ and $k^2 - ib$ vanish in the above domain, respectively, at

$$k = \sqrt{b}\, e^{3i\pi/4} = -\sqrt{b}\, e^{-i\pi/4} \quad \text{and} \quad k = \sqrt{b}\, e^{i\pi/4},$$

we find

$$u_2(x,t) = \frac{-2\pi i}{2\pi} \frac{i}{2} \left(e^{ibt - ix\sqrt{b/2}(1-i)} + e^{-ibt + ix\sqrt{b/2}(1+i)} \right)$$
$$= e^{-x\sqrt{b/2}} \cos\left(bt - x\sqrt{b/2} \right).$$

Hence,

$$u(x,t) = u_1(x,t) + e^{-x\sqrt{b/2}} \cos\left(bt - x\sqrt{b/2} \right), \tag{9.25}$$

where $u_1(x,t)$ denotes the first integral in the right-hand side of Eq. (9.24).

Derivation: The unified transform for the Black–Scholes equation
Here we derive the unified transform for the Black–Scholes equation which was introduced in the Application section of Chapter 8, see Section 8.8. The solution in terms of the unified transform is pertinent to point 8 made above.

The final version of the Black–Scholes equation reads exactly like the equation mentioned in point 8, namely $u_t = u_{xx} + \beta u_x$, see Eq. (8.143). We will focus on the call option, for which the range of values of the variable x for the option to be *"in the money"* is $x > 0$ (see Section 8.8). Thus, the corresponding domain is

$$\Omega = \{0 < x < \infty, \ t > 0\}.$$

In this case, the initial condition is as follows:

$$u(x,0) = K(e^{\sigma x/\sqrt{2}} - 1)\phi(e^{\sigma x/\sqrt{2}} - 1), \quad \sigma > 0.$$

We will assume that the relevant ranges of the parameters ρ and σ are such that $\beta > 0$. To derive the unified transform, we follow the three simple steps discussed earlier:

Step 1. Employing the Fourier transform on the half-line we obtain the global relation:

$$e^{(k^2 - i\beta k)t}\hat{u}(k, t) = \hat{u}_0(k) - \tilde{g}_1 - (ik + \beta)\tilde{g}_0, \quad \text{Im}[k] \leq 0, \tag{9.26}$$

where $\hat{u}_0$ is defined in Eq. (9.8) and $\tilde{g}_1$, $\tilde{g}_0$ are functions of $\omega(k) = k^2 - i\beta k$ and t. The above expression is consistent with the fact that $\exp[ikx - (k^2 - i\beta k)t]$ is a particular solution of Eq. (8.143).

Step 2. Taking into consideration that

$$\text{Re}[\omega(k)] < 0 \Rightarrow \left(k_R^2 - k_I^2 + \beta k_I\right) < 0,$$

it follows that

$$\mathcal{D} = \left\{k \in \mathbf{C}, \ k_R^2 - k_I^2 + \beta k_I < 0\right\}. \tag{9.27}$$

The equation $k_R^2 - k_I^2 + \beta k_I = 0$ implies that if $k_R = 0$ then $k_I = 0$ or $k_I = \beta$, as seen from the diagram below.

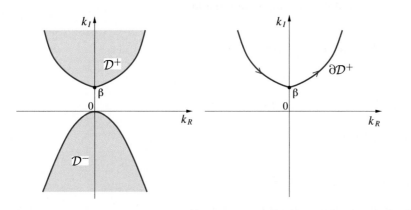

Thus,

$$u(x, t) = \int_{-\infty}^{\infty} e^{ikx - (k^2 - i\beta k)t}\hat{u}_0(k)\frac{dk}{2\pi}$$

$$- \int_{\partial \mathcal{D}^+} e^{ikx - (k^2 - i\beta k)t}\left[\tilde{g}_1 + (ik + \beta)\tilde{g}_0\right]\frac{dk}{2\pi}. \tag{9.28}$$

Step 3. Let $\nu(k)$ be the transformations which leave the expression $k^2 - i\beta k$ invariant, that is,

$$k^2 - i\beta k = \nu^2 - i\beta\nu, \quad \text{or} \quad (k - \nu)(k + \nu - i\beta) = 0.$$

Hence,

$$\nu(k) = -k + i\beta. \tag{9.29}$$

Thus, replacing k by $\nu(k)$ in the global relation Eq. (9.26) we find

$$e^{(k^2 - i\beta k)t}\hat{u}(-k + i\beta, t) = \hat{u}_0(-k + i\beta) - \tilde{g}_1 + ik\tilde{g}_0, \tag{9.30}$$

where we have used the computation

$$-[i\nu(k) + \beta] = -i(-k + i\beta) - \beta = ik.$$

We distinguish two cases, depending on the boundary conditions supplied:

(i) $u(0,t)$ is given
Solving Eq. (9.30) for $\tilde{g}_1$ and then substituting the resulting expression in Eq. (9.28) we find

$$u(x,t) = \int_{-\infty}^{\infty} e^{ikx-(k^2-i\beta k)t}\hat{u}_0(k)\frac{dk}{2\pi} + \int_{\partial\mathcal{D}^+} e^{ikx}\hat{u}(-k+i\beta,t)\frac{dk}{2\pi}$$
$$- \int_{\partial\mathcal{D}^+} e^{ikx-(k^2-i\beta k)t}\left[\hat{u}_0(-k+i\beta) + (2ik+\beta)\tilde{g}_0(k^2-i\beta k,t)\right]\frac{dk}{2\pi}.$$

We note that $\hat{u}(-k+i\beta,t)$ is given by

$$\hat{u}(-k+i\beta,t) = \int_0^{\infty} e^{-i(-k+i\beta)x}u(x,t)dx, \tag{9.31}$$

and its dependence on k is only through the complex exponential $\exp[ikx]$. Moreover,

$$\text{Re}[i(k-i\beta)] = -(k_I - \beta) < 0 \text{ in } \mathcal{D}^+,$$

and therefore the exponential that appears in the integral of Eq. (9.31) decays in $\mathcal{D}^+$ for all values of x in the domain Ω. These observations imply that $\hat{u}(-k+i\beta,t)$ decays exponentially in $\mathcal{D}^+$. Hence, Cauchy's theorem and Jordan's lemma in $\mathcal{D}^+$ shows that contribution of this integral is zero, leaving as the solution

$$u(x,t) = \int_{-\infty}^{\infty} e^{ikx-(k^2-i\beta k)t}\hat{u}_0(k)\frac{dk}{2\pi}$$
$$- \int_{\partial\mathcal{D}^+} e^{ikx-(k^2-i\beta k)t}\left[\hat{u}_0(-k+i\beta) + (2ik+\beta)\tilde{g}_0(k^2-i\beta k,t)\right]\frac{dk}{2\pi}. \tag{9.32}$$

(ii) $u_x(0,t)$ is given
Solving Eq. (9.30) for $\tilde{g}_0$ and then substituting the resulting expression in Eq. (9.28) we find

$$u(x,t) = \int_{-\infty}^{\infty} e^{ikx-(k^2-i\beta k)t}\hat{u}_0(k)\frac{dk}{2\pi} + \int_{\partial\mathcal{D}^+} e^{ikx}\left(\frac{i\beta}{k}-1\right)\hat{u}(-k+i\beta,t)\frac{dk}{2\pi}$$
$$- \int_{\partial\mathcal{D}^+} \frac{e^{ikx-(k^2-i\beta k)t}}{k}$$
$$\times \left[(i\beta-k)\hat{u}_0(-k+i\beta) + (2k-i\beta)\tilde{g}_1(k^2-i\beta k,t)\right]\frac{dk}{2\pi}.$$

We can establish that the second integral gives a vanishing contribution by the same considerations as in the previous case. The singularity in the integrand of this integral for $k = 0$ does not affect the result because this point is outside the domain of integration. The third integral also has a singularity at $k = 0$, but this point is not on the curve $\partial\mathcal{D}^+$. The solution takes the form:

$$u(x,t) = \int_{-\infty}^{\infty} e^{ikx-(k^2-i\beta k)t} \hat{u}_0(k) \frac{dk}{2\pi}$$

$$- \int_{\partial D^+} \frac{e^{ikx-(k^2-i\beta k)t}}{k}$$

$$\times \left[(i\beta - k)\hat{u}_0(-k+i\beta) + (2k - i\beta)\tilde{g}_1(k^2 - i\beta k, t)\right] \frac{dk}{2\pi}. \quad (9.33)$$

With regard to the solutions of Eqs. (9.32) and (9.33), we note the following:

1. Analyticity considerations imply that it is straightforward to rewrite the representations of Eqs. (9.32) and (9.33) in the Ehrenpreis form: $\tilde{g}_0(k^2 - i\beta k, t)$ and $\tilde{g}_1(k^2 - i\beta k, t)$ can be replaced by $\tilde{G}_0(k^2 - i\beta k)$ and $\tilde{G}_1(k^2 - i\beta k)$.

2. Both representations of Eqs. (9.32) and (9.33) involve the function $\hat{u}_0(-k+i\beta k)$ which is defined by

$$\hat{u}_0(-k+i\beta) = \int_0^{\infty} e^{-i(-k+i\beta)x} u_0(x)dx = \int_0^{\infty} e^{ik_R x} e^{(\beta-k_I)x} u_0(x)dx.$$

Unless $u_0(x)$ is exponentially decaying, the function $\hat{u}_0(-k+i\beta k)$ is well-defined only for $k_I \geq \beta$. Thus, in this case we *cannot* deform the curve ∂D^+ back to the real axis. This implies that in spite of the fact that the highest order spatial-derivative of Eq. (8.143) is only 2, there does *not* exist for this simple equation a traditional x-transform!

Example 9.1: Consider the heat equation on the positive half-line with $u_0(x) = e^{-x}$, $g_0(t) = u(0,t) = \cos(t)$. We note that $u(0,0) = 1 = u_0(0)$, thus the given initial and boundary conditions are compatible at $x = t = 0$. The unified transform is applicable even if the above compatibility is violated but in this case the solution is not smooth at the origin.

The above choice of $u_0(x)$ and $g_0(t)$ corresponds to the initial and boundary conditions of Eq. (9.22a) with $a = b = 1$. We first verify the validity of the initial and boundary conditions. Equation (9.25) evaluated at $x = 0$ yields

$$u(0,t) = u_1(0,t) + \cos(bt).$$

It suffices to show that $u_1(0,t) = 0$. Indeed, this is clear by deforming the contour C^+ to the real line and observing that the integrand is an odd function of k. As far as the initial condition at $t = 0$ is concerned, Eq. (9.24) evaluated at $t = 0$ yields

$$u(x,0) = \int_{C^+} e^{ikx} \left(\frac{1}{ik+a} + \frac{1}{ik-a}\right) \frac{dk}{2\pi}.$$

Applying the Residue theorem for the pole at $k = ia$ leads to

$$u(x,0) = \frac{2\pi i}{2\pi} e^{i(ia)x} \frac{1}{i} = e^{-ax},$$

which is indeed the initial condition $u_0(x) = e^{-x}$, since $a = 1$.

We can then use numerical integration to evaluate the integral $u_1(x,t)$, which for the present case takes the form

$$u_1(x,t) = \int_{C^+} V(k,x,t)\,dk,$$

with

$$V(k,x,t) = -\frac{ie^{ikx-k^2t}}{2\pi}\left[\left(\frac{1}{k-i} + \frac{1}{k+i}\right) - k\left(\frac{1}{k^2+i} + \frac{1}{k^2-i}\right)\right].$$

The contour C^+ can be any curve between the real axis and $\partial\mathcal{D}^+$. We choose C^+ to be the union of the rays $\mathrm{Arg}[k] = \pi/8$ and $\mathrm{Arg}[k] = 7\pi/8$, with the orientation shown in Fig. 9.1, which correspond to $\phi = \pi/8$. This choice is motivated by the fact that $\exp[ikx]$ oscillates on the real axis, whereas $\exp[-k^2t]$ oscillates on $\partial\mathcal{D}^+$, and the contour C^+ is "half-way" between the above two contours. Thus, splitting the integral into two parts, using $k = r\exp[i\pi/8]$, $r \in (0,+\infty)$ in the first integral and $k = r\exp[i7\pi/8]$, $r \in (+\infty,0)$ in the second, we express the integral as

$$u_1(x,t) = \int_0^\infty \left[V\left(re^{i\pi/8},x,t\right)e^{i\pi/8} - V\left(re^{i7\pi/8},x,t\right)e^{i7\pi/8}\right]dr$$

and the full solution is given by the expression:

$$u(x,t) = e^{-x/\sqrt{2}}\cos\left(t - \frac{x}{\sqrt{2}}\right) + u_1(x,t).$$

We note that the integral $u_1(x,t)$ takes real values, so we can take advantage of this fact to evaluate only the real contribution of the integrand $V(k,x,t)$, which reduces the computational time significantly.

The plots below show the behavior of $u(x,t)$ for $x \in [0,2]$ and various values of t, including the short-time behavior (left panel) and the longer-time behavior (right panel), with different colored lines corresponding to different t values as indicated by the legend; the initial–boundary condition $u(x,0) = u_0(x)$ is marked by a black dashed line.

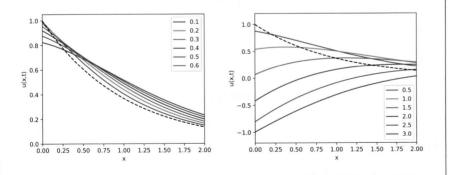

The solution $u(x,t)$ in the range $x \in [0,3]$ and $t \in [0,2\pi]$ is presented a three-dimensional perspective plot generated using Mathematica:

```
Plot3D [ exp(−x/√2) cos(t − x/√2)
  +NIntegrate [ Re[ V (re^{iπ/8}, x, t) e^{iπ/8} − V (re^{i7π/8}, x, t) e^{i7π/8} ],
    {r, 0, +∞} ], {x, 0, 3}, {t, 0, 2π} ]
```

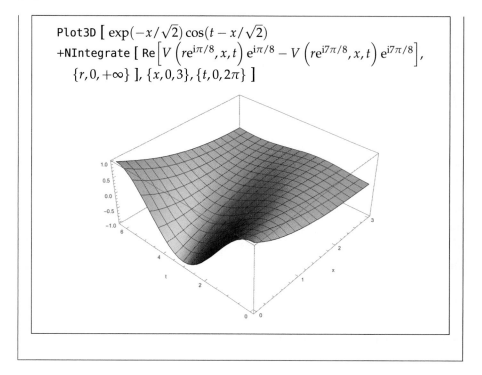

9.4   A PDE with a third-order spatial derivative

In this section, we take a closer look at evolution equations that involve a third-order spatial derivative. We first derive the domain $\mathcal{D}$ for several such equations, and then discuss the unified transform for one case in detail.

Derivation: Domain $\mathcal{D}$ for PDEs with a third-order spatial derivative

(A) As the first case, we consider the equation

$$u_t + u_{xxx} = 0. \tag{9.34}$$

Looking for a solution in the form $\exp[ikx − \omega(k)t]$, we find

$$\omega(k) = −ik^3. \tag{9.35}$$

The condition $\mathrm{Re}[\omega(k)] < 0$, which defines $\mathcal{D}$, implies

$$\mathrm{Re}[ik^3] > 0 \Rightarrow −|k|^3 \sin(3\phi) > 0 \Rightarrow \sin(3\phi) < 0,$$

where we have used the expression of k in terms of polar coordinates, $k = |k| \exp[i\phi]$. Hence,

$$\pi < 3\phi < 2\pi, \;\; 3\pi < 3\phi < 4\pi, \;\; 5\pi < 3\phi < 6\pi.$$

Thus, the allowed intervals for the argument ϕ are as follows:

$$\frac{\pi}{3} < \phi < \frac{2\pi}{3}, \;\; \pi < \phi < \frac{4\pi}{3}, \;\; \frac{5\pi}{3} < \phi < 2\pi.$$

These domains are shown as blue shaded areas in the diagram below. The diagram also shows the domain boundary $\partial\mathcal{D}^+$ as a thick blue line.

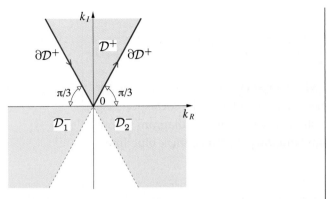

For future use we note that an alternative way to determine $\mathcal{D}$ is to use $k = k_R + ik_I$ instead of $k = |k| \exp[i\phi]$:

$$\text{Re}\left[i\left(k_R + ik_I\right)^3\right] > 0 \Rightarrow k_I\left(k_I^2 - 3k_R^2\right) > 0. \tag{9.36}$$

(B) For the second case, we consider the equation

$$u_t - u_{xxx} = 0. \tag{9.37}$$

In this case, looking for a solution in the form $\exp[ikx - \omega(k)t]$ we find

$$\omega(k) = ik^3. \tag{9.38}$$

By similar steps as in case (A), we find that the condition $\sin(3\phi) > 0$ yields

$$0 < \phi < \frac{\pi}{3}, \quad \frac{2\pi}{3} < \phi < \pi, \quad \frac{4\pi}{3} < \phi < \frac{5\pi}{3}.$$

These domains are shown as blue shaded areas in the diagram below. The diagram also shows the domain boundary $\partial \mathcal{D}^+$ as a thick blue line.

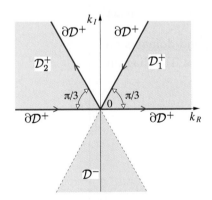

(C) For the next case we consider the equation

$$u_t + u_x + u_{xxx} = 0, \tag{9.39}$$

which is known as the "Stokes equation". In this case, looking for a solution in the form $\exp[ikx - \omega(k)t]$ leads to

$$\omega(k) = ik - ik^3. \tag{9.40}$$

Hence, using the representation $k = k_R + ik_I$,

$$\text{Re}\left[\mathrm{i}\left(k_R + ik_I\right)^3 - \mathrm{i}\left(k_R + ik_I\right)\right] > 0 \Rightarrow k_I(k_I^2 - 3k_R^2 + 1) > 0. \qquad (9.41)$$

If $k_I = 0$, then $k_I^2 - 3k_R^2 + 1 = 0$, which implies $k_R = \pm 1/\sqrt{3}$. Also, for large values of $|k|$, Eq. (9.41) becomes Eq. (9.36). Hence, the corresponding domains are those shown as blue shaded regions in the diagram below. The diagram also shows the domain boundary $\partial \mathcal{D}^+$ as a thick blue line.

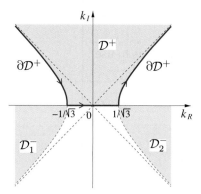

(D) As the final case, we consider the equation

$$u_t - u_x - u_{xxx} = 0. \qquad (9.42)$$

In this case, looking for a solution in the form $\exp[ikx - \omega(k)t]$ leads to

$$\omega(k) = -\mathrm{i}k + \mathrm{i}k^3. \qquad (9.43)$$

Hence, using the representation $k = k_R + ik_I$,

$$\text{Re}\left[-\mathrm{i}\left(k_R + ik_I\right)^3 + \mathrm{i}\left(k_R + ik_I\right)\right] > 0 \Rightarrow k_I(k_I^2 - 3k_R^2 + 1) < 0. \qquad (9.44)$$

By similar steps as in case (C), we find that the correspond domains are those shown as blue shaded regions in the diagram below. The diagram also shows the domain boundary $\partial \mathcal{D}^+$ as a thick blue line.

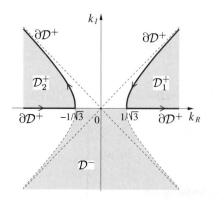

For Eqs. (9.34) and (9.39), there exist two subdomains, $\mathcal{D}_1^-$ and $\mathcal{D}_2^-$ in the lower half k-complex plane. Thus, for these PDEs an initial–boundary value problem defined for $0 < x < \infty$ requires only one boundary condition at $x = 0$. In contrast to the above situation, Eqs. (9.37) and (9.42) require two boundary conditions at $x = 0$, since for these PDEs there exists only one subdomain in the lower half k-complex plane.

For Eq. (9.37), in order to derive an integral representation in the k-complex plane, we use the fact that the integral of the relevant quantity vanishes in the union of the rays with $\text{Arg}[k] = \pi/3$ and $\text{Arg}[k] = 2\pi/3$. This follows from applying Cauchy's theorem in the domain bounded by the above rays.

We next discuss the unified transform for case (A), Eq. (9.34), namely the PDE

$$u_t + u_{xxx} = 0, \quad 0 < x < \infty, \quad t > 0.$$

The Stokes equation, Eq. (9.39), can be solved in a similar manner.

Step 1. Employing the Fourier transform, in analogy with Eq. (9.10) we find

$$e^{-ik^3 t}\hat{u}(k,t) = \hat{u}_0(k) + \tilde{g}_2(-ik^3, t) + ik\tilde{g}_1(-ik^3, t) - k^2\tilde{g}_0(-ik^3, t), \quad \text{Im}[k] \leq 0,$$
$$(9.45)$$

where $\hat{u}_0(k)$, as well as $\tilde{g}_0$ and $\tilde{g}_1$ are defined in Eq. (9.8) and

$$\tilde{g}_2(k,t) = \int_0^t e^{k\tau} u_{xx}(0,\tau) d\tau, \quad k \in \mathbb{C}. \tag{9.46}$$

Equation (9.45) is consistent with the fact that, for Eq. (9.34), $\omega(k) = -ik^3$, see Eq. (9.35).

Step 2.

$$u(x,t) = \int_{-\infty}^{\infty} e^{ikx + ik^3 t} \hat{u}_0(k) \frac{dk}{2\pi}$$
$$+ \int_{\partial\mathcal{D}^+} e^{ikx + ik^3 t} \left[\tilde{g}_2(-ik^3, t) + ik\tilde{g}_1(-ik^3, t) - k^2\tilde{g}_0(-ik^3, t) \right] \frac{dk}{2\pi},$$
$$(9.47)$$

where the domain $\mathcal{D}^+$ is the part of $\mathcal{D}$ that lies on the upper half k-complex plane and $\partial\mathcal{D}^+$ is its boundary. This is case (A) discussed in the derivation of the domain $\mathcal{D}$ for evolution equations with a third spatial derivative, where the explanation was given of the derivation of how $\mathcal{D}^+$ is defined.

Step 3. Let $\nu(k)$ be the transformations which leaves ik^3 invariant, that is, $ik^3 = i[\nu(k)]^3$. In the following, we will find it convenient to introduce the constant α, defined by

$$\alpha \equiv e^{i2\pi/3} \Rightarrow \alpha^2 = e^{i4\pi/3}, \quad \alpha^3 = 1, \tag{9.48}$$

for which the following identity holds:

$$1 + \alpha + \alpha^2 = 0, \tag{9.49}$$

as can be easily checked.

The transformations that leave ik^3 invariant are as follows:

$$\nu_1(k) = \alpha k, \quad \nu_2(k) = \alpha^2 k. \tag{9.50}$$

If k is in $\mathcal{D}^+$ then $\nu_1(k)$ is in $\mathcal{D}_1^-$ and $\nu_2(k)$ is in $\mathcal{D}_2^-$. Hence, if we replace in the global relation (9.45) k by $\nu_1(k)$ and by $\nu_2(k)$ we obtain two equations which are valid for k in $\mathcal{D}^+$:

$$e^{-ik^3t}\hat{u}(\alpha k, t) = \hat{u}_0(\alpha k) + \tilde{g}_2 + i\alpha k\tilde{g}_1 - \alpha^2 k^2 \tilde{g}_0, \quad k \in \mathcal{D}^+, \tag{9.51a}$$

$$e^{-ik^3t}\hat{u}(\alpha^2 k, t) = \hat{u}_0(\alpha^2 k) + \tilde{g}_2 + i\alpha^2 k\tilde{g}_1 - \alpha k^2 \tilde{g}_0, \quad k \in \mathcal{D}^+, \tag{9.51b}$$

where for convenience of notation we have suppressed the dependence of $\tilde{g}_0$, $\tilde{g}_1$ and $\tilde{g}_2$ on $-ik^3$ and t.

As with the earlier examples we expect that the contribution of the terms $\hat{u}(\alpha k, t)$ and $\hat{u}(\alpha^2 k, t)$ vanishes; thus, Eqs. (9.51a) and (9.51b) are two algebraic equations coupling $\tilde{g}_0$, $\tilde{g}_1$ and $\tilde{g}_2$. Hence, as stated in the derivation of the domain $\mathcal{D}$ for evolution equations with a third spatial derivative for a well-posed problem, one needs to specify only one boundary condition at $x = 0$.

For example, consider the initial and boundary conditions of Eq. (8.33). In this case, Eqs. (9.51a) and (9.51b) can be solved for the unknown functions $\tilde{g}_1$ and $\tilde{g}_2$, and then the resulting expressions can be substituted to the bracket of (9.47). Alternatively, we can compute directly this bracket as follows: we supplement Eqs. (9.51a) and (9.51b), with the equation

$$\tilde{g}(k, t) = \tilde{g}_2 + ik\tilde{g}_1 - k^2 \tilde{g}_0. \tag{9.52}$$

We multiply Eq. (9.51a) by α, Eq. (9.51b) by α^2 and add the resulting equations to Eq. (9.52):

$$\tilde{g}(k, t) = -e^{-ik^3t}\left[\alpha\hat{u}(\alpha k, t) + \alpha^2\hat{u}(\alpha^2 k, t)\right] + \alpha\hat{u}_0(\alpha k) + \alpha^2\hat{u}_0(\alpha^2 k) - 3k^2\tilde{g}_0, \tag{9.53}$$

where we have used the identity of Eq. (9.49). Substituting the above expression of $\tilde{g}(k, t)$ into (9.47) and noting that Cauchy's theorem in $\mathcal{D}^+$ implies that the contributions of the terms $\hat{u}(\alpha k, t)$ and $\hat{u}(\alpha^2 k, t)$ vanish, we find

$$u(x, t) = \int_{-\infty}^{\infty} e^{ikx + ik^3t}\hat{u}_0(k)\frac{dk}{2\pi}$$
$$+ \int_{\partial\mathcal{D}^+} e^{ikx + ik^3t}\left\{\left[\alpha\hat{u}_0(\alpha k) + \alpha^2\hat{u}_0(\alpha^2 k)\right] - 3k^2\tilde{g}_0(-ik^3, t)\right\}\frac{dk}{2\pi}. \tag{9.54}$$

The occurrence of the terms $\hat{u}_0(\alpha k)$ and $\hat{u}_0(\alpha^2 k)$ imply that the contour $\partial\mathcal{D}^+$ *cannot* be deformed back to the real axis, and hence there does *not* exist a traditional x-transform for this simple problem. Indeed, $\hat{u}_0(\alpha k)$ is bounded for $\text{Im}[\alpha k] \leq 0$; letting $k = |k|\exp(i\phi)$, this inequality becomes $\sin(2\pi/3 + \phi) \leq 0$, or $\pi \leq 2\pi/3 + \phi \leq 4\pi/3$. Hence, regarding $\hat{u}_0(\alpha k)$, it is possible to deform the ray $\text{Arg}[k] = 2\pi/3$ to the real axis but not the ray $\text{Arg}[k] = \pi/3$. Similar considerations are valid for $\hat{u}_0(\alpha^2 k)$.

Example 9.2: We consider Eq. (9.54) with the initial and boundary conditions of Eq. (9.22a) with $a = b = 1$. In order to achieve exponential decay in both $\exp[ikx]$ and $\exp[ik^3t]$ we deform both the real axis and $\partial\mathcal{D}^+$ to the contour C^+ defined by the union of the rays specified by $\text{Arg}[k] = \pi/6$ and $\text{Arg}[k] = 5\pi/6$, namely the path on the upper half of the k-complex plane corresponding to $\phi = \pi/6$ in Fig. 9.1. This choice is motivated by the fact that $\exp[ikx]$ oscillates on the real axis, whereas $\exp[ik^3t]$ oscillates on $\partial\mathcal{D}^+$, and the contour C^+ is "half-way" between the above two contours. With these considerations, and using the transforms of Eq. (9.22b), we find that Eq. (9.54) takes the form:

$$u(x,t) = \int_{C^+} V(k,x,t)dk - \frac{3i}{2}\int_{C^+} e^{ikx}k^2\left(\frac{e^{it}}{k^3-1} + \frac{e^{-it}}{k^3+1}\right)\frac{dk}{2\pi},$$

where

$$V(k,x,t) = -\frac{ie^{ikx+ik^3t}}{2\pi}\left[\left(\frac{1}{k-i} + \frac{\alpha}{\alpha k-i} + \frac{\alpha^2}{\alpha^2 k-i}\right) \right.$$
$$\left. -\frac{3}{2}k^2\left(\frac{1}{k^3-1} + \frac{1}{k^3+1}\right)\right],$$

with $\alpha = \exp[2\pi i/3]$. Employing the same procedure as in Example 9.1, the second integral in the above expression for $u(x,t)$ can be evaluated analytically. Indeed, observing the only poles in the domain enclosed by C^+ in the upper half-plane are

$$k_1 = e^{i2\pi/3}, \quad \text{and} \quad k_2 = e^{i4\pi/3},$$

and applying the Residue theorem we obtain

$$-\frac{3i}{2}\int_{C^+} e^{ikx}k^2\left(\frac{e^{it}}{k^3-1} + \frac{e^{-it}}{k^3+1}\right)\frac{dk}{2\pi} = e^{-\sqrt{3}x/2}\cos\left(t - \frac{x}{2}\right),$$

which leads to:

$$u(x,t) = \int_{C^+} V(k,x,t)dk + e^{-\sqrt{3}x/2}\cos\left(t - \frac{x}{2}\right).$$

We first verify that this expression is consistent with the initial and boundary conditions. Setting $x = 0$ in the above expression leads to

$$u(0,t) = \int_{C^+} V(k,0,t)dk + \cos(t).$$

It suffices to show that the first term of the left-hand side of the above expression vanishes. Indeed, the contribution of the integral on the ray $\text{Arg}[k] = \pi/6$ cancels the contribution of the integral on the ray $\text{Arg}[k] = 5\pi/6$. This follows from the change of variables $k = \alpha\lambda$, which maps the second ray to the first ray, and the property $\alpha V(\alpha k, 0, t) = V(k,0,t)$. On the other hand, evaluating the last expression for $u(x,t)$ at $t = 0$ yields

$$u(x,0) = -i\int_{C^+} e^{ikx}\left(\frac{1}{k-i} + \frac{\alpha}{\alpha k-i} + \frac{\alpha^2}{\alpha^2 k-i}\right)\frac{dk}{2\pi}.$$

Then, the Residue theorem applied to this contour integral yields

$$u(x,0) = -i\frac{2\pi i}{2\pi}e^{-ix} = e^{-x},$$

which is the initial condition $u_0(x)$.

For the numerical evaluation of the integral we follow similar steps as those in Example 9.1. Specifically, we split the integral into two parts, using $k = r\exp[i\pi/6]$, $r \in (0, +\infty)$ in the first integral and $k = r\exp[i5\pi/6]$, $r \in (+\infty, 0)$ in the second, so that the full solution is given by the expression:

$$u(x, t) = e^{-\sqrt{3}x/2} \cos\left(t - \frac{x}{2}\right)$$
$$+ \int_0^\infty \left[V\left(re^{i\pi/6}, x, t\right) e^{i\pi/6} - V\left(re^{i5\pi/6}, x, t\right) e^{i5\pi/6} \right] dr.$$

The numerical evaluation of the solution leads to the results shown below for $x \in [0, 2]$ and for various values of t for short timescales (left panel) and for longer timescales (right panel), with the initial–boundary condition shown by a black dashed line.

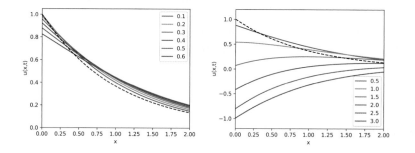

The solution $u(x, t)$ in the range $x \in [0, 3]$ and $t \in [0, 2\pi]$ is presented a three-dimensional perspective plot generated using Mathematica:

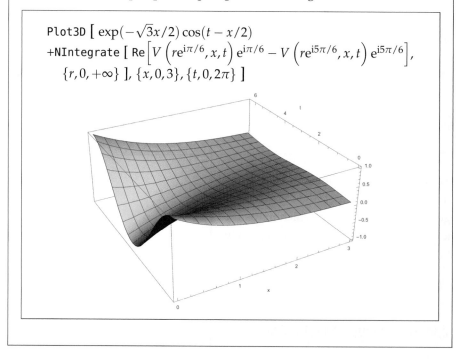

```
Plot3D [ exp(-√3x/2) cos(t − x/2)
  +NIntegrate [ Re[ V(re^{iπ/6}, x, t) e^{iπ/6} − V(re^{i5π/6}, x, t) e^{i5π/6} ],
    {r, 0, +∞} ], {x, 0, 3}, {t, 0, 2π} ]
```

9.5   Inhomogeneous PDEs and other considerations

The representations for the solutions of the homogeneous PDEs obtained in the Sections 9.2 and 9.4 can be immediately extended to the case of inhomogeneous

PDEs: the solution of an inhomogeneous PDE with forcing $f(x,t)$ is obtained from the representation of the corresponding homogeneous PDE by the following simple procedure:

replace $\hat{u}_0(k)$ with $\hat{v}(k,t) = \hat{u}_0(k) + \int_0^\infty e^{-ikx} \left[\int_0^t e^{\omega(k)\tau} f(x,\tau) d\tau \right] dx,$

where $\exp[ik - \omega(k)t]$ is a solution of the homogeneous PDE.

For example, consider the inhomogeneous heat equation

$$u_t - u_{xx} = f(x,t).$$

The first step of the unified transform yields

$$e^{k^2 t} \hat{u}(k,t) = \hat{u}_0(k) + \int_0^\infty e^{-ikx} \int_0^t e^{k^2 \tau} f(x,\tau) d\tau dx$$
$$- \tilde{g}_1(k^2,t) - ik\tilde{g}_0(k^2,t), \quad \text{Im}[k] \leq 0.$$

Then, since steps 2 and 3 are identical with those for the heat equation, we find

$$u(x,t) = \int_{-\infty}^\infty e^{ikx - k^2 t} \hat{v}(k,t) \frac{dk}{2\pi}$$
$$- \int_{\partial \mathcal{D}^+} e^{ikx - k^2 t} \left[\hat{v}(-k,t) + 2ik\tilde{G}_0(k^2) \right] \frac{dk}{2\pi}, \qquad (9.55a)$$

where

$$\hat{v}(k,t) = \hat{u}_0(k) + \int_0^\infty e^{-ik\xi} \left[\int_0^t e^{k^2 \tau} f(\xi,\tau) d\tau \right] d\xi. \qquad (9.55b)$$

Derivation

We will show that Eq. (9.55a) with $\hat{v}(k,t)$ defined by Eq. (9.55b) solves the forced heat equation with the initial and boundary conditions specified by the functions $u_0(x)$ and $g_0(t)$.

The only (x,t) dependence of $u(x,t)$ is through the exponential $\exp[ikx - k^2 t]$ (which solves the heat equation), as well as through $\hat{v}$. Hence, a simple computation yields

$$u_t - u_{xx} = \int_{-\infty}^\infty e^{ikx} \left(\int_0^\infty e^{-ik\xi} f(\xi,t) d\xi \right) \frac{dk}{2\pi}$$
$$- \int_{\partial \mathcal{D}^+} e^{ikx} \left(\int_0^\infty e^{ik\xi} f(\xi,t) d\xi \right) \frac{dk}{2\pi}.$$

Noting that $\exp[ik(x + \xi)]$ vanishes for large k in $\mathcal{D}^+$ (since $x + \xi > 0$), it follows that the second integral in the right-hand side of the above equation vanishes. Also, using the integral representation for the δ-function

$$\int_{-\infty}^\infty e^{ik(x-\xi)} \frac{dk}{2\pi} = \delta(x - \xi),$$

which was derived in Chapter 7, Eq. (7.53), we find that indeed u solves the inhomogeneous heat equation with the forcing $f(x,t)$.

Evaluating Eq. (9.55a) at $t = 0$ we find

$$u(x,0) = \int_{-\infty}^\infty e^{ikx} \hat{u}_0(k) \frac{dk}{2\pi} - \int_{\partial \mathcal{D}^+} e^{ikx} \hat{u}_0(-k) \frac{dk}{2\pi} - 2i \int_{\partial \mathcal{D}^+} e^{ikx} k \tilde{G}_0(k^2) \frac{dk}{2\pi}.$$

This is precisely the equation arising in the case of the homogeneous heat equation. Thus, as noted earlier, the first integral in the right-hand side of the above equation yields $u_0(x)$ (in view of the standard Fourier transform identity), the second integral vanishes (in view of the exponential decay of $\hat{u}_0(-k)$ for large k in $\mathcal{D}^+$), and the third integral vanishes in $\mathcal{D}^+$ (in view of integration by parts).

Evaluating Eq. (9.55a) at $x = 0$ we find

$$u(0, t) = \int_{-\infty}^{\infty} e^{-k^2 t} \hat{v}(k, t) \frac{dk}{2\pi} - \int_{\partial \mathcal{D}^+} e^{-k^2 t} \hat{v}(-k, t) \frac{dk}{2\pi}$$
$$- 2i \int_{\partial \mathcal{D}^+} e^{-k^2 t} k \tilde{G}_0(k^2) \frac{dk}{2\pi}.$$

Integration by parts implies that $e^{-k^2 t} \hat{v}(-k, t)$ decays for large k in the domain bounded by the real axis and $\partial \mathcal{D}^+$, thus the contour in the second integral above can be replaced by the real line. Then, replacing k with $-k$ we find an integral that cancels the first integral of the right-hand side of the above equation. Furthermore, as shown earlier, the third integral equals $g_0(t)$.

It is worth noting that it is straightforward to justify rigorously the procedure of showing that an integral decays for large k through the use of integration by parts. In what follows, we illustrate this approach for the case of $k \tilde{G}_0(k^2)$.

$$\tilde{G}_0(k^2) = \int_0^T e^{k^2 t} g_0(t) dt = \frac{1}{k^2} \left[e^{k^2 T} g_0(T) - g_0(0) \right] - \frac{1}{k^2} \int_0^T e^{k^2 t} \frac{dg_0(t)}{dt} dt.$$

The function $\exp[k^2 T]$ decays for k in the interior of $\mathcal{D}^+$ and oscillates on $\partial \mathcal{D}^+$. Thus,

$$|\tilde{G}_0(k^2)| \leqslant \frac{1}{|k|^2} \left[|g_0(T)| + |g_0(0)| + \left\| \frac{dg_0}{dt} \right\|_{L_1(0,T)} \right],$$

where $L_1(0, T)$ is the L_1-norm, defined in Eq. (1.55). As long as the $L_1(0, T)$ norm of dg_0/dt is bounded, $k \tilde{G}_0(k^2)$ is of order $1/k$, and $k \tilde{G}_0(k^2)$ vanishes as $|k| \to \infty$.

9.5.1   Robin boundary conditions

Steps 1 and 2 of the unified transform yield the global relation and the integral representation. Then, employing the transformations $\nu(k)$ which leave $\omega(k)$ invariant, it is straightforward to solve a variety of boundary value problems. For concreteness, we consider the case of the so-called Robin boundary condition, namely we consider the boundary condition

$$u_x(0, t) - \gamma u(0, t) = g_R(t), \quad t > 0, \tag{9.56}$$

where g_R is a given function with sufficient smoothness and γ is a real constant.

The t-transform of Eq. (9.56) yields

$$\tilde{g}_1(k, t) = \gamma \tilde{g}_0(k, t) + \tilde{g}_R(k, t); \quad \tilde{g}_R(k, t) = \int_0^t e^{k\tau} g_R(\tau) d\tau, \quad t > 0. \tag{9.57}$$

The above equations together with the GR Eq. (9.12) are two equations for the two unknown functions $\tilde{g}_0$ and $\tilde{g}_1$. Solving these two equations we find

$$\tilde{g}_0(k^2, t) = \frac{i}{k + i\gamma} \left[\hat{u}_0(-k) - \tilde{g}_R(k^2, t) - e^{k^2 t} \hat{u}(-k, t) \right], \tag{9.58a}$$

$$\tilde{g}_1(k^2, t) = \frac{i}{k + i\gamma} \left[\gamma \hat{u}_0(-k) - ik\tilde{g}_R(k^2, t) - \gamma e^{k^2 t} \hat{u}(-k, t) \right]. \tag{9.58b}$$

Substituting the above expressions in Eq. (9.14) we obtain

$$u(x, t) = \int_{-\infty}^{\infty} e^{ikx - k^2 t} \hat{u}_0(k) \frac{dk}{2\pi}$$
$$- \int_{\partial \mathcal{D}^+} \frac{e^{ikx - k^2 t}}{k + i\gamma} \left[2k\tilde{g}_R(k^2, t) - (k - i\gamma)\hat{u}_0(-k) \right] \frac{dk}{2\pi}$$
$$- \int_{\partial \mathcal{D}^+} e^{ikx} \left(\frac{k - i\gamma}{k + i\gamma} \right) \hat{u}(-k, t) \frac{dk}{2\pi}. \tag{9.59}$$

The factor $k + i\gamma$ vanishes at the point $k_0 = -i\gamma$, thus, we distinguish the following two cases:

(i) $\gamma > 0$
In this case, k_0 is outside $\mathcal{D}^+$, thus employing Cauchy's theorem in $\mathcal{D}^+$ we find that the contribution of $\hat{u}(-k, t)$ vanishes. Indeed, $(k - i\gamma)\hat{u}(-k, t)/(k + i\gamma) \sim \hat{u}(-k, t)$ as $k \to \infty$ and $\hat{u}(-k, t)$ decays exponentially for large $|k|$ in $\mathcal{D}^+$. Hence Eq. (9.59) becomes

$$u(x, t) = \int_{-\infty}^{\infty} e^{ikx - k^2 t} \hat{u}_0(k) \frac{dk}{2\pi}$$
$$- \int_{\partial \mathcal{D}^+} \frac{e^{ikx - k^2 t}}{k + i\gamma} \left[2k\tilde{g}_R(k^2, t) + (k - i\gamma)\hat{u}_0(-k) \right] \frac{dk}{2\pi}, \quad \gamma > 0. \tag{9.60}$$

(ii) $\gamma < 0$
In this case, taking into consideration that $(k - i\gamma)\hat{u}(-k, t)/(k + i\gamma)$ vanishes for large $|k|$ in $\mathcal{D}^+$, the Residue theorem implies that the third integral on the right-hand side of Eq. (9.59) equals $-2\gamma \exp(\gamma x)\hat{u}(i\gamma, t)$. Remarkably, $\hat{u}(i\gamma, t)$ can be computed explicitly. Indeed, multiplying either of Eqs. (9.58a) and (9.58b) by $(k + i\gamma)$ and then evaluating the resulting expression at $k = -i\gamma$ we find

$$\hat{u}(i\gamma, t) = e^{\gamma^2 t} \left[\hat{u}_0(i\gamma) - \tilde{g}_R(-\gamma^2, t) \right]. \tag{9.61}$$

Thus, Eq. (9.59) becomes

$$u(x, t) = \int_{-\infty}^{\infty} e^{ikx - k^2 t} \hat{u}_0(k) \frac{dk}{2\pi} - 2\gamma e^{\gamma x + \gamma^2 t} [\hat{u}_0(i\gamma) - \tilde{g}_R(-\gamma^2, t)]$$
$$- \int_{\partial \mathcal{D}^+} \frac{e^{ikx - k^2 t}}{k + i\gamma} \left[2k\tilde{g}_R(k^2, t) - (k - i\gamma)\hat{u}_0(-k) \right] \frac{dk}{2\pi}, \quad \gamma < 0, \tag{9.62}$$

where $\hat{u}_0$ and $\tilde{g}_R$ are the appropriate transforms of the given data $u_0(x)$ and $g_R(t)$.

Example 9.3: In order to illustrate the numerical advantage of the new method, and at the same time to avoid tedious analytical computations we take

$$u_0(x) = e^{-x}, \quad g_R(t) = -(\gamma + 1)\cos(t).$$

We note that since $u(x,0) = e^{-x}$, $u_x(x,0) = -e^{-x}$, it follows that

$$u_x(0,0) - \gamma u(0,0) = -(\gamma + 1) = g_R(0)$$

and the boundary condition Eq. (9.56) is satisfied at $t = 0$, for any $\gamma \in \mathbb{R}$.

The case $\gamma = 1$

As in Example 9.1, we first deform the contour C^+ defined by the union of the rays $\mathrm{Arg}[k] = \pi/8$ and $\mathrm{Arg}[k] = 7\pi/8$. For the above initial and boundary conditions, Eqs. (9.8b) and (9.57) yield

$$\hat{u}_0(k) = -\frac{i}{k-i}, \quad \text{and} \quad \tilde{g}_R(k^2, t) = \frac{2\left[k^2 - e^{-k^2 t}\left(k^2\cos t + \sin t\right)\right]}{k^4 + 1},$$

respectively. Using these expressions in Eq. (9.60), we obtain

$$u(x,t) = \int_{C^+} e^{ikx - k^2 t}\, \frac{4k\left[(k^2 + 1)\, e^{k^2 t}\left[k^2\cos(t) + \sin(t)\right] + (1 - k^2)\right]}{(k-i)(k+i)^2\,(k^4 + 1)}\,\frac{dk}{2\pi}.$$

This integral can be split into two contributions:

$$u(x,t) = \int_{C^+} e^{ikx - k^2 t}\, \frac{4k\left(1 - k^2\right)}{(k-i)(k+i)^2\,(k^4 + 1)}\,\frac{dk}{2\pi}$$

$$+ \int_{C^+} e^{ikx}\, \frac{4k\left(k^2 + 1\right)\left[k^2\cos(t) + \sin(t)\right]}{(k-i)(k+i)^2\,(k^4 + 1)}\,\frac{dk}{2\pi}.$$

Then, employing the same procedure as in Example 9.1, we can evaluate analytically the second integral in the above expression by calculating the residue contributions. In the domain which is over the curve C^+ there are two poles, namely $k_1 = e^{i\pi/4}$ and $k_2 = e^{3i\pi/4}$. The associated residues are

$$\frac{e^{it - (1+i)x/\sqrt{2}}}{1 + (1+i)/\sqrt{2}} \quad \text{and} \quad \frac{e^{-it - (1-i)x/\sqrt{2}}}{1 + (1-i)/\sqrt{2}},$$

respectively. Thus, the total contribution is the sum of these two terms and it is given by

$$v(x,t) = \frac{e^{-x/\sqrt{2}}}{\sqrt{2}}\left[\frac{e^{i\left(t - x/\sqrt{2}\right)}}{\sqrt{2} + 1 + i} + \frac{e^{-i\left(t - x/\sqrt{2}\right)}}{\sqrt{2} + 1 - i}\right].$$

Simplifying this expression we find

$$v(x,t) = e^{-x/\sqrt{2}}\left[\left(\sqrt{2} - 1\right)\sin\left(t - \frac{x}{\sqrt{2}}\right) + \cos\left(t - \frac{x}{\sqrt{2}}\right)\right],$$

and $u(x,t)$ becomes

$$u(x,t) = v(x,t) + \int_{C^+} V(k,x,t)dk,$$

where

$$V(k, x, t) = \frac{e^{ikx - k^2 t}}{2\pi} \frac{4k\left(1 - k^2\right)}{(k - i)(k + i)^2\left(k^4 + 1\right)}.$$

We first verify that this is compatible with the initial and boundary conditions. The last expression for $u(x, t)$ evaluated at $x = 0$ yields

$$u_x(0, t) - u(0, t) = \int_{C^+} \left[V_x(k, 0, t) - V(k, 0, t)\right] dk - 2\cos(t).$$

It suffices to show that the first term of the left-hand side of the above expression vanishes. Indeed, this is clear by deforming the contour C^+ to the real line and observing that the integrand

$$V_x(k, 0, t) - V(k, 0, t) = -\frac{e^{-k^2 t}}{2\pi} \frac{4ik\left(k^2 - 1\right)}{k^6 + k^4 + k^2 + 1}$$

is an odd function of k. On the other hand, the original expression for $u(x, t)$ as an integral over C^+ evaluated at $t = 0$ yields

$$u(x, 0) = \int_{C^+} e^{ikx} \frac{4k}{(k - i)(k + i)^2} \frac{dk}{2\pi}.$$

Then, application of the Residue theorem yields

$$u(x, 0) = \frac{2\pi i}{2\pi} e^{-ix} \frac{4i}{(2i)^2} = e^{-x},$$

which is the initial condition $u_0(x)$.

For the numerical evaluation of the integral we follow similar steps as those in Example 9.1. Specifically, we split the integral into two parts, using $k = r\exp[i\pi/8], r \in (0, +\infty)$ in the first integral and $k = r\exp[i7\pi/8]$, while r runs from ∞ to 0 in the second, so that the full solution is given by the expression:

$$u(x, t) = v(x, t) + \int_0^\infty \left[V\left(re^{i\pi/8}, x, t\right) e^{i\pi/8} - V\left(re^{i7\pi/8}, x, t\right) e^{i7\pi/8}\right] dr.$$

Plots of the numerical evaluation of the solution for $x \in [0, 1]$ at short (left panel) and longer (right panel) timescales with the initial–boundary condition shown as the black dashed line, as well as the three-dimensional perspective plot in the range $x \in [0, 1]$ and $t \in [0, 2\pi]$, generated using Mathematica, are shown below.

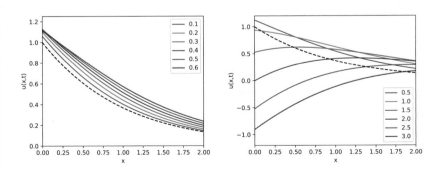

```
Plot3D [ v(x,t)
  +NIntegrate [ Re[ V (re^{iπ/8}, x, t) e^{iπ/8} − V (re^{i7π/8}, x, t) e^{i7π/8} ],
  {r, 0, +∞} ], {x, 0, 3}, {t, 0, 2π} ]
```

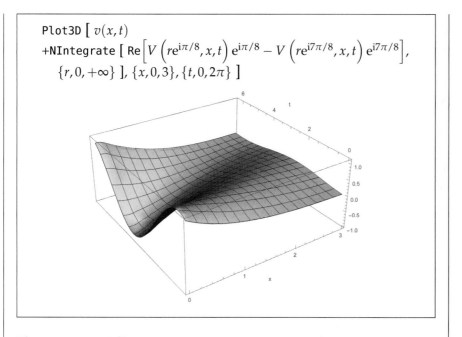

The case $\gamma = -1/2$

We follow similar steps as in the case $\gamma = 1$, using the same contour C^+, namely the one defined by the union of the rays $\text{Arg}[k] = \pi/8$ and $\text{Arg}[k] = 7\pi/8$. For the above initial and boundary conditions, Eqs. (9.8b) and (9.57) yield

$$\hat{u}_0(k) = -\frac{i}{k-i},$$

$$\tilde{g}_R(k^2, t) = \frac{1}{2(k^4+1)} \left[k^2 - e^{-k^2 t} \left(k^2 \cos(t) + \sin(t) \right) \right],$$

respectively. Using these expressions in Eq. (9.62) we obtain

$$u(x,t) = \int_{C^+} e^{ikx - k^2 t} \frac{2k \left[(k^2+1) e^{k^2 t} \left[k^2 \cos(t) + \sin(t) \right] + (1-k^2) \right]}{(k-i)(k+i)(2k-i)(k^4+1)} \frac{dk}{2\pi}$$
$$+ \frac{2}{51} e^{-x/2} \left[20 e^{t/4} + 12 \sin(t) - 3 \cos(t) \right].$$

The integral over C^+ can again be split into two parts:

$$u(x,t) = \int_{C^+} e^{ikx - k^2 t} \frac{2k \left(1 - k^2 \right)}{(k-i)(k+i)(2k-i)(k^4+1)} \frac{dk}{2\pi}$$
$$+ \int_{C^+} e^{ikx} \frac{2k \left(k^2 + 1 \right) \left[k^2 \cos(t) + \sin(t) \right]}{(k-i)(k+i)(2k-i)(k^4+1)} \frac{dk}{2\pi}$$
$$+ \frac{2}{51} e^{-x/2} \left[20 e^{t/4} + 12 \sin(t) - 3 \cos(t) \right].$$

We can evaluate analytically the second integral in the above expression by calculating the residue contributions. In the domain which is over the curve C^+ there are three poles, namely

$$k_1 = e^{i\pi/4}, \quad k_2 = e^{3i\pi/4}, \quad \text{and} \quad k_3 = \frac{i}{2}.$$

The associated residues are

$$\frac{e^{-it-(1-i)x/\sqrt{2}}}{-2+2\sqrt{2}(1-i)}, \quad \frac{e^{it-(1+i)x/\sqrt{2}}}{-2+2\sqrt{2}(1+i)}, \quad \text{and} \quad \frac{2}{17}e^{-x/2}\left(\cos(t)-4\sin(t)\right),$$

respectively. Then, following the exact same steps as in the case $\gamma = 1$, the contribution of the three residues is

$$\frac{e^{-x/\sqrt{2}}}{5-2\sqrt{2}}\left[\sqrt{2}\sin\left(t-\frac{x}{\sqrt{2}}\right)+\left(\sqrt{2}-1\right)\cos\left(t-\frac{x}{\sqrt{2}}\right)\right]$$
$$-\frac{2}{17}e^{-x/2}\left(4\sin t - \cos t\right).$$

The solution then takes the form

$$u(x,t) = v(x,t) + \int_{C^+} V(k,x,t)dk,$$

where

$$v(x,t) = \frac{40}{51}e^{(t-2x)/4}$$
$$-\frac{e^{-x/\sqrt{2}}}{2\sqrt{2}-5}\left[\sqrt{2}\sin\left(t-\frac{x}{\sqrt{2}}\right)+\left(\sqrt{2}-1\right)\cos\left(t-\frac{x}{\sqrt{2}}\right)\right],$$

$$V(k,x,t) = \frac{e^{ikx-k^2t}}{2\pi}\frac{2k\left(1-k^2\right)}{(k-i)(k+i)(2k-i)\left(k^4+1\right)}.$$

We first verify that these expressions are compatible with the initial and boundary conditions. The last expression for $u(x,t)$ evaluated at $x = 0$ yields

$$u_x(0,t) + \frac{1}{2}u(0,t) = \int_{C^+}\left[V_x(k,0,t)+\frac{1}{2}V(k,0,t)\right]\frac{dk}{2\pi}-\frac{1}{2}\cos(t).$$

It suffices to show that the first term of the left-hand side of the above expression vanishes. Indeed, this is clear by deforming the contour C^+ to the real line and observing that the integrand

$$V_x(k,0,t)+\frac{1}{2}V(k,0,t) = -\frac{ik\left(k^2-1\right)e^{-k^2t}}{k^6+k^4+k^2+1}$$

is an odd function of k. On the other hand, the previous expression for $u(x,t)$ evaluated at $t = 0$ yields

$$u(x,0) = \int_{C^+}e^{ikx}\frac{2k}{(k-i)(k+i)(2k-i)}\frac{dk}{2\pi}+\frac{2}{3}e^{-x/2}.$$

Then, applying the Residue theorem leads to

$$u(x,0) = \frac{2\pi i}{2\pi}e^{-x}\frac{4i}{(2i)^2}-\frac{2\pi i}{2\pi}e^{-x/2}\frac{\frac{i}{2}}{\left(\frac{i}{2}-i\right)\left(\frac{i}{2}+i\right)}+\frac{2}{3}e^{-x/2}=e^{-x}=u_0(x).$$

For the numerical evaluation of the integral we follow similar steps as above, namely we split the integral into two parts, using $k = r\exp[i\pi/8]$,

$r \in (0, +\infty)$ in the first integral and $k = r \exp[\mathrm{i} 7\pi/8]$, $r \in (+\infty, 0)$ in the second, so that the full solution is given by the expression:

$$u(x,t) = v(x,t) + \int_0^\infty \left[V\left(r e^{\mathrm{i}\pi/8}, x, t \right) e^{\mathrm{i}\pi/8} - V\left(r e^{\mathrm{i}7\pi/8}, x, t \right) e^{\mathrm{i}7\pi/8} \right] dr.$$

Plots of the numerical evaluation of the solution for $x \in [0,1]$ at short (left panel) and longer (right panel) timescales with the initial–boundary condition shown as the black dashed line, as well as the three-dimensional perspective plot in the range $x \in [0,1]$ and $t \in [0, 2\pi]$, generated using Mathematica, are shown below.

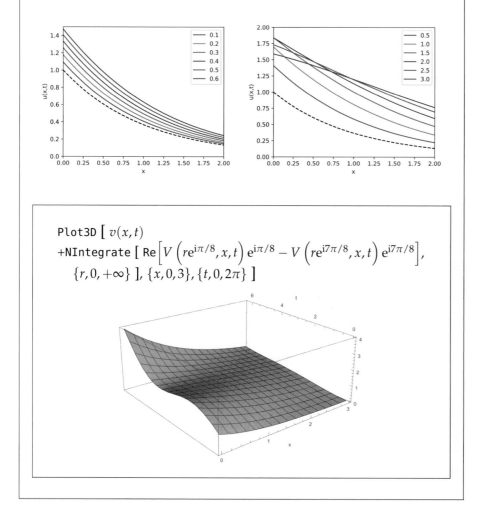

```
Plot3D [ v(x,t)
  +NIntegrate [ Re[ V( re^{iπ/8}, x, t ) e^{iπ/8} - V( re^{i7π/8}, x, t ) e^{i7π/8} ],
  {r, 0, +∞} ], {x, 0, 3}, {t, 0, 2π} ]
```

9.5.2   *From PDEs to ODEs*

The heat equation with Robin boundary conditions *can* be solved via a traditional transform pair, but the derivation of this pair cannot be guessed; specifically for the case of $\gamma < 0$, where the relevant transform pair involves an additional term (arising from the so-called "point spectrum"). Taking into consideration that the representations obtained via the unified transform have several advantages in comparison with the representations obtained via traditional transform pairs, it follows that the use of traditional transforms is obsolete. However, if, for whatever reason, one requires a traditional transform pair, the unified transform provides a much easier way for constructing this pair than the usual Green's function approach. This was already discussed in Section 9.2 for the simple cases when either $u_x(0,t)$ or $u(0,t)$ are given. In what follows, we discuss the case of the Robin boundary condition, Eq. (9.56) with $\gamma < 0$.

Evaluating Eq. (9.62) at $t = 0$ and deforming $\partial \mathcal{D}^+$ back to the real axis, we find

$$u_0(x) = \int_{-\infty}^{\infty} e^{ikx} \hat{u}_0(k) \frac{dk}{2\pi} + \int_{-\infty}^{+\infty} e^{ikx} \frac{k - i\gamma}{k + i\gamma} \hat{u}_0(-k) \frac{dk}{2\pi} - 2\gamma e^{\gamma x} \hat{u}_0(i\gamma).$$

Using the definition of $\hat{u}_0(k)$, the above equation becomes

$$u_0(x) = \int_{-\infty}^{\infty} e^{ikx} \left[\int_0^{\infty} \left(e^{-ik\xi} + \frac{k - i\gamma}{k + i\gamma} e^{ik\xi} \right) u_0(\xi) d\xi \right] \frac{dk}{2\pi}$$
$$- 2\gamma e^{\gamma x} \int_0^{\infty} e^{\gamma \xi} u_0(\xi) d\xi.$$

Thus, renaming $u_0(x)$ as $f(x)$, we obtain the following transform pair:

$$\hat{f}_R(k) = \int_0^{\infty} \left(e^{-ikx} + \frac{k - i\gamma}{k + i\gamma} e^{ikx} \right) f(x) dx, \tag{9.63a}$$

$$f(x) = \int_{-\infty}^{\infty} e^{ikx} \hat{f}_R(k) \frac{dk}{2\pi} - 2\gamma e^{\gamma x} \int_0^{\infty} e^{\gamma \xi} f(\xi) d\xi. \tag{9.63b}$$

An even more complicated transform pair will be derived in the last section.

It is worth noting that the unified transform also generates novel transform pairs. This will be elaborated on in Section 10.4.2.

9.5.3   Green's functions

After constructing the final integral representation, it is straightforward to obtain the associated Green's functions. For example, for the heat equation on the half-line with the initial and boundary conditions (8.33), Eq. (9.16) can be written in the following form:

$$u(x,t) = \int_0^{\infty} u_0(\xi) \left(\int_{-\infty}^{\infty} e^{ik(x-\xi)-k^2 t} \frac{dk}{2\pi} \right) d\xi$$
$$- \int_0^{\infty} u_0(\xi) \left(\int_{\partial \mathcal{D}^+} e^{ik(x+\xi)-k^2 t} \frac{dk}{2\pi} \right) d\xi$$
$$- \int_0^{T} g_0(\tau) \left(\int_{\partial \mathcal{D}^+} e^{ikx+k^2(\tau-t)} 2ik \frac{dk}{2\pi} \right) d\tau. \tag{9.64}$$

Hence,

$$u(x,t) = \int_0^{\infty} u_0(\xi) G(x,\xi,t) d\xi + \int_0^{T} g_0(\tau) G_L(x,\tau,t) d\tau, \tag{9.65}$$

where $G(x,\xi,t)$ and $G_L(x,\tau,t)$ are defined as follows:

$$G(x,\xi,t) = \int_{-\infty}^{\infty} e^{ik(x-\xi)-k^2 t} \frac{dk}{2\pi} - \int_{\partial \mathcal{D}^+} e^{ik(x+\xi)-k^2 t} \frac{dk}{2\pi}, \tag{9.66a}$$

$$G_L(x,\tau,t) = -2i \int_{\partial \mathcal{D}^+} k e^{ikx+k^2(\tau-t)} \frac{dk}{2\pi}, \quad t > \tau, \tag{9.66b}$$

with $0 \leq x < \infty, 0 \leq \xi < \infty, 0 \leq \tau \leq T$. It is straightforward to compute the above integrals in closed form.

Derivation: Green's functions for the heat equation on the half-line

The Green's functions defined in Eq. (9.66) can be computed in closed form, leading to

$$G(x, \xi, t) = \frac{1}{2\sqrt{\pi t}} \left[e^{-(x-\xi)^2/4t} - e^{-(x+\xi)^2/4t} \right], \qquad (9.67a)$$

$$G_L(x, \tau, t) = \frac{1}{2\sqrt{\pi}} \frac{x}{(t-\tau)^{3/2}} e^{-x^2/4(t-\tau)}, \quad t > \tau. \qquad (9.67b)$$

To derive Eq. (9.67a) we first "complete the square" associated with $ik(x - \xi) - k^2 t$:

$$-k^2 t + ik(x - \xi) = -t \left[k^2 + \frac{ik}{t}(x - \xi) \right] = -t \left[k + \frac{i}{2t}(x - \xi) \right]^2 - \frac{(x - \xi)^2}{4t}.$$

Hence,

$$\int_{-\infty}^{\infty} e^{ik(x-\xi)-k^2 t} dk = e^{-(x-\xi)^2/4t} \int_{-\infty}^{\infty} e^{-t(k^2 + i(x-\xi)/2t)^2} dk$$

$$= e^{-(x-\xi)^2/4t} \int_{-\infty+ia}^{\infty+ia} e^{-t\ell^2} d\ell, \quad a = \frac{x - \xi}{2t}.$$

The function $\exp[-t\ell^2]$ decays for ℓ in the domain between the real axis and the line $(-\infty + ia, \infty + ia)$ as $|\ell| \to \infty$. Thus, using Cauchy's theorem we can replace the above line with the real axis. Hence,

$$\int_{-\infty}^{\infty} e^{ik(x-\xi)-k^2 t} dk = e^{-(x-\xi)^2/4t} \int_{-\infty}^{\infty} e^{-t\ell^2} d\ell = \frac{\sqrt{\pi}}{\sqrt{t}} e^{-(x-\xi)^2/4t},$$

where we have used the identity we derived in Section 1.5.2, Eq. (1.57),

$$\int_{-\infty}^{\infty} e^{-t\ell^2} d\ell = \frac{1}{\sqrt{t}} \int_{-\infty}^{\infty} e^{-x^2} dx = \frac{\sqrt{\pi}}{\sqrt{t}}.$$

Similarly,

$$\int_{\partial \mathcal{D}^+} e^{ik(x+\xi)-k^2 t} dk = \frac{\sqrt{\pi}}{\sqrt{t}} e^{-(x+\xi)^2/4t},$$

and then Eq. (9.67a) follows.

To derive Eq. (9.67b), we first use the identity

$$k e^{ikx+k^2(\tau-t)} = \frac{e^{ikx+k^2(\tau-t)}}{2(\tau - t)} [2k(\tau - t) + ix] - \frac{ix e^{ikx+k^2(\tau-t)}}{2(\tau - t)}.$$

The first term above is the k-derivative of $\exp[ikx + k^2(\tau - t)]$, hence its integral along $\partial \mathcal{D}^+$ vanishes since the above exponential vanishes in $\mathcal{D}^+$ as $|k| \to \infty$. Thus,

$$\int_{\partial \mathcal{D}^+} k e^{ikx+k^2(\tau-t)} dk = \frac{ix}{2(t - \tau)} e^{-x^2/4(t-\tau)} \int_{-\infty}^{\infty} e^{-(t-\tau)\ell^2} d\ell$$

$$= \frac{ix\sqrt{\pi}}{2(t - \tau)^{3/2}} e^{-x^2/4(t-\tau)},$$

and Eq. (9.67b) follows.

9.6   Application: Heat flow along a solid rod

Consider a tank filled with liquid, open at the top and with a free surface area A_l. A horizontal semi-infinite solid rod is placed in direct contact with the tank on one of the vertical sides so that heat conduction takes place between the two. The liquid, of arbitrary initial temperature, absorbs heat from its surrounding which has a temperature $T(t)$, and is kept well stirred so that its temperature $r(t)$ is uniform throughout its volume. The lateral surfaces of the container and the rod are insulated so that there is no heat flow to the surrounding medium. The temperature u of the rod can also be regarded as uniform across the cross-section A_s, that is, we can assume that it is only a function of the distance x from the point of contact with the tank and of the time t, as shown in Fig. 9.2.

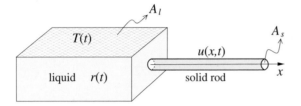

Figure 9.2: Heat conduction between a tank and a solid rod.

Hence, $u(x,t)$ obeys the one-dimensional heat equation

$$u_t = u_{xx}, \quad 0 < x < \infty, \quad t > 0. \tag{9.68}$$

At time $t = 0$ we have

$$u(x,0) = u_0(x), \quad 0 < x < \infty. \tag{9.69}$$

At the point of contact $x = 0$ we have the boundary condition

$$u(0,t) = r(t), \tag{9.70}$$

where $r(t)$ is the unknown temperature of the liquid.

The rate of accumulation of heat in the liquid can be expressed in terms of the incoming heat from the surrounding medium and the heat which flows from the tank to the rod:

$$\sigma \frac{dr(t)}{dt} = -u_x(0,t) + \lambda \left[T(t) - r(t) \right], \tag{9.71}$$

where the real constants σ and λ depend on the specific heat, the mass, and the conductivities of the rod and the liquid.

Combining Eqs. (9.70) and (9.71), we obtain the following boundary condition, known as an *oblique Robin* condition:

$$\sigma u_t(0,t) + u_x(0,t) + \lambda \left[u(0,t) - T(t) \right] = 0. \tag{9.72}$$

For simplicity of notation, we let

$$\alpha = \frac{1}{\sigma}, \quad \beta = \frac{\lambda}{\sigma}, \quad \gamma(t) = \frac{\lambda}{\sigma} T(t), \tag{9.73}$$

so that the boundary condition (9.72) becomes:

$$u_t(0,t) + \alpha u_x(0,t) + \beta u(0,t) = \gamma(t). \tag{9.74}$$

9.6.1   The conventional solution

The particular case of $\alpha = 1$, $\beta = \gamma = 0$, was solved by D.S. Cohen.[1] Even for this particular simpler case it is not possible to guess the correct transform. Actually, in this case even the derivation of the equation needed to be analyzed in order to derive the correct transform is quite complicated. In this connection, we note that for the heat equation formulated on the half-line with the initial condition Eq. (9.69) and the boundary condition $u_x(0,t) = \gamma(t)$, the formulation of the eigenvalue equation that needs to be analyzed in order to obtain the correct transform (which in this case is the cosine-transform) is straightforward:

[1] D.S. Cohen, An integral transform associated with boundary conditions containing an eigenvalue parameter. *SIAM Journal of Applied Mathematics* **14** (1966).

$$\frac{\partial^2 G(x,\xi,\lambda)}{\partial x^2} - \lambda G(x,\xi,\lambda) = \delta(x - \xi), \quad 0 < x, \xi < \infty, \ \lambda \in \mathbb{C}, \quad (9.75)$$

$$\frac{\partial G(0,\xi,\lambda)}{\partial x} = 0, \quad \lim_{x \to \infty} G(x,\xi,\lambda) = 0, \quad 0 < \xi < \infty. \quad (9.76)$$

However, if the above boundary condition is replaced by

$$u_t(0,t) + u_x(0,t) = 0, \quad (9.77)$$

using a series of elaborate arguments, D.S. Cohen showed (see footnote 1) that the analogue of Eqs. (9.75) and (9.76) are the following matrix and scalar equations:

$$\begin{pmatrix} -g_{1,xx}(x,\xi,\lambda) & -g_{2,xx}(x,\xi,\lambda) \\ g_{1,x}(0,\xi,\lambda) & g_{2,x}(0,\xi,\lambda) \end{pmatrix} - \lambda \begin{pmatrix} g_1(x,\xi,\lambda) & g_2(x,\xi,\lambda) \\ g_1(0,\xi,\lambda) & g_2(0,\xi,\lambda) \end{pmatrix}$$
$$= \begin{pmatrix} \delta(x-\xi) & 0 \\ 0 & e^{-\xi} \end{pmatrix}, \quad (9.78a)$$

$$\lim_{x \to \infty} g_j(x,\xi,\lambda) = 0, \quad j = 1, 2, \ \lambda \in \mathbb{C}. \quad (9.78b)$$

By analyzing these equations, it is shown that the proper transform is given by

$$f(x) = 4 \int_0^\infty \frac{\cos(kx) + k\sin(kx)}{k^2 + 1} \left[\int_0^\infty (\cos(k\xi) + k\sin(k\xi)) f(\xi)d\xi - f(0) \right] \frac{dk}{2\pi}$$
$$+ 2e^{-x} \left[f(0) - \int_0^\infty e^{-\xi} f(\xi)d\xi \right]. \quad (9.79)$$

Employing this transform it can be shown that

$$u(x,t) = 4 \int_0^\infty e^{-k^2 t} \frac{\cos(kx) + k\sin(kx)}{k^2 + 1}$$
$$\times \left[\int_0^\infty (\cos(k\xi) + k\sin(k\xi)) u_0(\xi)d\xi - u_0(0) \right] \frac{dk}{2\pi}$$
$$+ 2e^{-x+t} \left[u_0(0) - \int_0^\infty e^{-\xi} u_0(\xi)d\xi \right]. \quad (9.80)$$

As will be shown in the next subsection, the heat equation with the initial condition (9.69) and the boundary condition (9.74) can be solved via the unified transform in a straightforward manner. In the particular case of $\alpha = 1$, $\beta = \gamma = 0$, the form of the solution representation derived through the unified transform can be transformed to Eq. (9.80) (see the derivation below). The form derived through the unified transform has several advantages in comparison to Eq. (9.80). This suggests that the derivation of Eq. (9.79) is superseded. If, for whatever reason, one needs this transform for some other purpose, Eq. (9.79) can be deduced readily from the solution obtained through the unified transform.

9.6.2 The solution through the unified transform

The heat equation with the initial condition Eq. (9.69) and the boundary condition Eq. (9.74) was solved through the unified transform.[2] Multiplying the boundary condition Eq. (9.74) by $e^{k^2 s}$ and integrating with respect to s from 0 to t, we find

[2] D. Mantzavinos and A. Fokas, The unified method for the heat equation: I. non-separable boundary conditions and non-local constraints in one dimension. *European Journal of Applied Mathematics* **24**, pp. 857–886 (2013).

$$\int_0^t e^{k^2 s} u_s(0,s)\mathrm{d}s + \alpha \int_0^t e^{k^2 s} u_x(0,s)\mathrm{d}s + \beta \int_0^t e^{k^2 s} u(0,s)\mathrm{d}s = \int_0^t e^{k^2 s}\gamma(s)\mathrm{d}s.$$

Using integration by parts in the first integral above, we obtain the condition

$$e^{k^2 t}u(0,t) - u_0(0) - k^2 \tilde{g}_0(k^2,t) + \alpha \tilde{g}_1(k^2,t) + \beta \tilde{g}_0(k^2,t) = \tilde{\gamma}(k^2,t), \qquad (9.81)$$

where $\tilde{\gamma}$ is defined by

$$\tilde{\gamma}(k^2,t) = \int_0^t e^{k^2 s}\gamma(s)\mathrm{d}s. \qquad (9.82)$$

Equation (9.81) can be written in the form

$$\alpha \tilde{g}_1 + \left(\beta - k^2\right)\tilde{g}_0 = \tilde{\delta} - e^{k^2 t}u(0,t), \qquad (9.83)$$

where $\tilde{\delta}(k^2,t)$ is defined by

$$\tilde{\delta}(k^2,t) = \tilde{\gamma}(k^2,t) + u_0(0). \qquad (9.84)$$

Solving Eqs. (9.83) and (9.21), namely the global relation evaluated at $-k$ for $\tilde{g}_0$ and $\tilde{g}_1$, we find

$$\tilde{g}_0(k^2,t) = -\frac{1}{\Delta(k)}\left\{\tilde{\delta}(k^2,t) - \alpha \hat{u}_0(-k) + e^{k^2 t}\left[\alpha \hat{u}(-k,t) - u(0,t)\right]\right\}, \qquad (9.85)$$

$$\tilde{g}_1(k^2,t) = -\frac{1}{\Delta(k)}\left\{ik\tilde{\delta}(k^2,t) + \left(\beta - k^2\right)\hat{u}_0(-k)\right.$$
$$\left. + e^{k^2 t}\left[\left(k^2 - \beta\right)\hat{u}(-k,t) - iku(0,t)\right]\right\}, \qquad (9.86)$$

where

$$\Delta(k) = k^2 - i\alpha k - \beta. \qquad (9.87)$$

Inserting Eqs. (9.85) and (9.86) in the integral representation Eq. (9.14) for $u(x,t)$ we find

$$u(x,t) = \int_{-\infty}^{\infty} e^{ikx - k^2 t}\hat{u}_0(k)\frac{\mathrm{d}k}{2\pi}$$
$$+ \int_{\partial \mathcal{D}^+} \frac{e^{ikx - k^2 t}}{\Delta(k)}\left[2ik\tilde{\delta}(k^2,t) - \left(k^2 + i\alpha k - \beta\right)\hat{u}_0(-k)\right]\frac{\mathrm{d}k}{2\pi}$$
$$+ \int_{\partial \mathcal{D}^+} \frac{e^{ikx}}{\Delta(k)}\left[\left(k^2 + i\alpha k - \beta\right)\hat{u}(-k,t) - 2iku(0,t)\right]\frac{\mathrm{d}k}{2\pi}, \qquad (9.88)$$

where the contour $\partial \mathcal{D}^+$ is shown in Fig. 9.1.

The representation (9.88) is not yet effective, since it involves the unknown functions $\hat{u}(-k,t)$ and $u(0,t)$. However, the contributions of the corresponding k-integrals can be computed explicitly. Indeed, denoting the roots of the polynomial $k^2 - i\alpha k - \beta$ by

$$k_{1,2} = \frac{1}{2}\left(i\alpha \pm \sqrt{-\alpha^2 + 4\beta}\right), \tag{9.89}$$

and employing Cauchy's theorem along with Jordan's lemma in the region $\mathcal{D}^+$ depicted in Fig. 9.1, we can obtain the results below.

(i) If the roots $k_{1,2}$ do not lie in the region $\mathcal{D}^+$, then the heat equation on the half-line with the initial condition (9.69) and the boundary condition (9.74) admits the solution

$$u(x,t) = \int_{-\infty}^{\infty} e^{ikx - k^2 t} \hat{u}_0(k) \frac{dk}{2\pi}$$
$$+ \int_{\partial\mathcal{D}^+} \frac{e^{ikx - k^2 t}}{\Delta(k)} \left[2ik\,\tilde{\delta}(k^2,t) - \left(k^2 + i\alpha k - \beta\right)\hat{u}_0(-k)\right] \frac{dk}{2\pi}, \tag{9.90}$$

where $\hat{u}_0(k)$ denotes the Fourier transform of the initial data, the function $\tilde{\delta}(k^2,t)$ is defined by Eq. (9.84) and the contour $\partial\mathcal{D}^+$ is shown in Fig. 9.1. Indeed, in this case $\Delta(k)$ has no zeros in $\mathcal{D}^+$. Furthermore,

$$\frac{1}{\Delta(k)}\left[\left(k^2 + i\alpha k - \beta\right)\hat{u}(-k,t) - 2iku(0,t)\right] \sim -\hat{u}(-k,t) - \frac{2i}{k}u(0,t), \quad k \to \infty. \tag{9.91}$$

Thus, Jordan's lemma implies that the contribution of the second integral in Eq. (9.88) vanishes.

(ii) If $k_1 \in \mathcal{D}^+$ and $k_2 \notin \mathcal{D}^+$, then the solution is given by the expression (9.90) plus the function $u_1(x,t)$ defined by

$$u_1(x,t) = -\frac{2k_1}{2k_1 - i\alpha} e^{ik_1 x - k_1^2 t}\left[\alpha\,\hat{u}_0(-k_1) - \tilde{\delta}(k_1^2,t)\right]. \tag{9.92}$$

Indeed, noting that the expression defined by the left-hand side of Eq. (9.91) vanishes as $|k| \to \infty$ in $\mathcal{D}^+$, and employing Cauchy's theorem, we find that the contribution of the second integral in (9.88), denoted by u_1, equals

$$u_1 = ie^{ik_1 x} \frac{\left(k_1^2 + i\alpha k_1 - \beta\right)\hat{u}(-k_1,t) - 2ik_1 u(0,t)}{\Delta'(k_1)}. \tag{9.93}$$

Remarkably, the unknown quantities appearing above can be obtained explicitly: multiplying Eq. (9.85) by $\Delta(k)$ and evaluating the resulting equation at k_j we find

$$e^{k_j^2 t}\left[\alpha\hat{u}(-k_j,t) - u(0,t)\right] = \alpha\hat{u}_0(k_j) - \tilde{\delta}(k_j^2,t), \tag{9.94}$$

where $\Delta(k_j) = 0$, namely

$$k_j^2 = i\alpha k_j + \beta.$$

Using the above equations we find

$$\left(k_j^2 + i\alpha k_j - \beta\right)\hat{u}(-k_j,t) - 2ik_j u(0,t) = 2ik_j\left[\alpha\hat{u}(-k_j,t) - u(0,t)\right]$$
$$= 2ik_j e^{-k_j^2 t}\left[\alpha\hat{u}_0(k_j) - \tilde{\delta}(k_j^2,t)\right]. \tag{9.95}$$

By employing Eq. (9.95), Eq. (9.93) becomes Eq. (9.92).

(iii) If $k_2 \in \mathcal{D}^+$ and $k_1 \notin \mathcal{D}^+$, then the solution is given by the expression (9.90) plus the function $u_2(x,t)$ defined by

$$u_2(x,t) = -\frac{2k_2}{2k_2 - i\alpha}\, e^{ik_2 x - k_2^2 t}\left[\alpha\,\hat{u}_0(-k_2) - \tilde{\delta}(k_2^2, t)\right]. \qquad (9.96)$$

This situation is similar with the above.

(iv) If both k_1 and k_2 lie in the region $\mathcal{D}^+$, then the solution is given by the expression (9.90) plus the functions $u_1(x,t)$ and $u_2(x,t)$. Indeed, in this case both poles yield a contribution.

Derivation: Return to the real line and classical spectral representation
In the particular case of $\alpha = 1$, $\beta = \gamma = 0$, Eqs. (9.90) and (9.92) yield the solution expression

$$u(x,t) = \int_{-\infty}^{\infty} e^{ikx - k^2 t}\hat{u}_0(k)\frac{dk}{2\pi}$$
$$+ \int_{\partial\mathcal{D}^+} e^{ikx - k^2 t}\frac{2iu_0(0) - (k+i)\,\hat{u}_0(-k)}{k - i}\frac{dk}{2\pi}$$
$$+ 2e^{-x+t}\left[u_0(0) - \hat{u}_0(-i)\right]. \qquad (9.97)$$

In order to obtain the representation (9.80), we deform the contour of integration $\partial\mathcal{D}^+$ back to the real line. As stated earlier, for general boundary value problems such a deformation is *not* possible. However, in this particular case the terms involved in Eq. (9.97) are bounded and analytic for $k \in \mathbb{C}^+\backslash\mathcal{D}^+$, thus $\partial\mathcal{D}^+$ can indeed be deformed back to the real line, giving:

$$u(x,t) = \int_{-\infty}^{\infty} e^{ikx - k^2 t}\hat{u}_0(k)\frac{dk}{2\pi}$$
$$+ \int_{-\infty}^{\infty} e^{ikx - k^2 t}\left[\frac{2i}{k-i}u_0(0) - \frac{k+i}{k-i}\hat{u}_0(-k)\right]\frac{dk}{2\pi}$$
$$+ 2e^{-x+t}\left[u_0(0) - \hat{u}_0(-i)\right]. \qquad (9.98)$$

Using the definition of $\hat{u}_0(k)$ and rearranging, we find

$$u(x,t) = \int_0^{\infty} e^{-k^2 t}\left[e^{ikx}\int_0^{\infty} e^{-ik\xi}u_0(\xi)d\xi + e^{-ikx}\int_0^{\infty} e^{ik\xi}u_0(\xi)d\xi\right]\frac{dk}{2\pi}$$
$$- \int_0^{\infty} e^{-k^2 t}\left[e^{ikx}\frac{k+i}{k-i}\int_0^{\infty} e^{ik\xi}u_0(\xi)d\xi\right]\frac{dk}{2\pi}$$
$$- \int_0^{\infty} e^{-k^2 t}\left[e^{-ikx}\frac{k-i}{k+i}\int_0^{\infty} e^{-ik\xi}u_0(\xi)d\xi\right]\frac{dk}{2\pi}$$
$$- 2i\int_0^{\infty} e^{-k^2 t}\left[\frac{e^{-ikx}}{k+i} - \frac{e^{ikx}}{k-i}\right]u_0(0)\frac{dk}{2\pi}$$
$$+ 2e^{-x+t}\left[u_0(0) - \int_0^{\infty} e^{-\xi}u_0(\xi)d\xi\right].$$

Simplifying, we obtain the representation (9.80).

Example 9.4: We illustrate the solution for $u_0(x) = xe^{-4x}$, $\gamma(t) = \sin(5t)$, $\alpha = -2$ and $\beta = 1$. Then, the transforms become

$$\tilde{u}_0(k) = \frac{1}{(4 + ik)^2} \quad \text{and} \quad \tilde{\delta}(k^2, t) = \frac{e^{k^2 t} \left(k^2 \sin(5t) - 5\cos(5t)\right) + 5}{k^4 + 25}.$$

Following the same procedure as in Example 9.1 and with the same contour C^+, Eq. (9.90) takes the form

$$u(x, t) = v(x, t) + \int_C V(k, x, t) dk,$$

where

$$V(k, x, t) = \frac{e^{ikx - k^2 t}}{2\pi} \left[\frac{1}{(k + i)^2} \left(\frac{10ik}{k^4 + 25} + \frac{(k - i)^2}{(k + 4i)^2} \right) + \frac{1}{(4 + ik)^2} \right],$$

and

$$v(x, t) = \int_{C^+} e^{ikx} \frac{2ik \left(k^2 \sin(5t) - 5\cos(5t)\right)}{(k + i)^2 (k^4 + 25)} \frac{dk}{2\pi}.$$

The latter integral can be calculated analytically, by evaluating the residue contribution. In the domain which is over the curve C^+ there are two poles, namely

$$k_1 = e^{i\pi/4}\sqrt{5} \quad \text{and} \quad k_2 = e^{3i\pi/4}\sqrt{5}.$$

The associated residues are

$$-\frac{ie^{-\sqrt{5}(1-i)x/\sqrt{2}} \left(\cos(5t) - i\sin(5t)\right)}{\left(\sqrt{5} + i\left(\sqrt{2} + \sqrt{5}\right)\right)^2},$$

$$\frac{ie^{-\sqrt{5}(1+i)x/\sqrt{2}} \left(\cos(5t) + i\sin(5t)\right)}{\left(\sqrt{5} - i\left(\sqrt{2} + \sqrt{5}\right)\right)^2}.$$

Thus, the sum of the above expressions yields

$$v(x, t) = \frac{e^{-x\sqrt{5/2}}}{\left(\sqrt{10} + 6\right)^2} \left[\left(\sqrt{10} + 1\right) \sin\left(5t - \frac{\sqrt{5}x}{\sqrt{2}}\right) \right.$$

$$\left. - \left(\sqrt{10} + 5\right) \cos\left(5t - \frac{\sqrt{5}x}{\sqrt{2}}\right) \right].$$

Verification of the initial and boundary conditions is obtained as in Example 9.1.

Plots of the numerical evaluation of the solution for $x \in [0, 1]$ at short (left panel) and longer (right panel) timescales with the initial-boundary condition shown as the black dashed line, as well as the three-dimensional perspective plot in the range $x \in [0, 2]$ and $t \in [0, 2]$, generated using Mathematica, are shown below.

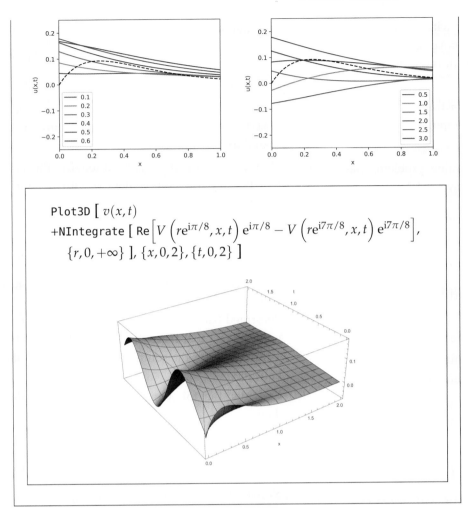

Further reading

1. A.S. Fokas, *A Unified Approach for Boundary Value Problems* (Society for Industrial and Applied Mathematics, 2008). This book provides a thorough account of the unified transform method with numerous examples. The level of treatment is more detailed than in the present book.

2. A.S. Fokas and B. Pelloni, Eds, *Unified Transform for Boundary Value Problems: Applications and Advances* (Society for Industrial and Applied Mathematics, 2015). This book is a state-of-the-art account of the applications of the unified transform to a wide variety of problems and includes new theoretical results as well as numerical applications.

Problems

1. By replacing k with $-k$ in the first term of the second integral of Eq. (9.13) show that this equation

becomes

$$
u(x,t) = 2i \int_{-\infty}^{\infty} e^{-k^2 t}[\sin(kx)\hat{u}_0(k) \\
- \tilde{g}_0(k^2,t)]\frac{dk}{2\pi}. \tag{9.99}
$$

Split the above integral to an integral along $(-\infty, 0)$ and an integral along $(0, +\infty)$. Then, by replacing k with $-k$ in the integral along $(-\infty, 0)$, show that Eq. (9.99) becomes the expression of Eq. (8.39) obtained through the sine-transform, Eq. (7.40).

2. Consider the heat equation on the half-line with the initial and boundary conditions

$$
u(x,0) = u_0(x), \quad 0 < x < +\infty,
$$
$$
u_x(0,t) = g_1(t), \quad t > 0. \tag{9.100}
$$

(a) Solve the above initial–boundary value problem using the cosine-transform defined in Eq. (7.42).

(b) Starting with Eq. (9.11), derive the analogue of Eq. (9.99).

(c) Then, following steps similar with those used in Problem 1, show that the above solution coincides with the one obtained through the cosine-transform.

3. Find the analogue of Eq. (9.15) for the heat equation defined on the half-line with the initial and boundary conditions of Eq. (9.100).

4. Consider the initial–boundary value problem analyzed in Problem 4 with the following data:

$$u_0(x) = e^{-ax}, \quad 0 < x < +\infty;$$
$$g_1(t) = -a\cos(bt), \quad a, b > 0. \quad (9.101)$$

Note that

$$\frac{du_0}{dx}(0) = -a = g_1(0),$$

which means that the data of Eq. (9.101) are compatible at $x = t = 0$. By employing the hybrid analytical-numerical technique used in Example 9.1, compute $u(x,t)$. For the numerical computation take $a = 1.6$, $b = 2.1$.

5. Consider the one-dimensional Schrödinger equation with a zero potential,

$$iu_t + u_{xx} = 0, \quad (x,t) \in \Omega, \quad (9.102)$$

where $u(x,t)$ is a complex-valued function and Ω is a convex two-dimensional domain.

(a) Suppose that Eq. (9.102) is defined on the half-line, $0 < x < +\infty$, with the initial and boundary conditions of Eq. (9.100). Using the unified transform, derive an integral representation of the solution $u(x,t)$ that involves a contour along the real axis and along the positive imaginary axis of the k-complex plane.

(b) Consider the following initial–boundary value data:

$$u_0(x) = xe^{-ax}, \quad 0 < x < +\infty;$$
$$g_1(t) = \cos(bt), \quad a, b > 0. \quad (9.103)$$

Note that

$$\frac{du_0}{dx}(0) = 1 = g_1(0),$$

which means that the data of Eq. (9.103) are compatible at $x = t = 0$. By employing the hybrid analytical–numerical technique used in Example 9.1, compute $u(x,t)$. For the numerical computation take $a = 2.2$, $b = 1.5$.

6. Suppose that $u(x,t)$ satisfies the PDE

$$u_t - u_{xxx} = 0, \quad (9.104)$$

defined on the half-line, $0 < x < +\infty$, with $t > 0$.

(a) By using the unified transform show that a well-posed problem for this equation requires two-boundary conditions.

(b) Derive an integral representation for the case that Eq. (9.104) is supplemented with the initial and boundary conditions

$$u(x,0) = u_0(x), \quad 0 < x < +\infty;$$
$$u(0,t) = g_0(t), \quad u_x(0,t) = g_1(t), \quad t > 0. \quad (9.105)$$

(c) Show that the contours associated with $\hat{u}_0(-\alpha k)$ and $\hat{u}_0(-\alpha^2 k)$, where $\alpha = \exp[i2\pi/3]$, cannot be deformed back to the real axis. This is consistent with the fact that there does not exist a standard x-transform for the solution of Eq. (9.104) on the half-line.

(d) Consider the following initial–boundary value data:

$$u_0(x) = e^{-ax}, \quad g_0(t) = \cos(bt),$$
$$g_1(t) = -a\cos(ct), \quad a, b, c > 0. \quad (9.106)$$

Note that

$$\frac{du_0}{dx}(0) = -a = g_1(0),$$

which means that the data of Eq. (9.106) are compatible at $x = t = 0$. By employing the hybrid analytical–numerical technique used in Example 9.1, compute $u(x,t)$. For the numerical computation take $a = 1.3$, $b = 2.4$, $c = 3.5$.

7. By evaluating Eq. (9.54) at $t = 0$, derive the following transform:

$$\hat{f}(k) = \int_0^\infty e^{-ikx} f(x)dx, \quad \text{Im}[k] \le 0, \quad (9.107a)$$

$$f(x) = \int_{-\infty}^\infty e^{ikx} \hat{f}(k)\frac{dk}{2\pi}$$
$$+ \int_{\partial D^+} e^{ikx}\left[\alpha\hat{f}(\alpha k) + \alpha^2\hat{f}(\alpha^2 k)\right]\frac{dk}{2\pi}, \quad (9.107b)$$

where, $0 < x < +\infty$, $\alpha = \exp[i2\pi/3]$ and ∂D^+ is the union of the rays making angles $\pi/3$ and $2\pi/3$ with the real axis, as shown in the diagram for case (A) of the derivation of the domain D for PDEs with a third-order spatial derivative [see Section 9.4].

(a) Prove directly the validity of Eqs. (9.107a) and (9.107b).

(b) Use the transform of Eqs. (9.107a) and (9.107b) to solve Eq. (9.34) with the initial and boundary conditions

$$u(x,0) = u_0(x), \ 0 < x < +\infty;$$

$$u(0,t) = g_0(t), \ t > 0. \tag{9.108}$$

8. Consider the equation

$$iu_t + u_{xx} = F(x,t), \quad 0 < x < +\infty; \quad t > 0, \tag{9.109}$$

with the initial and boundary condition of Eq. (9.108).

(a) Derive an integral representation for the solution $u(x,t)$.

(b) Compute the solution $u(x,t)$ in the particular case that

$$u_0(x) = 0, \ g_0(t) = \sin(bt), \ F(x,t) = e^{-ax+ict},$$

where the constants $a, b, c > 0$. For the numerical solution use $a = 2.1$, $b = 1.3$, $c = 2.4$.

Chapter 10

Unified Transform II: Evolution PDEs
on a Finite Interval

In this chapter, we turn our attention to the case of evolution PDEs on a finite interval and we proceed to derive their solution through the unified transform. We follow the same itinerary, namely we start with the heat equation, followed by an evolution PDE that contains a third-order spatial derivative, and we close with a discussion of inhomogeneous PDEs and other considerations. The unified transform on a finite interval for evolution PDEs involves three steps which are very similar with those used in Section 9.2.

10.1   The heat equation

For the heat equation on the finite interval $x \in [0, L]$ the three steps of the unified transform are as follows:

Step 1. *Obtain the global relation.*
By employing the Fourier transform on the finite interval $[0, L]$, namely, Eq. (9.6a), we find

$$e^{k^2 t}\hat{u}(k, t) = \hat{u}_0(k) - \tilde{g}_1(k^2, t) - ik\tilde{g}_0(k^2, t) + e^{-ikL}\left[\tilde{h}_1(k^2, t) + ik\tilde{h}_0(k^2, t)\right], \quad k \in \mathbb{C},$$

$$(10.1)$$

where $\tilde{g}_0$ and $\tilde{g}_1$ are defined in Eq. (9.8), and $\hat{u}, \hat{u}_0, \tilde{h}_0, \tilde{h}_1$ are defined as follows:

$$\hat{u}(k, t) = \int_0^L e^{-ikx}u(x, t)dx, \quad \hat{u}_0(k) = \int_0^L e^{-ikx}u_0(x)dx, \quad k \in \mathbb{C}, \qquad (10.2a)$$

$$\tilde{h}_0(k, t) = \int_0^t e^{k\tau}u(L, \tau)d\tau, \quad \tilde{h}_1(k, t) = \int_0^t e^{k\tau}u_x(L, \tau)d\tau, \quad k \in \mathbb{C}. \qquad (10.2b)$$

Indeed, differentiating the first of Eq. (10.2a) with respect to t, replacing u_t by u_{xx}, and using integration by parts, we find

$$\frac{\partial\hat{u}}{\partial t} + k^2\hat{u} = -u_x(0, t) - iku(0, t) + e^{-ikL}\left[u_x(L, t) + iku(L, t)\right].$$

Solving this ODE for $\hat{u}(k, t)$ we find Eq. (10.1).

We emphasize that the procedure for obtaining (10.1) is very similar with the procedure used for obtaining (8.49). However, Eq. (10.1) in contrast to Eq. (8.49) involves *all* possible boundary values, namely $u(0, t), u_x(0, t), u(L, t), u_x(L, t)$. On the other hand, Eq. (10.1) is valid for *all* complex values of k, whereas Eq. (8.49) is restricted to $k = n\pi/L$. The success of the unified transform for boundary value problems on a finite interval is based on the fact that Eq. (10.1) is valid for *all* complex values of k.

Step 2. *By using the inverse Fourier transform,*

$$u(x) = \int_{-\infty}^{\infty} e^{ikx} \hat{u}(k) \frac{dk}{2\pi}, \ 0 < x < L,$$

the global relation yields an integral representation on the real line. By deforming the real axis to a contour in the k-complex plane, it is possible to rewrite this expression as an integral along the contour $\partial \mathcal{D}$, where $\partial \mathcal{D}$ is the boundary of the domain $\mathcal{D}$ defined by

$$\partial \mathcal{D} : \{k \in \mathbb{C}, \mathrm{Re}[\omega(k)] = 0\},$$

and $\omega(k)$ is obtained by the requirement that $\exp[ikx - \omega(k)t]$ solves the given PDE. For the heat equation, Eq. (10.1) implies

$$\begin{aligned}
u(x,t) = &\int_{-\infty}^{\infty} e^{ikx - k^2 t} \hat{u}_0(k) \frac{dk}{2\pi} \\
&- \int_{\partial \mathcal{D}^+} e^{ikx - k^2 t} \left[\tilde{g}_1(k^2, t) + ik\tilde{g}_0(k^2, t) \right] \frac{dk}{2\pi} \\
&- \int_{\partial \mathcal{D}^-} e^{-ik(L-x) - k^2 t} \left[\tilde{h}_1(k^2, t) + ik\tilde{h}_0(k^2, t) \right] \frac{dk}{2\pi},
\end{aligned} \tag{10.3a}$$

where the contours $\partial \mathcal{D}^+$ and $\partial \mathcal{D}^-$ are depicted in Fig. 9.1. The justification for deforming the contour associated with $\tilde{h}_1$ and $\tilde{h}_0$ to $\partial \mathcal{D}^-$ is similar with the argument presented in Section 9.2. The only difference is that now, instead of the factor $\exp[ikx]$ there exists the factor $\exp[-ik(L - x)]$ which is bounded in the lower half of the k-complex plane. Also, by convention, the domain is to the *left* of the *increasing* direction of its boundary. Thus, we have changed the sign in front of $\partial \mathcal{D}^-$ to accommodate the fact that we have reversed the direction of integration.

By using the transformations which leave $\omega(k)$ invariant it is always possible to map $\partial \mathcal{D}^-$ to $\partial \mathcal{D}^+$. For the heat equation, using the transformation $k \to -k$ in the third integral on the right-hand side of Eq. (10.3a), we find that this equation becomes

$$\begin{aligned}
u(x,t) = &\int_{-\infty}^{\infty} e^{ikx - k^2 t} \hat{u}_0(k) \frac{dk}{2\pi} \\
&- \int_{\partial \mathcal{D}^+} e^{-k^2 t} \left\{ e^{ikx} \left[\tilde{g}_1(k^2, t) + ik\tilde{g}_0(k^2, t) \right] \right. \\
&\left. + e^{ik(L-x)} \left[-\tilde{h}_1(k^2, t) + ik\tilde{h}_0(k^2, t) \right] \right\} \frac{dk}{2\pi}.
\end{aligned} \tag{10.3b}$$

Step 3. *By using the global relation and by employing the transformations in the k-complex plane which leave $\omega(k)$ invariant, it is possible to eliminate from the integral representation of $u(x,t)$ the transforms of the unknown boundary values.*

In the case of boundary value problems on the half-line, the global relation is valid only for k in the lower half of the k-complex plane. Thus, one can only use those transformations $\nu(k)$ which have the property that for k in the upper half of the k-complex plane, $\nu(k)$ is in the lower half of the k-complex plane. However, now the global relation is valid for *all* complex values of k, thus one can use *all* transformations $\nu(k)$.

For example, for the heat equation, one can use *simultaneously* the global relation (10.1) together with the equation obtained from (10.1) via the transformation $k \to -k$, namely the equation

$$e^{k^2 t}\hat{u}(-k,t) = \hat{u}_0(-k) - \tilde{g}_1(k^2,t) + ik\tilde{g}_0(k^2,t)$$
$$+ e^{ikL}\left[\tilde{h}_1(k^2,t) - ik\tilde{h}_0(k^2,t)\right], \quad k \in \mathbb{C}. \qquad (10.4)$$

Equations (10.1) and (10.4) are two equations involving the four boundary values $u(0,t), u_x(0,t), u(L,t), u_x(L,t)$. Since we expect that $\hat{u}(k,t)$ and $\hat{u}(-k,t)$ will *not* contribute to $u(x,t)$, it follows that for a well-posed problem one needs to prescribe two boundary conditions. Actually, it can be shown that in order for the contribution of $\hat{u}(k,t)$ and $\hat{u}(-k,t)$ to be eliminated, it is *necessary* to prescribe one boundary condition at each end.

In the case that $u(0,t)$ and $u(L,t)$ are given, Eqs. (10.1) and (10.4) become, respectively,

$$e^{k^2 t}\hat{u}(k,t) = G(k,t) - \tilde{g}_1 + e^{-ikL}\tilde{h}_1, \qquad (10.5a)$$
$$e^{k^2 t}\hat{u}(-k,t) = G(-k,t) - \tilde{g}_1 + e^{ikL}\tilde{h}_1, \qquad (10.5b)$$

where the known function $G(k,t)$ is defined by

$$G(k,t) = \hat{u}_0(k) - ik\tilde{g}_0(k^2,t) + ike^{-ikL}\tilde{h}_0(k^2,t). \qquad (10.6)$$

Solving Eqs. (10.5a) and (10.5b) for $\tilde{g}_1$ and $\tilde{h}_1$ we find

$$\tilde{g}_1 = \frac{1}{\Delta(k)}\left[e^{ikL}G(k,t) - e^{-ikL}G(-k,t)\right] + e^{k^2 t}\check{g}_1, \qquad (10.7a)$$

$$\tilde{h}_1 = \frac{1}{\Delta(k)}\left[G(k,t) - G(-k,t)\right] + e^{k^2 t}\check{h}_1, \qquad (10.7b)$$

where $\Delta(k)$ is defined by

$$\Delta(k) = e^{ikL} - e^{-ikL}, \qquad (10.8)$$

and $\check{g}_1, \check{h}_1$ contain the terms $\hat{u}(k,t)$ and $\hat{u}(-k,t)$:

$$\check{g}_1 = -\frac{1}{\Delta(k)}\left[e^{ikL}\hat{u}(k,t) - e^{-ikL}\hat{u}(-k,t)\right], \qquad (10.9a)$$

$$\check{h}_1 = -\frac{1}{\Delta(k)}\left[\hat{u}(k,t) - \hat{u}(-k,t)\right]. \qquad (10.9b)$$

Inserting the expressions for $\tilde{g}_1$ and $\tilde{h}_1$ in Eq. (10.3b) and simplifying we find

$$u(x,t) = \int_{-\infty}^{\infty} e^{ikx - k^2 t}\hat{u}_0(k)\frac{dk}{2\pi}$$
$$- \int_{\partial\mathcal{D}^+} \frac{e^{-k^2 t}}{\Delta(k)}\left\{2\sin(kx)\left[ie^{ikL}\hat{u}_0(k) - 2k\tilde{h}_0(k^2,t)\right]\right.$$
$$\left. + 2\sin(k(L-x))\left[i\hat{u}_0(-k) - 2k\tilde{g}_0(k^2,t)\right]\right\}\frac{dk}{2\pi}, \qquad (10.10)$$

plus the following term which vanishes:

$$\int_{\partial\mathcal{D}^+}\left[e^{ikx}\check{g}_1(k,t) - e^{ik(L-x)}\check{h}_1(k,t)\right]\frac{dk}{2\pi}. \qquad (10.11)$$

Regarding the integral defined in Eq. (10.11) we note that as $k \to \infty$ in the upper half of the k-complex plane, the exponential term $\exp[ikL]$ decays, thus $\Delta(k) \sim \exp[-ikL]$. Hence,

$$\check{g}_1 \sim -\hat{u}(-k,t) + e^{ikL} \int_0^L e^{ik(L-x)} u(x,t) dx, \quad \text{for } k \to \infty.$$

The function $\hat{u}(-k,t)$, as well as the exponentials $\exp[ikx]$ and $\exp[ik(L-x)]$ decay in $\mathcal{D}^+$, thus $\check{g}_1$ decays in $\mathcal{D}^+$ and Cauchy's theorem applied in $\mathcal{D}^+$ implies that the contribution of $\check{g}_1$ vanishes.

Similarly,

$$\check{h}_1 \sim \int_0^L e^{ik(L-x)} u(x,t) dx - e^{ikL} \hat{u}(-k,t), \quad \text{for } k \to \infty,$$

and since the functions $\exp[ik(L-x)]$, $\exp[ikL]$, $\hat{u}(-k,t)$ decay for k in $\mathcal{D}^+$, Cauchy's theorem in the domain $\mathcal{D}^+$ implies that the contribution of the term $\exp[ik(L-x)]\check{h}_1$ vanishes.

The function $\Delta(k)$ vanishes for k on $\partial \mathcal{D}$ only for $k = 0$, but the numerators of the relevant ratios also vanish at $k = 0$; thus, $k = 0$ is a removable singularity.

Equation (10.10) can be rewritten in a form which is consistent with the Ehrenpreis fundamental principle: if the boundary condition is specified for $0 < t < T$, where T is a positive constant, then using Cauchy's theorem together with Jordan's lemma, we can rewrite Eq. (10.10) in the following form:

$$u(x,t) = \int_{-\infty}^{\infty} e^{ikx - k^2 t} \hat{u}_0(k) \frac{dk}{2\pi}$$

$$- \int_{\partial \mathcal{D}^+} \frac{2e^{k^2 t}}{\Delta(k)} \left\{ \sin(kx) \left[i e^{ikL} \hat{u}_0(k) - 2k \tilde{H}_0(k^2) \right] \right.$$

$$\left. + \sin(k(L-x)) \left[i \hat{u}_0(-k) - 2k \tilde{G}_0(k^2) \right] \right\} \frac{dk}{2\pi}, \qquad (10.12)$$

where $\tilde{G}_0(k^2)$, $\tilde{H}_0(k^2)$ are defined as $\tilde{g}_0(k^2,t)$ and $\tilde{h}_0(k^2,t)$ but with t replaced by T.

Example 10.1: We consider the heat equation for $0 < x < 1$, $t > 0$, with

$$u_0(x) = e^{-x}, \quad g_0(t) = \cos(t), \quad h_0(t) = e^{-1} \cos(t).$$

We note that

$$u(0,0) = 1 = u_0(0) \quad \text{and} \quad u(1,0) = e^{-1} = u_0(1),$$

thus the initial and boundary conditions are compatible both at $x = t = 0$ and at $x = 1, t = 0$. The transforms that correspond to these conditions are as follows:

$$\hat{u}_0(k) = \frac{-i\left(1 - e^{-1-ik}\right)}{k - i}, \quad \tilde{g}_0(k^2,t) = \frac{e^{k^2 t}\left(k^2 \cos(t) + \sin(t)\right)}{k^4 + 1},$$

$$\tilde{h}_0(k^2,t) = \frac{\tilde{g}_0(k^2,t)}{e}.$$

The first integral of Eq. (10.10) can now be rewritten as

$$-\int_{\partial \mathcal{D}^+} e^{ikx-k^2t} \frac{i}{k-i} \frac{dk}{2\pi} - \int_{\partial \mathcal{D}^-} e^{ikx-k^2t} \frac{ie^{-1-ik}}{k-i} \frac{dk}{2\pi}$$

$$= -i\int_{\partial \mathcal{D}^+} e^{-k^2t} \left[\frac{e^{ikx}}{k-i} + \frac{e^{ik(1-x)}e^{-1}}{k+i} \right] \frac{dk}{2\pi}.$$

Following the same logic as in Example 9.1, we deform $\partial \mathcal{D}^+$ to the contour C^+ defined by the union of the rays specified by $\text{Arg}[k] = \pi/8$ and $\text{Arg}[k] = 7\pi/8$.

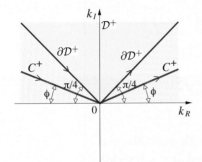

We define the quantity $Q(k, x)$ as

$$Q(k, x) = -4k \left\{ \frac{\sin(kx)}{e} + \sin(k(1-x)) \right\}.$$

Using this definition, Eq. (10.10) becomes

$$u(x,t) = \int_{C^+} \frac{e^{-k^2t}}{e^{ik} - e^{-ik}} \frac{Q(k,x)}{(k^2+1)(k^4+1)}$$

$$\times \left[k^2 - 1 - \left(k^2 + 1 \right) e^{k^2t} \left(k^2\cos(t) + \sin(t) \right) \right] \frac{dk}{2\pi}.$$

The above integral can be split into two parts as follows:

$$u(x,t) = \int_{C^+} \frac{e^{-k^2t}}{e^{ik} - e^{-ik}} \frac{Q(k,x)\,(k^2-1)}{(k^2+1)(k^4+1)} \frac{dk}{2\pi}$$

$$- \int_{C^+} \frac{1}{e^{ik} - e^{-ik}} \frac{Q(k,x)}{(k^4+1)} \left[k^2\cos(t) + \sin(t) \right] \frac{dk}{2\pi}.$$

The second integral of the above expression can be calculated analytically, by evaluating the residue contribution. In the domain which is above the curve C^+ there are two poles, namely

$$k_1 = e^{i\pi/4} \quad \text{and} \quad k_2 = e^{3i\pi/4}.$$

Evaluating the residue contribution of k_1 and k_2, and then summing them up we obtain the total residue contribution

$$u(x,t) = v(x,t) + \int_C V(k,x,t)dk,$$

where

$$v(x,t) = 2e^{x/\sqrt{2}} \left[\left(\frac{1}{2} - \alpha \right) \cos \left(t + \frac{x}{\sqrt{2}} \right) - \beta \sin \left(t + \frac{x}{\sqrt{2}} \right) \right]$$
$$+ 2e^{-x/\sqrt{2}} \left[\alpha \cos \left(t - \frac{x}{\sqrt{2}} \right) + \beta \sin \left(t - \frac{x}{\sqrt{2}} \right) \right],$$

and

$$V(k,x,t) = \frac{e^{-k^2 t}}{2\pi (e^{ik} - e^{-ik})} \frac{Q(k,x)\,(k^2 - 1)}{(k^2 + 1)(k^4 + 1)}.$$

The parameters α, β that appear in the expression for $v(x,t)$ are as follows:

$$\alpha = \frac{-e^{\sqrt{2}} + \cos \left(\sqrt{2} \right) + 2e^{-1} \cos \left(1/\sqrt{2} \right) \sinh \left(1/\sqrt{2} \right)}{4 \cos \left(\sqrt{2} \right) - 4 \cosh \left(\sqrt{2} \right)},$$

$$\beta = \frac{-\sin \left(\sqrt{2} \right) + 2e^{-1} \sin \left(1/\sqrt{2} \right) \cosh \left(1/\sqrt{2} \right)}{4 \cos \left(\sqrt{2} \right) - 4 \cosh \left(\sqrt{2} \right)}.$$

We next verify that these expressions are consistent with the initial and boundary conditions. The expression for $u(x,t)$ evaluated at $x = 0$ yields

$$u(0,t) = \cos(t) + \int_{C^+} V(k,0,t) dk.$$

It suffices to show that the integral in the above expression vanishes. From the definition of $V(k,x,t)$ we obtain

$$\int_{C^+} V(k,0,t) dk = - \int_{C^+} e^{-k^2 t} \frac{2ik\,(k^2 - 1)}{(k^2 + 1)(k^4 + 1)} \frac{dk}{2\pi} = 0.$$

The last equality is obtained by deforming the contour C^+ to the real line and observing that the integrand is an odd function of k. Furthermore, when the expression for $u(x,t)$ is evaluated for $x = 1$ it yields

$$u(1,t) = e^{-1} \cos(t) + \int_{C^+} V(k,1,t) dk,$$

and the integral again vanishes by using similar arguments as in the case $x = 0$.

The original expression for $u(x,t)$ evaluated at $t = 0$ yields

$$u(x,0) = - \int_{C^+} \frac{Q(k,x)}{2i \sin(k)\,(k^2 + 1)} \frac{dk}{2\pi},$$

which has a simple pole $k = i$; we recall that all other singularities are on the real line, with the $k = 0$ being a removable one. Hence, the Residue theorem yields

$$u(x,0) = -(2\pi i) \frac{1}{2\pi} \frac{(2i)(i)}{(2i)} \left(\frac{e^{-x-1} - e^{x-1} + e^{x-1} - e^{-x+1}}{e^{-1} - e} \right) = e^{-x}.$$

Plots of the numerical evaluation of the solution for $x \in [0,1]$ at short (left panel) and longer (right panel) timescales with the initial–boundary condition shown as the black dashed line, as well as the three-dimensional perspective plot in the range $x \in [0,1]$ and $t \in [0,2\pi]$, generated using Mathematica, are shown below.

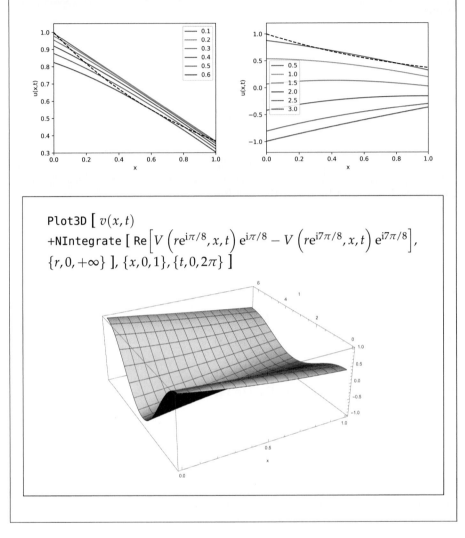

10.2   Advantages of the unified transform

The advantages of the unified transform mentioned in Section 9.3 in connection with boundary value problems on the half-line remain valid for the case that $x \in [0, L]$. In particular, it is straightforward to show that Eq. (10.12), or equivalently Eq. (10.10), satisfies the heat equation, as well as the initial and boundary conditions. Indeed, the only (x,t) dependence of the expression in Eq. (10.12) is in the form of $\exp[\pm ikx - k^2 t]$, thus this expression satisfies the heat equation. Evaluating the right-hand side of Eq. (10.12) at $x = 0$, we find

$$\int_{-\infty}^{\infty} e^{-k^2 t} \hat{u}_0(k) \frac{dk}{2\pi} - \int_{\partial \mathcal{D}^+} e^{-k^2 t} \left[\hat{u}_0(-k) + 2ik\tilde{G}_0(k^2) \right] \frac{dk}{2\pi}.$$

Deforming the integral involving $\hat{u}_0(-k)$ back to the real axis and then replacing k with $-k$ we find that this integral cancels with the integral involving

$\hat{u}_0(k)$. Moreover, using the same change of variables employed for the case of the half-line, namely $k = il$ the integral involving $\tilde{G}_0(k^2)$ yields $g_0(t)$. The proof that $u(L,t) = h_0(t)$ is even simpler since the relevant integral over $\partial\mathcal{D}^+$ involves $\hat{u}_0(k)$. Finally, evaluating the right-hand side of Eq. (10.10) at $t = 0$ we find $u(x,0) = u_0(x)$, since the integral along $\partial\mathcal{D}^+$ vanishes. This is the consequence of the fact that the functions

$$\sin(kx)e^{ikL}\frac{\hat{u}_0(k)}{\Delta(k)} \quad \text{and} \quad \sin(k(L-x))\frac{\hat{u}_0(-k)}{\Delta(k)}$$

are analytic and bounded in $\mathbb{C}^+$.

In the case of the heat equation we note that the zeros of $\Delta(k)$ are precisely the points $n\pi/L$ appearing in the sine series representation (the so-called "discrete spectra"):

$$e^{-ikL} - e^{ikL} = 0, \quad e^{2ikL} = 1 = e^{2in\pi}, \quad k = \frac{n\pi}{L}. \tag{10.13}$$

By deforming the contour $\partial\mathcal{D}^+$ back to the real axis, and by using the residue theorem to compute the contribution from the poles $k = n\pi/L$, it is possible to show that the representation of Eq. (10.10) yields the sine series representation of Eq. (8.53).

Actually, there exists a simpler way to connect the unified transform with the latter representation: subtracting Eqs. (10.1), (10.4) we find

$$e^{k^2t}[\hat{u}(-k,t) - \hat{u}(k,t)] = \hat{u}_0(-k) - \hat{u}_0(k) + \Delta(k)\tilde{h}_1(k^2,t) + 2ik\tilde{g}_0(k^2,t)$$
$$- ik\tilde{h}_0(k^2,t)\left(e^{ikL} + e^{-ikL}\right). \tag{10.14}$$

In order to eliminate the unknown term $\tilde{h}_1$, we evaluate the above equation at those values of k for which $\Delta(k) = 0$, that is, $k = n\pi/L$, and then Eq. (10.14) becomes Eq. (8.51).

We also note that the points $k = n\pi/L$ are those points in the k-complex plane where the integrands of the integrals involving $\partial\mathcal{D}^+$ are singular. Thus, in a sense, the standard series representations are associated with the "worst" possible choice of points.

Regarding the applicability of the unified transform to linear evolution PDEs involving a spatial derivative of order n, we note the following: there exist n unknown boundary values at each side, and the global relation together with the equations obtained via the transformation $k \to \nu(k)$ yield n algebraic relations. Thus, one must prescribe a total of n boundary conditions. The analysis of the associated half-line problem determines in a straightforward way the number of boundary conditions needed at $x = 0$: $n/2$ if n is even, and $(n-1)/2$ or $(n+1)/2$ if n is odd. Thus, the number of boundary conditions at $x = L$ is $n/2$ if n is even, and $(n-1)/2$ or $(n+1)/2$ if n is odd.

The unified transform indeed provides a *unified* treatment to initial boundary value problems for evolution equations: comparing the representations obtained for problems defined on the line, Eq. (8.26), on the half-line, Eq. (9.16), and on the interval, Eq. (10.12), one observes that for the half-line the solution contains an additional integral along $\partial\mathcal{D}^+$. Actually, this formulation provides a straightforward way for recovering the solution of the half-line from the solution of the finite interval by taking $L \to \infty$. Such limiting procedures do *not* exist for the traditional representations involving an infinite series.

10.3   A PDE with a third-order spatial derivative

In what follows, we implement the three steps outlined in Section 10.1 for the third-order PDE of Eq. (9.34), namely

$$u_t + u_{xxx} = 0. \tag{10.15}$$

We will find it convenient to employ the constant α, defined by

$$\alpha \equiv e^{i2\pi/3} \Rightarrow \alpha^2 = e^{i4\pi/3}, \quad \alpha^3 = 1.$$

Step 1. Employing the Fourier transform formula, Eq. (9.6a), we obtain the global relation

$$e^{-ik^3 t}\hat{u}(k,t) = \hat{u}_0(k) + \tilde{g}_2 + ik\tilde{g}_1 - k^2\tilde{g}_0 - e^{-ikL}(\tilde{h}_2 + ik\tilde{h}_1 - k^2\tilde{h}_0), \tag{10.16}$$

where for convenience we have suppressed the dependence of $\tilde{g}_j$, $\tilde{h}_j$, $j = 0, 1, 2$, on $(-ik^3, t)$. The functions $\tilde{g}_2$ and $\tilde{h}_2$ are defined by

$$\tilde{g}_2(k,t) = \int_0^t e^{k\tau} u_{xx}(0,\tau)d\tau, \quad \tilde{h}_2(k,t) = \int_0^t e^{k\tau} u_{xx}(L,\tau)d\tau, \quad k \in \mathbb{C}. \tag{10.17}$$

Step 2. Inserting the expression for $\hat{u}(k,t)$ in the inverse Fourier transform formula and then deforming the real axis to $\partial\mathcal{D}^+$ and $\partial\mathcal{D}^-$ we find

$$u(x,t) = \int_{-\infty}^{\infty} e^{ikx+ik^3 t}\hat{u}_0(k)\frac{dk}{2\pi}$$
$$+ \int_{\partial\mathcal{D}^+} e^{ikx+ik^3 t}\left[\tilde{g}_2(-ik^3,t) + ik\tilde{g}_1(-ik^3,t) - k^2\tilde{g}_0(-ik^3,t)\right]\frac{dk}{2\pi}$$
$$+ \int_{\partial\mathcal{D}^-} e^{-ik(L-x)+ik^3 t}\left[\tilde{h}_2(-ik^3,t) + ik\tilde{h}_1(-ik^3,t) - k^2\tilde{h}_0(-ik^3,t)\right]\frac{dk}{2\pi}, \tag{10.18a}$$

where $\partial\mathcal{D}^+$ and $\partial\mathcal{D}^-$ are depicted in Fig. 10.1; these are the boundaries of the domains $\mathcal{D}^+$ and $\mathcal{D}^-$, discussed in detail in relation to case (A) of the derivation of the domain $\mathcal{D}$ for evolution equations with a third spatial derivative.

If k is in $\mathcal{D}^+$, that is, $\pi/3 < \text{Arg}[k] < 2\pi/3$, then $\pi < \text{Arg}[\alpha k] < 4\pi/3$, that is, k is in the left subdomain of $\partial\mathcal{D}^-$. Similarly, if k is in $\mathcal{D}^+$, then $5\pi/3 < \text{Arg}[\alpha^2 k] < 2\pi$, that is, k is in the right subdomain of $\mathcal{D}^-$. Hence, replacing k by αk and by $\alpha^2 k$ maps the boundaries of the left and the right subdomains of $\mathcal{D}^-$ to the boundary of $\mathcal{D}^+$. Hence, Eq. (10.18a) can be rewritten in the following form:

$$u(x,t) = \int_{-\infty}^{\infty} e^{ikx+ik^3 t}\hat{u}_0(k)\frac{dk}{2\pi}$$

$$+ \int_{\partial\mathcal{D}^+} e^{ik^3 t}\left\{e^{ikx}\left[\tilde{g}_2(-ik^3,t) + ik\tilde{g}_1(-ik^3,t) - k^2\tilde{g}_0(-ik^3,t)\right]\right.$$

$$+ e^{-i\alpha k(L-x)}\left[\tilde{h}_2(-ik^3,t) + i\alpha k\tilde{h}_1(-ik^3,t) - \alpha^2 k^2\tilde{h}_0(-ik^3,t)\right]\alpha$$

$$\left. + e^{-i\alpha^2 k(L-x)}\left[\tilde{h}_2(-ik^3,t) + i\alpha^2 k\tilde{h}_1(-ik^3,t) - \alpha k^2\tilde{h}_0(-ik^3,t)\right]\alpha^2\right\}\frac{dk}{2\pi}. \tag{10.18b}$$

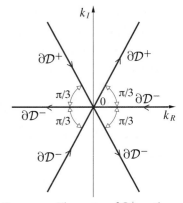

Figure 10.1: The contour $\partial\mathcal{D}^+$ on the upper half complex plane and the contour $\partial\mathcal{D}^-$ on the lower half complex plane (see also case (A) in the derivation of the domain $\mathcal{D}$ for evolution equations with a third spatial derivative for the corresponding domains $\mathcal{D}^+$ and $\mathcal{D}^-$).

Step 3. The global relation, Eq. (10.16), must be supplemented with the two equations obtained from (10.16) by replacing k with αk and $\alpha^2 k$.

Consider the following initial–boundary value problem:

$$u(x,0) = u_0(x), \ 0 < x < L; \ u(0,t) = g_0(t), \ u(L,t) = h_0(t),$$
$$u_x(L,t) = h_1(t), \ t > 0. \tag{10.19}$$

The global relation together with the equations obtained from the global relation via the transforms $k \to \alpha k$, $k \to \alpha^2 k$ yield the following three equations:

$$e^{-ik^3 t}\hat{u}(k,t) = G(k,t) + \tilde{g}_2 + ik\tilde{g}_1 - e^{-ikL}\tilde{h}_2,$$
$$e^{-ik^3 t}\hat{u}(\alpha k,t) = G(\alpha k,t) + \tilde{g}_2 + i\alpha k\tilde{g}_1 - e^{-i\alpha L}\tilde{h}_2,$$
$$e^{-ik^3 t}\hat{u}(\alpha^2 k,t) = G(\alpha^2 k,t) + \tilde{g}_2 + i\alpha^2 k\tilde{g}_1 - e^{-i\alpha^2 L}\tilde{h}_2,$$

where

$$G(k,t) = \hat{u}_0(k) - k^2\tilde{g}_0 + k^2 e^{-ikL}\tilde{h}_0 - ike^{-ikL}\tilde{h}_1.$$

Thus,

$$\begin{pmatrix} 1 & 1 & -e^{-ikL} \\ 1 & \alpha & -e^{-i\alpha kL} \\ 1 & \alpha^2 & -e^{-i\alpha^2 kL} \end{pmatrix} \begin{pmatrix} \tilde{g}_2 \\ ik\tilde{g}_1 \\ \tilde{h}_2 \end{pmatrix} = \begin{pmatrix} e^{-ik^3 t}\hat{u}(k,t) - G(k,t) \\ e^{-ik^3 t}\hat{u}(\alpha k,t) - G(\alpha k,t) \\ e^{-ik^3 t}\hat{u}(\alpha^2 k,t) - G(\alpha^2 k,t) \end{pmatrix}, \ k \in \mathbb{C}. \tag{10.20}$$

The unknown functions $\tilde{g}_2$, $ik\tilde{g}_1$, $\tilde{h}_2$ can be expressed in terms of the known function G as well as the unknown functions $\hat{u}(k,t)$, $\hat{u}(\alpha k,t)$, $\hat{u}(\alpha^2 k,t)$. It turns out that the terms involving these unknown functions times the associated exponentials involving x are bounded in $\mathcal{D}^+$, thus the contribution of these unknown functions vanishes provided that the determinant, $\Delta(k)$, of the 3×3 matrix in Eq. (10.20) does *not* have zeros in $\mathcal{D}$. This determinant is given by

$$\Delta(k) = (1 - \alpha)\left[\alpha e^{-ikL} - (1 + \alpha)e^{-i\alpha kL} + e^{-i\alpha^2 kL}\right]. \tag{10.21}$$

The zeros of $\Delta(k)$ can be found by making use of the following result:

Let $F(z)$ be defined by

$$F(z) = e^z + a_1 e^{k_1 z} + \cdots + a_n e^{k_n z},$$

where $\{a_j, k_j\}_1^n$, $a_j \neq 0$, are complex constants. Assume that the polygon with vertices at the points $\{1, k_1, \ldots, k_n\}$ is convex. Then, the zeros of $F(z)$ are clustered along the rays emanating from the origin with direction orthogonal to the sides of the polygon. Furthermore, these zeros can only accumulate at infinity on these rays.

Letting $-ikL = z$, it follows that the vertices $1, \alpha, \alpha^2$ yield the triangle and the lines depicted in Figs. 10.2 and 10.3, respectively. The relevant lines are at angles $\pi/3$, π and $5\pi/3$. Hence the corresponding lines in the k-complex plane are at angles $5\pi/6$, $3\pi/2$ and $\pi/6$. Thus, these lines are in the exterior of $\mathcal{D}$ and hence the contribution of the term involving $\hat{u}(k,t)$, $\hat{u}(\alpha k,t)$, $\hat{u}(\alpha^2 k,t)$ vanishes.

Recall that in the case of the heat equation, it is possible to deform $\partial\mathcal{D}$ back to the real axis and this yields the traditional representation in terms of an infinite sine series. Furthermore, in this case the zeros of $\Delta(k)$ can be computed *explicitly* and this is the reason that the sine series involves the explicit numbers

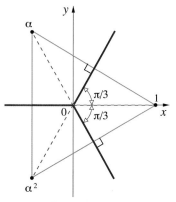

Figure 10.2: The triangle associated with $\Delta(k)$, defined in Eq.(10.21), in the z-complex plane.

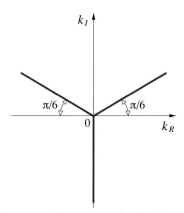

Figure 10.3: The rays associated with $\Delta(k)$ in the k-complex plane.

$n\pi/L$. However, it turns out that in the case of Eq. (10.15) it is *not* possible to deform $\partial \mathcal{D}$ on the three rays on which $\Delta(k)$ vanishes. Thus, it is *not* possible to express $u(x,t)$ in terms of an infinite series. But, even if this were possible, the associated series would involve the values k_n which would *not* be explicit but would be defined via the transcendental equation $\Delta(k) = 0$, with $\Delta(k)$ given by Eq. (10.21).

There exists a common misconception that boundary value problems defined on an infinite domain can be solved by integral transforms (continuous spectrum), whereas problems defined on an interval can be solved via infinite series (discrete spectrum). The analysis of this simple PDE involving a third-order derivative shows that there exist simple boundary value problems formulated on an interval for which there *does* exist an exact representation in terms of integrals in the k-complex plane, but for which there does *not* exist an infinite series representation.

10.4   Inhomogeneous PDEs and other considerations

The solution of an initial–boundary value problem for an inhomogeneous evolution PDE can be obtained from the solution of the homogeneous PDE through a similar procedure as the one used in Section 9.5:

$$\hat{u}_0(k) \text{ is replaced with } \hat{v}(k,t) = \hat{u}_0(k) + \int_0^L e^{-ikx}dx \int_0^t e^{\omega(k)\tau}f(x,\tau)d\tau,$$

$$(10.22)$$

where $f(x,t)$ is the forcing, and $\exp[ikx - \omega(k)t]$ solves the given homogeneous PDE.

10.4.1   Robin boundary conditions

As mentioned earlier, a major advantage of the unified transform is that it provides a "unified" approach for solving a given PDE with a variety of boundary conditions. Indeed, the integral representation and the global relation are the two basic equations needed for the solutions of any well-posed initial–boundary value problem.

For concreteness, we consider the heat equation with a Robin boundary condition at $x = 0$ and with u given at $x = L$, that is,

$$u_x(0,t) - \gamma u(0,t) = g_R(t), \quad u(L,t) = h_0(t), \quad t > 0. \qquad (10.23)$$

Also, for simplicity we assume that $\gamma > 0$.

Taking the t-transform of Eq. (10.23) we find $\tilde{g}_1 = \gamma\tilde{g}_0 + \tilde{g}_R$, where $\tilde{g}_R$ is the t-transform of the given function g_R defined in Eq. (9.57).

Both the global relation, Eq. (10.1) and the basic integral representation Eq. (10.3a) involve the combination $\tilde{g}_1 + ik\tilde{g}_0$. Using the above equation for $\tilde{g}_1$, we find

$$\tilde{g}_1(k^2,t) + ik\tilde{g}_0(k^2,t) = \tilde{g}_R(k^2,t) + (ik + \gamma)\tilde{g}_0(k^2,t). \qquad (10.24)$$

Using Eq. (10.24), the global relation Eq. (10.1) becomes

$$e^{k^2 t}\hat{u}(k,t) = G(k,t) - (ik + \gamma)\tilde{g}_0 + e^{-ikL}\tilde{h}_1, \qquad (10.25)$$

where the known function G is defined by

$$G(k,t) = \hat{u}_0(k) - \tilde{g}_R(k^2, t) + ike^{-ikL}\tilde{h}_0(k^2, t). \tag{10.26}$$

Replacing k with $-k$ in (10.25) we find

$$e^{k^2 t}\hat{u}(-k,t) = G(-k,t) - (-ik + \gamma)\tilde{g}_0 + e^{ikL}\tilde{h}_1. \tag{10.27}$$

Equations (10.25) and (10.27) are two algebraic equations for the unknown functions $\tilde{g}_0$ and $\tilde{h}_1$. Solving these equations we find

$$\tilde{g}_0 = \frac{1}{\Delta(k)}\left[e^{ikL}G(k,t) - e^{-ikL}G(-k,t)\right] + \check{g}_0, \tag{10.28}$$

and

$$\tilde{h}_1 = \frac{1}{\Delta(k)}\left[(\gamma - ik)G(k,t) - (\gamma + ik)G(-k,t)\right] + \check{h}_1, \tag{10.29}$$

where

$$\Delta(k) = (\gamma + ik)e^{ikL} - (\gamma - ik)e^{-ikL}, \tag{10.30}$$

and the unknown functions $\check{g}_0, \check{h}_1$ involve $\hat{u}(k,t)$ and $\hat{u}(-k,t)$:

$$\check{g}_0 = -\frac{e^{k^2 t}}{\Delta(k)}\left[e^{ikL}\hat{u}(k,t) - e^{-ikL}\hat{u}(-k,t)\right], \tag{10.31}$$

and

$$\check{h}_1 = -\frac{e^{k^2 t}}{\Delta(k)}\left[(\gamma - ik)\hat{u}(k,t) - (\gamma + ik)\hat{u}(-k,t)\right]. \tag{10.32}$$

The zeros of $\Delta(k)$ are *not* in $\mathcal{D}$, thus the contribution of $\check{g}_0$ and $\check{h}_1$ vanishes. Replacing in Eq. (10.3a) $\tilde{g}_1 + ik\tilde{g}_0$ by the right-hand side of Eq. (10.24), the functions $\tilde{g}_0$ and $\tilde{h}_1$ by the right-hand side of Eqs. (10.28), (10.29), simplifying, we find the following expression:

$$u(x,t) = \int_{-\infty}^{\infty} e^{ikx - k^2 t}\hat{u}_0(k)\frac{dk}{2\pi}$$
$$+ \int_{\partial\mathcal{D}^+}\frac{2e^{-k^2 t}}{\Delta(k)}\Big\{i\left[\gamma\sin(kx) + k\cos(kx)\right]e^{ikL}\hat{u}_0(k)$$
$$+ (i\gamma - k)\sin(k(L-x))\hat{u}_0(-k) + 2k\sin(k(L-x))\tilde{g}_R(k^2, t)$$
$$- 2k\left[\gamma\sin(kx) + k\cos(kx)\right]\tilde{h}_0(k^2, t)\Big\}\frac{dk}{2\pi}, \tag{10.33}$$

with $G(k,t)$ defined in Eq. (10.26). The above representation is similar with the representation Eq. (10.10) obtained for the case that u is given at both $x = 0$ and $x = L$.

This problem can also be solved through an infinite series representation, but as expected from Eq. (10.33), the relevant series involves the numbers $\{k_n\}_1^\infty$ which are the solutions of the equation $\Delta(k_n) = 0$, with $\Delta(k)$ defined in Eq. (10.30).

In summary, the traditional approach yields a representation in terms of an infinite series over $\{k_n\}_1^\infty$, where k_n are defined via a transcendental equation. Furthermore, this series is *not* uniformly convergent at $x = 0$ and $x = L$, thus it is *not* suitable for the numerical evaluation of the solution. On the other hand, the unified transform yields the explicit representation Eq. (10.33) that *is* uniformly convergent at both $x = 0$ and $x = L$, and *can* be evaluated numerically.

Example 10.2: The effectiveness of the new method in comparison to the use of the standard transforms and series becomes evident by solving the heat equation in $0 < x < 1, t > 0$, with a Robin boundary condition at $x = 0$:

$$u_0(x) = e^{-x}, \quad u_x(0,t) - 2u(0,t) = -3\cos(t), \quad u(1,t) = e^{-1}\cos(t).$$

We note that

$$u_x(0,0) - 2u(0,0) = -3 = u_0'(0) - 2u_0(0),$$

and

$$u(1,0) = e^{-1} = u_0(1),$$

thus, there is no jump at $x = t = 0$ and at $x = 1, t = 0$.

Following the same procedure as in Example 10.1 and with the same contour C^+ and the same poles k_1, k_2, we find that the solution of Eq. (10.33) is given by

$$u(x,t) = v(x,t) + \int_{C^+} V(k,x,t)dk,$$

where

$$v(x,t) = e^{-x/\sqrt{2}}\left[\alpha\cos\left(t - \frac{x}{\sqrt{2}}\right) + \beta\sin\left(t - \frac{x}{\sqrt{2}}\right)\right]$$
$$+ e^{x/\sqrt{2}}\left[\gamma\cos\left(t + \frac{x}{\sqrt{2}}\right) + \delta\sin\left(t + \frac{x}{\sqrt{2}}\right)\right],$$

and

$$V(k,x,t) = \frac{e^{-k^2t}}{2\pi}\left(\frac{2ik\,(k^2 - 1)\,[2\sin(kx) + 3e\sin(k(1-x)) + k\cos(kx)]}{e(k^2+1)(k^4+1)\,(2\sin(k) + k\cos(k))}\right).$$

The parameters $\alpha, \beta, \gamma, \delta$ that appear in the expression for $v(x,t)$ are as follows:

$$\alpha = \frac{1}{\varepsilon}\left[3(4+\sqrt{2})e^{1+\sqrt{2}} + 2(5-2\sqrt{2})e^{-1/\sqrt{2}}c_1 - 2e^{1/\sqrt{2}}\left(3c_1 - 2\sqrt{2}s_1\right)\right.$$
$$\left. + 3e\left((\sqrt{2}-4)c_2 + \sqrt{2}s_2\right)\right],$$

$$\beta = \frac{1}{\varepsilon}\left[3\sqrt{2}e^{1+\sqrt{2}} - 2(5-2\sqrt{2})e^{-1/\sqrt{2}}s_1 - 2e^{1/\sqrt{2}}\left(3s_1 + 2\sqrt{2}c_1\right)\right.$$
$$\left. - 3e\left((\sqrt{2}-4)s_2 - \sqrt{2}c_2\right)\right],$$

$$\gamma = \frac{1}{\varepsilon}\left[3(4-\sqrt{2})e^{1-\sqrt{2}} + 2(5+2\sqrt{2})e^{1/\sqrt{2}}c_1 - 2e^{-1/\sqrt{2}}\left(3c_1 - 2\sqrt{2}s_1\right)\right.$$
$$\left. - 3e\left((\sqrt{2}+4)c_2 - \sqrt{2}s_2\right)\right],$$

$$\delta = \frac{1}{\varepsilon}\left[3\sqrt{2}e^{1-\sqrt{2}} - 2(5+2\sqrt{2})e^{1/\sqrt{2}}s_1 - 2e^{-1/\sqrt{2}}\left(3s_1 + 2\sqrt{2}c_1\right)\right.$$

$$\left. + 3e\left((\sqrt{2}+4)s_2 + \sqrt{2}c_2\right)\right],$$

where

$$c_1 = \cos\left(\frac{1}{\sqrt{2}}\right), \quad c_2 = \cos(\sqrt{2}), \quad s_1 = \sin\left(\frac{1}{\sqrt{2}}\right), \quad s_2 = \sin(\sqrt{2}),$$

$$\varepsilon = 4e\left[-3\cos(\sqrt{2}) + 5\cosh(\sqrt{2}) + 2\sqrt{2}\left(\sin(\sqrt{2}) + \sinh(\sqrt{2})\right)\right].$$

Plots of the numerical evaluation of the solution for $x \in [0,1]$ at short (left panel) and longer (right panel) timescales with the initial–boundary condition shown as the black dashed line, as well as the three-dimensional perspective plot in the range $x \in [0,1]$ and $t \in [0,2\pi]$, generated using Mathematica, are shown below.

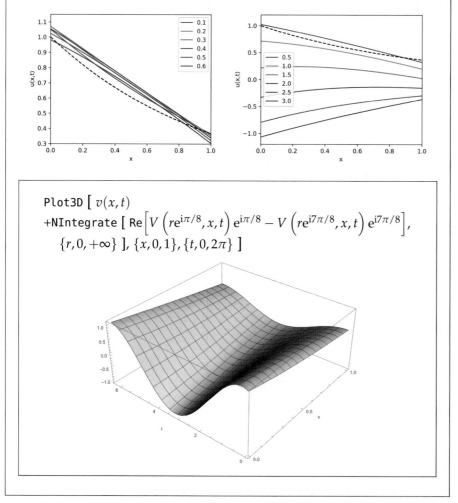

10.4.2   From PDEs to ODEs

As discussed in Section 9.5.2, although classical transforms appear to be superseded in view of the efficiency of the unified transform, if such transforms are needed the unified transform provides a simple way for their derivation:

evaluate the expression of the solution $u(x,t)$ obtained via the unified transform at $t = 0$. For this purpose, in order to obtain a classical transform, it is necessary to first deform the contours $\partial \mathcal{D}^+$ back to the real axis.

The evaluation of the expression of $u(x,t)$ at $t = 0$, *without* the deformation of $\partial \mathcal{D}^+$ back to the real axis, gives rise to novel transforms. A particular such transform is given below:

$$i\pi f(x) = \int_{-\infty}^{\infty} e^{ikx} \hat{f}_s(k) dk - \int_{\partial \mathcal{D}^+} \frac{e^{ikx} - e^{-ikx}}{e^{ikL} - e^{-ikL}} e^{ikL} \hat{f}_s(k) dk, \qquad 0 < x < L,$$

$$\text{(10.34)}$$

where the sine-transform $\hat{f}_s(k)$ is defined as

$$\hat{f}_s(k) = \int_0^L \sin(k\xi) f(\xi) d\xi, \qquad k \in \mathbb{C}. \qquad \text{(10.35)}$$

Given $f(x)$, Eq. (10.35) yields $\hat{f}_s(k)$, and given $\hat{f}_s(k)$, $f(x)$ can be computed through the right-hand side of Eq. (10.34).

Derivation of the novel transform

In order to derive Eq. (10.34) we evaluate Eq. (10.10) at $t = 0$ and rename $u_0(x)$ as $f(x)$:

$$f(x) = \int_{-\infty}^{\infty} e^{ikx} \hat{f}(k) \frac{dk}{2\pi}$$

$$- i \int_{\partial \mathcal{D}^+} \frac{2}{\Delta(k)} \left[\sin(kx) e^{ikL} \hat{f}(k) + \sin(k(L-x)) \hat{f}(-k) \right] \frac{dk}{2\pi}, \quad \text{(10.36)}$$

with $0 < x < L$, where $\hat{f}(k)$ denotes the finite Fourier transform:

$$\hat{f}(k) = \int_0^L e^{-ikx} f(x) dx. \qquad \text{(10.37)}$$

It turns out that Eq. (10.36) can be rewritten in the form of Eq. (10.34). Using the identities

$$e^{-ikx} = \cos(kx) - i\sin(kx), \quad e^{ikx} = \cos(kx) + i\sin(kx), \qquad \text{(10.38)}$$

it follows that

$$\hat{f}(k) = \hat{f}_c(k) - i\hat{f}_s(k), \quad \hat{f}(-k) = \hat{f}_c(k) + i\hat{f}_s(k), \qquad \text{(10.39)}$$

where the sine-transform, $\hat{f}_s(k)$, is defined in Eq. (10.35) and the cosine-transform, $\hat{f}_c(k)$, is defined by a similar equation with $\sin(k\xi)$ replaced with $\cos(k\xi)$. Using the expressions of Eq. (10.38) in Eq. (10.36), and deforming the part of the integral over $\partial \mathcal{D}^+$ involving $\hat{f}_c(k)$ back to the real axis, we find Eq. (10.34).

The novel pair of transforms (10.34), (10.35) has been used recently for the solution of an important problem in the area of controllability that had remained open for a long time.

10.4.3   Green's functions

As with the case of boundary value problems on the half-line, the solution representation obtained via the unified transform yields an *explicit* representation for the associated Green's functions.

For example, for the heat equation with $u(x,0)$, $u(0,t)$ and $u(L,t)$ given, employing Eq. (10.12) and recalling the definitions of $\hat{u}_0(k)$, $\tilde{G}_0(k^2)$, $\tilde{H}_0(k^2)$, we find the following representation:

$$u(x,t) = \int_0^L u_0(\xi)G(x,\xi,t)\mathrm{d}\xi + \int_0^T g_0(\tau)G_L(x,\tau,t)\mathrm{d}\tau$$
$$+ \int_0^T h_0(\tau)G_R(x,\tau,t)\mathrm{d}\tau, \tag{10.40}$$

where G, G_L, G_R are defined as follows:

$$G(x,\xi,t) = \int_{-\infty}^{\infty} e^{ik(x-\xi)-k^2 t}\frac{\mathrm{d}k}{2\pi}$$
$$- i\int_{\partial\mathcal{D}^+}\frac{e^{-k^2 t}}{\Delta(k)}\left[\sin(kx)e^{ik(L-\xi)} + \sin(k(L-x))e^{ik\xi}\right]\frac{\mathrm{d}k}{2\pi}, \tag{10.41a}$$

$$G_L(x,\tau,t) = \int_{\partial\mathcal{D}^+}\frac{2k}{\Delta(k)}\sin(k(L-x))e^{k^2(\tau-t)}\frac{\mathrm{d}k}{2\pi}, \quad t > \tau, \tag{10.41b}$$

$$G_R(x,\tau,t) = \int_{\partial\mathcal{D}^+}\frac{2k}{\Delta(k)}\sin(kx)e^{k^2(\tau-t)}\frac{\mathrm{d}k}{2\pi}, \quad t > \tau. \tag{10.41c}$$

10.5 Application: Detection of colloid concentration

Consider the situation depicted in Fig. 10.4: A clear tube of length $L > 0$ contains a colloidal suspension whose opacity is a known monotonic function of the concentration of the dispersed substance. The flux of the dispersed substance across the end of the tube vanishes. Assuming that the viscosity and temperature remain constant, and that there is no net flow of the liquid phase, the dispersed substance diffuses according to the one-dimensional heat equation. Therefore, the concentration of the dispersed substance, $u(x,t)$, satisfies the heat equation on the finite domain, $0 < x < L$, with the following initial and boundary conditions:

$$u(x,0) = u_0(x), \quad 0 < x < L, \quad u_x(L,t) = 0, \quad t > 0, \tag{10.42}$$

where it is assumed that the initial concentration profile, $u_0(x)$, is known.

In order to determine the concentration $u(x,t)$ it is necessary to supplement the above description with a measurement of the concentration of the dispersed substance at $x = 0$. For this purpose, one uses a lamp or a laser and a photodetector to measure the opacity of the colloidal suspension, from which the concentration can be deduced. In this apparatus, the photodetector must have some finite width l, such that $0 < l < L$. Therefore, the measurement will not yield $u(0,t)$, but rather the average concentration, namely,

$$\int_0^l u(x,t)\frac{\mathrm{d}x}{l}.$$

Actually, if $f(x) > 0$ is the known measure of the sensitivity of the photodetector at position x, then we have

$$\int_0^l f(y)u(y,t)\mathrm{d}y = g_N(t), \quad t > 0, \quad 0 < l < L, \tag{10.43}$$

where the subscript N denotes the fact that this boundary condition is non-local. Because of this non-locality, the problem is not separable, and hence it

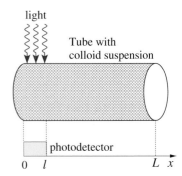

Figure 10.4: Illustration of the experimental setup for the detection of the concentration of a colloid suspension in a tube, using a photodetector of finite size. The length of the tube is L and the length of the photodetector is $l < L$.

cannot be solved by the standard transform methods. However, it is straight-forward to solve this class of problems using the unified transform. In this connection, we consider a slight generalization of the above, where we allow the flux at $x = L$ to be a given known function, $h_1(t)$. The solution of the heat equation with $0 < x < L$, and any boundary conditions is given by Eq. (10.3a), where the transforms of the boundary values satisfy the global relation of Eq. (10.1). However, this global relation involves boundary values at $x = 0$ and $x = L$, whereas in our case we have boundary values at $x = y$ and $x = L$. This suggests using an additional global relation obtained by defining the following transform:

$$\check{u}(k,t;y) = \int_y^L e^{-ikx} u(x,t) dx, \quad 0 < y < L, \quad k \in \mathbb{C}. \tag{10.44}$$

Then, in analogy with Eq. (10.1) we obtain

$$e^{k^2 t} \check{u}(k,t;y) = \check{u}_0(k;y)$$
$$- e^{-iky} \left[\int_0^t e^{k^2 \tau} u_y(y,\tau) d\tau + ik \int_0^t e^{k^2 \tau} u(y,\tau) d\tau \right]$$
$$+ e^{-ikL} \left[\tilde{h}_1(k^2,t) + ik\tilde{h}_0(k^2,t) \right], \quad k \in \mathbb{C}, \tag{10.45}$$

where $\tilde{h}_0$, and $\tilde{h}_1$ are defined as before [see Eq. (10.2)], whereas $\check{u}_0$ is defined by

$$\check{u}_0(k;y) = \int_y^L e^{-ikx} u_0(x) dx, \quad 0 < y < L, \quad k \in \mathbb{C}. \tag{10.46}$$

Multiplying Eq. (10.45) by $\exp[iky] f(y)$ and integrating from 0 to l, we find

$$e^{k^2 t} U(k,t) = U_0(k) - G_1(k^2,t) - ik\tilde{g}_N(k^2,t)$$
$$+ F(k) \left[\tilde{h}_1(k^2,t) + ik\tilde{h}_0(k^2,t) \right], \quad k \in \mathbb{C}, \tag{10.47}$$

where the known functions $U_0(k)$, $F(k)$ and the unknown functions $U(k,t)$, $G_1(k,t)$ are defined as follows:

$$U(k,t) = \int_0^l e^{iky} f(y) \left(\int_y^L e^{-ikx} u(x,t) dx \right) dy, \quad k \in \mathbb{C}, \; t > 0, \tag{10.48a}$$

$$U_0(k) = \int_0^l e^{iky} f(y) \left(\int_y^L e^{-ikx} u_0(x) dx \right) dy, \quad k \in \mathbb{C}, \tag{10.48b}$$

$$G_1(k,t) = \int_0^t e^{k\tau} \left(\int_0^l f(y) u_y(y,\tau) dy \right) d\tau, \quad k \in \mathbb{C}, \; t > 0, \tag{10.48c}$$

$$F(k) = \int_0^l e^{ik(y-L)} f(y) dy, \quad k \in \mathbb{C}. \tag{10.48d}$$

Equation (10.47) is a new global relation coupling known and unknown boundary values; G_1 and $\tilde{g}_N$ play the role of $\tilde{g}_1$ and $\tilde{g}_0$ in the usual global relations. Replacing in Eq. (10.47) k with $-k$ we find

$$e^{k^2 t} U(-k,t) = U_0(-k) - G_1(k^2,t) + ik\tilde{g}_N(k^2,t)$$
$$+ F(-k) \left[\tilde{h}_1(k^2,t) - ik\tilde{h}_0(k^2,t) \right], \quad k \in \mathbb{C}. \tag{10.49}$$

Solving Eqs. (10.47) and (10.49) for $\tilde{h}_0$ we find

$$k\tilde{h}_0(k^2,t) = \frac{G(k,t)}{\Delta(k)} + \frac{e^{k^2 t} U^+(k,t)}{\Delta(k)}, \tag{10.50}$$

where the functions $G(k,t)$, $\Delta(k)$ and $U^+(k,t)$ are defined as follows:

$$G(k,t) = k\tilde{g}_N(k^2,t) + \int_0^l \left(\int_y^L \sin(k(x-y))\, u_0(x)\mathrm{d}x \right) f(y)\mathrm{d}y$$

$$+ \left(\int_0^l \sin(k(L-y))\, f(y)\mathrm{d}y \right) \tilde{h}_1(k^2,t), \quad k \in \mathbb{C}, \ t > 0, \quad (10.51a)$$

$$\Delta(k) = \int_0^l \cos(k(L-y))\, f(y)\mathrm{d}y, \quad k \in \mathbb{C}, \quad\quad (10.51b)$$

$$U^+(k,t) = -\int_0^l \left(\int_y^L \sin(k(x-y))\, u(x,t)\mathrm{d}x \right) f(y)\mathrm{d}y, \quad k \in \mathbb{C}, \ t > 0.$$

$$(10.51c)$$

For the solution of this problem it is slightly more convenient to use Eq. (10.3a) instead of Eq. (10.3b). The global relation of Eq. (10.1) implies that

$$\tilde{g}_1 + ik\tilde{g}_0 = \hat{u}_0 + e^{-ikL}\left(\tilde{h}_1 + ik\tilde{h}_0 \right) - e^{-k^2 t}\hat{u}.$$

Using this expression in Eq. (10.3a), where $\tilde{h}_0$ is given by Eq. (10.50), it can be shown that the contribution of the combination of the functions $U^+(k,t)$ and $\hat{u}(k,t)$ vanishes. Hence, Eq. (10.3a) yields

$$u(x,t) = \int_{-\infty}^{\infty} e^{ikx-k^2t}\hat{u}_0(k)\frac{\mathrm{d}k}{2\pi}$$

$$- \int_{\partial\mathcal{D}^+} e^{ikx-k^2t} \left\{ \hat{u}_0(k) + e^{-ikL}\left[\tilde{h}_1(k^2,t) + \frac{iG(k,t)}{\Delta(k)} \right] \right\} \frac{\mathrm{d}k}{2\pi}$$

$$- \int_{\partial\mathcal{D}^-} e^{-ik(L-x)-k^2t} \left[\tilde{h}_1(k^2,t) + \frac{iG(k,t)}{\Delta(k)} \right] \frac{\mathrm{d}k}{2\pi}. \quad (10.52)$$

The function $\hat{u}_0(k)$ is bounded for $\mathrm{Im}[k] \leq 0$, thus, the term $\hat{u}_0(k)$ may not be bounded on $\partial\mathcal{D}^+$ (in the earlier examples we had $\hat{u}_0(-k)$ instead of $\hat{u}_0(k)$ which *is* bounded in $\partial\mathcal{D}^+$). However, as will be shown below, the entire integrand in the integrals over $\partial\mathcal{D}^+$ in Eq. (10.52) is bounded.

In what follows, in order to minimize calculations, we concentrate on the problem with the boundary conditions of Eq. (10.42), that is, we let $h_1 = 0$. Then, the curly bracket occurring in the second integral can be rewritten as follows:

$$\hat{u}_0(k) + e^{-ikL} \left[ik\frac{\tilde{g}_N(k^2,t)}{\Delta(k)} + \frac{1}{2\Delta(k)} \int_0^l \left(\int_y^L e^{ik(x-y)}u_0(x)\mathrm{d}x \right) f(y)\mathrm{d}y \right.$$

$$\left. - \frac{\hat{u}_0(k)}{2\Delta(k)} \int_0^l e^{iky}f(y)\mathrm{d}y + \frac{1}{2\Delta(k)} \int_0^l \left(\int_0^y e^{-ik(x-y)}u_0(x)\mathrm{d}x \right) f(y)\mathrm{d}y \right],$$

$$(10.53)$$

where, in the term in G involving the integral of $\exp[-ik(x-y)]$ we replaced the integral from y to L by an integral from 0 to L minus an integral from 0 to y. We note that for $k \in \partial\mathcal{D}^+$, the term $\exp[-ikL]/\Delta(k)$ behaves as follows:

$$\frac{e^{-ikL}}{\Delta(k)} = -2e^{-ikL}\left[\int_0^l e^{-ik(L-y)}\left[1 + e^{2ik(L-y)} \right] f(y)\mathrm{d}y \right]^{-1}$$

$$\sim -2\left[\int_0^l e^{iky}f(y)\mathrm{d}y \right]^{-1} \sim \frac{2ik}{f(0)}, \quad (10.54)$$

where we have used the fact that $\exp[2ik(L-y)]$ decays (since $L > y$) and we have estimated the last integral using integration by parts. Also, the exponentials $\exp[ik(x-y)]$ and $\exp[-ik(x-y)]$ are bounded for $x > y$ and $y > x$, respectively. Furthermore, using integration by parts, the leading order behavior for large $|k|$ of these integrals yields:

$$\int_y^L e^{ik(x-y)}u_0(x)dx \sim \frac{1}{ik}\left[e^{ik(L-y)}u_0(L) - u_0(y)\right], \qquad (10.55a)$$

$$\int_0^y e^{-ik(x-y)}u_0(x)dx \sim -\frac{1}{ik}\left[u_0(y) - e^{iky}u_0(0)\right]. \qquad (10.55b)$$

Hence, the leading behavior for large $|k|$ of the terms in Eq. (10.53) involving the double integrals is given by

$$-\frac{1}{ik}\int_0^l u_0(y)f(y)dy.$$

On the other hand,

$$ik\tilde{g}_N(k^2,t) = ik\int_0^t e^{k^2\tau}g_N(\tau)d\tau \sim \frac{1}{ik}g_N(0) = \frac{1}{ik}\int_0^l u_0(y)f(y)dy, \quad |k| \to \infty,$$

where we used integration by parts to estimate the above t-integral. Thus, the numerator of the ratios involving $\Delta(k)$ of the second, third and fourth terms in Eq. (10.53) combine to form a term of order $1/k^2$. Hence, Eq. (10.54) implies that this ratio is of order $1/k$ as $|k| \to \infty$. Finally, the combination of the terms containing $\hat{u}_0(k)$ simplifies as follows:

$$\hat{u}_0(k) - \frac{\hat{u}_0(k)}{2\Delta(k)}e^{-ikL}\int_0^l e^{iky}f(y)dy$$

$$= \hat{u}_0(k)\int_0^l e^{ik(L-y)}f(y)dy\left[\int_0^l e^{-ik(L-y)}\left[1 + e^{2ik(L-y)}\right]f(y)dy\right]^{-1}$$

$$\sim \hat{u}_0(k)e^{ikL}\int_0^l e^{ik(L-y)}f(y)dy\left[\int_0^l e^{iky}f(y)dy\right]^{-1}$$

$$\sim -\hat{u}_0(k)e^{ikL}\frac{e^{ik(L-l)}f(l) - e^{ikL}f(0)}{e^{ikl}f(l) - f(0)}, \quad |k| \to \infty,$$

where we used integration by parts to estimate the last two integrals. Thus, for large $|k|$ the leading behavior of the terms containing $\hat{u}_0(k)$ is given by $\hat{u}_0(k)\exp[ikL]/f(0)$, and integration by parts shows that, $\hat{u}_0(k)\exp[ikL]$ is of order $1/k$.

The above analysis implies that the integrand of $\partial\mathcal{D}^+$ consists of three terms each of which is bounded:

$$u(x,t) = \int_{-\infty}^{\infty} e^{ikx-k^2t}\hat{u}_0(k)\frac{dk}{2\pi}$$

$$-\int_{\partial\mathcal{D}^+} e^{ikx-k^2t}\left\{\frac{e^{-ikL}}{\Delta(k)}\left[ik\tilde{g}_N(k^2,t) + \frac{1}{2}\int_0^l\left(\int_y^L e^{ik(x-y)}u_0(x)dx\right.\right.\right.$$

$$\left.\left.\left. + \int_0^y e^{ik(y-x)}u_0(x)dx\right)f(y)dy\right] + \frac{\hat{u}_0(k)}{2\Delta(k)}\int_0^l e^{ik(L-y)}f(y)dy\right\}\frac{dk}{2\pi}$$

$$-\int_{\partial\mathcal{D}^-} e^{-ik(L-x)-k^2t}\frac{i}{\Delta(k)}\left[k\tilde{g}_N(k^2,t) + \int_0^l\left(\int_y^L \sin[k(x-y)]u_0(x)dx\right)f(y)dy\right]\frac{dk}{2\pi},$$

$$(10.56)$$

where $\tilde{g}_N$ is defined in terms of the given boundary condition $g_N(t)$,

$$\tilde{g}_N(k^2, t) = \int_0^t e^{k^2 \tau} g_N(\tau) d\tau, \quad k \in \mathbb{C}, \tag{10.57}$$

whereas the function $G(k, t)$ is defined by Eq. (10.51a) in terms of the above boundary conditions and the initial datum $u_0(t)$. The proof that the contribution of the combination of $U^+(k, t)$ and $\hat{u}(k, t)$ vanishes requires similar steps as those used for the simplification of the terms involving $\hat{u}_0(k)$. The proof that the expression obtained through the unified approach provides the unique solution of the above nonlocal initial-boundary value problem was given by P. Miller and D.A. Smith.[1]

[1] P. Miller and D.A. Smith, The diffusion equation with non-local data, *Journal of Mathematical Analysis and Applications* **466**, pp. 1119–1143 (2018).

Using the change of variables $k \to -k$ in the last integral in the right-hand side of Eq. (10.56), and splitting the term in G involving the integral of $\exp[-ik(x - y)]$ in terms of two integrals as was done above earlier, Eq. (10.56) takes the following form:

$$u(x, t) = \int_{-\infty}^{\infty} e^{ikx - k^2 t} \hat{u}_0(k) \frac{dk}{2\pi}$$

$$- \int_{\partial D^+} e^{-k^2 t} \left\{ \frac{2ik}{\Delta(k)} \cos(k(L - x)) \tilde{g}_N(k^2, t) + i \frac{e^{ikL}}{\Delta(k)} \hat{u}_0(k) \right.$$

$$\times \int_0^l \sin(k(x - y)) f(y) dy + \frac{\cos(k(L - x))}{\Delta(k)} \int_0^l \left(\int_y^L e^{ik(x-y)} u_0(x) dx \right.$$

$$\left. + \int_0^y e^{ik(y-x)} u_0(x) dx \right) f(y) dy \left. \right\} \frac{dk}{2\pi}. \tag{10.58}$$

Example 10.3: To give a specific numerical implementation of the final result for the solution $u(x, t)$, let

$$u(x, 0) = u_0(x), \qquad x \in (0, L), \tag{10.59a}$$

$$u_x(L, t) = 0, \qquad t > 0, \tag{10.59b}$$

$$\int_0^l f(x) u(x, t) dx = g_N(t), \qquad t > 0. \tag{10.59c}$$

Differentiating Eq. (10.59a), evaluating the resulting equation at $x = L$, and comparing with Eq. (10.59b) evaluated at $t = 0$ we conclude that

$$\left. \frac{du_0(x)}{dx} \right|_{x=L} = 0. \tag{10.60}$$

Also, evaluating Eq. (10.59c) at $t = 0$ we find

$$\int_0^l f(x) u_0(x) dx = g_N(0). \tag{10.61}$$

We next choose the specific boundary conditions

$$f(x) = e^{-ax}, \quad u_0(x) = \cosh(b(L - x)), \quad g_N(t) = A \cos(ct), \tag{10.62}$$

and let $l = L$. We note that with these choices Eq. (10.60) is satisfied. Also,

$$\int_0^L e^{-ax}\cosh[b(L-x)]dx = \frac{a\cosh(bL) - b\sinh(bL) - ae^{-aL}}{a^2 - b^2}.$$

We choose the value of the constant A in Eq. (10.62) to be equal to the right-hand side of the above equation, so that Eq. (10.61) is satisfied as well. We thus obtain

$$\hat{u}_0(k) = \frac{-ik\cosh(bL) + b\sinh(bL)}{b^2 + k^2} + \frac{ike^{-ikL}}{b^2 + k^2}.$$

We treat the first integral of Eq. (10.58) as follows:

$$\int_{-\infty}^{\infty} e^{ikx-k^2t}\hat{u}_0(k)\frac{dk}{2\pi}$$

$$= \int_{-\infty}^{\infty} e^{ikx-k^2t}\frac{-ik\cosh(bL) + b\sinh(bL)}{b^2 + k^2}\frac{dk}{2\pi} + \int_{-\infty}^{\infty} e^{ik(x-L)-k^2t}\frac{ik}{b^2 + k^2}\frac{dk}{2\pi}$$

$$= \int_{\partial D^+} e^{ikx-k^2t}\frac{-ik\cosh(bL) + b\sinh(bL)}{b^2 + k^2}\frac{dk}{2\pi} - \int_{\partial D^-} e^{ik(x-L)-k^2t}\frac{ik}{b^2 + k^2}\frac{dk}{2\pi}$$

$$= -\int_{\partial D^+} e^{-k^2t}\left[e^{ikx}\frac{ik\cosh(bL) - b\sinh(bL)}{b^2 + k^2} + e^{ik(L-x)}\frac{ik}{b^2 + k^2}\right]\frac{dk}{2\pi}.$$

$$(10.63)$$

Thus, Eq. (10.58) takes the form

$$u(x,t) = -\int_{\partial D^+} V(k,x,t)\frac{dk}{2\pi} = -\int_{C^+} V(k,x,t)\frac{dk}{2\pi},$$

where the function $V(k,x,t)$ consists of the integrand in the last integral of Eq. (10.63), as well as all the terms of the integrand in the second integral of Eq. (10.58). For the specific values of the parameters $L = 1$, $a = 2$, $b = 3$, $c = 4$, the numerical integration of the right-hand side of Eq. (10.63), employing the same steps as those used in the Examples 10.1 and 10.2, yields the result shown in the following figure.

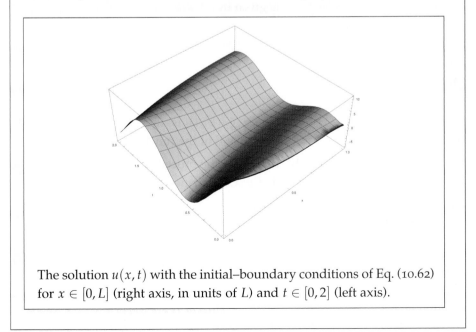

The solution $u(x,t)$ with the initial–boundary conditions of Eq. (10.62) for $x \in [0,L]$ (right axis, in units of L) and $t \in [0,2]$ (left axis).

Problems

1. Consider the heat equation in the finite interval, $0 < x < L$, $L > 0$, with the following initial and boundary conditions:

$$u(x,0) = u_0(x),\ 0 < x < L;$$
$$u_x(0,t) = g_1(t),\ u_x(L,t) = h_1(t),\ t > 0. \quad (10.64)$$

The global relation for this problem is given by [see Section 10.1]:

$$e^{k^2 t}\hat{u}(k,t) = \hat{u}_0(k) - \tilde{g}_1(k^2,t) - ik\tilde{g}_0(k^2,t)$$
$$+ e^{-ikL}\left[\tilde{h}_1(k^2,t) + ik\tilde{h}_0(k^2,t)\right], \quad (10.65)$$

for $k \in \mathbb{C}$, where $\hat{u}_0(k)$ is the Fourier transform of $u_0(x)$, the functions $g_0(t)$, $h_0(t)$ denote the unknown boundary values at $x = 0$, $x = L$, respectively, and $\tilde{f}(k,t)$ denotes the following t-transform of the function $f(t)$:

$$\tilde{f}(k,t) = \int_0^t e^{k\tau} f(\tau)\mathrm{d}\tau. \quad (10.66)$$

(a) By adding to Eq. (10.65) the equation obtained from Eq. (10.65) by replacing k with $-k$, show that $u(x,t)$ satisfies the equation

$$\int_0^\infty \cos(kx)u(x,t)\mathrm{d}x$$
$$= e^{-k^2 t}\int_0^\infty \cos(kx)u_0(x)\mathrm{d}x + \tilde{g}_1(k^2,t)$$
$$+ k\sin(kL)\tilde{h}_0(k^2,t) + \cos(kL)\tilde{h}_1(k^2,t).$$
$$(10.67)$$

(b) By choosing k so that $\sin(kL) = 0$, that is, $k = k_n$, where

$$e^{2ik_n L} = e^{2in\pi},\ k_n = \frac{n\pi}{L},\ n \text{ integer}, \quad (10.68)$$

show that Eq. (10.67) yields the solution of the problem obtained through the cosine-Fourier series.

(c) Obtain an integral representation of the solution $u(x,t)$ using the unified transform.

(d) Compute the solution $u(x,t)$ for the following data:

$$L = 1,\ u_0(x) = e^{-ax},\ 0 < x < 1,\ a > 0;$$
$$g_1(t) = -a\cos(bt),\ h_1(t) = -ae^{-a}\cos(ct),$$
$$(10.69)$$

where the constants $b,c > 0$. Note that the above data are compatible at $x = t = 0$ and at $x = 1$, $t = 0$. For the numerical evaluation take $a = 1.6$, $b = 2.3$, $c = 3.6$.

2. Suppose that $u(x,t)$ satisfies the equation

$$u_t = u_{xx} + u_x,\ 0 < x < L,\ t > 0, \quad (10.70)$$

with the initial and boundary conditions of Eq. (10.64).

(a) Obtain an integral representation of the solution $u(x,t)$.

(b) Compute the solution in the case that the initial–boundary value data are given by Eq. (10.69).

3. Consider the problem of the heat conduction in a finite rod of length L with insulated lateral surfaces, whose left end is held at the constant temperature T_0 and whose right end is exposed to convective heat transfer. Assume that the air temperature is T_1. This problem can be modeled by the heat equation, $0 < x < L$, $t > 0$, with the following boundary conditions:

$$u(0,t) = T_0,\ -\kappa u_x(L,t) = \lambda\left[u(L,t) - T_1\right], \quad (10.71)$$

where κ is the thermal conductivity coefficient and λ the convection coefficient.

(a) Show that the global relation associated with the boundary conditions

$$u(0,t) = g_0(t),$$
$$u_x(L,t) + au(L,t) = g_R(t),\ a > 0, \quad (10.72)$$

is given by

$$e^{k^2 t}\hat{u}(k,t) = \hat{u}_0(k) - \tilde{g}_1(k^2,t) - ik\tilde{g}_0(k^2,t)$$
$$+ e^{-ikL}(ik - a)\tilde{h}_0(k^2,t)$$
$$+ e^{-ikL}\tilde{g}_R(k^2,t),\ k \in \mathbb{C}, \quad (10.73)$$

where $\tilde{g}_R(k^2,t)$ is the t-transform [defined in Eq. (10.66)] of the function $g_R(t)$.

(b) Following the procedure used in Problem 1 show that $u(x,t)$ satisfies the following equation:

$$\int_0^\infty \sin(k_n x)u(x,t)\mathrm{d}x$$
$$= e^{-k^2 t}\left[\int_0^\infty \sin(k_n x)u_0(x)\mathrm{d}x + k\tilde{g}_0(k_n^2,t)\right.$$
$$\left. + \sin(k_n L)\tilde{g}_R(k_n^2,t)\right],$$

where k_n satisfies the *transcendental* equation

$$k_n \cos(k_n L) + a \sin(k_n L) = 0. \qquad (10.74)$$

This is consistent with the fact that the heat equation with the initial–boundary conditions of Eq. (10.71) can be solved by a sine-type Fourier series where the discrete values of k_n satisfy the condition of Eq. (10.74) [for more details see the textbook by D.L. Powers, mentioned in Further Reading of Chapter 8, pp. 170–175].

(c) Obtain an integral representation for $u(x,t)$, with the initial condition $u(x,0) = u_0(x)$, $0 < x < L$, and the boundary conditions of Eq. (10.72).

(d) Compute the solution $u(x,t)$ for the case of $u_0(x) = 0$, and of the boundary conditions of Eq. (10.71), with $L = 1.7$, $T_0 = 3.4$, $\kappa = 2.5$, $\lambda = 4.6$, $T_1 = 10$.

4. Obtain an integral representation of the following initial–boundary value problem:

$$u_t - u_{xxx} = 0, \quad 0 < x < L, \quad t > 0,$$
$$u(x,0) = u_0(x), \quad 0 < x < L,$$
$$u(0,t) = g_0(t), \quad u(L,t) = h_0(t), \quad t > 0,$$
$$u_x(0,t) = g_1(t), \quad t > 0.$$

Find an analytical formula for the determinant $\Delta(k)$ of the associated linear system [the analogue of Eq. (10.21)]. Show numerically that for large k, the values of $\Delta(k) = 0$ are consistent with those obtained by the general result in Section 10.3.

5. Obtain an integral representation of the solution of the following initial–boundary value problem:

$$iu_t + u_{xx} = F(x,t), \quad 0 < x < L, \quad t > 0,$$
$$u(x,0) = u_0(x), \quad 0 < x < L;$$
$$u(0,t) = g_0(t), \quad u(L,t) = h_0(t), \quad t > 0.$$

Chapter 11

Unified Transform III: The Wave Equation

In this chapter, we use the unified transform to derive the solution for one of the most important PDEs, namely the wave equation, both on the half-line and on a finite interval. Without loss of generality, we can let $c = 1$ in the wave equation, Eq. (8.60), and we obtain

$$u_{tt} - u_{xx} = 0. \tag{11.1}$$

Alternatively, Eq. (8.60) can be thought of as resulting from Eq. (11.1) by replacing t with ct.

As a first step in our discussion of the unified transform for the wave equation, we shall present an alternative derivation of the global relation for the familiar heat equation, which employs the important concept of the adjoint. This approach will also be useful in the implementation of the unified transform to the wave equation as well as to elliptic equations that will be discussed in Chapter 12.

11.1   An alternative derivation of the global relation

We first introduce the notion of the **adjoint**. Let $u(x_1, x_2)$ satisfy a given PDE; the adjoint $v(x_1, x_2)$ satisfies the PDE obtained from the given PDE by replacing ∂_{x_j} with $-\partial_{x_j}$, $j = 1, 2$. For example, the adjoint of the heat equation

$$u_t = u_{xx}, \tag{11.2}$$

is the equation

$$-v_t = v_{xx}. \tag{11.3}$$

Multiplying Eqs. (11.2) and (11.3) by v and u, respectively, and then subtracting the resulting equations, we find

$$(uv)_t = (vu_x - uv_x)_x. \tag{11.4}$$

Using Green's theorem, that is, Eq. (2.73), with $x = t, F_2 = uv, y = x, F_1 = (vu_x - uv_x)$, we find that if the heat equation is valid in the domain Ω with boundary $\partial\Omega$, then

$$\int_{\partial\Omega} [(uv)\mathrm{d}x + (vu_x - uv_x)\mathrm{d}t] = 0. \tag{11.5}$$

Noting that the expression $v = \exp[-ikx + k^2 t]$ is a particular solution of the adjoint equation, Eq. (11.3), the above equation becomes

$$\int_{\partial\Omega} e^{-ikx+k^2 t}[u(x,t)dx + (u_x(x,t) + iku(x,t))dt] = 0. \qquad (11.6)$$

Suppose that Ω is given by

$$\Omega = \{0 < x < L,\ 0 < t < T\}, \qquad (11.7)$$

as shown in Fig. 11.1. Then, integrating along $\partial\Omega$, Eq. (11.6) yields

$$-e^{k^2 T}\int_0^L e^{-ikx}u(x,T)dx + \int_0^L e^{-ikx}u(x,0)dx - \int_0^T e^{k^2 t}[u_x(0,t) + iku(0,t)]dt$$

$$+ e^{-ikL}\int_0^T e^{k^2 t}[u_x(L,t) + iku(L,t)]dt = 0, \quad k \in \mathbb{C}, \qquad (11.8)$$

which is equivalent to Eq. (10.1), evaluated at $t = T$.

It is important to note that using the adjoint it is always possible to rewrite a given PDE with constant coefficients in the form

$$\frac{\partial F_2}{\partial x_1} - \frac{\partial F_1}{\partial x_2} = 0, \qquad (11.9)$$

where F_1 and F_2 involve $u(x_1, x_2)$, derivatives of $u(x_1, x_2)$, and the complex parameter k used in the parameterization of a particular solution for $v(x_1, x_2)$. Thus, the introduction of the adjoint provides an algorithmic way for constructing a global relation for any linear PDE with constant coefficients, avoiding integration by parts.

11.2 The wave equation on the half-line

In order to derive a global relation for Eq. (11.1), we consider the function $v(x, t)$ satisfying the adjoint of the wave equation, which is clearly the wave equation itself,

$$v_{tt} - v_{xx} = 0. \qquad (11.10)$$

Multiplying Eqs. (11.1) and (11.10) by v and u, respectively, and then subtracting the resulting equations we find

$$(vu_t - uv_t)_t - (vu_x - uv_x)_x = 0.$$

Using Green's theorem, Eq. (2.73), with $x = t, y = x$, we find that if Eq. (11.1) is valid in a domain Ω, then

$$\int_{\partial\Omega}(vu_t - uv_t)dx + (vu_x - uv_x)dt = 0, \qquad (11.11)$$

where $\partial\Omega$ denotes the boundary of Ω.

Employing the particular solution $v = \exp[-ikx + ikt]$, Eq. (11.11) becomes the following global relation of the wave equation:

$$\int_{\partial\Omega} e^{-ikx+ik\tau}[(u_\tau - iku)dx + (u_x + iku)d\tau] = 0, \quad k \in \mathbb{C}. \qquad (11.12)$$

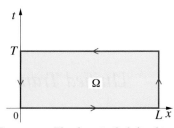

Figure 11.1: The domain Ω defined in Eq. (11.7).

For the domain $0 < x < \infty$, $0 < \tau < t$, shown in Fig. 11.2, Eq. (11.12) becomes

$$-\,e^{ikt} \int_0^\infty e^{-ikx}\left[u_t(x,t) - iku(x,t)\right]dx + \int_0^\infty e^{-ikx}\left[u_t(x,0) - iku(x,0)\right]dx$$

$$-\int_0^t e^{ik\tau}\left[u_x(0,\tau) + iku(0,\tau)\right]d\tau = 0, \quad \mathrm{Im}[k] \le 0.$$

Multiplying the above equation by $\exp[-2ikt]$ we obtain the equation

$$\int_0^\infty e^{-ikx}\frac{\partial}{\partial t}\left[e^{-ikt}u(x,t)\right]dx = e^{-2ikt}\Bigg[\left[\hat{u}_1(k) - ik\hat{u}_0(k)\right]$$

$$-\int_0^t e^{ik\tau}\left[u_x(0,\tau) + iku(0,\tau)\right]d\tau\Bigg], \quad \mathrm{Im}[k] \le 0,$$

$$(11.13)$$

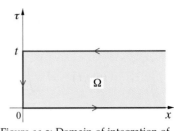

Figure 11.2: Domain of integration of Eq. (11.12).

where $\hat{u}_0(k)$ and $\hat{u}_1(k)$ denote the half-Fourier transforms of $u(x,0)$ and $u_t(x,0)$, respectively. In order to find the equation satisfied by the Fourier transform of $u(x,t)$, we need to integrate Eq. (11.13) with respect to t. For this purpose we will use the following identities:

$$\int_0^t \frac{\partial}{\partial \tau}\left[e^{-ik\tau}u(x,\tau)\right]d\tau = e^{-ikt}u(x,t) - u_0(x), \qquad (11.14a)$$

$$\int_0^t e^{-2ik\tau}d\tau = \frac{e^{-2ikt} - 1}{-2ik}, \qquad (11.14b)$$

$$\int_0^t e^{-2ik\tau}\left[\int_0^\tau f(\eta)d\eta\right]d\tau = \frac{e^{-2ikt}}{-2ik}\int_0^t f(\eta)d\eta + \frac{1}{2ik}\int_0^t e^{-2ik\tau}f(\tau)d\tau, \qquad (11.14c)$$

where in Eq. (11.14c) we have employed integration by parts, and we have used the fact that the integral of $\exp[-2ik\tau]$ with respect to τ is $\exp[-2ik\tau]/(-2ik)$. Integrating Eq. (11.13) with respect to t, employing the identities of Eqs. (11.14), multiplying the resulting expression by $\exp[ikt]$, and simplifying we find

$$\hat{u}(k,t) = \frac{\sin(kt)}{k}\hat{u}_1(k) + \cos(kt)\hat{u}_0(k) - \left[\check{g}_1(k,t) + ik\check{g}_0(k,t)\right], \quad \mathrm{Im}[k] \le 0, \qquad (11.15)$$

where $\hat{u}(k,t)$, $\hat{u}_0(k)$, $\hat{u}_1(k)$ denote the Fourier transforms on the half-line of $u(x,t)$, $u_0(x)$, $u_1(x)$, respectively, that is,

$$\hat{u}(k,t) = \int_0^\infty e^{-ikx}u(x,t)dx, \quad \mathrm{Im}[k] \le 0, \qquad (11.16a)$$

$$\hat{u}_0(k) = \int_0^\infty e^{-ikx}u_0(x)dx, \qquad (11.16b)$$

$$\hat{u}_1(k) = \int_0^\infty e^{-ikx}u_1(x)dx, \quad \mathrm{Im}[k] \le 0, \qquad (11.16c)$$

whereas $\check{g}_1$ and $\check{g}_0$ denote the following t-transforms of $u_x(0,t)$ and $u(0,t)$:

$$\check{g}_1(k,t) = \int_0^t \frac{1}{k}\sin(k(t-\tau))u_x(0,\tau)d\tau, \qquad (11.17a)$$

$$\check{g}_0(k,t) = \int_0^t \frac{1}{k}\sin(k(t-\tau))u(0,\tau)d\tau, \quad k \in \mathbf{C}, \ t > 0. \qquad (11.17b)$$

Interestingly, by employing Green's theorem, we have been able to obtain the global relation Eq. (11.15) without the need to solve a second-order ODE, similar to the one derived in Eq. (8.81).

Taking the inverse Fourier transform of Eq. (11.15), we find

$$u(x,t) = \int_{-\infty}^{\infty} e^{ikx} \left[\frac{\sin(kt)}{k} \hat{u}_1(k) + \cos(kt)\hat{u}_0(k) \right] \frac{dk}{2\pi}$$

$$- \int_{-\infty}^{\infty} e^{ikx} \left[\check{g}_1(k,t) + ik\check{g}_0(k,t) \right] \frac{dk}{2\pi}. \tag{11.18}$$

The global relation, Eq. (11.15), together with the integral representation (11.18) can be used for the solution of a variety of initial-boundary value problems. As an example, we next consider the Dirichlet problem, that is, the case that $u(0,t)$ is prescribed as a boundary datum.

11.2.1 The Dirichlet problem

Consider the boundary condition Eq. (8.79). Replacing in the global relation k with $-k$ and noting that $\check{g}_1(k,t)$ and $\check{g}_0(k,t)$ remain invariant, we find

$$\hat{u}(-k,t) = \frac{\sin(kt)}{k} \hat{u}_1(-k) + \cos(kt)\hat{u}_0(-k) - [\check{g}_1(k,t) - ik\check{g}_0(k,t)], \quad \text{Im}[k] \geq 0.$$

Solving this equation for $\check{g}_1(k,t)$ and substituting the resulting expression in (11.18), we find

$$u'(x,t) = \int_{-\infty}^{\infty} e^{ikx} \left[\frac{\sin(kt)}{k} \hat{u}_1(k) + \cos(kt)\hat{u}_0(k) \right] \frac{dk}{2\pi}$$

$$- \int_{-\infty}^{\infty} e^{ikx} \left[\frac{\sin(kt)}{k} \hat{u}_1(-k) + \cos(kt)\hat{u}_0(-k) + 2ik\check{g}_0(k,t) \right] \frac{dk}{2\pi}, \tag{11.19}$$

where we have used the fact that, as shown in Eq. (9.5), the contribution of $\hat{u}(-k,t)$ vanishes,

$$\int_{-\infty}^{\infty} e^{ikx} \hat{u}(-k,t) \frac{dk}{2\pi} = 0, \quad k \in \mathbb{R}, \quad x > 0.$$

d′ Alembert type representation

As discussed in Chapters 9 and 10, it is always possible to rewrite an integral representation from the Fourier plane to the physical plane. This involves the associated Green's functions, which in general are expressed through integrals in the Fourier plane. It turns out that, for the wave equation, these integrals can be computed explicitly, and this gives rise to d′ Alembert type representations.

Equation (11.19) can be simplified by using the integral representation of the δ-function, Eq. (7.53), which leads to the following expression:

$$\int_{-\infty}^{\infty} e^{ikx} \cos(kt)\hat{u}_0(k) \frac{dk}{2\pi} = \int_0^{\infty} \left[\int_{-\infty}^{\infty} e^{ik(x-\xi)} \left(\frac{e^{ikt} + e^{-ikt}}{2} \right) \frac{dk}{2\pi} \right] u_0(\xi) d\xi$$

$$= \frac{1}{2} \int_0^{\infty} \left[\delta(x - \xi + t) + \delta(x - \xi - t) \right] u_0(\xi) d\xi.$$

Similarly,

$$\int_{-\infty}^{\infty} e^{ikx} \cos(kt)\hat{u}_0(-k)\frac{dk}{2\pi} = \frac{1}{2}\int_0^{\infty} \left[\delta(x+\xi+t) + \delta(x+\xi-t)\right]u_0(\xi)d\xi.$$

Also, using integration by parts, we find

$$\int_{-\infty}^{\infty} e^{ikx}\frac{\sin(kt)}{k}\hat{u}_1(k)\frac{dk}{2\pi}$$

$$= \int_0^{\infty} \left[\int_{-\infty}^{\infty} \frac{1}{2ik}\left(e^{ik(x-\xi+t)} - e^{ik(x-\xi-t)}\right)\frac{dk}{2\pi}\right]u_1(\xi)d\xi$$

$$= \left[\int_0^{\xi} u_1(\xi')d\xi' \int_{-\infty}^{\infty} \frac{1}{2ik}\left(e^{ik(x-\xi+t)} - e^{ik(x-\xi-t)}\right)\frac{dk}{2\pi}\right]_0^{\infty}$$

$$+ \frac{1}{2}\int_0^{\infty}\left(\int_0^{\xi} u_1(\xi')d\xi'\right)\left[\delta(x-\xi+t) - \delta(x-\xi-t)\right]d\xi.$$

The boundary (first) term vanishes, since for $\xi = 0$, the integral of $u_1(\xi')$ vanishes, whereas for $\xi \to \infty$, the integral over k vanishes. Hence,

$$\int_{-\infty}^{\infty} e^{ikx}\frac{\sin(kt)}{k}\hat{u}_1(k)\frac{dk}{2\pi}$$

$$= \frac{1}{2}\int_0^{\infty}\left(\int_0^{\xi} u_1(\xi')d\xi'\right)\left[\delta(x-\xi+t) - \delta(x-\xi-t)\right]d\xi.$$

Similarly,

$$\int_{-\infty}^{\infty} e^{ikx}\frac{\sin(kt)}{k}\hat{u}_1(-k)\frac{dk}{2\pi}$$

$$= -\frac{1}{2}\int_0^{\infty}\left(\int_0^{\xi} u_1(\xi')d\xi'\right)\left[\delta(x+\xi+t) - \delta(x+\xi-t)\right]d\xi.$$

Using similar steps, the term involving $\check{g}_0(k,t)$ gives

$$\int_{-\infty}^{\infty} e^{ikx}k\check{g}_0(k,t)\frac{dk}{2\pi} = \int_0^t\left[\int_{-\infty}^{\infty} \frac{1}{2i}\left(e^{ik(x+t-\tau)} - e^{ik(x-t+\tau)}\right)\frac{dk}{2\pi}\right]g_0(\tau)d\tau$$

$$= \frac{1}{2i}\int_0^t\left[\delta(x+t-\tau) - \delta(x-t+\tau)\right]g_0(\tau)d\tau.$$

Using the above results, Eq. (11.19) becomes

$$u(x,t) = \frac{1}{2}u_0(x+t) + \begin{cases} \dfrac{1}{2}u_0(x-t) + \dfrac{1}{2}\displaystyle\int_{x-t}^{x+t} u_1(\xi)d\xi, & x > t, \\[3mm] g_0(t-x) - \dfrac{1}{2}u_0(t-x) + \dfrac{1}{2}\displaystyle\int_{t-x}^{t+x} u_1(\xi)d\xi, & x < t. \end{cases}$$

$$(11.20)$$

This equation provides the proper generalization of the celebrated formula of d'Alembert for the case that the purely initial value problem is replaced by a boundary value problem defined on the half-line. The expression in Eq. (11.20) depends only on $(x+t)$ and $(x-t)$, hence it satisfies the wave equation. Furthermore, it is straightforward to verify that it satisfies the given initial and boundary conditions:

(i) Using the expression of Eq. (11.20) valid for $x > t$, we find

$$u(x,0) = \frac{1}{2}u_0(x) + \frac{1}{2}u_0(x) = u_0(x).$$

(ii) Using the expression of Eq. (11.20) valid for $x < t$, we find

$$u(0,t) = \frac{1}{2}u_0(t) + g_0(t) - \frac{1}{2}u_0(t) = g_0(t).$$

(iii) Using the expression of Eq. (11.20) valid for $x > t$, and Eq. (1.48) to evaluate the contribution from the integral of $u_1(\xi)$, we find

$$u_t(x,0) = \frac{1}{2}u_0'(x) - \frac{1}{2}u_0'(x) + \frac{1}{2}\left[\frac{d(x+t)}{dt}u_1(x) - \frac{d(x-t)}{dt}u_1(x)\right]$$

$$= u_1(x).$$

It is worth noting that the continuity of u, u_x, and u_t at $x = t$ implies certain constraints between $u_0(0)$, $g_0(0)$, $u_1(0)$, and $g_0'(0)$:

$$\lim_{t \to x^+}[u(x,t)] = \lim_{t \to x^-}[u(x,t)] \Rightarrow u_0(0) = g_0(0), \tag{11.21a}$$

$$\lim_{t \to x^+}[u_x(x,t)] = \lim_{t \to x^-}[u_x(x,t)] \Rightarrow u_1(0) = g_0'(0), \tag{11.21b}$$

$$\lim_{t \to x^+}[u_t(x,t)] = \lim_{t \to x^-}[u_t(x,t)] \Rightarrow u_1(0) = g_0'(0). \tag{11.21c}$$

The constraint of Eq. (11.21a) is consistent with

$$\lim_{x \to 0}[u(x,0)] = \lim_{t \to 0}u(0,t),$$

while the constraint of Eq. (11.21b) is consistent with

$$\lim_{x \to 0}u_t(x,0) = \lim_{t \to 0}u_t(0,t).$$

Equation (11.20) immediately yields the unknown Neumann boundary value at $x = 0$. Indeed, differentiating the expression of Eq. (11.20) with respect to x for $x < t$ and then evaluating the resulting expression at $x = 0$, we find

$$u_x(0,t) = \frac{1}{2}u_0'(t) - g_0'(t) + \frac{1}{2}u_0'(t) + \frac{1}{2}u_1(t) + \frac{1}{2}u_1(t) = u_0'(t) - g_0'(t) + u_1(t).$$

Example 11.1: A wave with a decaying tail

Consider the wave equation, Eq. (11.1), with the following initial-boundary value conditions:

$$u_0(x) = e^{-ax}, \quad u_1(x) = xe^{-bx}, \quad g_0(t) = \cos(ct), \quad a,b,c > 0,$$

which satisfy the constraints

$$u_0(0) = g_0(0) = 1, \quad u_1(0) = \dot{g}_0(0) = 0.$$

The expression of Eq. (11.20) makes the solution of this problem remarkably simple. Employing this expression, and taking into consideration that

$$\int \xi e^{-b\xi}d\xi = -\frac{1}{b}\left(\xi + \frac{1}{b}\right)e^{-b\xi},$$

the solution is

$$u(x,t) = \frac{1}{2}e^{-a(x+t)} + \frac{1}{2}e^{-a(x-t)} - \frac{1}{2b}e^{-b(x+t)}\left(x+t+\frac{1}{b}\right)$$

$$+ \frac{1}{2b}e^{-b(x-t)}\left(x-t+\frac{1}{b}\right), \quad x > t,$$

and

$$u(x,t) = \frac{1}{2}e^{-a(x+t)} - \frac{1}{2}e^{-a(t-x)} - \frac{1}{2b}e^{-b(x+t)}\left(x+t+\frac{1}{b}\right)$$

$$+ \frac{1}{2b}e^{-b(t-x)}\left(t-x+\frac{1}{b}\right) + \cos(c(t-x)), \quad x < t.$$

It is easy to verify that this function satisfies the wave equation, Eq. (11.1), for all $t > 0$ and $0 < x < \infty$. In the following figure, we show examples of $u(x,t)$ for different moments in time (different color lines), including the initial condition, $t = 0$ (black dashed line), for the parameter values $a = 1, b = 1, c = \pi/2$. It is evident that this represents a traveling wave which decays exponentially for $x > t$.

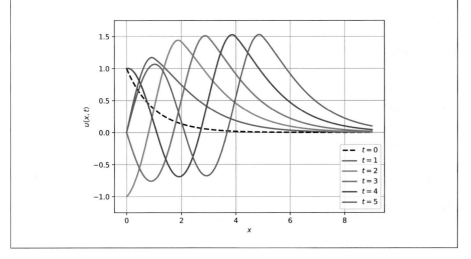

11.3   The wave equation on a finite interval

For the domain $0 < x < L$, $0 < \tau < t$, the global relation of Eq. (11.12), in addition to the terms similar to those appearing in Eq. (11.13), it also contains a new term arising from the boundary $\{x = L, 0 < \tau < t\}$:

$$-e^{ikt}\int_0^L e^{-ikx}\left[u_t(x,t) - iku(x,t)\right]dx + \int_0^L e^{-ikx}\left[u_t(x,0) - iku(x,0)\right]dx$$

$$-\int_0^t e^{ik\tau}\left[u_x(0,\tau) + iku(0,\tau)\right]d\tau$$

$$+e^{-ikL}\int_0^t e^{ik\tau}\left[u_x(L,\tau) + iku(L,\tau)\right]d\tau = 0, \quad k \in \mathbb{C}. \tag{11.22}$$

Multiplying this equation by $\exp[-2ikt]$, we find

$$e^{-ikt} \int_0^L e^{-ikx} \left[u_t(x,t) - iku(x,t) \right] dx$$

$$= e^{-2ikt} \left[\hat{u}_1(k) - ik\hat{u}_0(k) \right] - e^{-2ikt} \int_0^t e^{ik\tau} \left[u_x(0,\tau) + iku(0,\tau) \right] d\tau$$

$$+ e^{-2ikt - ikL} \int_0^t e^{ik\tau} \left[u_x(L,\tau) + iku(L,\tau) \right] d\tau, \quad k \in \mathbb{C}, \tag{11.23}$$

where $\hat{u}_1(k)$ and $\hat{u}_0(k)$ are defined by

$$\hat{u}_1(k) = \int_0^L e^{-ikx} u_1(x) dx, \qquad \hat{u}_0(k) = \int_0^L e^{-ikx} u_0(x) dx, \qquad k \in \mathbb{C}, \tag{11.24}$$

with $u_1(x) = u_t(x,0)$ and $u_0(x) = u(x,0)$. Using the same identities employed for the derivation of the global relation in the case of the half-line, and then integrating Eq. (11.23) with respect to t, we find

$$\hat{u}(k,t) = \frac{\sin(kt)}{k} \hat{u}_1(k) + \cos(kt)\hat{u}_0(k) - \check{g}_1(k,t) - ik\check{g}_0(k,t)$$

$$+ e^{-ikL} \left[\check{h}_1(k,t) + ik\check{h}_0(k,t) \right], \qquad k \in \mathbb{C}, \tag{11.25}$$

where

$$\hat{u}(k,t) = \int_0^L e^{-ikx} u(x,t) dx, \qquad k \in \mathbb{C}, \tag{11.26}$$

$\check{g}_0(k,t)$ and $\check{g}_1(k,t)$ are defined in Eqs. (11.17), while $\check{h}_0(k,t)$ and $\check{h}_1(k,t)$ are defined similarly but with $u(0,t)$, $u_x(0,t)$ replaced by $u(L,t)$, $u_x(L,t)$:

$$\check{h}_0(k,t) = \int_0^t \frac{1}{k} \sin\left(k(t-\tau)\right) u(L,\tau) d\tau, \quad k \in \mathbb{C}. \tag{11.27a}$$

$$\check{h}_1(k,t) = \int_0^t \frac{1}{k} \sin\left(k(t-\tau)\right) u_x(L,\tau) d\tau, \quad k \in \mathbb{C}. \tag{11.27b}$$

Taking the inverse Fourier transform of Eq. (11.25), we find

$$u(x,t) = \int_{-\infty}^{\infty} e^{ikx} \left[\frac{\sin(kt)}{k} \hat{u}_1(k) + \cos(kt)\hat{u}_0(k) \right] \frac{dk}{2\pi}$$

$$- \int_{-\infty}^{\infty} e^{ikx} \left[\check{g}_1(k,t) + ik\check{g}_0(k,t) \right] \frac{dk}{2\pi}$$

$$+ \int_{-\infty}^{\infty} e^{-ik(L-x)} \left[\check{h}_1(k,t) + ik\check{h}_0(k,t) \right] \frac{dk}{2\pi}. \tag{11.28}$$

Using Eqs. (11.28) and (11.25) it is possible to solve a variety of boundary value problems for the wave equation without the need to solve a second-order ODE, similar to the one derived in Eq. (8.81).

11.3.1   The Dirichlet–Dirichlet problem

The boundary conditions are

$$g_0(t) = u(0,t), \qquad h_0(t) = u(L,t), \qquad t > 0. \tag{11.29}$$

Let $G(k,t)$ denote the following known function:

$$G(k,t) = \frac{\sin(kt)}{k} \hat{u}_1(k) + \cos(kt)\hat{u}_0(k) - ik\check{g}_0(k,t) + e^{-ikL} ik\check{h}_0(k,t). \tag{11.30}$$

The global relation of Eq. (11.25) becomes

$$\hat{u}(k,t) = G(k,t) - \check{g}_1(k,t) + e^{-ikL}\check{h}_1(k,t), \qquad k \in \mathbb{C}. \tag{11.31}$$

The transformation $k \to -k$ leaves the functions $\check{g}_j(k,t)$ and $\check{h}_j(k,t)$, $j = 0,1$, invariant. Applying this transformation to Eq. (11.31), we find

$$\hat{u}(-k,t) = G(-k,t) - \check{g}_1(k,t) + e^{ikL}\check{h}_1(k,t), \qquad k \in \mathbb{C}. \tag{11.32}$$

Equations (11.31) and (11.32) can be solved for the unknown functions $\check{g}_1(k,t)$ and $\check{h}_1(k,t)$, as follows:

$$\check{h}_1(k,t) = \frac{1}{\Delta(k)} \left[G(k,t) - G(-k,t) + \hat{u}(-k,t) - \hat{u}(k,t) \right], \tag{11.33a}$$

$$\check{g}_1(k,t) = \frac{1}{\Delta(k)} \left[\left(e^{ikL}G(k,t) - e^{-ikL}G(-k,t) \right) + \left(e^{-ikL}\hat{u}(-k,t) - e^{ikL}\hat{u}(k,t) \right) \right], \tag{11.33b}$$

where we have defined the quantity $\Delta(k)$ as

$$\Delta(k) = e^{ikL} - e^{-ikL}. \tag{11.34}$$

The function $\Delta(k)$ vanishes for

$$k = \frac{n\pi}{L}, \qquad n = 0, \pm 1, \pm 2, \ldots \tag{11.35}$$

These points are referred to as the "discrete spectrum" and are of vital importance in the traditional approaches of solving the wave equation on the finite interval.

Since the functions $\check{h}_1(k,t)$ and $\check{g}_1(k,t)$ are well-defined for all k-complex values, it follows that the points $k = n\pi/L$ are removable singularities. Hence, before separating the two terms appearing in Eqs. (11.33a) and (11.33b), we first deform the contour along the real axis: for the contours of $\check{g}_1(k,t)$ and $\check{h}_1(k,t)$ we deformed above and below the points $k = n\pi/L$, respectively, and denoted these contours by $K^{(+)}$ and $K^{(-)}$, as shown in Fig. 11.3.

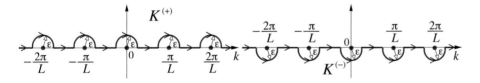

Figure 11.3: The contours $K^{(+)}$ (left panel) and $K^{(-)}$ (right panel).

With the above choice, the contribution of the terms involving $\hat{u}(-k,t)$ and $\hat{u}(k,t)$ vanish. Indeed, the contribution of the second term of $\check{h}_1(k,t)$ to the solution is given by

$$\int_{K^{(-)}} e^{-ik(L-x)} \frac{1}{\Delta(k)} \left[\hat{u}(-k,t) - \hat{u}(k,t) \right] dk. \tag{11.36}$$

The term $\exp[-ik(L-x)]$ is bounded and analytic for $\text{Im}[k] < 0$, whereas for large $|k|$ in the lower half of the k-complex plane, the term $\exp[-ikL]$ vanishes exponentially. Hence,

$$\frac{1}{\Delta(k)} \left[\hat{u}(-k,t) - \hat{u}(k,t) \right] \approx \frac{1}{e^{ikL}} \left[\hat{u}(-k,t) - \hat{u}(k,t) \right]$$

$$= \int_0^L e^{-ik(L-\xi)} u(\xi,t) d\xi + e^{-ikL}\hat{u}(k,t), \quad |k| \to \infty. \tag{11.37}$$

The first integral is bounded and analytic for $\text{Im}[k] < 0$, while both terms $\exp[-ikL]$ and $\hat{u}(k,t)$ are also bounded and analytic in the lower half of the k-complex plane. Furthermore, the integral of Eq. (11.37) decays as $1/|k|$ for large $|k|$. Hence, Jordan's lemma, applied to lower half of the k-complex plane, implies that the integral in Eq. (11.36) vanishes.

Similar considerations are valid for the contribution of the second term of $\check{g}_1(k,t)$, which is given by

$$\int_{K^{(+)}} e^{ikx} \frac{1}{\Delta(k)} \left[e^{-ikL} \hat{u}(-k,t) - e^{ikL} \hat{u}(k,t) \right] dk. \qquad (11.38)$$

For $\text{Im}[k] > 0$, $\exp[ikL]$ vanishes exponentially, hence, for $|k| \to \infty$,

$$\frac{1}{\Delta(k)} \left[e^{-ikL} \hat{u}(-k,t) - e^{ikL} \hat{u}(k,t) \right] \approx -\hat{u}(-k,t) + e^{ikL} \int_0^L e^{-ik(L-\xi)} u(\xi,t) d\xi, \quad |k| \to \infty.$$

The terms $\exp[ikL]$, $\hat{u}(-k,t)$, as well as the above integral are bounded and analytic for $\text{Im}[k] > 0$, thus the application of Jordan's lemma in the upper half of the k-complex plane implies that the integral appearing in Eq. (11.38) vanishes.

Replacing $\check{g}_1(k,t)$ and $\check{h}_1(k,t)$ in Eq. (11.28) in terms of their expressions from Eqs. (11.33), we obtain

$$u(x,t) = \int_{-\infty}^{\infty} e^{ikx} \left[\frac{\sin(kt)}{k} \hat{u}_1(k) + \cos(kt) \hat{u}_0(k) \right] \frac{dk}{2\pi}$$

$$- \int_{-\infty}^{\infty} e^{ikx} ik \check{g}_0(k,t) \frac{dk}{2\pi} + \int_{-\infty}^{\infty} e^{-ik(L-x)} ik \check{h}_0(k,t) \frac{dk}{2\pi}$$

$$- \int_{K^{(+)}} e^{ikx} \frac{1}{\Delta(k)} \left[e^{ikL} G(k,t) - e^{-ikL} G(-k,t) \right] \frac{dk}{2\pi}$$

$$+ \int_{K^{(-)}} e^{-ik(L-x)} \frac{1}{\Delta(k)} \left[G(k,t) - G(-k,t) \right] \frac{dk}{2\pi}. \qquad (11.39)$$

d' Alembert type representation

All terms that appear in Eq. (11.39) except for the last two integrals can be simplified using the same procedure employed for the analysis of the wave equation on the half-line. For the analysis of the last two integrals, we will use the following identities:

$$K^{(+)} : \frac{1}{\Delta(k)} = -\frac{e^{ikL}}{\left(1 - e^{2ikL}\right)} = -\sum_{n=0}^{\infty} e^{i(2n+1)kL}, \qquad (11.40a)$$

$$K^{(-)} : \frac{1}{\Delta(k)} = \frac{e^{-ikL}}{\left(1 - e^{-2ikL}\right)} = \sum_{n=0}^{\infty} e^{-i(2n+1)kL}. \qquad (11.40b)$$

Using these expressions, Eq. (11.39) takes the form

$$u(x,t) = \int_{-\infty}^{\infty} e^{ikx} G(k,t) \frac{dk}{2\pi}$$

$$+ \sum_{n=0}^{\infty} \int_{K^{(+)}} e^{ikx} \left[e^{i(2n+2)kL} G(k,t) - e^{i2nkL} G(-k,t) \right] \frac{dk}{2\pi}$$

$$+ \sum_{n=0}^{\infty} \int_{K^{(-)}} e^{ikx} \left[e^{-i(2n+2)kL} G(k,t) - e^{-i(2n+2)kL} G(-k,t) \right] \frac{dk}{2\pi},$$

$$(11.41)$$

with $G(k,t)$ defined in Eq. (11.30). Using the same procedure as for the case of the half-line, we obtain the solution in an explicit form. For brevity of presentation we analyze a typical case, namely the contribution of the last term of the third integral of Eq. (11.41), the term involving the integration of $G(-k,t)$ on the contour $K^{(-)}$. This term involves the four typical terms $\hat{u}_0(k,t)$, $\hat{u}_1(k,t)$, $\check{g}_0(k,t)$, $\check{h}_0(k,t)$. First, we consider the contribution of $\hat{u}_0(-k)$, namely

$$-\sum_{n=0}^{\infty}\int_{K^{(-)}} e^{ikx}e^{-i(2n+2)kL}\cos(kt)\hat{u}_0(-k)\frac{dk}{2\pi}$$

$$=-\sum_{n=0}^{\infty}\int_{K^{(-)}} e^{ik[x-(2n+2)L]}\frac{e^{ikt}+e^{-ikt}}{2}\left(\int_0^L e^{ik\xi}u_0(\xi)d\xi\right)\frac{dk}{2\pi}$$

$$=-\frac{1}{2}\sum_{n=0}^{\infty}\int_0^L \left(\int_{K^{(-)}}\left\{e^{ik[x+\xi-t-(2n+2)L]}+e^{ik[x+\xi+t-(2n+2)L]}\right\}\frac{dk}{2\pi}\right)u_0(\xi)d\xi.$$

The first term of the above integral yields zero contribution. Indeed, the inequality

$$x+\xi-t-(2n+2)L \leq x+\xi-t-2L = (x-L)+(\xi-L)-t < 0,$$

implies that the exponential is bounded in the lower k-complex plane. Then, Cauchy's theorem, supplemented with the Jordan's lemma, yields the desired result. The second term, yields the contribution

$$-\frac{1}{2}\sum_{n=0}^{\infty}\int_0^L \delta(x+\xi+t-(2n+2)L)u_0(\xi)d\xi$$

$$=-\frac{1}{2}u_0(-x-t+(2n+2)L), \qquad (2n+1)L \leq x+t \leq (2n+2)L.$$

Next, we consider the contribution of $\hat{u}_1(-k)$, namely

$$-\sum_{n=0}^{\infty}\int_{K^{(-)}} e^{ikx}e^{-i(2n+2)kL}\frac{\sin(kt)}{k}\hat{u}_1(-k)\frac{dk}{2\pi}$$

$$=-\sum_{n=0}^{\infty}\int_{K^{(-)}} e^{ik[x-(2n+2)L]}\frac{e^{ikt}-e^{-ikt}}{2ik}\left(\int_0^L e^{ik\xi}u_1(\xi)d\xi\right)\frac{dk}{2\pi}$$

$$=\frac{1}{2}\sum_{n=0}^{\infty}\int_0^L \left(\int_{K^{(-)}}\frac{1}{ik}\left\{e^{ik[x+\xi-t-(2n+2)L]}-e^{ik[x+\xi+t-(2n+2)L]}\right\}\frac{dk}{2\pi}\right)u_1(\xi)d\xi.$$

The first term of the above integral yields zero contribution. In order to compute the second term, we integrate by parts and obtain the contribution

$$\frac{1}{2}\sum_{n=0}^{\infty}\int_0^L \left[\int_{K^{(-)}} e^{ik[x+\xi+t-(2n+2)L]}\frac{dk}{2\pi}\right]\left(\int_0^\xi u_1(r)dr\right)d\xi$$

$$=\frac{1}{2}\sum_{n=0}^{\infty}\int_0^L \delta(x+\xi+t-(2n+2)L)\left(\int_0^\xi u_1(r)dr\right)d\xi$$

$$=\frac{1}{2}\int_0^{(2n+2)L-t-x} u_1(\xi)d\xi, \qquad (2n+1)L \leq x+t \leq (2n+2)L.$$

Finally, we consider the contribution of $\check{g}_0(k,t)$, namely

$$-\sum_{n=0}^{\infty}\int_{K(-)}e^{ikx}e^{-i(2n+2)kL}ik\check{g}_0(k,t)\frac{dk}{2\pi}$$

$$=-\sum_{n=0}^{\infty}\int_{K(-)}e^{ik[x-(2n+2)L]}\left(\int_0^t\frac{1}{2}\left[e^{ik(t-\tau)}-e^{-ik(t-\tau)}\right]g_0(\tau)d\tau\right)\frac{dk}{2\pi}$$

$$=-\frac{1}{2}\sum_{n=0}^{\infty}\int_0^t\left(\int_{K(-)}\left\{e^{ik[x+t-\tau-(2n+2)L]}-e^{ik[x-t+\tau-(2n+2)L]}\right\}\frac{dk}{2\pi}\right)g_0(\tau)d\tau.$$

The second term yields zero contribution, and the first term yields the contribution

$$-\frac{1}{2}\sum_{n=0}^{\infty}\int_0^t\delta(x+t-\tau-(2n+2)L)g_0(\tau)d\tau=-\frac{1}{2}\sum_{n=0}^{N}g_0(x+t-(2n+2)L),$$

where N is the largest integer satisfying the inequality $x+t-(2N+2)L>0$. The contribution of $\check{h}_0(k,t)$ is obtained in the same way as the one of $\check{g}_0(k,t)$, and is given by

$$\frac{1}{2}\sum_{n=0}^{N}h_0(x+t-(2n+1)L),$$

where N is the largest integer satisfying the inequality $x+t-(2N+1)L>0$.

Using a similar analysis to the other terms of Eq. (11.41), we obtain the solution in the following explicit form:

$$u(x,t)=B_1(x,t)-B_2(x,t)+\frac{1}{2}I_1(x,t)-\frac{1}{2}I_2(x,t),\qquad(11.42a)$$

where the terms $B_1(x,t)$ and $B_2(x,t)$ denote the contribution from the boundary conditions, whereas the terms $I_1(x,t)$ and $I_2(x,t)$ denote the contributions from the initial conditions. For compactness of notation in the expressions for these contributions we define the variables X,T as:

$$X=x+t,\quad T=t-x,$$

and the integers N,M as:

$$N=\left[\frac{t+x+L}{2L}\right]=\left[\frac{X+L}{2L}\right],\quad M=\left[\frac{t-x+L}{2L}\right]=\left[\frac{T+L}{2L}\right],\qquad(11.42b)$$

where $[q]$ denotes the integer part of the real number q. Then, the contributions from the boundary conditions are given by

$$B_1(x,t)=\sum_{n=0}^{N-1}h_0(X-(2n+1)L)-\sum_{n=0}^{[X/2L]-1}g_0(X-2(n+1)L),$$

$$B_2(x,t)=\sum_{n=0}^{M-1}h_0(T-(2n+1)L)-\sum_{n=0}^{[T/2L]}g_0(T-2nL),\qquad(11.42c)$$

and the contributions from the initial conditions are given by

$$I_1(x,t)=\text{sign}(X-2NL)u_0(X-2NL|)+\int_0^{|X-2NL|}u_1(\xi)\,d\xi,\quad|X-2NL|\le L,$$

$$I_2(x,t)=\text{sign}(T-2ML)u_0(|T-2ML|)+\int_0^{|T-2ML|}u_1(\xi)\,d\xi,\quad|T-2ML|\le L.$$

$$(11.42d)$$

The formulae for N and M introduced in Eq. (11.42b) can be obtained as follows:

$$|X - 2NL| < L \Leftrightarrow -L < 2NL - X < L \Leftrightarrow \frac{X - L}{2L} < N < \frac{X + L}{2L}$$

$$\Leftrightarrow N < \frac{X + L}{2L} = \frac{X - L}{2L} + 1 < N + 1 \Leftrightarrow N = \left[\frac{X + L}{2L} \right],$$

and similarly,

$$|T - 2ML| < L \Leftrightarrow M = \left[\frac{T + L}{2L} \right].$$

We note that $X = t + x \geq 0$, thus the first term of the expression (11.42d) yields no contribution for $N = 0$, in the interval $-L \leq t + x < 0$.

Example 11.2: We present an example illustrating how the limits of the summation, as well as the integers M and N can be computed.

Let $L = 1$, $x = 3/4$ and $t = 5/2$, from which we obtain $X = 13/4$ and $T = 7/4$. We first compute the limits of summation in the terms $B_1(x, t)$ and $B_2(x, t)$:

$$\left[\frac{X - L}{2L} \right] = \left[\frac{9}{8} \right] = 1, \quad \left[\frac{X - 2L}{2L} \right] = \left[\frac{5}{8} \right] = 0,$$

$$\left[\frac{T - L}{2L} \right] = \left[\frac{3}{8} \right] = 0, \quad \left[\frac{T}{2L} \right] = \left[\frac{7}{8} \right] = 0.$$

We next determine the integers N and M in the terms $I_1(x, t)$ and $I_2(x, t)$:

$$N = \left[\frac{X + L}{2L} \right] = \left[\frac{17}{8} \right] = 2, \quad M = \left[\frac{T + L}{2L} \right] = \left[\frac{11}{8} \right] = 1.$$

Hence, the solution (11.42) for $L = 1$, $x = 3/4$ and $t = 5/2$ takes the form:

$$u(x, t) = h_0(X - L) + h_0(X - 3L) - g_0(X - 2L) - h_0(T - L) + g_0(T)$$

$$- \frac{1}{2} u_0(4L - X) + \frac{1}{2} \int_0^{4L - X} u_1(\xi) d\xi$$

$$+ \frac{1}{2} u_0(2L - T) - \frac{1}{2} \int_0^{2L - T} u_1(\xi) d\xi,$$

namely

$$u \left(\frac{3}{4}, \frac{5}{2} \right) = h_0 \left(\frac{9}{4} \right) - h_0 \left(\frac{3}{4} \right) + h_0 \left(\frac{1}{4} \right) + g_0 \left(\frac{7}{4} \right) - g_0 \left(\frac{5}{4} \right)$$

$$+ \frac{1}{2} u_0 \left(\frac{1}{4} \right) - \frac{1}{2} u_0 \left(\frac{3}{4} \right) + \frac{1}{2} \int_0^{\frac{3}{4}} u_1(\xi) d\xi - \frac{1}{2} \int_0^{\frac{1}{4}} u_1(\xi) d\xi.$$

Verification of the initial and boundary conditions: In order to verify the initial and boundary conditions, we only need to keep track of which of the above terms are present when the solution of Eq. (11.42) is evaluated at $t = 0$, $x = 0$ and $x = L$, respectively. For $t = 0$, the terms $B_1(x, t)$ and $B_2(x, t)$ vanish, whereas the terms $I_1(x, t)$ and $I_2, (x, t)$ yield non-zero contributions only for

$N = M = 0$, namely,

$$I_1(x,0) = u_0(x) + \int_0^x u_1(\xi)\, d\xi,$$

$$I_2(x,0) = -u_0(x) - \int_0^x u_1(\xi)\, d\xi.$$

Hence, Eq. (11.42) yields

$$u(x,0) = \frac{1}{2}\left[u_0(x) + \int_0^x u_1(\xi)d\xi + u_0(x) - \int_0^x u_1(\xi)d\xi\right] = u_0(x), \ \ 0 \le x \le L.$$

Similarly,

$$u(x,0) = \frac{1}{2}[\dot{u}_0(x) + u_1(x) - \dot{u}_0(x) + u_1(x)] = u_1(x), \quad 0 \le x \le L.$$

Evaluating Eq. (11.42) at $x = 0$, *all* the terms appearing in $I_1(0,t)$ come in pairs with terms in $I_2(0,t)$, thus, they cancel. The same is true for the terms appearing in $B_1(0,t)$ and $B_2(0,t)$, apart from the first term of the second sum of $B_2(0,t)$, namely $g_0(t)$. Thus, $u(0,t) = g_0(t)$. Similarly, evaluating the solution at $x = L$, the only term that remains is the first term of the first sum of $B_1(L,t)$, namely $h_0(t)$. Thus, $u(L,t) = h_0(t)$.

Compatibility of the initial and boundary conditions — Continuity of the solution: By evaluating the solution of Eq. (11.42) as $x + t \to L^{(\pm)} = L \pm \epsilon$ where $\epsilon > 0$, we obtain the following two limits:

$$\lim_{x+t\to L^{(+)}} [u(x,t)] = \lim_{x+t\to L^{(+)}}\left[F(x-t) + h_0(0) + \frac{1}{2}\left(-u_0(L) + \int_0^L u_1(\xi)d\xi\right)\right],$$

and

$$\lim_{x+t\to L^{(-)}} u(x,t) = \lim_{x+t\to L^{(-)}}\left[F(x-t) + \frac{1}{2}\left(u_0(L) + \int_0^L u_1(\xi)d\xi\right)\right],$$

where F denotes all the contribution of the terms which depend on $(x-t)$. By definition

$$\lim_{x+t\to L^{(+)}} [F(x-t)] = \lim_{x+t\to L^{(-)}} [F(x-t)],$$

thus, the continuity of the solution at $x + t = L$, yields the condition

$$h_0(0) + \frac{1}{2}\left(-u_0(L) + \int_0^L u_1(\xi)d\xi\right) = \frac{1}{2}\left(u_0(L) + \int_0^L u_1(\xi)d\xi\right),$$

which is equivalent to $h_0(0) = u_0(L)$.

Similarly, continuity at $x - t = 0$ yields $g_0(0) = u_0(0)$. Furthermore, the continuity of the derivative of the solution at $x - t = 0$ and $x + t = L$ yields the conditions $\dot{g}_0(0) = u_1(0)$ and $\dot{h}_0(0) = u_1(L)$, respectively.

The formulae for u_x at the boundaries: Evaluating the derivative with respect to x of the solution of Eq. (11.42) at $x = 0$ we find

$$u_x(0,t) = -\dot{g}_0(t) - 2\sum_{n=1}^{[t/2L]}\dot{g}_0(t-2nL) + 2\sum_{n=0}^{[(t-L)/2L]}\dot{h}_0(t-(2n+1)L)$$

$$+ \begin{cases} \dot{u}_0(t-2nL) + u_1(t-2nL), & n \leq \dfrac{t}{2L} \leq n + \dfrac{1}{2} \\[2mm] \dot{u}_0((2n+2)L-t) + u_1((2n+2)L-t), & n+\dfrac{1}{2} \leq \dfrac{t}{2L} \leq n+1 \end{cases}$$

$$\text{for} \quad n = 0,1,2,\ldots, \left[\frac{t}{2L}\right] \qquad (11.43)$$

where $n = 0,1,2,\ldots,[t/2L]$, and the notation $[q]$ denotes the integer part of the real number q. Observe that for $0 < t < L$, the above expression simplifies to

$$u_x(0,t) = -\dot{g}_0(t) + \dot{u}_0(t) + u_1(t).$$

Evaluating the derivative with respect to x of the solution of Eq. (11.42) at $x = L$, we find

$$u_x(L,t) = \dot{h}_0(t) - 2\sum_{n=0}^{[(t-L)/2L]}\dot{g}_0(t-2nL) + 2\sum_{n=1}^{[t/2L]-1}\dot{h}_0(t-(2n+1)L)$$

$$+ \begin{cases} \dot{u}_0(t-(2n+1)L) + u_1(t-(2n+1)L), & n+\dfrac{1}{2} \leq \dfrac{t}{2L} \leq n+1 \\[2mm] \dot{u}_0((2n+1)L-t) + u_1((2n+1)L-t), & n \leq \dfrac{t}{2L} \leq n+\dfrac{1}{2} \end{cases}$$

$$\text{for} \quad n = 0,1,2,\ldots, \left[\frac{t}{2L}\right] \qquad (11.44)$$

where $n = 0,1,2,\ldots,[t/2L]$. Observe that for $0 < t < L$, the above simplifies to

$$u_x(L,t) = \dot{h}_0(t) + \dot{u}_0(L-t) + u_1(L-t).$$

Example 11.3: We give two examples of the solution $u(x,t)$ to the wave equation as expressed in Eq. (11.42).

Case 1: We use the initial conditions

$$u_0(x) = 1 - \cos\left(\frac{2\pi x}{L}\right), \quad u_1(x) = \cos\left(\frac{5\pi x}{3L}\right), \quad x \in (0,L), \quad (11.45a)$$

the boundary conditions

$$g_0(t) = \sin(t), \quad u(L,t) = h_0(t) = \sin\left(\frac{t}{2}\right), \quad t > 0. \qquad (11.45b)$$

Using the fact that

$$\int_0^x u_1(\xi)\,d\xi = \frac{3L}{5\pi}\sin\left(\frac{5\pi x}{3L}\right), \qquad (11.46)$$

we obtain the solution shown in the following plot.

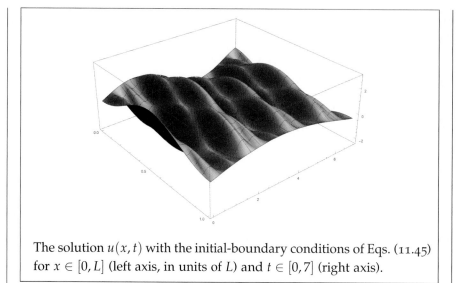

The solution $u(x,t)$ with the initial-boundary conditions of Eqs. (11.45) for $x \in [0,L]$ (left axis, in units of L) and $t \in [0,7]$ (right axis).

Case 2: We use the initial conditions

$$u_0(x) = \cos\left(\frac{\pi x}{L}\right) + 1, \quad u_1(x) = -\frac{5}{4}\cos\left(\frac{\pi x}{L}\right) - \frac{3}{4}, \quad x \in (0,L),$$
(11.47a)

the boundary conditions

$$g_0(t) = 2e^{-t}, \quad h_0(t) = 1 - e^{-t/2}, \quad t > 0.$$
(11.47b)

Using the fact that

$$\int_0^x u_1(\xi)\mathrm{d}\xi = -\frac{5L}{4\pi}\sin\left(\frac{\pi x}{L}\right) - \frac{3}{4}x,$$
(11.48)

we obtain the solution shown in the following plot.

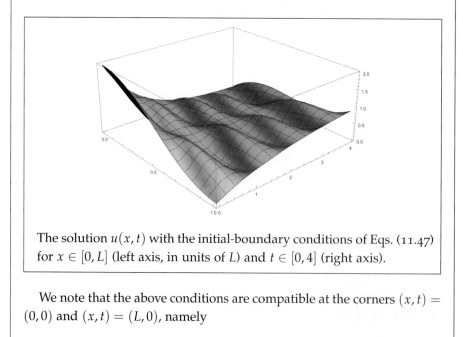

The solution $u(x,t)$ with the initial-boundary conditions of Eqs. (11.47) for $x \in [0,L]$ (left axis, in units of L) and $t \in [0,4]$ (right axis).

We note that the above conditions are compatible at the corners $(x,t) = (0,0)$ and $(x,t) = (L,0)$, namely

$$g_0(0) = u_0(0), \quad h_0(0) = u_0(L), \quad \dot{g}_0(0) = u_1(0), \quad \dot{h}_0(0) = u_1(L).$$

11.4   The forced problem

Let $u(x,t)$ satisfy the forced wave equation

$$u_{tt} - u_{xx} = f(x,t). \tag{11.49}$$

Defining the adjoint as earlier, we now find

$$(vu_t - uv_t)_t - (vu_x - uv_x)_x = vf.$$

Hence, Green's theorem implies

$$\int_{\partial\Omega} (vu_t - uv_t)\,dx + (vu_x - uv_x)\,dt = -\iint_\Omega vf\,dxdt.$$

Letting $v = \exp[-ikx + ik\tau]$, it follows that for the case of the half-line the global relation involves in the right-hand side the term

$$e^{-2ikt} \int_0^t \int_0^\infty e^{-ikx+ik\tau} f(x,\tau)\,dxd\tau.$$

Integrating this term with respect to t and using the identity

$$\int_0^t e^{-2ik\tau} \left(\int_0^\tau f(\eta)d\eta \right) d\tau = \frac{e^{-2ikt}}{-2ik} \int_0^t f(\eta)d\eta + \frac{1}{2ik} \int_0^t e^{-2ik\tau} f(\tau)d\tau, \tag{11.50}$$

we find that the global relation now has the additional term

$$\check{f}(k,t) = \int_0^\infty e^{-ikx} \left(\int_0^t \frac{\sin[k(t-\tau)]}{k} f(x,\tau)d\tau \right) dx. \tag{11.51}$$

For the case of the finite interval, we obtain a similar term where the integral with respect to x is replaced with the integral from 0 to L.

For the Dirichlet problem on the half-line, the integral representation of the solution is given by Eq. (11.18) with the additional term

$$F(x,t) = \int_{-\infty}^\infty e^{ikx} \left[\check{f}(k,t) - \check{f}(-k,t) \right] \frac{dk}{2\pi}. \tag{11.52}$$

This takes the form:

$$F(x,t) = -2i \int_{-\infty}^\infty e^{ikx} \left[\int_0^\infty \sin(k\xi) \left(\int_0^t \frac{\sin[k(t-\tau)]}{k} f(\xi,\tau)d\tau \right) d\xi \right] \frac{dk}{2\pi}.$$

Through integration by parts with respect to ξ, the above expression simplifies to

$$F(x,t) = 2i \int_{-\infty}^\infty e^{ikx} \left\{ \int_0^\infty \cos(k\xi) \left[\int_0^t \sin[k(t-\tau)] \left(\int_0^\xi f(\rho,\tau)d\rho \right) d\tau \right] d\xi \right\} \frac{dk}{2\pi},$$

which takes the form

$$F(x,t) = \frac{1}{2} \int_0^t \left\{ \int_0^\infty \left[\delta(x+\xi+t-\tau) + \delta(x-\xi+t-\tau) \right. \right.$$

$$\left. \left. - \delta(x+\xi-t+\tau) - \delta(x-\xi-t+\tau) \right] \left(\int_0^\xi f(\rho,\tau)d\rho \right) d\xi \right\} d\tau.$$

This expression can be evaluated explicitly as follows:

$$F(x,t) = \frac{1}{2} \int_0^t \left(\int_0^{x+t-\tau} f(\xi,\tau)d\xi \right) d\tau$$

$$-\frac{1}{2} \begin{cases} \int_0^t \left(\int_0^{-x+t-\tau} f(\xi,\tau)d\xi \right) d\tau, & t-\tau > x, \\ \int_0^t \left(\int_0^{x-t+\tau} f(\xi,\tau)d\xi \right) d\tau, & t-\tau < x. \end{cases}$$

Using the variables $X = (x+t)$ and $T = (t-x)$ for a more compact notation, $F(x,t)$ simplifies further to the following expression:

$$F(x,t) = \frac{1}{2} \int_0^t \left(\int_{|T-\tau|}^{X-\tau} f(\xi,\tau)d\xi \right) d\tau. \tag{11.53}$$

The material presented in this chapter, as well as in Chapters 9 and 10, shows that initial-boundary value problems for evolution PDEs and for the wave equation, can be solved as follows: if a problem is formulated on the half-line, then it can be analysed via the half-Fourier transform, whereas if it is formulated on a finite interval, it can be analyzed via the finite Fourier transform. This new approach to problems on a finite interval is clearly quite different that the traditional approach, which involves a Fourier series.

The unified transform for solving problems on a finite interval suggests a paradigm shift ,which may have far reaching implications beyond the solution of PDEs: if a function is defined on a finite interval, then this function should be represented via the finite Fourier transform and not via a Fourier series. This representation has two important advantages: First, a sum is replaced by an integral, which, in general, provides a more effective way for obtaining useful information, especially with respect to asymptotics. Second, the finite Fourier transform is an entire function, and this allows utilizing the powerful theory of analytic functions, including the use of Cauchy's theorem.

Problems

1. The behavior of the electromagnetic field $E = u(x,t)$ in some antennas (like those used in old cars) can be modeled by the equation

$$\frac{\partial^2 u}{\partial t^2} + \frac{\partial^4 u}{\partial x^4} = 0, \tag{11.54}$$

where, as usual, the speed of light c has been absorbed in the redefined time variable t. Rewriting Eq. (11.54) in the form

$$\frac{\partial}{\partial t}\left(\frac{\partial u}{\partial t}\right) + \frac{\partial^2}{\partial x^2}\left(\frac{\partial^2 u}{\partial x^2}\right) = 0,$$

motivates the introduction of the variable v through the equations

$$\frac{\partial u}{\partial t} = -\frac{\partial^2 v}{\partial x^2}, \quad \frac{\partial v}{\partial t} = \frac{\partial^2 u}{\partial x^2}.$$

By introducing the complex valued function $q = u + iv$, show that q satisfies a simple evolution equation that can be solved by the techniques of Chapters 9 and 10.

2. The so-called "water hammer" phenomenon is described by the equations

$$\frac{\partial u}{\partial t} = -\frac{\partial p}{\partial x}, \quad \frac{\partial p}{\partial t} = -c^2 \frac{\partial u}{\partial x}.$$

By expressing u and p in terms of an appropriately chosen function v, show that v satisfies the wave equation.

3. The "transmission-line equations," also known as the "telegraph equations," are given by

$$\frac{\partial I}{\partial x} + GV + C\frac{\partial V}{\partial t} = 0, \quad \frac{\partial V}{\partial x} + RV + L\frac{\partial I}{\partial t} = 0,$$

where $I(x,t)$ is the current and $V(x,t)$ is the voltage along the line described by the position variable x, and the time variable t. The constants that appear in these equations represent the following physical quantities: R is the resistance, L the inductance, G the leakage conductance, and C the capacitance. Obtain a single equation for the current and show that if the resistance and the leakage conductance vanish, then $I(x,t)$ satisfies the wave equation with velocity c given by $c = 1/\sqrt{LC}$.

4. It can be shown that the displacement, u, of a vibrating string in a medium that resists the motion satisfies the equation

$$\frac{1}{c^2}\frac{\partial^2 u}{\partial t^2} - \frac{\partial^2 u}{\partial x^2} + \alpha\frac{\partial u}{\partial t} = 0. \qquad (11.55)$$

Absorb c into t by redefining α. Suppose that

$$u(x,0) = u_0(x), \quad u_t(x,0) = u_1(x), \quad 0 < x < +\infty.$$

Denote the boundary values $u(0,t)$ and $u_x(0,t)$ by $g_0(t)$ and $g_1(t)$. The adjoint of Eq. (11.55) is given by

$$\frac{1}{c^2}\frac{\partial^2 v}{\partial t^2} - \frac{\partial^2 v}{\partial x^2} - a\frac{\partial v}{\partial t} = 0.$$

By searching for a particular solution of the adjoint equation in the form $\exp[-i\kappa x + \gamma t]$, construct the global relation in the case that $0 < x < +\infty$. Show that with $\alpha = 0$, the global relation becomes Eq. (11.15).

5. The equation satisfied by the voltage $V = u(x,t)$ of Problem 3 is of the form

$$\frac{1}{c^2}\frac{\partial^2 u}{\partial t^2} - \frac{\partial^2 u}{\partial x^2} + \alpha\frac{\partial u}{\partial t} - \beta\frac{\partial u}{\partial x} = 0.$$

Absorb the constant c into the variable t by redefining the value of α. Suppose that

$$u(x,0) = u_0(x), \quad u_t(x,0) = u_1(x), \quad 0 < x < L.$$

Denote the boundary values $u(0,t)$ and $u_x(0,t)$ by $g_0(t)$ and $g_1(t)$. Similarly, denote the boundary values $u(L,t)$ and $u_x(L,t)$ by $h_0(t)$ and $h_1(t)$. Search for

a particular solution of the associated adjoint equation in the form $\exp[-i\kappa x + \gamma t]$. Construct the global relation and show that if $\alpha = \beta = 0$, this equation becomes Eq. (11.25).

6. Consider the wave equation on the half-line with the initial and boundary conditions of Example 11.1. Obtain a representation of the solution in the form of Eq. (11.18). By computing the relevant integrals via the residue theorem, show that $u(x,t)$ coincides with the expressions obtained in Example 11.1.

7. Consider the wave equation on the half-line with initial and boundary conditions

$$u_0(x) = xe^{-ax}, \quad u_1(x) = e^{-bx}, \quad g_0(t) = \sin(t),$$

where the constants $a, b > 0$.

(a) Compute $u(x,t)$ by using Eq. (11.19).

(b) Verify directly that the solution obtained satisfies the wave equation as well as the given initial and boundary conditions.

8. Consider the wave equation on the finite-interval $0 < x < L$.

(a) Using the initial and boundary conditions of Eqs. (11.45), obtain a representation of the solution in the form of Eq. (11.39).

(b) Using the initial and boundary conditions

$$u_0(x) = Lx + x^2, \quad u_1(x) = 0,$$
$$g_0(t) = 0, \quad h_0(t) = 2L^2,$$

express the solution in the form of Eq. (11.42).

(c) Verify that the solution found in part (b) satisfies the wave equation, as well as given the initial and boundary conditions.

9. Starting with Eqs. (11.51) and (11.52) for the Dirichlet problem of the forced wave equation, provide the missing intermediate steps to arrive at the final expression for the term $F(x,t)$, given in Eq. (11.53).

Chapter 12

Unified Transform IV: Laplace, Poisson, and Helmholtz Equations

In this chapter, we discuss the implementation of the unified transform to the following elliptic equations which appear in many important applications:

$$\text{Laplace equation}: \quad u_{xx}(x,y) + u_{yy}(x,y) = 0,$$
$$\text{Poisson equation}: \quad u_{xx}(x,y) + u_{yy}(x,y) = f(x,y),$$
$$\text{Helmholtz equation}: \quad u_{xx}(x,y) + u_{yy}(x,y) + \kappa^2 u(x,y) = 0, \quad \kappa > 0,$$
$$\text{modified Helmholtz equation}: \quad u_{xx}(x,y) + u_{yy}(x,y) - \kappa^2 u(x,y) = 0, \quad \kappa > 0.$$

These equations were introduced in Chapter 8.

12.1 Introduction

It was shown in Chapters 9–11 that the implementation of the unified transform to evolution PDEs and to the wave equation is based on the proper analysis of the global relation. Actually, this relation was used in two different ways:

- First, it was employed for obtaining the Fourier transform of the solution in terms of appropriate transforms of the initial datum and all boundary values. Then, by inverting this Fourier transform and using contour deformation, the solution was expressed in terms of an integral in the k-complex plane. However, this expression did not provide an effective solution because some of the boundary values are not prescribed as boundary conditions, and hence the above integral representation contains transforms of unknown functions.

- Second, by applying to the global relation suitable transformations in the k-complex plane which leave the transforms of the boundary values invariant, it was possible to express the transforms of the unknown boundary values in terms of the given data. In this way, the integral representation obtained in the first two steps gave rise to a formula involving only transforms of the given data.

An important feature of evolution PDEs and of the wave equation is that, by definition, the boundary of the domain of interest contains the solution itself. Hence, since the global relation couples appropriate transforms of boundary values, the global relation of evolution PDEs and of the wave equation contains the Fourier transform of the solution, denoted by $\hat{u}(k,t)$. As explained in the first point above, this has the advantage of providing a simple way for computing $\hat{u}(k,t)$. On the other hand, it has the disadvantage that it obscures the essence of the global relation: in general, the global relation couples the initial and the boundary data with the unknown boundary values, and in contrast to $u(x,0)$, $u(0,t)$, $u_x(0,t)$, etc., the solution $u(x,t)$ is not a genuine boundary value.

The essence of the global relation becomes clear for elliptic PDEs: for the particular class of such PDEs that contain derivatives up to second order, the global relation couples the Dirichlet and the Neumann boundary values. As noted in Chapter 8, this terminology refers to the values of the function and of its outward normal derivative on the boundary, respectively. Since the global relation does not contain the solution itself, it is not possible to use the global relation to derive an integral representation of the solution. However, for a general *convex domain* it is still possible, using one of several different approaches, to construct an integral representation in the k-complex plane. This representation, just like the analogous one obtained for evolution equations and the wave equation, is not yet effective: it contains both the Dirichlet and Neumann boundary values, and for a given boundary value problem, only one of them or their combination is known. It is at this stage that the employment of the global relation becomes crucial. For the case of the Laplace, Poisson, Helmholtz, and modified Helmholtz equations defined in simple domains, including the interior of a quarter plane, or a half strip, or a rectangle, or a quarter plane, or an equilateral triangle, or an orthogonal isosceles triangle, the unified transform is as effective as the case of evolution equations: the use of the global relation allows one to eliminate the unknown boundary values and to obtain an effective solution. Furthermore, for the above domains this methodology can be implemented for a variety of complicated boundary conditions, including non-separable boundary conditions. Apparently, some of these problems cannot be solved by the standard methods, including the method of transforms or the method of images.

For an arbitrary domain the above methodology fails: although for a convex domain it is *always* possible to obtain an integral representation in the k-complex plane, in general it is *not* possible to make the above integral representations effective, in the sense that it is not possible to eliminate the transforms of the unknown boundary values. Despite the impossibility of constructing an effective solution in the case of an arbitrary domain, the employment of the global relation gives rise to a very simple *numerical algorithm* for solving a variety of physically significant problems.

In what follows we will concentrate in the case that the domain, denoted by Ω, is the interior of a closed, bounded, convex polygon, specified by the complex numbers $z_1, z_2, \ldots, z_n$, see Fig. 12.1.

It is straightforward to also handle unbounded domains. Furthermore, non-convex polygons can be analyzed by transforming them to convex domains via the use of "virtual boundaries".

Given the convex polygon Ω, the above discussion can be summarized as follows:

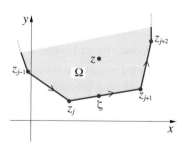

Figure 12.1: The interior, Ω, of a convex polygonal domain. The point z is in the interior of the polygonal domain and the point ζ is on the boundary.

(i) In analogy with the case of evolution PDEs and of the wave equation, it is always possible to express the solution in terms of an integral formula in the k-complex plane containing appropriate transforms of the Dirichlet and Neumann boundary values.

(ii) For very simple domains, by using the global relation as well as suitable transformations of the global relation, it is possible to eliminate the unknown boundary values from the formula of (i) and to obtain an effective solution involving only transforms of the given boundary data.

(iii) In the general case of an arbitrary convex polygon with different types of boundary conditions prescribed on different sides, the global relation gives rise to a very simple numerical algorithm, which determines the unknown boundary values in terms of the given data. We refer the reader to the original works for more details.[1]

[1] M.J. Colbrook, N. Flyer and B. Fornberg, On the Fokas method for the solution of elliptic problems in both convex and non-convex polygonal domains, *Journal of Computational Physics* 374, pp. 996–1016 (2018).

Taking into consideration the generality and simplicity of (iii) above, it is the implementation of this part that will provide the focus of this chapter. However, for completeness simple problems illustrating (ii) above will also be solved.

In order to emphasize the "unified" nature of the unified transform, we next illustrate the similarities of the integral representations of the heat equation and of the modified Helmholtz equation, which is a typical elliptic equation.

12.1.1   Integral representations in the k-complex plane

Before illustrating the general nature of the integral representations for elliptic PDEs obtained via the unified method by considering the particular case of the modified Helmholtz equation in the interior of the quarter plane, it is worth revisiting the integral representation of the heat equation in the finite interval, $0 < x < L$, with $0 < t < T$. The relevant domain is shown in Fig. 12.2.

It was argued in Chapters 9 and 10 that by using Cauchy's theorem and Jordan's lemma in $\mathcal{D}^+$ and $\mathcal{D}^-$ it is possible to replace $\tilde{g}_j(k^2, t)$ and $\tilde{h}_j(k^2, t)$ in Eq. (10.3a), which are appropriate t-transforms of the boundary values involving integrals up to t, by $\tilde{G}_j(k^2)$ and $\tilde{H}_j(k^2)$, $j = 0, 1$, which are the same transforms with integrals up to T. Thus, the integral representation of the solution takes the Ehrenpreis form

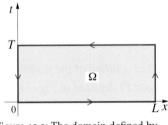

Figure 12.2: The domain defined by $0 < x < L$, with $0 < t < T$.

$$u(x,t) = \int_L e^{ikx - k^2 t} U(k) \frac{dk}{2\pi}, \tag{12.1}$$

where the integral along L consists of the union of three separate contours, and for each of these contours the function $\hat{u}(k)$ takes a different form:

$$\int_{L_1} = \int_{-\infty}^{\infty}, \quad \hat{u}_1(k) = \int_0^L e^{-ikx} u(x,0) dx, \tag{12.2a}$$

$$\int_{L_2} = \int_{\partial \mathcal{D}^+}, \quad \hat{u}_2(k) = -\int_0^T e^{k^2 t}[u_x(0,t) + iku(0,t)] dt, \tag{12.2b}$$

$$\int_{L_3} = \int_{\partial \mathcal{D}^-}, \quad \hat{u}_3(k) = -\int_0^T e^{k^2 t}[u_x(L,t) + iku(L,t)] dt. \tag{12.2c}$$

Thus, each part of the boundary of the domain under consideration gives rise to an integral transform, as well as to an appropriate contour in the k-complex plane: the boundary $\{t = 0,\, 0 < x < L\}$ gives rise to the transform $\hat{u}_1(k)$ and to the contour $(-\infty, \infty)$; the boundary $\{x = 0,\, 0 < t < T\}$ gives rise to the transform $\hat{u}_2(k)$ and to the contour $\partial \mathcal{D}^+$; and the boundary $\{x = L,\, 0 < t < T\}$ gives rise to the transform $\hat{u}_3(k)$ and to the contour $\partial \mathcal{D}^-$.

The situation for elliptic PDEs in the interior of the convex polygon Ω is similar. For example, it turns out that the integral representation for the solution of the modified Helmholtz equation in the interior of the quarter plane shown in Fig. 12.3 takes the following form:

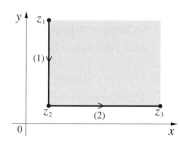

Figure 12.3: The interior of the quarter plane.

$$u(x,y) = \frac{1}{2i} \int_0^{i\infty} e^{i\kappa(kz - \bar{z}/k)} \hat{u}_1(k) \frac{dk}{2\pi} + \frac{1}{2i} \int_0^{\infty} e^{i\kappa(kz - \bar{z}/k)} \hat{u}_2(k) \frac{dk}{2\pi}, \qquad (12.3)$$

where $\hat{u}_1(k)$ and $\hat{u}_2(k)$ are defined by

$$\hat{u}_1(k) = \int_0^{\infty} e^{i\kappa(k+1/k)y} \left[u_x(0,y) + \frac{i\kappa}{2}\left(\frac{1}{k} - k\right) u(0,y) \right] dy, \qquad (12.4a)$$

$$\hat{u}_2(k) = \int_0^{\infty} e^{i\kappa(k-1/k)y} \left[-u_y(x,0) + \frac{\kappa}{2}\left(\frac{1}{k} + k\right) u(x,0) \right] dx. \qquad (12.4b)$$

Side (1) of the polygon of Fig. 12.3, namely, the side $\{x = 0, 0 < y < \infty\}$ gives rise to the transform $\hat{u}_1(k)$ and to the contour $(0, i\infty)$ along the positive imaginary axis of the k-complex plane. Similarly, side (2), namely the side $\{y = 0, 0 < x < \infty\}$ gives rise to the transform $\hat{u}_2(k)$ and to the contour $(0, \infty)$ along the positive real axis of the k-complex plane. The integral representation Eq. (12.3) is a particular case of the following general result derived by two different methods[2,3]:

Define the n rays $l_1, \ldots, l_n$ in the k-complex plane by

$$l_j = e^{-i\text{Arg}[z_{j+1} - z_j]} \rho, \quad 0 < \rho < \infty, \quad j = 1, 2, \ldots, n. \qquad (12.5)$$

The solution of the modified Helmholtz equation in the interior of the convex polygon Ω depicted in Fig. 12.1 is given by

$$u(x,y) = \frac{1}{2i} \sum_{j=1}^{n} \int_{l_j} \frac{1}{k} e^{i\kappa(kz - \bar{z}/k)/2} \hat{u}_j(k) \frac{dk}{2\pi}, \qquad (12.6)$$

where the functions $\hat{u}_j(k)$ are defined by

$$\hat{u}_j(k) = \int_{z_j}^{z_{j+1}} e^{i\kappa(\bar{z}/k - kz)/2} \left[\frac{\partial u_j}{\partial \mathcal{N}} ds + \frac{\kappa}{2} u_j \left(k dz + \frac{d\bar{z}}{k} \right) \right]. \qquad (12.7)$$

Equation (12.5) defining the rays l_j shows that the easiest way to construct these rays is to determine the directions of the unit vectors along each side of the polygon. For the case of the quarter plane depicted in Fig. 12.3, the unit vector along side (1) forms an angle $-\pi/2$ with the x-axis, thus the ray l_1 forms an angle $\pi/2$ with the real axis of the k-complex plane; hence, l_1 is along the positive imaginary axis. Similarly, the unit vector along side (2) makes a zero angle with the x-axis, hence the ray l_2 is along the real axis of the k-complex plane. Regarding $\hat{u}_1(k)$, we note that on side (1), $z = iy$, $\bar{z} = -iy$, $dz = idy$, $d\bar{z} = -idy$, the normal derivative is equal to $-u_x(0,y)$; hence, $\hat{u}_1(k)$ is given by Eq. (12.4a). Similarly, for side (2), $z = \bar{z} = x$, $dz = d\bar{z} = dx$, the normal derivative is equal to $-u_y(0,x)$; hence, $\hat{u}_1(k)$ is given by Eq. (12.4b).

[2] A.S. Fokas, Two-dimensional linear partial differential equations in a convex polygon, *Proceedings of the Royal Society London A* **457**, pp. 371–393 (2001).

[3] A.S. Fokas and M. Zyskin, The fundamental differential form and boundary value problems, *Quarterly Journal of Mechanics and Applied Mathematics* **55**, pp. 457–479 (2002).

12.2   The Laplace and Poisson equations

It was shown in Section 11.1 that the easiest way to derive the global relation is to employ the concept of the adjoint. The adjoint of the Laplace equation is itself,

$$v_{xx} + v_{yy} = 0. \tag{12.8}$$

Multiplying the Laplace equation by v, and its adjoint by u and then subtracting the resulting equations we find

$$(vu_x - uv_x)_x = (uv_y - vu_y)_y. \tag{12.9}$$

Thus, if the Laplace equation is valid in the domain Ω with the boundary $\partial\Omega$, then Green's theorem, Eq. (2.73), implies

$$\int_{\partial\Omega} [(vu_x - uv_x)\mathrm{d}y + (vu_y - uv_y)\mathrm{d}x] = 0. \tag{12.10}$$

Using the particular solutions of $\exp[-ikx + ky]$ and $\exp[ikx + ky]$ of Eq. (12.8), Eq. (12.10) implies the following two global relations:

$$\int_{\partial\Omega} \mathrm{e}^{-ikx+ky}[(u_x\mathrm{d}y - u_y\mathrm{d}x) + ku(\mathrm{d}x + i\mathrm{d}y)] = 0, \quad k \in \mathbb{C}, \tag{12.11}$$

and

$$\int_{\partial\Omega} \mathrm{e}^{ikx+ky}[(u_x\mathrm{d}y - u_y\mathrm{d}x) + ku(\mathrm{d}x - i\mathrm{d}y)] = 0, \quad k \in \mathbb{C}. \tag{12.12}$$

If u is real, then Eq. (12.12) can be obtained from Eq. (12.11) via the so-called Schwartz conjugation with respect to k, namely by taking the complex conjugate of Eq. (12.11) and then replacing $\bar{k}$ with k.

In what follows, we assume for simplicity that u is real, thus we concentrate on Eq. (12.11).

As noted earlier, the most well-known boundary value problems for Laplace equation involve boundary conditions for u or for the normal derivative of u which will be denoted by $\partial u/\partial\mathcal{N}$. In order to express the integral of Eq. (12.11) in terms of u and $\partial u/\partial\mathcal{N}$, we parameterize the boundary $\partial\Omega$ in terms of its arc length s. Then, recalling the relation

$$u_x\mathrm{d}y - u_y\mathrm{d}x = u_{\mathcal{N}}\mathrm{d}s, \tag{12.13}$$

we find that the global relation, Eq. (12.11), becomes

$$\int_{\partial\Omega} \mathrm{e}^{-ikx+ky}\left[\frac{\partial u}{\partial\mathcal{N}}\mathrm{d}s + ku\mathrm{d}z\right] = 0, \quad k \in \mathbb{C}. \tag{12.14}$$

It is important to emphasize that the global relation is *not* unique. In particular, since u satisfies $u_{z\bar{z}} = 0$, it follows that

$$\left(\mathrm{e}^{-ikz}u_z\right)_{\bar{z}} = 0, \quad z \in \Omega.$$

Hence, $\exp[-ikz]u_z$ is an analytic function. Thus, Cauchy's theorem implies

$$\int_{\partial\Omega} \mathrm{e}^{-ikz}u_z\mathrm{d}z = 0, \quad k \in \mathbb{C}. \tag{12.15}$$

The choice of which global relation to use depends on the particular boundary value problem: for a Dirichlet problem, Eq. (12.14) is preferable, while for a boundary value problem involving only derivatives, Eq. (12.15) is more convenient.

By utilizing the notion of the adjoint, it is also straightforward to obtain a global relation for the Poisson equation:

$$\int_{\partial\Omega} e^{-ikx+ky} \left[\frac{\partial u}{\partial \mathcal{N}} ds + ku dz \right] = -\iint_{\Omega} e^{-ikx+ky} f(x,y) dx dy. \qquad (12.16)$$

Indeed, multiplying the Poisson equation, Eq. (8.104), and the adjoint Laplace equation, Eq. (12.8), by v and u respectively, and then subtracting the resulting equations we find

$$(uv_x - vu_x)_x + (uv_y - vu_y)_y = fv.$$

Thus, Green's theorem, Eq. (2.73), with $F_2 = (uv_x - vu_x), F_1 = -(uv_y - vu_y)$ and $v = \exp[-ikx+ky]$, yields

$$\int_{\partial\Omega} e^{-ikx+ky}[(u_x dy - u_y dx) + ku(dx + idy)] = -\iint_{\Omega} e^{-ikx+ky} f(x,y) dx dy.$$

Simplifying the left-hand side of this equation using Eq. (12.13) we find Eq. (12.16). It is worth noting that, using the identity [see Eq. (8.95)]

$$\partial_z \partial_{\bar{z}} = \frac{1}{4}\left(\partial_{x^2} + \partial_{y^2}\right),$$

the Poisson equation can be written in the form

$$\left(e^{-ikz} u_z\right)_{\bar{z}} = \frac{e^{-ikz}}{4} f(x,y). \qquad (12.17)$$

Hence, employing the general form of Cauchy's theorem, Eq. (4.54), we find

$$\int_{\partial\Omega} e^{-ikz} u_z dz = \frac{i}{2}\iint_{\Omega} e^{-ikz} f(x,y) dx dy. \qquad (12.18)$$

12.2.1 Global relations for a convex polygon

In the case that Ω is the interior of the convex polygon of Fig. 12.1, Eq. (12.11) becomes

$$\sum_{j=1}^{n} \hat{u}_j(k) = 0, \quad k \in \mathbb{C}, \qquad (12.19)$$

where

$$\hat{u}_j(k) = \int_{z_j}^{z_{j+1}} e^{-ikx+ky} \left[(u_{j,x} dy - u_{j,y} dx) + ku_j dz\right], \quad j = 1,\ldots,n, \quad k \in \mathbb{C}. \qquad (12.20)$$

Equations (12.14) and (12.15) imply that Eq. (12.19) is also valid with $\hat{u}_j(k)$ defined by

$$\hat{u}_j(k) = \int_{z_j}^{z_{j+1}} e^{-ikz} \left[\frac{\partial u_j}{\partial \mathcal{N}} ds + ku_j dz\right], \quad j = 1,\ldots,n, \quad k \in \mathbb{C}, \qquad (12.21)$$

and

$$\hat{u}_j(k) = \int_{z_j}^{z_{j+1}} e^{-ikz}\frac{\partial u_j}{\partial z}dz, \quad j = 1,\ldots,n, \quad k \in \mathbb{C}. \tag{12.22}$$

For the Poisson equation, Eq. (12.16) yields

$$\sum_{j=1}^{n} \hat{u}_j(k) = -\iint_{\Omega} e^{-ikx+ky} f(x,y)dxdy, \tag{12.23}$$

where $\hat{u}_j(k)$ is given by Eq. (12.21).

Similarly, Eq. (12.18) yields

$$\sum_{j=1}^{n} \hat{u}_j(k) = -\frac{i}{2} \iint_{\Omega} e^{-ikx+ky} f(x,y)dxdy, \tag{12.24}$$

where $\hat{u}_j(k)$ is defined by Eq. (12.22).

Derivation: Global relation for a square

For a simple, closed convex polygon, consider the square with corners at

$$z_1 = -1+i, \quad z_2 = -1-i, \quad z_3 = 1-i, \quad z_4 = 1+i, \tag{12.25}$$

shown in Fig. 12.4. In this case, in order to compute $\hat{u}_j$, $j = 1,2,3,4$, it is convenient to use Eq. (12.20): on the side (1), which is the side between z_1 and z_2, the variable x takes the value $x = -1$ and the variable y takes values $-1 < y < 1$. Thus, $z = -1+iy$, $dx = 0$, and $dz = idy$, from which we obtain

$$\hat{u}_1(k) = -e^{ik}\int_{-1}^{1} e^{ky}[u_{1,x} + iku_1]dy \quad k \in \mathbb{C}. \tag{12.26}$$

Similarly,

$$\hat{u}_2(k) = e^{-k}\int_{-1}^{1} e^{-ikx}[-u_{2,y} + ku_2]dx, \tag{12.27}$$

$$\hat{u}_3(k) = e^{-ik}\int_{-1}^{1} e^{ky}[u_{3,x} + iku_3]dy, \tag{12.28}$$

$$\hat{u}_4(k) = -e^{k}\int_{-1}^{1} e^{-ikx}[-u_{4,y} + ku_4]dx, \quad k \in \mathbb{C}. \tag{12.29}$$

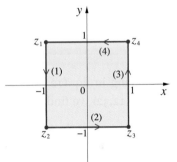

Figure 12.4: The square with corners $z_1 = -1+i$, $z_2 = -1-i$, $z_3 = 1-i$, $z_4 = 1+i$.

12.2.2 *An integral representation in the k-complex plane*

In order to construct an integral representation in the k-complex plane for the Laplace equation, we recall that u is harmonic if and only if u_z is analytic. Thus, it is sufficient to construct an integral representation for the analytic function u_z. In this connection, we recall Cauchy's formula for an analytic function $f(z)$ in the domain Ω with boundary $\partial\Omega$:

$$f(z) = \frac{1}{2\pi i}\int_{\partial\Omega} \frac{f(\zeta)}{\zeta - z}d\zeta. \tag{12.30}$$

If Ω is the interior of the convex polygon shown in Fig. 12.1, then the above equation becomes

$$f(z) = \frac{1}{2\pi i}\sum_{j=1}^{n}\int_{z_j}^{z_{j+1}} \frac{f(\zeta)}{\zeta - z}d\zeta. \tag{12.31}$$

Let l_j be a ray in the k-complex plane from the origin to infinity. Then,

$$\frac{1}{\zeta - z} = 2\pi \mathrm{i} \int_{l_j} \mathrm{e}^{\mathrm{i}k(z-\zeta)} \frac{\mathrm{d}k}{2\pi}, \tag{12.32}$$

provided that on this ray the term $\exp[\mathrm{i}k(z - \zeta)]$ vanishes as $|k| \to \infty$. Thus, we require that

$$0 < \mathrm{Arg}[k] + \mathrm{Arg}[z - \zeta] < \pi. \tag{12.33}$$

It becomes evident from Fig. 12.1 that convexity implies the inequality

$$\mathrm{Arg}[z_{j+1} - z_j] < \mathrm{Arg}[z - \zeta] < \mathrm{Arg}[z_{j-1} - z_j]. \tag{12.34}$$

Thus, Eq. (12.32) is valid provided that the ray l_j is defined by the requirement that

$$\mathrm{Arg}[k] = -\mathrm{Arg}[z_{j+1} - z_j]. \tag{12.35}$$

Replacing in Eq. (12.31) $f(z)$ with u_z and $1/(\zeta - z)$ with the right-hand side of Eq. (12.32) we find the following result:

Let l_j denoted the n rays in the k-complex plane defined by Eq. (12.5). The solution of the Laplace equation inside the convex polygon characterized by the corners $z_1, z_2, \ldots, z_n$, satisfies

$$\frac{\partial u}{\partial z} = \sum_{j=1}^{n} \int_{l_j} \mathrm{e}^{\mathrm{i}kz} \hat{u}_j(k) \frac{\mathrm{d}k}{2\pi}, \tag{12.36}$$

where

$$\hat{u}_j(k) = \int_{z_j}^{z_{j+1}} \mathrm{e}^{-\mathrm{i}k\zeta} \frac{\partial u}{\partial \zeta} \mathrm{d}\zeta. \tag{12.37}$$

Furthermore $\hat{u}_j(k)$, $j = 1, \ldots, n$, satisfy the global relation defined by Eq. (12.19).

Example 12.1: For the square defined by Eq. (12.25) (see Fig.12.4) the unit vectors along the sides (1), (2), (3), (4) make the angles $\pi/2$, 0, $-\pi/2$, $-\pi$ with the x-axis, respectively. Thus, the rays l_1, l_2, l_3, l_4 make the angles $\pi/2$, 0, $-\pi/2$, $-\pi$ with the real axis of the k-complex plane. Hence,

$$\frac{\partial u}{\partial z} = \int_0^{\mathrm{i}\infty} \mathrm{e}^{\mathrm{i}kz} \hat{u}_1(k) \frac{\mathrm{d}k}{2\pi} + \int_0^{\infty} \mathrm{e}^{\mathrm{i}kz} \hat{u}_2(k) \frac{\mathrm{d}k}{2\pi}$$
$$+ \int_0^{-\mathrm{i}\infty} \mathrm{e}^{\mathrm{i}kz} \hat{u}_3(k) \frac{\mathrm{d}k}{2\pi} + \int_0^{-\infty} \mathrm{e}^{\mathrm{i}kz} \hat{u}_4(k) \frac{\mathrm{d}k}{2\pi}.$$

The functions $\hat{u}_j(k)$ are defined by

$$\hat{u}_1(k) = -\frac{\mathrm{e}^{\mathrm{i}k}}{2} \int_{-1}^{1} \mathrm{e}^{ky}(\mathrm{i}u_{1,x} + u_{1,y})\mathrm{d}y,$$

$$\hat{u}_2(k) = \frac{\mathrm{e}^{-k}}{2} \int_{-1}^{1} \mathrm{e}^{-\mathrm{i}kx}(u_{2,x} - \mathrm{i}u_{2,y})\mathrm{d}x,$$

$$\hat{u}_3(k) = \frac{\mathrm{e}^{-\mathrm{i}k}}{2} \int_{-1}^{1} \mathrm{e}^{ky}(\mathrm{i}u_{3,x} + u_{3,y})\mathrm{d}y,$$

$$\hat{u}_4(k) = -\frac{\mathrm{e}^{k}}{2} \int_{-1}^{1} \mathrm{e}^{-\mathrm{i}kx}(u_{4,x} - \mathrm{i}u_{4,y})\mathrm{d}x.$$

Example 12.2: For the quarter plane, shown in Fig. 12.3, as discussed earlier, the rays l_1 and l_2 make the angles $\pi/2$ and 0 with the real axis of the k-complex plane. Thus,

$$\frac{\partial u}{\partial z} = \int_0^{i\infty} e^{ikz} \hat{u}_1(k) \frac{dk}{2\pi} + \int_0^{\infty} e^{ikz} \hat{u}_2(k) \frac{dk}{2\pi},$$

where

$$\hat{u}_1(k) = -\frac{i}{2} \int_0^{\infty} e^{ky}(u_{1,x} - iu_{1,y})dy, \quad \text{Re}[k] \leq 0,$$

$$\hat{u}_2(k) = \frac{1}{2} \int_0^{\infty} e^{-ikx}(u_{2,x} - iu_{2,y})dx, \quad \text{Im}[k] \geq 0.$$

In this case, because the domain is unbounded, $\hat{u}_1(k)$ is defined only in the second and third quadrants of the k-complex plane, whereas $\hat{u}_2(k)$ is defined only in the third and fourth quadrants of the k-complex plane. Thus, the global relation is valid in the overlap of the above domains, i.e., in the third quadrant:

$$\hat{u}_1(k) + \hat{u}_2(k) = 0, \quad \pi \leq \text{Arg}[k] \leq \frac{3\pi}{2}.$$

It was noted earlier that one of the most well-known boundary value problems for the Laplace equation is the Neumann problem where one prescribes the derivative of u normal to the boundary. There exist physical circumstances where one prescribes the derivative in some given direction to the boundary. This boundary condition is not separable, thus the classical transforms fail. However, the unified transform can be applied. Below we analyze such a problem for the interior of the first quadrant, as illustrated in Fig. 12.5:

The derivative of u in the direction making an angle α with the boundaries is prescribed:

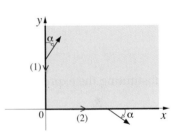

Figure 12.5: The interior of the first quadrant with oblique boundary conditions.

$$u_x(0,y)\sin\alpha + u_y(0,y)\cos\alpha = g_1(y), \quad 0 < y < \infty, \tag{12.38a}$$

$$-u_y(x,0)\sin\alpha + u_x(x,0)\cos\alpha = g_2(x), \quad 0 < x < \infty, \tag{12.38b}$$

where $0 \leq \alpha \leq \pi$ and the real function $u(x,y)$, decays as $|x| + |y| \to \infty$. We assume that the real functions $g_1(y)$ and $g_2(x)$ have appropriate smoothness and decay.

Equations (12.38a) and (12.38b) provide a relationship between the two unknown boundary values u_x and u_y on each side of the boundary. Instead of using these relations to express one of these unknown functions in terms of the other, we find it more convenient to introduce two new unknown functions $f_1(y)$ and $f_2(x)$ and to express the four unknown functions $u_x(0,y)$, $u_y(0,y)$, $u_x(x,0)$, $u_y(x,0)$ in terms of these two unknown functions. Let $f_1(y)$ and $f_2(x)$ denote the unknown derivatives in the direction normal to the directions of the given derivatives, that is,

$$u_y(0,y)\sin\alpha - u_x(0,y)\cos\alpha = f_1(y), \quad 0 < y < \infty, \tag{12.39a}$$

$$u_x(x,0)\sin\alpha + u_y(x,0)\cos\alpha = f_2(x), \quad 0 < x < \infty. \tag{12.39b}$$

Solving Eqs. (12.38a) and (12.39a) for $u_y(0,y)$ and $u_x(0,y)$, as well as Eqs. (12.38b) and (12.39b) for $u_x(x,0)$ and $u_y(x,0)$, we can express all four boundary values in terms of the two given functions and the two new unknown functions:

$$u_y(0,y) = g_1(y)\cos\alpha + f_1(y)\sin\alpha, \quad u_x(0,y) = g_1(y)\sin\alpha - f_1(y)\cos\alpha,$$

$$u_y(x,0) = -g_2(x)\sin\alpha + f_2(x)\cos\alpha, \quad u_x(x,0) = g_2(x)\cos\alpha + f_2(x)\sin\alpha.$$

Substituting these expressions in the definitions of $\hat{u}_1(k)$ and $\hat{u}_2(k)$, we find

$$\hat{u}_1(k) = e^{i\alpha}[\hat{g}_1(k) + i\hat{f}_1(k)], \tag{12.40}$$

$$\hat{u}_2(k) = e^{i\alpha}[\hat{g}_2(-ik) + i\hat{f}_2(-ik)], \tag{12.41}$$

where the known functions $\hat{g}_j(k)$, $j = 1, 2$, are defined as follows:

$$\hat{g}_1(k) = -\frac{1}{2}\int_0^\infty e^{ky}g_1(y)dy, \quad \text{Re}[k] \le 0,$$

$$\hat{g}_2(-ik) = \frac{1}{2}\int_0^\infty e^{-ikx}g_2(x)dx, \quad \text{Im}[k] \le 0,$$

whereas the unknown functions $\hat{f}_j(k)$, $j = 1, 2$, are defined by

$$\hat{f}_1(k) = \frac{1}{2}\int_0^\infty e^{ky}f_1(y)dy, \quad \text{Re}[k] \le 0,$$

$$\hat{f}_2(-ik) = -\frac{1}{2}\int_0^\infty e^{-ikx}f_2(x)dx, \quad \text{Im}[k] \le 0.$$

Substituting the expressions for $\hat{u}_1(k)$ and $\hat{u}_2(k)$ in the global relation, we find

$$\hat{g}_1(k) + i\hat{f}_1(k) + \hat{g}_2(-ik) + i\hat{f}_2(-ik) = 0, \quad \pi \le \text{Arg}[k] \le \frac{3\pi}{2}. \tag{12.42}$$

Taking the complex conjugate of this equation and replacing $\bar{k}$ with k we find

$$\hat{g}_1(k) - i\hat{f}_1(k) + \hat{g}_2(ik) - i\hat{f}_2(ik) = 0, \quad \frac{\pi}{2} \le \text{Arg}[k] \le \pi. \tag{12.43}$$

Equations (12.42) and (12.43) are two equations coupling the three unknown functions $\hat{f}_1(k)$, $\hat{f}_2(-ik)$ and $\hat{f}_2(ik)$. The pair $\hat{f}_2(-ik)$ and $\hat{f}_2(ik)$ remains invariant under the transformation $k \to -k$, thus we can supplement the above equations with the equations obtained from Eqs. (12.42) and (12.43) under the transformation $k \to -k$.

Equation (12.40) shows that the representation for u_1 involves integrals along the boundary of the first quadrant of the k-complex plane. Since in this domain the function $\hat{f}_2(ik)$ is analytic, we will express $\hat{u}_1$ and $\hat{u}_2$ in terms of this function with the expectation that its contribution to u_z vanishes [just as it happened in the case of evolution PDEs with the contribution of $\hat{u}(-k,t)$]. An alternative could have been to choose $\hat{f}_1(-k)$ instead of $\hat{f}_2(ik)$, since $\hat{f}_1(-k)$ is also analytic in the first quadrant. Solving Eq. (12.43) for $i\hat{f}_1(k)$ we find

$$i\hat{f}_1(k) = -i\hat{f}_2(ik) + \hat{g}_1(k) + \hat{g}_2(ik), \quad \frac{\pi}{2} \le \text{Arg}[k] \le \pi.$$

In order to determine $\hat{f}_2(-ik)$ we eliminate from Eqs. (12.42) and (12.43) the function $\hat{f}_1(k)$ and then we replace in the resulting equation k with $-k$:

$$\mathrm{i}\hat{f}_2(-ik) = \mathrm{i}\hat{f}_2(ik) + 2\hat{g}_1(-k) + \hat{g}_2(ik) + \hat{g}_2(-k), \quad 0 \le k \le \infty. \qquad (12.44)$$

The procedure of obtaining Eq. (12.44) is equivalent with the elimination of the function $\hat{f}_1(-k)$ from Eqs. (12.42) and (12.43) under the transformation $k \to -k$.

Substituting the expressions for $\mathrm{i}\hat{f}_1(k)$ and $\mathrm{i}\hat{f}_2(-ik)$ in the formulas for $\hat{u}_1(k)$ and $\hat{u}_2(k)$ we find

$$\hat{u}_1(k) = \mathrm{e}^{\mathrm{i}\alpha}[2\hat{g}_1(k) + \hat{g}_2(ik) - \mathrm{i}\hat{f}_2(ik)], \quad k \in k_I, \qquad (12.45a)$$

$$\hat{u}_2(k) = \mathrm{e}^{\mathrm{i}\alpha}[2\hat{g}_2(-ik) + \hat{g}_2(ik) + \mathrm{i}\hat{f}_2(ik) + 2\hat{g}_1(-k)], \quad k \in k_R. \qquad (12.45b)$$

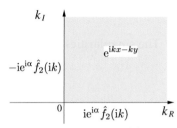

Figure 12.6: Contributions of the terms involving $\hat{f}_2(ik)$ in Eqs. (12.45a) and (12.45b).

The contribution of the terms involving $\hat{f}_2(ik)$ vanishes, see Fig. 12.6. Indeed, the real part of $\exp[ikx - ky]$ equals $\exp[-k_I x - k_R y]$ which is bounded in the first quadrant of the k-complex plane (since both x and y are positive). Furthermore, the function $\hat{f}_2(ik)$ is analytic in the first quadrant and integration by parts shows that it decays linearly as $|k| \to \infty$. Thus, Jordan's lemma implies that

$$\left(\int_{i\infty}^{0} + \int_{0}^{\infty} \right) \mathrm{i}\mathrm{e}^{ikz}\mathrm{e}^{\mathrm{i}\alpha} \hat{f}_2(ik) \frac{dk}{2\pi} = 0. \qquad (12.46)$$

In order to rewrite the solution in a more symmetric form we use Jordan's lemma to transform the contribution of the term involving $\hat{g}_2(ik)$ from the integral along k_R to an integral along k_I. Thus,

$$\frac{\partial u}{\partial z} = 2 \int_{0}^{i\infty} \mathrm{e}^{\mathrm{i}(\alpha+kz)}[\hat{g}_1(k) + \hat{g}_2(ik)] \frac{dk}{2\pi} + 2 \int_{0}^{\infty} \mathrm{e}^{\mathrm{i}(\alpha+kz)}[\hat{g}_1(-k) + \hat{g}_2(-ik)] \frac{dk}{2\pi},$$

$$0 < \mathrm{Arg}[z] < \frac{\pi}{2}, \qquad (12.47)$$

where $\hat{g}_1$ and $\hat{g}_2$ are the transforms of the functions g_1 and g_2.

Example 12.3: For a particular oblique Neumann problem on the quarter plane, we analyze further the case that

$$g_1(y) = \mathrm{e}^{-a_1 y}, \quad g_2(x) = \mathrm{e}^{-a_2 x}, \quad a_1, a_2 > 0.$$

Then,

$$\hat{g}_1(k) = -\tfrac{1}{2} \int_{0}^{\infty} \mathrm{e}^{(k-a_1)y} dy, \quad \hat{g}_2(k) = \tfrac{1}{2} \int_{0}^{\infty} \mathrm{e}^{(k-a_2)x} dx$$

and therefore

$$\hat{g}_1(k) = \frac{1}{2(k-a_1)}, \quad \hat{g}_2(k) = -\frac{1}{2(k-a_2)}.$$

Thus, Eq. (12.47) becomes

$$\frac{\partial u}{\partial z} = \int_{0}^{i\infty} \mathrm{e}^{\mathrm{i}(\alpha+kz)} \left(\frac{1}{k-a_1} - \frac{1}{ik-a_2} \right) \frac{dk}{2\pi}$$

$$- \int_{0}^{i\infty} \mathrm{e}^{\mathrm{i}(\alpha+kz)} \left(\frac{1}{k+a_1} - \frac{1}{ik+a_2} \right) \frac{dk}{2\pi}. \qquad (12.48)$$

The exponential $\exp[ikz]$ decays provided that

$$0 < \mathrm{Arg}[k] + \mathrm{Arg}[z] < \pi,$$

and since

$$0 < \mathrm{Arg}[z] < \frac{\pi}{2},$$

the exponential $\exp[ikz]$ decays on any ray such that

$$0 < \mathrm{Arg}[k] < \frac{\pi}{2}.$$

The singularities of the two integrands of Eq. (12.48) are shown below:

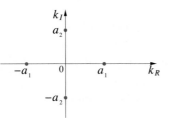

Thus, Eq. (12.48) takes the form

$$\frac{\partial u}{\partial z} = e^{i\alpha} \int_0^{\infty e^{i\phi}} e^{ikz} \left(\frac{1}{k - a_1} - \frac{1}{k + a_1} + \frac{i}{k + ia_2} - \frac{i}{k - ia_2} \right) \frac{dk}{2\pi}, \quad 0 < \phi < \frac{\pi}{2}.$$

Integrating the above equation and adding the complex conjugate of the resulting expression, we find

$$u(z) = -i e^{i\alpha} U(z) + i e^{-i\alpha} \overline{U(z)} + c, \tag{12.49}$$

where

$$U(z) = \lim_{\epsilon \to 0} \int_\epsilon^{\infty e^{i\phi}} \frac{e^{ikz}}{k} \left(\frac{1}{k - a_1} - \frac{1}{k + a_1} + \frac{i}{k + ia_2} - \frac{i}{k - ia_2} \right) \frac{dk}{2\pi} + \tilde{c}.$$

The integrand of the above integral has a pole singularity at $k = 0$. The constant $\tilde{c}$ must be chosen so that this singularity is canceled. Using partial fractions, we find that the integrand of the above integral is given by

$$\frac{1}{a_1} \left(\frac{1}{k - a_1} + \frac{1}{k + a_1} \right) - \frac{1}{a_2} \left(\frac{1}{k + ia_2} + \frac{1}{k - ia_2} \right) - \left(\frac{2}{a_1} - \frac{2}{a_2} \right) \frac{1}{k}.$$

The last term of the above expression yields the contribution

$$-\left(\frac{2}{a_1} - \frac{2}{a_2} \right) \int_0^{\infty e^{i\phi}} \frac{e^{ikz}}{k} \frac{dk}{2\pi}.$$

Ignoring for the moment the singularity of this integral, it is clear that it has the property that the change of variables $kz = t$ transforms it to an integral which is independent of z:

$$\int_0^{\infty e^{i\phi}} e^{it} \frac{dt}{t}, \qquad \phi \in (0, \pi).$$

Thus, we should subtract an integral which has two properties: first, cancels the singularity at $k = 0$, and second, is transformed to a z-independent integral under the change of variables $kz = t$. Hence

$$\tilde{c} = \left(\frac{2}{a_1} - \frac{2}{a_2} \right) c_1, \qquad c_1 = \int_0^{e^{i\phi}/z} \frac{1}{k} \frac{dk}{2\pi} + \frac{1}{2\pi} \ln z.$$

Thus,

$$-\int_0^{\infty e^{i\phi}} \frac{e^{ikz}}{k} \frac{dk}{2\pi} + c_1 = - \left(\int_0^{\infty e^{i\phi}} \frac{e^{ikz}}{k} \frac{dk}{2\pi} - \int_0^{e^{i\phi}/z} \frac{1}{k} \frac{dk}{2\pi} \right) + \frac{1}{2\pi} \ln z.$$

Using the change of variables $kz = t$, the above expression takes the form

$$\frac{1}{2\pi} \ln z - \left(\int_0^{e^{i\phi}} \frac{e^{it} - 1}{t} \frac{dt}{2\pi} + \int_{e^{i\phi}}^{\infty e^{i\phi}} \frac{e^{it}}{t} \frac{dt}{2\pi} \right) = \frac{1}{2\pi} \ln z + c_2,$$

where c_2 is a constant which can be absorbed in the undetermined constant associated with the Neumann problem of the Laplace equation. Hence, $U(z)$ is expressed as

$$U(z) = \int_0^{\infty e^{i\phi}} V(k, z) dk + v(z), \tag{12.50}$$

with

$$v(z) = \frac{1}{2\pi} \left(\frac{2}{a_1} - \frac{2}{a_2} \right) \ln z, \tag{12.51}$$

$$V(k, z) = \frac{e^{ikz}}{2\pi} \left[\frac{1}{a_1} \left(\frac{1}{k + a_1} + \frac{1}{k - a_1} \right) - \frac{1}{a_2} \left(\frac{1}{k - ia_2} + \frac{1}{k + ia_2} \right) \right]. \tag{12.52}$$

As explained in Chapter 9, the derivation of the solution assumes that the solution exists and it has sufficient smoothness and decay. In this sense, the derivation is formal. However, it is straightforward to make the derivation rigorous *a posteriori*.

For this example, U depends only on z, and its complex conjugate only on $\bar{z}$, hence, the final expression satisfies the Laplace equation. Thus, all that remains is to verify the boundary conditions. For simplicity let $\alpha = 0$. Then Eqs. (12.49) and (12.50) yield

$$u_z = -i U_z = \int_0^{\infty e^{i\phi}} e^{ikz} F(k) dk - \frac{i}{2\pi} \left(\frac{2}{a_1} - \frac{2}{a_2} \right) \frac{1}{z}, \qquad \phi \in (0, \pi/2),$$

where we have defined the auxiliary function

$$F(k) = \frac{k}{2\pi a_1}\left(\frac{1}{k-a_1}+\frac{1}{k+a_1}\right) - \frac{k}{2\pi a_2}\left(\frac{1}{k+ia_2}+\frac{1}{k-ia_2}\right).$$

Recalling that u_x is $1/2$ of the real part of u_z, and evaluating the expression of u_z at $x = 0$ we find:

$$u_x(x,0) = \int_0^{\infty e^{i\phi}} e^{ikx}F(k)dk + \overline{\int_0^{\infty e^{i\phi}} e^{ikx}F(k)dk}, \qquad \phi \in (0,\pi/2).$$

The second integral of the above expression takes the form

$$\overline{\int_0^{\infty e^{i\phi}} e^{ikx}F(k)dk} = \int_0^{\infty e^{i\phi}} e^{-i\bar{k}x}F(\bar{k})d\bar{k} = \int_{\infty e^{i(\pi-\phi)}}^0 e^{i\lambda x}F(-\lambda)d\lambda,$$

where we made the change of variables $\lambda = -\bar{k}$ in order to obtain the last equality. Using the fact that $F(-k) = F(k)$ the above equality becomes

$$\overline{\int_0^{\infty e^{i\phi}} e^{ikx}F(k)dk} = \int_{\infty e^{i(\pi-\phi)}}^0 e^{ikx}F(k)dk, \qquad \phi \in (0,\pi/2).$$

Then, it follows from the Residue theorem that

$$u_x(x,0) = 2\pi i \text{Residue}\left[e^{ikx}F(k)\right]_{k=ia_2} = ie^{iia_2 x}\left(-\frac{ia_2}{a_2}\right) = e^{-a_2 x}.$$

Similarly, by evaluating the expression of u_z at $y = 0$ we find that

$$u_y(0,y) = 2\pi \text{Residue}\left[e^{-ky}F(k)\right]_{k=a_1} = e^{-a_1 y}\frac{a_1}{a_1} = e^{-a_1 y}.$$

Because of the exponential decay for large k, the integral in Eq. (12.50) can be computed numerically very efficiently. For example, let $a_1 = 1$, $a_2 = 2$, $\alpha = 0$. This choice of parameter values corresponds to the boundary conditions

$$u_y(0,y) = e^{-y}, \qquad u_x(x,0) = e^{-2x}.$$

We choose $\phi = \pi/4$, and we evaluate $u(z)$ through Eq. (12.49), with the constant of integration $c = 0.5$. An example of such numerical evaluation using Mathematica, with $z = x + iy$ and the functions $v(z)$, $V(k,z)$ defined in Eqs. (12.51) and (12.52), respectively, is shown in the plot below for the range of the variables $x \in [0,4]$, $y \in [0,4]$.

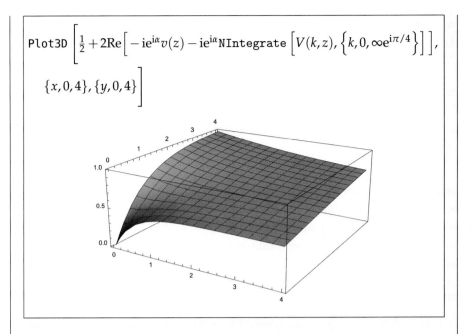

$$\texttt{Plot3D}\left[\frac{1}{2}+2\mathrm{Re}\left[-i e^{i\alpha}v(z)-i e^{i\alpha}\texttt{NIntegrate}\left[V(k,z),\left\{k,0,\infty e^{i\pi/4}\right\}\right]\right],\right.$$

$$\left.\{x,0,4\},\{y,0,4\}\right]$$

We note that the value of u at the boundary $y = 0$ is equal to the function $(1 - e^{-2x})/2$, while the value at the boundary $x = 0$ is equal to the function $1 - e^{-y}$. The behavior of these two functions is evident at the boundaries of the plot, with x along the bottom-labeled axis and y along the top-labeled axis.

12.2.3 The approximate global relation for a convex polygon

Since either u_j or $\partial u_j / \partial \mathcal{N}$ or a relationship between u_j and $\partial u_j / \partial \mathcal{N}$ is prescribed on each side of the polygon, it follows that each $\hat{u}_j$, $j = 1, \ldots, n$ contains one unknown function. Thus, the global relation (12.19) is a single equation for n unknowns.

Is it possible to use this single equation to determine the n unknowns? The answer is yes. This follows from the observation that the global relation is valid for all complex values of k. Thus, in a sense, the global relation gives us infinitely many equations (or rather, a *functional equation*). This observation yields a numerical procedure for evaluating the unknown boundary values. In its simplest form, this procedure involves three algorithmic steps discussed below:

(i) *Upon using a parameterization of each side with variable $x \in [-1, 1]$, expand all the unknown boundary values in terms of $N \in \mathbb{N}$ Legendre polynomials. For example, if a Neumann boundary condition is given along the jth side of the polygon, use*

$$u_j(x) \approx \sum_{l=0}^{N-1} a_l^{(j)} P_l(x), \tag{12.53}$$

where P_l denotes the lth Legendre polynomial. Similarly, if a Dirichlet boundary condition is given along the jth side of the polygon, use

$$\frac{\partial u_j}{\partial \mathcal{N}}(x) \approx \sum_{l=0}^{N-1} b_l^{(j)} P_l(x). \tag{12.54}$$

(ii) *Substitute the above approximation into the global relation to obtain an approximate global relation with nN unknown constants corresponding to the appropriate subset of the coefficients $\{a_l^{(j)}, b_l^{(j)} : j = 1, \ldots, n, l = 0, \ldots, N-1\}$.*

(iii) *Determine these unknown coefficients by collocating the approximate global relation at a sufficiently large number of well-chosen points in the k-complex plane. A convenient choice of such points, which leads to good numerical conditioning, will be discussed in the example below.*

Derivation: Approximate global relation for a square

We illustrate the application of this procedure for the case of the Dirichlet problem for Laplace's equation in the interior of the square defined by Eq. (12.25) (see Fig. 12.4). For this illustration, we will assume that the boundary conditions are as follows:

$$\text{side } (1): \quad u(-1,y) = \sin(\pi y), \quad 1 > y > -1, \tag{12.55a}$$

$$\text{side } (2): \quad u(x,-1) = 0, \quad -1 < x < 1, \tag{12.55b}$$

$$\text{side } (3): \quad u(1,y) = 0, \quad -1 < y < 1, \tag{12.55c}$$

$$\text{side } (4): \quad u(x,1) = 0, \quad 1 > x > -1, \tag{12.55d}$$

namely, all the Dirichlet data vanishes except on the side of the square labeled (1). Note that $u(-1,\pm1) = 0$ so that the boundary data are compatible. Equation (12.26) becomes

$$\hat{u}_1(k) = -e^{ik} \int_{-1}^1 e^{ky} u_{1,x} dy - ik e^{ik} \int_{-1}^1 e^{ky} \sin(\pi y) dy.$$

Noting that

$$\int_{-1}^1 e^{ky} \sin(\pi y) dy = \frac{\pi}{\pi^2 + k^2} \left(e^k - e^{-k} \right),$$

we find

$$\hat{u}_1(k) \approx -e^{ik} \sum_{l=0}^{N-1} b_l^{(1)} \hat{P}_l(k) - \frac{ik\pi e^{ik}}{\pi^2 + k^2} \left(e^k - e^{-k} \right). \tag{12.56a}$$

Similarly, Eqs. (12.27)–(12.29) become

$$\hat{u}_2(k) \approx -e^{-k} \sum_{l=0}^{N-1} b_l^{(2)} \hat{P}_l(-ik), \tag{12.56b}$$

$$\hat{u}_3(k) \approx e^{-ik} \sum_{l=0}^{N-1} b_l^{(3)} \hat{P}_l(k), \tag{12.56c}$$

$$\hat{u}_4(k) \approx e^k \sum_{l=0}^{N-1} b_l^{(4)} \hat{P}_l(-ik), \tag{12.56d}$$

where

$$\hat{P}_l(k) = \int_{-1}^1 e^{ky} P_l(y) dy. \tag{12.57}$$

Hence, the approximate global relation becomes

$$- e^{ik} \sum_{l=0}^{N-1} b_l^{(1)} \hat{P}_l(k) - e^{-k} \sum_{l=0}^{N-1} b_l^{(2)} \hat{P}_l(-ik)$$

$$+ e^{-ik} \sum_{l=0}^{N-1} b_l^{(3)} \hat{P}_l(k) + e^{k} \sum_{l=0}^{N-1} b_l^{(4)} \hat{P}_l(-ik) \approx \frac{ik\pi e^{ik}}{\pi^2 + k^2} \left(e^{k} - e^{-k} \right).$$

$$(12.58)$$

The functions $\hat{P}_l$ can be expressed in terms of the so-called modified Bessel functions of the first kind [these functions can be computed using standard routines in most modern computing environments]:

$$\hat{P}_l(k) = \left(\frac{2\pi}{k} \right)^{1/2} I_{l+1/2}(k).$$

It turns out that it is possible to collocate Eq. (12.58) at suitable complex values of k, such that the resulting linear system is diagonally dominant for large $|k|$. For this purpose, we choose k to be on the rays below:

$$L_1 = \{ -i\rho : 0 < \rho < \infty \}, \qquad (12.59a)$$

$$L_2 = \{ -\rho : 0 < \rho < \infty \}, \qquad (12.59b)$$

$$L_3 = \{ i\rho : 0 < \rho < \infty \}, \qquad (12.59c)$$

$$L_4 = \{ \rho : 0 < \rho < \infty \}. \qquad (12.59d)$$

These rays are the continuation of the rays $l_1, \dots, l_4$ appearing in the integral representation of u_z. In order to elucidate the significance of the above choice, we multiply Eq. (12.58) by $\exp[-ik]$ and evaluate at $k = -i\rho$:

$$\sum_{l=0}^{N-1} \left[b_l^{(1)} \hat{P}_l(-i\rho) + b_l^{(2)} e^{-\rho + i\rho} \hat{P}_l(-\rho) - b_l^{(3)} e^{-2\rho} \hat{P}_l(-i\rho) - b_l^{(4)} e^{-\rho - i\rho} \hat{P}_l(-\rho) \right]$$

$$\approx \frac{\rho\pi}{\pi^2 - \rho^2} \left(e^{i\rho} - e^{-i\rho} \right). \qquad (12.60a)$$

The coefficients of $b_l^{(1)}$ are the usual Fourier transforms of P_l, whereas the coefficients of $b_l^{(3)}$ decay exponentially as $\rho \to \infty$. Furthermore, the coefficients of $b_l^{(2)}$ and $b_l^{(4)}$ decay as $\mathcal{O}(\rho^{-1})$. Indeed, using integration by parts, we have

$$e^{-\rho} \hat{P}_l(-\rho) = \int_{-1}^{1} e^{-\rho(1+y)} P_l(y) dy$$

$$= \frac{1}{\rho} \left[P_l(-1) - P_l(1) e^{-2\rho} + \int_{-1}^{1} e^{-\rho(1+y)} P_l'(y) dy \right].$$

Note that if $y > -1$ then $\exp[-\rho(1+y)]$ decays exponentially as $\rho \to \infty$. Further integration by parts shows that the integral inside the brackets converges to zero as $\rho \to \infty$. One can then show that

$$e^{-\rho} \hat{P}_l(-\rho) = \frac{1}{\rho} P_l(-1) + \mathcal{O}(\rho^{-2}).$$

Similarly, multiplying Eq. (12.58) by $\exp[k]$, $\exp[ik]$ and $\exp[-k]$, and evaluating along L_2, L_3 and L_4 respectively, we find the following equations:

$$\sum_{l=0}^{N-1}\left[b_l^{(1)}e^{-\rho-i\rho}\hat{P}_l(-\rho)+b_l^{(2)}\hat{P}_l(i\rho)-b_l^{(3)}e^{-\rho+i\rho}\hat{P}_l(-\rho)-b_l^{(4)}e^{-2\rho}\hat{P}_l(i\rho)\right]$$

$$\approx\frac{i\rho\pi e^{-i\rho}}{\pi^2+\rho^2}\left(e^{-2\rho}-1\right), \tag{12.60b}$$

$$\sum_{l=0}^{N-1}\left[b_l^{(1)}e^{-\rho}\hat{P}_l(i\rho)+b_l^{(2)}e^{-\rho-i\rho}\hat{P}_l(\rho)-b_l^{(3)}\hat{P}_l(i\rho)-b_l^{(4)}e^{-\rho+i\rho}\hat{P}_l(\rho)\right]$$

$$\approx\frac{\rho\pi e^{-2\rho}}{\pi^2-\rho^2}\left(e^{i\rho}-e^{-i\rho}\right), \tag{12.60c}$$

$$\sum_{l=0}^{N-1}\left[b_l^{(1)}e^{-\rho+i\rho}\hat{P}_l(\rho)+b_l^{(2)}e^{-2\rho}\hat{P}_l(-i\rho)-b_l^{(3)}e^{-\rho-i\rho}\hat{P}_l(\rho)-b_l^{(4)}\hat{P}_l(-i\rho)\right]$$

$$\approx\frac{i\rho\pi e^{i\rho}}{\pi^2+\rho^2}\left(e^{-2\rho}-1\right). \tag{12.60d}$$

Example 12.4: Using Eqs. (12.60a)–(12.60d), we can determine the coefficients $\{b_l^{(j)}: j=1,\ldots,n, l=0,\ldots,N-1\}$ numerically as follows. We choose $M\in\mathbb{N}$ values of ρ given by $\rho=(R/M)\nu$, $\nu=1,\ldots,M$, where $R>0$ specifies the distance (given by R/M) between adjacent collocation points. Numerical experiments suggest the choice $M=Nn$ and $R=2M$. For real-valued u, we also evaluate the Schwartz conjugate of the global relation at the complex conjugates of the suggested collocation points. The result is an overdetermined linear system, which is inverted in the least squares sense.

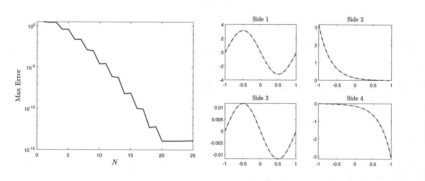

The plots above show the output of this procedure. The left panel shows the convergence of the method. The four panels on the right depict the computed solution in red dashed lines and the true solution in blue lines at the four sides of the square. We see exponential convergence, which is not surprising since this simple example has no corner singularities. To cope with corner singularities and maintain fast convergence, additional functions to the Legendre polynomials must be used (see Section 12.4).

12.2.4   The general case

In the general case, the jth side of the polygon, which is the side between z_j and z_{j+1}, can be parameterized as follows:

$$z = m_j + t\,h_j, \quad m_j = \frac{z_j + z_{j+1}}{2}, \quad h_j = \frac{z_{j+1} - z_j}{2}, \quad -1 < t < 1. \qquad (12.61)$$

It can be verified that if $t = -1$, $z = m_j - h_j = z_j$, and for $t = 1$, $z = m_j + h_j = z_{j+1}$. Also, for the jth side,

$$dz = h_j dt, \quad ds = |dz| = |h_j| dt. \qquad (12.62)$$

Hence, Eq. (12.21) becomes

$$\hat{u}_j(k) = e^{-ikm_j} \int_{-1}^{1} e^{-ikh_j t} \left(\frac{\partial u_j}{\partial \mathcal{N}} |h_j| + ku_j h_j \right) dt. \qquad (12.63)$$

For the square defined by Eq. (12.25) (see Fig. 12.4), we have

$$m_1 = -1, \ h_1 = -i; \ m_2 = -i, \ h_2 = 1; \ m_3 = 1, \ h_3 = i; \ m_4 = i, \ h_4 = -1.$$

For the Dirichlet boundary value problem, using

$$\frac{\partial u_j}{\partial \mathcal{N}}(t) \approx \sum_{l=0}^{N-1} b_l^j \hat{P}_l(t),$$

we find

$$\hat{u}_j(k) = e^{-ikm_j} |h_j| \sum_{l=0}^{N-1} b_l^j \hat{P}_l(-ikh_j) + kh_j e^{-ikm_j} U_j(-ikh_j), \qquad (12.64)$$

where

$$U_j(k) = \int_{-1}^{1} e^{kt} u_j(t) dt. \qquad (12.65)$$

Hence, the approximate global relation becomes

$$\sum_{j=1}^{n} e^{-ikm_j} |h_j| \sum_{l=0}^{N-1} b_l^j \hat{P}_l(-ikh_j) \approx G(k), \qquad (12.66)$$

where

$$G(k) = k \sum_{j=1}^{n} h_j e^{-ikm_j} \hat{u}_j(-ikh_j). \qquad (12.67)$$

In order to compute b_l^j we multiply Eq. (12.64) by $\exp[ikm_1]$, $\exp[ikm_2]$, $\ldots$, $\exp[ikm_n]$, respectively, and then evaluate the resulting n equations for k equal to

$$-\bar{h}_1\rho, -\bar{h}_2\rho, \ldots, -\bar{h}_n\rho, \quad \text{with } \rho = 2\nu, \ \nu = 1, 2, \ldots, nN.$$

12.3   The Helmholtz and modified Helmholtz equations

In order to derive a global relation for the Helmholtz equation, we consider the adjoint $v(x, y)$ of Eq. (8.116), which satisfies the Helmholtz equation itself,

$$v_{xx} + v_{yy} + \kappa^2 u = 0, \quad \kappa > 0. \qquad (12.68)$$

Multiplying Eq. (12.68) by v and Eq. (8.116) by u and then subtracting the resulting equations we find

$$(vu_x - uv_x)_x + (vu_y - uv_y)_y = 0. \tag{12.69}$$

Green's theorem, Eq. (2.73), with $F_2 = (vu_x - uv_x)$, $F_1 = -(vu_y - uv_y)$, implies that if the Helmholtz equation, Eq. (8.116), is valid in the domain Ω, then

$$\int_{\partial\Omega} [v(u_y dx - u_x dy) - u(v_y dx - v_x dy)] = 0, \tag{12.70}$$

where $\partial\Omega$ is the boundary of Ω and $v(x,y)$ is any solution of the Helmholtz equation.

Looking for a particular solution of the Helmholtz equation of the form $\exp[-i\alpha x - ky]$ we find $\alpha^2 = k^2 + \kappa^2$. Thus, a global relation of the Helmholtz equation is given by

$$\int_{\partial\Omega} e^{-i(\kappa^2+k^2)^{1/2}x - ky} [u_y dx - u_x dy + u(k dx + i(\kappa^2 + k^2)^{1/2} dy)] = 0, \quad k \in \mathbb{C}. \tag{12.71}$$

A more symmetric global relation can be obtained by seeking a particular solution in the form $\exp[-i\alpha_1 x + i\alpha_2 y]$, which yields $\alpha_1^2 + \alpha_2^2 = \kappa^2$. Thus, using the parameterizations

$$\alpha_1 = \kappa\cos(\phi), \quad \alpha_2 = \kappa\sin(\phi), \quad e^{i\phi} = k,$$

we find the particular solution $\exp[-i\kappa(kz + \bar{z}/k)/2]$. Hence, Eq. (12.71) yields an alternative global relation:

$$\text{H}: \int_{\partial\Omega} e^{-i\kappa(kz+\bar{z}/k)/2} \left[u_x dy - u_y dx + \frac{\kappa}{2} u \left(k dz - \frac{d\bar{z}}{k} \right) \right] = 0, \quad k \in \mathbb{C} \setminus \{0\}. \tag{12.72}$$

Letting $\kappa \to i\kappa$, the Helmholtz equation, Eq. (8.116), becomes the modified Helmholtz equation, Eq. (8.119). Thus, replacing in Eq. (12.72) κ with $i\kappa$ and k with $-ik$, we find the following global relation for the modified Helmholtz equation:

$$\text{MH}: \int_{\partial\Omega} e^{i\kappa(\bar{z}/k - kz)/2} \left[u_x dy - u_y dx + \frac{\kappa}{2} u \left(k dz + \frac{d\bar{z}}{k} \right) \right] = 0, \quad k \in \mathbb{C} \setminus \{0\}. \tag{12.73}$$

Employing the identity of Eq. (12.13), Eqs. (12.72) and (12.73) become the following global relations:

$$\text{H}: \int_{\partial\Omega} e^{-i\kappa(kz+\bar{z}/k)/2} \left[\frac{\partial u}{\partial \mathcal{N}} ds + \frac{\kappa}{2} u \left(k dz - \frac{d\bar{z}}{k} \right) \right] = 0, \quad k \in \mathbb{C} \setminus \{0\}. \tag{12.74}$$

$$\text{MH}: \int_{\partial\Omega} e^{i\kappa(\bar{z}/k - kz)/2} \left[\frac{\partial u}{\partial \mathcal{N}} ds + \frac{\kappa}{2} u \left(k dz + \frac{d\bar{z}}{k} \right) \right] = 0, \quad k \in \mathbb{C} \setminus \{0\}. \tag{12.75}$$

12.3.1 Global relations for a convex polygon

Let Ω be the interior of the polygonal domain of Fig. 12.1. Then the global relations become

$$\sum_{j=1}^{n} \hat{u}_j(k) = 0, \quad k \in \mathbb{C} \setminus \{0\}, \tag{12.76}$$

where $\hat{u}_j(k)$ are defined below. Equations (12.72) and (12.73) imply

H :

$$\hat{u}_j(k) = \int_{z_j}^{z_{j+1}} e^{-i\kappa(kz+\bar{z}/k)/2} \left[u_{j,x}dy - u_{j,y}dx + \frac{\kappa}{2}u_j\left(kdz - \frac{d\bar{z}}{k}\right) \right], \quad (12.77a)$$

$$\hat{u}_j(k) = \int_{z_j}^{z_{j+1}} e^{-i\kappa(kz+\bar{z}/k)/2} \left[\frac{\partial u_j}{\partial \mathcal{N}}ds + \frac{\kappa}{2}u_j\left(kdz - \frac{d\bar{z}}{k}\right) \right]. \quad (12.77b)$$

Similarly, Eqs. (12.74) and (12.75) imply

MH :

$$\hat{u}_j(k) = \int_{z_j}^{z_{j+1}} e^{i\kappa(\bar{z}/k-kz)/2} \left[u_{j,x}dy - u_{j,y}dx + \frac{\kappa}{2}u_j\left(kdz + \frac{d\bar{z}}{k}\right) \right], \quad (12.78a)$$

$$\hat{u}_j(k) = \int_{z_j}^{z_{j+1}} e^{i\kappa(\bar{z}/k-kz)/2} \left[\frac{\partial u_j}{\partial \mathcal{N}}ds + \frac{\kappa}{2}u_j\left(kdz + \frac{d\bar{z}}{k}\right) \right]. \quad (12.78b)$$

Derivation: Global relations for a square

We illustrate the application of these relations to the case of the square defined by Eq. (12.25) (see Fig. 12.4). With the aid of the new variables $\mathcal{K}_1, \mathcal{K}_2$ defined as

$$\mathcal{K}_1 = \frac{\kappa}{2}\left(k + \frac{1}{k}\right), \quad \mathcal{K}_2 = \frac{\kappa}{2}\left(k - \frac{1}{k}\right), \quad (12.79)$$

Eqs. (12.77a) and (12.78a) imply for the case of the Helmholtz equation we find:

$$\hat{u}_1(k) = -e^{i\mathcal{K}_1} \int_{-1}^{1} e^{\mathcal{K}_2 y} \left[u_{1,x} + i\mathcal{K}_1 u_1 \right] dy, \quad (12.80a)$$

$$\hat{u}_2(k) = e^{-\mathcal{K}_2} \int_{-1}^{1} e^{-i\mathcal{K}_1 x} \left[-u_{2,y} + \mathcal{K}_2 u_2 \right] dx, \quad (12.80b)$$

$$\hat{u}_3(k) = e^{-i\mathcal{K}_1} \int_{-1}^{1} e^{\mathcal{K}_2 y} \left[u_{3,x} + i\mathcal{K}_1 u_3 \right] dy, \quad (12.80c)$$

$$\hat{u}_4(k) = -e^{\mathcal{K}_2} \int_{-1}^{1} e^{-i\mathcal{K}_1 x} \left[-u_{4,y} + \mathcal{K}_2 u_4 \right] dx. \quad (12.80d)$$

Similarly, for the case of the modified Helmholtz equation:

$$\hat{u}_1(k) = -e^{i\mathcal{K}_2} \int_{-1}^{1} e^{\mathcal{K}_1 y} \left[u_{1,x} + i\mathcal{K}_2 u_1 \right] dy, \quad (12.81a)$$

$$\hat{u}_2(k) = e^{-\mathcal{K}_1} \int_{-1}^{1} e^{-i\mathcal{K}_2 x} \left[-u_{2,y} + \mathcal{K}_1 u_2 \right] dx, \quad (12.81b)$$

$$\hat{u}_3(k) = e^{-i\mathcal{K}_2} \int_{-1}^{1} e^{\mathcal{K}_1 y} \left[u_{3,x} + i\mathcal{K}_2 u_3 \right] dy, \quad (12.81c)$$

$$\hat{u}_4(k) = -e^{\mathcal{K}_1} \int_{-1}^{1} e^{-i\mathcal{K}_2 x} \left[-u_{4,y} + \mathcal{K}_1 u_4 \right] dx. \quad (12.81d)$$

12.3.2   Novel integral representations for the modified Helmholtz equation

Equation (12.6) shows that the integral representations of the modified Helmholtz equation in the k-complex plane involves the same rays $l_j, j = 1, \ldots, n$ as those appearing in the integral representation of the Laplace equation.

For instance, in the case of the quarter plane shown in Fig. 12.3, we find

$$u(x,y) = \frac{1}{2i} \int_0^{i\infty} \frac{1}{k} e^{i\kappa(kz - \bar{z}/k)/2} \hat{u}_1(k) \frac{dk}{2\pi} + \frac{1}{2i} \int_0^{\infty} \frac{1}{k} e^{i\kappa(kz - \bar{z}/k)/2} \hat{u}_2(k) \frac{dk}{2\pi},$$

(12.82)

where $\hat{u}_1(k)$ and $\hat{u}_2(k)$ are defined by

$$\hat{u}_1(k) = -i \int_0^{\infty} e^{\mathcal{K}_1 y} [u_{1,x} + i\mathcal{K}_2 u_1] \, dy, \quad \text{Re}[k] \leq 0,$$

(12.83a)

$$\hat{u}_2(k) = i \int_0^{\infty} e^{i\mathcal{K}_2 x} [-u_{2,y} + \mathcal{K}_1 u_2] \, dx, \quad \text{Im}[k] \leq 0,$$

(12.83b)

and the variables $\mathcal{K}_1, \mathcal{K}_2$ were defined in Eq. (12.79). Furthermore, $\hat{u}_1(k)$ and $\hat{u}_2(k)$ satisfy the global relation

$$\hat{u}_1(k) + \hat{u}_2(k) = 0, \quad \pi \leq \text{Arg}[k] \leq \frac{3\pi}{2}.$$

(12.84)

Indeed, it is straightforward to verify that for both $\hat{u}_1(k)$ and $\hat{u}_2(k)$, $\exp[ikz + i\bar{z}/k]$ is bounded in the same domains as $\exp[-ikz]$.

Similarly, in the case of a quarter plane with Neumann boundary conditions, in order to avoid computations, instead of considering the oblique Neumann boundary conditions (12.38), we consider the simpler case of Neumann boundary conditions:

$$u_x(0,y) = g_1(y), \quad 0 < y < \infty,$$

(12.85a)

$$u_y(x,0) = g_2(x), \quad 0 < x < \infty.$$

(12.85b)

Equations (12.83a) and (12.83b) yield

$$\hat{u}_1(k) = -i \int_0^{\infty} e^{\mathcal{K}_1 y} [u_x(0,y) + i\mathcal{K}_2 u(0,y)] \, dy, \quad \text{Re}[k] \leq 0,$$

(12.86a)

$$\hat{u}_2(k) = i \int_0^{\infty} e^{i\mathcal{K}_2 x} [-u_y(x,0) + i\mathcal{K}_1 u(x,0)] \, dx, \quad \text{Im}[k] \leq 0.$$

(12.86b)

Using Eqs. (12.85a) and (12.85b) in Eqs. (12.86a) and (12.86b) we find

$$\hat{u}_1(k) = \mathcal{K}_2 \hat{f}_1(k) - i\hat{g}_1(k), \quad \text{Re}[k] \leq 0,$$

(12.87a)

$$\hat{u}_2(k) = i\mathcal{K}_1 \hat{f}_2(-ik) - i\hat{g}_2(-ik), \quad \text{Im}[k] \leq 0,$$

(12.87b)

where $\hat{g}_1$ and $\hat{g}_2$ are the known functions

$$\hat{g}_1(k) = \int_0^{\infty} e^{\mathcal{K}_1 y} g_1(y) dy, \quad \text{Re}[k] \leq 0,$$

(12.88a)

$$\hat{g}_2(-ik) = \int_0^{\infty} e^{-i\mathcal{K}_2 x} g_2(x) dx, \quad \text{Im}[k] \leq 0,$$

(12.88b)

whereas $\hat{f}_1$ and $\hat{f}_2$ are the unknown functions

$$\hat{f}_1(k) = \int_0^{\infty} e^{\mathcal{K}_1 y} u(0,y) dy, \quad \text{Re}[k] \leq 0,$$

(12.89a)

$$\hat{f}_2(-ik) = \int_0^{\infty} e^{-i\mathcal{K}_2 x} u(x,0) dx, \quad \text{Im}[k] \leq 0.$$

(12.89b)

The global relation Eq. (12.84) together with the Schwartz conjugate yield

$$\mathcal{K}_2 \hat{f}_1(k) + i\mathcal{K}_1 \hat{f}_2(-ik) - i\hat{g}_1(k) - i\hat{g}_2(-ik) = 0, \quad k \in \text{III},$$

(12.90a)

$$\mathcal{K}_2 \hat{f}_1(k) - i\mathcal{K}_1 \hat{f}_2(ik) + i\hat{g}_1(k) + i\hat{g}_2(ik) = 0, \quad k \in \text{II},$$

(12.90b)

where I, II, III, IV denote the first, second, third and fourth quadrant of the k-complex plane. We will express $\hat{u}_1$ and $\hat{u}_2$ in terms of the unknown function $\hat{f}_2(ik)$: Eq. (12.90b) yields

$$\hat{f}_1(k) = i\frac{k^2+1}{k^2-1}\hat{f}_2(ik) - i\frac{\hat{g}_1(k)+\hat{g}_2(ik)}{\mathcal{K}_2}, \quad k \in \text{II}. \tag{12.91}$$

Eliminating $\hat{f}_1$ from Eqs. (12.90), replacing k with $-k$ in the resulting equation and then solving for $\hat{f}_2(-ik)$ we find

$$\hat{f}_2(-ik) = -\hat{f}_2(ik) - \frac{1}{\mathcal{K}_1}[2\hat{g}_1(-k) + \hat{g}_2(-ik) + \hat{g}_2(ik)], \quad k \in \mathbb{R}^+. \tag{12.92}$$

The unknown function $\hat{f}_2(ik)$ yields the contribution

$$-i\pi\kappa \int_{\partial \text{I}} \frac{1}{k}\left(k + \frac{1}{k}\right) \hat{f}_2(ik)e^{-i\kappa(k-1/k)x/2}e^{\kappa(k+1/k)y/2}\frac{dk}{2\pi}, \tag{12.93}$$

where ∂I denotes the boundary of I. The integrand of the above integral is bounded and analytic in I, thus its contribution vanishes. Hence,

$$u(x,y) = \frac{1}{2i}\int_0^{i\infty} \frac{1}{k}e^{i\kappa(kz-\bar{z}/k)/2}\hat{u}_1(k)\frac{dk}{2\pi} + \frac{1}{2i}\int_0^{\infty} \frac{1}{k}e^{i\kappa(kz-\bar{z}/k)/2}\hat{u}_2(k)\frac{dk}{2\pi}, \tag{12.94}$$

where

$$\hat{u}_1(k) = -2i\hat{g}_1(k) - i\hat{g}_2(ik), \quad \hat{u}_2(k) = -2i\hat{g}_2(-ik) - 2i\hat{g}_1(-k) - i\hat{g}_2(ik). \tag{12.95}$$

Using analyticity we can rewrite $\hat{u}_1$, $\hat{u}_2$ in the symmetric form

$$\hat{u}_1(k) = -2i[\hat{g}_1(k) + \hat{g}_1(-k)], \quad \hat{u}_2(k) = -2i[\hat{g}_2(ik) + \hat{g}_2(-ik)]. \tag{12.96}$$

Hence,

$$u(x,y) = -\int_0^{i\infty} \frac{1}{k}e^{i\kappa(kz-\bar{z}/k)/2}[\hat{g}_1(k) + \hat{g}_1(-k) + \hat{g}_2(ik) + \hat{g}_2(-ik)]\frac{dk}{2\pi}. \tag{12.97}$$

Equation (12.97) can also be derived by using the classical cosine-transform and then deforming in the k-complex plane. However, the boundary value problem with the boundary conditions (12.38) *cannot* be solved by a classical transform.

Example 12.5: We explore the following case for the modified Helmholtz equation with boundary conditions defined by

$$g_1(y) = e^{-a_1 y}, \qquad g_2(x) = e^{-a_2 x}, \qquad a_1 > 0, a_2 > 0.$$

Then,

$$\hat{g}_1(k) = \frac{1}{a_1 - \kappa(k+1/k)/2}, \qquad \hat{g}_2(k) = \frac{1}{a_2 - \kappa(k+1/k)/2}.$$

Furthermore, by steps similar to those discussed in Example 12.3 we find that Eq. (12.97) takes the form

$$u(x,y) = \int_0^{\infty e^{i\phi}} e^{i\kappa[k(x+iy)-(x-iy)/k]/2}V(k)dk, \qquad 0 < \phi < \frac{\pi}{2},$$

where we have defined the function $V(k)$ as

$$V(k) = \frac{1}{\pi}\left[\frac{1}{\kappa k^2 - 2a_1 k + \kappa} - \frac{1}{\kappa k^2 + 2a_1 k + \kappa}\right.$$

$$\left. - \frac{i}{\kappa k^2 + 2ia_2 k - \kappa} + \frac{i}{\kappa k^2 - 2ia_2 k - \kappa}\right].$$

For simplicity we take specific values for the parameters appearing in $V(k)$, namely $\kappa = 1$, $a_1 = 2$, $a_2 = 3$ and $\phi = \pi/4$. In order to verify the boundary conditions we perform the following evaluation:

$$u_y(x,0) = \int_0^{\infty e^{i\phi}} e^{i\kappa(k-1/k)x/2}\left(-\frac{\kappa}{2}\right)\left(k + \frac{1}{k}\right)V(k)dk$$

$$= \int_0^{\infty e^{i\pi/4}} e^{i(k-1/k)x/2}F(k)dk,$$

where we have defined the auxiliary function $F(k)$ as

$$F(k) = -\frac{1}{2\pi}\left(k + \frac{1}{k}\right)\left(\frac{1}{k^2 - 4k + 1} - \frac{1}{k^2 + 4k + 1}\right.$$

$$\left. - \frac{i}{k^2 + 6ik - 1} + \frac{i}{k^2 - 6ik - 1}\right).$$

Recalling that $u_y(x,0)$ is real we rewrite the last equation as

$$u_y(x,0) = \int_0^{\infty e^{i\pi/4}} e^{i(k-1/k)x/2}F(k)dk + \overline{\int_0^{\infty e^{i\pi/4}} e^{i(k-1/k)x/2}F(k)dk}.$$

Then, following the same procedure as in Example 12.3 for the Laplace equation, and using the fact that $F(-k) = F(k)$, we obtain the equation

$$u_y(x,0) = \int_0^{\infty e^{i\pi/4}} e^{i(k-1/k)x/2}F(k)dk + \int_{\infty e^{i3\pi/4}}^0 e^{i(k-1/k)x/2}F(k)dk.$$

Then, it suffices to compute the residue contribution of the above integrand. The only poles which occur in the upper half-plane come from the last term of $F(k)$, namely from the term

$$\frac{i}{k^2 - 6ik - 1},$$

which has two poles at $k_1 = i(3 + 2\sqrt{2})$ and $k_2 = i(3 - 2\sqrt{2})$. Applying the Residue theorem we find

$$u_y(x,0) = 2\pi i\text{Residue}\left[e^{i(k-1/k)x/2}F(k)\right]_{k=k_1}$$

$$+ 2\pi i\text{Residue}\left[e^{i(k-1/k)x/2}F(k)\right]_{k=k_2}$$

$$= \frac{e^{-3x}}{2} + \frac{e^{-3x}}{2} = e^{-3x}.$$

Similarly, we find that

$$u_x(0,y) = e^{-2y}.$$

Because of the exponential decay for large k, the integral representation of the solution can be computed numerically very efficiently. We employ the change of variables $k \to r \exp[i\pi/4]$, in order to compute the integral along the real line. Then it suffices to compute only the real part of the integrand. An example of such numerical evaluation using Mathematica, is shown in the plot below for the range of the variables $x \in [0,3]$, $y \in [0,5]$.

$$\texttt{Plot3D}\left[\texttt{NIntegrate}\left[\texttt{Re}\left[\texttt{Exp}\left[\tfrac{i}{2}\left(e^{i\pi/4}r(x+iy) - e^{-i\pi/4}(x-iy)/r\right)\right]\right.\right.\right.$$

$$\left.\left.\left. V\left(re^{i\pi/4}\right)e^{i\pi/4}\right], \{r,0,+\infty\}\right], \{x,0,3\}, \{y,0,5\}\right]$$

Derivation: Approximate global relations for a square

The three algorithmic steps discussed in Section 12.2.3 provide again an effective way for coupling numerically the unknown boundary values. We consider again the square defined by Eq. (12.25) (see Fig. 12.4), with the boundary conditions of Eqs. (12.55a)–(12.55d). We first solve the Helmholtz equation for this domain and boundary conditions. Employing the approximations Eq. (12.53) for the Neumann boundary values, and the variables $\mathcal{K}_1, \mathcal{K}_2$ defined in Eq. (12.79), it follows that Eq. (12.80) yield the relations

$$\hat{u}_1(k) \approx e^{i\mathcal{K}_1} \sum_{l=0}^{N-1} b_l^{(1)} \hat{P}_l(-\mathcal{K}_2) - i\pi\mathcal{K}_1 e^{i\mathcal{K}_1} \frac{e^{\mathcal{K}_1} - e^{-\mathcal{K}_1}}{\pi^2 + \mathcal{K}_1^2}, \qquad (12.98a)$$

$$\hat{u}_2(k) \approx e^{-\mathcal{K}_2} \sum_{l=0}^{N-1} b_l^{(2)} \hat{P}_l(-i\mathcal{K}_1), \qquad (12.98b)$$

$$\hat{u}_3(k) \approx e^{-i\mathcal{K}_1} \sum_{l=0}^{N-1} b_l^{(3)} \hat{P}_l(\mathcal{K}_2), \qquad (12.98c)$$

$$\hat{u}_4(k) \approx e^{\mathcal{K}_2} \sum_{l=0}^{N-1} b_l^{(4)} \hat{P}_l(i\mathcal{K}_1). \qquad (12.98d)$$

Substituting Eqs. (12.98a)–(12.98d) into the global relation (12.76), we obtain a single equation for the unknowns $\{b_l^{(j)} : l = 0, \dots, N-1, j = 1, \dots, 4\}$. We evaluate this equation along the rays defined in Eq. (12.59) to obtain four sets of equations. Using these four equations and the same choices of collocation points as in Example 12.4, we can numerically approximate the unknown coefficients.

We next solve the modified Helmholtz equation for the same domain and boundary conditions. With the variables $\mathcal{K}_1, \mathcal{K}_2$ defined in Eq. (12.79), the approximate global relation leads to

$$\hat{u}_1(k) \approx e^{i\mathcal{K}_2} \sum_{l=0}^{N-1} b_l^{(1)} \hat{P}_l(-\mathcal{K}_1) - i\pi\mathcal{K}_2 e^{i\mathcal{K}_2} \frac{e^{\mathcal{K}_1} - e^{-\mathcal{K}_1}}{\pi^2 + \mathcal{K}_1^2}, \tag{12.99a}$$

$$\hat{u}_2(k) \approx e^{-\mathcal{K}_1} \sum_{l=0}^{N-1} b_l^{(2)} \hat{P}_l(-i\mathcal{K}_2), \tag{12.99b}$$

$$\hat{u}_3(k) \approx e^{-i\mathcal{K}_2} \sum_{l=0}^{N-1} b_l^{(3)} \hat{P}_l(\mathcal{K}_1), \tag{12.99c}$$

$$\hat{u}_4(k) \approx e^{\mathcal{K}_1} \sum_{l=0}^{N-1} b_l^{(4)} \hat{P}_l(i\mathcal{K}_2). \tag{12.99d}$$

Following exactly the same procedure as in the case of the Helmholtz equation, we can approximate the unknown coefficients.

Example 12.6: We illustrate the numerical solution with two specific cases.

For the Helmholtz equation we take the value of the parameter $\kappa = \sqrt{17}\pi/4$. The plots below show the output of the procedure outlined above for finding the coefficients in Eqs. (12.98a)–(12.98d). The left panel shows the convergence of the method. We see exponential convergence, which is not surprising since this simple example has no corner singularities. The four panels on the right depict the computed solution in red dashed lines and the true solution in blue lines at the four sides of the square.

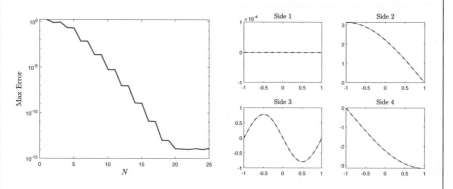

For the modified Helmholtz equation, we choose the value of the parameter $\kappa = \sqrt{3}\pi$. The plots below show the output of the procedure outlined above for finding the coefficients in Eqs. (12.99a)–(12.99d). The left panel shows the convergence of the method. Again, we see exponential convergence, which is not surprising since this simple example has no corner singularities. The four panels on the right depict the computed solution in red dashed lines and the true solution in blue lines at the four sides of the square.

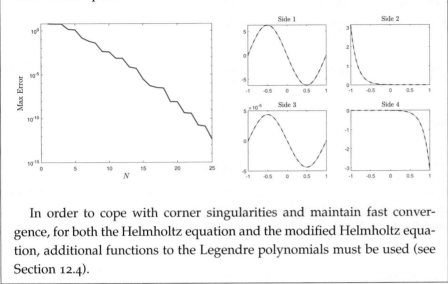

In order to cope with corner singularities and maintain fast convergence, for both the Helmholtz equation and the modified Helmholtz equation, additional functions to the Legendre polynomials must be used (see Section 12.4).

12.3.3   The general case

Recall that the jth side of a polygon, which is the side between z_j and z_{j+1}, can be parameterized through Eq. (12.61). For the following discussion, we will find it convenient to define the variables

$$\mathcal{K}_1^j = \frac{\kappa}{2}\left(km_j + \frac{\bar{m}_j}{k}\right), \quad \mathcal{K}_2^j = \frac{\kappa}{2}\left(km_j - \frac{\bar{m}_j}{k}\right),$$

$$\Lambda_1^j = \frac{\kappa}{2}\left(kh_j + \frac{\bar{h}_j}{k}\right), \quad \Lambda_2^j = \frac{\kappa}{2}\left(kh_j - \frac{\bar{h}_j}{k}\right).$$

Then Eqs. (12.77b) and (12.78b) become

$$\text{H}: \ \hat{u}_j(k) = e^{-i\mathcal{K}_1^j}\int_{-1}^{1} e^{-i\Lambda_1^j t}\left[\frac{\partial u_j}{\partial \mathcal{N}}|h_j| + u_j\Lambda_2^j\right]dt. \tag{12.100}$$

$$\text{MH}: \ \hat{u}_j(k) = e^{-i\mathcal{K}_2^j}\int_{-1}^{1} e^{-i\Lambda_2^j t}\left[\frac{\partial u_j}{\partial \mathcal{N}}|h_j| + u_j\Lambda_1^j\right]dt. \tag{12.101}$$

For example, for the Dirichlet problem using Eq. (12.54) to approximate the Neumann boundary values, we find

$$\text{H}: \ \hat{u}_j(k) = e^{-i\mathcal{K}_1^j}\left\{\sum_{l=0}^{N-1} b_l^j|h_j|\hat{P}_l\left(-i\Lambda_1^j\right) + a_l^j\Lambda_2^j U_j\left(-i\Lambda_1^j\right)\right\}, \tag{12.102}$$

$$\text{MH}: \ \hat{u}_j(k) = e^{-i\mathcal{K}_2^j}\left\{\sum_{l=0}^{N-1} b_l^j|h_j|\hat{P}_l\left(-i\Lambda_2^j\right) + a_l^j\Lambda_1^j U_j\left(-i\Lambda_2^j\right)\right\}, \tag{12.103}$$

where $\hat{P}_l(k)$ and $U_j(k)$ are defined in Eq. (12.57) and Eq. (12.65), respectively. Substituting Eqs. (12.102) and (12.103) in the general relation (12.76) we obtain the approximate global relations for Helmholtz and modified Helmholtz equations. These equations can be evaluated at the same set of discrete values as those used for the Laplace equation.

12.3.4   *Computing the solution in the interior of a convex polygon*

After determining the unknown boundary values, the conventional expressions presented in Chapter 8 for the solutions of the Helmholtz and modified Helmholtz equations can be used for computing the solution $u(x, y)$. Indeed, using the identities $ds = |h_j| dt$ and $d\zeta = h_j dt$, Eq. (8.126) becomes

$$u(x, y) = \sum_{j=1}^{n} \int_{-1}^{1} \left[G \frac{\partial u_j}{\partial \mathcal{N}} |h_j| + iu \left(\frac{\partial G}{\partial \zeta} h_j - \frac{\partial G}{\partial \bar{\zeta}} \bar{h}_j \right) \right] dt. \tag{12.104}$$

In order to compute $\partial G / \partial \zeta$ we can use the following identities:

$$\frac{dH_0^{(1)}(\zeta)}{d\zeta} = -H_1^{(1)}(\zeta), \quad \frac{dK_0(\zeta)}{d\zeta} = -K_1(\zeta), \tag{12.105}$$

where $H_\alpha^{(1)}(x)$ are the Hankel functions, see Eq. (8.130), and $K_\alpha(x)$ are the modified Bessel functions of the second kind, see Eq. (8.129). These identities imply the following equations:

$$\text{H}: \quad \frac{\partial G}{\partial \zeta} = -\frac{i\kappa}{8} \left(\frac{\bar{z} - \bar{\zeta}}{z - \zeta} \right)^{\frac{1}{2}} H_1^{(1)}(\kappa |z - \zeta|). \tag{12.106}$$

$$\text{MH}: \quad \frac{\partial G}{\partial \zeta} = -\frac{\kappa}{2\pi} \left(\frac{\bar{z} - \bar{\zeta}}{z - \zeta} \right)^{\frac{1}{2}} K_1(\kappa |z - \zeta|). \tag{12.107}$$

Replacing in Eq. (12.104) the term $\partial G / \partial \zeta$ through Eqs. (12.106) and (12.107) and then letting $\zeta = m_j + t\, h_j$, we find the following expressions:

$$\text{H}: \quad u(x, y) = \frac{i}{4} \sum_{j=1}^{n} |h_j| \int_{-1}^{1} H_0^{(1)}\left(\kappa |z - m_j - th_j|\right) \frac{\partial u_j}{\partial \mathcal{N}} dt$$

$$- \frac{\kappa}{8} \sum_{j=1}^{n} h_j \int_{-1}^{1} \left(\frac{\bar{z} - \bar{m}_j - t\bar{h}_j}{z - m_j - th_j} \right)^{\frac{1}{2}} H_1^{(1)}\left(\kappa |z - m_j - th_j|\right) u_j dt$$

$$+ \frac{\kappa}{8} \sum_{j=1}^{n} \bar{h}_j \int_{-1}^{1} \left(\frac{z - m_j - th_j}{\bar{z} - \bar{m}_j - t\bar{h}_j} \right)^{\frac{1}{2}} H_1^{(1)}\left(\kappa |z - m_j - th_j|\right) u_j dt. \tag{12.108}$$

$$\text{MH}: \quad u(x, y) = \frac{1}{2\pi} \sum_{j=1}^{n} |h_j| \int_{-1}^{1} K_0\left(\kappa |z - m_j - th_j|\right) \frac{\partial u_j}{\partial \mathcal{N}} dt$$

$$+ \frac{i\kappa}{4\pi} \sum_{j=1}^{n} h_j \int_{-1}^{1} \left(\frac{\bar{z} - \bar{m}_j - t\bar{h}_j}{z - m_j - th_j} \right)^{\frac{1}{2}} K_1\left(\kappa |z - m_j - th_j|\right) u_j dt$$

$$- \frac{i\kappa}{4\pi} \sum_{j=1}^{n} \bar{h}_j \int_{-1}^{1} \left(\frac{z - m_j - th_j}{\bar{z} - \bar{m}_j - t\bar{h}_j} \right)^{\frac{1}{2}} K_1\left(\kappa |z - m_j - th_j|\right) u_j dt. \tag{12.109}$$

It should be emphasized that *after* evaluating the expansion coefficients a_l^j and b_l^j occurring in the Legendre expansions of u_j or $\partial u_j / \partial \mathcal{N}$, it is straightforward to evaluate numerically the right-hand side of Eqs. (12.108) and (12.109).

12.4   Generalizations

It is worth pointing out a few recently developed generalizations of the above procedure:

(a) In many applications, the solution of the PDE has singularities at the corners of the domain. In this case, the type of singular behavior can often be predicted analytically and suitable functions that capture such behavior can be added to the expansion to supplement the Legendre polynomials.[4] This improves the speed of convergence as N increases.

(b) One can extend the method to problems defined in the exterior domains.[5] Such problems appear in many applications including acoustic scattering.[6,7] In this case, the allowed values of k are determined by the boundary condition at infinity.

(c) One can use more general boundary conditions than Neumann, Dirichlet or a mixture of the two (Robin boundary conditions). For example, one can couple the PDE with an elastic boundary condition to model flexural waves.[8]

(d) The global relation can be extended to separable PDEs with variable coefficients. Furthermore, the numerical procedure can be extended to separable PDEs with variable coefficients and to curved boundaries.[9]

12.5   Application: Water waves

It is important to emphasize that the global relation provides a useful tool for analyzing boundary value problems with complicated boundary conditions, which may even be nonlinear. Such a problem arises in the study of free surface water waves, which has been at the forefront of applied mathematics and engineering for over 200 years. This problem impacts a variety of important practical problems including harbor design, shipping, and tsunami prediction. The problem formulated directly from the governing equations derived from physical principles is prohibitively difficult, because it requires the solution of Laplace's equation in an unknown domain, which is itself determined by nonlinear boundary conditions which depend on the solution. The reformulation of this problem in terms of the global relation[10] reduces the problem to the solution of a global relation and a Bernoulli-like equation, vastly reducing its complexity. This result has had a significant impact in this classical area spearheaded by Bernard Deconinck and collaborators. In particular, it has led to new computations of surface water waves, and the discovery of new instabilities of waves in shallow water. In addition, the employment of the global relation has inspired the incorporation of large amplitude effects in the reconstruction of the surface wave profile using pressure data measured at the bottom, as well as the solution of the inverse water wave problem, namely the determination of the bottom topography from surface wave data.

In what follows, rather than presenting the formulation in three-dimensions, for simplicity we consider waves in two-dimensions (x, y). Specifically, we consider the two-dimensional free surface waves, under the assumption that the flow is inviscid and irrotational.

[4] M.J. Colbrook, A.S. Fokas, and P. Hashemzadeh, A hybrid analytical-numerical technique for elliptic PDEs. *SIAM Journal on Scientific Computing* **41**(2), pp. A1066–1090 (2019).

[5] M.J. Colbrook, L.J. Ayton, and A.S. Fokas, The unified transform for mixed boundary condition problems in unbounded domains. *Proceedings of the Royal Society A* **475**(2222):20180605 (2019).

[6] L.J. Ayton, M.J. Colbrook, and A.S. Fokas. The unified transform: A spectral collocation method for acoustic scattering. *25th AIAA/CEAS Aeroacoustics Conference*, p. 2528 (2019).

[7] M.J. Colbrook and M.J. Priddin. Fast and spectrally accurate numerical methods for perforated screens (with applications to Robin boundary conditions). *IMA Journal of Applied Mathematics* **85**(5), pp. 790–821 (2020).

[8] M.J. Colbrook and L.J. Ayton. A spectral collocation method for acoustic scattering by multiple elastic plates. *Journal of Sound and Vibration* **461**:114904 (2019).

[9] M.J. Colbrook. Extending the unified transform: Curvilinear polygons and variable coefficient PDEs. *IMA Journal of Numerical Analysis* **40**(2), pp. 976–1004 (2020).

[10] M.J. Ablowitz, A.S. Fokas, A.S. and Z.H. Musslimani, On a new non-local formulation of water waves. *Journal of Fluid Mechanics* **562**, pp. 313–343 (2006).

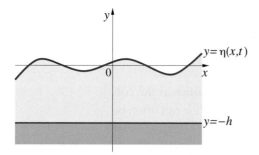

Figure 12.7: Illustration of the setup for the study of water waves in a two-dimensional problem: $y = \eta(x,t)$ is the free surface of the fluid where the waves occur and $y = -h$ is the position of the fixed flat bottom.

We shall assume an unbounded domain with a fixed flat bottom at $y = -h$, see Fig. 12.7. The free surface is denoted by $y = \eta(x,t)$ and the domain by $\Omega(t)$:

$$\Omega(t) = \{-\infty < x < \infty, \; -h < y < \eta(x,t)\}. \tag{12.110}$$

The velocity of the flow will be denoted by

$$\tau(x,y,t) = u(x,y,t)\hat{x} + v(x,y,t)\hat{y}.$$

The mass continuity equation implies

$$\nabla_{\mathbf{r}} \cdot \tau = 0 \Rightarrow u_x + v_y = 0, \quad \text{in } \Omega(t). \tag{12.111}$$

Since the flow is irrotational, the vorticity vanishes, hence

$$\nabla_{\mathbf{r}} \times \tau = 0 \Rightarrow v_x - u_y = 0 \quad \text{in } \Omega(t). \tag{12.112}$$

This equation implies the existence of a function Ψ, called the velocity potential, such that

$$\tau = u(x,y,t)\hat{x} + v(x,y,t)\hat{y} = \nabla_{\mathbf{r}}\Psi \Rightarrow u(x,y,t) = \Psi_x, \quad v(x,y,t) = \Psi_y. \tag{12.113}$$

Replacing in Eq. (12.111) the above expressions for u and v, we find the Laplace equation

$$\Psi_{xx} + \Psi_{yy} = 0, \quad \text{in } \in \Omega(t). \tag{12.114}$$

The fact that the same particles always form the free water surface is captured by the following kinematic boundary condition, which is valid on the free surface:

$$v = \eta_t + \eta_x u \quad \text{on } y = \eta(x,t).$$

Using Eq. (12.113), this last equation becomes

$$\Psi_y = \eta_t + \eta_x \Psi_x, \quad \text{on } y = \eta(x,t). \tag{12.115}$$

The normal velocity vanishes on the flat bottom, hence

$$\Psi_y = 0, \quad \text{on } y = -h. \tag{12.116}$$

Equations (12.114)–(12.116) define a Neumann boundary value problem for the Laplace equation, involving the unknown boundary $\eta(x,t)$. The type of fluid flow considered here, namely inviscid and irrotational, is governed by the so-called Euler equations. Since the external pressure is given by the constant atmospheric pressure, by taking the density of the fluid to be equal to 1 (which merely sets the unit system we adopt for this problem), the Euler equations for fluid flow result in the following condition:

$$\Psi_t + \frac{1}{2}\left(\Psi_x^2 + \Psi_y^2\right) + g\eta = 0, \quad \text{on } y = \eta(x,t), \qquad (12.117)$$

where g is the acceleration of gravity. Eq. (12.117) is referred to as "Bernoulli's law".

We denote the value of $\Psi(x,y,t)$ on the free surface by $q(x,t)$, that is,

$$q(x,t) = \Psi(x,\eta(x,t),t). \qquad (12.118)$$

Our aim is to find the equations satisfied by $\eta(x,t)$ and $q(x,t)$. In this respect, differentiating Eq. (12.118) with respect to x, we obtain

$$q_x = \Psi_x + \Psi_y \eta_x, \quad \text{on } y = \eta(x,t). \qquad (12.119)$$

Equation (12.119) along with Eq. (12.115) are two equations that can be solved for Ψ_x and Ψ_y at $y = \eta$ in terms of derivatives of q and η:

$$\Psi_x = \frac{q_x - \eta_x \eta_t}{1 + \eta_x^2}, \quad \Psi_y = \frac{\eta_t + \eta_x q_x}{1 + \eta_x^2}, \quad \text{on } y = \eta(x,t). \qquad (12.120)$$

In order to simplify Eq. (12.117), we differentiate Eq. (12.118) with respect to t,

$$q_t = \Psi_t + \Psi_y \eta_t, \quad \text{on } y = \eta(x,t). \qquad (12.121)$$

Using Eq. (12.121) and Ψ_y from Eq. (12.120), we can express Ψ_t in terms of derivatives of q and η:

$$\Psi_t = q_t - \frac{\eta_t + \eta_x q_x}{1 + \eta_x^2}\eta_t, \quad \text{on } y = \eta(x,t). \qquad (12.122)$$

Using Eq. (12.122) together with Eq. (12.120) in condition Eq. (12.117), we obtain the first equation coupling q and η:

$$q_t + g\eta + \frac{1}{2}q_x^2 - \frac{1}{2}\frac{(\eta_t + \eta_x q_x)^2}{1 + \eta_x^2} = 0. \qquad (12.123)$$

In order to obtain a second equation relating q and η, we introduce the complex variable $z = x + iy$, and we use the global relation of Eq. (12.15), which is the more convenient relation in this case, since the problem involves only derivatives:

$$\int_{\partial\Omega(t)} e^{-ikz}\Psi_z dz = \frac{1}{2}\int_{\partial\Omega(t)} e^{-ikz}\left[\Psi_x - i\Psi_y\right] dz = 0, \quad k \in \mathbb{R}, \qquad (12.124)$$

where we have used Eq. (4.31a) for the derivative with respect to z. The restriction on k in the above equation is due to the fact that the domain is unbounded. On the free surface,

$$z = x + i\eta(x,t) \Rightarrow dz = (1 + i\eta_x)dx.$$

Hence, replacing Ψ_x, Ψ_y with their expressions from Eq. (12.120), we find that the relevant contribution in the global relation is given by the expression

$$\int_{-\infty}^{\infty} e^{-ik(x+i\eta)} \left[\frac{q_x - \eta_x \eta_t}{1 + \eta_x^2} - i\frac{\eta_t + \eta_x q_x}{1 + \eta_x^2}\right] (1 + i\eta_x)dx, \quad k \in \mathbb{R}. \qquad (12.125)$$

This equation, remarkably, simplifies to the expression

$$\int_{-\infty}^{\infty} e^{-ik(x+i\eta)} \left(q_x - i\eta_t\right) dx, \quad k \in \mathbb{R}. \tag{12.126}$$

On the bottom, we have from the boundary condition (12.116),

$$z = x - ih, \quad \Psi_y(x, -h, 0) = 0,$$

so the contribution to the global relation is

$$\int_{-\infty}^{\infty} e^{-ikx - kh} \Psi_x(x, -h, 0) dx. \tag{12.127}$$

Hence the global relation becomes

$$\int_{-\infty}^{\infty} e^{-ikx + k\eta} \left(q_x - i\eta_t\right) dx + e^{-kh} \int_{-\infty}^{\infty} e^{-ikx} \Psi_x(x, -h, 0) dx = 0, \quad k \in \mathbb{R}. \tag{12.128a}$$

We can eliminate the second integral in this equation by the following steps: Taking the complex conjugate and then replacing k with $-k$ we find:

$$\int_{-\infty}^{\infty} e^{-ikx - k\eta} \left(q_x + i\eta_t\right) dx + e^{kh} \int_{-\infty}^{\infty} e^{-ikx} \Psi_x(x, -h, 0) dx = 0, \quad k \in \mathbb{R}. \tag{12.128b}$$

Multiplying Eq. (12.128a) by $\exp[kh]$ and Eq. (12.128b) by $\exp[-kh]$, subtracting the resulting equations, and finally replacing again k with $-k$ we find:

$$\int_{-\infty}^{\infty} e^{ikx} \left\{ q_x \sinh\left[k(\eta + h)\right] + i\eta_t \cosh\left[k(\eta + h)\right] \right\} dx = 0, \quad k \in \mathbb{R}. \tag{12.129}$$

This nonlocal equation together with the first-order PDE, Eq. (12.123), are two equations characterizing the two unknown functions $q(x, t)$ and $\eta(x, t)$.

We apply this result to a type of familiar and important waves, namely waves traveling with constant speed c. In this case, the functions η and q are functions of the same variable w which contains both the spatial dependence, x, and temporal dependence, t, in a specific linear combination, namely $w = x - ct$, and the functions η and q take the form

$$\eta(x, t) = \eta(x - ct), \quad q(x, t) = q(x - ct).$$

Since both η and q are functions of w, we will introduce their derivatives η' and q' with respect to this variable, namely

$$\eta'(w) = \frac{\partial \eta}{\partial w}, \quad q'(w) = \frac{\partial q}{\partial w},$$

and express the derivatives with respect to x and t in terms of η', q', using the chain rule:

$$q_x = \frac{\partial q}{\partial w} \frac{\partial w}{\partial x} = q', \quad q_t = \frac{\partial q}{\partial w} \frac{\partial w}{\partial t} = -cq',$$

$$\eta_x = \frac{\partial \eta}{\partial w} \frac{\partial w}{\partial x} = \eta', \quad \eta_t = \frac{\partial \eta}{\partial w} \frac{\partial w}{\partial t} = -c\eta',$$

where we have used $w_x = 1, w_t = -c$. Then Eq. (12.123) becomes

$$-cq' + \frac{1}{2} \left(q'\right)^2 + g\eta - \frac{\left(-c\eta' + q'\eta'\right)^2}{2\left[1 + (\eta')^2\right]} = 0. \tag{12.130}$$

This simplifies to the quadratic equation [see B. Deconinck and K. Oliveras, *The instability of periodic surface gravity waves*. *Journal of Fluid Mechanics*, **675**, pp. 141–167 (2011)] in the unknown quantity q',

$$(q')^2 - 2cq' + 2g\eta \left[1 + (\eta')^2\right] - c^2 (\eta')^2 = 0, \qquad (12.131)$$

which is easily solved to yield:

$$q' = c + \sqrt{\left[1 + (\eta')^2\right](c^2 - 2g\eta)},$$

where we have kept the solution with the positive sign in front of the square root, as the physically meaningful one. This immediately implies the classical estimate

$$(c^2 - 2g\eta) \geq 0 \Rightarrow \eta \leq \frac{c^2}{2g}.$$

In terms of the derivatives q', η' with respect to the variable w the non-local equation, Eq. (12.129), takes the form

$$\int_{-\infty}^{\infty} e^{ikw} \left\{ q' \sinh\left[k(\eta + h)\right] - ic\eta' \cosh\left[k(\eta + h)\right] \right\} dw = 0, \ k \in \mathbb{R}. \quad (12.132)$$

Then, Eqs. (12.131) and (12.132) can be solved to obtain the functions q and η.

Example 12.7: Waves with small amplitude

We consider waves with small amplitude a and long wavelength l, in comparison to the depth h. In this situation, we can take advantage of the two small parameters

$$\epsilon = \frac{a}{h}, \quad \delta = \frac{h}{l},$$

to expand the various terms in the basic equations, namely Eqs. (12.123) and (12.129), and obtain simpler expressions. We begin by replacing all variables by ones that contain the relevant scales, so that the resulting quantities are dimensionless. Specifically, we define the parameter

$$c_0 = \sqrt{gh},$$

which has the dimensions of velocity, and replace the original variables as follows:

$$x \to lx, \quad k \to \frac{k}{l}, \quad t \to \frac{l}{c_0}t, \quad q \to \frac{gla}{c_0}q, \quad \eta \to a\eta.$$

With these substitutions the basic equations take the form:

$$q_t + \eta + \frac{1}{2}\epsilon q_x^2 - \frac{1}{2}\epsilon\delta^2 \frac{(\eta_t + \epsilon\eta_x q_x)^2}{1 + \epsilon^2\delta^2\eta_x^2} = 0, \qquad (12.133a)$$

$$\int_{-\infty}^{\infty} e^{ikx} \left\{ q_x \sinh\left[k\delta(1 + \epsilon\eta)\right] + i\delta\eta_t \cosh\left[k\delta(1 + \epsilon\eta)\right] \right\} dx = 0, \ k \in \mathbb{R}.$$

$$(12.133b)$$

In what follows, we neglect terms of $\mathcal{O}(\epsilon^2)$, $\mathcal{O}(\delta^4)$, and $\mathcal{O}(\epsilon\delta^2)$. Then, using the Taylor expansions of the cosh and sinh functions, see Eqs. (3.19),

(3.20), and keeping the lowest order terms we obtain:

$$\cosh\left[k\delta(1+\epsilon\eta)\right] \approx 1 + \delta^2 k^2, \quad \sinh\left[k\delta(1+\epsilon\eta)\right] \approx \delta k + \frac{1}{6}\delta^3 k^3 + \epsilon\delta k\eta.$$

With these approximations, Eqs. (12.133a) and (12.133b) become:

$$\eta = -q_t - \frac{1}{2}\epsilon q_x^2, \qquad (12.134a)$$

$$\int_{-\infty}^{\infty} e^{ikx}\left\{ q_x\left(\delta k + \frac{1}{6}\delta^3 k^3 + \epsilon\delta k\eta\right) + i\delta\eta_t\left(1 + \delta^2 k^2\right)\right\} dx = 0,\ k \in \mathbb{R}. \qquad (12.134b)$$

Using integration by parts, Eq. (12.134b) gives rise to a PDE. Indeed, integration by parts leads to the general result

$$\int_{-\infty}^{\infty} e^{ikx}\frac{\partial f}{\partial x}dx = \left[e^{ikx}f(x)\right]_{-\infty}^{\infty} - i\int_{-\infty}^{\infty} ke^{ikx}f(x)dx$$

$$\Rightarrow \quad \int_{-\infty}^{\infty} e^{ikx}kf(x)dx = \int_{-\infty}^{\infty} e^{ikx}if_x dx,$$

where we have assumed that the function $f(x)$ vanishes for $|x| \to \infty$. We can take advantage of the last equation to replace every term of the form $kf(x)$ in the Fourier transform integral by the term if_x. With the aid of this substitution, Eq. (12.134b) becomes

$$\int_{-\infty}^{\infty} e^{ikx}\left\{ iq_{xx} + \frac{1}{6}\delta^2 i^3 q_{xxxx} + i\epsilon\left(\eta q_x\right)_x + i\eta_t + i\delta^2\left(i^2\eta_{txx}\right)\right\} dx = 0.$$

Thus, the inverse Fourier transform implies

$$\eta_t - \delta^2\eta_{txx} + q_{xx} - \frac{\delta^2}{6}q_{xxxx} + \epsilon\left(\eta q_x\right)_x = 0. \qquad (12.135)$$

Replacing in Eq. (12.135) η by the right-hand side of Eq. (12.134a) and neglecting terms of $\mathcal{O}(\epsilon^2)$ and $\mathcal{O}(\epsilon\delta^2)$ we find

$$q_{tt} - q_{xx} + \frac{\epsilon}{2}\left(q_x^2\right)_t + \epsilon\left(q_x q_t\right)_x - \delta^2 q_{ttxx} + \frac{\delta^2}{6}q_{xxxx} = 0. \qquad (12.136)$$

In the case of a traveling wave, $q(x,t) = q(w)$, where the dimensionless variable w is given in terms of the dimensionless variables x and t as

$$w = x - \frac{c}{c_0}t = x - \tilde{c}t, \quad \text{where } \tilde{c} = \frac{c}{c_0},$$

Eq. (12.136) becomes

$$\tilde{c}^2 q'' - q'' - \frac{\epsilon}{2}\tilde{c}\left(q'^2\right)' - \epsilon\tilde{c}\left(q'^2\right)' - \delta^2\tilde{c}^2 q'''' + \frac{\delta^2}{6}q'''' = 0. \qquad (12.137)$$

We will integrate this equation assuming that q' and q'' vanish as $|w| \to \infty$. Integrating Eq. (12.137) and then letting $q' = Q$, we find

$$(\tilde{c}^2 - 1)Q - \frac{3\epsilon\tilde{c}}{2}Q^2 + \delta^2\left(\frac{1}{6} - \tilde{c}^2\right)Q'' = 0. \qquad (12.138)$$

Multiplying by Q' and integrating we find the following first-order ODE for Q:

$$(\tilde{c}^2 - 1)Q^2 - \epsilon\tilde{c}Q^3 + \delta^2 \left(\frac{1}{6} - \tilde{c}^2\right)(Q')^2 = 0.$$

Dividing by $\epsilon\tilde{c}$ this equation becomes

$$\left(\frac{\delta}{\sqrt{\epsilon}}\sqrt{\tilde{c} - \frac{1}{6\tilde{c}}}\frac{dQ}{dw}\right)^2 = \frac{1}{\epsilon}\left(\tilde{c} - \frac{1}{\tilde{c}}\right)Q^2 - Q^3. \quad \tilde{c} > 1.$$

We next define the parameter λ and the new variable X as

$$\lambda = \frac{1}{\epsilon}\left(\tilde{c} - \frac{1}{\tilde{c}}\right), \quad X = \frac{\sqrt{\epsilon}}{\delta}\frac{1}{\sqrt{\tilde{c} - 1/(6\tilde{c})}}w,$$

in terms of which the last equation takes the form

$$\left(\frac{dQ}{dX}\right)^2 = \lambda Q^2 - Q^3. \tag{12.139}$$

The general solution of Eq. (12.139) which vanishes for $|X| \to \infty$ is given by

$$Q(X) = \lambda \left[\cosh\left(\frac{\sqrt{\lambda}}{2}(X - X_0)\right)\right]^{-2}, \tag{12.140}$$

where X_0 is an arbitrary constant. This is the same expression as the Korteweg–de Vries (KdV) soliton [see Section 8.2, Eq. (8.4c)]. From the solution for $Q(X)$ we can reconstruct the profile of the wave, $\eta(x,t)$, using Eq. (12.134a), which yields

$$\eta(x,t) = a\tilde{c}Q(X)\left[1 - \frac{\epsilon}{2\tilde{c}}Q(X)\right]. \tag{12.141}$$

In this last expression we have restored the dimensions of $\eta(x,t)$ by recalling that all quantities starting with Eqs. (12.133a) and (12.133b) are dimensionless and that the amplitude of the wave is $a = \epsilon h$. The plots below show the functions $Q(X)$ (left panel) and $\eta(x,t)$ (right panel) in terms of the variables x (in units of $l = h/\delta$) and t (in units of l/c_0). The values of the parameters are $\epsilon = \delta^2 = 0.1$, $\tilde{c} = 1.5$; the height of the soliton is in units of a.

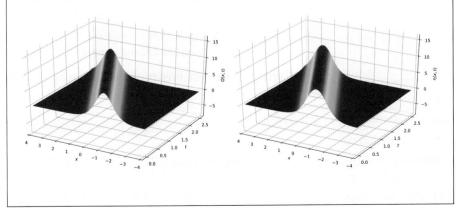

Problems

1. Suppose that $q(z,\bar{z})$ satisfies the biharmonic equation,

$$q_{zz\bar{z}\bar{z}} = 0,$$

in the interior of a closed convex polygon with corners at z_i, $i = 1,\ldots,n$. Using the fact that $q_{zz\bar{z}}$ is an analytic function deduce the formulas

$$q_{zz\bar{z}} = \sum_{j=1}^{n} \int_{l_j} e^{ikz}\tilde{q}_j(k)\frac{dk}{2\pi},$$

$$\tilde{q}_j(k) = \int_{z_j}^{z_{j+1}} e^{-ik\zeta}q_{\zeta\zeta\bar{\zeta}}d\zeta,$$

$$\sum_{j=1}^{n} \tilde{q}_j(k) = 0, \quad k \in \mathbb{C}.$$

Derive analogous formulas involving $q_{zz} - \bar{z}q_{zz\bar{z}}$.

2. The derivation of the global relation is based on Green's theorem, which is a special case of the Poincaré lemma, Eq. (2.80)

$$\int_{\partial\Omega} \mathcal{F} = \int_{\Omega} d\mathcal{F},$$

where $\mathcal{F}$ is a differential form (see the discussion in Section 2.4). If $\mathcal{F}$ is exact, that is, $d\mathcal{F}$ vanishes, then the left-hand side of the above equation yields a global relation.

(a) Using the fact that

$$d\left(e^{-ikz}u\right)$$

is an exact form, derive the following global relation for Laplace's equation:

$$\int_{\partial\Omega} e^{-ikz}\left\{[(1+\alpha)u_z - i\alpha ku]\,dz + \alpha u_{\bar{z}}d\bar{z}\right\},$$

where α is an arbitrary constant.

(b) By using the exact form

$$d\Big(e^{-ikz}\alpha_1 u_{z\bar{z}} + \alpha_2 u_{zz} + \alpha_3 u_{\bar{z}\bar{z}}$$

$$+\alpha_4 ku_z + \alpha_5 ku_{\bar{z}} + \alpha_6 k^2 u\Big),$$

for $k \in \mathbb{C}$, where α_j, $j = 1,\ldots,6$ are complex constants, form a new global relation associated with $u_{z\bar{z}\bar{z}}$ involving six arbitrary constants.

In a similar way, it is possible to derive a global relation associated with the basic entity $u_{zz} - \bar{z}u_{zz\bar{z}}$, which involves 16 constants [see M. Dimakos and A. Fokas, The Poisson and the biharmonic equations in the interior of a convex polygon, *Studies in Applied Mathematics* **134**(4), pp. 456–498 (2015)].

3. The real function $u(x,y)$ satisfies Laplace's equation in the semistrip $0 < x < \infty$, $-L < y < L$. Using Eqs. (12.36) and (12.37) show that

$$u_z(x,y) = \int_0^{-\infty} e^{ikz}\hat{u}_1(k)\frac{dk}{2\pi}$$

$$+ \int_0^{i\infty} e^{ikz}\hat{u}_2(k)\frac{dk}{2\pi}$$

$$+ \int_0^{\infty} e^{ikz}\hat{u}_3(k)\frac{dk}{2\pi}, \qquad (12.142)$$

where

$$\hat{u}_1(k) = \frac{i}{2}e^{kL}U_1(-ik) - \frac{1}{2}e^{kL}G_1(-ik), \quad \mathrm{Im}[k] \le 0,$$

$$\hat{u}_2(k) = -\frac{i}{2}U_2(k) - \frac{1}{2}G_2(k), \quad k \in \mathbb{C},$$

$$\hat{u}_3(k) = -\frac{i}{2}e^{-kL}U_3(-ik) + \frac{1}{2}e^{kL}G_3(-ik), \quad \mathrm{Im}[k] \le 0,$$

and U_j, G_j, $j = 1,2,3$, are integrals to be determined. The subscripts $1,2,3$ refer to the sides $\{0 < x < \infty, y = L\}$, $\{x = 0, -L < y < L\}$, and $\{0 < x < \infty, y = -L\}$, respectively.

4. Consider Laplace's equation in the semistrip domain of Problem 3 with Dirichlet boundary conditions. In this case, the functions G_j, $j = 1,2,3$, can be computed in terms of the boundary conditions.

(a) Use the global relation and the equation obtained from the global relation through complex conjugation and the transformation $k \to -\bar{k}$ to express $U_1(-ik)$ and $U_3(-ik)$ in terms of $U_2(k)$, $U_2(-k)$ and known functions.

(b) By deforming the contours involving $U_2(k)$ and $U_2(-k)$ from the real axis to the positive imaginary axis, show that the contribution of $U_2(k)$ and $U_2(-k)$ to the solution $u(x,y)$ can be computed explicitly. Indeed, this contribution can be expressed in terms of the residues associated with poles at $k = n\pi/2L$, $n = 0,1,2,\ldots$, and these residues can be computed explicitly.

5. Consider the Dirichlet problem of Laplace's equation in the semistrip domain of Problem 3, with

$$u(x,L) = u(x,-L) = 0, \quad 0 < x < \infty,$$

$$u(0,y) = \sin\left(\frac{\pi y}{L}\right), \quad -L < y < L.$$

Follow the approach introduced in Problem 4 to show that

$$u_z = -\frac{i\pi}{2L}e^{-\pi z/L}.$$

Integrating this expression and recalling that u is real, deduce that

$$u(x,y) = e^{-\pi x/L} \sin\left(\frac{\pi y}{L}\right).$$

6. Let $\hat{u}_j(k)$ be defined by Eq. (12.37). Using integration by parts show that $\hat{u}_j(k)$ can be expressed in terms of a formula which involves boundary values at the corners, as well as the following entity:

$$\tilde{u}_j(k) = \int_{z_j}^{z_{j+1}} e^{-ikz} \left[u_x dy - u_y dx + kudz\right]$$

$$= \int_{z_j}^{z_{j+1}} e^{-ikz} \left[\frac{\partial u}{\partial \mathcal{N}} ds + kudz\right]. \quad (12.143)$$

Assuming that the boundary values are continuous at the corners show that

$$u(x,y) = -\mathrm{Re}\left[\sum_{j=1}^{n} \int_{l_j} e^{ikz} \tilde{u}_j(k) \frac{dk}{2\pi}\right] + \text{const.,}$$

$$(12.144)$$

where the rays l_j, $j = 1,\ldots,n$ are defined in Eq. (12.5). The constant can be fixed by using the boundary conditions. Show that $\tilde{u}_j(k)$ follows from the expression defined in part (a) of Problem 2 with the choice of $\alpha = -1/2$. Hence, deduce that Eqs. (12.143) and (12.144) are valid without the assumption of the continuity of the boundary values at the corners.

7. The famous Saffman–Taylor problem in fluid mechanics can be mapped to the Laplace equation in the semistrip domain of Problem 3 with the Dirichlet boundary conditions

$$u(x,\pm L) = 0, \quad 0 < x < \infty;$$

$$u(0,y) = \alpha y, \quad -L < y < L.$$

The above Dirichlet data are discontinuous at the two corners of the domain. This implies that u_z has a pole singularity at these corners and hence the formulas of Eqs. (12.36), (12.37) cannot be used for the solution of this problem. However, this problem can be solved using the formulas derived in Problem 6.

(a) Show that, for $k \in \mathbb{C}$,

$$\hat{u}_j(k) = e^{kL} U_j(-ik), \quad j = 1,3, \quad \mathrm{Im}[k] \leq 0,$$

$$\hat{u}_2(k) = -U_2(k) - 2i\alpha\left[L\cosh(kL) - \frac{1}{k}\sinh(kL)\right].$$

(b) Using the global relation and the equation obtained from the global relation under complex conjugation and the transformation $k \to -\bar{k}$, show that

$$\Delta\, U_1(-ik) = e^{kL} U_2(k) - e^{-kL} U_2(-k) + h(k),$$

$$\Delta\, U_3(-ik) = e^{-kL} U_2(k) - e^{-kL} U_2(-k) + h(k),$$

where $\Delta = 4\cosh(kL)\sinh(kL)$ and the function $h(k)$ is defined as

$$h(k) = 4i\alpha\left[L\cosh^2(kL) - \frac{\Delta}{4k}\right].$$

(c) Taking into consideration that $\cosh^2(kL) = 0$ at the zeros of Δ, deduce that $U_2(ik)$ yields a zero contribution to $u(x,y)$.

(d) Show that the contribution of the known part of $\tilde{u}_2(k)$ is due to the residues associated with $k = i\pi n/L$ and is given by

$$\frac{i\alpha L}{\pi} \sum_{n=1}^{\infty} \frac{(-1)^n}{n} e^{-\pi nz/L}.$$

(e) Deduce that

$$u(x,y) = \frac{\alpha L}{\pi} \mathrm{Im}\left[\ln\left(1 + e^{-\pi z/L}\right)\right] + \text{const.}$$

8. Let $u(x,y)$ satisfy the Laplace equation in the interior of the orthogonal isosceles triangle on the (x,y)-plane with corners at $(iL,0,L)$, $L > 0$, shown in the diagram.

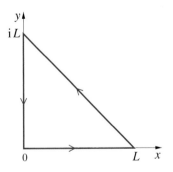

Suppose that $u(x,y)$ satisfies the following continuous Dirichlet boundary conditions:

side 1 : $x = 0$, $L > y > 0$, $u(0,y) = 0$, $\quad$ (12.145a)

side 2 : $0 < x < L$, $y = 0$, $u(x,0) = \sin\left(\frac{\pi x}{L}\right)$,

$$(12.145b)$$

side 3 : $L > x > 0$, $y = -x + L$, $u(x,-x+L) = 0$.

$$(12.145c)$$

Use the global relation (12.11) to derive the equations needed for the approximate evaluation of the Neumann boundary values on the three sides of the triangle. Namely, implement the following steps:

(a) Show that for $k \in \mathbb{C}$,

$$\hat{u}_1(k) = -\int_0^L e^{ky} u_x(0, y) dy,$$

$$\hat{u}_2(k) = -\int_0^L e^{-ikx} u_y(x, 0) dx + g(k),$$

$$\hat{u}_3(k) = 2e^{kL} \int_0^L e^{-(i+1)kx} u_y(x, -x + L) dx,$$

where the function $g(k)$ in the expression for $\hat{u}_2(k)$ is defined as

$$g(k) = \frac{k\pi}{2} \left(\frac{1 + e^{-ikL}}{k^2 - \pi^2/4} \right).$$

[Hint: For the derivation of $\hat{u}_3(k)$ use the fact that on the hypotenuse of the triangle, $y = -x + L$, thus $dy = -dx$. Also $u(x, -x + L) = 0$, thus, $u_x - u_y = 0$.]

(b) Letting

$$u_x(0, y) \approx \sum_{l=0}^N a_l^{(1)} P_l(y),$$

$$u_y(x, 0) \approx \sum_{l=0}^N a_l^{(2)} P_l(x),$$

$$u_y(x, -x + L) \approx \sum_{l=0}^N a_l^{(3)} P_l(x),$$

where $P_l(x)$ are the Legendre polynomials [see Section 3.2.4], show that for $k \in \mathbb{C}$ the approximate global relation takes the form

$$\sum_{l=0}^N \left\{ a_l^{(1)} \hat{P}_l(k) + a_l^{(2)} \hat{P}_l(-ik) + a_l^{(3)} \hat{P}_l(-ik - k) \right\}$$

$$= g(k), \qquad (12.146)$$

where

$$\hat{P}_l(k) = \int_0^L e^{kx} P_l(x) dx, \quad k \in \mathbb{C}.$$

(c) Show that convenient collocation points are points of the following rays:

$$L_1 = \{-i\rho, 0 < \rho < \infty\},$$
$$L_2 = \{-\rho, 0 < \rho < \infty\},$$
$$L_3 = \{-\rho e^{i\pi/4}, 0 < \rho < \infty\}.$$

(d) Verify the following:

 i. If $k = -i\rho$, then the terms of the first sum in Eq. (12.146) oscillate, whereas the terms of the second and third sums decay as $\rho \to \infty$.

 ii. If $k = -\rho$, then the terms of the second sum in Eq. (12.146) oscillate, whereas the terms of the first and the third sums decay as $\rho \to \infty$.

 iii. After multiplying Eq. (12.146) by $\exp[-kL]$ and evaluating the resulting equation at $k = -\rho \exp[i\pi/4]$, show that the terms of the third sum oscillate, whereas the terms of the first two sums decay as $\rho \to \infty$.

9. Let $u(x, y)$ satisfy the modified Helmholtz equation in the interior of the orthogonal isosceles triangle specified in Problem 8, with the Dirichlet boundary conditions of Eq. (12.145). Find the analogue of the approximate global relation of Eq. (12.146).

10. Repeat Problem 9 for the Helmholtz equation.

11. Use the numerical technique introduced in Examples 12.4 and 12.6 to solve numerically the approximate global relations derived in Problems 8, 9 and 10.

12. Starting with Bernoulli's law, Eq. (12.117), use Eqs. (12.120) and (12.122) to obtain the first equation coupling q and η, Eq. (12.123).

13. Provide the steps to derive Eq. (12.126) from Eq. (12.125).

14. Verify that the expression of Eq. (12.140) is a solution to Eq. (12.139), and that the profile of the wave $\eta(x, t)$ is described by Eq. (12.141).

Part IV

Probabilities, Numerical, and Stochastic Methods

Chapter 13

Probability Theory

Probability theory is nothing more than common sense confirmed by calculation.

Pierre-Simon Laplace (French mathematician, 1749–1827)

Although people have been fascinated for many centuries with the problem of calculating the probability of some event to occur, the methods that we use today toward this goal were only developed in the 19th century. The theory of probabilities has become a major branch of statistics and data analysis, fields whose importance in many technical and societal problems is steadily increasing. In this chapter, we introduce the main concepts of probability theory and explore their applications in a few simple problems.

We also make the connection between probabilities and entropy, a powerful concept for understanding many physical phenomena. This concept underlies the theory of statistical mechanics which provides the conceptual basis for thermodynamics, a phenomenological theory used broadly to describe physical phenomena that involve heat transfer. The concept of entropy is closely related to random sampling, a topic explored in Chapter 15.

13.1 Probability distributions

13.1.1 General concepts

We begin the discussion of probability theory by defining a few key concepts:

1. **Event space:** it consists of all possible outcomes of an experiment. For example, if an experiment involves the tossing of a fair die, the event space consists of the numbers 1, 2, 3, 4, 5, 6, which are all the possible outcomes for the top face of the labeled die. Similarly, if an experiment involves flipping a fair coin, the event space is H,T (for "heads" and "tails"), which are all the possible outcomes for the face of the coin that is up when the coin lands.

2. **Probability of event A:** it is the probability that event A will occur. This is equal to

$$P(A) = \frac{m}{N},$$ (13.1)

where m is the number of outcomes in the event space that corresponds to event A and N is the total number of all possible outcomes in the event space. For example, if we define as event A that a fair die gives 1, then $P(A) = 1/6$. Similarly, the probability that the die gives an even number, is $P(A) = 3/6 = 0.5$; that a fair coin gives heads, is $P(A) = 1/2$.

3. **Permutations:** it is the number of ways to arrange n different entities. There are $n! = n(n-1)(n-2)\cdots 1$ permutations of n objects. This is easy to justify: there are n possible choices for the first object, $(n-1)$ possible choices for the second object, etc. For example, if three objects are labeled as a, b, c, then the possible permutations are: abc, acb, bac, bca, cab, cba, that is, $3 \times 2 \times 1 = 6$ permutations.

4. **Combinations:** it is the number of ways of taking k entities out of a group of n entities, without specifying their order or identity. The number of combinations of k entities out of n are given by

$$\frac{n(n-1)\cdots(n-k+1)}{k(k-1)\cdots 1} = \frac{n!}{(n-k)!k!}. \tag{13.2}$$

This expression is also easy to justify: there are n ways of picking the first object, $n-1$ ways of picking the second, $\ldots$ $(n-k+1)$ ways of picking the kth object out of a total of n. But since we do not care in which order we pick these k objects, we have to divide the total number by the number of permutations among them, which is $k!$. Interestingly, these values enter in the binomial expansion:

$$(p+q)^n = \sum_{k=0}^{n} \frac{n!}{(n-k)!k!} p^k q^{n-k}, \quad p,q \in \mathbb{R}. \tag{13.3}$$

This expression has been encountered several times before [see, for example, Eq. (3.9)].

We give next some examples of how these simple concepts and definitions are very useful in calculating probabilities.

Example 13.1: A standard deck of playing cards contains 52 cards, consisting of 4 sets (clubs, diamonds, hearts, spades) of 13 cards in each set, with numbers from 2 to 10, the Ace (also counting as number one) and three figures (Jack, Queen, King). What is the probability of drawing a pair of cards with the same number or figure?

Answer: We start by using simple logic. Since we don't care which pair we draw, the probability of the first card being part of a pair is 1. The probability of the second card being part of a pair is the number of cards of the same type as the first, remaining in the deck, which is $(4-1) = 3$, divided by the total number of remaining cards, which is $(52-1) = 51$, giving as a result:

$$\frac{3}{51} = \frac{1}{17}.$$

We next use the concepts defined above. First, since we don't care which same number or figure the two cards have, there are 13 possibilities. Next, we need to account for how many ways we can draw two of the four cards that have the same number or figure. This is given by the number of combinations of 2 out of 4 objects:

$$\frac{4!}{2!(4-2)!} = \frac{1 \times 2 \times 3 \times 4}{1 \times 2 \times 1 \times 2} = 6.$$

But there are many possible ways of picking two cards out of the deck, and so far we have only accounted for the combinations that give the right cards (they have the same number or figure). In other words, we have figured out how many ways we can get a successful outcome. To calculate the actual probability, we need to divide this number by the total number of ways of picking any 2 cards from the deck, which is the total event space. This is given by the number of combinations of 2 objects out of 52:

$$\frac{52!}{2!(52-2)!} = \frac{51 \times 52}{1 \times 2} = 51 \times 26.$$

Putting these three numbers together, we find that the probability of getting a pair of cards with the same number or figure, of which there are 13 different possibilities, is given by

$$13 \times 6 \times \frac{1}{51 \times 26} = \frac{1}{17},$$

which is the same result found earlier.

Example 13.2: Three fair dice are thrown together. What is the probability of getting a total sum equal to some value N, say $N = 6$? What is the probability of getting a total sum larger than a given value, say $N = 15$?

Answer: Note that the possible values of N range from 3 (all ones) to 18 (all sixes). For each case, we have to count the total number of ways (multiplicity $m(N)$) to get a specific value and divide by the total number of all possible outcomes, which is $6 \times 6 \times 6 = 216$. There is only one way to get $N = 3$, namely all three dice being 1, denoted as $[1, 1, 1]$, so $m(3) = 1$. There are 3 ways to get $N = 4$, by two dice being 1 and the third one being 2, which can happen in 3 different ways: $[1, 1, 2]$, $[1, 2, 1]$, $[2, 1, 1]$, so $m(4) = 3$. The distribution of the number of ways $m(N)$ of getting each value of $N = 3$ to 18 is shown in the diagram below.

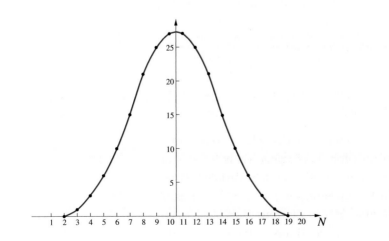

The table below gives each possible way of getting a specific sum from the three dice and its multiplicity $m(N)$ for $N = 5$ to 10. For instance, one can get $N = 5$ from one of the dice being 1 and the other two being 2, shown in the table as $[1, 2, 2]$ under the heading 5; there are three different ways to get this result, namely $[1, 2, 2]$, $[2, 1, 2]$, $[2, 2, 1]$ (only one is listed in the table, the others being easily obtained by permutations). The total number of ways of getting $N = 5$ is then $m(5) = 6$: 3 from the outcome $[1, 2, 2]$ and its permutations plus 3 from the outcome $[1, 1, 3]$ and its permutations.

$N = 5$		$N = 6$		$N = 7$		$N = 8$		$N = 9$		$N = 10$	
$[1, 2, 2]$	3	$[2, 2, 2]$	1	$[2, 2, 3]$	3	$[2, 3, 3]$	3	$[3, 3, 3]$	1	$[2, 4, 4]$	3
$[1, 1, 3]$	3	$[1, 2, 3]$	6	$[3, 3, 1]$	3	$[1, 3, 4]$	6	$[1, 4, 4]$	3	$[1, 4, 5]$	6
		$[1, 1, 4]$	3	$[4, 2, 1]$	6	$[1, 2, 5]$	6	$[2, 3, 4]$	6	$[3, 3, 4]$	3
				$[1, 1, 5]$	3	$[2, 2, 4]$	3	$[1, 3, 5]$	6	$[2, 3, 5]$	6
						$[1, 1, 6]$	3	$[1, 2, 6]$	6	$[2, 2, 6]$	3
								$[2, 2, 5]$	3	$[1, 3, 6]$	6
	6		10		15		21		25		27

We also note that in order to get a number N higher that 10, we have exactly the same permutations as for the corresponding number $(21 - N)$ with the substitutions $1 \leftrightarrow 6$, $2 \leftrightarrow 5$, $3 \leftrightarrow 4$, as can be easily checked. So we only need to find all the ways of getting a number N ranging from 3 to 10, since the numbers 11 to 18 have the same multiplicities as the numbers 10 to 3:

$$m(N) = m(21 - N), \quad \text{for } N > 10.$$

Using this table we can calculate any desired probability. For instance, the probability of getting a sum of $N = 6$ can be obtained from the corresponding entry in the table for $N = 6$:

$$P(6) = \frac{m(6)}{216} = \frac{10}{216}.$$

The probability of getting a number larger that 15 can be obtained by summing the probabilities of getting $N = 16, 17, 18$, or equivalently, the probabilities of getting a number smaller than $21 - 15 = 6$, that is, getting $N = 5, 4, 3$:

$$P(N > 15) = P(N < 6) = \frac{m(3) + m(4) + m(5)}{216} = \frac{1 + 3 + 6}{216} = \frac{10}{216}.$$

13.1.2   Probability distributions and their features

We define a general **random variable** x, which can be discrete or continuous. The **probability distribution function** (pdf for short), denoted here as $f(x)$, represents the probability of finding the value x of the random variable. There is a subtle but important difference in the meaning of $f(x)$ for the cases of a discrete or a continuous random variable: If x is discrete, then $f(x)$ is the actual *probability* of finding the value x. If x is continuous, then $f(x)$ is the probability *density*. It is important to distinguish these two quantities because in the case of a continuous variable x for which $f(x) \neq 0$ for $x \in [a, b]$, where $a < b$, there is an infinite number of values in the interval $[a, b]$ and therefore the probability of finding any of them must be zero. Thus, the proper way to define the probability in this case is to consider an interval $[x, x + dx]$, so that the probability of finding a value in this interval is $f(x)dx$, with dx an infinitesimal quantity.

For $f(x)$ to be a proper pdf, it must be normalized to unity, namely,

$$\sum_{j=1}^{N} f(x_j) = 1, \quad \text{or} \quad \int_{-\infty}^{\infty} f(x)dx = 1, \tag{13.4}$$

where the first expression holds for a discrete variable with possible values x_j identified by the discrete index j (taking the values $1, \ldots, N$), and the second expression holds for a continuous variable. We also define the **cumulative probability function** $F(x)$: the value of $F(x)$ gives the probability of finding any value of the random variable smaller than or equal to the value x.

$$F(x) = \sum_{j, x_j \leq x} f(x_j), \quad \text{or} \quad F(x) = \int_{-\infty}^{x} f(x')dx', \tag{13.5}$$

for a discrete or for a continuous variable, respectively. Note that from this definition, we conclude that $F(x_{\max}) = 1$, where $x_{\max}$ is the maximum value that x can take (in either the discrete or continuous version).

We define the **median** m as the value of the variable x such that the total probability of all values of x below m is equal to $1/2$:

$$F(m) = \frac{1}{2}. \tag{13.6}$$

From the definition of $F(x)$ we conclude that the total probability of all values of x above m is also $1/2$. Thus, the median m splits the range of values of x in two parts, each containing a set of values that have total probability $1/2$.

The **mean value** μ and the **variance** σ^2 of the the pdf are defined as follows:

$$\mu = \sum_{j=1}^{N} x_j f(x_j), \quad \text{or} \quad \mu = \int_{-\infty}^{\infty} x f(x)dx, \tag{13.7}$$

$$\sigma^2 = \sum_{j=1}^{N} (x_j - \mu)^2 f(x_j), \quad \text{or} \quad \sigma^2 = \int_{-\infty}^{\infty} (x - \mu)^2 f(x)dx, \tag{13.8}$$

for the discrete or the continuous cases, respectively. From the definitions of the mean and the variance, we can easily prove the following relation:

$$\sigma^2 = \sum_{j=1}^{N} x_j^2 f(x_j) - \mu^2, \quad \text{or} \quad \sigma^2 = \int_{-\infty}^{\infty} x^2 f(x)dx - \mu^2, \tag{13.9}$$

thus, *the variance is equal to the second moment minus the mean squared.*

We define the **expectation** E of a function $g(x)$ of a random variable x as

$$E(g(X)) = \sum_{j=1}^{N} g(x_j) f(x_j), \quad \text{or} \quad E(g(X)) = \int_{-\infty}^{\infty} g(x) f(x)dx, \tag{13.10}$$

for the discrete or the continuous cases, respectively. The notation $E(g(X))$ signifies that the expectation involves a function g of the random variable, but the expectation itself is not a function of x, since in the process of determining the expectation the values of x are summed (or integrated). With this definition, the mean and the variance can be expressed as follows:

$$\mu = E(X), \quad \sigma^2 = E((X - \mu)^2) \Rightarrow \sigma^2 = E(X^2) - \mu^2. \tag{13.11}$$

The above identities imply the relation

$$E(X^2) = \sigma^2 + [E(X)]^2 = \sigma^2 + \mu^2. \tag{13.12}$$

13.2   Common probability distributions

In this section, we discuss the most common probability distributions and calculate the mean and the variance in each case.

13.2.1   Uniform distribution

Definition: The simplest probability distribution is the one with a uniform probability density $p_u(x)$, that is, the one in which every possible event occurs with the same probability, which is a constant,

$$p_u(x) = c. \tag{13.13}$$

The value of this constant is determined by the normalization condition,

$$\sum_{i=1}^{N} p_u(x_i) = Nc = 1 \Rightarrow c = \frac{1}{N}, \tag{13.14}$$

for a discrete random variable x_i taking N different values, or

$$\int_a^b p_u(x; a, b) \mathrm{d}x = 1 \Rightarrow c = \frac{1}{b - a}, \tag{13.15}$$

for a continuous random variable x taking values in the interval $x \in [a, b]$. Thus, for the case of a continuous variable x the probability distribution is entirely determined by the range of values that x takes, $[a, b]$. In Fig. 13.1, we show some examples of uniform distributions.

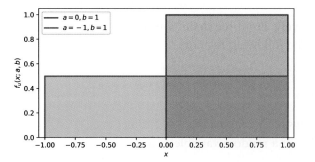

Figure 13.1: Examples of uniform probability distributions: The first (red) has $a = 0$, $b = 1$ and $c = 1$; the second (green) has $a = -1$, $b = 1$ and $c = 0.5$.

The uniform probability density is important because many simple ways of getting a random variable are based on natural events with uniform probability densities. For example, the probability of getting a number from 1 to 6 by throwing a fair die is the constant value $1/6$.

The mean value of a uniform distribution is, for the case of a discrete variable

$$\mu_u = \sum_{i=1}^{N} p_u(x_i)x_i = \frac{1}{N}\sum_{i=1}^{N}x_i,$$

and for the case of a continuous variable

$$\mu_u = \int_a^b p_u(x;a,b)x\mathrm{d}x = \frac{1}{b-a}\int_a^b x\mathrm{d}x = \frac{1}{2(b-a)}(b^2-a^2) = \frac{1}{2}(a+b).$$

$$(13.16)$$

The variance for the discrete case is

$$\sigma_u^2 = \sum_{i=1}^{N} p_u(x_i)(x_i-\mu_u)^2 = \frac{1}{N}\sum_{i=1}^{N}x_i^2 - \mu_u^2,$$

and for the continuous case

$$\sigma_u^2 = \int_a^b p_u(x;a,b)(x-\mu_u)^2\mathrm{d}x = \frac{1}{b-a}\int_a^b x^2\mathrm{d}x - \mu_u^2 = \frac{(b^3-a^3)}{3(b-a)} - \frac{(a+b)^2}{4}$$

$$\Rightarrow \quad \sigma_u^2 = \frac{(b-a)^2}{12}.$$

$$(13.17)$$

13.2.2 Binomial distribution

Definition: The binomial distribution applies when we know the total number of trials n, the probability of success for each trial p (the probability of failure is $q = 1 - p$) and we are interested in the probability of x successful outcomes. Both n and x are integers, both p and q are real ($0 < p,q < 1$). The mean is $\mu_b = np$ and the variance is $\sigma_b^2 = npq$. The binomial distribution is given by

$$f_b(x;n,p) = \frac{n!}{(n-x)!x!}p^x q^{n-x}, \quad n \text{ integer}, \quad 0 < p < 1, \, q = 1-p, \quad (13.18)$$

which is a function of the random variable x and its definition involves two parameters, the integer n and the real value $p \in (0,1)$. In Fig. 13.2 we show the binomial probability distribution as defined in Eq. (13.18) for $p = 0.5$ and $n = 6, 10, 20, 30$, for all the integer values of x in the range $0 \le x \le n$. In the following, for simplicity in notation we omit the parameters n and p in the expression of this probability distribution.

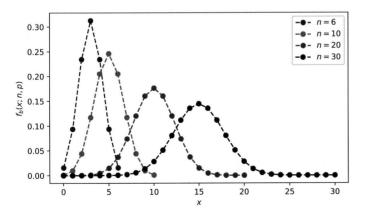

Figure 13.2: The binomial probability $f_b(x;n,p)$ for all $x \in [0,n]$, for different values of n. Note that the shape of $f_b(x;n,p)$ evolves closer to that of a normalized Gaussian function of x for increasing n and is very close to the Gaussian form already for the largest value of $n = 30$ shown here.

Note that this is a properly normalized pdf because according to Eq. (13.3) we have

$$\sum_{x=0}^{n} f_b(x) = \sum_{x=0}^{n} \frac{n!}{(n-x)!x!}p^x q^{n-x} = (p+q)^n = 1.$$

Derivation of the binomial distribution

Consider an event that occurs with probability p. The probability that this event does not occur is $q = 1 - p$. The probability that the event occurs x times in n trials (x here is by definition an integer, thus it will be treated as a discrete variable) is given by

$$pqqq \cdots qp = p^x q^{n-x},$$

where we have assumed that it occurred in the first trial, it did not occur on the next three trials, ..., it did not occur on the $(n-1)$th trial and it occurred in the nth trial. But this expression corresponds to a particular sequence of trials. If we want to compute the probability of x successful outcomes out of n trials, without specifying the sequence in which the successes and failures occurred, then we have to consider all possible combinations of the above sequence of outcomes, which is given by Eq. (13.2). Combining the two results, we conclude that the desired probability distribution is given by Eq. (13.18).

The mean and the variance of the binomial distribution can be calculated as follows. For the mean, we use the expression of Eq. (13.7) for the discrete case,

$$\mu_{\mathrm{b}} = \sum_{x=0}^{n} x \frac{n!}{(n-x)!x!} p^x q^{n-x}. \tag{13.19}$$

To calculate the value of μ_{b} we employ the Fourier transform: The Fourier transform of $f_{\mathrm{b}}(x)$ is given by

$$\hat{f}_{\mathrm{b}}(k) = \int_{-\infty}^{\infty} f_{\mathrm{b}}(x) e^{ikx} dx = \sum_{x=0}^{n} \frac{n!}{(n-x)!x!} p^x q^{n-x} e^{-ikx}$$

$$= \sum_{x=0}^{n} \frac{n!}{(n-x)!x!} \left(pe^{-ik}\right)^x q^{n-x} = \left(pe^{-ik} + q\right)^n, \tag{13.20}$$

where we have turned the integral over x to a sum over the discrete values that this variable takes, from 0 to n, and we have used the binomial expansion Eq. (3.9) to obtain the last step. Since the mean is the first moment of the variable x with respect to the function $f(x)$ [see definition in Eq. (13.7)], we can evaluate the mean from the general expression derived earlier for the first moment using FT's, namely Eq. (7.25):

$$\mu_{\mathrm{b}} = i \left[\frac{d\hat{f}_{\mathrm{b}}}{dk} \right]_{k=0} = np(-i)i(p+q)^{n-1} \Rightarrow \mu_{\mathrm{b}} = np, \tag{13.21}$$

where we have also used the fact that $p + q = 1$ to obtain the last expression. We can also use the expression for the second moment that involves FT's, Eq. (7.26), to obtain

$$i^2 \left[\frac{d^2 \hat{f}_{\mathrm{b}}}{dk^2} \right]_{k=0} = n(n-1)p^2 + np = np - np^2 + n^2 p^2. \tag{13.22}$$

Recalling that the variance is equal to the second moment minus the mean squared, Eq. (13.9), and using the above result yields for the variance of the binomial distribution the following:

$$\sigma_b^2 = i^2 \left[\frac{d^2 \hat{f}_b}{dk^2}\right]_{k=0} - \mu_b^2 \Rightarrow \sigma_b^2 = npq, \qquad (13.23)$$

where we have used $q = 1 - p$ to obtain the last expression.

Example 13.3: A fair coin is tossed n times: (a) What is the probability of getting x heads, where $x = 0, \ldots, n$? (b) What is the probability of getting at least 3 heads in 10 tosses?

Answer: Here the binomial distribution applies because we need to select x successful outcomes out of n tries, with the probability of success in each try $p = 0.5$ (and probability of failure $q = 1 - p = 0.5$), and we do not care about the order in which the successes or failures occur.
(a) The general case for n tosses yields:

$$f_b(x) = \frac{n!}{x!(n-x)!} p^x q^{1-x}, \quad p = q = \frac{1}{2} \Rightarrow f_b(x) = \frac{n!}{x!(n-x)!} \frac{1}{2^n}.$$

(b) For $n = 10$, the probabilities are as follows:

$$f_b(0) = \frac{10!}{0!10!} \left(\frac{1}{2}\right)^0 \left(\frac{1}{2}\right)^{10} = \frac{1}{1024}, \quad f_b(1) = \frac{10!}{1!9!} \left(\frac{1}{2}\right)^1 \left(\frac{1}{2}\right)^9 = \frac{10}{1024},$$

$$f_b(2) = \frac{10!}{2!8!} \left(\frac{1}{2}\right)^2 \left(\frac{1}{2}\right)^8 = \frac{45}{1024}, \quad f_b(3) = \frac{10!}{3!7!} \left(\frac{1}{2}\right)^3 \left(\frac{1}{2}\right)^7 = \frac{120}{1024},$$

$$f_b(4) = \frac{10!}{4!6!} \left(\frac{1}{2}\right)^4 \left(\frac{1}{2}\right)^6 = \frac{210}{1024}, \quad f_b(5) = \frac{10!}{5!5!} \left(\frac{1}{2}\right)^5 \left(\frac{1}{2}\right)^5 = \frac{252}{1024}.$$

Note that from the general expression for $f_b(x)$ we have $f_b(n-x) = f_b(x)$ for $0 \le x \le n/2$, for the special case $p = q = 0.5$, so the remaining probabilities are the same as the ones we have already calculated. In Fig. 13.2 we showed the values of $f_b(x)$ for all values of $0 \le x \le n$ for several values of n, namely $n = 6, 10, 20, 30$. The probability of getting at least 3 heads in 10 tosses is 1 minus the probability of getting any number of heads that is smaller than 3, that is, 0, 1 and 2 heads:

$$1 - f_b(0) - f_b(1) - f_b(2) = 1 - \frac{56}{1024} = 0.9453125.$$

13.2.3   Poisson distribution

Definition: The Poisson distribution applies when we know the mean value λ, and we are interested in the probability of x successful outcomes, in the limit when the number of trials n is much larger than the number of successes, $x \ll n$, the probability of success for an individual trial is very small ($p \ll 1$), but the mean is finite ($np = \lambda$). Hence, x is an integer and λ is a real positive number. The Poisson distribution is given by

$$f_P(x; \lambda) = \frac{\lambda^x e^{-\lambda}}{x!}, \quad x \text{ integer}, \quad \lambda > 0. \qquad (13.24)$$

This expression is a properly normalized pdf, because

$$\sum_{x=0}^{\infty} \frac{\lambda^x}{x!} = e^{\lambda}, \tag{13.25}$$

and therefore

$$\sum_{x=0}^{\infty} f_{\mathrm{P}}(x) = \sum_{x=0}^{\infty} \frac{\lambda^x}{x!} e^{-\lambda} = 1. \tag{13.26}$$

Also, note that for $\lambda < 1$ the Poisson pdf is peaked at $x = 0$, thus, the highest value is $f_{\mathrm{P}}(0)$, whereas for $\lambda > 1$ it is peaked at some value of $x > 0$, as illustrated in Fig. 13.3; for $\lambda = 1$, the values at $x = 0$ and $x = 1$ are equal, $f_{\mathrm{P}}(0) = f_{\mathrm{P}}(1) = e^{-1}$.

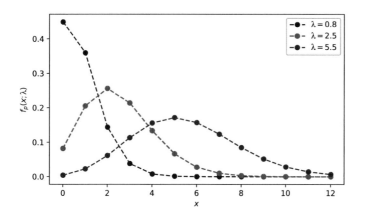

Figure 13.3: The Poisson probability distribution function $f_{\mathrm{P}}(x)$, defined in Eq. (13.24), for three different values of λ.

Derivation of the Poisson distribution from the binomial distribution

To derive the Poisson distribution, we start from the binomial distribution. We first note that

$$\frac{n!}{(n-x)!} = (n-x+1)(n-x+2)\cdots(n-1)n \approx n^x,$$

since for $x \ll n$ each term in the above product is approximately equal to n and there are x such terms. Next, we note that

$$q^{n-x} = \frac{(1-p)^n}{(1-p)^x} \Rightarrow \ln(q^{n-x}) = n\ln(1-p) - x\ln(1-p) \approx -np + xp,$$

where we have used $q = 1 - p$ and $\ln(1-\epsilon) \approx -\epsilon$ for $\epsilon \to 0$. From this last result, taking exponentials of both sides, we obtain

$$q^{n-x} \approx e^{-np},$$

where we have neglected xp compared to np, since we assumed $x \ll n$. Combining this estimate with the previous relation between $n!/(n-x)!$ and n^x, we find

$$f_{\mathrm{P}}(x) \approx \frac{(np)^x e^{-np}}{x!},$$

which is the desired expression for the Poisson distribution for $\lambda = np$.

The mean of the Poisson distribution can be calculated as follows: From the general definition of the mean, we have

$$\mu_{\rm p} = \sum_{x=0}^{\infty} f_{\rm p}(x)x = \sum_{x=0}^{\infty} x \frac{{\rm e}^{-\lambda}\lambda^x}{x!} = {\rm e}^{-\lambda} \sum_{x=0}^{\infty} x \frac{\lambda^x}{x!}.$$

We will use the identity:

$$\frac{\partial}{\partial \lambda}\left(\sum_{x=0}^{\infty} \frac{\lambda^x}{x!}\right) = \sum_{x=0}^{\infty} x \frac{\lambda^{x-1}}{x!}.$$

Multiplying both sides of this equation by $\lambda {\rm e}^{-\lambda}$ we find

$$\lambda {\rm e}^{-\lambda} \frac{\partial}{\partial \lambda}\left(\sum_{x=0}^{\infty} \frac{\lambda^x}{x!}\right) = \sum_{x=0}^{\infty} x \frac{{\rm e}^{-\lambda}\lambda^x}{x!} = \mu_{\rm p}.$$

But we also have

$$\left(\sum_{x=0}^{\infty} \frac{\lambda^x}{x!}\right) = {\rm e}^{\lambda} \Rightarrow \lambda {\rm e}^{-\lambda} \frac{\partial}{\partial \lambda}\left(\sum_{x=0}^{\infty} \frac{\lambda^x}{x!}\right) = \lambda {\rm e}^{-\lambda}{\rm e}^{\lambda} \Rightarrow \mu_{\rm p} = \lambda. \qquad (13.27)$$

For the variance of the Poisson distribution we have

$$\sigma_{\rm p}^2 = \sum_{x=0}^{\infty} f_{\rm p}(x)(x - \mu_{\rm p})^2 = \sum_{x=0}^{\infty} (x - \mu_{\rm p})^2 \frac{{\rm e}^{-\lambda}\lambda^x}{x!} = \sum_{x=0}^{\infty} x^2 \frac{{\rm e}^{-\lambda}\lambda^x}{x!} - \mu_{\rm p}^2. \quad (13.28)$$

We will use the identity:

$$\frac{\partial^2}{\partial \lambda^2}\left(\sum_{x=0}^{\infty} \frac{\lambda^x}{x!}\right) = \sum_{x=0}^{\infty} x(x-1) \frac{\lambda^{x-2}}{x!}.$$

Multiplying both sides of this equation by $\lambda^2 {\rm e}^{-\lambda}$ we find

$$\lambda^2 {\rm e}^{-\lambda} \frac{\partial^2}{\partial \lambda^2}\left(\sum_{x=0}^{\infty} \frac{\lambda^x}{x!}\right) = \sum_{x=0}^{\infty} x(x-1) \frac{{\rm e}^{-\lambda}\lambda^x}{x!} = \sum_{x=0}^{\infty} x^2 \frac{{\rm e}^{-\lambda}\lambda^x}{x!} - \mu_{\rm p} = \lambda^2,$$

where the last step is obtained by using the same steps as in Eq. (13.27). Then, substituting this result in Eq. (13.28) and taking advantage of $\mu_{\rm p} = \lambda$, we obtain

$$\sigma_{\rm p}^2 = \sum_{x=0}^{\infty} x^2 \frac{{\rm e}^{-\lambda}\lambda^x}{x!} - \mu_{\rm p}^2 = \lambda^2 + \mu_{\rm p} - \mu_{\rm p}^2 = \lambda. \qquad (13.29)$$

Example 13.4: During a storm, a field is hit by lightning at a rate of 20 hits per hour. A lightning can knock to the ground the grazing cattle. The farmer whose herd has been grazing on the field needs about five minutes to collect the cattle in a shelter. Once lightning activity has begun, what is the probability that the farmer will loose three animals before collecting them in the shelter? It may be assumed that the animals are dispersed and if a lightning strikes the field it will hit one animal.

Answer: In this case we can calculate the average number of lightning hits in the given time interval $t = 5$ min, denoted by λ, from the rate, $r = 20$ hits/hour as follows:

$$\lambda = t \times r = 5 \text{ min} \times \frac{20}{60 \text{ min}} = \frac{5}{3}.$$

We want to calculate the probability of $x = 3$ animals being struck during this given interval, so we can use the Poisson distribution to get the probability of this event:

$$f_p(x = 3) = \frac{\lambda^x e^{-\lambda}}{x!} = \frac{(5/3)^3 e^{-5/3}}{3!} = 0.145737, \qquad (13.30)$$

or about 14.6%. Alternatively, we can look at this problem from the point of view of the binomial distribution. In this case we must subdivide the total time interval of $t = 5$ min into more elementary units $\tau = t/n$, where n is the number of these units in the entire time interval t, in order to have a well-posed problem; eventually we want to take the limit of the number of these units being very large $n \to \infty$ and therefore their size going to zero. For example, if we take the elementary time unit to be $\tau = 1$ min then $\tau = t/5$, so there are $n = 5$ such subdivisions in the total time interval of interest (for $t = 5$ min); the probability of success in one subdivision is

$$p = \tau \times r = 1 \text{ min} \times \frac{20}{60 \text{ min}} = \frac{1}{3},$$

and the probability of failure is $q = 1 - p = 2/3$, while the total number of successful outcomes is $x = 3$, which can occur, one at a time, in any of the total $n = 5$ tries. From the binomial distribution we obtain the probability of this happening, as follows:

$$n = 5, p = \frac{1}{3}, q = \frac{2}{3}, x = 3 \Rightarrow f_b(3) = \frac{5!}{3!2!}\left(\frac{1}{3}\right)^3\left(\frac{2}{3}\right)^2 = 0.1646.$$

Similarly, for $\tau = 0.5$ min $= t/10$, we have

$$n = 10, p = \frac{1}{6}, q = \frac{5}{6}, x = 3 \Rightarrow f_b(3) = \frac{10!}{3!7!}\left(\frac{1}{6}\right)^3\left(\frac{5}{6}\right)^7 = 0.1550,$$

and for $\tau = 0.25$ min $= t/20$, we have

$$n = 20, p = \frac{1}{12}, q = \frac{11}{12}, x = 3 \Rightarrow f_b(3) = \frac{20!}{3!17!}\left(\frac{1}{12}\right)^3\left(\frac{11}{12}\right)^{17} = 0.1503.$$

We see that the result keeps changing and gets closer and closer to the answer we obtained with the Poisson distribution as the size of the subdivision τ is decreased or equivalently the number of trials n is increased, since $n\tau = t$. It is evident that in the limit $\tau \to 0$ we find $n \to \infty, p \to 0$, but $np = 5/3 = \lambda$, as in the case of the Poisson distribution; in this way we recover the result of Eq. (13.30).

13.2.4 *Normal distribution*

Definition: The normal distribution applies when the number of trials n is very large. There are two parameters needed to completely specify this distribution, namely, the real numbers μ and σ; x is a real variable that can range from $-\infty$ to $+\infty$, the mean value is μ and the standard deviation is σ. Connecting the normal distribution with the binomial as the limit of $n \to \infty$, we obtain for the mean $\mu = np$ and for the variance $\sigma^2 = npq$. The normal distribution is given by a normalized Gaussian:

$$f_g(x; \mu, \sigma) = \frac{1}{\sqrt{2\pi}\sigma} e^{-(x-\mu)^2/2\sigma^2}. \tag{13.31}$$

In Fig. 13.4 we show two examples of a normal distribution for different values of the parameters μ and σ that define it. In the following, for simplicity of notation we omit the explicit dependence in the expression of the normal or Gaussian distribution on the parameters μ and σ.

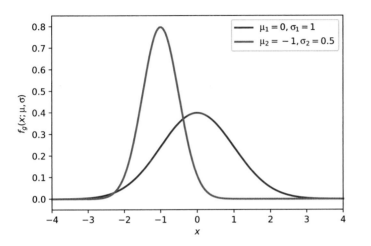

Figure 13.4: The normal probability distribution function $f_g(x; \mu, \sigma)$, defined in Eq. (13.31), for different values of the mean and variance, namely $\mu_1 = 0$, $\sigma_1 = 1$ (red curve) and $\mu_2 = -1$, $\sigma_2 = 0.5$ (green curve).

Derivation of the normal distribution from the binomial distribution

To derive the expression for the normal distribution, we will start again with the binomial distribution. In this derivation it will prove useful to express the random variable x in terms of the mean μ and its difference from this value, which we call δ, because the normal distribution is vanishingly small when x is far from μ, and therefore δ must be a small quantity compared to μ. We can then take advantage of this fact to use Taylor expansions in the small quantities that involve δ. We thus begin by setting

$$\mu = np, \quad x = \mu + \delta, \quad \text{where } n \to \infty. \tag{13.32}$$

Since p and q are finite and n is very large, we can write the following inequalities:

$$\delta \ll np = \mu, \quad \delta \ll nq = n(1 - p) = n - \mu, \tag{13.33}$$

as well as the following relations, using the fact that $q = 1 - p$:

$$n - x = n - \mu - \delta = n - np - \delta = nq - \delta. \tag{13.34}$$

We will use Stirling's formula

$$n! \approx (2\pi n)^{1/2} n^n e^{-n}, \tag{13.35}$$

which is valid in the limit of large n, to rewrite the factorials that appear in the binomial expression, Eq. (13.18). From this formula, we find that the binomial expression takes the following form:

$$
\frac{n!}{(n-x)!x!}p^{x}q^{n-x} \approx \frac{n^{n}}{x^{x}(n-x)^{n-x}}p^{x}q^{n-x}\left(\frac{n}{x(n-x)}\right)^{1/2}\frac{1}{\sqrt{2\pi}}
$$
$$
= \frac{1}{\sqrt{2\pi}}\left(\frac{np}{x}\right)^{x}\left(\frac{nq}{n-x}\right)^{n-x}\left(\frac{x(n-x)}{n}\right)^{-1/2}.
$$

(13.36)

We will deal with each term in the parentheses in the above expression separately, where we will substitute n, p and q through the variables μ, x and δ, using Eqs. (13.32), (13.33) and (13.34). The first term in the parentheses on the right-hand side of Eq. (13.36) yields

$$
\left(\frac{np}{x}\right)^{x} = \left(\frac{np}{np+\delta}\right)^{\mu+\delta} = \left(1+\frac{\delta}{np}\right)^{-(\mu+\delta)}.
$$

(13.37)

Taking the logarithm of this expression, we find

$$
-(np+\delta)\ln\left(1+\frac{\delta}{np}\right) = -(np+\delta)\left(\frac{\delta}{np}-\frac{\delta^{2}}{2(np)^{2}}+\cdots\right)
$$
$$
= -\delta - \frac{\delta^{2}}{2np}\cdots,
$$

where we have used the fact that $\delta \ll np$ and we have employed the Taylor expansion of the logarithm $\ln(1-\epsilon)$ for $\epsilon \ll 1$, Eq. (3.14), with $\epsilon = \delta/(np)$, keeping the first two terms in the expansion. Similarly, the second term in the parentheses on the right-hand side of Eq. (13.36) yields:

$$
\left(\frac{nq}{n-x}\right)^{n-x} = \left(\frac{nq}{nq-\delta}\right)^{nq-\delta} = \left(1-\frac{\delta}{nq}\right)^{-(nq-\delta)}.
$$

(13.38)

Taking the logarithm of this expression we find

$$
-(nq-\delta)\ln\left(1-\frac{\delta}{nq}\right) = -(nq-\delta)\left(-\frac{\delta}{nq}-\frac{\delta^{2}}{2(nq)^{2}}+\cdots\right)
$$
$$
= \delta - \frac{\delta^{2}}{2nq}\cdots,
$$

where we have used the fact that $\delta \ll nq$ and we have employed again the Taylor expansion of the logarithm with $\epsilon = \delta/nq$, and kept the first two terms in the expansion. The product of the two terms in Eqs. (13.37), (13.38), which upon taking logarithms is the sum of their logarithms, yields the following:

$$
\ln\left[\left(\frac{np}{x}\right)^{x}\left(\frac{nq}{n-x}\right)^{n-x}\right] = -\frac{\delta^{2}}{2np}-\frac{\delta^{2}}{2nq} = -\frac{\delta^{2}}{2n}\left(\frac{1}{p}+\frac{1}{q}\right)
$$
$$
= -\frac{\delta^{2}(p+q)}{2npq} = -\frac{\delta^{2}}{2\sigma^{2}},
$$

since $p+q=1$ and $\sigma^{2}=npq$. The third term in the parentheses on the right-hand side of Eq. (13.36) yields:

$$\left(\frac{x(n-x)}{n}\right)^{-1/2} = \left[npq\left(\frac{np+\delta}{np}\right)\left(\frac{nq-\delta}{nq}\right)\right]^{-1/2}$$

$$\approx (npq)^{-1/2}\left(1 - \frac{\delta}{2np} + \frac{\delta}{2nq} + \cdots\right)$$

$$\approx (npq)^{-1/2} = (\sigma^2)^{-1/2}.$$

Putting all the partial results together, we arrive at

$$f_g(x; \mu, \sigma) \approx e^{-\delta^2/2\sigma^2}(2\pi\sigma^2)^{-1/2},$$

which, with $\delta = x - \mu$ gives the desired result of Eq. (13.31).

The mean of the normal distribution is given by

$$(\bar{x})_g = \int_{-\infty}^{\infty} x f_g(x; \mu, \sigma)\mathrm{d}x = \frac{1}{\sqrt{2\pi}\sigma}\int_{-\infty}^{\infty} x e^{-(x-\mu)^2/2\sigma^2}\mathrm{d}x = \mu. \qquad (13.39)$$

To arrive at this result we first change variables from x to $x' = (x - \mu)$ and then take advantage of the fact that $\exp[-x'^2/2\sigma^2]$ is an even function of x', and use the result for the normalized Gaussian, Eq. (1.57). For the calculation of the variance of the normal distribution, we take advantage of the properties of the normalized Gaussian function to obtain

$$\left(\overline{(x-\mu)^2}\right)_g = \int_{-\infty}^{\infty} (x-\mu)^2 f_g(x; \mu, \sigma)\mathrm{d}x = \sigma^2. \qquad (13.40)$$

The derivation of the results in Eqs. (13.39) and (13.40) is straightforward, so we leave them as an exercise [see Problem 1].

We note that if we scale the variable x by a factor s, namely, we define the new variable $\bar{x} = sx$, we can rewrite the Gaussian probability as

$$f_g(x; \mu, \sigma) = s\frac{1}{\sqrt{2\pi}(s\sigma)}e^{-(sx-s\mu)^2/2(s\sigma)^2} = s\frac{1}{\sqrt{2\pi}\bar{\sigma}}e^{-(\bar{x}-\bar{\mu})^2/2\bar{\sigma}^2},$$

where we have defined the new parameters

$$\bar{\mu} = s\mu, \quad \bar{\sigma} = s\sigma. \qquad (13.41)$$

Usually, we employ the probability distributions to evaluate the probability of finding a value x within a small range $[x, x + \mathrm{d}x]$, which is given by the pdf multiplied by this small range, that is, we are interested in the quantity $f_g(x; \mu, \sigma)\mathrm{d}x$. With this in mind, we find that the scaling by the factor s gives

$$f_g(x; \mu, \sigma)\mathrm{d}x = f_g(\bar{x}; \bar{\mu}, \bar{\sigma})\mathrm{d}\bar{x}, \qquad (13.42)$$

where we have used $s\mathrm{d}x = \mathrm{d}(sx) = \mathrm{d}\bar{x}$. The relation of Eq. (13.42) shows that scaling the variable x by the factor s gives rise to another Gaussian distribution in the scaled variable $\bar{x}$, with a mean and a standard deviation that are scaled by the same factor.

Standard normal distribution: In the final form of the Poisson distribution derived above, the variable x ranges from 0 to n and its mean value is $\mu = np$.

When n is very large, it is difficult to work with the original values of the parameters that enter in the definition of $f_g(x; \mu, \sigma)$; it is more convenient to scale and shift the origin of x, so that the mean of the scaled and shifted variable is zero and its standard deviation is 1; this is referred to as the "standard normal distribution", $\varphi(x) = f_g(x; 0, 1)$:

$$\varphi(x) = \frac{1}{\sqrt{2\pi}} e^{-x^2/2}. \tag{13.43}$$

Clearly, any normal distribution can be expressed as the standard one, after shifting and scaling, which simplifies calculations because the values and integrals of the standard normal distribution can be tabulated. We explain these ideas in detail below.

We first define the scaled variables

$$\tilde{x} = \frac{x}{n}, \quad \tilde{\mu} = \frac{\mu}{n}, \quad \tilde{\sigma} = \frac{\sigma}{n}. \tag{13.44}$$

Next, we define the variable y through

$$y = \frac{\tilde{x} - \tilde{\mu}}{\tilde{\sigma}} \Rightarrow dy = \frac{d\tilde{x}}{\tilde{\sigma}}, \tag{13.45}$$

that is, y is related to $\tilde{x}$ through a shift by $-\tilde{\mu}$ and a rescaling by $1/\tilde{\sigma}$. Since the limits of x are $[0, n]$, the limits of the variable $\tilde{x}$ are $[0, 1]$ and the limits of the variable y are $[-\tilde{\mu}/\tilde{\sigma}, (1 - \tilde{\mu})/\tilde{\sigma}]$. Note that

$$\mu = pn \Rightarrow \tilde{\mu} = \frac{\mu}{n} = p \Rightarrow (1 - \tilde{\mu}) = 1 - p = q.$$

We also have

$$\sigma^2 = npq \Rightarrow \tilde{\sigma} = \frac{\sigma}{n} = \frac{\sqrt{pq}}{\sqrt{n}}.$$

With these results, the limits of the variable y become

$$-\frac{\tilde{\mu}}{\tilde{\sigma}} = -\sqrt{n}\sqrt{\frac{p}{q}}, \quad \frac{(1 - \tilde{\mu})}{\tilde{\sigma}} = \sqrt{n}\sqrt{\frac{q}{p}}.$$

Since we are taking the limit of very large n, when p and q have finite values between 0 and 1, we conclude that the limits of the variable y are essentially $\pm\infty$. Thus, the mean and variance of the variable y have now become

$$\mu_y = \frac{\tilde{\mu} - \tilde{\mu}}{\tilde{\sigma}} = 0, \quad \sigma_y^2 = \frac{\tilde{\sigma}^2}{\tilde{\sigma}^2} = 1.$$

The final expression is a properly normalized probability distribution function, since

$$\int_{-\infty}^{\infty} \frac{1}{\sqrt{2\pi}} e^{-y^2/2} dy = 1,$$

as was shown in Eq. (1.57).

The function $F_g(w; \mu, \sigma)$, defined as

$$F_g(w; \mu, \sigma) = \frac{1}{\sqrt{2\pi}\sigma} \int_{-\infty}^{w} e^{-(y-\mu)^2/2\sigma^2} dy, \tag{13.46}$$

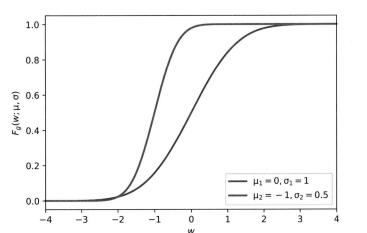

Figure 13.5: The cumulative distribution function $F_g(w; \mu, \sigma)$ for the two examples of the normal distribution shown in Fig. 13.4, namely $\mu_1 = 0, \sigma_1 = 1$ (red curve), and $\mu_2 = -1, \sigma_2 = 0.5$ (green curve).

is the cumulative distribution function of the normal distribution with mean μ and variance σ^2. In Fig. 13.5 we show the cumulative distribution function as defined in Eq. (13.46) for some values of the mean and the variance. This function is very useful in calculating probabilities as we demonstrate in the next example. However, the values of $F_g(w; \mu, \sigma)$ cannot be obtained analytically, so they have to be obtained numerically or from tables. Instead of calculating this function for the given values of μ and σ, it is more convenient to define the cumulative distribution function for the standard normal distribution and then shift and scale the variables of each problem appropriately, as explained above. Because of its importance, this function is denoted by a special symbol, $\Phi(w) = F_g(w; 0, 1)$, and was first introduced in Chapter 1, Eq. (1.60a):

$$\Phi(w) = \frac{1}{\sqrt{2\pi}} \int_{-\infty}^{w} e^{-y^2/2} dy = \frac{1}{\sqrt{2\pi}} \int_{-w}^{\infty} e^{-y^2/2} dy. \qquad (13.47)$$

The function $\Phi(w)$ satisfies the following relations:

$$\Phi(+\infty) = 1, \quad \Phi(-\infty) = 0, \quad \Phi(-w) = 1 - \Phi(w), \quad \Phi(0) = 0.5, \qquad (13.48)$$

and its values can be found in tables (for convenience we provide such a table in Appendix B for $w \in [0, 2.5]$ in steps of 0.01). $\Phi(w)$ can also be calculated by a Taylor expansion near $w_0 = 0$, see Example 3.2 in Chapter 3.

Example 13.5: We consider a fair coin tossed a million times. What is the probability that the number of tails is between 49.9% and 50.1% of the total?

<u>Answer:</u> Here we are obviously dealing with a Gaussian distribution since $n = 10^6$ is a very large number of trials. We could, in principle, use the binomial distribution, but this would lead to extremely tedious and difficult calculations, which would give essentially the same result. We will first calculate the probability that the number of tails is between 0 and 49.9% of the total.

In the context of the Gaussian distribution, we have

$$p = 0.5, \ q = 0.5, \ \Rightarrow \mu = np = 0.5 \times 10^6,$$

$$\sigma^2 = npq = 0.25 \times 10^6 \Rightarrow \sigma = 0.5 \times 10^3.$$

The probability that we get any number of tails up to 49.9% of the total is given by the integral of the Gaussian probability distribution over all values of x from $x_1 = 0$ (no tails) to $x_2 = 0.499n$ (49.9% tails) with μ and σ determined above:

$$\int_0^{0.499n} \frac{1}{\sqrt{2\pi}\sigma} e^{-(x-\mu)^2/2\sigma^2} dx.$$

This, however, is an awkward integral to calculate. Instead, we change variables as indicated in the general discussion of the Gaussian distribution:

$$\tilde{x} = \frac{x}{n}, \quad \tilde{\sigma} = \frac{\sigma}{n} = 0.5 \times 10^{-3}, \quad \tilde{\mu} = \frac{\mu}{n} = 0.5,$$

so that the range of the new variable $\tilde{x}$ is $[0, 1]$. We express the limits of the integral in terms of $\tilde{\mu}$ and $\tilde{\sigma}$:

$$\tilde{x}_1 = \frac{x_1}{n} = 0 = 0.5 - 10^3 \times 0.5 \times 10^{-3} = \tilde{\mu} - 10^3 \tilde{\sigma},$$

$$\tilde{x}_2 = \frac{x_2}{n} = 0.499 = 0.5 - 2 \times 0.5 \times 10^{-3} = \tilde{\mu} - 2\tilde{\sigma}.$$

In this way, the desired integral becomes

$$\int_{\tilde{\mu}-10^3\tilde{\sigma}}^{\tilde{\mu}-2\tilde{\sigma}} \frac{1}{\sqrt{2\pi}\tilde{\sigma}} e^{-(\tilde{x}-\tilde{\mu})^2/2\tilde{\sigma}^2} d\tilde{x}.$$

We next shift the origin of the variable $\tilde{x}$ by $-\tilde{\mu}$ and scale it by $\tilde{\sigma}$, defining a new variable y:

$$y = \frac{\tilde{x} - \tilde{\mu}}{\tilde{\sigma}} \Rightarrow dy = \frac{d\tilde{x}}{\tilde{\sigma}}.$$

Using this new variable, the above integral becomes

$$\int_{-10^3}^{-2} \frac{1}{\sqrt{2\pi}} e^{-y^2/2} dy \approx \int_{-\infty}^{-2} \frac{1}{\sqrt{2\pi}} e^{-y^2/2} dy = \Phi(-2) = 1 - \Phi(2) = 0.02275,$$

or 2.275%. In the last expression, we have replaced the lower limit of $y = -10^3$ with $-\infty$, since for such large negative values of y the integrand is zero for all practical purposes, which is the value that it takes for $y \to -\infty$. The numerical value is obtained from the tabulated values of the normalized Gaussian integral $\Phi(w)$, Eq. (13.47), see Appendix B. The probability that the number of tails is between 50.1% and 100% of the total is the same as for being between 0 and 49.9% of the total. Thus, the probability of the number of tails being between 49.9% and 50.1% of the total is: $1 - 2 \times 0.02275 = 0.95450$ (95.45%).

Derivation of the normal distribution from the autoconvolution of uniform distribution

There is another interesting way to obtain a normal distribution starting with a *uniform* distribution: by exploiting the properties of the convolution. We start with the uniform probability distribution $g_0(x)$ defined in the range $x \in [0, 1]$:

$$g_0(x) = 0, \quad |x| > \frac{1}{2},$$
$$= 1, \quad |x| \leq \frac{1}{2}, \tag{13.49}$$

which is shown in the diagrams below in red lines. We next calculate the convolution of $g_0(x)$ with itself, called the "autoconvolution", which is a new function $g_1(x)$:

$$g_1(x) = \int_{-\infty}^{\infty} g_0(x - y) \, g_0(y) \mathrm{d}y. \tag{13.50}$$

The two different possibilities of arranging the functions $g_0(y)$ and $g_0(x - y)$, in order to calculate their convolution are shown below:

$$\text{(i)} \ -1 \leq x \leq 0, \quad \text{(ii)} \ 0 \leq x \leq 1.$$

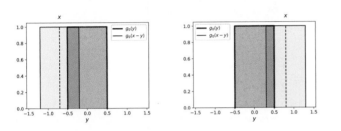

These give the following integrals:

$$\text{(i)} : \ -1 \leq x \leq 0 \Rightarrow g_1(x) = \int_{-\frac{1}{2}}^{x+\frac{1}{2}} \mathrm{d}y = x + 1,$$

$$\text{(ii)} : \ 0 \leq x \leq 1 \Rightarrow g_1(x) = \int_{x-\frac{1}{2}}^{\frac{1}{2}} \mathrm{d}y = 1 - x.$$

Thus, the result is the following function:

$$g_1(x) = 0, \quad |x| > 1,$$
$$= 1 - |x|, \quad |x| \leq 1. \tag{13.51}$$

$g_1(x)$ is a proper probability distribution: it is *linear* in the variable x, its values lie in the range $0 \leq g_1(x) \leq 1$ for all x and its integral over all x is equal to 1, as can be readily checked. The function $g_1(x)$ is shown in the diagrams below in blue lines.

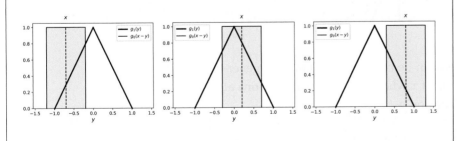

We can repeat the convolution step as follows: we consider the convolution of $g_1(x)$ with $g_0(x)$:

$$g_2(x) = \int_{-\infty}^{\infty} g_0(x-y)\, g_1(y)\mathrm{d}y. \tag{13.52}$$

There are three different possibilities of arranging the functions $g_1(y)$ and $g_0(x-y)$ in order to calculate their convolution, as shown in the three diagrams above:

$$\text{(i)} \ -\frac{3}{2} \le x \le -\frac{1}{2}, \quad \text{(ii)} \ -\frac{1}{2} \le x \le \frac{1}{2}, \quad \text{(iii)} \ \frac{1}{2} \le x \le \frac{3}{2}.$$

These give the following integrals:

$$\text{(i)} \ -\frac{3}{2} \le x \le -\frac{1}{2} \Rightarrow g_2(x) = \int_{-1}^{x+\frac{1}{2}} (1+y)\mathrm{d}y = \frac{1}{2}\left(x+\frac{3}{2}\right)^2,$$

$$\text{(ii)} \ -\frac{1}{2} \le x \le \frac{1}{2} \Rightarrow g_2(x) = \int_{x-\frac{1}{2}}^{0} (1+y)\mathrm{d}y + \int_{0}^{x+\frac{1}{2}} (1-y)\mathrm{d}y$$

$$= \frac{3}{4} - x^2,$$

$$\text{(iii)} \ \frac{1}{2} \le x \le \frac{3}{2} \Rightarrow g_2(x) = \int_{x-\frac{1}{2}}^{1} (1-y)\mathrm{d}y = \frac{1}{2}\left(\frac{3}{2}-x\right)^2.$$

Thus, the result is the following function:

$$g_2(x) = \begin{cases} 0, & |x| > \dfrac{3}{2}, \\[2mm] \dfrac{1}{2}\left(\dfrac{3}{2}-|x|\right)^2, & \dfrac{1}{2} \le |x| \le \dfrac{3}{2}, \\[2mm] \dfrac{3}{4} - x^2, & |x| \le \dfrac{1}{2}. \end{cases} \tag{13.53}$$

$g_2(x)$ is also a proper probability distribution: it is *quadratic* in the variable x, its values lie in the range $0 \le g_2(x) \le 1$ for all x and its integral over all x is equal to 1, as can be readily checked. The function $g_2(x)$ is included in the diagram below (cyan), along with $g_0(x)$ (red) and $g_1(x)$ (blue).

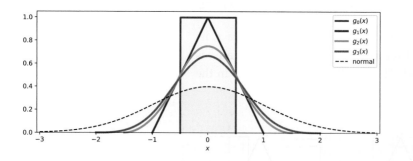

It is easily seen that this process can be continued indefinitely, generating functions that are polynomials of x whose order increases by one at each step. For instance, the next function $g_3(x)$ defined as

$$g_3(x) = \int_{-\infty}^{\infty} g_0(x-y)\, g_2(y)\mathrm{d}y \tag{13.54}$$

turns out to be

$$g_3(x) = \begin{cases} 0, & |x| > 2, \\ \dfrac{1}{6}\left(2 - |x|\right)^3, & 1 \le |x| \le 2, \\ \dfrac{2}{3} - x^2 + \dfrac{1}{2}|x|^3, & |x| \le 1. \end{cases} \tag{13.55}$$

This function is also shown in the diagram above (green line). Each of these functions is a proper probability distribution. Taking the limit of an infinite number of steps generates the normal distribution, shown in the diagram as the dashed black line for comparison. This approach emphasizes the fact that the normal distribution is the result of a very large number of samplings, which in the present case is based on a uniform distribution. We provide a more complete proof of this fact in Section 13.5 in the context of the Central Limit Theorem.

13.2.5 Arbitrary probability distributions

It is often useful to consider how to generate probability distributions that are *not* uniform, through the use of an easily accessible random variable with uniform probability density. This becomes an important consideration in the case of random (or pseudo-random) numbers, for which we can easily produce a uniform distribution; this distribution can then become the basis for generating random numbers with any other desirable distribution. We revisit these topics in the next chapter.

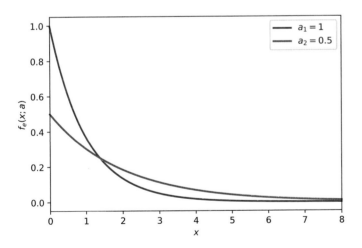

Figure 13.6: The exponential probability distribution function $f_e(x)$, defined in Eq. (13.56), for two different values of the parameter a that defines it.

For a concrete example, suppose we wish to work with a random variable y that has exponential probability density, namely, the probability of finding the value y is proportional to e^{-ay}, with $a > 0$ and $y \in [0, \infty)$

$$p_e(y; a) = A e^{-ay}, \quad a > 0, \ y \ge 0, \tag{13.56}$$

where A is the normalization constant; integrating $p_e(y; a)$ over all values of y and setting this integral equal to unity gives

$$\int_0^\infty A e^{-ay} dy = A \left[-\frac{1}{a} e^{-ay} \right]_0^\infty = 1 \Rightarrow A = a.$$

However, instead of having access to random values y distributed with probability $p_e(y; a)$, suppose we only have access to a random variable x with a uniform probability density in the range $x \in [0, 1]$, namely $p_u(x) = 1$. For the new variable y, we want to generate a probability density that obeys

$$p_e(y; a)\mathrm{d}y = p_u(x)\mathrm{d}x. \tag{13.57}$$

In this way, no probability density will be lost in switching from the variable x to the variable y. In other words, since we only know how densely distributed are the values of the new random variable for a given value y, which is given by $p_e(y; a)\mathrm{d}y$, Eq. (13.57) will ensure that the overall number of values contains the same amount of random values as the original one. The result of this analysis is that the two probability densities must be related by the equation

$$p_e(y; a) = p_u(x) \left| \frac{\mathrm{d}x}{\mathrm{d}y} \right|, \tag{13.58}$$

where we have inserted an absolute value for the derivative $\mathrm{d}x/\mathrm{d}y$, since we are not *a priori* certain that this derivative is a positive quantity but we do know that all other quantities involved in Eq. (13.57) are positive.

For the case of an exponential probability density in the variable y, we require

$$p_e(y; a) = ae^{-ay} = p_u(x) \left| \frac{\mathrm{d}x}{\mathrm{d}y} \right| = \left| \frac{\mathrm{d}x}{\mathrm{d}y} \right| \Rightarrow x(y) = e^{-ay},$$

where we solved for x this equation viewing it as a function of y. We can invert this last relation to express y as a function of x by taking the logarithm of both sides:

$$y(x) = -\frac{1}{a} \ln(x).$$

Thus, the new random values y are given in terms of the original random values x, which have uniform probability distribution, through the relation $y(x) = -\ln(x)/a$. In other words, by taking the natural logarithm of the original random values (including a minus sign, which makes sure that the new numbers are positive) we generate random values y in the range $[0, \infty)$ with a probability distribution $p_e(y; a) = ae^{-ay}$. Examples of this distribution for different values of a are shown in Fig. 13.6.

This exponential probability distribution is encountered in stochastic optimization problems, as we discuss in the next chapter. In physics and chemistry it is often referred to as the "Boltzmann distribution" and plays a key role in describing the equilibrium properties of physical systems consisting of a large number of microscopic particles, like gases, liquids and solids.

It is straightforward to calculate the mean, μ_e, and the variance, σ_e^2, of the exponential distribution. An elegant way to calculate these quantities is by the use of the properties of the Fourier transform. We showed in Chapter 7 that the FT of this distribution is (see Example 7.3):

$$\hat{f}(k) = \frac{a}{ik + a}.$$

From this result, using the expression for the FT of the moments of the function $f(x)$, Eq. (7.26), we obtain

$$\mu_e = \int_0^\infty xae^{-ax}\mathrm{d}x = i\left[\frac{\mathrm{d}\hat{f}}{\mathrm{d}k}\right]_{k=0} = i\left[\frac{-ia}{(ik + a)^2}\right]_{k=0} = \frac{1}{a},$$

$$\sigma_e^2 = \int_0^\infty x^2 a e^{-ax} dx - \mu_e^2 = i^2 \left[\frac{d^2 \hat{f}}{dk^2} \right]_{k=0} - \mu_e^2 = i^2 \left[\frac{-i(-2i)a}{(ik+a)^3} \right]_{k=0} - \frac{1}{a^2} = \frac{1}{a^2}.$$

We summarize the key features of the common probability distributions in the following table:

Distribution	Expression	Parameters	Variable	Mean	Variance
Uniform	$\frac{1}{(b-a)}$	$a, b \in \mathbb{R}$ $b > a$	$x \in [a, b]$	$\frac{(a+b)}{2}$	$\frac{(b-a)^2}{12}$
Binomial	$\frac{n! p^x q^{n-x}}{x!(n-x)!}$	$n \in \mathbb{Z}^{0+}; p, q \in \mathbb{R}$ $0 < p, q < 1$ $q = 1 - p$	$x \in [0, n]$ $x \in \mathbb{Z}^{0+}$	np	npq
Poisson	$\frac{\exp[-\lambda]\lambda^x}{x!}$	$\lambda > 0$	$x \in [0, \infty)$ $x \in \mathbb{Z}^{0+}$	λ	λ
Normal	$\frac{\exp[-(x-\mu)^2/2\sigma^2]}{\sqrt{2\pi}\sigma}$	$\mu, \sigma \in \mathbb{R}$	$x \in (-\infty, \infty)$	μ	σ^2
Exponential	$a e^{-ax}$	$a > 0$	$x \in [0, \infty)$	$\frac{1}{a}$	$\frac{1}{a^2}$

13.3   Probabilities, information and entropy

An important concept in discussing probabilistic events is the amount of information contained in a statement about the occurrence of an event whose probability is known [for a thorough discussion see *"Pattern Recognition and Machine Learning"*, by C.M. Bishop, Chapter 1 (Springer, 2006)]. Specifically, if we are told that a low-probability event has occurred, this statement contains very valuable information; in contrast the occurrence of a high-probability event contains very little valuable information, and the occurrence of an event of probability 1 contains no useful information. We wish to quantify the associated information by the concept of **entropy**, which is expected to depend on the probability distribution $p(x)$. As is evident from the general idea we want to convey, the information content, which we denote by s, must be a function of the probability distribution $p(x)$. Hence, we are dealing with the function $s(p)$; actually, $s(p)$ is a functional, but we deal with it as simple function here, since we use as the primary variable the probability p. The function $s(p)$ must be always positive, it must take large positive values when the probability p is low, and must tend to zero when the probability $p \to 1$. Moreover, for two independent events x_1, x_2 with probabilities $p_1 = p(x_1)$ and $p_2 = p(x_2)$, the probability $p_{1,2} = p(x_1, x_2)$ of the two events occurring is given by

$$p_{1,2} = p(x_1, x_2) = p(x_1)p(x_2) = p_1 p_2.$$

This is intuitively an obvious result; we shall examine the meaning of this result in the following section, when we discuss multivariate probabilities. We expect the information contained in the statement that the two independent events have occurred to be the sum of the information contained in the two events occurring separately, that is,

$$s(p_{1,2}) = s(p_1) + s(p_2).$$

A function that captures all the desired aspects of this behavior is

$$s(p) = -\ln(p).$$

The overall minus sign ensures that the function $s(p)$ takes positive values for $0 \leq p \leq 1$; moreover, the property

$$\ln(ab) = \ln(a) + \ln(b)$$

ensures the additive property of independent events. We define the entropy functional $S[p(x)]$ in terms of $p(x)$ as

$$S[p(x)] = \int_{-\infty}^{\infty} p(x)s(p)\, dx = -\int_{-\infty}^{\infty} p(x)\ln[p(x)]\, dx. \qquad (13.59)$$

The probability distribution function that corresponds to the maximum entropy, for a given variance σ^2 and mean μ, is the Gaussian probability distribution with this variance and mean, as shown in the derivation below.

This result for the entropy can also be established as follows: consider a large number of objects N that can be distributed in m bins. These objects can in general be different, and have different affinities to be placed in the various bins, but for simplicity we consider here the case that all objects are the same and have the same affinity for any bin. In a particular distribution of the objects there are n_i objects in the bin labeled i, with $i = 1, 2, \ldots, m$, so that

$$\sum_{i=1}^{m} n_i = N.$$

The probability of finding an object in bin i, which is referred to as the "state" of the system, is given by

$$p_i = \frac{n_i}{N}.$$

This is a proper probability distribution since

$$\sum_{i=1}^{m} p_i = \sum_{i=1}^{m} \frac{n_i}{N} = \frac{1}{N}\sum_{i=1}^{m} n_i = 1. \qquad (13.60)$$

The number of ways to obtain a particular distribution is given by

$$W = \frac{N!}{n_1! n_2! \cdots n_m!}.$$

The different ways of distributing the objects in the bins contain information about the specific state of this system which is characterized by how many objects are in each bin. We define the entropy S of the system with the logarithm of W for each state of the system [this argument is essentially the one used by Ludwig Boltzmann in his derivation of the entropy of a system of identical particles, a cornerstone of statistical physics for the description of matter]. The logarithm of W must be normalized by the number of objects N, where N can be very large, $N \to \infty$:

$$S = \frac{1}{N} \ln(W) = \frac{1}{N} \ln\left(\frac{N!}{n_1! n_2! \cdots n_m!}\right).$$

Using Stirling's formula for the logarithm of large numbers, Eq. (13.35), and neglecting factors of order $\ln(n)/n$ which tend to zero for n very large, we obtain

$$S = \lim_{N \to \infty} \left\{ \frac{1}{N} \left[N \ln(N) - \sum_{i=1}^{m} n_i \ln(n_i) \right] \right\} = \ln(N) - \sum_{i=1}^{m} p_i [\ln(p_i) + \ln(N)]$$

$$\Rightarrow \quad S = - \sum_{i=1}^{m} p_i \ln(p_i), \tag{13.61}$$

where we have used the fact that the p_i's represent probabilities, in other words, that Eq. (13.60) holds. The last expression is simply a discrete form of the general expression, Eq. (13.59).

Derivation of the maximum entropy probability distribution

We wish to find the probability distribution function $f(x)$ with given average μ and variance σ^2, that maximizes the entropy functional, Eq. (13.59), namely

$$S[f(x)] = - \int_{-\infty}^{\infty} f(x) \ln[f(x)] dx \quad : \quad \text{maximum}.$$

Since the function $f(x)$ is a probability distribution, it is subject to the constraints that the zeroth moment is equal to 1 (normalization condition), the first moment is equal to μ (the mean of the distribution), and the second moment relative to μ is equal to σ^2 (the variance of the distribution):

$$\int_{-\infty}^{\infty} f(x) dx = 1, \quad \int_{-\infty}^{\infty} x f(x) dx = \mu, \quad \int_{-\infty}^{\infty} (x - \mu)^2 f(x) dx = \sigma^2. \tag{13.62}$$

We will implement these constraints with the help of the Lagrange multipliers $\lambda_0, \lambda_1, \lambda_2$, that is, we will minimize a functional which, in addition to $S[(f(x)]$, also contains the following terms:

$$- \lambda_0 \left[\int_{-\infty}^{\infty} f(x) dx - 1 \right] - \lambda_1 \left[\int_{-\infty}^{\infty} x f(x) dx - \mu \right]$$

$$- \lambda_2 \left[\int_{-\infty}^{\infty} (x - \mu)^2 f(x) dx - \sigma^2 \right].$$

We will use the methods developed in the section on calculus of variations (see Section 2.6): the above expression is a functional of the general form given by Eq. (2.88) with

$$G(f, f', x) = -f \ln[f] - \lambda_0 f - \lambda_1 x f - \lambda_2 (x - \mu)^2 f.$$

To find the stationary solution which corresponds to an extremum of this functional, we apply the Euler–Lagrange equation, Eq. (2.89):

$$\frac{\partial G}{\partial f'} = 0 \Rightarrow \frac{\partial G}{\partial f} = -\ln(f) - f\frac{1}{f} - \lambda_0 - \lambda_1 x - \lambda_2 (x - \mu)^2 = 0$$

$$\Rightarrow \ln(f) = -(\lambda_0 + 1) - \lambda_1 x - \lambda_2 (x - \mu)^2 \Rightarrow f(x) = e^{-(\lambda_0+1)} e^{-\lambda_1 x} e^{-\lambda_2 (x-\mu)^2}.$$

In order to satisfy the constraints, we must take $\lambda_1 = 0$, because if $\lambda_1 \neq 0$ the function $f(x)$ blows up for either $x \to +\infty$ (if $\lambda_1 > 0$) or for $x \to -\infty$ (for $\lambda_1 < 0$). For the same reason, we must have $\lambda_2 > 0$. The other two constraints give

$$\int_{-\infty}^{\infty} f(x)\mathrm{d}x = 1 \Rightarrow e^{-(\lambda_0+1)}\sqrt{\pi}\lambda_2^{-1/2} = 1,$$

$$\int_{-\infty}^{\infty} (x - \mu)^2 f(x)\mathrm{d}x = \sigma^2 \Rightarrow e^{-(\lambda_0+1)}\frac{\sqrt{\pi}}{2}\lambda_2^{-3/2} = \sigma^2,$$

where we have used the properties of the Gaussian function, Eq. (1.66). Dividing the first of these relations by the second, we find

$$\lambda_2 = \frac{1}{2\sigma^2} > 0,$$

as desired; substituting this result in the first expression we find

$$\lambda_0 = \ln(\sqrt{2\pi}\sigma) - 1 \Rightarrow e^{-(\lambda_0+1)} = \frac{1}{\sqrt{2\pi}\sigma}.$$

Thus, the function which satisfies the constraints of Eq. (13.62) and maximizes the entropy is the normalized Gaussian or normal probability distribution is given by

$$f(x) = \frac{1}{\sqrt{2\pi}\sigma}e^{-(x-\mu)^2/2\sigma^2}.$$

13.4 Multivariate probabilities

We next consider cases where we have more than one random variable. We will deal explicitly with two random variables, but the discussion is easily generalized to any finite number of variables. We will use the symbol $P(X, Y)$ to denote the value of the probability that depends on two variables, without specifying the values of the variables (this is what the capital symbols X, Y mean). When these variables take specific values the quantity $P(X, Y)$ is equal to the probability of these specific values actually occurring.

We define the probability distribution $f(x, y)$ for the two random variables x, y as the probability of obtaining the first variable with value x and the second variable with value y:

$$f(x, y) = P(X = x, Y = y). \tag{13.63}$$

With this definition, we can define the probability of obtaining the value x for the first variable independent of what the value of the second variable is, which is given by

$$f_1(x) = P(X = x, Y: \text{ arbitrary}) = \sum_y f(x, y), \qquad (13.64)$$

where the symbol $\sum_y$ implies a summation over all the possible values of y, whether this is a discrete or a continuous variable. The quantity $f_1(x)$ is known as the "marginal probability" for the random variable x. Similarly, the probability of obtaining the value y for the second variable independent of what the value of the first variable is, is given by

$$f_2(y) = P(X: \text{ arbitrary}, Y = y) = \sum_x f(x, y), \qquad (13.65)$$

which is the marginal probability for the random variable y.

If the two variables are **independent** of each other, then the following relation is true:

$$f(x, y) = f_1(x) f_2(y). \qquad (13.66)$$

A consequence of this equation is that the corresponding cumulative probability distributions $F_1(x), F_2(y)$ obey a similar relation:

$$F(x, y) = F_1(x) F_2(y). \qquad (13.67)$$

By analogy with the case of single-variable probability distributions we define the expectation value of a function of two variables $g(x, y)$ as

$$E(g(X, Y)) = \int_{-\infty}^{\infty} \int_{-\infty}^{\infty} g(x, y) f(x, y) dx dy. \qquad (13.68)$$

Using this definition, we can derive the relation

$$E(X + Y) = E(X) + E(Y), \qquad (13.69)$$

which holds for any pair of random variables because it is a consequence of the following relation:

$$\int_{-\infty}^{\infty} \int_{-\infty}^{\infty} (x + y) f(x, y) dx dy = \int_{-\infty}^{\infty} \int_{-\infty}^{\infty} x f(x, y) dx dy$$
$$+ \int_{-\infty}^{\infty} \int_{-\infty}^{\infty} y f(x, y) dx dy.$$

We also have

$$E(XY) = \int_{-\infty}^{\infty} \int_{-\infty}^{\infty} (xy) f(x, y) dx dy,$$

which for **independent** variables becomes

$$\int_{-\infty}^{\infty} \int_{-\infty}^{\infty} (xy) f(x, y) dx dy = \int_{-\infty}^{\infty} x f_1(x) dx \int_{-\infty}^{\infty} y f_2(y) dy,$$

giving the result

$$\text{independent variables}: \quad E(XY) = E(X) E(Y). \qquad (13.70)$$

We can express the mean and variance of each of the random variables as

$$\mu_x = E(X), \quad \sigma_x^2 = E((X - \mu_x)^2), \qquad (13.71a)$$
$$\mu_y = E(Y), \quad \sigma_y^2 = E((Y - \mu_y)^2). \qquad (13.71b)$$

A consequence of these relations is

$$\sigma_x^2 = E(X^2) - \mu_x^2 = E(X^2) - [E(X)]^2, \qquad (13.71c)$$
$$\sigma_y^2 = E(Y^2) - \mu_y^2 = E(Y^2) - [E(Y)]^2. \qquad (13.71d)$$

Derivation: Normal distributions from uniform distributions

Here we shall take advantage of the concept of independent random variables to generate variables with normal distribution using as input variables that have uniform probability distributions. The latter type of random variables are much easier to generate, whereas the former are more challenging.

Starting with two random variables of uniform distribution x_1, x_2, both in the interval $[0, 1]$, we define the new variables y_1, y_2 through

$$y_1 = \sqrt{-2\ln(x_1)}\cos(2\pi x_2), \quad y_2 = \sqrt{-2\ln(x_1)}\sin(2\pi x_2).$$

From these definitions, we find

$$\frac{y_2}{y_1} = \frac{\sin(2\pi x_2)}{\cos(2\pi x_2)} = \tan(2\pi x_2) \Rightarrow x_2(y_1, y_2) = \frac{1}{2\pi}\tan^{-1}\left(\frac{y_2}{y_1}\right),$$

$$y_1^2 + y_2^2 = -2\ln(x_1)\left[\cos^2(2\pi x_2) + \sin^2(2\pi x_2)\right]$$

$$\Rightarrow \quad x_1(y_1, y_2) = e^{-(y_1^2 + y_2^2)/2}.$$

We can then use the general formula of Eq. (13.58) to relate the two new distributions:

$$p(y_1, y_2)\mathrm{d}y_1\mathrm{d}y_2 = p(x_1, x_2)\left|\frac{\partial(x_1, x_2)}{\partial(y_1, y_2)}\right|\mathrm{d}y_1\mathrm{d}y_2,$$

where the quantity inside the absolute value is the generalization of the partial derivative to the case of a two-variable system [called the "Jacobian", see Eq. (2.33)]:

$$\frac{\partial(x_1, x_2)}{\partial(y_1, y_2)} = \left[\frac{\partial x_1}{\partial y_1}\frac{\partial x_2}{\partial y_2} - \frac{\partial x_2}{\partial y_1}\frac{\partial x_1}{\partial y_2}\right]. \tag{13.72}$$

Using the relations between x_1, x_2, and y_1, y_2 from above, we can calculate the Jacobian to obtain

$$p(y_1, y_2) = \frac{1}{\sqrt{2\pi}}e^{-y_1^2/2}\frac{1}{\sqrt{2\pi}}e^{-y_2^2/2}. \tag{13.73}$$

The last expression is actually two independent normal probability distributions in the variables y_1, y_2. Thus, the process described above is quite efficient: we start with *two* independent random variables of *uniform* distribution, x_1, x_2 and generate *two* independent random variables of *normal* (Gaussian) distribution. No computation is wasted in this process; by "computation" here we refer to the actual cost of computing time to generate the two uniform random values x_1, x_2, which can be quite demanding for high-quality pseudo-random numbers. These topics are discussed in detail in Chapter 14, see Section 14.8.

Example 13.6: Buffon's needle problem and the value of π

We shall take advantage of the notion of independent random variables to prove an interesting relation which is known as "Buffon's problem" [this problem was first solved by Georges-Louis Leclerc, Compte de Buffon].

The problem can be stated as follows:

If a needle of length l is dropped at random onto a surface with parallel stripes of width $W > l$, what is the probability that it intersects one of the lines that separate the stripes?

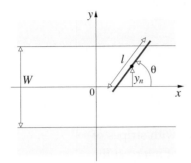

The geometry of the problem is shown in the diagram, with the needle in red and its center marked by a black dot. To describe the placement of the needle we consider two variables, the position of its center along the axis perpendicular to the stripe direction, denoted as y_n, and its orientation relative to the direction parallel to the stripes, denoted by the angle θ. Since the needle is dropped randomly on the surface, the probability p_y of its center being at any point within a stripe is a uniform distribution. Similarly, the probability p_θ of its orientation relative to the direction parallel to the stripes is also a uniform distribution. Placing the origin at the middle of the stripe in which the needle has landed, the range of the variable y_n is $-W/2 \leq y_n \leq W/2$, and the range of the angle θ is $0 \leq \theta \leq \pi$. We therefore have

$$p_y = \frac{1}{W}, \quad \text{for} \quad \frac{W}{2} \leq y_n \leq \frac{W}{2}, \quad \text{and} \quad p_\theta = \frac{1}{\pi}, \quad \text{for} \ 0 \leq \theta \leq \pi.$$

Moreover, the position of the center and the orientation are independent variables. Thus, the probability of finding a given position of the center y_n and a given orientation θ is given by

$$p(y_n, \theta) = p_y p_\theta = \frac{1}{W\pi}, \quad \text{for} \quad \frac{W}{2} \leq y_n \leq \frac{W}{2}, \text{ and } 0 \leq \theta \leq \pi.$$

From the symmetry of the problem, we can focus our attention to the center being on the top half of the stripe and the orientation being between the horizontal (parallel to the stripes) and the vertical (perpendicular to the stripes) directions, and compensate by multiplying by a factor of 2 for each restriction, which leads to

$$p(y_n, \theta) = \frac{4}{W\pi}, \quad \text{for} \ 0 \leq y_n \leq \frac{W}{2}, \text{ and } 0 \leq \theta \leq \frac{\pi}{2}.$$

From the diagram above, we easily see that the condition for the needle to intersect the line between stripes is given by

$$y_n + \frac{l}{2}\sin(\theta) \geq \frac{W}{2} \Rightarrow y_n \geq \frac{1}{2}\left[W - l\sin(\theta)\right].$$

Thus, the total probability P that the needle intersects a line is obtained by summing $p(y_n, \theta)$ over all the allowed values of its two variables, which leads to

$$P = \int_0^{\pi/2} \int_{[W-l\sin(\theta)]/2}^{W/2} p(y_n, \theta) dy_n d\theta = \int_0^{\pi/2} \left(\int_{[W-l\sin(\theta)]/2}^{W/2} \frac{4}{W\pi} dy_n \right) d\theta$$

$$= \frac{4}{W\pi} \int_0^{\pi/2} \frac{l}{2} \sin(\theta) d\theta = \frac{2l}{W\pi} \left[-\cos(\theta) \right]_0^{\pi/2} = \frac{2l}{W\pi}.$$

This result can be used to obtain the value of π as

$$\pi = \frac{2l}{WP},$$

by dropping N needles of length l randomly on a surface with stripes of width $W > l$, and simply finding the fraction of needles that intersect the lines between stripes, since for large N this fraction is equal to P.

13.5   Composite random variables

An important class of random variables are the so-called "composite random variables", that is, variables that are combinations of several other random variables. Here we study some aspects of composite random variables in their simplest version, that is when the composite variable is the sum of two or more random variables. We first introduce the general behavior of composite variables, next we present some examples, and finally we discuss an important theorem that refers to the sum of a large number of independent variables, the so-called "Central Limit Theorem".

As the simplest case, consider a random variable z which is the sum of two other random variables, x and y:

$$z = x + y. \tag{13.74}$$

For the new random variable z, we can calculate the mean value μ_z in terms of the corresponding values for the original random variables μ_x, μ_y, as follows:

$$\mu_z = E(Z) = E(X + Y) = \mu_x + \mu_y. \tag{13.75}$$

If, however, we try to express the variance σ_z^2 in terms of σ_x, σ_y, for the general case, we discover that this is not possible. For this purpose, we need to introduce an important new concept, the "covariance". The variance of z is given by

$$\sigma_z^2 = E((Z - \mu_z)^2) = E(Z^2 - 2Z\mu_z + \mu_z^2) = E(Z^2) - \mu_z^2.$$

The first term in the last expression can be rewritten as

$$E(Z^2) = E(X^2 + 2XY + Y^2) = E(X^2) + 2E(XY) + E(Y^2).$$

From the definition of $E(Z)$ we have

$$[E(Z)]^2 = [E(X) + E(Y)]^2 = [E(X)]^2 + [E(Y)]^2 + 2E(X)E(Y).$$

Subtracting the second expression from the first and using Eqs. (13.71c) and (13.71d), we obtain

$$\sigma_z^2 = \sigma_x^2 + \sigma_y^2 + 2\sigma_{xy}, \tag{13.76}$$

where we have introduced the **covariance** σ_{xy}, defined as

$$\sigma_{xy} = E(XY) - E(X)E(Y). \tag{13.77}$$

For independent variables, using Eq. (13.70) we conclude that the covariance must be equal to zero. This leads to the following relation for independent variables:

$$\sigma_z^2 = \sigma_x^2 + \sigma_y^2. \tag{13.78}$$

In the case of independent variables x and y, an important consequence of the definition of the Eq. (13.74) is that the probability distribution $p_z(Z)$ is the *convolution* of the probabilities $p_x(X), p_y(Y)$. To show this, we note that if we do not care which value of x we have for a given value of z, then we can sum over all possible values of x that give the desired value of z, with the value of y being given by $y = z - x$. Since the variables x and y are independent, we have

$$Y = z - x \Rightarrow p_z(z) = \sum_x p_x(x) p_y(z - x), \tag{13.79a}$$

with the sum on the right-hand side of the last equation being the definition of the convolution of the functions $p_x(x), p_y(y)$. An equivalent expression can be obtained if we do not care which value of y we have for a given value of z, with $x = z - y$:

$$X = z - y \Rightarrow p_z(z) = \sum_y p_x(z - y) p_y(y). \tag{13.79b}$$

Thus, the results of Eqs. (13.75) and (13.78) can also be established by taking advantage of the properties of the convolution and its Fourier transform (see Problem 4).

Example 13.7: Consider two yards, which we will label A and B, and a big tree which sits right in the middle of the fence separating the two yards. The leaves from the tree fall randomly to the yards below, with probabilities p and q, respectively. The values of p and q are not necessarily equal (for example, a slight breeze may lead to $p > q$) but they always add up to unity: $p + q = 1$. Let x be the random variable associated with the fate of leaf 1, which may land either in yard A, in which case we assign the value $x_0 = 0$ to it, or in yard B, in which case we assign the value $x_1 = 1$. Similarly, the random variable y describes the fate of leaf 2, and it can also have values $y_0 = 0$ or $y_1 = 1$, depending on where leaf 2 lands. The random variables x, y take on the values 0 with probability p each, that is

$$p_x(X = 0) = p, \quad p_y(Y = 0) = p,$$

and they take on the values 1 with probability q each, that is

$$p_x(X = 1) = q, \quad p_y(Y = 1) = q.$$

We define a new random variable z as the sum of x and y. From the way we have indexed the possible values of the random variables x, y, we have

$$x_i = i, \quad y_j = j, \quad i, j = 0, 1.$$

The values that the random variable z can take are as follows:

$$z_0 = 0 \rightarrow (X, Y) = (x_0, y_0),$$
$$z_1 = 1 \rightarrow (X, Y) = (x_1, y_0) \text{ or } (x_0, y_1),$$
$$z_2 = 2 \rightarrow (X, Y) = (x_1, y_1).$$

From the way we have indexed the possible values of z we deduce

$$z_k = x_i + y_{k-i}.$$

Regarding the probabilities associated with the three possible values of z, from the above relation we find that

$$p_z(Z_k) = \sum_i p_x(X_i) p_y(Y_{k-i}).$$

This expression is exactly what we have defined as the convolution of the two functions $p_x(x)$ and $p_y(y)$, where we can think of the subscript of the variables X, Y as the integration variable in the general definition of the convolution. From the definition of the probabilities for $p_x(X_i)$ and $p_y(Y_j)$ we then find the following:

$$p_z(0) = p_x(0) p_y(0) = p^2,$$
$$p_z(1) = p_x(1) p_y(0) + p_x(0) p_y(1) = 2pq,$$
$$p_z(2) = p_x(1) p_y(1) = q^2.$$

It is easy to check that

$$\sum_k p_z(Z_k) = p^2 + 2pq + q^2 = (p + q)^2 = 1,$$

that is, $p_z(Z)$ is normalized to unity, for any values of p and q, provided that $p + q = 1$; thus, $p_z(Z)$ is a proper pdf.

We next prove the following theorem.

Theorem: *The probability distribution of the composite variable $z = x_1 + x_2$, where x_1 and x_2 are two independent variables with normal distributions, is a normal distribution with mean equal to the sum of the means and variance equal to the sum of the variances of x_1 and x_2.*

Proof: The probability distributions of the variables x_1, x_2, with means μ_1, μ_2 and variances σ_1, σ_2, respectively, are as follows:

$$f_g(x_1; \mu_1, \sigma_1) = \frac{1}{\sqrt{2\pi}\sigma_1} e^{-(x_1-\mu_1)^2/2\sigma_1^2}, \quad f_g(x_2; \mu_2, \sigma_2) = \frac{1}{\sqrt{2\pi}\sigma_2} e^{-(x_2-\mu_2)^2/2\sigma_2^2}.$$

Since the two variables are independent, the probability of finding any pair of values (x_1, x_2) is given by the probability distribution:

$$g(x_1, x_2) = f_g(x_1; \mu_1, \sigma_1) \, f_g(x_2; \mu_2, \sigma_2) = \frac{1}{2\pi\sigma_1\sigma_2} e^{-(x_1-\mu_1)^2/2\sigma_1^2} e^{-(x_2-\mu_1)^2/2\sigma_2^2}.$$

For the probability distribution $f(z)$ of the composite variable $z = x_1 + x_2$, since for any value of x_1 we have $x_2 = z - x_1$, we obtain

$$f(z) = \int_{-\infty}^{\infty} g(x_1, z - x_1) dx_1 = \frac{1}{2\pi\sigma_1\sigma_2} \int_{-\infty}^{\infty} e^{-(x_1-\mu_1)^2/2\sigma_1^2} e^{-(z-x_1-\mu_2)^2/2\sigma_2^2} dx_1.$$

The exponential in the variables x_1 and z, after a few steps of algebra, becomes

$$\exp\left[-\frac{(x_1 - \mu_1)^2}{2\sigma_1^2} - \frac{(z - x_1 - \mu_1)^2}{2\sigma_2^2}\right] = \exp\left[-\frac{(x_1 - \mu')^2}{2\sigma'^2} - \frac{(z - \bar\mu)^2}{2\bar\sigma^2}\right],$$

(13.80)

where the new parameters μ', σ', and $\bar\mu, \bar\sigma$ are defined by

$$\mu' = \frac{(z + \mu_2)\sigma_1^2 + \mu_1\sigma_2^2}{\sigma_1^2 + \sigma_2^2}, \quad \sigma' = \frac{\sigma_1\sigma_2}{\sqrt{\sigma_1^2 + \sigma_2^2}}, \quad \bar\mu = \mu_1 + \mu_2, \quad \bar\sigma = \sqrt{\sigma_1^2 + \sigma_2^2}.$$

(13.81)

With these parameters, the dependence on the variable x_1 becomes a normalized Gaussian integral with mean μ' and variance σ'^2, which integrates to 1; the remaining part is

$$f(z) = \frac{1}{\sqrt{2\pi}\bar\sigma}e^{-(z-\bar\mu)^2/2\bar\sigma^2} = f_g(z; \bar\mu, \bar\sigma),$$

(13.82)

that is, a normal distribution in the variable z with mean $\bar\mu$ and variance $\bar\sigma^2$. This completes the proof of the theorem. ∎

Example 13.8: Consider the random variables x, y, z, all of which have standard normal distributions, and are independent. We define the new variables: $u = 3x + 4y$, and $v = 2x + z$. The two new variables are *not* independent, because they both depend on the same variable x. We will consider three different questions about the behavior of the variables u and v.

(i) Find the probability that $u \geq u_0$, where u_0 is some fixed value.
First, from the statement of the problem we have

$$\mu_x = \mu_y = \mu_z = 0, \quad \sigma_x = \sigma_y = \sigma_z = 1.$$

We define the variables $x' = 3x$ and $y' = 4y$, which, according to our earlier results, Eqs. (13.41), (13.42), have Gaussian distributions with means $\mu_{x'} = \mu_{y'} = 0$ and variances $\sigma_{x'}^2 = (3\sigma_x)^2 = 9$ and $\sigma_{y'}^2 = (4\sigma_y)^2 = 16$, respectively. Then, from the results derived earlier, the variable $u = x' + y'$ will have a Gaussian distribution with mean and variance:

$$\mu_u = \mu_{x'} + \mu_{y'} = 0, \quad \sigma_u^2 = \sigma_{x'}^2 + \sigma_{y'}^2 = 25.$$

We next express u_0 in terms of the mean μ_u and variance σ_u as: $u_0 = \mu_u + m\sigma_u$, where m is a real value: The probability that the variable u has values greater than or equal to u_0 is given by

$$P_u(u \geq u_0) = \int_{\mu_u + m\sigma_u}^{\infty} \frac{1}{\sqrt{2\pi}\sigma_u}e^{-(u-\mu_u)^2/2\sigma_u^2}du = 1 - \Phi(m).$$

For example, if we choose $u_0 = 5 = 0 + 1 \times 5$, that is, $m = 1$, we find from the table of values of $\Phi(w)$ (see Appendix B) that $P_u(u \geq 5) = 1 - \Phi(1) = 0.158655$.

(ii) Find the covariance σ_{uv} of the variables u and v.
We know that the two variables are not independent, because they both contain the variable x, so we expect a covariance $\sigma_{uv} \neq 0$. To calculate the

covariance from Eq. (13.77), we need to find the expectation values $E(UV)$ and $E(U)E(V)$:

$$E(U) = E(3X + 4Y) = 3E(X) + 4E(Y) = 0,$$
$$E(V) = E(2X + Z) = 2E(X) + E(Z) = 0,$$
$$E(UV) = E(6X^2 + 8XY + 3XZ + 4YZ)$$
$$= 6E(X^2) + 8E(XY) + 3E(XZ) + 4E(YZ).$$

Now each pair of variables are independent, so $E(XY) = E(X)E(Y) = 0$, and similarly for the other two pairs, XZ and YZ. To compute $E(X^2)$ we use Eq. (13.12) and obtain $E(X^2) = \sigma_x^2 + \mu_x^2 = 1$. Combining these results we find

$$\sigma_{uv} = E(UV) - E(U)E(V) = 6.$$

(iii) Find the variance σ_w^2 of the variable w which is defined as: $w = u + v$. We can use the fact that both u and v are variables with normal distributions, as proven earlier. We have also found that the variance of u is $\sigma_u^2 = 25$; by similar steps we can calculate the variance of v which turns out to be $\sigma_v^2 = 2^2 + 1^2 = 5$. Now we can use the general formula for the variance of a random variable which is the sum of two other random variables, Eq. (13.76), to obtain

$$\sigma_w^2 = \sigma_u^2 + \sigma_v^2 + 2\sigma_{uv} = 25 + 5 + 2 \times 6 = 42.$$

Alternatively, we have $w = u + v = 5x + 4y + z$ and since the variables x, y, z are independent, with zero covariances, and the mean of each is 0 and the variance 1, the variance of w is given by

$$\sigma_w^2 = (5\sigma_x)^2 + (4\sigma_y)^2 + \sigma_z^2 = 42.$$

13.5.1   Central Limit Theorem

An important theorem in statistics is the "Central Limit Theorem", which deals with the limit of a composite variable consisting of large number of independent variables. In this limit, the theorem asserts that the composite variable obeys a normal distribution, independent of what the distribution of the original variables is.

Theorem: *Consider the variable z which is defined as*
$$z = x_1 + x_2 + \cdots + x_n,$$

where $x_1, x_2, \ldots, x_n$ are independent random variables, each one drawn from a probability distribution with mean μ and variance σ^2. In the limit $n \to \infty$, the variable z obeys a normal probability distribution with mean $\bar{\mu} = n\mu$ and variance $\bar{\sigma}^2 = n\sigma^2$.

Proof: The theorem applies for an arbitrary probability distribution of each random variable $x_i, i = 1, \ldots, n$. We will prove it here for the simplest distribution, namely a uniform probability distribution with $x_i \in [-1/2, 1/2]$, which has mean $\mu = 0$ and variance $\sigma^2 = 1/12$, as discussed in Section 13.2.1. In other words, the probability distribution from which every x_i variable is drawn is the function $g_0(x)$ defined in Eq. (13.49).

We define the following variables:

$$y_j = y_{j-1} + x_{j+1}, \quad j = 1, \ldots, n-1, \quad y_0 = x_1,$$

from which it follows that $y_{n-1} = z$. For the first of those variables we have

$$y_1 = y_0 + x_2 = x_1 + x_2 \Rightarrow x_2 = y_1 - x_1,$$

and therefore the probability distribution for this variable is given by

$$g_1(y_1) = \int_{-\infty}^{\infty} g_0(x_1) g_0(x_2) dx_1 = \int_{-\infty}^{\infty} g_0(x_1) g_0(y_1 - x_1) dx_1.$$

This function was obtained earlier, see Eq. (13.51). Similarly, for y_2,

$$y_2 = y_1 + x_3 \Rightarrow x_3 = y_2 - y_1,$$

and its probability distribution is given by

$$g_2(y_2) = \int_{-\infty}^{\infty} g_1(y_1) g_0(x_3) dy_1 = \int_{-\infty}^{\infty} g_1(y_1) g_0(y_2 - y_1) dy_1,$$

which is also a function obtained earlier, see Eq. (13.53). We note that each probability distribution $g_j(y_j)$ is a convolution of the probability distribution $g_0(x)$ with the previous one $g_{j-1}(y_{j-1})$. Using the Fourier transforms of these probability distributions, and the fact that the Fourier transform of the convolution is the product of the Fourier transforms of the two functions appearing in the convolution integral, Eq. (7.28), we obtain

$$\hat{g}_j(k) = \hat{g}_0(k)\hat{g}_{j-1}(k) = \hat{g}_0(k)\hat{g}_0(k)\hat{g}_{j-2}(k) = \cdots = [\hat{g}_0(k)]^{j+1},$$

since for the Fourier transform of the first distribution we have

$$\hat{g}_1(k) = \hat{g}_0(k)\hat{g}_0(k) = [\hat{g}_0(k)]^2.$$

Therefore the FT, $\hat{f}(k)$, of the probability distribution of the variable $z = y_{n-1}$ is given by

$$\hat{f}(k) = \hat{g}_{n-1}(k) = [\hat{g}_0(k)]^n.$$

The FT $\hat{g}_0(k)$ of the function $g_0(x)$ was calculated in Example 7.4 of Chapter 7, see Eq. (7.34). Using that result with $a = 1/2$ we obtain

$$\hat{f}(k) = \left[\frac{\sin(k/2)}{k/2}\right]^n.$$

The expression inside the square brackets is a function of k that decays as k^{-1}, its value at $k = 0$ is 1, and it is smaller than 1 in absolute value for all $k \neq 0$. Therefore, raising this quantity to a large positive power n implies that only the values near $k = 0$ are relevant and the rest are vanishingly small. Using the Taylor expansion of $\sin(k/2)$ near $k = 0$ we obtain

$$\hat{f}(k) \approx \left[1 - \frac{1}{6}\left(\frac{k}{2}\right)^2\right]^n \approx 1 - \frac{n}{24}k^2 \approx e^{-k^2(n/24)} = e^{-k^2\bar{\sigma}^2/2},$$

where we have defined $\bar{\sigma}^2 = n/12$. Recall that we calculated in Section 7.3.1 the FT of a normalized Gaussian, the result of which was the same as the last expression above as a function of k. By applying the Inverse Fourier transform we find the probability distribution of the variable z, namely

$$f(z) = \frac{1}{\sqrt{2\pi}\bar{\sigma}}e^{-z^2/2\bar{\sigma}^2}.$$

Thus, we have proven that the variable z obeys a normal distribution with mean $\bar{\mu} = n\mu = 0$ and variance $\bar{\sigma}^2 = n/12 = n\sigma^2$, where $\sigma^2 = 1/12$ is the variance of the uniform probability distribution $g_0(x)$ of the independent variables x_i; this completes the proof of the theorem. ■

13.6 Conditional probabilities

Consider two sets of events, one denoted by A, the other by B. The events common to both A and B comprise the intersection $A \cap B$. The events that can be either in A or in B comprise the union $A \cup B$. If all events belong to a superset S that includes both A and B and possibly other events that do not belong to either A or B, then the following relations will hold for the probabilities $P(A), P(B), P(S)$, associated with the above sets:

$$P(S) = 1, \quad 0 \le P(A) \le 1, \quad 0 \le P(B) \le 1, \tag{13.83}$$

$$P(A \cup B) = P(A) + P(B) - P(A \cap B), \tag{13.84}$$

where the last relation is justified by the fact that the events that belong to both A and B are counted twice in $P(A) + P(B)$. The probability $P(A \cap B)$ is also known as the **joint probability**.

We define as **mutually exclusive** the sets of events that have no common elements, that is, $A \cap B = 0$; in that case we have

$$\text{mutually exclusive}: P(A \cap B) = 0 \Rightarrow P(A \cup B) = P(A) + P(B). \tag{13.85}$$

We further define the complement of A, denoted by A^c, as the set of all events that do not belong to A, and similarly for the complement of B, denoted by B^c; the complements satisfy the relations:

$$A^c = S - A, \quad A \cup A^c = S, \quad P(A^c) = 1 - P(A). \tag{13.86}$$

We define the **conditional probability** $P(B|A)$ as the probability that an event in B will occur under the condition an event in A has occurred:

$$P(B|A) = \frac{P(A \cap B)}{P(A)}. \tag{13.87}$$

Conditional probabilities play a very important role in probability theory and in many applications. The conditional probability refers to the occurrence of an event B, under the assumption that we are also in the space of event A, in which case we are not concerned (or we do not know) the probability of A occurring; the joint probability characterizes the occurrence of an event B and *simultaneously* the occurrence of an event A, in which case the probability of each event occurring does matter.

We call the events in A and B **independent** if the probability of an event in A occurring has no connection to the probability of an event in B occurring. This is expressed by the relation:

$$\text{independent events}: P(A \cap B) = P(A)P(B). \tag{13.88}$$

In the case of independent events, when Eq. (13.88) holds, the conditional probabilities become

$$P(A|B) = \frac{P(A \cap B)}{P(B)} = P(A), \quad P(B|A) = \frac{P(B \cap A)}{P(A)} = P(B). \qquad (13.89)$$

This simply says that the probability of an event in A occurring under the condition that an event in B has occurred, is the same as the probability of an event in A occurring, $P(A)$, since this has no connection to an event in B (the same holds with the roles of A and B interchanged).

Note that the following relationship always holds:

$$P(A \cap B) = P(B \cap A), \qquad (13.90)$$

whereas for the conditional probabilities in general we have

$$P(A|B) \neq P(B|A), \qquad (13.91)$$

and the conditional probabilities $P(A|B)$, $P(B|A)$ are equal only if $P(A) = P(B)$.

Example 13.9: Consider two fair dice that are thrown simultaneously. We define two events A and B as follows:

- event A: the sum of values of the two dice is equal to 7;

- event B: only one of the two dice has the value 2.

Our goal is to calculate the conditional probabilities $P(A|B)$ and $P(B|A)$. We can construct the table of all possible outcomes and identify the events A and B in this table, as shown below (the line of values $i = 1 - 6$ represents the outcome of the first die, the column of values $j = 1 - 6$ represents the values of the second die):

$j \mid i$	1	2	3	4	5	6
1		B				A
2	B		B	B	A, B	B
3		B		A		
4		B	A			
5		A, B				
6	A	B				

The total number of outcomes $6 \times 6 = 36$ is the event space S. The boxes labeled with A represent the set of values corresponding to event A and the boxes labeled with B the set of values corresponding to event B. There are several boxes carrying both labels, and this is the intersection of the two sets, $A \cap B$; all the labeled boxes comprise the union of A and B, $A \cup B$. From this table, it is straightforward to calculate the probabilities:

$$P(A) = 6 \times \frac{1}{36} = \frac{1}{6}, \quad P(B) = 10 \times \frac{1}{36} = \frac{10}{36}.$$

Indeed, the probability of event A is the total number of boxes containing the label A, which is 6, divided by the total number of possible outcomes which is 36; similarly for the probability of event B. We also have

$$P(A \cup B) = P(A) + P(B) - P(A \cap B) = \frac{6}{36} + \frac{10}{36} - \frac{2}{36} = \frac{14}{36}.$$

Indeed, the probability of either event occurring is the probability of the event A occurring plus the probability of event B occurring minus the probability of the intersection which is counted twice. There are 14 occupied entries in the entire table, two of which have two labels (representing the intersection $A \cap B$).

Finally, the conditional probabilities are obtained as follows: the probability that event A will occur (sum of values is 7) under the condition that event B has occurred (only one die has value 2) is

$$P(A|B) = \frac{P(A \cap B)}{P(B)} = \frac{2/36}{10/36} = \frac{2}{10}.$$

The probability that event B will occur (only one die has value 2) under the condition that event A has occurred (sum of values is 7) is

$$P(B|A) = \frac{P(B \cap A)}{P(A)} = \frac{2/36}{6/36} = \frac{2}{6}.$$

These conditional probabilities are readily obtained from the table of events shown earlier in the present example: out of the 10 boxes containing a label B, that is, all the cases that only one die has value 2, there are only two boxes that also contain the label A, that is, the sum of values of the two dice is 7: thus, the probability that the sum of values is 7 under the condition that only of the dice has value 2 is 2/10, which is precisely the conditional probability $P(A|B)$. We can justify the value obtained for the conditional probability $P(B|A)$ by a similar argument.

13.6.1 Bayes' theorem

An important application of the concept of conditional probabilities is the case when some of the marginal probabilities can be easily established or measured while others are difficult to obtain, yet they are more interesting. These situations can be handled through an ingenious scheme, based on Bayes' theorem, which we discuss next.

Theorem: *The conditional probability $P(A|B)$ for an event A to occur given that event B has occurred, is equal to the conditional probability $P(B|A)$ for the event B to occur given that event A has occurred, divided by the probability of event B occurring, $P(B)$, and multiplied by the probability of event A occurring, $P(A)$:*

$$P(A|B) = \frac{P(B|A)}{P(B)} P(A). \tag{13.92}$$

Proof: Based on the definition or conditional probabilities, Eq. (13.87), we find that the probability of an event which belongs to both A and B to occur, that is, the joint probability $P(A \cap B)$, can be obtained from the conditional probabilities:

$$P(A \cap B) = P(B|A)P(A). \tag{13.93}$$

In other words, the joint probability $P(A \cap B)$ is equal to the probability of an event in B occurring under the condition that an event in A has occurred, times the probability of an event in A occurring. The equivalent expression holds with the roles of A and B interchanged, for the joint probability $P(B \cap A)$:

$$P(B \cap A) = P(A|B)P(B). \tag{13.94}$$

But the two joint probabilities must be equal, $P(A \cap B) = P(B \cap A)$, giving

$$P(A|B)P(B) = P(B|A)P(A) \Rightarrow P(A|B) = \frac{P(B|A)}{P(B)}P(A),$$

which proves Bayes' theorem.   ∎

This is an important result, because we typically can obtain the probability $P(A)$ easily, which is called the **prior**, but we are really interested in the conditional probability $P(A|B)$, called the **posterior**, which is often hard to calculate. We can then obtain the posterior, by calculating another conditional probability $P(B|A)$, which is presumably easier to obtain or known, and weigh it by $1/P(B)$, also presumably known. The ratio $P(B|A)/P(B)$ is called the **support**.

Example 13.10: A television game played between a host and a player consists of finding a prize which is hidden behind one of three doors. The rules are as follows:

1. The player chooses one of the three doors without opening it.

2. The host opens one of the two remaining doors, which does not have the prize behind it.

3. The player has a chance to change her choice of door or stick with the original choice.

What is the right strategy for the player in order to maximize her chances of finding the prize: switch choice of doors or stick to the original choice?
<u>Answer:</u> We label the door that the player chooses initially as door 1 and the other two doors as 2 and 3. We have to calculate probabilities from the player's perspective, that is, taking into account only the information that the player has. We define the following events:
A_i: the prize is behind door $i = 1, 2, 3$.
B_i: the host opens door $i = 1, 2, 3$.
From the player's perspective, the prize can be behind any one of the three doors, hence $P(A_1) = P(A_2) = P(A_3) = 1/3$; these are the priors for the present case. Also, the host will not open door 1 since the player picked it first, but can open door 2 or 3 with equal probability as far as the player is concerned, hence $P(B_1) = 0, P(B_2) = P(B_3) = 1/2$. We can also calculate the following conditional probabilities:

if the prize is behind door $1 : P(B_2|A_1) = 1/2, \quad P(B_3|A_1) = 1/2,$

since the host is equally likely (from the player's perspective) to open door 2 or 3. Also

if the prize is behind door $2 : P(B_2|A_2) = 0, \quad P(B_3|A_2) = 1,$

if the prize is behind door $3 : P(B_2|A_3) = 1, \quad P(B_3|A_3) = 0,$

since the host will not open the door with the prize. In all cases, since $P(B_1) = 0$, we have

$$P(B_1 \cap A_i) = P(A_i|B_1)P(B_1) = 0 \Rightarrow P(A_i \cap B_1) = 0, \quad i = 1,2,3.$$

With this information and the marginal probabilities for A_i and B_i, we can now calculate the table of joint probabilities $P(A_i \cap B_j) = P(B_j|A_i)P(A_i)$ for $i = 1,2,3$ and $j = 2,3$:

	B_1	B_2	B_3	$P(A_i)$
A_1	0	1/6	1/6	1/3
A_2	0	0	1/3	1/3
A_3	0	1/3	0	1/3
$P(B_i)$	0	1/2	1/2	

The entries in the 3×3 table formed by B_1, B_2, B_3 and A_1, A_2, A_3, are the joint probabilities $P(A_i \cap B_j)$; the entries in the far-right column are the marginal probabilities $P(A_i)$ and the entries in the bottom row are the marginal probabilities $P(B_i)$. From this table, we can then construct the posteriors, which is what the player actually wants to know; these are given by the following conditional probabilities:

$$P(A_1|B_2) = \frac{P(A_1 \cap B_2)}{P(B_2)} = \frac{1}{3}, \quad P(A_1|B_3) = \frac{P(A_1 \cap B_3)}{P(B_3)} = \frac{1}{3},$$

$$P(A_2|B_3) = \frac{P(A_2 \cap B_3)}{P(B_3)} = \frac{2}{3}, \quad P(A_3|B_2) = \frac{P(A_3 \cap B_2)}{P(B_2)} = \frac{2}{3}.$$

Thus, the probability that the prize is behind door 1 (the original choice) under the condition that the host opens door 2 or 3 is 1/3 in each case, but the probability that the prize is behind door 2 if the host opens door 3 or behind door 3 if the host opens door 2 is 2/3 in each case. Therefore, the player should always switch choice of doors to maximize her chances.

Equivalently, we can solve the problem by applying Bayes' theorem. We know that the marginal probabilities from the player's perspective are $P(A_1) = P(A_2) = P(A_3) = 1/3$, since the player does not know behind which door the prize is hidden; these are the priors. The player also figures out that $P(B_2) = P(B_3) = 1/2$, since the host only opens one of the doors labeled 2 and 3 under any circumstances and with equal probability from the player's perspective (according to the rules of the game, the host never opens door 1, thus, $P(B_1) = 0$). The probabilities that the player wants to know are the conditional probabilities $P(A_1|B_2)$ and $P(A_3|B_2)$ if the host opens door 2, or $P(A_1|B_3)$ and $P(A_2|B_3)$ if the host opens door 3. We can calculate them by using Bayes' theorem. The player figures out that if the prize is behind door 1, then the host will open door 2 or 3 with equal probability, that is, the player calculates $P(B_2|A_1) = P(B_3|A_1) = 1/2$. Therefore,

$$P(A_1|B_2) = \frac{P(B_2|A_1)P(A_1)}{P(B_2)} = \frac{(1/2)(1/3)}{(1/2)} = \frac{1}{3},$$

$$P(A_1|B_3) = \frac{P(B_3|A_1)P(A_1)}{P(B_3)} = \frac{(1/2)(1/3)}{(1/2)} = \frac{1}{3}.$$

The player also calculates that if the prize is behind door 2, the host will open door 2 with probability $P(B_2|A_2) = 0$ and door 3 with probability $P(B_3|A_2) = 1$. Therefore,

$$P(A_2|B_3) = \frac{P(B_3|A_2)P(A_2)}{P(B_3)} = \frac{1(1/3)}{(1/2)} = \frac{2}{3}.$$

Similarly, if the prize is behind door 3, the host will open door 2 with probability $P(B_2|A_3) = 1$ and door 3 with probability $P(B_3|A_3) = 0$. Therefore,

$$P(A_3|B_2) = \frac{P(B_2|A_3)P(A_3)}{P(B_2)} = \frac{1(1/3)}{(1/2)} = \frac{2}{3}.$$

From these results, the player figures out that her chances of finding the prize increase from $1/3$ if she sticks to the original choice to $2/3$ if she switches her choice of doors.

13.6.2   The Fokker–Planck equation

The concept of conditional probabilities plays a central role in understanding physical processes like Brownian motion; this type of motion describes the evolution of systems with dynamics governed by random forces. The effect of the random forces is captured by introducing *probabilistic* rather than *deterministic* dynamics for the evolution of the system. The probability distribution involved in this description, which depends on the time variable t and the position variable x, obeys a partial differential equation which is known as the "Fokker–Planck equation". We derive this important equation next[1] (the mathematics behind this derivation is an elegant combination of conditional probabilities, the Taylor series expansion and integration by parts).

As a first step we introduce the so-called "Smoluchowski equation" which involves the conditional probability that a particle at position x_0 at the moment $t = 0$ moves to another position x after a time $t + \Delta t$ (Δt is a small time interval). Denoting this conditional probability by as $P(x, t + \Delta t|x_0, 0)$, we find the Smoluchowski equation

$$P(x, t + \Delta t|x_0, 0) = \int P(x, \Delta t|y, 0)P(y, t|x_0, 0)\mathrm{d}y. \tag{13.95}$$

This equation can be justified from simple physical arguments: The step $x_0 \rightarrow x$ can be thought of as occurring through many possible intermediate two-step events, namely $x_0 \rightarrow y$ followed by $y \rightarrow x$. Since these are independent events each happening with the corresponding conditional probability, the probability of one such two-step process is given by

$$P(x, t + \Delta t|y, t)P(y, t|x_0, 0).$$

[1] The derivation presented here follows closely the classic work of M.C. Wang and G.E. Uhlenbeck, *Reviews of Modern Physics*, **17**, pp. 323–342 (1945).

This is illustrated in Fig. 13.7.

We observe that due to translational time invariance, that is, the physical process is independent of the time we start the clock and only depends on the time elapsed, we have

$$P(x, t + \Delta t | y, t) = P(x, \Delta t | y, 0).$$

The Smoluchowski equation then follows from using the last two expressions and the fact that we have to sum over all possible intermediate positions y, which is expressed with the integration over the variable y.

To derive the Fokker–Planck equation we begin by considering the integral

$$\int f(x) \frac{\partial}{\partial t} [P(x, t | x_0, 0)] \, dx, \tag{13.96}$$

where $f(x)$ is an arbitrary function of x such that $f(x) \to 0$ sufficiently fast for $|x| \to \infty$. We then write the time derivative inside the integral of Eq. (13.96) as the difference between two terms at the time moments $t + \Delta t$ and t, divided by Δt, in the limit $\Delta t \to 0$, namely,

$$\int f(x) \frac{\partial}{\partial t} [P(x, t | x_0, 0)] \, dx$$
$$= \lim_{\Delta t \to 0} \frac{1}{\Delta t} \int f(x) [P(x, t + \Delta t | x_0, 0) - P(x, t | x_0, 0)] \, dx.$$

We use the Smoluchowski equation, Eq. (13.95), to replace the first conditional probability in the square bracket of the last expression to arrive at:

$$\int f(x) \frac{\partial}{\partial t} [P(x, t | x_0, 0)] \, dx$$
$$= \lim_{\Delta t \to 0} \frac{1}{\Delta t} \int f(x) \left[\int P(x, \Delta t | y, 0) P(y, t | x_0, 0) dy - P(x, t | x_0, 0) \right] dx$$
$$= \lim_{\Delta t \to 0} \frac{1}{\Delta t} \left[\int \left(\int f(x) P(x, \Delta t | y, 0) dx \right) P(y, t | x_0, 0) dy - \int f(x) P(x, t | x_0, 0) dx \right].$$
$$\tag{13.97}$$

The next step in the derivation is to introduce the Taylor expansion of $f(x)$ around y, namely,

$$f(x) \approx f(y) + (x - y) f'(y) + \frac{1}{2} (x - y)^2 f''(y),$$

where $f'(y)$ and $f''(y)$ are the first and second derivatives, respectively, of $f(x)$, evaluated at $x = y$. Replacing this expression in the first integral over x of Eq. (13.97) we obtain

$$\int f(x) P(x, \Delta t | y, 0) dx = f(y) \int P(x, \Delta t | y, 0) dx$$
$$+ f'(y) \int (x - y) P(x, \Delta t | y, 0) dx$$
$$+ \frac{1}{2} f''(y) \int (x - y)^2 P(x, \Delta t | y, 0) dx$$
$$= f(y) + f'(y) \int (x - y) P(x, \Delta t | y, 0) dx$$
$$+ \frac{1}{2} f''(y) \int (x - y)^2 P(x, \Delta t | y, 0) dx,$$

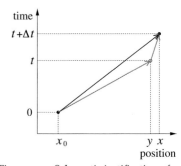

Figure 13.7: Schematic justification of the Smoluchowski equation, Eq. (13.95): the probability for making the step $x_0 \to x$ can be considered as the sum of the probabilities of all possible intermediate, independent steps from $x_0 \to y$ followed by $y \to x$.

where in the last step we have taken advantage of the fact that the conditional probability distribution $P(x, \Delta t|x_0, 0)$ is properly normalized over the variable x, namely,

$$\int P(x, \Delta t|y, 0)dx = 1.$$

We next define the following two functions:

$$G_1(y) \equiv \lim_{\Delta t \to 0} \left[\frac{1}{\Delta t} \int (x - y)P(x, \Delta t|y, 0)dx \right], \qquad (13.98a)$$

$$G_2(y) \equiv \lim_{\Delta t \to 0} \left[\frac{1}{\Delta t} \int (x - y)^2 P(x, \Delta t|y, 0)dx \right]. \qquad (13.98b)$$

We note that $G_1(y)$ and $G_2(y)$ represent the time average of the first and second moments, respectively, of the probability distribution $P(x, \Delta t|x_0, 0)$ with respect to the variable x, in the limit $\Delta t \to 0$. Substituting the above results in the right-hand side of Eq. (13.97) we find

$$\int f(x)\frac{\partial}{\partial t}[P(x, t|x_0, 0)]\,dx = \int P(y, t|x_0, 0)\left[f'(y)G_1(y) + \frac{1}{2}f''(y)G_2(y)\right]dy.$$

Using integration by parts and taking into consideration the fact that $f(x) \to 0$ for $|x| \to \infty$, the right-hand side of the above equation is replaced by the expression

$$\int f(y) \left\{ -\frac{\partial}{\partial y}[P(y, t|x_0, 0)G_1(y)] + \frac{1}{2}\frac{\partial^2}{\partial y^2}[P(y, t|x_0, 0)G_2(y)] \right\} dy.$$

We can switch the name of the variable of integration on the right-hand side from y to x, and bring all terms to the same side of the equation, which leads to the result that the integral

$$\int f(x) \left\{ \frac{\partial P(x, t|x_0, 0)}{\partial t} + \frac{\partial [P(x, t|x_0, 0)G_1(x)]}{\partial x} - \frac{1}{2}\frac{\partial^2 [P(x, t|x_0, 0)G_2(x)]}{\partial x^2} \right\} dx$$

must be zero. Finally, noting that this result must be valid for an *arbitrary* function $f(x)$, we conclude that

$$\frac{\partial P(x, t|x_0, 0)}{\partial t} + \frac{\partial [P(x, t|x_0, 0)G_1(x)]}{\partial x} - \frac{1}{2}\frac{\partial^2 [P(x, t|x_0, 0)G_2(x)]}{\partial x^2} = 0. \quad (13.99)$$

This is the celebrated Fokker–Planck equation.

An interesting case is that of the conditional probability $P(x, \Delta t|x_0, 0)$ being "translationally invariant", that is, this probability does not depend on the specific values of x and x_0 but only on their difference $(x - x_0)$. In this case $P(x, \Delta t|x_0, 0)$ takes the form

$$P(x, \Delta t|x_0, 0) = P((x - x_0), \Delta t),$$

and the integral involved in the definition of $G_1(y)$, Eq. (13.98a), becomes

$$\int (x - x_0)P((x - x_0), \Delta t)dx.$$

If $P((x - x_0), \Delta t)$ is an even function of $(x - x_0)$ this integral vanishes, leading to $G_1(y) = 0$. Moreover, in this case $G_2(y)$ becomes a constant, $G_2(y) = 2D$, and the Fokker–Planck equation reduces to

$$\frac{\partial P((x - x_0), t)}{\partial t} = D\frac{\partial^2 [P((x - x_0), t)]}{\partial x^2}. \qquad (13.100)$$

This equation is equivalent to the diffusion or heat equation, Eq. (8.15), whose solution is a normalized Gaussian,

$$P((x - x_0), t) = \frac{1}{\sqrt{4\pi Dt}} e^{-(x-x_0)^2/4Dt},$$

as was discussed in detail in Chapter 8 (see Section 8.3, Example 8.1).

13.7   Application: Hypothesis testing

A problem that arises in many situations, including engineering, social, medical, economic and other contexts, is the following: how can we choose between two events, A_1 and $A_2 = A_1^c$, having at our disposal only a finite number of N sampling outcomes. For large N, the distributions for each event separately will be normal:

$$g_1(x) = \frac{\alpha_1}{\sqrt{2\pi}\sigma_1} e^{-(x-\mu_1)^2/2\sigma_1^2}, \quad g_2(x) = \frac{\alpha_2}{\sqrt{2\pi}\sigma_2} e^{-(x-\mu_2)^2/2\sigma_2^2}, \qquad (13.101)$$

but the fractions α_1 and α_2 of each type of event will not necessarily be equal to each other, although their sum must add up to unity:

$$\alpha_1 + \alpha_2 = 1.$$

The two choices will have different means μ_1, μ_2 and different standard deviations σ_1, σ_2. We define the average of the two means, κ, which is called the **decision boundary**,

$$\kappa = (\mu_1 + \mu_2)/2.$$

We will assume without loss of generality that

$$\mu_2 < \kappa < \mu_1,$$

since this simply means that we assign the index "2" to the event with lower mean. The usual strategy is to use κ as a criterion for deciding: if most samples give a value smaller than κ, we conclude that we are dealing with event A_2; because the samples indicate that we are below the decision boundary and closer to the mean of event A_2 which is therefore more likely. Conversely, if most samples give a value larger than κ we assume that we are dealing with event A_1. This is referred to as **hypothesis testing**. We emphasize that this strategy of defining the decision boundary as the average of μ_1 and μ_2 is not unique. Other considerations may dictate a different choice of its value, so that one type of errors is minimized (the types of errors are discussed below). Figuring out the proper choice of the decision boundary can be complicated, and is the subject of "decision theory", a subject beyond the scope of the present treatment.

An important issue is how to define a quantitative measure of the errors involved in making the decision between the two choices. We analyze this problem next. The source of the errors is related to making decisions based on sampling from an overall distribution that is the sum of two underlying distributions. This is illustrated in Fig. 13.8, for two different choices of the means and variances of the two distributions: in the first choice, the two underlying

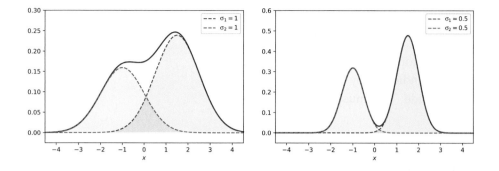

Figure 13.8: Illustration of hypothesis testing: The blue line shows the total distribution. The underlying Gaussian distributions for events A_1 (red dashed curve) and A_2 (green dashed curve) can have large overlap of the Gaussian tails near $x = 0$ as in the left panel, or they may have essentially zero overlap of the tails as in the right panel. The parameter values for the Gaussian of Eq. (13.101) are: $\alpha_1 = 0.6, \mu_1 = 1.5$ and $\alpha_2 = 0.4, \mu_2 = -1$, in both cases, with only the values of σ_1 and σ_2 differing, which are both equal to 1 in the first case and to 0.5 in the second case. Note the different scale of values on the vertical axis for the two cases.

distributions exhibit large overlap of the tails of the two Gaussians in the region close to κ; in the second choice of parameters, the overlap between the two underlying distributions is negligible. The amount of overlap is a measure of the error, because it corresponds to getting sampling values that are above or below the critical value κ but come from the "wrong" Gaussian distribution. Specifically, sampling values *larger* than κ can be obtained from the tail of the distribution for A_2, whose mean is $\mu_2 < \kappa$, which according to our criterion would count as corresponding to the event A_1; these are referred to as **false positives** or **false acceptances**. Conversely, sampling values *smaller* than κ can be obtained from the tail of the distribution for A_1, whose mean is $\mu_1 > \kappa$, which according to our criterion would count as corresponding to the event A_2; these are referred to as **false negatives** or **false rejections**. At each value of the variable x the probability of a false call represent the difference between the total probability and the underlying probability of event A_1, to the right of the critical value κ, or of event A_2, to the left of κ. The total probability of false calls is the integral over these two tails that lead to wrong calls. The false positives and false negatives contributions to the error, denoted by $\mathcal{E}^{(+)}$ and $\mathcal{E}^{(-)}$, respectively, are given by

$$\mathcal{E}^{(+)} = \int_{\kappa}^{\infty} g_2(x) dx, \quad \mathcal{E}^{(-)} = \int_{-\infty}^{\kappa} g_1(x) dx,$$

with the total probability of error given by the sum of these two expressions. As discussed in Section 13.2.4, we can rescale and shift the two Gaussians so that we can obtain the values of the integrals from the tabulated values of $\Phi(w)$. We illustrate these notions in the next example.

Example 13.11: Drawing gold and silver coins from two boxes

Here we work out in detail a problem which is at the core of hypothesis testing. Our example is a variation on a problem first discussed by Jacob Bernoulli in his book *Ars Conjectandi*, published posthumously in 1713 (for more information see the book by Diaconis and Skyrms mentioned in Further Reading):

Consider two boxes containing gold and silver coins: in the first box the gold/silver coin ratio is 70%/30%, whereas in the second box the ratio is 30%/70%. The total number of coins in each box is the same and the two kinds of coins, as well as the two boxes, are indistinguishable by weight or other features. The objective is to find, with a given degree of confidence, which box contains more gold coins by drawing n coins from one box, with replacement.

We will employ conditional probabilities to solve this problem. We define the following two random variables, which are *not* independent:

- The variable A identifies the box choice from which we decide to draw the coins; it can take only two values, indicated by A_1, A_2. Since the boxes look identical from the outside we must assign $P(A_1) = P(A_2) = 0.5$.

- The variable B counts the number of gold coins; it can take several values, which we denote as B_j, depending on the total number of coin draws we are allowed to make. For example if n draws are allowed, we can find from $j = 0$ (B_0) to $j = n$ (B_n) gold coins, and the rest, $n - j$, will be silver coins.

The difficulty of this problem lies in the fact that, when we perform an experiment that involves drawing n coins from one box, the probabilities of getting a certain number of gold coins, $P(B_j), j = 0, \ldots, n$, cannot be directly calculated, because their values depend on the box from which we are drawing coins, and this information is not known.

We first calculate the conditional probabilities for a single draw, $n = 1$, in which case B can take the values B_1 (one gold coin) or B_0 (one silver coin). The conditional probabilities in this case are determined by

$$P(B_1|A_1) = 0.7, \quad P(B_1|A_2) = 0.3, \quad P(B_0|A_1) = 0.3, \quad P(B_0|A_2) = 0.7.$$

These numbers are the result of the definitions we have made: The probability of getting a gold coin under the condition that we have chosen the first box is 70%, or $P(B_1|A_1) = 0.7$, etc. From these results we can calculate the joint probabilities $P(A_1 \cap B_1), P(A_2 \cap B_1), P(A_1 \cap B_0), P(A_2 \cap B_0)$; using the general formulas of Eqs. (13.93), (13.94), we find

$$P(A_1 \cap B_1) = P(B_1|A_1)P(A_1) = 0.35,$$

$$P(A_1 \cap B_0) = P(B_0|A_1)P(A_1) = 0.15,$$

$$P(A_2 \cap B_1) = P(B_1|A_2)P(A_2) = 0.15,$$

$$P(A_2 \cap B_0) = P(A_2|B_0)P(A_2) = 0.35.$$

We summarize these joint probabilities in the table below:

	B_1	B_0	
A_1	0.35	0.15	0.50
A_2	0.15	0.35	0.50
	0.50	0.50	

The four entries under B_1, B_0 and across A_1, A_2 are the corresponding joint probabilities $P(A_1 \cap B_1) = P(B_1 \cap A_1)$, etc. The extra entries are the column and row sums. These are interesting quantities: The first row is the sum of the probabilities $P(A_1 \cap B_1) + P(A_1 \cap B_0)$, namely, the probability of getting either a gold coin (event B_1) or a silver coin (event B_0)

from box 1 (event A_1); this is the same as the probability of choosing box 1 itself, which must be equal to $P(A_1) = 0.5$, as indeed it is. Similarly for the second row, $P(A_2) = 0.5$. The first column is the probability of choosing a gold coin (event B_1) from either box 1 (event A_1) or box 2 (event A_2), which turns out to be $P(B_1) = 0.5$. Similarly for the second column, $P(B_0) = 0.5$. These are the marginal probabilities for B_1 and B_0: the first column represents the probability of getting a gold coin ($P(B_1)$) whether we picked it from box 1 (event A_1) or from box 2 (event A_2); similarly for the second column.

We next draw $n > 1$ coins from one box. We illustrate what can happen for $n = 3$. If the box is the one containing 70% gold and 30% silver coins, then we can figure out using the binomial distribution the probabilities of picking three gold coins, two gold and one silver coins, one gold and two silver coins or three silver coins. We know that the probability of getting a gold coin from box 1 is $p = 0.7$ and the probability of getting a silver coin is $q = 1 - p = 0.3$; each time we pick a coin these probabilities are unchanged, because of replacement. Therefore, the probabilities of getting, 3, 2, 1, 0 gold coins (and hence 0, 1, 2, 3 silver coins) in $n = 3$ tries, on the condition that we are picking coins from box 1 (event A_1), are as follows:

$$P(B_3|A_1) = \frac{3!}{3!0!}(0.7)^3(0.3)^0 = 0.343 : \quad 3 \text{ gold}, \quad 0 \text{ silver},$$

$$P(B_2|A_1) = \frac{3!}{2!1!}(0.7)^2(0.3)^1 = 0.441 : \quad 2 \text{ gold}, \quad 1 \text{ silver},$$

$$P(B_1|A_1) = \frac{3!}{1!2!}(0.7)^1(0.3)^2 = 0.189 : \quad 1 \text{ gold}, \quad 2 \text{ silver},$$

$$P(B_0|A_1) = \frac{3!}{0!3!}(0.7)^0(0.3)^3 = 0.027 : \quad 0 \text{ gold}, \quad 3 \text{ silver},$$

where the events B_j denote getting j gold coins ($j = 0, 1, 2, 3$) in the three picks. These probabilities add up properly to 1. We can use these conditional probabilities to obtain the joint probabilities $P(A_1 \cap B_j)$ by the same procedure as before, through the relation

$$P(A_1 \cap B_j) = P(B_j|A_1)P(A_1),$$

with $P(A_1) = 0.5$. Similarly for the joint probabilities $P(A_2 \cap B_j)$, with $P(A_2) = 0.5$. We summarize these joint probabilities in the table below:

	B_0	B_1	B_2	B_3	
A_1	0.0135	0.0945	0.2205	0.1715	0.50
A_2	0.1715	0.2205	0.0945	0.0135	0.50
	0.185	0.315	0.315	0.185	

The numbers under both entries A_1 or A_2 and B_j are the joint probabilities $P(B_j \cap A_1)$ or $P(B_j \cap A_2)$, while the last column is the sum of individual rows of joint probabilities which, as expected, sum up to 0.5;

the numbers in the last row are the column sums of the $P(B_j \cap A_1)$ or $P(B_j \cap A_2)$ entries. Note that the column sums represent the marginal probabilities

$$P(B_j) = P(B_j \cap A_1) + P(B_j \cap A_2)$$

of getting j coins, whether we chose box 1 (event A_1) or box 2 (event A_2). Thus, the numbers of the bottom row in the above table are the marginal probabilities $P(B_0)$, $P(B_1)$, $P(B_2)$, $P(B_3)$, in this order.

Having determined the joint and marginal probabilities, we can use them to calculate another set of conditional probabilities, which are actually more interesting. Specifically, we can calculate the probability that we have chosen box 1, under the condition that we drew three gold coins in our three trials; this is given by the expression

$$P(A_1|B_3) = \frac{P(B_3 \cap A_1)}{P(B_3)} = \frac{0.1715}{0.185} = 0.927.$$

In other words, if in our three tries we get three gold coins, then the probability that we have found the box with the most gold coins (box 1 or event A_1) is 92.7%. If our goal were to identify the box with the most gold coins by picking coins from one of the two identical-looking boxes, this would be very useful information. Conversely, if we had obtained three silver coins (and hence zero gold coins) in our three picks, the probability that we had found box 1 would be

$$P(A_1|B_0) = \frac{P(B_0 \cap A_1)}{P(B_0)} = \frac{0.0135}{0.185} = 0.073,$$

or 7.3%, a rather small value, and equal to $1 - P(A_1|B_3)$.

Let us also calculate the average number of gold coins in the $n = 3$ picks: it is given by

$$\mu = \sum_{j=0}^{n} jP(B_j) = 0 \times 0.185 + 1 \times 0.315 + 2 \times 0.315 + 3 \times 0.185 = 1.5.$$

We note that because the ratio of gold and silver coins in box 1 is the reverse of that in box 2, the average number scaled by the number of picks is given by

$$\frac{\mu}{n} = \frac{1.5}{3} = 0.5.$$

This will not change if the value of n is changed.

What would be the right strategy for finding the box with the most gold coins? A logical strategy would be to choose as the box with the most gold coins the one from which we get a larger than average number of gold coins in three picks. Since the average is 1.5, getting more than average gold coins means getting two or three gold coins. We should then compare the probability of getting two or three gold coins to the probability of getting zero or one gold coins. We note that the events of getting i and j gold coins are mutually exclusive for $i \neq j$, therefore:

$$P(B_i \cup B_j) = P(B_i) + P(B_j), \quad i \neq j.$$

The same relation will hold for the joint probabilities:

$$P((B_i \cup B_j) \cap A_1) = P(B_i \cap A_1) + P(B_j \cap A_1), \quad i \neq j,$$

$$P((B_i \cup B_j) \cap A_2) = P(B_i \cap A_2) + P(B_j \cap A_2), \quad i \neq j.$$

From the table of joint probabilities for the individual B_j events, we can then calculate the joint probabilities $P((B_0 \cup B_1) \cap A_1)$, $P((B_2 \cup B_3) \cap A_1)$, $P((B_0 \cup B_1) \cap A_2)$ and $P((B_2 \cup B_3) \cap A_2)$:

	$B_0 \cup B_1$	$B_2 \cup B_3$	
A_1	0.108	0.392	0.50
A_2	0.392	0.108	0.50
	0.50	0.50	

From these entries, we can next obtain the conditional probabilities of getting more than average gold coins ($B_2 \cup B_3$) or less than average gold coins ($B_0 \cup B_1$), on the condition that we are picking from box 1 (event A_1) or box 2 (event A_2):

$$P((B_2 \cup B_3)|A_1) = 0.784, \quad P((B_0 \cup B_1)|A_1) = 0.216,$$

$$P((B_2 \cup B_3)|A_2) = 0.216, \quad P((B_0 \cup B_1)|A_2) = 0.784.$$

These results show that if we get more than average gold coins there is a 78.4% probability that we have indeed identified the box containing the larger percentage of gold coins. There is also a non-negligible 21.6% probability that we have the wrong box, that is, the one with smaller percentage of gold coins, even though we got more than average gold coins in our three picks. Similarly, if we had gotten fewer than average gold coins, there is a 78.4% probability that we are picking from box 2, the box with the smaller percentage of gold coins, but there is also a 21.6% probability that we were picking from the box with the larger percentage of gold coins. Assuming that our strategy is to keep as the right choice the box for which we obtain more than the average gold coins and to reject the box for which we obtain fewer than the average gold coins, then we would be making the correct choice with a probability 78.4%, and the wrong choice with probability 21.6%.

It is evident that if we increase the number of samplings n, the odds of finding the right box with this strategy will increase, but there will always be a non-zero chance of making the wrong choice. Essentially, with n increasing, the distribution of probabilities for each case evolves into a Gaussian, and for large enough n the Gaussians corresponding to A_1 and A_2 will have very little overlap.

We next pose the problem in a more interesting manner: Suppose that we desire a degree of certainty that we pick the right box, namely, the one that has a majority of gold coins. How many trials should we make in order to be sure we have the right box with the desired degree of certainty? If the number of trials n is large enough we know that we will have Gaussian distributions. If we are picking from box 1, then

$$A_1 \; : \; \mu_1 = np_1, \; p_1 = 0.7 \Rightarrow \frac{\mu_1}{n} = 0.7.$$

If we are picking from box 2, then

$$A_2 : \; \mu_2 = np_2, \; p_2 = 0.3 \Rightarrow \frac{\mu_2}{n} = 0.3.$$

Since the two boxes look identical from the outside, we may be picking coins from either box, and therefore the normalization of each Gaussian representing the probabilities of getting any combination of gold and silver coins must be equal to $P(A_1) = P(A_2) = 0.5$. For both curves, the variance will be given by

$$\sigma_1^2 = np_1q_1, \quad \sigma_2^2 = np_2q_2 \Rightarrow \frac{\sigma_1^2}{n} = \frac{\sigma_2^2}{n} = \frac{\sqrt{0.21}}{\sqrt{n}}.$$

The first curve will be peaked at $\mu_1/n = 0.7$, the second at $\mu_2/n = 0.3$, and for large enough n the overlap between them will be small.

The tails of the Gaussians have interesting meaning: we define a common random variable which represents the number of gold coins divided by n:

- The tail of the first curve for values of the random variable less than 0.5 corresponds to situations where we would get less than half gold coins by sampling from box 1; these would lead to erroneous rejection of the box (false negatives).

- The tail of the second curve for values of the random variable greater than 0.5 corresponds to situations where we would get more than half gold coins by sampling from box 2; these would lead to erroneous acceptance of the box (false positives).

The only freedom we have in order to reach the desired degree of certainty is to increase the number of picks n. To pose a specific question:

How big should n be to guarantee that the probability of making either type of mistake is less than 1%?

To answer this question, we have to sum up all the probabilities that correspond to the tails of the two Gaussian curves above and below the average number of gold coins. In the present example, the variance and normalization of the two Gaussians is the same, and their position with respect to the average is symmetrical, therefore the contribution of each tail is the same. Thus, we simply calculate the contribution of one of the tails and multiply that by a factor of 2; this turns one of the Gaussians into a properly normalized one. We will choose the Gaussian that corresponds to event A_1, centered at $\mu_1 = 0.7n$. Thus, we must calculate the Gaussian integral

$$\Phi(-w) = \frac{1}{\sqrt{2\pi}} \int_{-\infty}^{-w} e^{-y^2/2} dy,$$

so that it gives a value of 0.01. From the tabulated values of $\Phi(w)$ we find that this value occurs for $w = 2.326$ (see Appendix B). This is of course a value expressed in units of $\sigma = 1$ for the integrand $\varphi(x)$ of $\Phi(w)$, measured with respect to the mean $\mu = 0$ for $\varphi(x)$. All that remains to do is to express this result in terms of the actual number of coin draws n. Accordingly, we express the mean and variance in terms of the scaled variables

$$\tilde{\mu}_1 = \frac{\mu_1}{n} = 0.7, \quad \tilde{\sigma}_1 = \frac{\sigma_1}{n} = \frac{\sqrt{0.21}}{\sqrt{n}},$$

and then express the cutoff point of the tail relative to $\tilde{\mu}_1$ in units of $\tilde{\sigma}_1$:

$$0.5 = \tilde{\mu}_1 - 2.326\,\tilde{\sigma}_1.$$

Plugging into this expression the values of $\tilde{\mu}_1$ and $\tilde{\sigma}_1$ from above (the latter containing a factor of $\sqrt{n}$) and solving for n, we find $n = 28.4$. Taking the value of n to be the nearest integer larger than this calculated value, guarantees that using our strategy for choosing the right box, the overall error will not exceed 1%. This includes both false negatives and false positives.

Further reading

The mathematical-methods textbooks mentioned under "General Sources" in the Preface include brief introductions to probability theory. For more extensive treatments the following books are recommended:

1. Dimitri P. Bertsekas and John N. Tsitsiklis, *Introduction to Probability*, Second Ed. (Athena Scientific, 2008). This is a remarkably thorough and clear introduction to all the important concepts in probability theory.

2. Joseph K. Blitzstein and Jessica Hwang, *Introduction to Probability*, Second Ed. (Taylor and Francis/CRC Press, 2019). This is an inspired, modern introduction to probability theory and its applications.

3. Persi Diaconis and Brian Skyrms, *Ten Great Ideas About Chance* (Princeton University Press, 2018). This is a wonderful exposition of the basic ideas of probability, the history of their development, and their connection to other fields like those of psychology and philosophy.

Problems

1. (a) Prove the result of Eq. (13.39) by following the steps outlined in the text.

   (b) Using the moments of the normalized Gaussian function derived in Chapter 1, Eq. (1.66), and the result of Eq. (13.39), prove the expression for the variance of the normal distribution, Eq. (13.40).

   (c) Prove the relation of Eq. (13.80) where the values of the parameters in the two exponentials are related by the expressions in Eq. (13.81).

2. (a) A fair coin is tossed five times and comes up head each time. What is the probability of this occurring? What is the probability of heads on the sixth toss?

   (b) A student takes a true–false test containing 10 problems, and guesses the answers. What is the probability that the student gets at least seven right?

   (c) During a thunderstorm, lightning flashes are seen at a rate of 12 per minute. What is probability that during a 30 sec. interval fewer than 2 flashes will be seen?

(d) Water contains a certain kind of bacteria in the density of 1 per cubic inch. What is the probability that a glass of water contains no bacteria? (estimate the volume of a typical glass in cubic inches).

(e) A novice basketball player is able to score from the three-point line only once in 20 tries on average. What is the probability that the player will score a three-pointer at least once in the 20 tries. With practice the player is able to improve his average three-point scoring. How many three-pointers must he be able to score on average, so that in one game where he gets to shoot four times from the three-point line, he has the probability of scoring at least one?

(f) Two fair dice are tossed. What is the probability that the sum of numbers that come up is at least 9, if one of the dice comes up even?

3. The Harvard admissions committee aims to have 1667 acceptances to its admission offers each year, because this is the number of beds in the Harvard Yard dormitories where all first-year students live. From the viewpoint of the admissions committee, the acceptance or refusal of each offer is a binary random process. Assign the value 1 to acceptance and 0 to refusal, and assume that the probability of refusal is 20%.

(a) What is the mean and variance of a single event.

(b) If 2020 admissions were offered one year, find the mean and variance of the sum of choices, assuming the single events are independent.

(c) The admissions committee needs to know the probability that the target will be exceeded. Describe the principle underlying the estimation process using a diagram, assuming a normal distribution of events. From the table of the normal probability distribution (see Appendix B) estimate the probability that the target will be exceeded.

4. Consider the *independent* random variables x and y with probability distributions $p_x(X)$ and $p_y(Y)$, with means μ_x, μ_y, and variances σ_x^2, σ_y^2, respectively. We define a random variable $z = x + y$, as in Eq. (13.74); its probability distribution $p_z(Z)$, can be expressed as the convolution of $p_x(X)$ and $p_y(Y)$,

see Eqs. (13.79a) and (13.79b). We wish to obtain the mean, μ_z, and the variance, σ_z^2, of the variable z, in terms of the means and variances of x and y, by taking advantage of the convolution properties, namely the fact that the Fourier transform of the convolution of two functions is the product of the Fourier transforms of the two functions:

$$\hat{p}_z(k) = \hat{p}_x(k)\hat{p}_y(k). \qquad (13.102)$$

(a) From the first three moments of the probabilities $p_x(X)$, $p_y(Y)$, using the Fourier transforms and their derivatives evaluated for $k = 0$, show that:

zeroth moment : $\hat{p}_x(0) = 1 = \hat{p}_y(0)$,
first moment : $i\hat{p}_x'(0) = \mu_x$, $i\hat{p}_y'(0) = \mu_y$,
second moment : $i^2\hat{p}_x''(0) = \sigma_x^2 + \mu_x^2$,
$i^2\hat{p}_y''(0) = \sigma_y^2 + \mu_y^2$.

(b) Using the FT of the probability $p_z(Z)$ and the linearity of the FT show that

$$\mu_z = \mu_x + \mu_y.$$

(c) Using the above results and the linearity of the FT, show that

$$\sigma_z^2 = \sigma_x^2 + \sigma_y^2.$$

5. Suppose that x, y and z are three independent continuous random variables, each of them having a uniform probability $p_x(x), p_y(y), p_z(z)$ over the interval $[0,1]$, and zero everywhere else. What is the value of the three uniform probabilities $p_x(x), p_y(y), p_z(z)$ in the interval $[0,1]$?

(a) Calculate the zeroth, first, and second moments of $p_x(x)$, and from those the mean and variance of x.

(b) Calculate the Fourier transform $\hat{p}_x(k)$ of $p_x(x)$ and show that:

$$i^n \frac{d^n \hat{p}_x(k)}{dk^n} = n\text{th moment of } p(x),$$

by explicitly comparing to the previous results for $n = 0, 1, 2$ [the same results hold for the variables y and z].

(c) Consider the variable $u = x + y$: find its probability function $p_u(u)$ making sure that it is properly normalized.

(d) Calculate the mean and variance of the variable u, and its Fourier transform $\hat{p}_u(k)$. Relate the behavior of $\hat{p}_u(k)$ to the features of $p_u(u)$.

(e) Consider the variable $v = x + y + z$: find its probability function $p_v(v)$, the mean and the variance of v, and the Fourier transform $\hat{p}_v(k)$ of $p_v(v)$. Compare the three probability functions $p_x(x)$, $p_u(u)$, $p_v(v)$, and comment on their relation to the normal probability function.

6. A pair of unfair dice always come up with the same number. If we consider the two dice as random variables x, y, and the probability of any event as given by $p(x, y)$, what are the values that $p(x, y)$ takes for all the possible values of x and y?

(a) Calculate the mean of x and $y, \mu_x = E[x]$ and $\mu_y = E[y]$.

(b) Calculate the variance of x and $y, \sigma_x^2 = E[(x - \mu_x)^2]$ and $\sigma_y^2 = E[(y - \mu_y)^2]$.

(c) Calculate the mixed moment $E[xy]$ and the covariance of x and $y, \sigma_{xy} = E[(x - \mu_x)(y - \mu_y)]$. From this obtain the correlation coefficient, defined as $\sigma_{xy}/\sigma_x\sigma_y$.

(d) Consider the random variable $z = x + y$. Calculate the mean and variance of z, given by $\mu_x = E[(x + y)]$ and $\sigma_z^2 = E[(z - \mu_z)^2]$ respectively.

7. Let X be a continuous random variable with finite mean.

(a) If the probability distribution function of X, denoted as $f(x)$, is symmetric about some real value c, that is, $f(c + x) = f(c - x)$, show that $E(x) = c$.

(b) Consider $f(x = l) = 2^{-l}$ for $l = 1, 2, \ldots$. Can this be a probability distribution function of a discrete random variable X? If so, determine $E(X)$, $E(X^2)$ and σ^2 of X.

8. An electronic device can fail at any time t after operating properly up to this point. Its lifetime T (in hours) can be viewed as a random variable whose cumulative distribution function $F(t)$ is given by

$$F(t) = P(T \le t) = \begin{cases} 1 - k/t, & t > 1000, \\ 0, & t \le 1000, \end{cases}$$

where k is a constant.

(a) Determine the value of k and find the probability distribution function $f(t)$ of this random variable.

(b) What is the probability that T is greater than 2000 hours given that T exceeded 1500 hours?

9. 90% of Harvard students take a math course in the Fall semester and 60% take a math course in the spring semester. If a randomly chosen student takes a math course in the spring semester, the probability that this student attended a math course in the last fall semester is 30%. Given that a random student took a math course last fall, what is the probability that she attends a math course in the spring?

10. Consider the probability distribution function $f(x)$ defined as

$$f(x) = \begin{cases} A/\sqrt{1 - x^2}, & \text{for } |x| < 1, \\ 0, & \text{otherwise.} \end{cases}$$

(a) Find the value of A. [Hint: See Problem 13 of Chapter 1.]

(b) Find the corresponding cumulative density function $F(x)$.

11. Consider the following joint probability density function of two continuous random variables X, Y:

$$f(x, y) = \begin{cases} c(x^2 + y^2), & 0 < x < 1, \ 0 < y < 1, \\ 0, & \text{elsewhere,} \end{cases}$$

where c is a constant.

(a) Calculate the value of c.

(b) Calculate the probability $P(X + Y > 1)$.

12. In a country there is a referendum on an important issue. Voters cast ballots by punching a YES or NO hole in a card, using a machine which does not always function properly. Define as A the event that the machine punches a YES hole (A^c is the event that the machine punches a NO hole). Define as B the event that the voter intends to vote YES (B^c is the event that the voter intends to vote NO). The machine works correctly 70% of the time when trying to vote YES, and 90% of the time when trying to vote NO; when the machine does not work correctly it punches the wrong hole.

(a) An independent and reliable poll has found that 70% of the voters intended to vote YES. Construct a table of probabilities of intersections $[P(A \cap B), P(A^c \cap B), P(A \cap B^c), P(A^c \cap B^c)]$, and from that determine the marginal probabilities that the ballots cast will be counted as YES, $P(A)$, or they will be counted as NO, $P(A^c)$.

(b) Four voters go into a booth to vote. If all these voters intended to vote the same (either all YES or all NO), what is the probability that the results will show 0, 1, 2, 3 or 4 YES votes? Make a rough plot of these two sets of probabilities as a function of the percentage of YES votes, on the same graph. What do you expect the two sets of results to look like for a very large number of voters? Indicate carefully where the characteristic features of the curves (that is, the extrema) will be in the large sample limit and what their normalization is.

(c) 100 voters cast votes in the referendum. Find what is the probability that the outcome will be NO despite the intention of the voters to vote YES. The referendum needs two-thirds majority to pass. Determine what type of error is more likely.

Chapter 14

Numerical Methods

The use of numerical methods to solve scientific and engineering problems has become widespread with the advent of easy access to computational resources. These methods are quite interesting in their own right and present their distinct challenges. In fact, thorough accounts of numerical methods are rather extensive works, not to mention the associated scientific literature. Thus, a proper treatment of the field is beyond the scope of the present book. However, it would be remiss not to mention here a few basic ideas in this field.

We have selected for this chapter only certain numerical methods that are directly related to material covered earlier in the book. The methods discussed here are relatively simple to implement, typically in a few lines of computer code. These numerical approaches can serve as checks on the analytical answers, or as tools to obtain a quick solution for problems too difficult to solve analytically. For more advanced applications, the reader should consult one of the more extensive treatments (see Further Reading at the end of this chapter for suggestions).

14.1 Calculating with digital computers

In previous chapters, we dealt mostly with real and complex numbers, which are represented by the "real axis" (the x-axis) or the "complex plane" (the xy-plane), and take all possible values in these domains, including integer, rational and irrational values. Based on this picture we could safely assume that a given real value could be approached arbitrarily close. Unfortunately, this luxury is not available when working with a digital computer where numbers are represented with an accuracy that depends on the computer architecture. To understand the limitations implied by this fact, we first explore this topic in some detail.

Currently, digital computers are based on the electric diode, a physical device that either lets current pass or not, when subjected to an external electric potential. This means that the natural way to represent numbers in a digital computer is a "binary" system, that is, a system with only two possible states corresponding to the physical states of "current on" or "current off" in the diode. To achieve this, a computer stores the values 0 or 1 in a "bit", which is another physical unit, typically with magnetic properties. A set of 8 bits is called the "byte". By convention, computer memory is measured in multiples of the byte. Early commercial computers had a 1-byte (8-bit) architecture and until the late 1990s widely available computers were based on a 4-byte (32-bit) architecture. Most modern computers are based on an 8-byte (64-bit) architecture. Accordingly, in the following we give examples assuming a 64-bit computer architecture.

The total number of bytes on which the computer architecture is based is called the "word". In making numerical operations, the computer's processor must find the "address" of the word, that is, find where the number "resides" in memory, then bring this number to the processor to perform the required operations with other numbers (addition, multiplication, division, etc.), and return the resulting number to another address in memory.

Thus, to understand the limitations of working with numbers represented in a computer we must familiarize ourselves with the binary system of numbers. In the standard "decimal" number system, that is, a system with the number 10 as base, a given number is represented, for instance, as

$$4,728.597 = 4 \times 10^3 + 7 \times 10^2 + 2 \times 10^1 + 8 \times 10^0$$
$$+ 5 \times 10^{-1} + 9 \times 10^{-2} + 7 \times 10^{-3},$$

that is, each digit *before* the decimal point is multiplied by the power 10^{n-1} where n is its position to the *left* of the decimal point, and each digit *after* the decimal point is multiplied by the power 10^{-m} where m is its position to the *right* of the decimal point. The same rules apply to the representation of numbers in the number system with base 2, the so-called "binary" system; for example, an integer represented by a 7-bit binary number with some of the bits being 0 and some being 1, could be [we use a subscript on the right side of the number to indicate the base when it is different than 10]:

$$1010011_2 = 1 \times 2^6 + 0 \times 2^5 + 1 \times 2^4 + 0 \times 2^3 + 0 \times 2^2 + 1 \times 2^1 + 1 \times 2^0$$
$$= 64 + 16 + 2 + 1 = 83, \tag{14.1}$$

where we also provided its decimal-system value for comparison. Similarly, a number smaller than 1 represented by a 7-bit binary number with the bit before the point being set to 0 (since the number is of magnitude smaller than 1), and the other six bits being either 0 or 1, could be:

$$0.101011_2 = 1 \times 2^{-1} + 0 \times 2^{-2} + 1 \times 2^{-3} + 0 \times 2^{-4} + 1 \times 2^{-5} + 1 \times 2^{-6}$$
$$= \frac{1}{2} + \frac{1}{8} + \frac{1}{32} + \frac{1}{64} = \frac{43}{64} = 0.671875, \tag{14.2}$$

where we provided again the corresponding decimal-system value.

Notice that in each case we have N different symbols, including the number 0, for the values of each digit: for the base-10 case, we have the ten symbols 0, 1, 2, 3, 4, 5, 6, 7, 8, 9, whereas for the base-2 case we only have the two symbols 0, 1. Another number system sometimes encountered in computations is the so-called "hexadecimal" system with base 16, using the 10 digits of the decimal system plus the letters A, B, C, D, E, F to represent the values 10, 11, 12, 13, 14, 15, respectively. As an amusing example, the magnitude of the largest *signed* integer that can be represented in a 32-bit computer architecture, after saving one bit for the sign, corresponds to the remaining 31 bits being equal to 1, which gives

$$\underbrace{11111111111111111111111111111111_2}_{\text{31 digits}} = \sum_{n=0}^{30} 2^n = 2^{31} - 1 = 2,147,483,647.$$

This is a prime number as was first proven by Euler; its hexadecimal representation is a 7 followed by seven F's:

$$7\text{FFFFFFF}_{16} = 7 \times 16^7 + 15 \times \sum_{n=0}^{6} 16^n = 2,147,483,647.$$

Byte $\downarrow$	Bit$\rightarrow$ 7	6	5	4	3	2	1	0
7	S	X_{10}	X_9	X_8	X_7	X_6	X_5	X_4
6	X_3	X_2	X_1	X_0	M_{51}	M_{50}	M_{49}	M_{48}
5	M_{47}	M_{46}	M_{45}	M_{44}	M_{43}	M_{42}	M_{41}	M_{40}
4	M_{39}	M_{38}	M_{37}	M_{36}	M_{35}	M_{34}	M_{33}	M_{32}
3	M_{31}	M_{30}	M_{29}	M_{28}	M_{27}	M_{26}	M_{25}	M_{24}
2	M_{23}	M_{22}	M_{21}	M_{20}	M_{19}	M_{18}	M_{17}	M_{16}
1	M_{15}	M_{14}	M_{13}	M_{12}	M_{11}	M_{10}	M_9	M_8
0	M_7	M_6	M_5	M_4	M_3	M_2	M_1	M_0

Table 14.1: Assignment of bits in 64-bit architecture machine: one bit is assigned to the sign (S), 11 bits are assigned to the exponent ($X_0 - X_{10}$), and 52 bits are assigned to the mantissa ($M_0 - M_{51}$). The locations are organized in 8 bytes with 8 bits each.

In a 64-bit machine, the bits are used to represent a real *signed* number according to the scheme shown in Table 14.1: The bit labeled as "7" of the byte labeled "7", with the labeling of both bits and bytes starting at 0, is reserved for the sign (S). Of the remaining 63 bits, 11 are allocated to the representation of the exponent and 52 to the representation of the "mantissa". The mantissa is the actual value of the number, excluding the sign and the order of magnitude, the latter being provided by the exponent value. By convention, the mantissa value is taken to be smaller than 1 in magnitude, and is subsequently multiplied by the value obtained from the exponent to obtain the full value of the number represented by a given 64-bit word.

In what concerns the mantissa, the largest *integer* that can be represented with its 52 bits is a value corresponding to all bits set to 1:

$$\underbrace{1111 \ldots 1111_2}_{\text{52 digits}} = \sum_{n=0}^{51} 2^n = 2^{52} - 1 = 4.5035996 \times 10^{15}.$$

The smallest integer (other than zero) that can be represented by the mantissa bits has only the right-most bit equal to 1 and the rest set to zero, which is equal to 1. Taking into consideration that by convention the mantissa represents a number smaller than 1, this means that the *range* of numbers that can be represented by the mantissa extends over 15 orders of magnitude, the smallest being of order 10^{-15} and the largest being of order 1. This determines the "precision" with which the 64-bit architecture can handle numbers, namely 1 part in 10^{15}. In other words, a number known with precision higher than 15 significant digits simply *cannot* be handled by a 64-bit machine, and must therefore be *truncated* to 15 digits or fewer. The loss of accuracy in doing calculations under this limitation is referred to as "truncation error".

Derivation: Largest and smallest values in 64-bit computer

When dealing with numbers in a computer, there are two other important questions, beyond the precision issue:

(a) What is the largest possible value (in magnitude) that can be represented in a computer? This is essentially the value of "infinity" that the computer can handle.

(b) What is the smallest possible value (in magnitude), other than zero, that can be represented in a computer? This is essentially the value of the infinitesimal quantity we called ϵ, which was used in many analytical arguments; in those cases we treated ϵ as *arbitrarily* small, but this is not possible when working with a computer.

In order to answer these questions we need to take into account not only the values handled by the mantissa but also the values handled by the exponent. We concentrate again on a 64-bit architecture (see Table 14.1). The 11 bits reserved for the exponent give as the highest accessible integer value

$$2^{11} - 1 = 2047.$$

However, one bit needs to be reserved for the sign of the exponent, which can be positive or negative. Instead of reserving a bit for the exponent sign, the following trick is employed: the range is shifted by -1023, to give the range of exponents $[-1023, 1024]$. This results in slightly larger range, since saving one bit for the exponent sign and combining it with the remaining 10 bits would have produced a range of $[-1023, 1023]$. Thus, the largest and smallest numbers (in magnitude) that can be represented by the exponent are as follows:

$$\text{exponent}: \quad \text{largest} \to 2^{1024} \approx 10^{308}, \quad \text{smallest} \to 2^{-1023} \approx 10^{-308}.$$

As already discussed, the largest number represented with the mantissa is ~ 1, so the largest number that can be represented in a 64-bit machine is $\sim 10^{308}$. As far as the smallest (in magnitude) number, other than zero, that can be represented, we have a contribution from the exponent of $\sim 10^{-308}$ and from the mantissa of $\sim 10^{-15}$, giving a number of $\sim 10^{-308} \times 10^{-15} = 10^{-323}$. This quantity is sometimes referred to as the "machine epsilon", or "machine precision".

Examples of interesting questions that can be addressed with numerical methods are as follows:

- What are the roots of the function within the computational domain?

- What are the values of the various derivatives of the function in this domain?

- What is the integral of the function over this domain (or over some subdomains within it)?

- How can the values of the function be employed in the computation of other quantities like the solution of differential equations in which it appears as a source or a boundary condition?

In the following sections, we discuss specific computational methods of how these types of questions can be addressed.

14.2 Numerical representation of functions

It is evident from the discussion of the previous section that when working with a computer a real variable x is limited to a *finite* number of values which are by necessity also *discrete*, as opposed to the continuous nature of the full set of real values. Moreover, there is an intrinsic limit to how close these values can be, dictated by the "precision" imposed by the computer architecture. For example, in a 64-bit architecture machine, the successive values of a real variable cannot be closer than one part in 10^{15}. The question we wish to consider next is: What are the limitations that these facts impose on the representation of functions of real variables? To address this question, we first describe the two basic ways of representing functions of real variables, namely grid methods and spectral methods.

14.2.1 Grid and spectral methods

A common approach is to employ a regularly spaced set of $N + 1$ values of the real variable x, known as the "computational grid" and defined through the following relations:

$$x_n = x_0 + n \, \Delta x, \quad n = 0, 1, 2, \ldots, N, \tag{14.3a}$$

$$\Delta x \equiv \frac{x_N - x_0}{N} = x_{n+1} - x_n \quad \forall n \in [0, N-1]. \tag{14.3b}$$

In other words, the domain extends from the value x_0 to the value x_N in N equal intervals between successive values of the variable. Grids of non-uniformly spaced points are also possible, and are used for problems where the non-uniformity of the grid makes it more efficient or accurate. For simplicity, we will only concern ourselves with uniform grids here. The values of the function $f(x)$ provided on such a grid of $N + 1$ points x_n and are denoted as:

$$f_n = f(x_n), \quad n = 0, 1, 2, \ldots, N. \tag{14.3c}$$

In many computational methods the goal is to determine the function $f(x)$ itself by the conditions it obeys, for example, when it is the solution of a differential equation. Typically, each value of the function f_n depends on previous values, f_m where $m < n$, as well as the values of other known functions evaluated on the same grid, so that the determination of unknown the function is an iterative process. In these cases, the interval Δx between successive values of x can play a crucial role in the accuracy of the solution. We return to this issue below.

A different way of representing a function $f(x)$ is through its expansion in other *known* functions, $g_n(x)$,

$$f(x) = \sum_{n=0}^{N} a_n g_n(x). \tag{14.4}$$

In this case, the objective is to determine the values of the coefficients a_n, which makes it possible to evaluate $f(x)$ at any value of x that can be represented numerically. For $N \to \infty$, Eq. (14.4) is the familiar series expansion discussed in detail in Chapter 3 for the case of special polynomials and in Chapter 7 for trigonometric functions and complex exponentials. In numerical applications, N is by necessity finite, and therefore the expression of Eq. (14.4) is an approximation of the function. The larger the value of N, the better the approximation becomes, but of course this comes at the cost of having to determine the value of a larger set of coefficients a_n.

The functions $g_n(x)$ typically involve a variable k_n in inverse space and their argument is the product $(k_n x)$, a dimensional quantity. The values that the variable k_n takes are called the "spectrum" of the function, and therefore the numerical methods using the expression of Eq. (14.4) are known as "spectral methods". In numerical applications, it is often the case that spectral methods are much more accurate in representing the unknown function $f(x)$, given the constraints that this function obeys. In fact, the constraints obeyed by $f(x)$ are important in choosing the proper set of functions $g_n(x)$ that should be used in the expansion. The reasons for the superior performance of spectral methods are beyond the scope of what can be covered here; we suggest at the end of this chapter several textbooks where this subject is treated in detail (see Further Reading).

14.2.2 Accuracy and stability

When using a numerical grid and the evaluation of a function involves differential operators or integrals, the finite size of the interval Δx between grid points implies that the numerical solution is *not* the same as the analytical solution, since the latter is defined in the limit $\Delta x \to 0$. Thus, the numerical solution is only an *approximation* to the analytical solution. It is expected that the numerical solution will approach the analytical solution when Δx becomes sufficiently small, but a sufficiently small value is subject to the limitations imposed by the cost of the computation and the computer architecture. Specifically, the number N of grid points, which determines the grid spacing Δx, must be such that an answer can be obtained with a reasonable amount of computer calculations, referred to as the "computational cost", which scales with some power of N. In complicated algorithms this power of N can be rather large [for instance, the cost of matrix diagonalization typically scales as N^3]. Thus, there is an incentive in numerical solutions to keep N as small as possible, but this makes Δx large. In some cases, either when N is very large or the total interval $(x_N - x_0)$ is very small, the spacing between successive numbers can become comparable to the inherent limit due to the computer architecture (the machine epsilon), and this situation has to be handled carefully.

It is therefore important to quantify the magnitude of the numerical error that is implied by the numerical approximation to the analytical solution. This is referred to as the "numerical accuracy". This error is given in terms of the dominant power of Δx, which is typically obtained by invoking the relevant Taylor series expansion. In the discussion of the various computational methods we will pay particular attention to determining the magnitude of the error, referred to as the "order" of the numerical error, which refers to the power of Δx in the dominant error term. We will discuss several examples of this related to specific numerical methods.

A different issue that affects the numerical solution is the so-called "stability" of the algorithm employed. To give an illustrative example of an *unstable* method, suppose that at each step of an iterative evaluation of the unknown function, the previous value is multiplied by a factor larger than 1. Then, it is plausible that after a large number of iterations the values will become unboundedly large, leading to the explosive failure of the algorithm. Thus, the issue of numerical stability can be of crucial importance, particularly in methods that employ a regular grid to represent an unknown function. Moreover, an important theorem, known as the "Lax equivalence theorem", states that stability is a requirement for the numerical approximation to converge to the correct solution, at least in the case of partial differential equations with well-posed initial conditions. In essence, establishing the stability of a numerical method also guarantees that it will converge to the correct answer. This is important because establishing the convergence of the numerical solution itself is more difficult than establishing its stability. Typically, requiring a method to be stable leads to constraints on the values of the parameters involved, including the size of the numerical grid interval. It is also possible to devise methods that are *unconditionally* stable, that is, their stability is guaranteed without having to impose any conditions on the parameters involved. In the following discussion of numerical methods, we provide comments relevant to the method stability, as appropriate.

14.3   Numerical evaluation of derivatives

In calculating numerically the derivative of a function on a computational grid as defined in Eq. (14.3a), the grid spacing Δx plays an important role. We recall that the analytical definition of the derivative, Eq. (1.23), involves the limit of the difference between successive points being infinitesimally small, $dx \to 0$. However, in actual applications the number of points N where the function can be evaluated is relatively small, and therefore the grid spacing Δx is *not* infinitesimally small. Thus, it is important to take advantage of all the available information in order to obtain the best possible numerical results.

From the analytical expression for the derivative, there are two intuitive definitions of the numerical value at a point x_n. These formulas are referred to as the "forward" and "backward" difference, respectively:

$$\text{forward}: \quad f'(x_n) \quad = \quad \frac{f_{n+1} - f_n}{x_{n+1} - x_n}, = \frac{f_{n+1} - f_n}{\Delta x}, \qquad (14.5)$$

$$\text{backward}: \quad f'(x_n) \quad = \quad \frac{f_{n-1} - f_n}{x_{n-1} - x_n} = \frac{f_n - f_{n-1}}{\Delta x}. \qquad (14.6)$$

In the limit $\Delta x \to 0$, or equivalently $N \to \infty$, the two numerical values become identical to the analytical expression, Eq. (1.23). However, for finite Δx the choice of the expression may affect the accuracy, depending on the behavior of the function. Thus, a better approximation would be to take the *average* of the two expressions, which gives the symmetric, or "central" difference expression:

$$\text{central}: f'(x_n) = \frac{1}{2}\left[\frac{f_{n+1} - f_n}{x_{n+1} - x_n} + \frac{f_{n-1} - f_n}{x_{n-1} - x_n}\right] = \frac{f_{n+1} - f_{n-1}}{2\Delta x}. \qquad (14.7)$$

We note that this expression does *not* involve the value of the function at the point x_n, in contrast to the forward and backward derivatives, but it involves the values of the function on either side of x_n.

Of particular interest is the accuracy with which derivatives can be evaluated numerically, which depends on the size of the grid spacing Δx. It turns out that in the case of the forward-difference or the backward-difference formula, in order to obtain a reliable numerical approximation of the derivative, Δx must be approximately equal to the *square root* of the machine precision. However, for the central difference formula, the size of Δx must be approximately equal to the *cubic root* of the machine precision.[1] Since the machine precision is a small quantity, its cubic root is larger than its square root, indicating the superior performance of the central-difference formula since it provides similar accuracy to the other two formulas but with larger Δx.

It is intriguing to try to extend the finite-difference scheme to higher-order derivatives. For example, for the second derivative, using the notation $f'_n = f'(x_n)$ and the forward- and backward-difference expressions, we obtain the symmetric expression:

$$f''(x_n) = \frac{f'_{n+1} - f'_n}{x_{n+1} - x_n} = \frac{1}{\Delta x}\left[\frac{f_{n+1} - f_n}{\Delta x} - \frac{f_n - f_{n-1}}{\Delta x}\right]$$
$$= \frac{f_{n+1} - 2f_n + f_{n-1}}{(\Delta x)^2}. \tag{14.8}$$

Thus, working with the values of the function on either side of x_n we obtain symmetric expressions for both the first and the second derivative.

To illustrate some of the pitfalls of simply re-using previously derived numerical expressions to calculate higher derivatives, we mention the following interesting case: If we apply the central difference expression, Eq. (14.7), to obtain the second derivative in terms of first derivative, we obtain

$$f''(x_n) = \frac{f'_{n+1} - f'_{n-1}}{2\Delta x} = \frac{1}{2\Delta x}\left[\frac{f_{n+2} - f_n}{2\Delta x} - \frac{f_n - f_{n-2}}{2\Delta x}\right]$$
$$= \frac{f_{n+2} - 2f_n + f_{n-2}}{4(\Delta x)^2}, \tag{14.9}$$

which is different than the formula of Eq. (14.8): it involves the values f_{n+2} and f_{n-2}, which are at a distance of $\pm 2\Delta x$ from the point x_n, rather than the points f_{n+1} and f_{n-1} which are at a distance $\pm \Delta x$. Thus, this last expression is less accurate, as it uses information about the function farther away from the point where we are trying to evaluate its derivative: it is effectively the same as having a grid spacing of $2\Delta x$, or equivalently, half the number of points in the grid.

The expressions derived for the first and second derivatives in Eqs. (14.7) and (14.8) are advantageous in the sense that they include information about the function both to the right (at x_{n+1}) and to the left (at x_{n-1}) of the point x_n where we are trying to evaluate the derivatives. Of course this is not possible to do at the endpoints of the interval. To compensate for this deficiency, we use the following expression based on evaluating the derivative at the point x_n as the *average* of its values at nearest points:

$$f'_n = \frac{1}{2}\left(f'_{n+1} + f'_{n-1}\right) \Rightarrow f'_{n-1} = 2f'_n - f'_{n+1}, \quad \text{and} \quad f'_{n+1} = 2f'_n - f'_{n-1}.$$

[1] See Philip E. Gill, Walter Murray and Margaret H. Wright, *Practical Optimization* (Classics in Applied Mathematics, Society of Industrial and Applied Mathematics, 2019).

Applying these expressions to the endpoints, we obtain

$$f_0' = 2f_1' - f_2', \quad \text{and} \quad f_N' = 2f_{N-1}' - f_{N-2}'. \tag{14.10}$$

A similar argument can be applied to evaluate the second derivative at the endpoints by the following expressions:

$$f_0'' = 2f_1'' - f_2'', \quad \text{and} \quad f_N'' = 2f_{N-1}'' - f_{N-2}''. \tag{14.11}$$

These expressions incorporate information about the function from several points, which provides a more reliable estimate of the value of the derivatives.

Example 14.1: Numerical evaluation of derivatives of polynomials

For an example of how the above formulas perform in the evaluation of derivatives we consider the polynomial

$$f(x) = x^5 - 2x^4,$$

whose first and second derivatives can be easily evaluated,

$$f'(x) = 5x^4 - 8x^3, \quad f''(x) = 20x^3 - 24x^2.$$

The values of the analytical expressions can then be compared to those obtained from the numerical expressions.

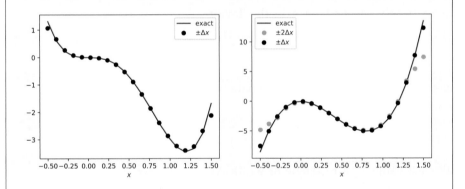

In the plots above we show results for the first (left panel) and the second (right panel) derivatives in the range $x \in [-0.5, 1.5]$, as obtained from numerical evaluation using the function values on a discrete grid with $N = 20$ in the above interval. These are compared to the values of the analytical expressions (red lines).

For the first derivative, the numerical values were obtained with the use of the central derivative formula, Eq. (14.7), with the endpoints obtained from the extrapolated formula of Eq. (14.10).

For the second derivative, one set of the numerical results was obtained by employing the formula of Eq. (14.8) (labeled $\pm\Delta x$, blue dots), and a second set of results by employing the formula of Eq. (14.9) (labeled $\pm 2\Delta x$, yellow dots). In both cases the endpoints obtained from the extrapolated formula of Eq. (14.11). These results are consistent with our expectations from the preceding discussion, namely, the results of the formula with effective spacing $2\Delta x$ are less accurate throughout the range of x values, with the difference being more pronounced at the endpoints of the interval.

14.4 Numerical evaluation of integrals

For the numerical evaluation of integrals we assume that the values of the integrand $f(x)$ are known for a discrete grid of values of its argument, as defined in Eq. (14.3a). The basic idea is to apply the formula in the definition of the definite integral discussed in Chapter 1, namely, Eq. (1.47). The main difficulty is that instead of using the limit $N \to \infty$, we typically have only a relatively small number of point in the computational grid. In other words, N is not sufficiently large for the general formula of the definition to apply so that this expression may give a poor approximation of the actual value of definite integral. The goal then is to devise methods that can give a reasonable approximation with the available information. In the following, we denote the endpoints of the domain as $a = x_0$ and $b = x_N$, so that the integral $I(a,b)$ (viewed as a function of the values a and b) is expressed as:

$$I(a,b) = \int_a^b f(x)dx = \lim_{N \to \infty}\left[\sum_{n=0}^{N} f(x_n)\Delta x\right],$$

$$x = a + n\,\Delta x, \quad \Delta x = \frac{b-a}{N}. \tag{14.12}$$

The simplest approximation is to keep only the first term in the Taylor expansion of the function,

$$f(x) = f(x_n) + (x - x_n)f'(x_n) + \frac{1}{2}(x - x_n)^2 f''(x_n) + \cdots,$$

namely $f(x_n) = f_n$, to which we refer as the "zeroth-order approximation". This gives for the integral in the interval $[x_n, x_{n+1}]$

$$\int_{x_n}^{x_n+\Delta x} f(x)dx = f_n \Delta x. \tag{14.13}$$

This is equivalent to associating with each value x_n a rectangle of width Δx and height f_n, as shown in Fig. 14.1.

The dominant contribution to the error comes from the first term in the Taylor expansion which has been *neglected*, which in the present case is the term proportional to the first derivative $f'(x)$. This leads to

$$\int_{x_n}^{x_n+\Delta x} f'(x_n)(x - x_n)dx = \frac{f'(x_n)}{2}\left[(x - x_n)^2\right]_{x_n}^{x_n+\Delta x} = \frac{f'(x_n)}{2}(\Delta x)^2.$$

However, this is the contribution of only one interval of size Δx, while the integral $I(a,b)$ involves the summation over all such intervals. Defining the average value of the first derivative as

$$\bar{f}' \equiv \frac{1}{N}\sum_1^N f'(x_n),$$

we can obtain the *accumulated* error $\mathcal{E}_0(\Delta x)$ for the integral as the sum of all the associated contributions:

$$\mathcal{E}_0(\Delta x) = \sum_{n=1}^{N} \frac{f'(x_n)}{2}(\Delta x)^2 = \frac{\bar{f}'}{2}N(\Delta x)^2 = \frac{\bar{f}'}{2}\frac{b-a}{\Delta x}(\Delta x)^2 = \frac{\bar{f}'}{2}(b-a)\Delta x,$$

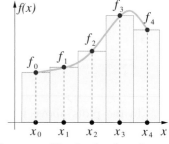

Figure 14.1: The simplest (zeroth-order) approximation to an integral, Eq. (14.12).

where we have used the last relation from Eq. (14.12) to replace N in terms of Δx. This result shows that the total error for applying the zeroth-order formula, Eq. (14.13), is of order Δx. Thus, our final result for the zeroth-order approximation of the integral is

$$0^{\text{th}} \text{ order}: \quad I(a,b) \approx \sum_{n=0}^{N} f_n \Delta x + \mathcal{O}(\Delta x). \tag{14.14}$$

The estimate of the error generalizes in the sense that the nth-order approximation of the function in terms of its Taylor expansion leads to an error in the integral of $\mathcal{O}\left[(\Delta x)^{n+1}\right]$, as we mention is the examples below.

To improve on this result we must keep higher order terms in the Taylor expansion. The next two terms give the methods referred to as the "trapezoid formula" and "Simpson's 1/3-rule".

14.4.1   The trapezoid formula

Keeping the first two terms in the Taylor expansion at each point, namely those containing $f(x_n)$ and $f'(x_n)$, which is the first-order approximation, and performing an integral between successive points we find:

$$\int_{x_n}^{x_{n+1}} f(x)\mathrm{d}x \approx \int_{x_n}^{x_{n+1}} \left[f(x_n) + (x - x_n)f'(x_n)\right] \mathrm{d}x$$

$$= f_n \left[x\right]_{x_n}^{x_{n+1}} + \frac{1}{2} f'(x_j)\left[(x - x_n)^2\right]_{x_n}^{x_{n+1}} = f_n \Delta x + \frac{1}{2} f'(x_n)(\Delta x)^2.$$

Next, we substitute for the derivative $f'(x_j)$ the numerical approximation that involves only the points x_n and x_{n+1}, namely

$$f'(x_n) \approx \frac{f(x_{n+1}) - f(x_n)}{\Delta x}.$$

Then we obtain

$$\int_{x_n}^{x_{n+1}} f(x)\mathrm{d}x \approx f_n \, \Delta x + \frac{1}{2}\left[f_{n+1} - f_n\right] \Delta x = \frac{1}{2}\left[f_{n+1} + f_n\right] \Delta x.$$

This expression is equivalent to the integral of the function $f(x)$ between the points x_n and x_{n+1} being equal to the area of the *trapezoid* that has two parallel vertical sides equal to $f(x_n)$ and $f(x_{n+1})$, respectively, and a horizontal side equal to Δx, as shown in Fig. 14.2. Hence the resulting formula is known as the "trapezoid rule".

The error in the trapezoid rule, obtained from the first term that has been *neglected*, namely $(f''(x_n)/2)(x - x_n)^2$, is of order $(\Delta x)^3$. Thus, applying this rule to all the points in the interval from $x_0 = a$ to $x_N = b$ we obtain

$$1^{\text{st}} \text{ order}: \quad I(a,b) \approx \frac{\Delta x}{2}\left[f_0 + 2\sum_{n=1}^{N-1} f_n + f_N\right] + \mathcal{O}\left[(\Delta x)^2\right], \tag{14.15}$$

where the estimate of the order of the error is obtained by the same procedure as in case of the zeroth-order approximation.

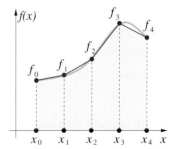

Figure 14.2: The trapezoid (first-order) approximation, to an integral Eq. (14.15). In both plots, the spacing of x values is $x_{n+1} - x_n = \Delta x$, and the corresponding values of the function are f_n, $n = 0, \ldots, 4$.

14.4.2   Simpson's 1/3-rule formula

The next level of approximation is to include one more term in the Taylor expansion, namely the one proportional to $f''(x_n)$, and repeat the process of integrating this approximate function. However, in order to be able to use the symmetric rule for the first and second derivatives, in this case we integrate over the interval, $[x_{n-1}, x_{n+1}]$. These steps give

$$\int_{x_{n-1}}^{x_{n+1}} f(x)dx \approx \int_{x_{n-1}}^{x_{n+1}} \left[f(x_n) + (x - x_n)f'(x_n) + \frac{1}{2}f''(x_n)(x - x_n)^2 \right] dx$$

$$= f_n [x]_{x_{n-1}}^{x_{n+1}} + \frac{1}{2}f'(x_n) [(x - x_n)^2]_{x_{n-1}}^{x_{n+1}} + \frac{1}{6}f''(x_n) [(x - x_n)^3]_{x_{n-1}}^{x_{n+1}}$$

$$= \Delta x \left[2f_n + \frac{1}{3}f''(x_n)(\Delta x)^2 \right].$$

In this expression the first derivative has vanished. For the second derivative we use the symmetric formula of Eq. (14.8), and substituting this in the above expression for the integral we find

$$\int_{x_{n-1}}^{x_{n+1}} f(x)dx \approx \Delta x \left[2f_n + \frac{1}{3}(f_{n+1} - 2f_n + f_{n-1}) \right] = \frac{\Delta x}{3} [f_{n+1} + 4f_n + f_{n-1}].$$

There are three points involved in this result, x_{n-1}, x_n, x_{n+1}, and the corresponding values of the function, f_{n-1}, f_n, f_{n+1}. Since we used the first three terms in the Taylor expansion to obtain this formula, it is equivalent to fitting a second-order (quadratic) polynomial to each segment of the function that contains three points. These segments must not overlap, and therefore the selection of triplets of points consists of the set f_0, f_1, f_2 first, the set f_2, f_3, f_4 second, and so on, as illustrates in Fig. 14.3. Fitting each triplet of points by a quadratic polynomial, which has three coefficients, means that the polynomial passes through all values of the function, f_{n-1}, f_n, f_{n+1}.

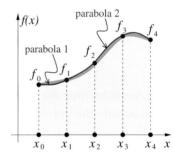

Figure 14.3: The Simpson's "1/3-rule" approximation for calculating an integral numerically, illustrated for $N = 4$. One quadratic polynomial ("parabola 1") is fit to the three values of the function f_0, f_1, f_2, and a different quadratic polynomial ("parabola 2") to the next three values of the function f_2, f_3, f_4.

Applying the above three-point formula repeatedly as described above, we obtain

$$\int_{x_0}^{x_N} f(x)dx \approx \frac{\Delta x}{3} [f_0 + 4f_1 + f_2 + f_2 + 4f_3 + f_4 + \cdots].$$

The generalization of this formula to a number of points N is given by the following expression:

$$\text{2}^{\text{nd}} \text{ order}: \quad I(a,b) \approx \frac{\Delta x}{3} \left[f_0 + 4 \sum_{j=1}^{N/2} f_{2j-1} + 2 \sum_{j=1}^{(N/2)-1} f_{2j} + f_N \right] + \mathcal{O}\left[(\Delta x)^3 \right],$$

$$(14.16)$$

where the first sum includes all the *odd* values of the index,

$$n = 2j - 1, \;\; j = 1, 2, \ldots, \frac{N}{2} \Rightarrow n = 1, 3, \ldots, N - 1,$$

and the second sum includes all the *even* values of the index

$$n = 2j, \;\; n = 1, 2, \ldots, \frac{N}{2} - 1 \Rightarrow n = 2, 4, \ldots, N - 2.$$

Evidently, for the formula of Eq. (14.16) to work, N must be an even number. The estimate of the order of the error is again obtained by the same procedure as in case of the zeroth-order approximation.

Example 14.2: Numerical evaluation of a definite integral

To demonstrate the efficacy of the different methods, we apply them to obtain numerically the value of the following integral:

$$I(-\pi, \pi) = \int_{-\pi}^{\pi} \frac{\cos(\pi x)}{\sqrt{4-x}} dx = -0.231761303.$$

In the table below we show a set of results for the zeroth-order, first-order, and second-order approximations, Eqs. (14.14), (14.15), (14.16), respectively, and for different number of grid points, N. For the zeroth-order result, we also give the error $\mathcal{E}_0(\Delta x)$ and the grid spacing $\Delta x = 2\pi/N$.

	Zeroth order			First order	Second order
N	$I(-\pi, \pi)$	$\mathcal{E}_0(\Delta x)$	Δx	$I(-\pi, \pi)$	$I(-\pi, \pi)$
16	-0.471059	-0.239298	0.392699	-0.213433	-0.232160
64	-0.295026	-0.063265	0.098175	-0.230620	-0.231762
256	-0.247792	-0.016031	0.024544	-0.231690	-0.231761
1024	-0.235782	-0.004021	0.006136	-0.231757	-0.231761
4096	-0.232767	-0.001006	0.001534	-0.231761	-0.231761
16,384	-0.232013	-0.000252	0.000383	-0.231761	-0.231761
65,536	-0.231824	-0.000063	0.000096	-0.231761	-0.231761
262,144	-0.231777	-0.000016	0.000024	-0.231761	-0.231761
1,048,576	-0.231765	-0.000004	0.000006	-0.231761	-0.231761

The second-order method, Simpson's 1/3-rule formula of Eq. (14.16), converges to the exact answer to six significant digits for $N = 64$, whereas the zeroth- and first-order methods require many more points to converge to the exact answer with the same accuracy: over four thousand points are required for the first-order method and over one million points for the zeroth-order method. It is worth pointing out that in the three numerical evaluations of the integral, for a given N we use exactly the same values of the function, $f_n = f(x_n)$, but with different coefficients in each case, as determined by the corresponding formulas.

The convergence rate with the number of points N, or equivalently the size of the interval Δx between successive points, is consistent with our estimate of the error. For the zeroth-order case, the calculated error $\mathcal{E}_0(\Delta x)$, namely the difference between the numerical value for the chosen Δx and the actual value of the integral, tracks closely the value of Δx, by a factor of $\sim 2/3$. This is expected from the expression we postulated for this case, $\mathcal{O}(\Delta x)$. Note that the sign of the error *cannot* predicted by our analysis, which provides only its order of magnitude, unless we include more information like the average value of the derivative.

14.5 Numerical solution of ODEs

14.5.1 First-order ODEs

A general first-order ODE in the unknown function $u(x)$ is

$$\frac{du}{dx} = f(u, x), \tag{14.17}$$

where $f(u, x)$ is a given function of $u(x)$ and the independent variable is x. To solve this type of equation we also need to know the relevant boundary conditions (for more details see Chapter 8, Section 8.2). The numerical solution of ODEs is based on the same ideas as those introduced for the numerical evaluation of derivatives.

Euler's method: The simplest numerical solution is to use an expression for the derivative of the function $u(x)$ like the forward derivative, Eq. (14.5):

$$\frac{u(x_{n+1}) - u(x_n)}{\Delta x} = f(u(x_n), x_n) \Rightarrow u_{n+1} = u_n + \Delta x \, f(u_n, x_n), \qquad (14.18)$$

where we have adopted the notation $u_n = u(x_n)$. This already provides a formula for constructing a numerical solution if the value of the function $u(x)$ is known at some initial value of the variable $x = x_0$: we iterate the above expression using a regularly-spaced grid of points x_n, see Eq. (14.3a). This is known as "Euler's method". It turns out that Euler's method is not very accurate, because the numerical solution quickly diverges from the exact solution.

For a simple example, we consider the following equation:

$$\frac{du}{dx} = a \, u(x) \Rightarrow f(u, x) = a \, u(x), \qquad (14.19)$$

whose solution is

$$u(x) = Ae^{ax}. \qquad (14.20)$$

The value of the constant A is determined by the boundary condition $u(x_0) = u_0$:

$$u(x_0) = Ae^{ax_0} = u_0 \Rightarrow A = u_0 e^{-ax_0}.$$

The numerical solution using Euler's method, obtained by applying Eq. (14.18) from the first value u_0 at x_0 to the last value u_N at x_N, is shown in Fig. 14.4. The limitations of Euler's method are evident from the comparison of the numerical solution to the exact result, Eq. (14.20).

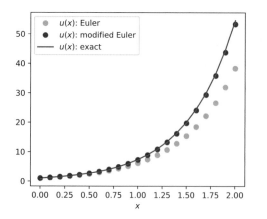

Figure 14.4: Example of numerical solution of the first-order ODE of Eq. (14.19), using Euler's method, Eq. (14.18) (yellow dots) and the modified Euler's method, Eq. (14.21). For this example the values of the parameters are: $a = 2$, $x_0 = 0$, $u_0 = 1$, $N = 20$, $x_N = 2$ and therefore $\Delta x = (x_N - x_0)/N = 0.1$. The exact solution, from Eq. (14.20), is shown as a red line.

Modified Euler's method: To improve on Euler's method, we use the approximate solution for $u(x)$ at x_{n+1} obtained from Eq. (14.18) to evaluate the right-hand side of the ODE at x_{n+1}, namely

$$f(x_{n+1}, u_{n+1}) \approx f(x_{n+1}, u_n + \Delta x\, f(x_n, u_n)).$$

This provides information on the *derivative* of the function $u(x)$ at x_{n+1}. Then, we take advantage of this information to obtain a better approximation for u_{n+1} by using the *average* of the two values of the function $f(x, u)$, at the points (x_n, u_n) and (x_{n+1}, u_{n+1}):

$$u_{n+1} = u_n + \frac{1}{2}\Delta x\left[f(x_n, u_n) + f(x_{n+1}, u_{n+1})\right]$$

$$\Rightarrow \quad u_{n+1} \approx u_n + \frac{1}{2}\Delta x\left[f(x_n, u_n) + f(x_{n+1}, u_n + \Delta x\, f(x_n, u_n))\right]. \quad (14.21)$$

The last expression provides an improved formula for propagating the values of the unknown function from the boundary x_0 where its value is given as u_0, to the desired final value, and is known as the "modified Euler's method". The improvement in accuracy is demonstrated in Fig. 14.4.

Example 14.3: The SIR epidemic model

We present here a simple yet powerful application of the numerical methods described so far to a non-trivial case of first-order ODEs. Many physical phenomena can be described by models that involve several ODEs, each with several unknown functions that also appear in other equations. This type of problem is referred to as "coupled ODEs". For example, a simple mathematical description of the spread of a disease in a population is the so-called "SIR model", which divides the total population of N individuals into three groups: the "susceptible" individuals denoted by S, that is, those not yet infected, the "infected" (or "infectious") individuals denoted by I, and the "recovered" individuals denoted by R, that is, those who were infected and then cured of the disease, now with immunity. All three populations are functions of time t, so that at any moment

$$S(t) + I(t) + R(t) = N.$$

The SIR model as originally developed to describe the spread of epidemics [see W. O. Kermack and A. G. McKendrick, A contribution to the mathematical theory of epidemics, *Proceedings of the Royal Society A*, **115**, 772 (1927)] involves two parameters, β and γ:

- β describes the effective contact rate of the disease: an infected individual comes into contact with β other individuals per unit time (the fraction that are susceptible to contracting the disease is S/N);

- γ is the mean recovery rate; equivalently, $(1/\gamma)$ is the mean period of time during which an infected individual can pass it on.

The evolution of the three population in the SIR model is captured by the following coupled first-order differential equations:

$$\frac{dI}{dt} = \beta I \frac{S}{N} - \gamma I, \quad \frac{dS}{dt} = -\beta I \frac{S}{N}, \quad \frac{dR}{dt} = \gamma I.$$

An analytical solution for these equations is not easy to derive. The numerical solution on the other hand, using the methods discussed above, is rather straightforward. In the plot below we show the numerical solution of these equations for the three populations as functions of time using the modified Euler's method. The values of the parameters for this numerical solution are: $\beta = 0.3$, $\gamma = 1/10$, $N = 1000$, and the initial conditions $I(0) = 1$, $S(0) = N - I(0)$.

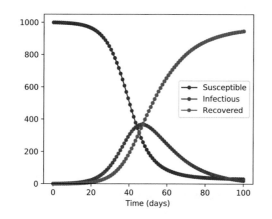

The solution describes quite accurately the observed evolution of the populations as functions of time, under the assumption that no measures to contain the epidemic are imposed and no cure or vaccine exists. This predicted behavior can then be used to decide the necessity of imposing measures on the allowed activities of the population, with the aim of shortening the duration of the epidemic. Such severe measures, known as "lockdown", were imposed during the COVID-19 pandemic.

14.5.2   Second-order ODEs

A general second-order ODE in the unknown function $u(x)$ is

$$\frac{d^2 u}{dx^2} = g(x, u, u'), \tag{14.22}$$

where $g(x, u, u')$ is a given function of $u(x)$, its first derivative is

$$u'(x) = \frac{du}{dx},$$

and the independent variable is x. This ODE must also be accompanied by appropriate boundary conditions. These can be values of the function and the first derivative at the same point, or values of the function at two different points, or some other combination of values of the function and its derivative. To solve this type of equation we follow steps similar to those applied for the numerical solution of first-order ODEs.

Euler's method: An efficient numerical method for second-order ODEs is to use the central formula for the derivative twice, first for the function itself, and then for the first derivative:

$$\frac{u(x_{n+1}) - u(x_{n-1})}{2\Delta x} = u'(x_n) \Rightarrow u_{n+1} = u_{n-1} + 2\,\Delta x\,u'_n, \qquad (14.23a)$$

$$\frac{u'(x_{n+1}) - u'(x_{n-1})}{2\Delta x} = u''(x_n) = g(x_n, u(x_n), u'(x_n))$$

$$\Rightarrow \quad u'_{n+1} = u'_{n-1} + 2\,\Delta x\,g(x_n, u_n, u'_n). \qquad (14.23b)$$

The set of Eqs. (14.23a) and (14.23b) allows the evaluation of the function and its derivative at the point x_{n+1} assuming that we know the values of the function and its derivative at the two previous points x_n and x_{n-1}. The equations involve the small quantity Δx only to the first power, even though the *original* ODE, Eq. (14.22), involves a second derivative of the function, and therefore we might expect to have to deal with the second power of Δx, as in the formula for the second derivative of an ordinary function, Eq. (14.8).

Starting with just the two initial values $u_0 = u(x_0)$ and $u'_0 = u'(x_0)$, and using Eqs. (14.23a) and (14.23b) to find successive values of the function and its derivative, we encounter the obvious problem that for each point we need the values of u and u' at the previous *two* points, which cannot be applied at the very beginning of the process. To circumvent this difficulty, we can use the two given values, u_0, u'_0 to find the values $u(x_1) = u_1$ and $u'(x_1) = u'_1$ by extrapolation:

$$\frac{u(x_1) - u(x_0)}{\Delta x} = u'(x_0) \Rightarrow u_1 = u_0 + \Delta x\,u'_0,$$

$$\frac{u'(x_1) - u'(x_0)}{\Delta x} = u''(x_0) = g(x_0, u_0, u'_0) \Rightarrow u'_1 = u'_0 + \Delta x\,g(x_0, u_0, u'_0).$$

Having obtained these values we can iterate the formulas of Eqs. (14.23a) and (14.23b) to find the numerical solution for $u(x_n)$ at values of $x_n > x_0$. This is known as "Euler's method" for the second-order ODE. An example is shown in Fig. 14.5 for the simple case of the equation

$$\frac{d^2u}{dx^2} = -a^2 u(x), \quad u_0 = 0, \quad u'_0 = 1, \qquad (14.24)$$

whose solution is $u(x) = \sin(ax)$. Evidently, this approach is not very accurate as the numerical solution deviates from the exact solution significantly.

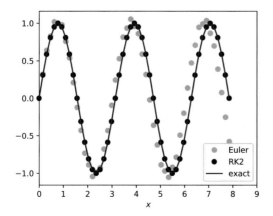

Figure 14.5: Example of numerical solution of the second-order ODE of Eq. (14.24), using Euler's method, Eqs. (14.23a) and (14.23b) (yellow dots) and the Runge–Kutta method of second order (RK2), Eqs. (14.27a)–(14.27c). For this example we used the values of the parameters $a = 2$, $x_0 = 0$, $u_0 = 0$, $u'_0 = 2$, $N = 50$, $x_N = 2.5\pi$ and therefore $\Delta x = (x_N - x_0)/N = 0.05\pi$. The analytical (exact) solution, $u(x) = \sin(2x)$, is shown as a red line.

The Runge–Kutta method: An improvement of Euler's method is the so-called "Runge-Kutta" method. This method relies on evaluating the derivatives at the mid-point of the interval h between successive values of x, and using this information to propagate from x_n to the next value x_{n+1}. We illustrate the method for the case of a second-order ODE: we begin by reducing the original second-order ODE to a set of coupled first-order ODEs, by the following process:

$$v \equiv \frac{\mathrm{d}u}{\mathrm{d}x} \Rightarrow \frac{\mathrm{d}^2 u}{\mathrm{d}x^2} = \frac{\mathrm{d}v}{\mathrm{d}x} = g(x, u, u'). \tag{14.25}$$

We then have to solve the following two coupled first-order ODEs:

$$\frac{\mathrm{d}u}{\mathrm{d}x} = v(x), \quad \frac{\mathrm{d}v}{\mathrm{d}x} = g(x, u, v). \tag{14.26}$$

This is accomplished by the following steps:

$$k_n^{(1)} = v_n, \quad l_n^{(1)} = g(x_n, u_n, v_n) \tag{14.27a}$$

$$k_n^{(2)} = v_n + \frac{h}{2}l_n^{(1)}, \quad l_n^{(2)} = g\left(x_n + \frac{h}{2}, u_n + \frac{h}{2}k_n^{(1)}, v_n + \frac{h}{2}l_n^{(1)}\right), \tag{14.27b}$$

with the final expressions for u_{n+1} and v_{n+1} being

$$u_{n+1} = u_n + hk_n^{(2)}, \quad v_{n+1} = v_n + hl_n^{(2)}, \tag{14.27c}$$

where h is the length of the interval $[x_n, x_{n+1}]$, namely $h = (x_{n+1} - x_n)$. This scheme is referred to as the "second order" Runge-Kutta method, or RK2.

The geometric interpretation of the RK2 method is illustrated in Fig.14.6: The quantity $k_n^{(1)}$ is simply the value of v at x_n, where we assume the values u_n and v_n have already been calculated. The quantity $l_n^{(1)}$, which depends on x_n as well as on u_n and v_n, is equal to the *derivative* of the function v at this point. Therefore, if we linearly extrapolate the value of $k_n^{(1)}$ to the *midpoint* of the interval by using this value of the derivative, we obtain the new value $k_n^{(2)}$. This can be used as an estimate of the value of the function v at the mid-point. By the same logic, we use the derivative of the function u at x_n, which is the value of v_n, to extrapolate its value to the *midpoint*, which produces the value

$$u_n + \frac{h}{2}k_n^{(1)}.$$

With these values of v and u at the midpoint, we can evaluate the function $g(x, u, v)$, which gives the value of $l_n^{(2)}$, the derivative of the function v at the midpoint. We now have improved estimates of the derivatives at the midpoint of the interval, which we employ to update the values of the functions u and v from the current values u_n and v_n to the new values u_{n+1} and v_{n+1}.

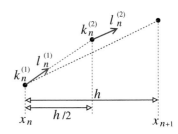

Figure 14.6: Geometric interpretation of the second-order Runge-Kutta method, RK2, defined by Eqs (14.27). The colored dashed lines have the same slope as the corresponding colored arrows, which represent the values of the derivative of the function v, at the starting point x_n (red), and at the midpoint $x_n + h/2$ (blue).

As usual, we want to establish the order of the error in the RK2 method. It turns out that the evaluation of the derivatives at the midpoint eliminates the first-order error terms and leads to a local error of order $(\Delta x)^3$, or, an overall error of second order, hence the designation of this method as "second order".

The Runge–Kutta method is versatile in the sense that it can be readily generalized to higher-order schemes. In fact, the most commonly used scheme of this type is the so-called "fourth-order" Runge–Kutta (RK4) method. This method consists of the following steps, based again on the two coupled first-order ODEs of Eq. (14.26):

$$k_n^{(1)} = v_n, \qquad l_n^{(1)} = g\left(x_n, u_n, k_n^{(1)}\right), \tag{14.28a}$$

$$k_n^{(2)} = v_n + \frac{h l_n^{(1)}}{2}, \qquad l_n^{(2)} = g\left(x_n + \frac{h}{2}, u_n + \frac{h k_n^{(1)}}{2}, k_n^{(2)}\right), \tag{14.28b}$$

$$k_n^{(3)} = v_n + \frac{h l_n^{(2)}}{2}, \qquad l_n^{(3)} = g\left(x_n + \frac{h}{2}, u_n + \frac{h k_n^{(2)}}{2}, k_n^{(3)}\right), \tag{14.28c}$$

$$k_n^{(4)} = v_n + h l_n^{(3)}, \qquad l_n^{(4)} = g\left(x_{n+1}, u_n + h k_n^{(3)}, k_n^{(4)}\right), \tag{14.28d}$$

and the updated valued of u_{n+1} and v_{n+1} are given by the expressions

$$u_{n+1} = u_n + \frac{h}{6}\left[k_n^{(1)} + 2\,k_n^{(2)} + 2\,k_n^{(3)} + k_n^{(4)}\right], \tag{14.28e}$$

$$v_{n+1} = v_n + \frac{h}{6}\left[l_n^{(1)} + 2\,l_n^{(2)} + 2\,l_n^{(3)} + l_n^{(4)}\right]. \tag{14.28f}$$

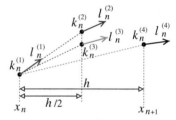

Figure 14.7: Geometric interpretation of the fourth-order Runge–Kutta method, RK4, defined by Eqs. (14.28a)–(14.28f). The colored dashed lines have the same slope as the corresponding colored arrows, which represent the values of the derivative of the function v, at the starting point x_n (red), and at the midpoint $x_n + h/2$ (blue and green).

A geometric interpretation of the steps involved in the RK4 scheme is shown in Fig. 14.7: The first step defines the value of the function v at x_n as $k_n^{(1)}$ and its derivative as $l_n^{(1)}$. In the second step, the value of v, called $k_n^{(2)}$, is obtained by linear extrapolation to the *midpoint* of the interval $[x_n, x_{n+1}]$, that is, at $x = x_n + h/2$, using the derivative $l_n^{(1)}$. At the same time, the new value of the derivative, called $l_n^{(2)}$, is also determined using this new value of v at the midpoint. In the third step, the new value of the derivative at the midpoint, $l_n^{(2)}$, is used to extrapolate anew, starting from the value v_n at x_n to a new midpoint value, called $k_n^{(3)}$. As in the previous steps, the value of the derivative at the midpoint, called $l_n^{(3)}$, is also updated with the $k_n^{(3)}$ estimate of v at the midpoint. In the fourth step the value of v at the *endpoint*, namely x_{n+1}, called $k_n^{(4)}$, is obtained by linear extrapolation from v_n by using the last estimate of the derivative at the midpoint, $l_n^{(3)}$, and the new estimate of the value of the derivative of v at the endpoint is obtained, called $l_n^{(4)}$, with the latest value of v, namely $k_n^{(4)}$. Finally, to obtain the values of u_{n+1} and v_{n+1}, a *weighted average* is used, which includes the quantities at the two endpoints and at the midpoint, given in Eqs. (14.28e) and (14.28f). The advantage of this rather elaborate updating scheme is that the resulting method is fourth order in the grid spacing Δx, and therefore quite accurate.

14.6 Numerical solution of PDEs

Solving PDEs with numerical methods involves steps similar to those discussed in the context of the numerical evaluation of derivatives and the solution of ODEs. However, the more complicated nature of PDEs requires commensurately more sophisticated numerical methods, the detailed discussion of which is beyond the scope of the present book. Instead, we will discuss only one evolution equation to illustrate some of the common themes encountered in the numerical solution of PDEs.

Let us consider the familiar heat equation for the function $u(x,t)$, namely $u_t = \alpha u_{xx}$, which was discussed at length in Chapters 8–10. For the computational grid associated with the spatial variable x, we will use

$$x_0 = 0, \ x_N = L, \quad x_n = n\,\Delta x, \quad \Delta x = \frac{L}{N}, \quad n = 0, 1, \ldots, N, \qquad (14.29)$$

while for the grid associated with the temporal variable t, we will use

$$t_0 = 0, \ t_N = T, \quad t_m = m\,\Delta t, \quad \Delta t = \frac{T}{M}, \quad m = 0, 1, \ldots, M. \qquad (14.30)$$

The initial–boundary conditions we will assume for the following discussion are

$$u(x,0) = u_0(x), \text{ for } 0 \leq x \leq L, \quad u(0,t) = u(L,t) = 0, \text{ for } 0 \leq t \leq T, \quad (14.31)$$

which imply that the initial function must also satisfy $u_0(0) = 0$ and $u_0(L) = 0$.

For the second-order spatial derivative we will use the central formula, namely Eq. (14.8), which in the present case takes the form

$$u_{xx}(x_n, t_m) = \frac{u_{n+1}^{(m)} - 2u_n^{(m)} + u_{n-1}^{(m)}}{(\Delta x)^2}. \qquad (14.32)$$

The choice of the numerical formula used for the first-order time derivative, makes a difference in the numerical scheme, so we consider the various possibilities separately.

Forward time derivative: As our first choice we adopt the forward formula for the time derivative, Eq. (14.5), which here takes the form:

$$u_t(x_n, t_m) = \frac{u_n^{(m+1)} - u_n^{(m)}}{\Delta t}. \qquad (14.33)$$

This gives the following discrete form for the heat equation:

$$u_t(x_n, t_m) = \alpha u_{xx}(x_n, t_m) \Rightarrow \frac{u_n^{(m+1)} - u_n^{(m)}}{\Delta t} = \alpha \frac{u_{n+1}^{(m)} - 2u_n^{(m)} + u_{n-1}^{(m)}}{(\Delta x)^2},$$

which we solve for $u_n^{(m+1)}$ to obtain

$$u_n^{(m+1)} = \gamma u_{n-1}^{(m)} + (1 - 2\gamma)u_n^{(m)} + \gamma u_{n+1}^{(m)}, \qquad (14.34)$$

where we have defined the dimensionless constant γ as

$$\gamma \equiv \alpha \frac{\Delta t}{(\Delta x)^2}. \tag{14.35}$$

Equation (14.34) has a useful structure: it gives the value of the unknown function u at the spatial point x_n and at time t_{m+1} in terms of its values at the *previous* time step t_m and the positions x_n and $x_{n\pm 1}$, which are the neighbors of x_n. From Eq. (14.31), the boundary conditions on the computational grid become:

$$u_n^{(0)} = u_0(x_n), \ \forall n \in [0, N], \quad \text{and} \quad u_0^{(m)} = 0, \ u_N^{(m)} = 0, \ \forall m \in [0, M].$$

From the first set of boundary conditions at $t = 0$ we realize that we have all the information needed, namely $u_n^{(0)}$ for all values of $n \in [0, N]$, to calculate the values $u_n^{(1)}$ using Eq. (14.34). The only problematic points are the endpoints of the interval, that is, the values $n = 0$ and $n = N$ on the left-hand side of Eq. (14.34) for which the indices $n \pm 1$ of u on the right-hand side would exceed the allowed range; for these cases we can simply ignore the terms whose index falls outside the allowed range. Thus, we can propagate the values of the unknown function u, starting with the initial set of values $u_n^{(0)}$ which are provided by the initial conditions, all the way to the end of the time interval, $m = M$. For this reason, this scheme is referred to as an "explicit propagation" method.

The error involved in this solution at each step is of first order in Δt and second order in Δx, as is evident from Eq. (14.34) and the definition of γ, Eq. (14.35).

An important consideration is the reliability of the propagation scheme, that is, whether the values it produces are a reasonable approximations to the exact solution. This depends on the values of Δt and Δx. As a check, the numerical solution must be consistent with known constraints on the values of the solution. An important condition that must hold for the solution of the heat equation is the following:

$$\min_{0 \le x' \le L} \{u(x', t')\} \le u(x, t) \le \max_{0 \le x' \le L} \{u(x', t')\}, \quad \forall x \in [0, L], \text{ for } t \ge t'.$$
$$\tag{14.36}$$

This condition is known as the "maximum principle" for the heat equation and applies when the boundary condition is of the type specified in Eq. (14.31), namely that $u(x, t)$ vanishes at the boundary for all t (see the related discussion for the case of $u_0(x)$ being a δ-function, in Eq. (8.30), Chapter 8, Section 8.3.2). Eq. (14.36) implies the relation

$$\max_{0 \le x \le L} \{|u(x, t)|\} \le \max_{0 \le x \le L} \{|u(x, t')|\}, \quad \text{for } t \ge t'. \tag{14.37}$$

When translated to the computational grids for the x and t variables, Eqs. (14.29) and (14.30), the last expression takes the form

$$\max_n \{|u_n^{(m+1)}|\} \le \max_n \{|u_n^{(m)}|\}, \quad \forall m \in [0, M]. \tag{14.38}$$

In order for this condition to be satisfied, we must choose values for Δx and Δt such that

$$\Delta t \le \frac{1}{2\alpha}(\Delta x)^2 \Rightarrow \gamma \le \frac{1}{2}. \tag{14.39}$$

Indeed, when this last condition is satisfied, and therefore $(1 - 2\gamma) \geq 0$, using the relation of Eq. (14.34), taking the absolute value of both sides and making use of the triangle inequality, Eq. (1.89), we obtain

$$\left| u_n^{(m+1)} \right| \leq \gamma \left| u_{n-1}^{(m)} \right| + (1 - 2\gamma) \left| u_n^{(m)} \right| + \gamma \left| u_{n+1}^{(m)} \right|.$$

This last inequality remains valid if we replace each term on the right-hand side by the maximum value of $\left| u_n^{(m)} \right|$ over all values of n, which gives

$$\left| u_n^{(m+1)} \right| \leq \max_n \left\{ \left| u_n^{(m)} \right| \right\}.$$

Since this holds for any value of n on the left-hand side, we can replace the left-hand side term by its maximum value, which leads to the desired result, namely Eq. (14.38).

The condition of Eq. (14.39) imposes severe constraints on the allowed values of Δt, which must be very small since Δx itself must be a small quantity.

This relation is also referred to as the "stability condition", because when it is obeyed the numerical method is stable, that is, it produces a reasonable approximation to the exact solution. When the condition is *not* obeyed, the numerical method quickly diverges from the exact solution (the method is "unstable").

Backward time derivative: As our second choice we adopt the backward formula for the time derivative, Eq. (14.6), which here takes the form:

$$u_t(x_n, t_{m+1}) = \frac{u_n^{(m+1)} - u_n^{(m)}}{\Delta t}. \tag{14.40}$$

This gives the following discrete form for the heat equation:

$$u_t(x_n, t_{m+1}) = \alpha u_{xx}(x_n, t_{m+1})$$

$$\Rightarrow \quad \frac{u_n^{(m+1)} - u_n^{(m)}}{\Delta t} = \alpha \frac{u_{n+1}^{(m+1)} - 2u_n^{(m+1)} + u_{n-1}^{(m+1)}}{(\Delta x)^2},$$

from which we obtain

$$-\gamma u_{n-1}^{(m+1)} + (1 + 2\gamma) u_n^{(m+1)} - \gamma u_{n+1}^{(m+1)} = u_n^{(m)}, \tag{14.41}$$

with the same definition of the constant γ as before, namely the one given in Eq. (14.35). The structure of the above equations is the reverse of what we had in Eq. (14.34). Specifically, in Eq. (14.41) we have three values of the function at the time step t_{m+1}, $u_n^{(m+1)}$ and $u_{n\pm1}^{(m+1)}$, related to a single value at time step t_m, $u_n^{(m)}$. We can view this as a system of $N + 1$ equations that can be written in matrix form as

$$\mathbf{G} \cdot \mathbf{u}^{(m+1)} = \mathbf{u}^{(m)},$$

where $\mathbf{u}^{(m)}$ is a vector of length $N + 1$ with entries $u_n^{(m)}, n = 0, \ldots, N$, and $\mathbf{G}$ is a square matrix of size $(N + 1) \times (N + 1)$ with the following form:

$$\mathbf{G} = \begin{bmatrix} 1 + 2\gamma & -\gamma & 0 & 0 & \cdots & 0 & 0 & 0 & 0 \\ -\gamma & 1 + 2\gamma & -\gamma & 0 & \cdots & 0 & 0 & 0 & 0 \\ 0 & -\gamma & 1 + 2\gamma & -\gamma & \cdots & 0 & 0 & 0 & 0 \\ \vdots & \vdots & \vdots & \vdots & \ddots & \vdots & \vdots & \vdots & \vdots \\ 0 & 0 & 0 & 0 & \cdots & -\gamma & 1 + 2\gamma & -\gamma & 0 \\ 0 & 0 & 0 & 0 & \cdots & 0 & -\gamma & 1 + 2\gamma & -\gamma \\ 0 & 0 & 0 & 0 & \cdots & 0 & 0 & -\gamma & 1 + 2\gamma \end{bmatrix}.$$

$\mathbf{G}$ is called a "tri-diagonal matrix" because all its elements are zero except the diagonal elements which are all equal to $(1 + 2\gamma)$ and the elements just below and just above the diagonal, which are all equal to $-\gamma$. It is straightforward to calculate the inverse of this tri-diagonal matrix, denoted as $\mathbf{G}^{-1}$, using standard matrix inversion routines, which leads to the solution

$$\mathbf{u}^{(m+1)} = \mathbf{G}^{-1} \cdot \mathbf{u}^{(m)}.$$

Starting with the vector $\mathbf{u}^{(0)}$, which represents the initial condition $u_n^{(0)} = u_0(x_n)$, we can then use the above matrix equation to obtain the values of $u_n^{(1)}$ for the next time moment, and so on. Each multiplication of the matrix $\mathbf{G}^{-1}$ with the vector $\mathbf{u}^{(m)}$ produces the entire vector $\mathbf{u}^{(m+1)}$. This approach is called an "implicit propagation" method. The error is again first order in Δt and second order in Δx.

We examine under what condition the implicit propagation method obeys the maximum principle of the heat equation, Eq. (14.38). Starting with Eq. (14.41), we obtain

$$(1 + 2\gamma)u_n^{(m+1)} = \gamma \left(u_{n+1}^{(m+1)} + u_{n-1}^{(m+1)} \right) + u_n^{(m)}.$$

Taking the absolute value of both sides and using the triangle inequality, we find

$$(1 + 2\gamma)\left| u_n^{(m+1)} \right| \leq \gamma \left(\left| u_{n+1}^{(m+1)} \right| + \left| u_{n-1}^{(m+1)} \right| \right) + \left| u_n^{(m)} \right|.$$

Substituting the maximum value of the terms on the right-hand side of the inequality we arrive at

$$(1 + 2\gamma)\left| u_n^{(m+1)} \right| \leq 2\gamma \max_n \left\{ \left| u_n^{(m+1)} \right| \right\} + \max_n \left\{ \left| u_n^{(m)} \right| \right\}.$$

This inequality is true for all values of n on the left-hand side, so we can replace the term on the left-hand side by its maximum value, which gives, after eliminating the common terms on the two sides, the inequality

$$\max_n \left\{ \left| u_n^{(m+1)} \right| \right\} \leq \max_n \left\{ \left| u_n^{(m)} \right| \right\},$$

which is the desired relation, Eq. (14.38). Interestingly, we arrive at this result without having to impose any conditions on the relative magnitude of Δt and Δx, as was necessary in the case of the forward time derivative. Thus, we conclude that the implicit propagation method is *unconditionally stable*.

The Crank–Nicolson method: Finally, we consider a method that improves the performance in terms of the order of the Δt term in the error. The basic idea is to mimic the modified Euler's method in obtaining the derivative, which produces a method of higher order than the simple Euler's method, the latter being the equivalent of the simple forward or backward derivative evaluation. Specifically, we set the time derivative equal to the average of the second spatial derivative, evaluated at the time moments t_m and t_{m+1}. This produces the following expression:

$$\frac{u_n^{(m+1)} - u_n^{(m)}}{\Delta t} = \frac{\alpha}{2} \left[\frac{u_{n+1}^{(m)} - 2u_n^{(m)} + u_{n-1}^{(m)}}{(\Delta x)^2} + \frac{u_{n+1}^{(m+1)} - 2u_n^{(m+1)} + u_{n-1}^{(m+1)}}{(\Delta x)^2} \right].$$

Rearranging terms in the above expression and using the definition of the parameter γ from Eq. (14.35), we find

$$-\gamma u_{n-1}^{(m+1)} + 2(1+\gamma)u_n^{(m+1)} - \gamma u_{n+1}^{(m+1)} = \gamma u_{n-1}^{(m)} + 2(1-\gamma)u_n^{(m)} + \gamma u_{n+1}^{(m)}.$$
$$(14.42)$$

This expression relates three values of u at the time moment t_{m+1} (those on the left-hand side) to three values of u at the time moment t_m (those on the right-hand side). As in the case of the implicit propagation method, Eq. (14.42) can be expressed as a matrix equation, involving two square tri-diagonal matrices, one on each side of the equation. Thus, the solution can be found by using matrix inversion and following the same process as in the implicit propagation method, namely starting with the first vector at t_0 which represents the initial condition to obtain the vectors for $t_m, m > 0$. This method is referred to as the "Crank–Nicolson" method.

The error in the Crank–Nicolson method is second order in Δt as well as in Δx, which is a significant advantage of the method. The Crank–Nicolson method is unconditionally stable, not surprisingly, due to its similarity to the implicit propagation method.

Comparison of the three methods: The three methods for the numerical solution of PDEs have different errors but also different stability conditions and different computational cost. The explicit propagation method has error of $\mathcal{O}\left[\Delta t, (\Delta x)^2\right]$ and, as discussed above, its stability condition is

$$\Delta t \leq \frac{(\Delta x)^2}{2\alpha},$$

which imposes a very stringent limit on the size of the time step, Δt. This limit implies that a very small time step must be used to ensure adequate accuracy: for instance, if Δx is of order 10^{-3} and α is taken to be 1 (which simply sets the units to be used in the expressing the solution), then Δt is of order 10^{-6}, in other words a million time steps are required to cover the entire range of the time variable, 0 to T. On the other hand, the implicit propagation method has the same error but unconditional stability, implying that the time step can be larger than in the explicit propagation method. Thus, the implicit propagation method may seem as a more efficient approach to find a solution for $u(x,t)$ over the given range of time. However, each step of the implicit propagation method involves the multiplication of a square matrix of size $(N+1) \times (N+$

1) with a vector of size $(N+1)$, a computationally costly operation for large N that must be performed *at each time step*. As a result, the latter method is not more efficient than the former one. Finally, the Crank–Nicolson method is certainly of superior performance, since it has an error of $\mathcal{O}\left[(\Delta t)^2, (\Delta x)^2\right]$ and is unconditionally stable. But this method also involves the computationally costly matrix multiplication at each time step, as in the implicit propagation method.

14.7 Solving differential equations with neural networks

A rather recent approach for solving differential equations, is based on machine learning techniques and in particular on the concept of artificial neural networks (ANNs), which was introduced in Chapter 2 (see Section 2.7). ANNs have the property of universal approximation, that is, they can approximate as closely as desired any finite function no matter how complex it may be. Hence they can also approximate the solution of a differential equation.

In the following, we will use the symbol $\hat{\mathcal{L}}$ to denote the differential operator which applies to a function u on the left-hand side of a differential equation, and the symbol $\hat{\mathcal{B}}$ to denote the corresponding differential operator for the boundary conditions. For simplicity, we will denote the independent variable of the function u by x, even though we may be interested in a function with more than one independent variables; in other words, we are using a shorthand notation in which x in the expression $u(x)$ represents *all* the relevant independent variables in the problem. The domain of interest for the values of x will be denoted as Ω and the boundary where the boundary conditions apply will be denoted as $\partial\Omega$. With these conventions for the notation, the differential equation, including the boundary conditions will be represented as

$$\hat{\mathcal{L}}u(x) = f(x), \text{ for } x \in \Omega, \quad \text{and} \quad \hat{\mathcal{B}}u(x) = g(x), \text{ for } x \in \partial\Omega. \quad (14.43)$$

The functions $f(x)$ and $g(x)$ that appear on the right-hand side of the differential equation and the boundary conditions, respectively, can be non-zero in the general case of an inhomogeneous differential equation, or for different types of boundary conditions.

The main idea is to model $u(x)$ in terms of a neural network,[2] that is, by considering the trial solution $u_{\text{trial}}(x)$ given by

$$u_{\text{trial}}(x) \approx \tilde{u}(x; w), \quad (14.44)$$

[2] I. E. Lagaris, A. Likas and D. I. Fotiadis, Artificial neural networks for solving ordinary and partial differential equations, *IEEE Transactions: Neural Networks*, **9**, pp. 987–1000 (1998).

where $\tilde{u}(x; w)$ is the output of the ANN with weights $\{w\}$. Substituting the approximate solution for $u(x)$ from Eq. (14.44) in Eq. (14.43) we obtain

$$\hat{\mathcal{L}}\tilde{u}(x; w) = f(x), \text{ for } x \in \Omega, \quad \text{and} \quad \hat{\mathcal{B}}\tilde{u}(x; w) = g(x), \text{ for } x \in \partial\Omega, \quad (14.45)$$

where the only unknowns are the ANN parameters, namely the weights $\{w\}$. Therefore, the original problem stated in Eq. (14.43) has been transformed to a parameter estimation problem described by Eq. (14.45).

To obtain a numerical solution, we introduce a discretization of the domain Ω and its boundary $\partial\Omega$ by picking a set of M_d domain points $x_i \in \Omega, i = 1, 2, \ldots, M_d$ and a set of M_b boundary points $x_j \in \partial\Omega, j = 1, 2, \ldots, M_b$. For an approximate numerical solution, we then require

$$\hat{\mathcal{L}}\tilde{u}(x_i, w) - f(x_i) \approx 0 \quad \text{for } i = 1, 2, \ldots, M_d, \quad x_i \in \Omega, \tag{14.46a}$$

$$\hat{\mathcal{B}}\tilde{u}(x_j, w) - g(x_j) \approx 0 \quad \text{for } j = 1, 2, \ldots, M_b, \quad x_j \in \partial\Omega. \tag{14.46b}$$

This may be accomplished by minimizing with respect to w (colloquially referred to as "training of the ANN"), the following error function, $\mathcal{E}_1(w)$, which depends only of the ANN weights $\{w\}$:

$$\mathcal{E}_1(w) = \frac{1}{M_d}\sum_{i=1}^{M_d}\left[\hat{\mathcal{L}}\tilde{u}(x_i; w) - f(x_i)\right]^2 + \frac{\lambda}{M_b}\sum_{j=1}^{M_b}\left[\hat{\mathcal{B}}\tilde{u}(x_j; w) - g(x_j)\right]^2. \tag{14.47}$$

In the above expression for the error we have introduced the parameter $\lambda > 0$, which may be used to regulate how accurately the boundary conditions are satisfied. If the minimum error value $\mathcal{E}_1^* \equiv \mathcal{E}_1(w^*)$ is close to zero, then an approximate numerical solution $u(x) \approx \tilde{u}(x; w^*)$ has been obtained. The values of the weights $\{w^*\}$ for which the error is close to zero can be determined by using standard numerical optimization techniques.

We note that the function $\tilde{u}(x, w)$ and its derivatives, which depend on the parameters $\{w\}$, are used directly to calculate the error $\mathcal{E}_1(w)$. Therefore, the expression for the error used to optimize the values of the parameters does not need to involve the values of the solution $u(x)$, which is *not* known; this is one of the advantages of the method.

Another advantage of the method is that once the error-minimizing weights $\{w^*\}$ have been determined, the value of the approximate solution $\tilde{u}(x, w^*)$ can be obtained for *any* value of x. In this sense, the approximate solution represents an essentially analytical expression since its derivatives can be obtained analytically by the chain rule of differentiation.

Example 14.4: To illustrate how the ANN solution of a differential equation works, we consider the following simple case of a second-order ODE:

$$\frac{d^2u}{dx^2} - u(x) = 0, \quad x \in [0, 2\pi],$$

with boundary conditions

$$u(0) = 0, \quad u_x(0) = 1.$$

The solution for this ODE was discussed in Chapter 8 (see Section 8.2), so we can compare the numerical solution to the exact analytical result.

From a numerical point of view, it is more convenient to write this ODE as a system of two coupled first-order ODEs by defining the function $v(x)$:

$$v(x) = \frac{du}{dx} \Rightarrow u(x) = \frac{d^2u}{dx^2} = \frac{dv}{dx}$$

[although this step circumvents the above formalism of breaking the problem into the minimization of the $(\hat{\mathcal{L}}\tilde{u} - f)$ and $(\hat{\mathcal{B}}\tilde{u} - g)$ errors]. Then the corresponding boundary condition for $v(x)$ becomes

$$v(0) = u_x(0) = 1.$$

The exact solution of this system of ODEs is given by

$$u(x) = C\sin(x), \quad v(x) = C\cos(x),$$

with C an arbitrary constant, which we choose here to be $C = 1$.

The results of the numerical solutions with the ANN method, and its comparison to the exact expressions for $u(x)$ and $v(x)$ are shown in the plot below [this solution was produced with the software package "NeuroDiffEq", see F. Chen, D. Sondak, P. Protopapas, M. Mattheakis, S. Liu, D. Agarwal, and M. Di Giovanni, NeuroDiffEq: A Python package for solving differential equations with neural networks, *Journal of Open Source Software* **5**, p. 46 (2020)]. The dots are values where the ANN solution is obtained for values of the variable x that were *not* part of the training set, and can therefore be considered as "test" points. The agreement between the numerical and the exact result is quite satisfactory.

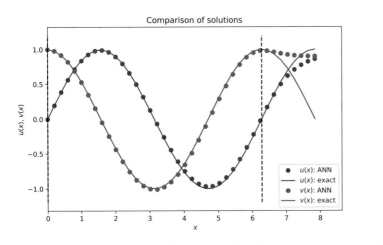

We note that the input to the ANN model, in principle, can be values of x *outside* the range used to "train" the network. However, as was also discussed in Section 2.7, the ANN solution is acceptable only *within* this range of values (in the present case $x \in [0, 2\pi]$), and typically fails for values of x outside this range. To demonstrate this point, we have plotted the numerical solutions for $u(x)$ and $v(x)$ for the range of $x \in [0, 5\pi/2]$. As expected, for $x > 2\pi$ (values to the right of the vertical dashed line) the ANN solution quickly and unpredictably diverges from the exact result.

A slightly different trial solution may sometimes be feasible: Consider expressing the trial solution as

$$u_{\text{trial}}(x; w) = A(x) + F(x, \tilde{u}(x; w)), \tag{14.48a}$$

where again $\tilde{u}(x; w)$ is the output of the ANN with weights $\{w\}$, $A(x)$ is a function without any parameters that satisfies the boundary conditions, that is,

$$\hat{\mathcal{B}}A(x) = g(x), \quad \text{for } x \in \partial\Omega, \tag{14.48b}$$

and $F(x, \tilde{u}(x; w))$ satisfies the equation

$$\hat{\mathcal{B}}F(x, \tilde{u}(x; w)) = 0, \quad \text{for } x \in \partial\Omega. \tag{14.48c}$$

If $\hat{B}$ is a linear operator then the trial solution of Eq. (14.48a) satisfies the boundary conditions by construction, and the training of the ANN amounts to minimizing the following error:

$$\mathcal{E}_2(w) = \frac{1}{M_d} \sum_{i=1}^{M_d} \left[\hat{\mathcal{L}} \left[A(x_i) + F(x_i, \tilde{u}(x_i; w)) \right] - f(x_i) \right]^2, \quad x_i \in \Omega. \quad (14.49)$$

The advantage of this approach is that the boundary conditions are automatically satisfied by construction through the functions $A(x)$ and $F(x, \tilde{u}(x; w))$, which can prove rather useful when the boundaries are irregular.[3] Then, what remains to do is to find the weights that minimize the error $\mathcal{E}_2(w)$ over *only* the domain points $x_i \in \Omega$, a task that can be considerably simpler than minimizing the error $\mathcal{E}_1(w)$ over *both* the domain points $x_i \in \Omega$ and the boundary points $x_j \in \partial\Omega$.

[3] I. E. Lagaris, A. Likas and D. G. Papageorgiou, Neural-network methods for boundary value problems with irregular boundaries, *IEEE Transactions: Neural Networks and Learning Systems* **11**, pp. 1041–1049 (2000).

We illustrate how this approach works for different types of boundary conditions, by considering the case of a second-order ODE, with the independent variable x in the range $x \in [a, b]$.

(a) For Dirichlet boundary conditions:

$$u(a) = u_0 \text{ and } u(b) = u_1,$$

$A(x)$ may be constructed as follows:

$$A(x) = u_0 \frac{b - x}{b - a} + u_1 \frac{x - a}{b - a}.$$

Two alternative choices for the function $F(x, \tilde{u}(x; w))$ are given by

$$F_1(x, \tilde{u}(x; w)) = (x - a)(x - b)\tilde{u}(x; w),$$

$$F_2(x, \tilde{u}(x; w)) = \tilde{u}(x; w) - \tilde{u}(b; w)\frac{x - a}{b - a} - \tilde{u}(a; w)\frac{b - x}{b - a}.$$

(b) For Dirichlet boundary conditions at $x = a$ and Neumann boundary conditions at $x = b$, that is,

$$u(a) = u_0 \text{ and } u_x(b) = u_1',$$

$A(x)$ may be constructed as

$$A(x) = u_0 + (x - a)u_1',$$

while two alternative choices for the function $F(x, \tilde{u}(x, ; w))$ are given by

$$F_1(x, \tilde{u}(x; w)) = (x - a)[\tilde{u}(x; w) - \tilde{u}(b; w) - (b - a)\tilde{u}_x(b; w)],$$

$$F_2(x, \tilde{u}(x; w)) = \tilde{u}(x; w) - \tilde{u}(a; w) - (x - a)\tilde{u}_x(b; w).$$

(c) For Cauchy (initial) conditions, defined as

$$u(a) = u_0 \text{ and } u_x(a) = u_0',$$

we can choose the function $A(x)$ as follows:

$$A(x) = u_0 + (x - a)u_0',$$

while two alternative choices for the function $F(x, \tilde{u}(x, ; w))$ are given by

$$F_1(x, \tilde{u}(x; w)) = (x - a)^2 \tilde{u}(x; w),$$

$$F_2(x, \tilde{u}(x; w)) = \tilde{u}(x; w) - \tilde{u}(a; w) - (x - a)\tilde{u}_x(a; w).$$

14.8 Computer-generated random numbers

Many applications of probability theory depend on the availability of long sequences of uncorrelated, uniformly distributed random numbers. Such sequences make it possible to construct reliable samples by choosing among several equally likely events. As we have seen in Chapter 13, given a uniform probability distribution we can also generate probability distributions of any other desirable type.

Due to the importance of random numbers, we present here a short introduction to the concept of computer-generated "pseudo-random numbers" which mimic the behavior of true random numbers. Before the advent of fast and widely available computers, random numbers were actually generated in practical ways, and lists of such numbers were published as reference books.[4] Later studies revealed that these lists did not qualify as truly random. In modern day applications, random numbers are generated by computer algorithms. Since they are generated by a deterministic algorithm they are by definition *not* truly random, and this the reason they are called "pseudo-random".[5]

A standard computational algorithm for generating pseudo-random numbers is the so-called "modulo algorithm". This algorithm consists of the following operations: given the positive integers A, B, M, R_0, generate in sequence the values R_i and r_i:

$$R_i = \mathrm{mod}[(AR_{i-1} + B), M], \quad r_i = \frac{R_i}{M}, \quad i = 1, \ldots, N \ll M. \quad (14.50)$$

The values r_i are the pseudo-random numbers uniformly distributed in the range between 0 and 1. R_0 is usually referred to as the **seed**. A different choice of seed with the same values for the rest of the parameters generates a different sequence of random numbers. Note that with this algorithm we can generate a number N of pseudo-random numbers where $N \ll M$. Without this restriction, the numbers generated are not really random, because the numbers generated by this algorithm repeat after a period M. For example, with

$$M = 4, \quad A = 5, \quad B = 3, \quad R_0 = 1,$$

the first five values generated are

$$R_1 = 0, \quad R_2 = 3, \quad R_3 = 2, \quad R_4 = 1, \quad R_5 = 0,$$

and then the numbers keep repeating, $R_6 = R_1$, *etc*, because at R_5 we have hit the same value that R_0 had. Actually this is to be expected, since $M = 4$ and all the possible outcomes of the modulo are $0, 1, 2, 3$. These numbers appear to occur in a "random" sequence in the first period, until the range of possible outcomes is exhausted, and then the sequence starts over. Thus, if N is allowed to reach M, the numbers generated do not have any resemblance of randomness anymore!

As this simple example shows, a careful choice of the variables A, B, M, R_0 is important to avoid a short period. In particular, M should be very large. An example is shown in Fig. 14.8, with $M = 197$, resulting in a sequence of $N = 100$ values in the interval $[0, 1]$, which appear random even for this very modest value of M. Actually, the sequence of pseudo-random numbers

[4] The first such book, *Random Sampling Numbers* (Cambridge University Press, 1927), was compiled by L.H.C. Tippett, who used numbers chosen at random from census registers.

[5] The first list of random numbers generated in this manner, *A Million Random Digits with 100,000 Normal Deviates*, was published by the RAND Corporation in 1955, for use with the Monte Carlo integration method (see Section 14.9) and related stochastic methods (see Chapter 15).

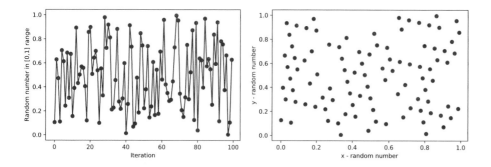

Figure 14.8: **Left:** 100 successive pseudo-random numbers generated with the modulo algorithm, Eq. (14.50), using $A = 1000$, $B = 400$, $M = 197$. **Right:** 100 points on a two-dimensional (x, y) plane, generated by calling the modulo algorithm with the same parameters for the x and y values, but with different seeds. The hyperplanes (lines in this case) where the pseudo-random numbers lie are evident.

generated by the modulo algorithm always involves correlations for any finite value of M. Specifically, if we were to use D such numbers to define a random variable in an D-dimensional space ($D = 2$ for a plane, $D = 3$ for a cube, etc.), then these points do not fill uniformly the D-dimensional space but lie on $(D-1)$-dimensional hyper-planes. There are at most $M^{1/D}$ such hyper-planes, if A, B, M, R_0 are carefully chosen. In fact, the number of such hyper-planes can be smaller than $M^{1/D}$, meaning poorer quality pseudo-random numbers, if the values of the parameters are not well chosen; an example is shown in Fig. 14.8.

One way to improve the randomness of pseudo-random numbers is to create a shuffling algorithm. This works as follows: we first create an array of N random numbers $r_j, j = 1, \ldots, N$, with N large. Next, we choose at random a value j between 1 and N and use the random number r_j, but at the same time we take it out of the array, and replace it by a new random number. This continuous shuffling of the random numbers reduces the correlations between them. Elaborate algorithms have been produced to make this procedure, known as Random Number Generators (RNG's), very efficient.

14.9   Application: Monte Carlo integration

As an application of random sampling we explore a computational method that uses random numbers to perform familiar mathematical operations like the calculation of an area, or more generally a volume in D-dimensional space, which is closely related to the computation of definite integrals. Because of the random nature of this approach, as in real-life processes like throwing dice or picking cards from a deck, it is often referred to as the "Monte Carlo" method from one of the most famous casinos in the world.

We first discuss the calculation of the value of π by using a geometric argument. Consider pairs of random number, (x, y), each with uniform probability distribution in the interval $[-1, 1]$. These can be easily generated by using a standard RNG for uniform random numbers r in the interval $[0, 1]$, and then shifting and scaling these numbers to obtain the numbers $r' = 2(r - 0.5)$, which have the desired distribution. If we use N pairs of such numbers, $(r'_1, r'_2) = (x, y)$ to represent points on the xy-plane that cover the square of side $a = 2$ centered at the origin, then the fraction N_0 that lies within the circle of radius $\rho = 1$ will correspond to an area equal to π, as is easily seen in Fig. 14.9. Therefore, in the limit of N very large the ratio N_0/N will be equal to the ratio of the area of the circle to area of the square, namely

$$\frac{N_0}{N} = \frac{\text{Area of circle } (\rho = 1)}{\text{Area of square } (a = 2)} = \frac{\pi \rho^2}{a^2} = \frac{\pi}{4}.$$

Thus, by simply generating a large number of random pairs of numbers (r'_1, r'_2) and simply keeping track of how many of them correspond to points (x, y) within the circle of radius $\rho = \sqrt{x^2 + y^2} = 1$ we can find the value of π.

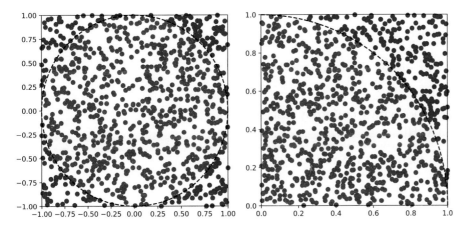

Figure 14.9: Example of Monte Carlo evaluation of π: red dots are random points within the range of the circle and blue dots are outside this region; the dashed black line indicates the perimeter of the circle of radius 1. **Left**: Using the ranges $x \in [-1, 1]$ and $y \in [-1, 1]$ which encompass the entire circle inscribed in a square of side 2. **Right**: Using the ranges $x \in [0, 1]$ and $y \in [0, 1]$ which encompass only the first quadrant of the circle within a square of side 1 with one corner at the origin.

Clearly, the same logic can be applied to the first quadrant of the circle of radius $\rho = 1$, as shown in Fig. 14.9. In this case, we use the numbers $(x, y) = (r_1, r_2)$ with uniform distribution in the range $[0, 1]$, and calculate the ratio of the number of points N_0 inside the first quadrant, to the number of points N inside the square of side $a = 1$, to obtain the same result, namely:

$$\frac{N_0}{N} = \frac{\text{Area of circle quadrant } (\rho = 1)}{\text{Area of square } (a = 1)} = \frac{(1/4)\pi\rho^2}{a^2} = \frac{\pi}{4}.$$

The geometric problem of calculating the area of the first quadrant of the circle can be cast as a definite integral, applying the concept of the integral being the "area under the curve" (see Section 1.4). The points (x, y) in the first quadrant that define the circle of radius $\rho = 1$ centered at the origin, obey the following relation:

$$x^2 + y^2 = \rho^2 = 1 \Rightarrow y = \sqrt{1 - x^2} = f(x).$$

The area under the curve $y = f(x)$ is the area of the first quadrant of the circle of radius $\rho = 1$, thus

$$I_0 = \int_0^1 f(x)dx = \int_0^1 \sqrt{1 - x^2}dx. \tag{14.51}$$

This integral is easy to evaluate explicitly by the change of variables $x = \cos(\theta)$, which implies $dx = -\sin(\theta)d\theta$ and $\sqrt{1 - x^2} = \sin(\theta)$, giving

$$I_0 = \int_0^1 \sqrt{1 - x^2}dx = -\int_{\pi/2}^0 \sin^2(\theta)d\theta = \int_0^{\pi/2} \frac{1}{2}\left[\sin^2(\theta) + \cos^2(\theta)\right] d\theta = \frac{\pi}{4}.$$

This concept can be extended to the numerical evaluation of any definite integral

$$I = \int_a^b f(x)dx,$$

for finite values of the integration limits a, b. The method consists of the following steps:

1. Find the extrema of the function $f(x)$ for $x \in [a, b]$, defined as

$$f_1 \equiv \min_{x \in [a,b]} \left[f(x) \right], \qquad f_2 \equiv \max_{x \in [a,b]} \left[f(x) \right].$$

Shift the integrand by f_1, by defining $\tilde{f}(x) = f(x) - f_1$, which satisfies $\tilde{f}(x) \geq 0$ since $f(x) \geq f_1$. The original integral can be written as

$$I = \tilde{I} + f_1(b - a), \quad \text{where } \tilde{I} = \int_a^b \tilde{f}(x) \mathrm{d}x.$$

2. Construct a rectangle of width $W = (b - a)$ on the x-axis and height $H \geq (f_2 - f_1)$ on the y-axis.

3. Generate N pairs of random values, (x, y), with x uniformly distributed in $[a, b]$, and y uniformly distributed in $[0, H]$.

4. Count the number N_0 of pairs (x, y) for which the relation $y \leq \tilde{f}(x)$ is satisfied.

5. The approximate value of the integral $\tilde{I}$ is given by

$$\tilde{I} \approx \frac{N_0}{N} WH.$$

This approach, referred to as "Monte Carlo integration", can be especially useful for cases when an analytic solution is difficult or impossible. We have assumed above that, in addition to the integration limits a and b being finite, the values of f_1, f_2 are also finite. However, even if the latter condition is not satisfied, in which case H is not finite, it may still be possible to approximate the integral by choosing a finite H and studying the convergence of the numerical result as a function of H (see Problem 7).

Example 14.5: We give two more examples of using the Monte Carlo integration method to approximate definite integrals:
(i) Consider the integral of Eq. (5.27), calculated in Section 5.7 by using contour integration, namely

$$I_1 = \int_{-1}^{1} \frac{\sqrt{1 - x^2}}{1 + x^2} \mathrm{d}x = 2 \int_0^1 \frac{\sqrt{1 - x^2}}{1 + x^2} \mathrm{d}x = \pi(\sqrt{2} - 1) = 1.301290.$$

$$(14.52)$$

In the above expression we took advantage of the fact that the integrand is an even function to restrict the integration range from the original $x \in [-1, 1]$ to $x \in [0, 1]$.

To compute this integral numerically, we generate random values in the range $[0, 1]$, which will be the values of the variable x, and for each x compare the value of the integrand $f(x)$ to a second random value, y, which is also in the range $[0, 1]$. If $y \leq f(x)$, then the point (x, y) would be under the curve $f(x)$. Since the integral represents the area under the curve described by $y = f(x)$ (see Section 1.4), we may compare the number of points that satisfy the relation $y \leq f(x)$ to the total number of points generated within the square, and this ratio will give the result for the integral of Eq. (14.52). Conveniently, the area of the square on the xy-plane that contains all the points generated is 1, so we need not worry about normalization of the result by the area of the square. In this example, this choice of the x and y random points, each within the interval $[0,1]$, is adequate because the integrand $f(x) = y$ takes values in the range $[0,1]$ for $x \in [0,1]$. We show below the distribution of $N = 10^3$ random points for this computation.

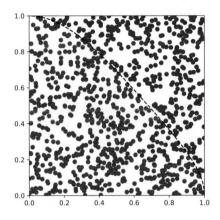

(ii) Consider the following integral:

$$I_2(w) = \int_{-w}^{w} \frac{1}{\sqrt{2\pi}} e^{-x^2/2} dx = 2 \int_{0}^{w} \frac{1}{\sqrt{2\pi}} e^{-x^2/2} dx, \qquad (14.53)$$

for finite values of w. Integrals of this type often arise in problems that involve normal probability distributions (see the discussion in Section 13.2.4 and the Example in Section 13.7). The value of the integral can be obtained from the tabulated values of the Gaussian integral $\Phi(w)$. Specifically, using the definition of $\Phi(w)$, Eq. (13.47), and its properties, Eq. (13.48), we find:

$$I_2(w) = \Phi(w) - \Phi(-w) = \Phi(w) - [1 - \Phi(w)] = 2\,\Phi(w) - 1.$$

For example, for $w = 2$, we find from the tabulated values of the Gaussian integral,

$$I_2(2) = 2\,\Phi(2) - 1 = 2 \times 0.977250 - 1 = 0.954500,$$

where the value of $\Phi(2)$ comes from the table in Appendix B. This result represents the often quoted fact that 95% of the normal probability distribution is contained within ± 2 standard deviations from the mean.

To compute the integral $I_2(w)$ numerically for a given value of w, we must generate values of the random variable in the interval $[-w, w]$, or, due to the fact that the integrand is an even function of x, we can simply integrate in the interval $[0, w]$ and multiply the result by a factor of 2. We must also make sure that the result is scaled by the total area which would be covered by random points. In order to make this computation efficient, we can arrange the total area to be a rectangle of width $2w$ which is the range of the variable x, (or w if we use the half range $[0, w]$), and a height that facilitates the comparison of the area under the Gaussian curve to the total area of the rectangle. For the height of the rectangle we can choose the value 0.5, so that its total area is 1 since $w = 2$. The computation then proceeds by generating random numbers to represent points (x, y) inside this rectangle, identifying those that lie under the Gaussian curve, namely those for which $y < \exp[-x^2/2]/\sqrt{2\pi}$, and comparing the number of points that satisfy this inequality to the total number of points inside the rectangle. We show below the distribution of $N = 10^3$ random points for this computation.

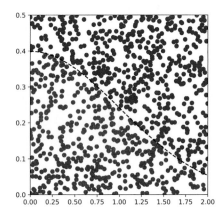

In the table below we collect the results of the calculation of the three integrals, I_0 of Eq. (14.51), I_1 of Eq. (14.52) and $I_2(w)$ of Eq. (14.53) for $w = 2$, using Monte Carlo integration with different numbers of random points N (from one thousand to one billion). $\Delta_i, i = 0, 1, 2$ are the differences from the true result given to a precision of six decimal places. As is evident from this table, the larger the number of points used, the more accurate the result (closer to the true value).

N	$4I_0$	$|\Delta_0|$	I_1	$|\Delta_1|$	$I_2(2)$	$|\Delta_2|$
10^3	3.164000	0.022407	1.312000	0.010710	1.002143	0.047643
10^4	3.121200	0.020393	1.315800	0.014510	0.952834	0.001666
10^5	3.142560	0.000967	1.302040	0.000750	0.957605	0.003105
10^6	3.143352	0.001759	1.301556	0.000266	0.954688	0.000188
10^7	3.142856	0.001263	1.301114	0.000176	0.954613	0.000113
10^8	3.141668	0.000075	1.301180	0.000110	0.954552	0.000052
10^9	3.141624	0.000031	1.301266	0.000024	0.954511	0.000011
true	3.141593		1.301290		0.954500	

Further reading

Numerical methods is a subject of growing relevance to applications, and the number of textbooks devoted to

it is commensurately large and growing. We offer here some suggestions for more detailed treatments:

1. William H. Press, Saul A. Teukolsky, William T. Vetterling, Brian P. Flannery, *Numerical Recipes—*

The Art of Scientific Computing, Third Ed., (Cambridge University Press, 2007). This is the "Bible" of scientific computing. It contains a thorough and extensive discussion of almost any topic that may arise in the context of scientific computation. It also includes sample code for each topic, and extensive references to the original literature.

2. Wen Shen, *An Introduction to Numerical Computation*, Second Ed. (World Scientific Publishing Co., 2020). This is a modern and concise description of the basic algorithms encountered in scientific computation, at an approachable level, both numerically and mathematically. It also contains code in Matlab for most topics discussed.

3. Michael T. Heath, *Scientific Computing, An Introductory Survey* (McGraw-Hill, 1997). This is an overview of methods used in scientific computing presented more from a computer science perspective, with emphasis on algorithms and containing many useful examples and exercises.

4. Michael L. Overton, *Numerical Computing with IEEE Floating Point Arithmetic* (Society for Industrial and Applied Mathematics, 2001). This book is the classic reference on floating point arithmetic.

5. Nicholas J. Higham, *Accuracy and Stability of Numerical Algorithms*, Second Ed. (Society for Industrial and Applied Mathematics, 2002). This book offers a careful study of error analysis and stability issues related to numerical algorithms.

6. M.J.D. Powell, *Approximation Theory and Methods* (Cambridge University Press, 1981). This book is a classic in approximation theory.

7. Arieh Iserles, *A First Course in the Numerical Analysis of Differential Equations*, Second Ed. (Cambridge University Press, 2009). This is a comprehensive treatment of the field of numerical solution of differential equations. It contains detailed explanation of the numerical methods and careful analysis of their convergence and stability.

8. Bengt Fornberg, *A Practical Guide to Pseudospectral Methods* (Cambridge University Press, 1996). This is a very useful reference for spectral methods.

9. Lloyd N. Trefethen, *Spectral Methods in MATLAB* (Society for Industrial and Applied Mathematics, 2000). This is a short book with very useful examples of the implementation of spectral methods using Matlab.

Problems

1. Show that the trapezoid formula for calculating definite integrals, Eq. (14.15), gives rise to an error of $\mathcal{O}\left[(\Delta x)^2\right]$.

2. Show that the Simpson's 1/3-rule formula for calculating definite integrals, Eq. (14.16), gives rise to an error of $\mathcal{O}\left[(\Delta x)^3\right]$.

3. Implement the trapezoid and the Simpson's 1/3-rule formulas for calculating definite integrals and calculate the value of the integrals defined in Eqs. (14.52) and (14.53) (see Example 14.4). Use different numbers of points for the computational grid and compare the efficiency of these methods to the method of Monte Carlo integration for these integrals (the results given in the table of Example 14.4).

4. Show that the total error in the fourth-order Runge–Kutta method, defined by Eqs. (14.26), and (14.28a)–(14.28f), is third order in the grid spacing Δx.

5. Show that the fourth-order Runge–Kutta method, defined by Eqs. (14.26), and (14.28a)–(14.28f), is equivalent to the Simpson's 1/3-rule for calculating integrals, Eq. (14.16).

6. Implement the simple Euler's method, Eq. (14.23), and the fourth-order Runge–Kutta method, Eqs. (14.28a)–(14.28f), to solve the second-order ODE of Eq. (14.24), with parameters: $a = 2$, $x_0 = 0$, $u_0 = 1$, $u_0' = 0$, $N = 50$, $x_N = 2.5\pi$. Compare the results of the two methods.

7. Consider the following numerical methods for estimating the value of π:

(a) Use the integral
$$\int_0^1 \frac{1}{\sqrt{1-x^2}}dx = \frac{\pi}{2},$$
calculated in Problem 13 of Chapter 1. Calculate numerically the value of this integral by using Monte Carlo integration. In this case the integrand $f(x)$ is not bounded for $x \to 1$. Choose a height H on the y-axis and study the convergence of the integral as a function of H.

(b) Another geometric method for estimating π is Buffon's needle problem, discussed in Example 13.6 of Chapter 13. In this method, the value of π given by the relation
$$\pi = \frac{2l}{WP},$$

where l is the needle length, W is the stipe width (with $W > l$), and P is the probability that randomly dropped needles intersect the lines separating the parallel stripes. The probability P is given by

$$P = \frac{N_X}{N},$$

where N is the total number of needles dropped and N_X is the number of needles that intersect the lines. How does the estimate of π with this method depend on the values of the parameters l and W?

(c) Compare your results to those presented in Section 14.9 for the estimation of π by generating (x, y) points within a square of side equal to 1. Which of the three methods converges faster, that is, with the fewest points N sampled?

Chapter 15

Stochastic Methods

When you come to a fork in the road, take it.

Yogi Berra (American professional baseball player, 1925–2015)

The concept of random events, and more generally of randomness, extends beyond calculating the probabilities of different possible outcomes. In fact, random sampling of possible outcomes, combined with efficient computational algorithms for searching through a vast range of such outcomes, has become one of the most powerful approaches for solving complex problems. In this chapter we explore some computational methods for taking advantage of randomness to tackle such problems. These methods are referred to as "stochastic methods".

We distinguish stochastic methods into two broad categories: The first category consists of methods that *simulate* the system of interest by randomly chosen events or processes, and then sample these events to obtain the properties of interest. We refer to such a method as "stochastic simulation". The second consists of methods that rely on random sampling to find an optimal solution to a complex problem. We refer to such a method as "stochastic optimization". The usefulness of these methods becomes evident in problems where traditional approaches fail due to the complexity or the high dimensionality of the space that must be explored for the optimal solution. A major drawback of stochastic optimization methods is that they do not guarantee that the *best* solution will be found; they only offer a way of finding a solution, which may be close to the optimal one. However, for many practical purposes this is adequate, and in some cases despite their limitations, these methods do offer the best available solution.

In the following, we present examples of stochastic simulation and stochastic optimization methods applied to representative problems. Specifically, we examine the physical process of diffusion which is described by the familiar heat (or diffusion) equation, and the solution of mathematical puzzles, and the traveling salesman problem which is relevant to many practical applications.

15.1 Stochastic simulation: Random walks and diffusion

An important process that can be efficiently modeled by using stochastic simulation is diffusion. As discussed in Chapter 8, diffusion is a process encountered in many physical contexts (see, for example, Sections 8.3 and 8.6). Here we shall first derive the diffusion equation in the simplest, one-dimensional case, and then explore how its solution can be modeled by employing the technique of "random walks". Each random walk bears close resemblance to actual physical processes, for instance, the so-called "Brownian motion" of microscopic particles, or stock prices as a function of time.

Derivation of the diffusion equation

The diffusion equation can be derived by considering how the concentration of the diffusing particles evolves in a small interval of length Δx, extending from x to $x + \Delta x$.

The number of particles in this interval at some moment in time t is equal to $\rho(x,t)\Delta x$, where $\rho(x,t)$ is the density at position x and time t, defined as the number particles per unit length. The change in this number of particles with time is then given by

$$\Delta n_1 = \frac{\partial \rho(x,t)}{\partial t}\Delta x.$$

This change must be equal to the total number of particles entering or leaving this region from its endpoints (whether the particles are entering or leaving at each end depends on the sign of the velocity at that point). Denoting the particle velocity at point x as $v(x)$, the number of particles entering or leaving from the unit surface at $x + \Delta x$ per unit time, that is, the flux through the unit area on the right end, will be given by

$$\rho(x + \Delta x, t)v(x + \Delta x, t),$$

since the outward-pointing surface unit vector at this point is $\hat{x}$ as shown in the diagram below. Similarly, the number of particles entering or leaving from the unit surface at x per unit time will be given by

$$-\rho(x,t)v(x,t),$$

since the outward-pointing surface unit vector at this point is $-\hat{x}$.

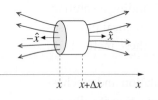

The sum of these two quantities, which equals the net change of the particle number in the region between x and $x + \Delta x$, is given by

$$\Delta n_2 = \rho(x + \Delta x, t)v(x + \Delta x, t) - \rho(x,t)v(x,t) = J(x + \Delta x, t) - J(x,t),$$

where we have defined a new quantity $J(x,t)$, the "mass current",

$$J(x,t) \equiv \rho(x,t)v(x,t).$$

Conservation of the number of particles in the small region of size Δx implies that the two changes must cancel out, that is

$$\Delta n_1 + \Delta n_2 = 0 \Rightarrow \frac{\partial \rho(x,t)}{\partial t} \Delta x + [J(x + \Delta x, t) - J(x,t)] = 0.$$

Dividing through by Δx and taking the limit $\Delta x \to 0$ we find

$$\frac{\partial \rho(x,t)}{\partial t} + \frac{\partial J(x,t)}{\partial x} = 0.$$

This equation is referred to as the "continuity equation", since it was derived by claiming conservation of the total number of particles at each point, in other words, by asserting that particles do not appear or disappear at any point in space (the density does not have discontinuities). Under diffusive motion, the current is proportional to the gradient of the local concentration of particles; this fact is known as "Fick's law", and is expressed by the equation:

$$J(x,t) = -D \frac{\partial \rho(x,t)}{\partial x},$$

where D is the diffusion constant. Substituting this expression in the continuity equation we obtain the diffusion equation:

$$\frac{\partial \rho(x,t)}{\partial t} = -\frac{\partial J(x,t)}{\partial x} \Rightarrow \frac{\partial \rho(x,t)}{\partial t} = D \frac{\partial^2 \rho(x,t)}{\partial x^2}. \tag{15.1}$$

Identifying the particle density $\rho(x,t)$ with the function $u(x,t)$ and the diffusion constant D with the parameter α, the diffusion equation becomes identical to the heat equation, Eq. (8.15).

The generalization of the above derivation to higher dimensions is straightforward. For simplicity and ease of visualization, in the following we discuss the diffusion process in two spatial dimensions (x,y); in this case the diffusion equation takes the form

$$\frac{\partial \rho(x,y,t)}{\partial t} = D_x \frac{\partial^2 \rho(x,y,t)}{\partial x^2} + D_y \frac{\partial^2 \rho(x,t)}{\partial y^2}. \tag{15.2}$$

Since the diffusion process, in general, is *anisotropic* we have allowed for different diffusion coefficients D_x and D_y in the x and y directions (the case $D_x = D_y$ corresponds to *isotropic* diffusion, that is, motion with equal probability in the two directions).

At the microscopic level, diffusion consists of a series of random moves by the diffusing particles, which leads to changes in their concentration that are governed by the diffusion equation. The process begins with an initial concentration of N particles at some point (x_0, y_0). We can model the initial concentration as a δ-function in each coordinate:

$$\rho(x,y,0) = N\delta(x - x_0)\delta(y - y_0).$$

Using the general solution of the heat equation, Eq. (8.31), with the above initial condition, we obtain the value of the density at any point (x,y) for time $t > 0$:

$$\rho(x,y,t) = \frac{N}{4\pi\sqrt{D_x D_y}t}e^{-(x-x_0)^2/4D_xt}e^{-(y-y_0)^2/4D_yt}. \qquad (15.3)$$

From the full solution for the density $\rho(x,y,t)$, we can calculate the average distances $\langle(x-x_0)\rangle$ and $\langle(y-y_0)\rangle$, as well as the average square distances $\langle(x-x_0)^2\rangle$ and $\langle(y-y_0)^2\rangle$, traveled by the diffusing particles in the x and y directions, after time t. To calculate these quantities we use the first and second moments of the Gaussian distribution [see Example 1.5 in Chapter 1 and Eq. (1.66)]. With the aid of these expressions, we obtain

$$\langle(x-x_0)\rangle = 0, \quad \langle(x-x_0)^2\rangle = 2D_xt, \quad \langle(y-y_0)\rangle = 0, \quad \langle(y-y_0)^2\rangle = 2D_xt.$$

These results show that while the average displacement relative to original point (x_0, y_0) in both the x and y directions is zero, the average *square* displacement in the x and y directions are proportional to the time t. The profile of the density is a Gaussian with increasing variance in both x and y, namely $\sigma_x^2 = 2D_xt$ and $\sigma_y^2 = 2D_yt$. From these results, we can also calculate the average distance squared from the origin:

$$\langle r^2 \rangle = \langle(x-x_0)^2\rangle + \langle(y-y_0)^2\rangle = 2(D_x + D_y)t. \qquad (15.4)$$

For simplicity, in the following we choose $(x_0, y_0) = (0,0)$.

The process of diffusion can be simulated by a particle executing a "random walk" on the (x,y)-plane. This is simply implemented by drawing pairs of random numbers (n_x, n_y) with uniform distribution, $0 < n_x, n_y < 1$ and moving the particle (or "walker") from its current position to a new one differing by $(\Delta x, \Delta y) = (n_x\delta x, n_y\delta y)$, where $\delta x, \delta y$ are predetermined step sizes. The motion of the walker is then monitored to obtain the average displacements and the average squared displacements in the x and y directions. In order to obtain meaningful averages, the walker must undertake a very large number of steps, and the whole process must be repeated many times, starting at the origin each time, and averaging over all the paths taken. Alternatively, we can use a large number N of walkers and follow their evolution simultaneously as they execute independent random walks, and average over their displacements. In other words, at each step we draw N pairs of random numbers and update all N walker positions simultaneously. An example of such a simulation is shown in Fig. 15.1, for the case of anisotropic diffusion, implemented here by choosing the step sizes in the x and y directions to be different, which is equivalent to moving in the two directions with different probabilities.

The analysis of the results is consistent with our expectations from the solution of the diffusion equation, Eq. (15.2). Specifically, the averages of the displacements in the two directions, $\langle x\rangle$ and $\langle y\rangle$, are fluctuating very close to zero, while the averages of the square displacements $\langle x^2\rangle$ and $\langle y^2\rangle$, increase linearly with time, as shown in Fig. 15.2. We can also calculate the average square distance from the origin, $\langle r^2\rangle$, defined in Eq. (15.4), which also increases linearly with time. In fact, since we are dealing with anisotropic diffusion in

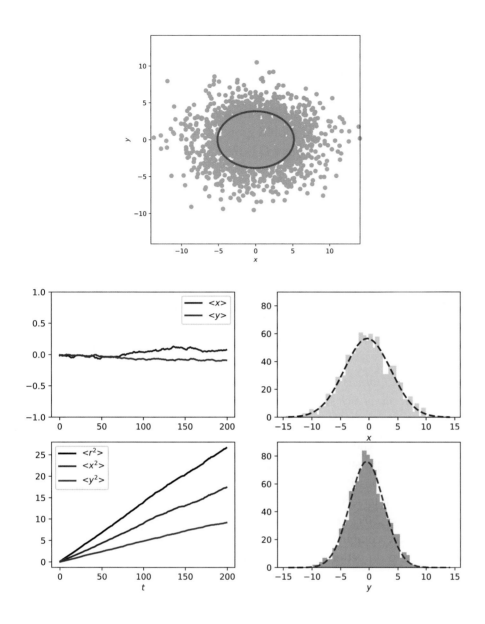

Figure 15.1: Simulation of diffusion with $N = 2000$ walkers in 2D: Each light-colored dot represents the position of a walker after $\Delta t = 200$ time steps. The diffusion is anisotropic, with the sizes of steps in the x and y directions given by $\delta x = 1$, $\delta y = 0.75$. The dark-green ellipse, representing the average displacement of particles along different directions, is defined by the set of points $(\bar{x}, \bar{y})$ given in Eq. (15.5).

Figure 15.2: Statistics of the simulation of diffusion with walkers in two dimension, shown in Fig. 15.1. **Left**: The average displacements $\langle x \rangle$, $\langle y \rangle$ (top panel), and $\langle x^2 \rangle$, $\langle y^2 \rangle$, $\langle r^2 \rangle$ (bottom panel), as functions of the time t (measured in time steps). Notice the large difference in the vertical scale of the two panels. **Right**: The histograms of the distribution of x (top panel) and y (bottom panel) displacements at the end of the simulation. The dashed lines in each case are Gaussian fits to the histogram data.

this example, we can define an ellipse described by the set of points

$$\bar{x}(\phi) = \sqrt{\langle r^2 \rangle} \delta x \cos(\phi), \quad \bar{y}(\phi) = \sqrt{\langle r^2 \rangle} \delta y \sin(\phi), \quad \text{with } 0 \leq \phi < 2\pi,$$

$$(15.5)$$

where $\langle r^2 \rangle$ is the average square distance. This ellipse represents the average displacement of particles along different directions in the xy-plane.

Finally, we can use all the x and y displacements of the walkers at some moment t to produce a histogram of their values, as shown in Fig. 15.2. From the solution of the diffusion equation, Eq. (15.2), we expect these distances to have a Gaussian distribution. Indeed, assuming that we have a large number of walkers, the histograms in the x and y displacements resemble closely the shape of a Gaussian. A fit of a Gaussian function to the actual data of the histograms gives the values of the variance in each direction, $\sigma_x^2 = 2D_x t$ and $\sigma_y^2 = 2D_y t$.

A process very closely resembling the simulation of diffusion can actually be used to measure the diffusion constant experimentally: One can monitor the motion of microscopic particles undergoing diffusion (for instance, viruses in solution, pollen in water, dust particles in air), plot the histograms of their displacements, and obtain the time-dependent variance of the Gaussians fitted to the data, for each independent spatial direction. The variance should grow linearly with time t, with the coefficient of t being equal to $2D$, where D is the diffusion constant in this direction.

15.2   Stochastic optimization

There are many problems of practical importance in which finding the optimal solution by conventional methods is impossible. One difficulty can be that enumerating all possible answers in order to find the optimal one may involve a truly large number of possibilities. To illustrate this type of situation we consider the following apparently simple problem: Two squares intersect as shown in Fig. 15.3, so that there 8 corner sites, 4 from each square, and 8 intersection sites. The question is:

How can the integers from 1 to 16 be placed at the 8 corners and the 8 intersection sites so that the sum of the 4 numbers along each side of each square is the same?

Note that the placement of numbers shown in Fig. 15.3 is *not* an optimal solution, that is, not all sides give the same sum. The brute force approach to answer this question would be to construct all possible combinations and find which ones satisfy the desired constraint. The possible ways of arranging the 16 consecutive integers in the 16 sites of the intersecting squares is 16!, a very large number, just short of 21 trillion [$16! = 2.09 \times 10^{13}$]. Arguably, one could use a computer to go through all possible combinations and find the right answer, but one would expect to be able to do better for such a simple problem!

The intersecting squares problem may seem like a rather contrived mathematical puzzle, but it is not too different from more realistic problems. Another type of mathematical puzzle is shown in Fig. 15.4: consider a triangular lattice on which we place identical magnets that can point North (up) or South (down). The optimal arrangement of the magnets consists of each magnet being surrounded by neighbors pointing in the *opposite* direction, because this minimizes the magnetic energy and makes the system stable. In Fig. 15.4, we show 25 magnets in a triangular lattice, with the ones pointing up colored blue and the ones pointing down colored red. It is easily seen that in the configuration shown the magnets point *opposite* to their immediate neighbors along the horizontal direction (labeled "a" in Fig. 15.4), and along one of the slanted directions (labeled "c"), but point in the *same* direction as their neighbors along the other slanted direction (labeled "b"). Upon some reflection it becomes evident that it is *impossible* to satisfy the requirement that each magnet points in direction opposite to all its neighbors. This is a property of the underlying lattice, which for this reason is called a "frustrated lattice". The question then becomes:

What are the optimal configurations of the magnets that have as many as possible pairs of neighbors pointing in opposite directions?

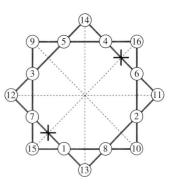

Figure 15.3: Two intersecting squares, colored red and blue; the 8 corners and the 8 points of intersection are highlighted by black circles; the 16 integers placed at these sites add up to the same number along each side of the blue and the red square except in the two sides marked by black crosses.

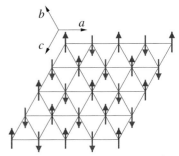

Figure 15.4: A triangular lattice with 25 magnets pointing up ("North", shown as blue arrows) and down ("South", shown as red arrows). The arrows are thought to be pointing in a direction perpendicular to the plane of the lattice.

The number of possible arrangements of N magnets on the lattice, is 2^N, since each can point in two directions (North or South). This number grows very fast with N. For instance, a systems of $N = 100$ magnets has $2^{100} = 1.27 \times 10^{30}$ possible configurations, a very large number indeed. Although we expect quite a few of these configurations will not satisfy the criterion of having as many as possible neighbor pairs pointing in opposite directions, many others will satisfy this criterion. Simply searching through all the possibilities to find the optimal solutions is a daunting task. The magnets on a triangular lattice problem may also sound like a contrived mathematical puzzle, but it turns out to be a reasonable model for describing the magnetic properties of many real solids.

Many other problems that arise in science and engineering are even more complicated. There are two reasons for this. The first is that, instead of one optimal solution, as in the first example mentioned above, there may be several (often a large number) of solutions close to the optimal as in the second example, and finding all of them is very difficult. A case in point is the so-called "traveling salesman" problem, discussed in the Application section of this chapter (see Section 15.4). A second reason is that the space spanned by the variables in the problem can be very large, so the search for even one of the nearly optimal solutions is very difficult.

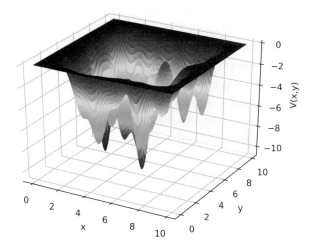

Figure 15.5: Illustration of a cost function, $V(x, y)$, with many local minima, in a two-dimensional space of the independent variables x and y.

A useful picture to bear in mind is shown in Fig. 15.5. In this figure, we show a "topographic map" of a function $V(x, y)$ of two independent variables, x and y. This function represents the values of the quality of the solution, often referred to as the "cost function". In this example, the optimal solution would have a very low cost, represented by the deep wells (or valleys), relative to some reference point taken to be zero. There are many deep wells, competing for the optimal solution, and it is not clear which one is the best choice. Moreover, if we devise a strategy for getting close to a well by starting at some point and following the local gradient to ever lower values, it is clear that we could be trapped in a well depending on where our search started, which may not be the deepest well. Therefore, our strategy must include the possibility of climbing out of a well and looking at other nearby, or even farther wells. Only if we have this ability we are guaranteed that we will be able to identify the deepest well. But this ability implies that for some parts of the search we will be going uphill, to values of *higher* cost, rather than monotonically descending to lower cost values. Then the obvious question is: "How far uphill should we allow the search to move?" Evidently, the strategy must be quite complicated. The problem is compounded by the fact that in typical cases the space we need to search involves many independent variables, often in the range of thousands or millions. To address problems of this difficulty, interesting computational methods have been developed which are based on the use of computer generated pseudo-random numbers.

In the following subsections, we discuss some key ideas and methods for numerical stochastic optimization. Before we delve into these topics we mention some concepts that are useful in understanding how and why these methods work. Specifically, we first introduce the concepts that underlie how to perform sampling in a computationally efficient way, and then we discuss a key algorithm, the Metropolis algorithm. An application of this is given in Section 15.3, namely the method of simulated annealing, that actually can deliver optimal solutions of complicated problems.

15.2.1 The method of "importance sampling"

To provide a more concrete picture, rather than abstract formalism, we adopt a physically motivated model like the example of Fig. 15.5. Specifically, we consider the quantity we want to optimize to be the potential energy of a system consisting of many particles that can move relative to each other. A "configuration" of the system is defined as a specific set of positions of the particles that comprise it, and corresponds to a particular value of the potential energy. In the picture of Fig. 15.5, we show a simple example with only two variables, x and y; each set of values of these two independent variables, gives a value of the potential, $\mathcal{V}(x, y)$. This potential energy "landscape" would describe a system of two independent particles each existing in a one-dimensional space, or one particle in a two-dimensional space.

To keep the discussion more general, we denote a configuration by **R** which represents the values of all the variables that define the state of the system, like the positions of the N particles comprising the system,

$$\mathbf{R} = \{\mathbf{r}_1, \mathbf{r}_2, \ldots, \mathbf{r}_N\}, \quad \mathbf{r}_k = (x_k, y_k, \ldots), \quad k = 1, \ldots, N,$$

with each particle, labeled by k, existing in a D-dimensional space. We emphasize that the adoption of this model does not restrict the applicability of the method. Namely, we can use any set of variables that determine the state of the system, and we can define any cost function in terms of these variables, whose value we want to optimize. Optimization may involve finding a minimum or a maximum value, as the problem of interest dictates.

We next define the concept of ergodicity, which is helpful in capturing the complexity of the task we are facing.

- The independent variables in the problem are referred to as **degrees of freedom**. The number of such independent degrees of freedom is called the **dimensionality** of the system. In our adopted model of N particles, each existing in a D-dimensional space, the dimensionality is $N \times D$.

- The range of values that all of the degrees of freedom can take is called the **phase space**. In the example of Fig. 15.5 the phase space consist of the range $x \in [0, 10]$, $y \in [0, 10]$, which is a finite region of the set $\mathbb{R}^2$.

- Covering all of phase space is referred to as achieving **ergodicity**. The time that it takes to achieve ergodicity is referred to as the **ergodic time**, symbolized by τ_{erg}. When the number of degrees of freedom $N \to \infty$ the ergodic time also tends to infinity, $\tau_{\text{erg}} \to \infty$. A process that allows the exploration of all regions of phase space is called an **ergodic** process.

- A sequence of discrete steps through points in phase space visited with probabilities that depend only on the previously visited point is called a **Markov chain**. This can be expressed formally by using conditional probabilities as:

$$p(\mathbf{R}_n | \mathbf{R}_{n-1}, \ldots, \mathbf{R}_0) = p(\mathbf{R}_n | \mathbf{R}_{n-1}), \tag{15.6}$$

namely, the probability of visiting the state $\mathbf{R}_n$ after having visited the states $\mathbf{R}_{n-1}, \ldots, \mathbf{R}_0$ in reverse sequence, depends only only on the probability of having visited the immediately preceding state $\mathbf{R}_{n-1}$; in other words, the process of visiting the different states has no memory beyond the immediately preceding one.

From the above definitions, we can see that in order to identify the optimal configuration of a system with N degrees of freedom we need to employ ergodic processes that allow the exploration of all interesting regions of its phase space. In the example of Fig. 15.5, our search for the optimal solution must make it possible to visit all the wells since only then we can claim that we have identified the one with the lowest cost. When N is very large, the time that this exploration takes can be exceedingly large; therein lies the main challenge for this type of problems.

As a final general concept, we need to define a parameter T to which we will compare changes in the value of the function $\mathcal{V}(\mathbf{R})$. This parameter sets the scale of the allowed change of values of $\mathcal{V}(\mathbf{R})$ that we will use to explore the phase space in order to find the optimal value. In other words, we will be comparing the values of $\mathcal{V}_{ij}$ changes to T in order to determine if those changes are large or small. Note that the units of T are the same as the units that we use to measure $\mathcal{V}(\mathbf{R})$. The parameter T is often referred to as the **temperature** because of its connection to actual temperature values when dealing with systems composed of many microscopic particles, as is the case for all matter.

The overall goal is to search the phase space of $\mathbf{R}$ *stochastically* in order to find a configuration with optimal value of $\mathcal{V}(\mathbf{R})$. But if the search is completely blind and random then we will be wasting most of the time in irrelevant parts of the phase space with very small chance of getting to the correct answer. The crucial idea is to sample the phase space using a probability distribution that captures the most important configurations. It turns out that the theory of statistical mechanics provides the tool for this type of sampling. This theory, based on ideas of Ludwig Boltzmann about entropy, shows that if a system is in equilibrium with its environment which has fixed temperature T, the probability of finding a configuration $\mathbf{R}$ with energy $\mathcal{V}(\mathbf{R})$, denoted as $\tilde{p}(\mathbf{R}; T)$, is proportional to the exponential of $-\mathcal{V}(\mathbf{R})/T$:

$$\tilde{p}(\mathbf{R};T) = \frac{1}{\mathcal{Z}(T)}e^{-\mathcal{V}(\mathbf{R})/T}, \tag{15.7}$$

where we have written the constant of proportionality in front of the exponential as $1/\mathcal{Z}(T)$, a quantity that depends on the temperature T. Since this is a probability distribution, summing over all values must give 1, therefore:

$$\sum_{\mathbf{R}} \tilde{p}(\mathbf{R};T) = \frac{1}{\mathcal{Z}(T)}\sum_{\mathbf{R}}e^{-\mathcal{V}(\mathbf{R})/T} = 1 \Rightarrow \mathcal{Z}(T) = \sum_{\mathbf{R}}e^{-\mathcal{V}(\mathbf{R})/T}. \tag{15.8}$$

$\mathcal{Z}$ is called the "partition function", and plays a central role in evaluating thermodynamic properties. In the following discussions, for simplicity, we will employ the shorthand notation

$$\tilde{p}_k = \tilde{p}(\mathbf{R}_k;T), \quad \forall k \text{ integer}.$$

The expression for $\mathcal{Z}(T)$ defines a probability distribution that corresponds to the equilibrium situation of the system of interest at temperature T, which is referred to as the "Boltzmann distribution". We note that in the general case what enters in the Boltzmann distribution is the *total energy* of the system in the configuration described by $\mathbf{R}$, which includes, in addition to the potential energy $\mathcal{V}(\mathbf{R})$, also the kinetic energy. In order to illustrate the key ideas, we confine ourselves only to the potential energy.

Derivation of the Boltzmann distribution

We can derive the equilibrium Boltzmann distribution by using the method of "ensembles", namely by considering $\mathcal{N}$ copies of a system, which exist in states with energy E_j, $j = 1, 2, \ldots, J$. The number of identical copies corresponding to the state with energy E_j is defined to be n_j, so that we will have:

$$\mathcal{N} = \sum_{j=1}^{J} n_j, \quad \mathcal{E} = \sum_{j=1}^{J} n_j E_j, \quad \bar{E} = \frac{\mathcal{E}}{\mathcal{N}},$$

where $\mathcal{E}$ is the total energy and $\bar{E}$ the average energy. The equilibrium condition is defined by the average energy $\bar{E}$ being fixed and the number of states W in which the system exists being a maximum. This corresponds to the special set of occupation numbers of each state denoted as $\tilde{n}_j$. The number of states W available to the system is given by the number of combinations of states, namely

$$W = \frac{\mathcal{N}!}{n_1! n_2! \cdots n_J!} \Rightarrow \ln(W) = \ln(\mathcal{N}!) - \sum_{j=1}^{J} \ln(n_j!),$$

under the condition that the average energy is equal to $\bar{E}$. We will use Stirling's formula for approximating the factorial of a large number n, Eq. (13.35):

$$n! \approx (2\pi n)^{1/2} n^n e^{-n} \Rightarrow \ln(n!) \approx n \ln(n) - n + \frac{1}{2}\ln(2\pi n).$$

The last term is much smaller than the other two contributions, so we will neglect it. Applying Stirling's formula to the factorials in W we find

$$\ln(W) = \mathcal{N}\ln(\mathcal{N}) - \mathcal{N} - \sum_{j=1}^{J}[n_j\ln(n_j) - n_j] = \mathcal{N}\ln(\mathcal{N}) - \sum_{j=1}^{J}[n_j\ln(n_j)].$$

Taking the differential of this expression we obtain

$$d\ln(W) = d\mathcal{N}\ln(\mathcal{N}) + \mathcal{N}\frac{1}{\mathcal{N}}d\mathcal{N} - \sum_{j=1}^{J}\left[dn_j\ln(n_j) + n_j\frac{1}{n_j}dn_j\right]$$

$$= -\sum_{j=1}^{J}\left[\ln(n_j) + 1\right]dn_j,$$

where we have used the fact that $d\mathcal{N} = 0$ since $\mathcal{N}$ is a fixed number; we will invoke this fact again below. At equilibrium, the number of states W is maximized, which is expressed as the condition that the differential vanishes:

$$W: \text{ maximum} \Rightarrow d\ln(W) = 0 \Rightarrow \sum_{j=1}^{J}dn_j\ln(n_j) + \sum_{j=1}^{J}dn_j = 0.$$

But the second summation is equal to the differential of $\mathcal{N}$, so it vanishes:

$$\mathcal{N}: \text{ fixed} \Rightarrow d\mathcal{N} = 0 \Rightarrow \sum_{j=1}^{J}dn_j = 0.$$

Therefore, the condition that we are dealing with an extremum of W reduces to

$$d\ln(W) = 0 \Rightarrow \sum_{j=1}^{J}dn_j\ln(n_j) = 0.$$

A second constraint is that the average energy of the system $\bar{E}$ is fixed, also expressed through the relation that the differential $d\bar{E}$ vanishes,

$$\bar{E}: \text{ fixed} \Rightarrow d\bar{E} = 0 \Rightarrow \sum_{j=1}^{J}E_j dn_j = 0.$$

Since all three conditions must be satisfied separately, if we multiply the constraint $d\mathcal{N} = 0$ by an arbitrary constant α and the constraint $d\bar{E} = 0$ by an arbitrary constant β, and add those two to $d\ln(W) = 0$, we will still obtain zero, that is

$$d\ln(W) + \alpha d\mathcal{N} + \beta d\bar{E} = 0 \Rightarrow \sum_{j=1}^{J}[\ln(n_j) + \alpha + \beta E_j]dn_j = 0.$$

The infinitesimal quantities dn_j are arbitrary. Therefore, the last condition can only be satisfied if there exist a set of special values of the variables n_j such that the square bracket in the summation vanishes. We call $\tilde{n}_j$ the special values of the variables n_j, hence

$$\ln(\tilde{n}_j) = -\alpha - \beta E_j \Rightarrow \tilde{n}_j = e^{-\alpha}e^{-\beta E_j}.$$

These are the values that correspond to equilibrium. In order to satisfy the condition of fixed $\mathcal{N}$ we must also have

$$\sum_{j=1}^{J} \tilde{n}_j = \mathcal{N} \Rightarrow e^{-\alpha} \sum_{j=1}^{J} e^{-\beta E_j} = \mathcal{N} \Rightarrow e^{-\alpha} = \mathcal{N} \left[\sum_{j=1}^{J} e^{-\beta E_j} \right]^{-1} = \frac{\mathcal{N}}{\mathcal{Z}},$$

where the partition function $\mathcal{Z}$ is defined here as

$$\mathcal{Z} = \sum_{j=1}^{J} e^{-\beta E_j},$$

and is the analog of the definition given in Eq. (15.8). We now set the parameter $\beta = 1/T$; in this way we obtain for the occupation number $\tilde{n}_j$ of the state with energy E_j at equilibrium:

$$\tilde{n}_j = \frac{\mathcal{N}}{\mathcal{Z}} e^{-E_j/T}.$$

From this, we obtain the *probability* $\tilde{p}_j$ of finding a state with energy E_j at equilibrium:

$$\tilde{p}_j = \frac{\tilde{n}_j}{\mathcal{N}} = \frac{1}{\mathcal{Z}} e^{-E_j/T}.$$

This last expression is the equilibrium Boltzmann distribution, consistent with the expression of Eq. (15.7) for a system with energy states $\mathcal{V}(\mathbf{R})$.

The conclusion from the above discussion is that if we want to sample the phase space in an efficient way, we should pick samples according to the equilibrium Boltzmann distribution. This distribution allows us to focus the search to configurations that have the largest probabilities, thus enhancing the efficiency of the sampling. At the same time, this does *not* exclude the configurations of low probability, which guarantees the ergodic nature of the search; the low-probability configurations just play a less important role (they come up less frequently, as they should). This process is referred to as **importance sampling**.

15.2.2   *The Metropolis algorithm*

In a seminal paper Metropolis, Rosenbluth, Rosenbluth, Teller and Teller,[1] proposed an algorithm for the sampling of large numbers of possible configurations of a system. This method has become known as the **Metropolis algorithm**; it is an efficient practical implementation of importance sampling. We first outline the basic steps of the Metropolis algorithm and then explain why it works.

The Metropolis algorithm consists of the following steps:

1. Choose a state of the system $\mathbf{R}_i$ that we call the "current" state, denoted as $\mathbf{R}_{cur}$, $\mathbf{R}_{cur} = \mathbf{R}_i$, and calculate the value of the potential energy (or more generally the "cost function") $\mathcal{V}(\mathbf{R}_i)$.

2. Make a move from the current state of the system to a new state characterized by the value $\mathbf{R}_j$. In practice, for the model adopted above this means

[1] N. Metropolis, A.W. Rosenbluth, M.N. Rosenbluth, A.H. Teller and E. Teller, Equation of state calculations by fast computing machines, *Journal of Chemical Physics* **21**, pp. 1087–1092 (1953).

that we can change the values of one or several of the variables $x_k, y_k, \ldots$ to go from state $\mathbf{R}_i$ to state $\mathbf{R}_j$.

3. Calculate the difference in the cost function:

$$\Delta \mathcal{V}_{ji} = \mathcal{V}(\mathbf{R}_j) - \mathcal{V}(\mathbf{R}_i).$$

4. If $\Delta \mathcal{V}_{ji} \leq 0$ accept the move, by which we mean make the current state of the system $\mathbf{R}_{\text{cur}} = \mathbf{R}_j$.
   Go to step 2.

5. If $\Delta \mathcal{V}_{ji} > 0$ accept the move with probability

$$\tilde{p}_{ji} = e^{-\Delta \mathcal{V}_{ji}/T}. \tag{15.9}$$

The acceptance of the move with probability $\tilde{p}_{ji}$ is accomplished by the following practical implementation: Choose a random number r of uniform distribution in the range $(0,1)$. If $r < \exp[-\Delta \mathcal{V}_{ji}/T]$ accept the move, $\mathbf{R}_{\text{cur}} = \mathbf{R}_j$; otherwise, keep the old state, $\mathbf{R}_{\text{cur}} = \mathbf{R}_i$.
Go to step 2.

It can be shown that the Metropolis algorithm is an ergodic Markov chain, that is, it allows for an ergodic search of the phase space.

We elaborate briefly on two important points related to step 5:

(i) *Why can the quantity $\tilde{p}_{ji}$ defined in Eq. (15.9) be considered a probability?*
   For $\Delta \mathcal{V}_{ji} > 0$ the quantity $y = \Delta \mathcal{V}_{ji}/T$ takes all possible values in the interval $(0,\infty)$ and therefore for $\tilde{p}_{ij} = \exp[-y]$ we have $0 < \tilde{p}_{ij} < 1$ and

$$\sum_{\{i,j\}} \tilde{p}_{ji} = \sum_{\{i,j\}} e^{-[\mathcal{V}(\mathbf{R}_j) - \mathcal{V}(\mathbf{R}_i)]/T} \longrightarrow \int_0^\infty e^{-y} dy = \left[-e^{-y} \right]_0^\infty = 1.$$

Thus, the quantity $\tilde{p}_{ij}$ has the right attributes of a probability distribution, namely it takes values in the interval $(0,1)$, and the sum of all its values is equal to 1.

(ii) *Why is the practical implementation in terms of drawing random numbers $r \in (0,1)$ and comparing them to the Boltzmann exponential equivalent to accepting the move with probability $\tilde{p}_{ji} = \exp[-\Delta \mathcal{V}_{ji}/T]$?*
   This is most readily seen by considering the limiting cases of $\Delta \mathcal{V}_{ji} \ll T$ and $\Delta \mathcal{V}_{ji} \gg T$. Recall that in step 5 we are dealing with the case $\Delta \mathcal{V}_{ji} > 0$. In the first limiting case we have

$$0 < \Delta \mathcal{V}_{ji} \ll T \Rightarrow 0 < \frac{\Delta \mathcal{V}_{ji}}{T} \ll 1 \Rightarrow \frac{\Delta \mathcal{V}_{ji}}{T} \approx 0 \Rightarrow e^{-\Delta \mathcal{V}_{ji}/T} \approx 1,$$

and because the random number $r \in (0,1)$, the condition $r < \exp[-\Delta \mathcal{V}_{ji}/T]$ is almost always satisfied; in other words, for $\Delta \mathcal{V}_{ji} \ll T$ the move is almost always accepted. By contrast, in the second limiting case we have

$$\Delta \mathcal{V}_{ji} \gg T \Rightarrow \frac{\Delta \mathcal{V}_{ji}}{T} \gg 1 \Rightarrow e^{-\Delta \mathcal{V}_{ji}/T} \approx 0,$$

and because the random number $r \in (0,1)$, the condition $r < \exp[-\Delta \mathcal{V}_{ji}/T]$ is almost never satisfied; in other words, for $\Delta \mathcal{V}_{ji} \gg T$ the move is almost never accepted.

Thus, starting from an arbitrary initial configuration $\mathbf{R}_i$, after a certain number of steps, the Metropolis algorithm produces configurations that occur with probabilities according to the equilibrium distribution. Once this is accomplished, any subsequent samples generated by the algorithm can be used as part of the importance sampling process. The number of steps required to arrive at the equilibrium distribution depends on the nature of the system and on the choice of initial configuration; the steps before we arrive at the equilibrium distribution are simply discarded from consideration.

We next provide a heuristic argument of why the Metropolis algorithm leads to an equilibrium distribution and therefore allows for importance sampling. Suppose we are selecting states from an arbitrary distribution in which the state denoted by $\mathbf{R}_i$ occurs with probability p_i and another state denoted by $\mathbf{R}_j$ occurs with probability p_j. We define the "transition probability" $w(\mathbf{R}_i \to \mathbf{R}_j)$ to go from state i to state j, that is, to start from the current state $\mathbf{R}_i$ and accept the move to state $\mathbf{R}_j$. According to the Metropolis algorithm, we have

$$\Delta V_{ji} \leq 0 \Rightarrow w(\mathbf{R}_i \to \mathbf{R}_j) = 1, \quad \text{and} \quad \Delta V_{ji} > 0 \Rightarrow w(\mathbf{R}_i \to \mathbf{R}_j) = e^{-\Delta V_{ji}/T}.$$

Now let us assume that for these two states, $V(\mathbf{R}_j) > V(\mathbf{R}_i)$ so that $\Delta V_{ji} > 0$; then the change in the probability p_i due to the acceptance of the move out of state $\mathbf{R}_i$ and into state $\mathbf{R}_j$ will be given by

$$\delta p_i^{(1)} = -p_i w(\mathbf{R}_i \to \mathbf{R}_j) = -p_i e^{-\Delta V_{ji}/T}.$$

But at some later time, starting at state $\mathbf{R}_j$, there will be a transition to state $\mathbf{R}_i$, and in this case we will have

$$\Delta V_{ij} = V(\mathbf{R}_i) - V(\mathbf{R}_j) < 0 \Rightarrow w(\mathbf{R}_j \to \mathbf{R}_i) = 1.$$

Thus, the change in the probability p_i due to the acceptance of the move out of state $\mathbf{R}_j$ and into state $\mathbf{R}_i$ will be given by

$$\delta p_i^{(2)} = p_j w(\mathbf{R}_j \to \mathbf{R}_i) = p_j.$$

The net change in the probability p_i will then be given by

$$\delta p_i = \delta p_i^{(1)} + \delta p_i^{(2)} = -p_i \left(e^{-\Delta V_{ji}/T} - \frac{p_j}{p_i} \right)$$

$$= p_i \left(\frac{p_j}{p_i} - \frac{e^{-V(\mathbf{R}_j)/T}}{e^{-V(\mathbf{R}_i)/T}} \right) = p_i \left(\frac{p_j}{p_i} - \frac{\tilde{p}_j}{\tilde{p}_i} \right).$$

From this result we see that if the probability distribution for p_i, p_j is the equilibrium one, $p_k = \tilde{p}_k$ for both $k = i, j$, then $\delta p_i = 0$, in other words, once the equilibrium distribution has been reached, then it remains unchanged. If the probability distribution from which we are selecting configurations is *not* the equilibrium one, then we can have two cases:

$$\text{(i)} \quad \frac{p_j}{p_i} > \frac{\tilde{p}_j}{\tilde{p}_i} \Rightarrow \left(\frac{p_j}{p_i} - \frac{\tilde{p}_j}{\tilde{p}_i} \right) > 0 \Rightarrow \delta p_i > 0,$$

which means that p_i *increases* and therefore the ratio (p_j/p_i) *decreases* so that it gets closer to the ratio $(\tilde{p}_j/\tilde{p}_i)$; or,

$$\text{(ii)} \quad \frac{p_j}{p_i} < \frac{\tilde{p}_j}{\tilde{p}_i} \Rightarrow \left(\frac{p_j}{p_i} - \frac{\tilde{p}_j}{\tilde{p}_i} \right) < 0 \Rightarrow \delta p_i < 0,$$

which means that p_i *decreases* and therefore the ratio (p_j/p_i) *increases* so that it gets closer to the ratio $(\tilde{p}_j/\tilde{p}_i)$. The net result is that the arbitrary probability distribution with which we start is driven by the algorithm toward the equilibrium probability distribution, and once the latter is reached it remains unchanged, and therefore sampling from it is equivalent to importance sampling.

The above results are formally captured by the so-called **master equation** that the probability distribution obeys during an ergodic Markov process, with the probability $p(\mathbf{R}, t)$ viewed as a time-dependent quantity:

$$\frac{\mathrm{d}p(\mathbf{R}, t)}{\mathrm{d}t} = -\sum_{\mathbf{R}'} w(\mathbf{R} \to \mathbf{R}')p(\mathbf{R}, t) + \sum_{\mathbf{R}'} w(\mathbf{R}' \to \mathbf{R})p(\mathbf{R}', t), \qquad (15.10)$$

where $w(\mathbf{R} \to \mathbf{R}')$ is the transition probability from state $\mathbf{R}$ to state $\mathbf{R}'$; the equilibrium probability distribution $\tilde{p}(\mathbf{R}, t)$ satisfies

$$\frac{\mathrm{d}\tilde{p}(\mathbf{R}, t)}{\mathrm{d}t} = 0.$$

15.3   The "simulated annealing" method

In the process of finding the optimal state of a complex system, the last stage is to combine importance sampling with an algorithm that allows the systematic evaluation of the various likely states. There are several ways to perform this last stage of optimization. We discuss here one such method which is very versatile so that it can be applied to a broad range of problems; furthermore, it is relatively transparent in how it works. This method is referred to as "simulated annealing"[2] because it mimics the physical process of creating high-quality crystals; this consists of repeated cycles of heating and cooling the sample so that the crystal defects are "annealed". The simulated annealing method is particularly useful for problems with a large number of degrees of freedom and very large phase space, which require evolution over long time scales in order to achieve ergodicity.

We assume that in the simulation of the system of interest, we can define a cost function, such as the total energy for a system consisting of microscopic particles. We also assume that it possible to generate new configurations from the current one that describes the state of the system, and that we can introduce the equivalent of a temperature T; recall that this scale allows comparisons of the differences in the value of the cost function when going from one configuration to the next. Finally, we assume that the process of generating new configurations does not *a priori* exclude any regions of phase space, which ensures that we are dealing with a Markov process.

In the following, by analogy to the preceding discussion, we shall use "energy" to denote the cost function, but this notion can be extended to any relevant quantity that characterizes the system of interest.

With the above assumptions, the simulated annealing algorithm consists of the following steps:

1. Start a "cooling" cycle from a high temperature: Select a random state of the system and a temperature value T_1 which is of the same order or larger

[2] S. Kirkpatrick, C.D. Gelatt Jr, and M.P. Vecchi, Optimization by simulated annealing, *Science* **220**, pp. 671–680 (1983).

than the energy of this random state. Set the current temperature of the system $T = T_1$.

2. Choose a temperature value T_0 down to which the system will be "cooled", which is much lower than T_1, typically by several orders of magnitude.

3. Use the Metropolis algorithm for a sufficient number of steps to arrive at the equilibrium distribution at temperature T.

4. Sample the equilibrium distribution for M steps at temperature T to find a low-energy state; this is not necessarily the *lowest* energy state, rather, it is a state of sufficiently low energy at this temperature value.

5. Lower the current temperature T by a predetermined factor $\lambda < 1$.
   If $T \geq T_0$ go to step 3.

6. Save the state with the lowest energy identified through this cycle.
   Go to step 1 and start a new cycle of cooling.

7. Compare the energies of the various states saved from each cycle to identify the one with the lowest energy.

Remark: The schedule for lowering the temperature in step 5 is quite important for finding a global extremum. Specifically, the temperature must be lowered at a slow enough rate to guarantee that all points in state-space have a finite probability of being visited; if this rate is too fast this condition is not satisfied and many important states may be missed.[3]

[3] For details see S. Geman and D. Geman, Stochastic relaxation, Gibbs distributions and the Bayesian restoration of images, *IEEE Proceedings Pattern Analysis and Machine Intelligence* **6**, pp. 721–741 (1984).

Example 15.1: To illustrate how the simulated annealing method works we address the problem of the two intersecting squares mentioned at the beginning of this section. We first label the sites according to some scheme of our choice, like the one shown in the figure below.

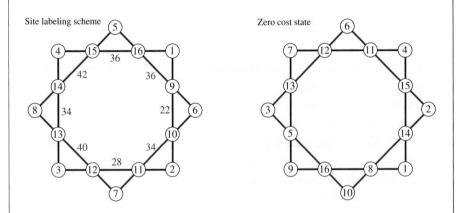

Next we define the sums along each side of the blue square (referred to as $B_i, i = 1, 2, 3, 4$) and the red square (referred to as $R_j, j = 1, 2, 3, 4$):

$$B_1 = s(1) + s(9) + s(10) + s(2), \quad B_2 = s(2) + s(11) + s(12) + s(3),$$
$$B_3 = s(3) + s(13) + s(14) + s(4), \quad B_4 = s(4) + s(15) + s(16) + s(1),$$

$$R_1 = s(5) + s(16) + s(9) + s(6), \quad R_2 = s(6) + s(10) + s(11) + s(7),$$
$$R_3 = s(7) + s(12) + s(13) + s(8), \quad R_4 = s(8) + s(14) + s(15) + s(5).$$

The problem we are seeking to solve is to place the first 16 integers in the 16 sites so that the sums along each side of the blue and red squares are equal, in other words $B_i = B_j$, $R_i = R_j$ and $B_i = R_j$, for any pair of indices, $i = 1, 2, 3, 4$ and $j = 1, 2, 3, 4$. It is easy to see that each integer is involved in two sides: the corner sites of the squares are involved in the two sides that share a corner, and the sites in the interior of each side are involved in one blue and one red side. Therefore, summing all the sums B_i and R_i will involve each integer twice. But the sum of the first N integers is a well known result, namely it is equal to $N(N+1)/2$; to obtain the sum of each side we must then multiply this number by 2 and divide it by 8, the number of all sides. From these considerations we conclude that the sum of integers on each side must be 34:

$$\sum_{n=1}^{N} n = \frac{N(N+1)}{2} \Rightarrow \frac{2}{8} \sum_{n=1}^{16} n = 34.$$

We can then define the "cost function" for a given state of the system as the sum of the absolute values of the *difference* between the sum on each side from the number 34:

$$C = \sum_{i=1}^{4} |B_i - 34| + \sum_{i=1}^{4} |R_i - 34|.$$

Finally, we need to define the "moves" that allow us to go from one state of the system, equivalent to a particular way of placing the 16 integers in the 16 sites, to another state. There are several ways to define such a move: for instance, we can select k sites and flip their order, placing the integers in these sites in the reverse order from the current state. This generates a new state whose cost we can calculate and compare to the cost of the old state; this will make it possible to perform stochastic sampling of configuration space using the Metropolis algorithm.

To implement optimization by simulated annealing, we used the following steps and parameters:

1. We can start the process with any random state of the system, in other words, any random order of placing the 16 integers in the 16 sites of the intersecting squares. As our random initial state we used the state defined by the labeling of the sites; the cost of this state is 36, as shown in the figure above.

2. We chose the initial temperature to be $T_1 = 20$ and the final temperature to be $T_0 = 0.5$. These choices are reasonable, since cost differences between states at the beginning of the search can be high, of order 10, and cost differences near the end of the optimization should be low, of order 1.

3. We chose the factor by which the temperature is reduced at each cycle of the annealing process to be $\lambda = 0.7$. This means that, starting with T_1, a temperature below T_0 is reached after 10 cycles.

4. We used a mixture of moves, 4–site flips, 3–site flips and 2–site flips, for 10%, 20% and 70% of the total moves, respectively. We found this mixture to be effective at identifying a zero cost state; moves of only one type are considerably less efficient and they tend to lead to local minima, typically only reaching states of cost 4 or 2. Such a low (but non-zero) cost state is shown in Fig. 15.4: the two sides for which the sum is not 34 are marked by black crosses.

With these choices, we could identify a zero-cost state by a single annealing run, with $M = 10^4$ Metropolis steps at each value of the temperature; since we have $K = 10$ temperature values from our choices above, this corresponds to sampling a total of $K \times M = 10^5$ configurations to find an optimal (zero-cost) one, which is an insignificant fraction of the total number of 21 trillion possible configurations.

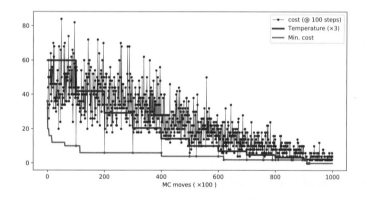

To gain some insight on how the search proceeds, we show in the figure above the cost of the different states that were sampled by the Metropolis algorithm at each temperature T step in our example (for clarity, we show only one out of every 100 states sampled). The cost associated with these sates, and therefore their cost *differences* which are of the same order, can be quite high, comparable to about $3T$. This is to be expected, since $e^{-3} \approx 0.05$, so even states with cost equal to $3T$ occur with a probability of about 5%. As the temperature is reduced, the range of cost of the states sampled is significantly reduced. The optimal value of the cost keeps decreasing steadily and reaches the desired value of zero by the last step in the temperature.

Why does the simulated annealing method work? A key feature is that it is not confined in the neighborhood of any single well near which the search starts. Because of the way the Metropolis algorithm is structured, the method explores many possible states with energy in the range of T. When T is large (the initial part of each cycle), the process explores the gross features of the energy landscape, while when T becomes smaller and smaller (later parts of the cycle), features at a smaller energy are explored. In this way, the process moves closer and closer to low-energy states over a wide range of configurations and is not confined ("trapped") in the immediate neighborhood of a shallow well. This is referred to as a **hierarchical** approach.

A key limitation of the simulated annealing method is that there is no guarantee we will find the "global" optimal solution. The method guarantees that we will identify one or more "local" optimal solutions, the best among all the low-cost wells visited by the process of importance sampling. An example of such a local optimal solution was mentioned for the two-intersecting-squares problem discussed in Example 15.1. A well-designed algorithm consists of finding moves that allow the system to escape from low-cost wells; this cannot be always guaranteed, so a careful choice of moves is important.

In real-life problems, the global optimal solution is usually not known, and there is no simple way to establish that it has been reached at the end of any annealing run. Hence, it is important to perform several annealing runs starting from initial configurations that are as different as possible; this at least gives some hope that a very broad range of the configuration space has been explored and the most important low-cost regions have been identified. As with many real-life situations, one can only hope that one has obtained a reasonable solution, given the finite amount of resources and time available. True ergodicity has infinite cost and takes forever!

15.4 Application: The traveling salesman problem

Suppose we are faced with the following challenge: The Rolling Stones are planning their last concert tour, and want to visit the 50 largest cities of the continental United States of America, shown in Fig. 15.6, starting in New York City and returning to it at the end of the tour. Their manager wants to find the path of shortest distance connecting all these cities. Evidently, the order in which the cities are visited will play a crucial role in the total distance traveled. The possible combinations for this case is a truly large number [$50! = 3.04 \times 10^{64}$], which far exceeds even astronomically large numbers like the number of all stars in all the galaxies of the Universe, estimated to be of order 10^{24}. Thus, finding all the combinations of cities in order to determine which one corresponds to the shortest distance traveled is clearly impossible. This problem is referred to as the "traveling salesman" problem, and has been studied extensively for various groups of cities in different parts of the world.

Figure 15.6: Map indicating the major cities by population in the United States of America.

We apply the simulated annealing method to solve this problem, which can be stated as follows:

Find the shortest path that connects a set of N cities, at the given positions (x_i, y_i), $i = 1, \ldots, N$ on a map, starting at one city and returning to it after visiting once only each other city.

The steps that can lead to an optimal solution using the concepts of the Metropolis algorithm and the simulated annealing method are the following:

1. Choose a random permutation of the numbers $1, 2, \ldots, N$ which defines the order in which the cities are visited; this constitutes the initial random configuration (state) of the system.

2. A change in the configuration to produce a new one is accomplished by choosing the path between two cities selected at random and reversing the order in which these cities are visited.

3. The "cost function" (equivalent to the "energy" in the examples of the preceding discussion) is the length L of the distance traveled:

$$L = \sum_{i=1}^{N} \left\{ (x_i - x_{i+1})^2 + (y_i - y_{i+1})^2 \right\}^{1/2},$$

where the order in which the cities are visited is determined by steps 1 and 2. The city with index $N + 1$ in the above summation is the same as the city with index 1, so that the path closes (returns to the original city).

4. Create an annealing schedule: for each move generated according to step 2 calculate the change in length ΔL and choose an initial temperature $T_1 \gg \Delta L$. Decrease the temperature by a factor λ at each step for K steps, where a step consists of M moves in the Metropolis algorithm.

The result of a few cycles of annealing is the identification of an optimal or near-optimal path. As mentioned earlier, it is not guaranteed that the process will identify the absolutely shortest path, but it *will* identify a path of reasonably short distance. The number of cycles needed depends on the number of cities N in the specific set, and the difficulty of finding a good solution given the geographic distribution of cities in this set.

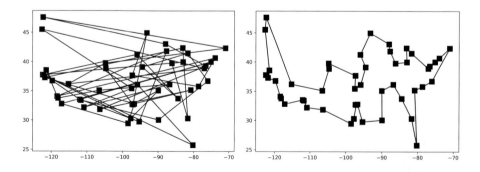

Figure 15.7: Two paths connecting the fifty major cities in the continental USA. The horizontal and vertical axes are labeled by the longitude and latitude in degrees. The path on the left corresponds to a random order of visiting the cities and has total length of about 62,100 miles. The path on the right is optimized by simulated annealing and has total length of about 10,300 miles.

As a case in point, we illustrate the traveling salesman problem applied to the $N = 50$ largest cities (in terms of population) in the continental USA. Using $M = 10^4$ moves in the Metropolis algorithm, $\lambda = 0.7$ for the scaling of the temperature at each simulated annealing step, and $K = 20$ such steps, we obtain an excellent solution in just one cooling cycle. The result is shown in Fig. 15.7. Given that the continental USA has an East–West length of about 3000 miles and a North–South length of about 1500 miles, or a perimeter of about 9000 miles, the optimal path found is rather short (it is shorter by a factor of 6 compared to the initial random path), and by visual inspection reasonably close to optimal (compare to the map in Fig. 15.6). The total number of paths explored to arrive at this result is $M \times K = 2 \times 10^5$, a truly insignificant fraction of the total number of possible combinations mentioned earlier, which is of order 10^{64}!

Further reading

1. James C. Spall, *Introduction to Stochastic Search and Optimization: Estimation, Simulation, and Control* (John Wiley & Sons, 2003). This book offers a comprehensive treatment of stochastic optimization methods. It includes in-depth discussions of all the topics mentioned in this chapter.

2. Mykel J. Kochenderfer and Tim A. Wheeler, *Algorithms for Optimization: Illustrated Edition* (MIT Press, 2019). This is a useful compilation of algorithms for a wide range of numerical methods and their applications, including stochastic optimization methods covered in the present chapter. For each algorithm, computer code in the Julia programming language is also provided.

Problems

1. The values of stocks resemble random walks as a function of time. Create a hypothetical portfolio of 20 stocks and simulate its worth for a period of 10 years, with a time-step of 1 day; you may assume that there are 248 trading days per year. Evaluate the annual yield of the portfolio, that is, the increase or decrease of its value relative to that at the beginning of the year, for each quarter (62 trading sessions). What is the yield on an annual basis? What is the yield for the 10-year period?

2. The European Union (EU) consists of 27 member states. The President of the EU Council intends to visit the capitals of all member states to present to their parliaments an important treaty. The President starts the trip in Brussels (Belgium), the seat of the EU Council, and returns there at the end. The number of combinations for this set of cities is $27! = 1.09 \times 10^{28}$, making it challenging to find the shortest-distance path. Solve the traveling salesman problem for this case using the method of simulated annealing. For the position of each capital on a map you may use its latitude and longitude in the geographic coordinate system, in degrees, minutes and seconds. In the North–South direction, the size of a latitudinal second is 30.7 m, while in the East–West direction, for the size of a longitudinal second you may take that of $45°$ latitude (approximately the average for Europe), which is 21.9 m.

3. Use the simulated annealing method to find solutions as close to the optimal as possible for up-or-down pointing magnets on a triangular lattice, as described in Section 15.2 (see Fig. 15.4). Use $N = 100$ magnets, and apply periodic boundary conditions so that each magnet on the lattice has the same number of neighbors ["periodic boundary conditions" imply that a site at the boundary of the lattice with some missing neighbors has as neighbors those in the corresponding position at the opposite boundary of the lattice].

Appendix A

Solution of the Black–Scholes Equation

We provide here the derivation of the solution to the Black–Scholes equation for the call option, given by Eq. (8.146), starting from the expression

$$v_c(x, \tau) = \int_{-\infty}^{\infty} v_{c,0}(x') \frac{1}{\sqrt{4\pi\tau}} e^{-(x-x')^2/4\tau} dx',$$

with

$$v_{c,0}(x) = K(e^{\sigma x/\sqrt{2}} - 1)\theta(e^{\sigma x/\sqrt{2}} - 1)e^{\beta x/2},$$

and

$$\beta = \left(\frac{\sqrt{2}}{\sigma}\rho - \frac{\sigma}{\sqrt{2}}\right).$$

We introduce the following change of variables:

$$y = \frac{\sigma x'}{\sqrt{2}} \Rightarrow x' = \frac{\sqrt{2}y}{\sigma}, \quad dx' = \frac{\sqrt{2}}{\sigma}dy.$$

Substituting the value of the parameter β in the expression for $v_{c,0}(x')$ we find

$$v_{c,0}(x') = 2K \sinh\left(\frac{y}{2}\right) e^{\rho y/\sigma^2}\theta(e^y - 1),$$

which, when inserted in the integral for $v_c(x, \tau)$ gives

$$
\begin{aligned}
v_c(x, \tau) &= \frac{2K}{\sqrt{2\pi\tau\sigma^2}} \int_{-\infty}^{\infty} \sinh\left(\frac{y}{2}\right) e^{\rho y/\sigma^2}\theta(e^y - 1)e^{-(\sqrt{2}y/\sigma - x)^2/4\tau}dy \\
&= \frac{2K}{\sqrt{2\pi\tau\sigma^2}} \int_{-\infty}^{\infty} \sinh\left(\frac{y}{2}\right) e^{\rho y/\sigma^2}\theta(e^y - 1)e^{-(y - x\sigma/\sqrt{2})^2/2\tau\sigma^2}dy \\
&= \frac{Ke^{-x^2/4\tau}}{\sqrt{2\pi\tau\sigma^2}} \int_{-\infty}^{\infty} \theta(e^y - 1)\left[e^{\rho y/\sigma^2 + y/2} - e^{\rho y/\sigma^2 - y/2}\right]e^{-(y^2 - \sqrt{2}x\sigma y)/2\tau\sigma^2}dy.
\end{aligned}
$$

We write the above expression as $v_c(x, \tau) = K(I_1 - I_2)$ where I_1, I_2 are two integrals defined by the expressions:

$$I_1 = \frac{e^{-x^2/4\tau}}{\sqrt{2\pi\tau\sigma^2}} \int_{-\infty}^{\infty} \theta(e^y - 1)e^{\rho y/\sigma^2 + y/2}e^{-(y^2 - \sqrt{2}x\sigma y)/2\tau\sigma^2}dy,$$

$$I_2 = \frac{e^{-x^2/4\tau}}{\sqrt{2\pi\tau\sigma^2}} \int_{-\infty}^{\infty} \theta(e^y - 1)e^{\rho y/\sigma^2 - y/2}e^{-(y^2 - \sqrt{2}x\sigma y)/2\tau\sigma^2}dy.$$

For the computation of each integral, we complete the square of the exponential. For I_1, completing the square gives

$$-\frac{y^2 - \sqrt{2}\sigma xy}{2\tau\sigma^2} + \left(\frac{\rho}{\sigma^2} + \frac{1}{2}\right)y = -\frac{1}{2\tau\sigma^2}(y - y_1)^2 + \frac{1}{2\tau\sigma^2}y_1^2,$$

where

$$y_1 = \left(\frac{\sigma x}{\sqrt{2}} + \tau\rho + \frac{\tau\sigma^2}{2}\right).$$

This leads to the following result for the first integral:

$$\begin{aligned}I_1 &= \frac{e^{-x^2/4\tau}}{\sqrt{2\pi\tau\sigma^2}}e^{y_1^2/2\tau\sigma^2}\int_{-\infty}^{\infty}\theta(e^y - 1)e^{-(y-y_1)^2/2\tau\sigma^2}dy\\ &= \frac{1}{\sqrt{2\pi\tau\sigma^2}}e^{\tau(\rho+\sigma^2/2)^2/2\sigma^2}e^{x(\rho+\sigma^2/2)/\sqrt{2}\sigma}\int_{-y_1}^{\infty}e^{-y^2/2\tau\sigma^2}dy\end{aligned}$$

and with the change of variables

$$w = \frac{y}{\sigma\sqrt{\tau}}, \quad \tilde{w}_1 = \frac{y_1}{\sigma\sqrt{\tau}},$$

we arrive at the final result

$$I_1 = \Phi(\tilde{w}_1)e^{\tau(\rho+\sigma^2/2)^2/2\sigma^2}e^{x(\rho+\sigma^2/2)/\sqrt{2}\sigma}, \quad \tilde{w}_1 = \frac{1}{\sigma\sqrt{\tau}}\left(\frac{\sigma x}{\sqrt{2}} + \tau\rho + \frac{\tau\sigma^2}{2}\right),$$

where $\Phi(w)$ is the normal cumulative distribution function, defined in Eq. (1.60a). For the second integral, following identical steps we find the result

$$I_2 = \Phi(\tilde{w}_2)e^{\tau(\rho-\sigma^2/2)^2/2\sigma^2}e^{x(\rho-\sigma^2/2)/\sqrt{2}\sigma}, \quad \tilde{w}_2 = \frac{1}{\sigma\sqrt{\tau}}\left(\frac{\sigma x}{\sqrt{2}} + \tau\rho - \frac{\tau\sigma^2}{2}\right).$$

Combining the two results, we can calculate the function $v_c(x, \tau)$. We can then use this, and the expression of Eq. (8.144), to obtain an expression for $u_c(x, \tau)$, namely:

$$\begin{aligned}u_c(x, \tau) &= v_c(x, \tau)e^{-\beta^2\tau/4 - \beta x/2} = K(I_1 - I_2)e^{-\beta^2\tau/4 - \beta x/2}\\ &= K\left[e^{\xi+\tau\rho}\Phi(\tilde{w}_1) - \Phi(\tilde{w}_2)\right],\end{aligned}$$

with the last step involving only trivial algebraic manipulations after inserting the value of the parameter β in terms of ρ and σ, and the definition $\xi = x\sigma/\sqrt{2}$. The last expression is the desired result, Eq. (8.146).

Appendix B

Gaussian Integral Table

In the following table we give values of the integral

$$\Phi(w) = \frac{1}{\sqrt{2\pi}} \int_{-\infty}^{w} e^{-y^2/2} dy = \frac{1}{\sqrt{2\pi}} \int_{-w}^{\infty} e^{-y^2/2} dy, \qquad (B.1)$$

for values of w ranging from 0 to 2.5 in steps of 0.01. Recall also the following useful relations:

$$\Phi(-w) = 1 - \Phi(w), \quad \Phi(0) = 0.5. \qquad (B.2)$$

w	$\Phi(w)$	w	$\Phi(w)$	w	$\Phi(w)$	w	$\Phi(w)$	w	$\Phi(w)$
0.01	0.503989	0.26	0.602568	0.51	0.694974	0.76	0.776373	1.01	0.843752
0.02	0.507978	0.27	0.606420	0.52	0.698468	0.77	0.779350	1.02	0.846136
0.03	0.511966	0.28	0.610261	0.53	0.701944	0.78	0.782305	1.03	0.848495
0.04	0.515953	0.29	0.614092	0.54	0.705401	0.79	0.785236	1.04	0.850830
0.05	0.519939	0.30	0.617911	0.55	0.708840	0.80	0.788145	1.05	0.853141
0.06	0.523922	0.31	0.621720	0.56	0.712260	0.81	0.791030	1.06	0.855428
0.07	0.527903	0.32	0.625516	0.57	0.715661	0.82	0.793892	1.07	0.857690
0.08	0.531881	0.33	0.629300	0.58	0.719043	0.83	0.796731	1.08	0.859929
0.09	0.535856	0.34	0.633072	0.59	0.722405	0.84	0.799546	1.09	0.862143
0.10	0.539828	0.35	0.636831	0.60	0.725747	0.85	0.802337	1.10	0.864334
0.11	0.543795	0.36	0.640576	0.61	0.729069	0.86	0.805105	1.11	0.866500
0.12	0.547758	0.37	0.644309	0.62	0.732371	0.87	0.807850	1.12	0.868643
0.13	0.551717	0.38	0.648027	0.63	0.735653	0.88	0.810570	1.13	0.870762
0.14	0.555670	0.39	0.651732	0.64	0.738914	0.89	0.813267	1.14	0.872857
0.15	0.559618	0.40	0.655422	0.65	0.742154	0.90	0.815940	1.15	0.874928
0.16	0.563559	0.41	0.659097	0.66	0.745373	0.91	0.818589	1.16	0.876976
0.17	0.567495	0.42	0.662757	0.67	0.748571	0.92	0.821214	1.17	0.879000
0.18	0.571424	0.43	0.666402	0.68	0.751748	0.93	0.823814	1.18	0.881000
0.19	0.575345	0.44	0.670031	0.69	0.754903	0.94	0.826391	1.19	0.882977
0.20	0.579260	0.45	0.673645	0.70	0.758036	0.95	0.828944	1.20	0.884930
0.21	0.583166	0.46	0.677242	0.71	0.761148	0.96	0.831472	1.21	0.886861
0.22	0.587064	0.47	0.680822	0.72	0.764238	0.97	0.833977	1.22	0.888768
0.23	0.590954	0.48	0.684386	0.73	0.767305	0.98	0.836457	1.23	0.890651
0.24	0.594835	0.49	0.687933	0.74	0.770350	0.99	0.838913	1.24	0.892512
0.25	0.598706	0.50	0.691462	0.75	0.773373	1.00	0.841345	1.25	0.894350

w	$\Phi(w)$	w	$\Phi(w)$	w	$\Phi(w)$	w	$\Phi(w)$	w	$\Phi(w)$
1.26	0.896165	1.51	0.934478	1.76	0.960796	2.01	0.977784	2.26	0.988089
1.27	0.897958	1.52	0.935745	1.77	0.961636	2.02	0.978308	2.27	0.988396
1.28	0.899727	1.53	0.936992	1.78	0.962462	2.03	0.978822	2.28	0.988696
1.29	0.901475	1.54	0.938220	1.79	0.963273	2.04	0.979325	2.29	0.988989
1.30	0.903200	1.55	0.939429	1.80	0.964070	2.05	0.979818	2.30	0.989276
1.31	0.904902	1.56	0.940620	1.81	0.964852	2.06	0.980301	2.31	0.989556
1.32	0.906582	1.57	0.941792	1.82	0.965620	2.07	0.980774	2.32	0.989830
1.33	0.908241	1.58	0.942947	1.83	0.966375	2.08	0.981237	2.33	0.990097
1.34	0.909877	1.59	0.944083	1.84	0.967116	2.09	0.981691	2.34	0.990358
1.35	0.911492	1.60	0.945201	1.85	0.967843	2.10	0.982136	2.35	0.990613
1.36	0.913085	1.61	0.946301	1.86	0.968557	2.11	0.982571	2.36	0.990863
1.37	0.914657	1.62	0.947384	1.87	0.969258	2.12	0.982997	2.37	0.991106
1.38	0.916207	1.63	0.948449	1.88	0.969946	2.13	0.983414	2.38	0.991344
1.39	0.917736	1.64	0.949497	1.89	0.970621	2.14	0.983823	2.39	0.991576
1.40	0.919243	1.65	0.950529	1.90	0.971283	2.15	0.984222	2.40	0.991802
1.41	0.920730	1.66	0.951543	1.91	0.971933	2.16	0.984614	2.41	0.992024
1.42	0.922196	1.67	0.952540	1.92	0.972571	2.17	0.984997	2.42	0.992240
1.43	0.923641	1.68	0.953521	1.93	0.973197	2.18	0.985371	2.43	0.992451
1.44	0.925066	1.69	0.954486	1.94	0.973810	2.19	0.985738	2.44	0.992656
1.45	0.926471	1.70	0.955435	1.95	0.974412	2.20	0.986097	2.45	0.992857
1.46	0.927855	1.71	0.956367	1.96	0.975002	2.21	0.986447	2.46	0.993053
1.47	0.929219	1.72	0.957284	1.97	0.975581	2.22	0.986791	2.47	0.993244
1.48	0.930563	1.73	0.958185	1.98	0.976148	2.23	0.987126	2.48	0.993431
1.49	0.931888	1.74	0.959070	1.99	0.976705	2.24	0.987455	2.49	0.993613
1.50	0.933193	1.75	0.959941	2.00	0.977250	2.25	0.987776	2.50	0.993790

Index